# Spon's Civil Engineering and Highway Works Price Book

# Spon's Civil Engineering and Highway Works Price Book

*Edited by*
DAVIS LANGDON & EVEREST
*Chartered Quantity Surveyors*

**1991**

*Fifth Edition*

E. & F.N. SPON
An imprint of Chapman and Hall
London · New York · Tokyo · Melbourne · Madras

| | |
|---|---|
| UK | Chapman and Hall, 2–6 Boundary Row, London SE1 8HN |
| USA | Van Nostrand Reinhold, 115 5th Avenue, New York NY10003 |
| JAPAN | Chapman and Hall Japan, Thomson Publishing Japan, Hirakawacho Nemoto Building, 7F, 1–7–11 Hirakawa-cho, Chiyoda-ku, Tokyo 102 |
| AUSTRALIA | Chapman and Hall Australia, Thomas Nelson Australia, 480 La Trobe Street, PO Box 4725, Melbourne 3000 |
| INDIA | Chapman and Hall India, R. Seshadri, 32 Second Main Road, CIT East, Madras 600 035 |

First edition 1984
Fifth edition 1990



Printed in Great Britain by
Richard Clay Ltd, Bungay, Suffolk

ISBN 0 419 16810 9    0 442 31300 4 (USA)

ISSN 0957–171X



**British Library Cataloguing in Publication Data**

Spon's civil engineering and highway works price book. – 5th ed.:
1991 (Spon's price book series)
I. Civil engineering – Estimates – Great
Britain – Periodicals
624 TA181.G7
ISBN 0–419–16810–9
ISSN 0957–171X

# Preface

:his edition follows the now established format covering Civil Engineering Works ınd Highway Works, structured around the Civil Engineering Standard Method of leasurement (2nd edition) and the Method of Measurement for Highway Works, but ıll prices, data and other information have been thoroughly reviewed and revised ıp to June/July 1990. Consequently the scope, range and detail of information ıas again expanded, which enables readers to compare or adjust any net rates with 'espect of outputs or costs.

ates, prices and outputs included in the resources and unit cost calculations ncluding allowances for wastage productivity and efficiency are all based on edium sized civil engineering schemes of about £6 million in value, with no cute access or ground condition problems, but are equally applicable for a wide ange of construction projects from £0.5 million to £50 million with little or no djustments. Where adjustments to outputs or costs can be made simply and uickly, we have detailed tables of multipliers within the appropriate sections.

he unit costs sections (Parts 4 and 5) now cover a wider range of items than ver before, where appropriate we have included notes and details on assumptions ade such as labour and plant resources, and materials waste factors. Part 3 of he book (Resources section) allows these unit rates to be adjusted to suit ndividual requirements by the substitution of alternative labour, materials or lant costs. Unit costs are obviously dependent upon the outputs or manhours sed to calculate them, the outputs used in this work have been compiled in etail from the editors' wide ranging experience and are based almost exclusively ı time and motion studies and records reported from a large number of recent ivil engineering schemes. A number of prices and outputs are based upon etailed specialist advice, acknowledgements to the main contributors are ıcluded in the book.

ıe same basis of information as used in the unit costs has been used to compile ıe Approximate Estimates section (Part 10). This has also been expanded and eviewed for this edition, but it must be emphasised that the data in this ection is only approximate and if more details are known then reference ought to made to the unit cost sections.

ıe expected reduction of United Kingdom civil engineering workload is just eginning to bite at the time of writing (July 1990) and this has obviously fected the pricing levels used in the book. With this in mind attention is awn to the free price updating service which will advise of any significant anges to the published information. The Update is circulated free every 3 nths to those readers who have registered with the publishers, in order to do readers should complete and return the coloured card bound in with this lume.

ilst all efforts are made to ensure the accuracy of the data and information ed in the book neither the editors nor the publishers can in any way accept ability for loss of any kind resulting from the use made by any person of such formation.

Davis Langdon & Everest
Chartered Quantity Surveyors
Princes House
39 Kingsway
London WC2B 6TP

# Contents

## PART 7 : COST AND TENDER PRICE INDICES



## PART 8 : DAYWORK



## PART 9 : PROFESSIONAL FEES



## PART 10 : APPROXIMATE ESTIMATES



## PART 11 : OUTPUTS

## PART 12 : TABLES AND MEMORANDA



## PART 13 : UPDATES

# Abbreviations

| | |
|---|---|
| BS/BSS | British Standard/British Standard Specification |
| cwt | hundredweight |
| DL&E | Davis Langdon and Everest |
| DERV | Diesel engine road vehicle |
| DoE | Department of Environment |
| DTp | Department of Transport |
| ft | foot |
| $ft^3$ | cubic foot |
| $ft^3$/min | cubic foot per minute |
| ha | hectare |
| HMSO | Her Majesty's Stationery Office |
| hp | horsepower |
| hr | hour |
| in | inch |
| kg | kilogramme |
| km | kilometre |
| kVA | kilovolt ampere |
| kw | kilowatt |
| m | metre |
| $m^2$ | square metre |
| $m^3$ | cubic metre |
| micron | $10^{-6}$ metre |
| mm | millimetre |
| $mm^2$ | square millimetre |
| N | newton |
| ne | not exceeding |
| NEDO | National Economic Development Office |
| nr | number |
| pa | per annum |
| PC | Prime Cost |
| sq | square |
| t | tonne |
| wk | week |
| yd | yard |
| $yd^3$ | cubic yard |

# Acknowledgements

The Editors wish to record their appreciation of the assistance given by many individuals and organisations in the compilation of this publication.

Abacus Municipal
Oddicroft Lane
Sutton in Ashfield
Notts
NG17 5FT
**STREET FURNITURE**

ACO Polymer Products Ltd
Hitchin Road
Shefford
Beds
SG17 5JS
**DRAINAGE SYSTEMS**

Anda Crib
Oaklands House
Solartron Road
Farnborough
Hants GU14 7LQ
**CRIBWALLS**

ARC Conbloc
Mill Lane
Higher Ongar
Essex
CM5 9RH
**BLOCKWORK**

ARC Pipes
Mells Road
Mells
Frome
Somerset
**CONCRETE MANHOLES**

Asset International
Stephenson Street
Newport
Gwent
NP9 0XH
**STEEL CULVERTS**

Bachy (UK) Limited
Foundation Court
Godalming Business Centre
Catteshall Lane
Gadalming
Surrey GU17 1XW
**DIAPHRAGM WALLING**

Badderidge Brick plc
Gospel End
Sedyley
Dudley
West Midlands
DY3 4AA
**BRICKS**

Baldwin Durafencing Limited
18 Park Road
Worsley
Manchester
M28 5EA
**FENCING PRODUCTS**

Barfab Reinforcements
68 First Avenue
Mortlake
London
SW14 8SR
**REINFORCEMENT**

Barnard Pipeline Technology
Lower Tower Street
Birmingham
B19 3NL
**COATED STEEL PIPES AND FITTINGS**

Beacon Hill Brick Company
Corfe Mollen
Wimborne
Dorset
BH21 3RX
**BRICKWORK**

Beaver 84 Ltd
Beaver House
Crompton Close
Basildon
Essex
SS14 3AY
**SCAFFOLDING**

Bell & Webster Ltd
Essex Road
Hoddesdon
Herts
EN11 ODR
**PRECAST CONCRETE RETAINING WALLS**

Blaw Knox Construction Equipment
Shorts Way
Rochester
Kent
ME1 3AP
**PLANT MANUFACTURERS**

Blockleys PLC
Sommerfield Road
Trench Lock
Telford
Shropshire
TF1 4RY
**BRICKS AND PAVERS**

BM Stainless
Station Road Industrial Estate
Tadcaster
North Yorkshire
LS24 9SG
**STAINLESS STEEL PIPES AND FITTINGS**

BM Plant Limited
38 Castlefields Industrial Estate
Bridgewater
Somerset
TA6 4DH
**PLANT AGENTS**
**(HITACHI EXCAVATORS)**

BRC Engineering Co Ltd
Lichfield Road
Stafford
ST17 4NN
**REINFORCEMENT AND**
**CONSTRUCTION EQUIPMENT**

British Alcan Building Products Ltd
Blackpole Trading Estate
Worcester
WR3 8TJ
**PROFILED ALUMINIUM ROOFING**
**AND CLADDING**

British Seed Houses Ltd
Bewsey Industrial Estate
Pitt Street
Warrington
WA5 5LE
**HORTICULTURAL PRODUCTS**

British Steel - General Steels
Piling Sales Dept
PO Box 1
Scunthorpe
South Humberside
DN16 1BP
**STEEL PILES**

British Steel - General Steel
Welded Tubes
PO Box 101
Corby
Northants
**STRUCTURAL STEEL**

Brixey Engineering Ltd
Fancy Road
Newtown
Poole
Dorset
BH17 7NZ
**EXCAVATOR EQUIPMENT**

BSS (UK) Ltd
Fleet House
Lee Circle
Leicester
LE1 3QQ
**STEEL PIPES AND FITTINGS**

Capital Demolition
(New Haw Enterprises Ltd)
Woodham Park Road
Woodham
Nr Weybridge
Surrey KT15 3TG
**DEMOLITION CONTRACTORS**

Caradon
24 Kingsland Grange
Woolston
Warrington
Cheshire
**UPVC PIPES**

Cementation Piling Limited
Cementation House
Maple Cross
Denham Way
Rickmansworth
WD3 2SW
**PILING CONTRACTORS**

Colas Building Products Limited
Riverside
Saltney
Chester
CH4 8RS
**WATERPROOFING**

Concrete Utilities
Lower Road
Great Anwell
Ware
Herts
SG12 9TA
**STREET FURNITURE**

Costain Dow Mac
Tallington
Stamford
Lincolnshire
PE9 4RL
**PRE STESSED CONCRETE BEAMS**

Crown Protective Coatings
Hadrian Works
Haltwistle
Northumberland
NE49 0HF
**PROTECTIVE COATINGS**

Crown Berger Trade Paints
PO Box 37
Crown House
Hollins Road
Darwen
Lancashire
BB3 0BG
**PAINTS/FINISHES**

C U Lighting Ltd
Charles House
Great Amwell
Ware
Herts
SG12 9TA
**STREET FURNITURE**

Duckbill Anchors
Vallance-by-Ways
Lowfield Heath Road
Charlwood
Horley
Surrey
RH6 0BT
**EARTH ANCHORS**

Drainage Systems Ltd
Thames Bank
Lodge Road
St Johns Wood
London
NW8 8NU
**CAST IRON PIPES AND FITTINGS**

Economic Forestry Group
Market Street
Fordham
Ely
Cambridgeshireshire
CB7 5LQ
**BARK PRODUCTS**

ECC Building Products Ltd
Hulland Ward
Derby
DE6 3ET
**KERBS, EDGINGS AND PAVINGS**

EFCO UK Ltd
35 Stilebrook Road
Yardley Road Industrial Estate
Olney
Bucks
MK46 7EA
**STEEL FORMS**

Expanded Metal Co Ltd
28 Rufford Court
Hardwich Grange
Woolston
Warrington
WA1 4RF
**EXPANDED METAL PRODUCTS**

Expandite
1-9 Chase Road
London
NW10 6PS
**WATERPROOFING & EXPANSION JOINTS**

Finning Limited
Cannock
Staffordshire
**PLANT AGENTS (CATERPILLAR)**

Fischer Arthur (UK) Ltd
Hithercroft Road
Wallingford
Oxon
OX10 9AT
**FIXINGS**

Greenham Trading Ltd
Greenham House
671 London Road
Iselworth
Middlesex
TW7 4EX
**CONTRACTORS SITE EQUIPMENT**

Glynwed Foundries
PO Box 3
Sinclair Works
Ketley
Shropshire
TF1 4AD
**MANHOLE COVERS & STREET FURNITURE**

GEM Professional
Brookside Lane
Oswaldtwistle
Accrington
Lancashire
BB5 3NY
**LANDSCAPING MATERIALS**

E G Harris Ltd
Dock Road
Brentford
Middlesex
TW8 8AQ
**DREDGING (INLAND WATER)**

Hepworth Building Products Ltd
Hazlehead
Stocksbridge
Sheffield
S30 5HG
**CLAYWARE PIPES**

Hercules Piling Ltd
Norman House
Heritage Gate
Friar Gate
Derby
DE1 1NU
**PILING CONTRACTORS**

P J Hopkins Railway Contractors Ltd
Monastery Lane
Sutton
St Helens
Merseyside
WA9 3SP
**RAIL TRACK**

Hill & Smith
PO Box 4
Canal Street
Brierly Hill
West Midlands
DY5 1JL
**SAFETY FENCING**

Hume Pipe Ltd
PO Box 28
Alton
Hants
GU34 1SU
**CONCRETE PIPES**

Independant Formwrok Ltd
Convent Drive
Waterbeach
Cambridge
CB5 9PB

JCB Sales Ltd
Rochester
Staffordshire
ST14 5JP
**LANDSCAPING MATERIALS**

Joseph Metcalf
Brookside Lane
Oswaldtwistle
Accrington
Lancashire
BB5 3NY
**LANDSCAPING MATERIALS**

Klargester Environmental Engineering Ltd
College Road
Aston Clinton
Aylesbury
Bucks
HP22 5EW
**CESSPOOLS, SEPTIC TANKS AND INTERCEPTORS**

Lathams, James PLC
Leeside Wharf
Mount Pleasant Hill
Clapton
London
E5 9NG
**TIMBER MERCHANTS**

London Brick Co Ltd
Stewartby
Bedford
Beds
MK43 9LZ
**BRICKS**

L & R Roadlines Ltd
24-32 Forth Street
Liverpool
L20 8JW
**ROAD MARKINGS**

Marshalls Mono Ltd
Southowram
Halifax
HX3 9SY
**KERBS, EDGINGS AND PAVINGS**

Marley Extrusions Ltd
Lenham
Maidstone
Kent
ME17 2DE
**UPVC DRAINAGE AND RAINWATER SYSTEMS**

Milton Pipes Ltd
Milton Regis
Sittingbourne
Kent
ME10 2QF
**PRECAST CONCRETE CULVERTS, MANHOLES AND RETAINING WALLS**

Marubeni-Komatsu Limited
Padgets Lane
Redditch
Worcs
B98 ORT
**PLANT MANUFACTURERS**

Mallinson-Denny (Northern) Ltd
Chaddock Lane
Worsley
Manchester
M28 4DS
**TIMBER**

MMG Civil Engineering Systems Ltd
Vermuyden House
Wiggenhall St Germans
Kings Lynn
Norfolk
PE34 3ES
**SCOUR PROTECTION, GEOTEXTILES**

Naylor Bros (Clayware) Ltd
Clough Green
Cawthorne
Barnsley
S75 4AD
**CLAYWARE PIPES AND FITTINGS**

Oldroyd Bros Ltd
Chester Road
Sutton Weaver
Runcorn
Cheshire
WA7 3EG
**STRUCTURAL DESIGN**

Frank Parker & Co Ltd
John O'Gaunt's Industrial Estate
Rothwell
Leeds
**CONTRACTORS SITE EQUIPMENT**

PSE Machinery Ltd
Thorney Mill Road
West Drayton
Middlesex
UB7 7E2
**EXCAVATOR ATTACHMENTS**

Perma Holdings Ltd
72/74 Bath Road
Cheltenham
Glos
GL5 37J
**RETAINING WALLS**

Rapid Metal Developments Ltd
Stubbers Green Road
Aldridge
Walsall
WS9 8BW
**TEMPORARY WORKS EQUIPMENT, CONSTRUCTION EQUIPMENT**

Rawlplug Company Limited
Skibo Drive
Thornsliebank Industrial Estate
Glasgow
G46 8JR
**FIXINGS**

RMC (N.W.) Limited
RMC House
St. Marys Road
Widnes
Cheshire
WA8 0DL
**CONCRETE (READY MIXED)**

Redland Bricks Ltd
York House
31 - 36 York Place
Leeds
LS1 2ED
**BRICKS**

Reinforced Earth Company Ltd
Walker House
Telford
Shropshire
TS3 4HA
**REINFORCED EARTH SYSTEMS**

Rees/Hough (Civil Engineering) Ltd
Nowhurst Lane
Broadbridge Heath
Horsham
Sussex
RH12 3PL
**PIPE JACKING**

Reinforced Steel Services
PO Box 41
Meadowhall Road
Sheffield
S9 1ED
**TEMPORARY WORKS EQUIPMENT**
**CONSTRUCTION EQUIPMENT**

River and Sea Gabions (London) Ltd
Tricorn House
51-53 Hagley Road
Birmingham
B16 8TP
**GABIONS**

RIW Ltd
Arc House
Terrace Road South
Binfield
Bracknell
Berkshire
RG12 5PZ
**WATERPROOFING**

RMD
Unit H2
Lynwell Road
Lyntown Trading Estate
Eccles
Manchester
M30 9QG
**TEMPORARY WORKS, STEEL FORMS**

Rowlawn (Turf Growers) Limited
Elvington
York
YO4 5AR
**TURF**

Sanderson Group
Croft
Skegness
Lincs
PE24 4BR
**PLANT AGENTS (PRIESTMAN, WINGET)**

David Roberts Son & Co (Bootle) Ltd
Coburg Sawmills
Coburg Dock
Liverpool
L3 4AP
**HARDWOOD**

Sadolin (UK) Ltd
Unit 13
Tower Close
St Peters Industrial Estate
Huntingdon
Cambs
PE18 7DR
**TIMBER TREATMENT**

Selwood Limited
Bournemouth Road
Eastleigh
Hampshire
SO5 3ZL
**PLANT HIRE**

Servicised Ltd
Ajax Avenue
Slough
Berkshire
SL1 4BH
**WATERPROOFING/EXPANSION JOINTS**

Solignum Ltd
Thames Road
Craywood
Kent
DA1 4QJ
**PROTECTIVE FINISHES**

Square Grip Ltd
11 Mulberry Business Park
Fishponds Road
Wokingham
Berkshire
RG11 2FH
**RETAINING WALL SYSTEM**

Stanton plc
PO Box 72
Nottingham
NG10 5AA
**CAST IRON PIPES AND FITTINGS**
**POLYMER CONCRETE CHANNELS**

Stent Foundations Ltd
Osbourn Way
Hook
Hampshire
RG27 9EX
**PILING AND FOUNDATIONS**

Stocksigns Limited
Ormside Way
Holmethorpe Industrial Estate
Redhill
Surrey
RH1 2LG
**ROAD SIGNS AND POSTS**

Sykes Construction Services
Bath Road
Padworth
Reading
Berkshire
RG7 5HR
**GROUNDWATER CONTROLS**

Tarmac Structural Concrete Ltd
Francon House
Ford Airfield
Ford
Arundel
West Sussex
BN18 OBN
**PRECAST CONCRETE**

Tarmac Topblock Ltd
Wergs Hall
Wergs Hall Road
Wolverhampton
WV8 2HZ
**BLOCKWORK**

Terry's Timber Ltd
35 Regent Road
Liverpool
L5 9SR
**TIMBER**

Tilcon Ltd
Unit 4
Hammond Industrial Estate
Reddish
Stockport
Greater Manchester
**READY MIXED CONCRETE**

Varley and Gulliver
57-70 Alfred Street
Sparkbrook
Birmingham
B12 8JR
**BRIDGE PARAPETS**

Volvo BM (UK) Ltd
Duxford
Cambridge
CB2 4QX
**PLANT MANUFACTURERS**

Wavin Building Products Ltd
Parsonage Way
Chippenham
Wiltshire
SN15 5PN
**UPVC DRAIN PIPES AND FITTINGS**

Weatherwise (Roofing and Cladding) Ltd
Manor Lane
Hawarden
Clwyd
CM5 3PP
**ROOFING AND CLADDING CONTRACTORS**

Wimpey Geotech Ltd
Chester Road
Birmingham
B35 7AH
**GEOTECHNICAL INVESTIGATIONS AND PROCESSES**

Wells Spiral Tubes Ltd
Pilsworth Way
Pilsworth Road
Bury
Greater Manchester
**BL9 8RE**
**STEEL CULVERTS**

Winn & Coales (Denso) Ltd
Denso House
Chapel Road
London
SE27 0TR
**DENSO PRODUCTS**

W & J Leigh
Tower Works
Kester Street
Bolton
BL2 2AL
**PAINT/FINISHES**

PART 1

# General

**PURPOSE AND CONTENT OF THE BOOK**

For many years the Editors have compiled a price book for use in the building industry with, more recently, companion volumes for use in connection with engineering services contracts and landscaping work. All of these price books take their reliability from the established practice within these sectors of the construction industry of pricing work by the application of unit rates to quantities measured from the designer's drawings. This practice is valid because most building work can be carried out under similar circumstances regardless of site location; a comparatively low proportion of contract value is subject to the risks that attend upon work below ground level; and once the building envelope is complete most trades can proceed without serious disruption from the weather.

This is not, however, the general method of pricing Civil Engineering work: the volume of work below ground, increased exposure to weather and the tremendous variety of projects, in terms of type, complexity and scale, makes the straightforward use of unit rates less reliable. So, whilst even in building work similar or identical measured items attract a fairly broad range of prices, the range is much greater in Civil Engineering bills. This uncertainty is compounded by the lower number of bill items generated by the Civil Engineering Method of Measurement, so that the precise nature of the work is less apparent from the bill descriptions and the statistical effect of 'swings and roundabouts' has less scope to average out extremes of pricing.

To prepare a price for a Civil Engineering project, then, it is necessary to have regard to the method to be adopted in executing the work, draw up a detailed programme and then cost out the resources necessary to prosecute the chosen method. Because the first part of this process is the province of the Contractor's planner, there has been a tendency to postpone detailed estimating until the tendering stage itself, with the employer relying up to that point upon an estimate prepared on a 'broad brush' basis.

The result has been a growing pressure on the part of project sponsors for an improvement in budgetary advice, so that a decision to commit expenditure to a particular project is taken on firmer grounds. The absence of a detailed pricing method during the pre-contract phase also inhibits the accurate costing of alternative designs and regular cost checking to ensure that the design is being developed within the employer's budget.

This book therefore seeks to draw together the information appropriate to two methods of pricing: the cost of resources for use where an operational plan has been outlined, and unit rates for use where quantities can be taken from available drawings.

To take some note of the range of unit rates that might apply to an item, the rates themselves are in some cases related to working method - for example by identifying the different types of plant that would suit varying circumstances. Nonetheless, it would be folly to propose that all types of Civil Engineering work could be covered by a price book such as this. The Editors have therefore had in mind the type and scale of work commissioned by a local authority, a public corporation or a large private company.

This does embrace the great majority of work undertaken by the industry each year. Although almost all projects will have individual features that require careful attention in pricing, and there will be some projects that are so specialist that they will not conform to standard pricing information at all. But for most projects, within the range of work covered, this book should provide a firm foundation of cost information upon which a job-specific estimate can be built.

The contents of the book are therefore set out in a form that permits the user to follow the estimating process through in a structured way, as follows:

**PURPOSE AND CONTENT OF THE BOOK** - *continued*

**Part 1: General**

The balance of this section describes in narrative form the work stages normally followed in a Contractor's office from receipt of the tender documents through to the submission of the tender.

**Part 2: Preliminaries and General Items**

Containing a checklist of items to be priced with Preliminaries and General Items (or 'Method Related Charges') and a worked example containing specific cost information.

**Part 3: Resources**

This deals with the basic cost of resources, so that a resource-based system of estimating can be adopted where it is possible to develop an outline programme and method statement. Reference to this section will also assist the user to make adjustments to unit rates where different labour or material costs are thought to apply and to calculate analogous rates for work based on the hypothetical examples given. It is stressed that all of the costs given in this section are exclusive of the items costed with the preliminaries and of financing charges, head office overheads and profit.

**Part 4 and 5: Unit Costs**

This section is structured around the two main methods of measurement for Civil Engineering Work and gives 'trade by trade' unit rates for those circumstances where the application of unit rates to measured quantities is possible and practical. Again, it is stressed that the rates are exclusive of the items costed with preliminaries and of financing charges, head office overheads and profit.

**Part 6: Oncosts and Profit**

Having produced an estimate for the predicted cost of the work, being the sum of the preliminaries and the measured work, the estimate must be converted to a tender by the application of any adjustment made by management (which follows the Management Appraisal described later in this Part of the book) and by additions for financing charges, head office overheads and profit. These additions are discussed in Part 5, which includes a worked example of a tender summary.

**Part 7: Costs and Tender Prices Indices**

The cost and tender price indices included in this part of the book provide a basis for updating historical cost or price information, by presenting changes in the indices since 1975.

**PURPOSE AND CONTENT OF THE BOOK**

**Part 8: Daywork**

Including details of the FCEC daywork schedule and advice on costing excluded items.

**Part 9: Professional Fees**

These contain reproductions of standard documentation relating to professional fees for Consulting Engineers and Quantity Surveyors.

**Part 10: Approximate Estimates**

The prices in this section have been assembled from a number of sources, including the relevant items in the unit costs section and recovered data from recent projects. They are intended to give broad price guides or to assist in comparison exercises.

**Part 11: Outputs**

Scheduled here are various types of operations and the outputs expected of them. Also listed are man hours for various trades found in Civil Engineering.

**Part 12: Tables and Memoranda**

These include conversion tables, formulae, and a series of reference tables structured around a trade heading. Also included here is a schedule of names and addresses of Trade associations, professional institutions, Government departments who have assisted with technical information.

**Part 13: Updates**

This section is a pro-forma update which allows the user of the book to insert the adjustments to the published prices as and when they are advised through the free quarterly updating bulletin.

## OUTLINE OF THE ESTIMATING AND TENDERING PROCESS

This section of the book outlines the nature and purpose of Civil Engineering estimating and provides background information for users. It comprises an outline of the estimating and tendering process with supporting notes and commentaries on particular aspects. Some worked examples on tender preparation referred to in this part are included at the end of Part 5.

It must be emphasised that the main purpose of this book is to aid the estimating process, thus it is concerned more with the predicted cost of Civil Engineering work than with the prices in a bill of quantities. To ensure the correct interpretation of the information provided it is important to distinguish clearly between estimating and tendering; the following definitions are followed throughout:

The *Estimate* is the prediction of the cost of a project to the Contractor.
The *Tender* is the price submitted by the Contractor to the Employer.

The tender is based on the estimate with adjustments being made after review by management; these include allowances for risk, overheads, profit and finance charges. As discussed later in this section, prices inserted against individual items in a bill of quantities may not necessarily reflect the true cost of the work so described due to the view taken by the Contractor on the risks and financial aspects involved in executing the work.

The following list summarises the activities involved in the preparation of a tender for a typical Civil Engineering project. The diagram that follows illustrates the relationships between the activities and, after the diagram, notes on factors affecting each stage of the process are given.

1. An overall appraisal of the project is made.

2. Material requirements are abstracted from the tender documents and prices are obtained from suppliers. Details of work normally sub-contracted are abstracted, together with relevant extracts from the tender documents, and prices are obtained from sub-contractors.

3. The site of the works and the surrounding area is visited and studied. Local information is obtained on factors affecting the execution of the contract.

4. A programme and a detailed method of working for the execution of the contract is prepared. This will include details of plant requirements, temporary works, unmeasured work, appropriate supervisory staff, etc.

5. Designs are made for temporary works and other features left to be designed by the Contractor, and quantities are taken off for costing.

6. Major quantities given in the tender documents are checked.

7. The cost estimate for the project is prepared by costing out all of the resources identified by Stages 2 to 6. An appraisal is made on the conditions of contract, financial requirements, etc., and an assessment of risk/opportunity is prepared.

8. The tender documents are priced.

9. Management reviews the estimate, evaluates the risks and makes allowances for overheads, profit and finance.

10. The tender is completed and submitted.

**OUTLINE OF THE ESTIMATING AND TENDERING PROCESS**

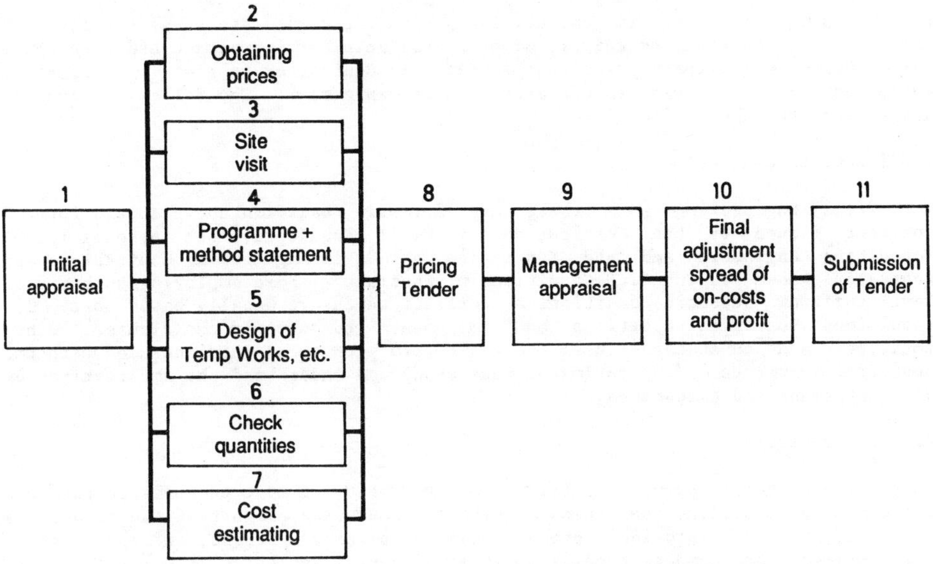


1. **INITIAL APPRAISAL**

The purpose of the initial overall appraisal is to highlight any high value areas or any particular problems which could require specialist attention, it can also identify possible alternative methods of construction or temporary works.

Tender lists are usually very selective and Contractors will have earned their place on them, therefore tenders are rarely declined or returned, a client who has an enquiry declined may not invite that Contractor to price for work again.

Other points to be considered:

the location and type of project and its suitability to the tenderer's particular expertise

the size of project, its financing requirement and the proportion of annual turnover it would represent

the availability of resources

the size of the tender list and the nature of the competition

the identity of the employer and his professional consultants

the adequacy of the tender documents

an initial appraisal of risk and opportunity (see Section 8)

## OUTLINE OF THE ESTIMATING AND TENDERING PROCESS

At the same time, it is essential for the estimator to study the contract documents issued with the enquiry or made available for inspection and note those parts which will affect pricing or involve the Contractor in risks which must be evaluated later and drawn to the attention of management. The following comments are indicative only:

### Conditions of Contract

For Civil Engineering work these are normally based on the standard form of contract issued by the Institution of Civil Engineers, the Association of Consulting Engineers and the Federation of Civil Engineering Contractors and generally known as the 'ICE Conditions of Contract'. Some employing authorities have their own standard conditions of contract which are usually based on the ICE conditions but with additions and amendments to some of the clauses. Other employers and consultants use the basic form with amendments included with the contract documents. The estimator must study and understand the implications of any variations and amendments.

### Bill of Quantities

This serves three purposes: first and foremost it must be prepared with the objective of providing the estimator with as accurate a picture of the project as possible, so as to provide a proper basis for pricing. Second, it should enable the employer to compare tenders on an equal basis, and third it will be used to evaluate the work executed for payment purposes. Individual items in the Bill do not necessarily describe in detail the work to be carried out under a particular item; reference must be made to the specification and the drawings to ascertain the full scope of the work involved.

The method of preparing the Bill may be based on the 'Civil Engineering Standard Method of Measurement' issued by the Institution of Civil Engineers or the 'Method of Measurement for Highway Works' issued by the Department of Transport, but some employing authorities have evolved their own methods and it is important for the estimator to study the Bill and its preambles to ensure that his rates and prices are comprehensive.

### Specification

This gives a detailed description of the workmanship, finish and materials to be used in the construction of the work. It may also give completion periods for sections and/or the whole of the work together with details of the requirements for the employer and/or the Consulting Engineer in connection with their supervision on site.

Water Authority and Highway Works in particular are based around a standard specification.

### Drawings

These give details of the work to be carried out and must be read in conjunction with the specification. It is important for the estimator to study the notes and descriptions given on the drawings as these amplify the specification.

Should the estimator discover any conflict between the various documents, it is important to have such discrepancies clarified by the Employer or Engineer prior to submission of any offer.

## OUTLINE OF THE ESTIMATING AND TENDERING PROCESS

### 2. OBTAINING PRICES

### (a) Materials

When pricing materials, the following points must be noted:

checks must be made to ensure that the quality of materials to be supplied meets with the requirements of the specification. If necessary, samples should be obtained and tested for compliance

checks must be made to ensure that the rates of delivery fabrication periods can meet the demands of the programme. It is sometimes necessary to use more than one supplier, with differing prices, to ensure a sufficient flow of materials

tests should be carried out to ascertain allowances to be made for operations such as compaction of soils and aggregates. Records of past contracts using similar materials can give this information, providing such records are accurate and reliable

### (b) Sub-contractors

It is common practice among Civil Engineering Contractors to sub-contract a significant proportion of their work. This is done for two main reasons:

to avoid the necessity of holding large amounts of special plant (earthmoving equipment for example) or specialised labour which, without continuity of work, generally have low utilisation and therefore provide a low return on capital employed

to spread the risks inherent in work of this nature

Sub-contracted work can form up to 65-70% of the value of measured work.

When utilising sub-contractors' prices it is extremely important to ensure that the rates given cover the full extent of the work described in the main contract, and that the sub-contractors quotation allows for meeting the main Tender programme and methods of working.

### 3. SITE VISIT

Factors to check during the site visit include:

access
limitations of working space
existing overhead and underground services
nearby public transport routes
availability of services - water, gas, electricity, telephone, etc.
availability of labour and sub-contractors
availability of materials - particularly aggregates and fill materials
location of nearest tipping facility
nature of the ground, including inspection of trial bores/ pits if dug
ground water level
presence of other Contractors working on or adjacent to the site.

## OUTLINE OF ESTIMATING AND TENDERING PROCESS

### 4. METHOD STATEMENT AND PROGRAMME

Whilst an estimate is being prepared it is preferable that a detailed method of working and a programme for the execution of the works is drawn up; the latter can take the form of a bar chart or, for large and more complex projects, may be prepared by more sophisticated means such as a precedence or critical path network. Compliance with the employer's target completion dates is, of course, essential. The method of working will depend on this programme in so far as the type and size of plant and the gang sizes to be used must be capable of achieving the output necessary to meet the programmed times. Allowance must be made for delays due to adverse weather and other hazards particular to the site.

A method statement is prepared in conjunction with the programme, setting out the resources required, outputs to be achieved and the requirements in respect of temporary works, etc.

At the same time, separate bar charts may be produced giving:

plant requirements

staff and site supervision requirements

labour requirements.

Where practical and economic, the method of working will be based on the use of Contractor-owned plant, but Contractors will also generally use hired plant especially where their overall utilisation of an item of plant is less than about 70% of its available working time.

These programmes and method statements will form the basis of the actual contract programme should the tender be accepted. They will also enable the Contractor to assure himself that he has available or can gain access to the necessary resources in plant, labour, materials and supervision to carry out the work should he be awarded the contract.

### 5. DESIGN OF TEMPORARY WORKS, ETC

Normally the period of time allowed for tendering is relatively short and therefore it is important that those aspects requiring design work are recognised as early as possible in the tender period. Design can be carried out either by the Contractor using his own engineers or by utilising the services of a Consulting Engineer. There are three aspects of design to be considered:

**Temporary works to the Contractor's requirements to enable the works to be constructed**

Design of temporary works covers the design of those parts of the work for which the Contractor accepts full responsibility in respect of stability and safety during construction. Such parts include support structures, cofferdams, temporary bridges, jetties and river diversions, special shuttering, scaffolding, etc. Design must be in sufficient detail to enable materials, quantities and work content to be assessed and priced by the estimator. In designing such work, it is important that adequate attention is given to working platforms and access for labour and plant and also to ease of dismantling and re-erection for further uses without damage.

It should be noted that many specialist sub-contractors will provide a design service when submitting quotations. For example, scaffolding Contractors will design suitable support work for soffit shuttering, etc.

## OUTLINE OF ESTIMATING AND TENDERING PROCESS

### 5. DESIGN OF TEMPORARY WORKS, ETC (continued)

**Specific items of the permanent works to meet a performance specification set out by the client or the Consulting Engineer**

It is common practice for certain parts of the work to be specified by means of a performance specification. For example, concrete is specified by strength only, piles by load carrying capacity, etc. It is then left to the Contractor to use those materials, workmanship and design which he feels are most suited to the particular site and conditions.

In many cases such design will be carried out by specialist suppliers or sub-contractors.

**Alternative designs for sections of the permanent works where the Contractor's experience leads him to consider that a more economical design could be used.**

It is possible for a Contractor to use his expertise and experience to design and submit alternative proposals and prices for complete sections of the permanent work without, of course, altering the basic requirements of the original design. Examples would be foundations in difficult ground conditions, bridge superstructures, use of precast in place of in situ concrete, etc. Such designs may be carried out by the Contractor's own staff or in conjunction with Consulting Engineers or specialist Contractors.

Obviously this is only done when the Contractor can offer considerable savings in cost and/or a reduced construction period. It is necessary to include a price for the original design in the tender, but the decision to submit a keen tender may be underpinned by the hope of sharing such savings with the Employer.

In all cases, the Contractor's designs and calculations must be checked and approved by the Employer's Consulting Engineer, but such checks and approvals do not relieve the Contractor of his responsibilities for the safety and stability of the work.

Where any design work is undertaken it is important to ensure that adequate design insurance is maintained.

### 6. QUANTITIES CHECK

Working within obvious time constraints, the estimating team will endeavour to complete a quantities check on at least the major and high price quantities of the Bill, as this could affect his pricing strategy, for example, in pricing provisional items. If any major discrepancies are noted these should be referred to the management appraisal agenda.

### 7. COST ESTIMATE

At this stage the estimator draws together the information and prepares a cost estimate made up of:

preliminaries and general items (see Part 2)

temporary works

**OUTLINE OF ESTIMATING AND TENDERING PROCESS**

7. **COST ESTIMATE** (continued)

labour, costed by reference to the method statement, with appropriate allowances for labour oncosts (non-productive overtime, travelling time/fares, subsistence, guaranteed time, bonus, Employer's Liability Insurance, training, statutory payments, etc.) (see page 67)

material costs taken from price lists or suppliers' quotations, with appropriate allowances for waste

plant, whether owned or hired, with appropriate allowances for transport to/from site, erection/dismantling, maintenance, fuel and operation. Heavy static plant (batching plant, tower cranes, etc.) will normally be priced with general items; the remainder will normally be allocated to the unit rates

sublet work, as quoted by specialist sub-contractors, with appropriate allowances for all attendances required to be provided by the main Contractor

At the same time, a preliminary assessment of risk/opportunity will be made for consideration with the Management Appraisal (see Section 8 below). This will include a look at:

weather conditions - costs not recoverable by claims under the Conditions of Contract, ground becoming unsuitable for working due to the effect of weather, etc.

flooding - liability of site to flooding and the consequent costs

suitability of materials - particular risk can arise if prices are based on the use of borrow pits or quarries and inadequate investigation has been carried out or development is refused by the local authorities

reliability of sub-contractors - failure of a sub-contractor to perform can result in higher costs through delays to other operations and employing an alternative sub-contractor at higher cost

non-recoverable costs - such as excesses on insurance claims

estimator's ability - eg. outputs allowed. This can only be gauged from experience

cost increase allowances for fixed price contract

terms and conditions contained in the contract documents

ability to meet specification requirements for the prices allowed

availability of adequate and suitable labour

Risk is, of course, balanced by opportunity and consideration needs to be given to areas for which particular expertise possessed by the Contractor will lead to a price advantage against other tenderers.

## OUTLINE OF ESTIMATING AND TENDERING PROCESS

### 8. PRICING THE TENDER

Once the cost estimate is complete, the estimator prices the items in the Bill. The rate to be entered against the items at this stage should be the correct rate for doing the job, whatever the quantity shown.

Where the overall operation covers a number of differing Bill items, the estimator will allocate the cost to the various items in reasonable proportions; the majority of the work is priced in this manner. The remaining items are normally priced by unit rate calculation.

Resources schedules, based on the programme and giving details of plant, labour and staff, perform an important role in enabling the estimator to check that he has included the total cost of resources for the period they are required on site. It is not unusual for an item of plant to be used intermittently for more than one operation. A check on the total time for which the cost has been included in the estimate against the total time the item is needed on site as shown on the programme, gives a period of non-productive time. The cost of this is normally included in site oncosts and preliminaries. A similar situation can arise in the cases of skilled labour, craftsmen and plant drivers.

Having priced the Bill at cost, there will remain a sum to be spread over the Bill items. The way in which this is done depends on the view taken by the Contractor of the project; for example:

sums can be put against the listed general items in the Preliminaries Bill

for work measured in accordance with the Civil Engineering Standard Method of Measurement 2nd Edition (CESMM2) a fixed 'Adjustment Item' will be included in the Bill for the convenience of the estimator. This can be used for the adjustment made following the Management Appraisal, and for taking advantage of any late but favourable quotations received from sub-contractors or suppliers

the balance can be spread over the rates by equal percentage on all items, or by unequal percentages to assist in financing the contract or to take advantage of possible contract variations, or expected quantity changes (see notes against 7)

The Contractor will normally assess the financial advantages to be gained from submitting his bid in this manner and possibly enabling him to submit a more competitive offer.

After completing the pricing of all aspects of the tender, the total costs are summarised and profit, risk, etc., added to arrive at the total value of the Tender. A suggested form of this summary is set out in Part 5: Oncosts and Profit.

Finally, reasonable forecasts of cash flow and finance requirements are essential for the successful result of the project. Preliminary assessments may have been made for the information of management, but contract cash flow and the amount of investment required to finance the work can now be estimated by plotting income against expenditure using the programme of work and the priced Bill of Quantities. Payment in arrears and retentions, both from the Employer and to the suppliers and sub-contractors, must be taken into account.

**OUTLINE OF ESTIMATING AND TENDERING PROCESS**

8. **PRICING THE TENDER** (continued)

It is unlikely that sufficient time will be available during the tender period to produce such information accurately, but an approximate figure, for use as a guide for finance requirements, can be assessed. The assumption is made that all work which is sub-contracted is self financing - i.e. that the sub-contractor is paid after the main Contractor receives payment and that the same contract conditions and terms of payment apply. A worked example is set out in Part 5: Oncosts and Profit.

9. **MANAGEMENT APPRAISAL**

Clearly, as far as the detail of the tender build-up is concerned, management must rely upon its established tendering procedures and upon the experience and skill of its estimators. The Management Appraisal will therefore concentrate upon matters of policy and factors having a significant impact upon cost or price, including a review of:

the major quantities

the programme and method statement

plant usage

major suppliers and/or sub-contractors, and discounts

the nature of the competition

risk and opportunity

contract conditions, including in particular the level of damages for late completion, the minimum amount of certificates, retention and bonding requirements

cash flow and finance

margin for head office overheads and profit

the weighting and spreading of the cost estimate over the measured items in the Bill

10. **SUBMISSION OF TENDER**

On completion of the tender, the documents are read over and then despatched to the employer in accordance with the conditions set down in the invitation letter.

A complete copy of the tender as submitted should be retained by the Contractor. Drawings on which the offer has been based should be clearly marked 'Tender Copy', their numbers recorded and the drawings filed for future reference. These documents will then form the basis for price variations should the design be amended during the currency of the contract.

**OUTLINE OF ESTIMATING AND TENDERING PROCESS**

**10. SUBMISSION OF TENDER** (continued)

The Contractor may wish to qualify his offer to clarify the basis of his price. Normally such qualifications are included in a letter accompanying the tender. Legally the Form of Tender constitutes the offer and it is important that reference to such a letter is made on the Form of Tender to ensure that it forms part of the offer. Wording such as 'and our letter dated..., reference...' should be added prior to quoting the Tender Sum.

Before any qualifications are quoted, careful note must be taken of the 'instructions to tenderers' as qualifications; or at least qualifications submitted without a conforming tender, may be forbidden.

PART 2

# Preliminaries and General Items

*This part deals with that portion of Civil Engineering costs not, or only indirectly, related to the actual quantity of work being carried out. It comprises a definition of Method Related Charges a checklist of items to be accounted for on a typical Civil Engineering contract; and a worked example illustrating how the various items on the checklist can be dealt with.*

**METHOD RELATED CHARGES**

The concept of Method Related charges was first defined in *Measurement in Contract Control* by Martin Barnes, published by The Institution of Civil Engineers, London, as a guide to the 1976 Civil Engineering Standard Method of Measurement, and is now embodied in the updated version of that book by the same author entitled *The CESMM2 Handbook* and can be summarised as follows :-

> In commissioning Civil Engineering work the Employer buys the materials left behind, but only hires from the Contractor the men and machines which manipulate them, and the management skills to manipulate them effectively. It is logical to assess their values in the same terms as the origin of their costs. It is illogical not to do so if the Employer is to retain the right at any time to vary what is left behind and if the financial uncertainties affecting Employer and Contractors are to be minimised.
>
> Tenderers have the option to define a group of bill items and insert charges against them to cover those expected costs which are not proportional to the quantities of Permanent Works. To distinguish these items they are called Method Related Charges. They are themselves divided into charges for recurrent or time related cost elements, such as maintaining site facilities or operating major plant, and charges for elements which are neither recurrent nor directly related to quantities, such as setting up, bringing plant to site and Temporary Works.'

Although the more familiar terminology of 'Preliminaries and General Items' is used in this book the principle of Method Related Charges - separating non quantity related charges from quantity related charges - is adopted. Generally the former are dealt within this part, while the latter are dealt with in Part 4: Unit Costs. In this part non quantity related charges are further subdivided into those that are *time related* and those that are *non time related.*

Another hope expressed with the introduction of Method Related Charges was that they should accurately reflect the work described in the item and that they should not, as had become the practice with some of the vague general items frequently included in Civil Engineering Bills, be used as a home for lump sum tender adjustments quite unrelated to the item. Where cost information is given in the worked example presented at the end of this part of the book, therefore, it must be stressed that only direct and relevant costs are quoted.

Where no detailed information is available, it is suggested that when preparing a preliminary estimate, an addition of between 7½%-15% of nett contract value is made to cover Contractor's Site Oncosts both time and non time related.

CHECKLIST OF ITEMS

The following checklist is representative but not exhaustive. It lists and describes the major preliminary and general items which are included, implicitly or explicitly, in a typical Civil Engineering contract and, where appropriate, gives an indication of how they might be costed. Generally contract documents give detailed requirements for the facilities and equipment to be provided for the Employer and for the Engineer's representative and Bills of Quantities provide items against which these may be priced; no such items are provided for Contractor's site oncosts or, usually, for temporary works and general purpose plant. For completeness a checklist of both types of item is given here under the following main headings :

CONTRACTOR'S SITE ONCOSTS
- time related
- non time related

EMPLOYER'S AND CONSULTANTS' SITE REQUIREMENTS
- time related
- non time related

OTHER SERVICES, CHARGES AND FEES

TEMPORARY WORKS OTHER THAN THOSE INCLUDED IN UNIT COSTS

GENERAL-PURPOSE PLANT OTHER THAN THAT INCLUDED IN UNIT COSTS

**CONTRACTOR'S SITE ONCOSTS - TIME RELATED**

Site staff salaries
all non-productive supervisory staff on site including; agent, sub-agent, engineers, general foremen, non-productive section foremen, clerks, typists, timekeepers, checkers, quantity surveyors, cost engineers, etc. Cost includes salaries, subsistence allowance, National Health Insurance and Pension Scheme contributions, etc. Average cost approximately 3% to 5% of contract value.

Site staff expenses
travelling, hotel and other incidental expenses incurred by staff. Average cost approximately 1% of staff salaries

Attendant labour
chainmen, storemen, drivers for staff vehicles, watchmen, cleaners, etc.

General yard labour
labour employed on loading and offloading stores, general site cleaning, removal of rubbish, etc.

Plant maintenance
fitters, electricians, and assistants engaged on general plant maintenance on site. This excludes drivers and banksmen

Site transport for staff and general use
vehicles provided for use of staff and others including running costs, licence and insurance and maintenance if not carried out by site fitters

CHECKLIST OF ITEMS

**CONTRACTOR'S SITE ONCOSTS - TIME RELATED** - *continued*

Transport for labour to and from site
buses or coaches provided for transporting employees to and from site including cost of drivers and running costs, etc., or charges by coach hire company for providing this service

Contractor's office rental
rental charges for provision of offices for Contractor's staff
main office
section offices
timekeepers and checkers
laboratory, etc.
an allowance of approximately 8 $m^2$ per staff member should be made

Contractor's site huts
rental charges for stores and other general-use site huts

Canteen and welfare huts
rental charges for canteen and huts for other welfare facilities required under Rule XVI of the Working Rule Agreement

Rates
chargeable by local authorities on any site, temporary buildings or quarry

General office expenditure
provision of postage, stationery and other consumables for general office use

Telephone and calls
rental charges and charges for calls

Telex and calls
rental charges and charges for calls

Furniture and equipment
rental charges for office furniture and equipment including photocopying, calculators, typewriters, etc.

Surveying equipment
rental

Canteen and welfare equipment
rental charges for canteen and other welfare equipment

Radio communication equipment
rental

Testing and laboratory equipment
rental

Lighting and heating for offices and huts
electricity, gas or other charges in connection with lighting and heating site offices and hutting

Site lighting electrical consumption
electricity charges in connection with general external site lighting

CHECKLIST OF ITEMS

**CONTRACTOR'S SITE ONCOSTS - TIME RELATED** - *continued*

Water consumption
water rates and charges

Canteen operation
labour, consumables and subsidy costs in operating site canteens

Carpenter's shop equipment
rental of building and mechanical equipment

Fitter's shop equipment
rental of building and equipment

Small tools
provision of small tools and equipment for general use on site. Average cost 5% of total labour cost

Protective clothing
provision of protective clothing for labour including boots, safety helmets, etc. Average cost of total labour cost

Traffic control
hire and operation of traffic lights

Road lighting
hire and operation of road lighting and traffic warning lights

Cleaning vehicles
equipment and labour cleaning vehicles before entering public roads

Cleaning roads and footpaths
equipment and labour cleaning public roads and footpaths

Progress photographs for Contractor's records

Rent of additional land
for Contractor's use for erection of huts, storage of soil and other materials, etc.

**CONTRACTOR'S SITE ONCOSTS - NON TIME RELATED**

Staff removal expenses
costs of staff moving house to new location. Generally only applies on longer-term contracts.

Erection of offices including drainage, paths, etc.
construction of foundations, drainage, footpaths and parking areas, erection of huts, installation of electric wiring, in situ fittings and decorating, etc.

Dismantle offices and restore site on completion

Erection of site huts
as previous item but for general site huts

Dismantle site huts

CHECKLIST OF ITEMS

**CONTRACTOR'S SITE ONCOSTS - NON TIME RELATED** - *continued*

Erection of canteen and welfare huts
as previous item but for canteen and welfare huts

Dismantle canteen and welfare huts

Caravan site construction and clearance
construction of site for employees' caravans including provision for water, electricity and drainage, and clear away and restoration of site on completion, allow credit for any charges to be levied

Telephone installation
charges for initial installation and removal

Furniture and equipment
purchase costs of furniture and equipment, allow for residual sale value

Survey equipment
purchase costs of survey equipment including pegs, profiles, paint, etc., for setting out

Canteen and welfare equipment
purchase costs of equipment

Testing and laboratory equipment
purchase cost of equipment

Radio communication
installation costs

Electrical connection and installation
initial charges for connections to mains supply

Electrical connection site plant

Electrical connection site lighting
installation costs and final dismantling

Water supply
installation on site and connection charges

Haulage plant
cost of transport of plant and equipment to and from site

Progress photographs

Depot loading and unloading charges

Carpenter's shop
erection of building, installation of equipment including electrical installation, etc.

Carpenter's shop
dismantle and clear away on completion

Fitter's shop
erection of building, installation of equipment including electrical installation, etc.

CHECKLIST OF ITEMS

**CONTRACTOR'S SITE ONCOSTS - NON TIME RELATED** - *continued*

Fitter's shop
    dismantle and clear away on completion

Stores compound
    erect and dismantle stores compound

Notice boards and signs
    supply, erect and remove Contractor's signboards, traffic control signs, etc.

Insurances
    premiums for all Contractor's insurance obligations (see separate section on insurances and bond below)

Bond
    charges for provision of bond (see separate section on insurances and bond below)

Plant erection
    cost of erection of Contractor's plant on site including foundations, hardstandings, drainage, etc.

Plant dismantling

Clear site on completion including removal of rubbish and reinstatement

**EMPLOYER'S AND CONSULTANTS' SITE REQUIREMENTS - TIME RELATED**

Office and other huts
    rental of office accommodation, sub-offices, laboratory, etc.

Office and site attendant labour
    office cleaning, chainmen, laboratory assistants, etc.

Site transport
    rental for vehicles for use of client and engineer

Telephone and calls
    rental and cost of calls (if to be borne by Contractor)

Furniture and equipment
    rental of office furniture and equipment

Survey equipment
    rental of surveying equipment

Testing and laboratory equipment
    rental of testing and laboratory equipment

Radio communication equipment
    rental and maintenance

Office lighting and heating
    cost of heating and lighting all offices and huts

CHECKLIST OF ITEMS

**EMPLOYER'S AND CONSULTANTS' SITE REQUIREMENTS - TIME RELATED** - *continued*

Office consumables
cost of office consumables to be provided by the Contractor

**EMPLOYER'S AND CONSULTANTS' SITE REQUIREMENTS - NON TIME RELATED**

Erection of huts and offices
client's and engineer's offices and other huts including foundations, pathways, parking area, electrical installation and drainage, etc.

Dismantling huts and offices
restoration of site on completion

Telephone installation charges

Furniture and equipment purchase
purchase cost for furniture and equipment

Survey equipment purchase

Testing and laboratory equipment purchase

Radio communication equipment installation

Progress photographs
cost of professional photographer and supplying prints as required

**OTHER SERVICES, CHARGES AND FEES**

Design fees for alternative designs for permanent works
design and drawing office costs and charges for preparing alternative designs and specifications and bill of quantities for alternative designs for permanent works

Design and design office charges for temporary works
design and drawing office costs and charges for preparing designs and drawings for temporary works

Preparation of bending schedules
drawing office charges for preparation of bending schedules

Fees to local authorities

Legal advice and fees
fees and charges from legal adviser

**TEMPORARY WORKS OTHER THAN THOSE INCLUDED IN UNIT COSTS**

Fencing

Traffic diversions

Lighting

Traffic signs

CHECKLIST OF ITEMS

**TEMPORARY WORKS OTHER THAN THOSE INCLUDED IN UNIT COSTS** - *continued*

Traffic control

Footpath diversions

Stream or river diversion

Cofferdam installation

Cofferdam removal

Support works

Jetties

Bridges

De-watering

General pumping
including construction of collecting sumps, etc.

Site access roads and maintenance

Scaffolding

**GENERAL PURPOSE PLANT OTHER THAN THAT INCLUDED IN UNIT COSTS**

Lorries and dumpers for general transport around site

Tractors and trailers

Craneage for general use

Compressed air plant

Pumps

Bowsers for fuelling plant

Bowsers for water supply

Non productive time for plant on site
obtained by comparing plant requirements as on programme with the plant time included in the build up of bill rates

*Note*: For all items of plant listed above the cost of drivers and other attendants must be allowed for together with consumables and other operating costs

**INSURANCES AND BOND**

Contractors are legally required to insure against liability which may be incurred when employees are injured at work and when individuals are injured by owners' vehicles. There is also a statutory requirement for certain types of machinery to be inspected at regular intervals. In addition to legal requirements, companies insure against possible loss due to fire, explosion, fraud, liability incurred as a result of damage to the property of others and through serious injury to individuals.

CHECKLIST OF ITEMS

**INSURANCES AND BOND** - *continued*

Certain risks are excluded from insurers' policies; these include war, revolution, etc, contamination by radioactivity and risks which arise from bad management.

Generally insurance companies take into account the claims record of a Contractor when assessing premiums payable on a particular contract or policy. Premiums are related to the risks involved and, on large Civil Engineering contracts, the insurers will require full details of the work, the methods of construction plant used and risks involved due to flood, ground conditions, etc. Insurance companies or brokers should be consulted before submitting a tender for major Civil Engineering work.

The following gives an outline of the items to be allowed for in a tender.

**Employer's liability insurance**

This provides indemnity to the Employer against legal liability for death of or injury to employees sustained in the course of their employment. The cost is normally allowed for in the build-up of the 'all-in' labour rate as a percentage addition to the gross cost. This will vary, depending on the Contractor's record. An allowance of 2% has been made in this book.

**Vehicle insurance**

This can cover individual vehicles or fleets. The cost is normally covered in the rate charged to the contract for the use of the vehicles.

**All risks insurance**

This provides for loss of or damage to permanent and/or temporary works being executed on the contract. It also covers plant, materials, etc. The cost is allowed for in the tender as a percentage of the total contract value. This will vary depending on the Contractor's record and type of contract undertaken and also on the value of excesses included in the policy.

**Public liability insurance**

This provides indemnity against legal liability which arises out of business activities resulting in bodily injury to any person (other than employees), loss of or damage to property (not owned or under the control of the company), obstruction, trespass or the like.

Such insurance can be extended to include labour-only sub-contractors and self-employed persons if required. The cost is generally included with head office overheads.

**Professional indemnity insurance**

This provides against liability arising out of claims made against the conduct and execution of the business. This covers such items as design liability, etc. The cost may be with head office overheads where such insurance is considered desirable.

CHECKLIST OF ITEMS

**INSURANCES AND BOND** - *continued*

**Loss of money insurance**

This covers loss of money and other negotiable items and loss or damage to safes and strong rooms as a result of theft. It is necessary to cover cash in transit for wages, etc. The cost is included with head office overheads.

**Fidelity guarantee insurance**

This covers loss by reason of any act of fraud or dishonesty committed by employees. The cost is included with head office overheads.

**Other insurances**

Other insurances which may be carried by a Contractor include fire insurance on his permanent premises and contents, consequential loss insurance in relation to his permanent premises, personal accident insurance on a 24 hour per day basis for employees.

**Contract bond**

Where the contract calls for a bond to be provided, this is normally given by either banks or insurance companies. The total value of these guarantees available to any company is limited, depending on the goodwill and assets of the company. It will also affect the borrowing facilities available to the company and therefore, to some extent, can restrict his trading. An allowance should be made at a rate of 1½% per annum on the amount of the bond for the construction period plus a rate of ½% per annum for the maintenance period.

WORKED EXAMPLE

The example is of a contract for the construction of an Airport extension. The contract includes concrete surfaced aprons/runways, surface water drainage, construction of two concrete structures, minor accommodation works and two culverts. The Conditions of contract are ICE Standard conditions; the contract period is 12 months; and the approximate value is £5.8 million. It is assumed that the main Contractor will sub-let:

bulk earthworks and landscape, fencing, concrete surfacing, signs and lighting, waterproofing to structures, and that all materials are obtained off site (including concrete).

The worked example demonstrates a method of assessing preliminary costs and is based on the programme below together with an assessment of general purpose plant (a plant reconciliation is also given).

**CONTRACTOR'S SITE ONCOSTS - TIME RELATED**

| | | £ | £ |
|---|---|---|---|
| Site staff salaries (see programme) | | | |
| Agent | 54 wks at £480 | 25 920 | |
| Senior Engineer | 54 wks at £400 | 21 600 | |
| Engineers | 76 wks at £325 | 24 700 | |
| General foreman | 52 wks at £400 | 20 800 | |
| Office manager | 52 wks at £385 | 20 020 | |
| Timekeeper | 44 wks at £245 | 10 780 | |
| Storeman/checker | 52 wks at £270 | 14 040 | |
| Typist/telephonist | 52 wks at £155 | 8 060 | |
| Cost clerk | 52 wks at £240 | 12 480 | |
| Quantity Surveyor | 54 wks at £471 | 25 434 | |
| Fitter | 42 wks at £355 | 14 910 | 198 744 |
| Site staff expenses (1% of staff salaries) | | say | 1 980 |
| Attendant labour | | | |
| Chainman | 128 wks at £190 | 24 320 | |
| Driver | 52 wks at £220 | 11 440 | |
| Office cleaner (part-time) | 52 wks at £ 80 | 4 160 | 39 920 |
| General yard labour (part-time) (loading and offloading, clearing site rubbish, etc.) | | | |
| 1 ganger | 10 wks at £290 | 2 900 | |
| 4 labourers | 40 wks at £260 | 10 400 | 13 300 |
| Plant maintenance (Contractor's own plant) | | | |
| Fitter (included in staff above) | | | |
| Fitter's mate | 32 wks at £300 | 9 600 | 9 600 |
| Site transport - Staff | | | |
| QS/Agent's cars | 54 wks at £180 | 9 720 | |
| Engineers' cars | 130 wks at £ 80 | 10 400 | |
| Land Rover | 54 wks at £160 | 8 640 | 28 760 |
| Site transport - Labour | 48 wks at £150 | | 7 200 |
| Contractor's offices | | | |
| Mobile offices (11 staff x 8 $m^2$) $88m^2$ for | 52 wks at £125 | 6 500 | |
| Section offices (2nr at 10 $m^2$) 20 $m^2$ for | 52 wks at £ 30 | 1 560 | 8 060 |
| | Carried forward | | £307 564 |

WORKED EXAMPLE

**CONTRACTOR'S SITE ONCOSTS - TIME RELATED** - *continued*

Programme of Activities, Staffing and Plant

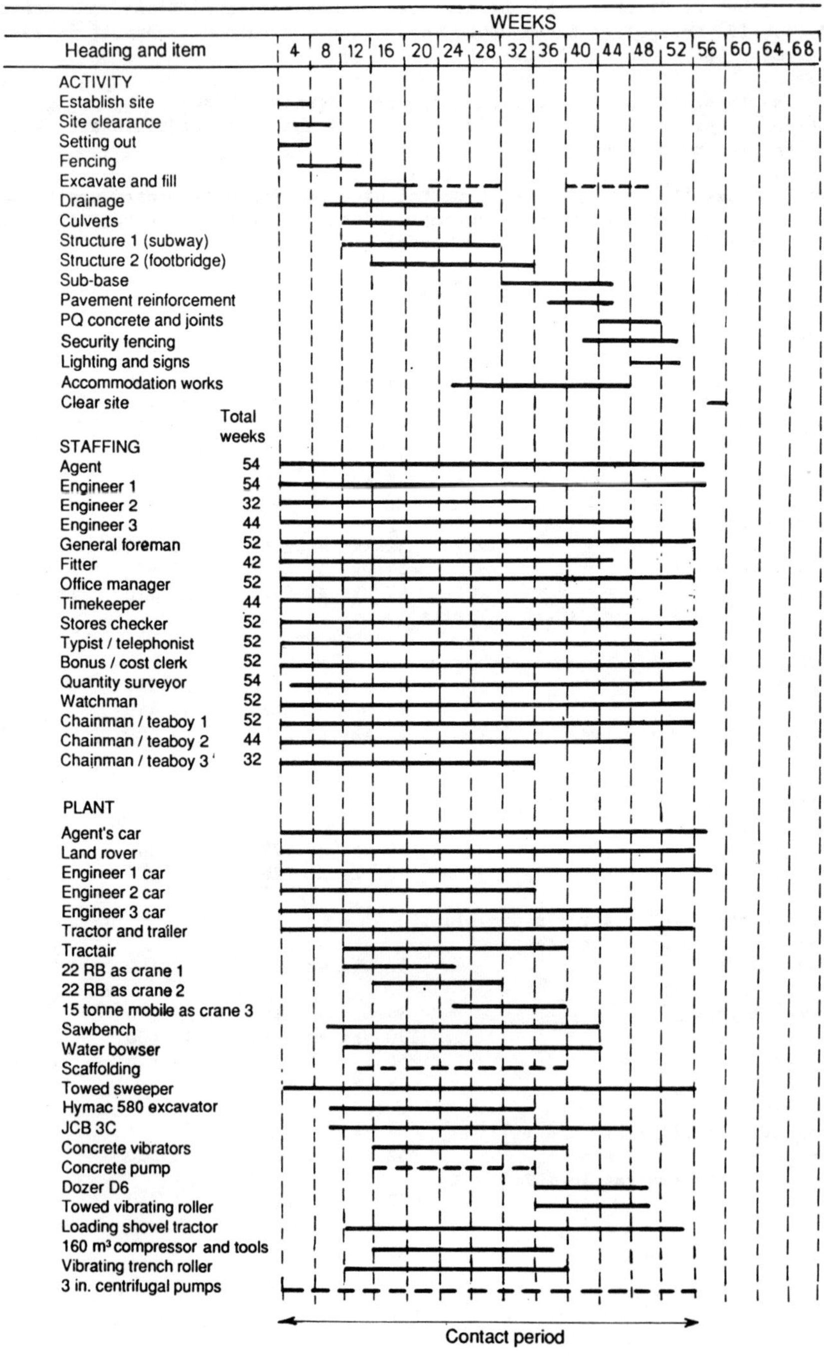

WORKED EXAMPLE

**CONTRACTOR'S SITE ONCOSTS - TIME RELATED** - *continued*

| | | | £ | £ |
|---|---|---|---|---|
| | Brought forward | | | 307 564 |
| Contractor's site huts | | | | |
| Stores hut 30 $m^2$ | 52 wks at £ 25 | | 1 300 | |
| Canteen 70 $m^2$ (assume 70 men) | 52 wks at £ 75 | | 3 900 | |
| Washroom 30 $m^2$ | 52 wks at £ 33 | | 1 716 | |
| Staff toilets | | say | 1 050 | |
| Site toilets | | say | 1 300 | 9 266 |
| Rates | | | | NIL |
| General office expenditure | 52 wks at £ 46 | | | 2 392 |
| Telephone calls and rental | 52 wks at £ 53 | | | 2 756 |
| Furniture rental | 52 wks at £ 35 | | | 1 820 |
| Canteen and welfare rental | 52 wks at £ 40 | | | 2 080 |
| Surveying equipment rental | 52 wks at £ 28 | | | 1 456 |
| Testing equipment rental | 52 wks at £ 17 | | | 884 |
| Lighting and heating offices 200 $m^2$ 52 wks at | £ 46 | | | 2 392 |
| Water supply | | | | |
| 2 250 000 litres at £1.52 per 5 000 litres | | | 684 | |
| Small tools | | | | |
| 1% on labour costs of say £1 000 000 | | | | 10 000 |
| Protective clothing | | | | |
| ½% on labour costs of say £1 000 000 | | | | 5 000 |
| Road lighting | | say | | 2 100 |
| Cleaning roads | | | | |
| Towed sweeper (tractors elsewhere) | 40 wks at £ 68 | | 2 720 | |
| Brushes | | say | 860 | |
| Labour | 40 wks at £180 | | 7 200 | 10 780 |
| Progress photographs | | say | | 630 |
| **Total Contractor's site oncosts - Time related** | | | | £359 804 |

**CONTRACTOR'S SITE ONCOSTS - NON TIME RELATED**

| | | | £ | £ |
|---|---|---|---|---|
| Erect and dismantle offices | | | | |
| Mobile | 108 $m^2$ at £ 6.30 | | 680 | |
| Site works | | say | 683 | |
| Toilets | | say | 168 | |
| Wiring, water, etc. | | say | 400 | 1 931 |
| Erect and dismantle other buildings | | | | |
| Stores and welfare | 130 $m^2$ at £11.55 | | 1 500 | |
| Site works | | say | 788 | |
| Toilets | | say | 567 | 2 855 |
| Telephone installation | | | | 200 |
| Furniture and office equipment | | | | |
| Purchase cost less residual value | | | | 1 470 |
| | | | Carried forward £ | 6 456 |

WORKED EXAMPLES

**CONTRACTOR'S SITE ONCOSTS - NON TIME RELATED** - *continued*

| | | £ | £ |
|---|---|---|---|
| | Brought forward | | 6 456 |
| Survey equipment and setting out | | | |
| Purchase cost including pegs, profiles, paint, ranging rods, etc. | | 780 | |
| Canteen and welfare equipment | | | |
| Purchase cost less residual value | | | 1 150 |
| Electrical installation | | | 1 780 |
| Water supply | | | |
| Connection charges | | 890 | |
| Site installation | | 1 000 | 1 890 |
| Transport of plant and equipment | | | 2 700 |
| Stores compound and huts | | | 400 |
| Sign boards and traffic signs | | | 580 |
| Insurances (dependent on Contractor's policy and record) | | | |
| Contractor's all risks | 1½% on £5.75 | 86 250 | |
| Allow for excesses | million | 12 500 | 98 750 |
| General site clearance | | | 2 625 |
| **Total Contractor's site oncosts - Non time related** | | | £117 111 |

**EMPLOYER'S AND CONSULTANTS' REQUIREMENTS ON SITE - TIME RELATED**
(details of requirements will be defined in the contract documents)

| | | £ |
|---|---|---|
| Attendant labour | 150 man wks at £205 | 30 750 |
| Offices 50 $m^2$ | 52 wks at £85 | 4 420 |
| Transport 2 Land Rovers | 52 wks at £315 | 16 380 |
| Telephone calls and rental | 52 wks at £26 | 1 352 |
| Furniture and equipment | 52 wks at £29 | 1 508 |
| Survey equipment | 52 wks at £29 | 1 508 |
| Heating and lighting 50 $m^2$ | 52 wks at £13 | 676 |
| Office consumables | 52 wks at £13 | 676 |
| Office cleaning | 52 wks at £40 | 2 080 |
| **Total employer's and consultants' requirements - time related £** | | 59 350 |

**EMPLOYER'S AND CONSULTANTS' REQUIREMENTS ON SITE - NON TIME RELATED**

(details of requirements will be as defined in the contract documents)

| | | £ | £ |
|---|---|---|---|
| Erection and dismantling of engineer's offices | | | |
| 50 $m^2$ at £6.30 | | 315 | |
| Site works, toilets, etc | say | 680 | 995 |
| Telephone installation | | | 200 |
| Electrical installation | | | 860 |
| Furniture and equipment | | | |
| Purchase cost less residual value | | | 575 |
| Progress photographs | | | |
| 100 sets at £31.50 | | | 3 150 |
| **Total employer's and consultants' requirements - Non time related** | | £ | 5 780 |

WORKED EXAMPLES

**TEMPORARY WORKS NOT INCLUDED IN UNIT COSTS**

| | | £ | £ |
|---|---|---|---|
| Temporary Fencing | | | |
| 1000m chestnut fencing | | | |
| Materials | 1000 m at £3.25 | 3 250 | |
| Labour | 500 man hours at £5.70 | 2 850 | 6 100 |
| Traffic diversions | | | |
| Structure No. 1 | | 5 040 | |
| Structure No. 2 | | 3 990 | 9 030 |
| Footpath diversion | | | 950 |
| Stream diversion | | | 2 625 |
| Site access roads | 400 m at £39.50 | | 15 800 |
| **Total temporary works** | | | £ 34 505 |

**GENERAL PURPOSE PLANT AND PLANT NOT INCLUDED IN UNIT COSTS**

| *Description* | | *Labour* £ | *Plant* £ | *Consumables* £ |
|---|---|---|---|---|
| Wheeled tractor | | | | |
| hire charge | 52 wks at £240 | | 12 480 | |
| driver | 50 hrs for 52 wks at £5.96 | 15 500 | | |
| consumables | 52 wks at £ 63.20 | | | 3 286 |
| Trailer | | | | |
| hire charge | 52 wks at £26 | | 1 352 | |
| Tractair | | | | |
| hire charge | 28 wks at £278 | | 7 784 | |
| driver | (use above driver)* | | | |
| consumables | 28 wks at £ 76 | | | 2 128 |
| 22 RB as crane | | | | |
| hire charge | 50 hrs for 40 wks at £15.50 | | 31 000 | |
| driver | 50 hrs for 40 wks at £ 8.15 | 16 300 | | |
| consumables | 50 hrs for 40 wks at £ 1.20 | | | 2 400 |
| Sawbench (diesel) | | | | |
| hire charge | 30 wks at £30.00 | | 900 | |
| consumables | 30 wks at £ 6.40 | | | 192 |
| Hymac 580D | | | | |
| hire charge | 50 hrs for 6 wks at £ 9.10 | | 2 730 | |
| driver | 50 hrs for 6 wks at £ 5.96 | 1 788 | | |
| banksman | 50 hrs for 6 wks at £ 5.70 | 1 710 | | |
| consumables | 50 hrs for 6 wks at £ 2.09 | | | 627 |
| Concrete vibrators | | | | |
| hire charge | 2 for 12 wks at £25.25 | | 606 | |
| D6 Dozer | | | | |
| hire charge | 50 hrs for 4 wks at £12.90 | | 2 580 | |
| driver | 50 hrs for 4 wks at £ 8.15 | 1 630 | | |
| consumables | 50 hrs for 4 wks at £ 4.52 | | | 904 |
| Towed roller BW15 | | | | |
| hire charge | 6 wks at £323.20 | | 1 939 | |
| Loading shovel Cat 931 | | | | |
| hire charge | 50 hrs for 16 wks at £10.07 | | 8 056 | |
| driver | 50 hrs for 16 wks at £ 5.96 | 4 768 | | |
| consumables | 50 hrs for 16 wks at £ 1.90 | | | 1 520 |
| | Carried forward | £41 696 | £69 427 | £11 057 |

WORKED EXAMPLES

**GENERAL PURPOSE PLANT AND PLANT NOT INCLUDED IN UNIT COSTS** - *continued*

| Description | | Labour £ | Plant £ | Consumables £ |
|---|---|---|---|---|
| | Brought forward | 41 696 | 69 427 | 11 057 |
| Compressor $21m^3$/min (silenced) | | | | |
| hire charge | 12 wks at £268.95 | | 3 227 | |
| consumables | 50 hrs for 12 wks at £5.95 | | | 3 570 |
| Plate Compactor (20x18") | | | | |
| hire charge | 12 wks at £32.00 | | 384 | |
| 2in. pumps | | | | |
| hire charge | 2 for 25 wks at £24.75 | | 1 238 | |
| consumables | 15 wks at £6.80 | | | 102 |
| Totals: labour plant and consumables | | £41 696 | £74 276 | £14 729 |

**Total general purpose plant and plant not included in unit costs** £130 701

**SUMMARY OF PRELIMINARIES AND GENERAL ITEMS**

| | £ |
|---|---|
| Contractor's site oncosts - Time related | 359 804 |
| Contractor's site oncosts - Non time related | 117 111 |
| Employer's and consultants' requirements on site - Time related | 59 350 |
| Employer's and consultants' requirements on site - Non Time related | 5 780 |
| Other services, charges and fees | NIL |
| Temporary works not included in unit costs | 34 505 |
| General purpose plant and plant not included in unit costs | 130 701 |
| **Total of Preliminaries and General Items** | £707 251 |

PART 3

# Resources

*This part comprises sections on labour, materials and plant for civil engineering work. These resources form the basis of the unit costs in Part 4 and are given here so that users of the book may:*

*calculate rates for work similar to but differing in detail from the unit costs given in Parts 4 and 5.*

*compare the costs given here with those used in their own organisation.*

*calculate the effects of changes in wage rates, material prices, etc.*

LABOUR COSTS

**NOTES FROM THE WORKING RULE AGREEMENT FOR THE CIVIL ENGINEERING INDUSTRY**

The following are brief details of the Working Rule Agreement for Civil Engineering operatives and of the 1990 Wage Settlement, which is effective from 25 June 1990.

Rates of Pay (Rule I)

The basic rates of pay are:-

| | |
|---|---|
| General Operative | £2.88 per hour |
| Craftsman | £3.38 per hour |

London and Liverpool differentials shall remain at 0.5p per hour.

Guaranteed Bonus (Rule IVA)

The maximum guaranteed bonus is:-

| | |
|---|---|
| General Operative | £14.04 per week |
| Craftsman | £16.38 per week |

Rates for Skill (Rule IVA)

This rule details the plus rates per hour applicable to various skills.

Working Hours (Rule VI)

| | | |
|---|---|---|
| Normal working hours are | Monday to Thursday | - 8 hours per day |
| | Friday | - 7 hours per day |
| | Total | - 39 hours per week |

Annual Holiday with Pay and Benefits Scheme (Rule VIA)

The weekly rate is £17.70

Overtime Rates (Rule VII)

Monday to Friday fisrt 4 hours after normal working day is paid at time and a half, after 4 hours double time shall be paid. Saturday time and a half until completion of the first four hours or until 12 noon, whichever occurs first. Remainder of Saturday and all Sunday at double time.

Travelling and Subsistence Allowances (Rule XIIIA)

This rule details the public transport and nightly subsistence allowances.

Plus Rates for Conditions (Rule XIV)

Civil Engineering opertives are eligible for plus rates per hour for conditions detailed in this rule.

Full copies of the booklet containing the Working Rule Agreement are available from the Federation of Civil Engineering Contractors, publications department.

LABOUR COSTS

## CALCULATION OF LABOUR COSTS

This section sets out a method of calculating 'all-in' labour costs generally - and the labour costs used elsewhere in this book specifically. The calculation is largely based on the wage rates, plus rates and other conditions of the Working Rule Agreement (WRA); important points are discussed below. The calculation has been revised in this edition to enable adjustments to be made to suit individual job conditions with respect to plus rate bonus, subsistence, etc and with respect to working hours and winter summer duration.

Average 'all-in' labour costs are calculated for five categories of labour:

Category 1 - Unskilled labour with rate additions 17p average for dumper driver, excavator banksman
Category 2 - Skilled labour with an average addition of 36p per hour to basic labour rate as plus rate for skill, ganger or conditions
Category 3 - Heavy plant operators with an average addition of 60p per hour to basic labour rates as plus rate for skill.
Category 4 - Light plant operators with an average addition of 43p per hour to basic labour rates as plus rate for skill.
Category 5 - Craftsmen (eg. formwork carpenters, bricklayers)

Costs are based on Class 1, general areas.

An average working week has been calculated to an equal balance between winter and summer (46.20 working weeks per year).
Subsistence payments are included for key men only: for a stated percentage of the workforce excluding category 1 and 2 labourers.
Periodic travel time and fares allowance are based on a fares allowance of £19.25 return every 6 weeks and a travelling time of 4 hours one way; again this applies to only a percentage of the workforce excluding category 1 and 2 labourers.

Allowances for bonus payments reflect current payments in the industry.
Working hours and non-productive overtime are calculated thus:

SUMMER (hours worked 8.00 am - 6.00pm)

| | SUN | MON | TUE | WED | THU | FRI | SAT | TOTAL | DEDUCT LOST & WET TIME | DEDUCT PAID BREAKS | EFFECTIVE HOURS |
|---|---|---|---|---|---|---|---|---|---|---|---|
| | - | 9.50 | 9.50 | 9.50 | 9.50 | 9.50 | 4.00 | 51.50 | 0.5 | 1.5 | 50 |
| Overtime | - | .75 | .75 | .75 | .75 | 1.25 | 2.00 | 6.25 | | | |
| | | | | total | hours | paid | = | 57.75 | | | |

WINTER (hours worked 7.30 am - 4.30pm)

| | SUN | MON | TUE | WED | THU | FRI | SAT | TOTAL | DEDUCT LOST & WET TIME | DEDUCT PAID BREAKS | EFFECTIVE HOURS |
|---|---|---|---|---|---|---|---|---|---|---|---|
| | - | 8.50 | 8.50 | 8.50 | 8.50 | 7.50 | 4.00 | 45.50 | 1.5 | - | 44 |
| Overime | - | - | .25 | .25 | .25 | .25 | - | 2.00 | 3.00 | | |
| | | | | total | hours | paid | = | 48.50 | | | |

So average of hours paid = 57.75 + 48.50 - 2 = 53.125 per week
hours worked = 50.00 + 44.00 - 2 = 47 per week

LABOUR COSTS

**CALCULATION OF LABOUR COSTS** - *continued*

| | Categories of labour | | | | |
|---|---|---|---|---|---|
| Payment details (based on 1990 settlement) | 1<br>£ | 2<br>£ | 3<br>£ | 4<br>£ | 5<br>£ |
| Basic rate per hour (Class 1 areas) | 2.88 | 2.88 | 2.88 | 2.88 | 3.38 |
| Allowance for skills and conditions | 0.17 | 0.36 | 0.60 | 0.44 | - |
| Total rate per hour (ordinary rate) | 3.05 | 3.24 | 3.485 | 3.32 | 3.38 |
| | £ | £ | £ | £ | £ |
| Total hours paid 53.125 @ ordinary rate | 162.03 | 172.15 | 184.88 | 176.38 | 179.56 |
| Bonus paid (based on averages advised for 1991) | 37% | 37% | 52% | 51% | 60% |
| | 221.98 | 235.85 | 281.01 | 266.33 | 287.30 |
| Rule XII | | | | | |
| LPO servicing time (6 hours) | - | - | - | 12.92 | - |
| HPO servicing time (11 hours) | - | - | 38.28 | - | - |
| Craftsman Tool allowance (joiner) | - | - | - | - | 1.68 |
| Total weekly earnings | 221.98 | 235.85 | 319.29 | 279.25 | 288.98 |
| Rule XIIIA | | | | | |
| Travelling allowances | | | | | |
| 14km = £2.69 x 6 days | | | | | |
| Category 1 & 2 = 100% | 16.14 | 16.14 | - | - | - |
| Category 3 & 4 = 80% | - | - | 12.91 | 12.91 | - |
| Category 5 = 90% | - | - | - | - | 14.53 |
| Total earnings and travel | 238.12 | 252.00 | 332.20 | 292.16 | 303.51 |
| Subsistence allowance | | | | | |
| Equivalent of 6.66 nights x £14.05 + 7% to cover periodic travel | | | | | |
| Category 1 & 2 = 0% | - | - | - | - | - |
| Category 3 & 4 = 20% | - | - | 20.02 | 20.02 | - |
| Category 5 = 10% | - | - | - | - | 10.01 |
| Total weekly earnings and allowances | 238.12 | 252.00 | 352.22 | 312.18 | 313.52 |
| Social Security contributions at current rates on weekly earnings | 23.20 | 24.65 | 33.37 | 29.00 | 30.02 |
| Annual holiday with pay contribution | 16.15 | 16.15 | 16.15 | 16.15 | 16.15 |
| Statutory holiday with pay 8 days @ 8 hours @ total rate plus 8/39 of guaranteed bonus (equal to 1.55 hours of total rate) | 4.73 | 4.96 | 5.38 | 5.15 | 5.23 |
| CITB levy | 0.39 | 0.39 | 0.53 | 0.53 | 1.62 |
| Sickness payment/death benefit | 1.55 | 1.55 | 1.55 | 1.55 | 1.55 |
| Employer's liability, third party insurance, safety officer time and all costs not overheads:allow 2.5% | 7.27 | 7.61 | 9.82 | 8.67 | 8.71 |
| Total weekly cost (47 hours) | £291.41 | 307.31 | 419.02 | 373.23 | 376.80 |
| Cost per hour | 6.20 | 6.54 | 8.92 | 7.95 | 8.01 |

LABOUR COSTS

From these a total gang cost per hour for particular operations can be assessed for use in the unit costs build ups (including those normally sublet). (An hourly paid Foreman or Chargehand has a + rate over the skilled labour rate where noted). All outputs used in unit cost calculations are based upon relevant gang's.

A. EXCAVATION/EARTHWORKS GANG

| Foundations and Trenches | Scraper | Excavator | Hand | |
|---|---|---|---|---|
| 1 | 0.5 | 0 | 1 | Ganger or Chargehand |
| 1 | 0 | 0 | 0 | Skilled operative |
| 2 | 1 | 1 | 2 | Unskilled operative |
| 0 | 2 | 0 | 0 | Heavy plant operator |
| 1 | 2 | 1 | 0 | Light plant operator |
| 33.71 | 43.35 | 14.15 | 19.22 | : Total hourly costs |

B. DRAINAGE GANG

| Manhole | Pipes (small bore) | Pipes (large bore) | |
|---|---|---|---|
| 1 | 0.5 | 0.50 | Chargehand pipelayer (+28p) |
| 1 | 1 | 1 | Skilled operative |
| 1 | 2 | 2 | Unskilled operative |
| 0 | 0 | 1 | Heavy plant operator |
| 1 | 1 | 0 | Light plant operator |
| 30.45 | 32.51 | 33.48 | : Total hourly costs |

C. FORMWORK/SHUTTERING AND STRIPPING

| Small Area F3 | Large Area F1 | |
|---|---|---|
| 1 | 1 | Foreman joiner (+28p) |
| 1 | 2 | Joiners |
| 0 | 0 | Skilled operative |
| 1 | 1 | Unskilled operative |
| 0.5 | 0.25 | Light plant operator |
| 29.20 | 27.21 | : hourly costs |

D. STEEL FIXING GANG

| Minor Fix | Major Fix | |
|---|---|---|
| 1 | 1 | Foreman steel fixer (+28p) |
| 2 | 4 | Steel fixer (skilled labour) |
| 1 | 1 | Unskilled operative |
| 0 | 0.25 | Heavy plant operator |
| 0.25 | 0 | Light plant operator |
| 28.09 | 41.41 | : hourly costs |

E. KERBING/PAVING GANG

| large areas | small areas | |
|---|---|---|
| 3 | 1 | Paver (skilled labour) |
| 1 | 1 | Unskilled operative |
| 0 | 0 | Heavy plant operator |
| 0.25 | 0 | Light plant operator |
| 27.81 | 12.74 | : Total hourly costs |

LABOUR COSTS

F. PILING/SPECIAL FOUNDATIONS GANG

| | |
|---|---|
| 1 | Foreman (+28p) |
| 2 | Skilled operative |
| 1 | Unskilled operative |
| 1 | Heavy plant operator |
| 0 | Light plant operator |
| 35.02 | : Total hourly costs |

G. ROAD SURFACING GANG (ASPHALT OR CONCRETE)

| Hand | Machine | |
|---|---|---|
| 1 | 1 | Chargehand (+28p) |
| 2 | 2 | Skilled operative |
| 5 | 4 | Unskilled operative |
| 0 | 1 | Heavy plant operator |
| 0.5 | 1 | Light plant operator |
| 54.88 | 61.57 | : Total hourly costs |

H. GENERAL LABOUR GANGS

| | | | |
|---|---|---|---|
| 1 | 0 | 0 | Ganger or Chargehand |
| 2 | 1 | 1 | Skilled operative |
| 5 | 1 | 1 | Unskilled operative |
| 0 | 0 | 0 | Heavy plant operator |
| 1 | 1 | 0 | Light plant operator |
| 58.85 | 20.69 | 12.74 | : Total hourly costs |

I. SPECIALIST ERECTION GANG (ROAD SIGNS/LIGHTING, PARAPETS, BARRIER ETC)

| Minor Work | Major Work | |
|---|---|---|
| 0 | 1 | Ganger or Chargehand |
| 1 | 1 | Unskilled operative |
| 2 | 2 | Skilled operative |
| 0 | 0.5 | Heavy plant operator |
| 1 | 1 | Light plant operator |
| 27.23 | 38.51 | : Total hoyurly costs |

J. BRICKWORK GANG

| Manholes like | Walls etc large | Walls etc small | |
|---|---|---|---|
| 0.5 | 1 | 0 | Foreman Bricklayer (+28p) |
| 2 | 4 | 2 | Bricklayer |
| 1 | 1 | 1 | Unskilled operative |
| 0 | 0 | 0 | Heavy plant operator |
| 0.25 | 0.5 | 0 | Light plant operator |
| 28.35 | 50.51 | 22.22 | : Total hourly costs |

K. CONCRETE GANG

| large area | small area | |
|---|---|---|
| 1 | 0 | Ganger |
| 2 | 1 | Skilled operative |
| 4 | 1 | Unskilled operatives |
| 0.25 | 0 | Heavy plant (crane operator) |
| 46.93 | 12.74 | : Total hourly costs |

## BASIC MATERIAL PRICES

This section comprises a price list for materials which assumes that the materials would be in quantities as required for a medium sized civil engineering project of, say, £5-6 million and the location of the works would be neither city centre nor excessively remote.
Material prices have been obtained from manufacturers and suppliers and incorporate appropriate trade discounts; generally they are those in force in June 1989. They are given for the units in which the materials are sold which may not necessarily be the unit used in the Unit Costs section. There is no allowance for waste or loss or any offloading or distribution charges, unless specifically noted.

All prices quoted are exclusive of Value Added Tax.

The section comprises:

BASIC MATERIAL PRICES

**ALUMINIUM CLADDING**

| | Unit | $1000m^2$ £ | $2000m^2$ £ |
|---|---|---|---|
| British Alcan building sheets | | | |
| Type A900, 0.55 mm thick, Duralcote finish | $m^2$ | 7.43 | 7.06 |
| Type A900, 0.70 mm thick, Duralcote finish | $m^2$ | 9.13 | 8.68 |
| Type WA6, 0.55 mm thick, Duralcote finish | $m^2$ | 7.26 | 6.90 |
| Type WA6, 0.70 mm thick, Duralcote finish | $m^2$ | 8.88 | 8.44 |
| Type C5, 0.90 mm thick, Duralcote finish | $m^2$ | 12.61 | 11.98 |
| Type C5, 1.20 mm thick, Duralcote finish | $m^2$ | 17.13 | 16.27 |
| Type M900, liner sheet, 0.50 mm thick | $m^2$ | 5.76 | 5.47 |
| Type M900, liner sheet, 0.70 mm thick | $m^2$ | 7.60 | 7.22 |
| British Alcan insulated sheets | | | |
| Type A900FL, 40 mm polyisocyanurate insulation, Duralcote finish | $m^2$ | 15.06 | 14.68 |
| British Alcan composite panels | | | |
| Type WA6, polyisocyanurate insulation, Duralcote finish | $m^2$ | 24.62 | 24.03 |
| Type SF50, polyisocyanurate insulation, Duralcote finish | $m^2$ | 24.60 | 24.01 |

| | Unit | £ |
|---|---|---|
| Flowclad curved sheets | | |
| Type A900 sheets | curve | 12.00 |
| Type WA6 sheets | curve | 12.00 |
| Type C5 sheets | curve | 12.00 |
| Type M900 sheets | curve | 12.00 |

| | Unit | Outer Panel £ | Lining Panel £ |
|---|---|---|---|
| British Alcan flashings and accessories | | | |
| Ridges/hips | m | 7.31 | 4.96 |
| Eaves gutters | m | 3.79 | 2.53 |
| Gables | m | 5.24 | 3.79 |
| Drips | m | 5.24 | 3.79 |
| External corners | m | 5.24 | 3.79 |
| Internal corners | m | 5.24 | 3.79 |
| Jambs | m | 5.10 | - |
| Roller shutter heads | m | 5.10 | 3.79 |
| Fillers | m | 2.20 | - |

BASIC MATERIAL PRICES

**BAR AND FABRIC REINFORCEMENT**

Grade 250 plain round reinforcing bars

Grade 460 type 2 'Barbond'deformed reinforcing bars

| | | Mild Steel to BS 4449 | | | | High Tensile to BS4449 | |
|---|---|---|---|---|---|---|---|
| | Unit | Straight £ | Bent £ | | Unit | Straight £ | Bent £ |
| R 6 | t | 461.75 | 546.75 | T 6 | t | 465.75 | 550.75 |
| R 8 | t | 422.75 | 502.75 | T 8 | t | 426.75 | 506.75 |
| R10 | t | 411.75 | 471.75 | T10 | t | 415.75 | 475.75 |
| R12 | t | 401.75 | 446.75 | T12 | t | 405.75 | 450.75 |
| R16 | t | 380.75 | 415.75 | T16 | t | 384.75 | 419.75 |
| R20 | t | 379.75 | 409.75 | T20 | t | 383.75 | 413.75 |
| R25 | t | 382.75 | 412.75 | T25 | t | 386.75 | 416.75 |
| R32 | t | 389.75 | 414.75 | T32 | t | 393.75 | 418.75 |
| R40 | t | 401.75 | 426.75 | T40 | t | 405.75 | 430.75 |

| Welded fabric to BS 4483 in sheets 4.80 x 2.40 metres | | | Unit | £ |
|---|---|---|---|---|
| BS Ref | A393 | (6.16kg per $m^2$) | $m^2$ | 2.96 |
| | A252 | (3.95kg per $m^2$) | $m^2$ | 1.87 |
| | A193 | (3.02kg per $m^2$) | $m^2$ | 1.44 |
| | A142 | (2.22kg per $m^2$) | $m^2$ | 1.06 |
| | A98 | (1.54kg per $m^2$) | $m^2$ | 0.81 |
| | B1131 | (10.90kg per $m^2$) | $m^2$ | 5.30 |
| | B785 | (8.14kg per $m^2$) | $m^2$ | 3.93 |
| | B503 | (5.93kg per $m^2$) | $m^2$ | 2.87 |
| | B385 | (4.53kg per $m^2$) | $m^2$ | 2.23 |
| | B283 | (3.73kg per $m^2$) | $m^2$ | 1.85 |
| | B196 | (3.05kg per $m^2$) | $m^2$ | 1.56 |
| | C785 | (6.72kg per $m^2$) | $m^2$ | 3.28 |
| | C636 | (5.55kg per $m^2$) | $m^2$ | 2.71 |
| | C503 | (4.34kg per $m^2$) | $m^2$ | 2.11 |
| | C385 | (3.41kg per $m^2$) | $m^2$ | 1.71 |
| | C283 | (2.61kg per $m^2$) | $m^2$ | 1.33 |
| | D49 | (0.77kg per $m^2$) | $m^2$ | 0.67 |

The prices shown above include delivery, but a charge of £25 is applicable for each delivery of under 8 tonnes.

| | Unit | £ |
|---|---|---|
| Black anealed tying wire | | |
| 16 swg tying wire (coil) | 25kg | 17.30 |
| 16 swg tying wire (coil) | 2kg | 1.99 |
| Stainless steel tying wire | | |
| 18 swg tying wire (coil) | 25kg | 226.50 |
| 18 swg tying wire (coil) | 2kg | 20.76 |

BASIC MATERIAL PRICES

**BRICKWORK**

| | Unit | £ |
|---|---|---|
| **ENGINEERING BRICKS** | | |
| Class A | | |
| Perforated/solid | 1000 | 335.00 |
| Mechanical offload charge | | inc |
| Number of bricks in pack | 400 | |
| Standard size load | 5600 | |
| Class B | | |
| Perforated/solid | 1000 | 203.00 |
| Mechanical offload charge | | inc |
| Number of bricks in pack | 400 | |
| Standard size load | 5600 | |
| **FACING BRICKS** | | |
| Haunchwood smooth blue | 1000 | 433.00 |
| Mechanical offload charge | | 5.00 |
| Number of bricks in pack | 300 | |
| Standard size load | 6000 | |
| Ashdown Crowborough multicoloured stock brick | 1000 | 388.00 |
| Mechanical offload charge | | 5.00 |
| Number of bricks in pack | 370 | |
| Standard size load | 8140 | |
| Otterham London stock | 1000 | 419.00 |
| Mechanical offload charge | | 5.00 |
| Number of bricks in pack | 392 | |
| Standard size load | 9016 | |
| **PAVER BRICKS** | | |
| Wealden stock flex paver (215 x 102.5 x 65 mm) | 1000 | 433.00 |
| Mechanical offload charge | | 5.00 |
| Number of bricks in pack | 300 | |
| Standard size load | 6300 | |
| Haunchwood wirecut blue paver (215 x 102.5 x 50 mm) | 1000 | 415.00 |
| Mechanical offload charge | | 5.00 |
| Number of bricks in pack | 390 | |
| Standard size load | 3900 | |
| Haunchwood smooth blue paver (215 x 102.5 x 50 mm) | 1000 | 505.00 |
| Mechanical offload charge | | 5.00 |
| Number of bricks in pack | 420 | |
| Standard size load | 4200 | |
| Holbrook wirecut red paver (215 x 102.5 x 35 mm) | 1000 | 325.00 |
| Mechanical offload charge | | 5.00 |
| Number of bricks in pack | 500 | |
| Standard size load | 8000 | |

BASIC MATERIAL PRICES

**BRICKWORK** - *continued*

CALCIUM SILICATE COMMON BRICKS

| | Unit | £ |
|---|---|---|
| Class 3 | 1000 | 122.11 |
| Mechanical offload charge | | 3.97 |
| Number of bricks in pack | 390 | |
| Standard size load | 9360 | |
| Class 4 | 1000 | 132.11 |
| Mechanical offload charge | | 3.97 |
| Number of bricks in pack | 390 | |
| Standard size load | 9360 | |
| Class 5 | 1000 | 150.24 |
| Mechanical offload charge | | 3.97 |
| Number of bricks in pack | 390 | |
| Standard size load | 9360 | |

LIGHTWEIGHT AGGREGATE BLOCKS

| | | |
|---|---|---|
| Solid (3.5 N/mm$^2$) | | |
| 75 mm | m$^2$ | 5.11 |
| 100 mm | m$^2$ | 4.80 |
| 140 mm | m$^2$ | 7.21 |
| 190 mm | m$^2$ | 10.46 |
| 215 mm | m$^2$ | 13.62 |

DENSE AGGREGATE BLOCKS

| | | |
|---|---|---|
| Cellular (3.5 N/mm$^2$) | | |
| 140 mm | m$^2$ | 6.61 |
| Hollow (3.5 N/mm$^2$) | | |
| 190 mm | m$^2$ | 7.59 |
| 215 mm | m$^2$ | 7.82 |
| Solid (3.5 or 7 N/mm$^2$) | | |
| 75 mm | m$^2$ | 4.08 |
| 100 mm | m$^2$ | 4.02 |
| 140 mm | m$^2$ | 6.67 |
| 190 mm | m$^2$ | 8.74 |
| 215 mm | m$^2$ | 13.82 |

PRESTRESSED CONCRETE LINTELS

| | | |
|---|---|---|
| 100 x 65 mm | m | 2.48 |
| 100 x 150 mm | m | 5.09 |
| 100 x 225 mm | m | 7.36 |
| 140 x 65 mm | m | 3.18 |
| 150 x 100 mm | m | 5.91 |
| 150 x 150 mm | m | 6.87 |
| 150 x 225 mm | m | 10.73 |
| 220 x 65 mm | m | 4.46 |
| 225 x 100 mm | m | 7.36 |
| 225 x 150 mm | m | 9.97 |

BASIC MATERIAL PRICES

**CAST IRON PIPES AND FITTINGS**

| **Diameter in mm** | Unit | **75** £ | **100** £ | **150** £ | **225** £ |
|---|---|---|---|---|---|
| Cast and ductile iron pipes & fittings to BS437 | | | | | |
| Plain ended cast iron pipes | m | 16.59 | 16.03 | 30.19 | 76.27 |
| Cast iron couplings complete with stainless steel nuts and bolts and synthetic rubber gaskets | nr | 8.67 | 10.25 | 12.42 | 28.44 |
| Bends 45° | nr | 14.66 | 16.30 | 33.06 | 133.85 |
| Bends 87.5° | nr | 15.56 | 19.16 | 38.68 | 133.85 |
| Access bends, side | nr | - | 50.36 | 93.52 | 200.04 |
| Bends, long radius | nr | - | 23.49 | 50.36 | - |
| Branches, 45°, equal | nr | 21.63 | 25.45 | 55.06 | 192.71 |
| Branches, access side, 45° (equal) | nr | - | 58.67 | 108.48 | 268.80 |
| Double branches, 87.5° (equal) | nr | - | 38.32 | - | 278.97 |
| Diminishing pipes | nr | 9.32 | 16.30 | 24.99 | 64.07 |

| **Diameter in mm** | Unit | **100** £ | **150** £ | **250** £ | **350** £ | **400** £ | **600** £ |
|---|---|---|---|---|---|---|---|
| Ductile spun iron pipe to BS 4772 with Tyton joints | | | | | | | |
| Spigot and socket pipes | m | 7.00 | 10.50 | 19.00 | 29.50 | 41.50 | 64.00 |
| Tyton joint | nr | 2.09 | 2.31 | 3.96 | 10.01 | 13.64 | 18.15 |
| Tyton tee (equal) | nr | 38.72 | 61.06 | 123.14 | 400.94 | 731.98 | 892.15 |
| Flange on Tyton tee (equal) | nr | 54.19 | 82.49 | 212.30 | 304.05 | 622.23 | - |
| Tyton bend, 90° | nr | 23.10 | 39.95 | 134.29 | 290.00 | 473.24 | 948.68 |
| Tyton bend, 45° | nr | 20.93 | 39.95 | 92.48 | 255.84 | 323.79 | 569.02 |
| Tyton duckfoot bend, 90° | nr | 54.12 | 91.04 | 266.07 | 538.41 | - | - |
| Tyton taper | nr | 26.73 | 46.99 | 116.41 | 243.47 | - | - |

| **Diameter in mm** | Unit | **700** £ | **800** £ | **900** £ | **1000** £ | **1100** £ |
|---|---|---|---|---|---|---|
| Ductile spun iron pipe to BS 4772 with Stantyte joints | | | | | | |
| Spigot and socket pipes | m | 80.00 | 100.00 | 120.00 | 145.00 | 172.00 |
| Stantyte joint | nr | 29.92 | 33.00 | 37.04 | 56.48 | 73.52 |
| Stantyte tee (equal) | nr | 1477.53 | 2048.46 | 3032.28 | 5012.27 | - |
| Stantyte bend, 90° | nr | 1726.41 | 2145.49 | 2717.51 | 4328.21 | - |
| Stantyte bend, 45° | nr | 856.70 | 906.63 | 1504.28 | 2260.11 | - |

BASIC MATERIAL PRICES

**CAST IRON PIPES AND FITTINGS** - *continued*

| **Diameter in mm** | Unit | **100** £ | **150** £ | **250** £ | **400** £ | **450** £ | **600** £ |
|---|---|---|---|---|---|---|---|
| Ductile spun iron pipe to BS 4772 with Stanlock joints | | | | | | | |
| Spigot and socket pipes | m | 10.41 | 15.63 | 29.54 | 57.08 | 64.64 | 97.72 |
| Stanlock joints | nr | 8.03 | 11.55 | 21.56 | 42.02 | 46.20 | 78.76 |
| Stanlock tee (equal) | nr | 37.79 | 56.33 | 164.50 | 468.11 | 639.88 | 895.25 |
| Flange on Stanlock tee (equal) | nr | 55.28 | 89.18 | 201.20 | 514.03 | 601.75 | 1005.78 |
| Stanlock bend, 90° | nr | 24.43 | 39.50 | 142.19 | 398.42 | 433.88 | 891.12 |
| Stanlock bend, 45° | nr | 23.35 | 34.39 | 93.69 | 313.76 | 349.95 | 604.53 |
| Stanlock and flange socket | nr | 21.79 | 32.18 | 74.12 | 159.06 | 182.02 | 334.23 |
| Stanlock cap | nr | 13.03 | 21.16 | 49.40 | 130.59 | 159.17 | 223.04 |

| **Diameter in mm** | Unit | **100** £ | **150** £ | **250** £ | **350** £ | **450** £ |
|---|---|---|---|---|---|---|
| Couplings on BS 4772 ductile spun iron pipe | | | | | | |
| Viking Johnson couplings | nr | 23.47 | 33.73 | 45.51 | 88.57 | 100.39 |
| Stanseal couplings | nr | 22.83 | 32.60 | - | - | - |

| **Diameter in mm** | Unit | **600** £ | **800** £ | **1000** £ | **1200** £ |
|---|---|---|---|---|---|
| Couplings on BS 4772 ductile spun iron pipe | | | | | |
| Viking Johnson couplings | nr | 118.60 | 216.76 | 348.82 | 488.62 |

BASIC MATERIAL PRICES

**CLAYWARE PIPES AND FITTINGS**

| **Diameter in mm** | | **100** | **150** | **225** | **300** | **375** | **400** | **450** |
|---|---|---|---|---|---|---|---|---|
| **Standard pipe length in m** | | **1.0** | **1.25 /1.4** | **1.6** | **1.6** | **1.5** | **1.5** | **1.5** |
| | Unit | £ | £ | £ | £ | £ | £ | £ |
| Vitrified clay spigot & socket pipes and fittings to BS65 (Denseal) | | | | | | | | |
| Straight Pipes | | | | | | | | |
| zone 1 | m | 3.14 | 4.50 | 8.93 | 14.19 | 27.52 | 27.83 | 35.9[illegible] |
| zone 2 | m | 3.21 | 4.59 | 9.05 | 14.47 | 28.11 | 28.28 | 36.6[illegible] |
| zone 3 | m | 3.21 | 4.59 | 9.25 | 14.81 | 28.60 | 28.79 | 37.3[illegible] |
| zone 4 | m | 3.21 | 4.59 | 9.45 | 15.11 | 29.20 | 29.24 | 38.0[illegible] |
| Short length pipes (0.6 m) or | | **0.9** | **0.9** | **0.9** | **0.9** | **1.0** | **1.0** | **1.0** |
| zone 1 | nr | 3.53 | 5.06 | 10.05 | 15.96 | 30.96 | 31.31 | 40.5[illegible] |
| zone 2 | nr | 3.62 | 5.16 | 10.18 | 16.28 | 31.63 | 31.82 | 41.2[illegible] |
| zone 3 | nr | 3.62 | 5.16 | 10.41 | 16.66 | 32.17 | 32.39 | 42.0[illegible] |
| zone 4 | nr | 3.62 | 5.16 | 10.63 | 17.01 | 32.85 | 32.90 | 42.7 |
| Bends (all zones) | | | | | | | | |
| $90^{o}$ $45^{o}$ $22.5^{o}$ $11.25^{o}$ | nr | 4.66 | 8.28 | 17.02 | 33.53 | 63.67 | 75.56 | 99.4 |
| Rest bends (all zones) | nr | 5.80 | 9.64 | 23.67 | - | - | - | - |
| Saddles (all zones) | | | | | | | | |
| oblique and square | nr | 4.62 | 7.52 | 19.64 | 40.21 | - | - | - |
| Junctions (all zones) | | | | | | | | |
| oblique/square/curved square | nr | 6.16 | 10.07 | 23.67 | 50.21 | 89.43 | 99.10 | 118.5[illegible] |
| tumbling bay/drop back | nr | 6.16 | 10.23 | 26.34 | 56.74 | 113.20 | 134.38 | 155.8[illegible] |
| Tapers (all zones) | | | | | | | | |
| reducers | nr | - | 12.47 | 19.64 | 40.21 | 59.58 | - | - |
| increasers | nr | 12.47 | 19.64 | 40.21 | 59.58 | - | - | - |
| Stoppers | | | | | | | | |
| all zones | nr | 1.83 | 2.71 | 6.13 | 13.50 | 18.75 | - | - |
| Double sockets | nr | 4.62 | 7.52 | 16.20 | 26.34 | 46.35 | - | - |

| **Diameter in mm** | | **100** | **150** | **200** | **225** |
|---|---|---|---|---|---|
| **Standard pipe length in m** | | **1.3** | **1/1.4** | **1.6** | **1.6** |
| | Unit | £ | £ | £ | £ |
| Plain ended vitrified clay sewer pipes and fittings with sleeve joints, to BS65, including 1 Nr coupling per standard pipe length (Densleeve) | | | | | |
| Straight pipes | | | | | |
| zone 1 | nr | 1.83 | 4.05 | 8.76 | 8.93 |
| zone 2 | nr | 1.88 | 4.12 | 8.92 | 9.05 |
| zone 3 | nr | 1.88 | 4.12 | 9.13 | 9.25 |
| zone 4 | nr | 1.88 | 4.12 | 9.32 | 9.45 |

BASIC MATERIAL PRICES

**CLAYWARE PIPES AND FITTINGS** - *continued*

| **Diameter in mm** | | **100** | **150** | **200** | **225** |
|---|---|---|---|---|---|
| | | £ | £ | £ | £ |
| Additional couplings all zones | nr | 0.80 | 1.87 | 3.43 | 3.43 |
| Bends (all zones) 90° 45° 22.5° 11.25° | nr | 2.42 | 5.93 | 16.85 | 16.85 |
| Rest bends all zones | nr | 3.48 | 7.11 | - | 18.03 |
| Saddles (all zones) oblique and square | nr | 4.03 | 7.59 | - | 21.96 |

| | Unit | £ |
|---|---|---|
| Taper pipes (all zones) | | |
| 100 x 150 mm | nr | 5.72 |
| 150 x 200 mm | nr | 19.47 |
| 150 x 225 mm | nr | 19.47 |
| Junctions (all zones) | | |
| Barrel diam x Arm diam | | |
| 100 x 100mm | nr | 4.21 |
| 150 x 100mm | nr | 7.32 |
| 150 x 150mm | nr | 7.84 |
| 225 x 100mm | nr | 23.44 |
| 225 x 150mm | nr | 23.44 |
| 225 x 225mm | nr | 23.44 |

| | Unit | Plain £ | Socketed £ |
|---|---|---|---|
| Low back traps | | | |
| 100 mm diameter 'P' trap | nr | 4.05 | 4.50 |
| 100 mm diameter 'Q' or 'S' trap | nr | 4.77 | 5.18 |

| **Diameter in mm** | | **100** | **150** | **225** | **300** | **375** | **400** | **450** |
|---|---|---|---|---|---|---|---|---|
| **Standard pipe length 0.9 m or** | | **0.9** | **0.9** | **0.9** | **0.9** | **0.9** | **0.9** | **0.9** |
| | Unit | £ | £ | £ | £ | £ | £ | £ |
| Vitrified clay channel pipes and fittings to BS 65 (Naylor) | | | | | | | | |
| Pipes all zones | nr | 1.56 | 2.58 | 5.75 | 11.38 | 17.24 | 18.12 | 22.23 |
| Bends (all zones) 90° 45° 22.5° 11.25° | nr | 1.59 | 2.63 | 8.78 | 17.43 | 54.23 | 55.80 | 75.49 |
| Half section bends all zones | nr | 3.26 | 5.35 | 17.67 | 34.98 | - | - | - |
| Three quarter section bends all zones | nr | 3.60 | 6.08 | 22.03 | 43.69 | - | - | - |

BASIC MATERIAL PRICES

**CLAYWARE PIPES AND FITTINGS** - *continued*

| **Diameter in mm** | | **100** | **150** | **225** | **300** | **375** | **400** | **450** |
|---|---|---|---|---|---|---|---|---|
| **Standard pipe length 0.9 m or:** | | **0.6** | **0.6** | - | - | - | - | - |
| | Unit | £ | £ | £ | £ | £ | £ | £ |
| Junctions (all zones) | | | | | | | | |
| single | nr | 3.03 | 5.02 | 14.39 | 34.24 | 54.63 | 70.87 | 76.15 |
| double | nr | 4.54 | 7.58 | 23.00 | 51.44 | 81.84 | - | - |
| breeches | nr | 4.54 | 7.58 | 23.00 | 51.44 | - | - | - |

| | Unit | £ |
|---|---|---|
| Tapers (all zones) | | |
| straight, 150 - 100 mm | nr | 6.66 |
| straight, 225 - 150 mm | nr | 14.72 |
| straight, 300 - 225 mm | nr | 29.12 |
| straight, 375 - 300 mm | nr | 45.10 |
| straight, 450 - 375/400 mm | nr | 60.79 |
| bend, 150 - 100 mm | nr | 10.14 |
| bend, 225 - 150 mm | nr | 28.91 |
| bend, 300 - 225 mm | nr | 57.21 |
| bend, 375 - 300 mm | nr | 154.92 |
| bend, 450 - 375/400 mm | nr | 207.39 |

| **Diameter in mm** | | **100** | **125** | **150** | **225** | **300** |
|---|---|---|---|---|---|---|
| **Standard pipe length in m** | | **1.3** | **1.0** | **1.0/1.4** | **1.6** | **1.6** |
| | Unit | £ | £ | £ | £ | £ |
| Dense vitrified clay plain ended conduits, to BS65 (Denduct) | | | | | | |
| Straight conduits | | | | | | |
| zone 1 | nr | 1.78 | 3.04 | 3.76 | 7.27 | 13.63 |
| zone 2 | nr | 1.82 | 3.19 | 3.86 | 7.48 | 14.04 |
| zone 3 | nr | 1.88 | 3.26 | 3.99 | 7.64 | 14.37 |
| zone 4 | nr | 1.94 | 3.32 | 4.17 | 7.88 | 14.78 |
| Bends/Bellmouths | | | | | | |
| all zones | nr | 1.96 | 3.19 | 4.07 | 9.69 | 16.01 |

| **Diameter in mm** | | **75** | **100** | **150** | **225** |
|---|---|---|---|---|---|
| **Standard pipe length in mm** | | **300** | **300** | **300** | **300** |
| | Unit | £ | £ | £ | £ |
| Land drain to BS 1196 (Hepworth) | | | | | |
| Straight pipes | | | | | |
| zone 1 | nr | 0.21 | 0.38 | 0.80 | 2.12 |
| zone 2 | nr | 0.23 | 0.41 | 0.84 | 2.22 |
| zone 3 | nr | 0.24 | 0.45 | 0.88 | 2.29 |
| zone 4 | nr | 0.24 | 0.45 | 0.88 | 2.29 |
| Junctions | | | | | |
| all zones | nr | 3.47 | 4.01 | 5.28 | 7.99 |

BASIC MATERIAL PRICES

**CLAYWARE PIPES AND FITTINGS** - *continued*

| **Diameter in mm** | | **100** | **150** | **225** | **300** | **400** | **450** |
|---|---|---|---|---|---|---|---|
| **Standard pipe length in m** | | **1.0** | **1.5** | **1.75** | **2.5** | **2.5** | **2.5** |
| | Unit | £ | £ | £ | £ | £ | £ |
| Vitrified clay pipes with rubber sealing rings (Hepseal) | | | | | | | |
| Straight pipes | | | | | | | |
| zone 1 | m | 4.87 | 6.32 | 12.37 | 19.30 | 37.84 | 48.96 |
| zone 2 | m | 4.96 | 6.42 | 12.57 | 19.68 | 38.46 | 49.84 |
| zone 3 | m | 5.04 | 6.59 | 12.86 | 20.17 | 39.16 | 50.78 |
| zone 4 | m | 5.23 | 6.77 | 13.16 | 20.56 | 39.78 | 51.73 |
| Bends (all zones) | | | | | | | |
| $90^{\circ}$ | nr | 6.83 | 11.28 | 23.12 | 45.60 | 143.89 | 189.46 |
| $45^{\circ}$ | nr | 6.83 | 11.28 | 23.12 | 45.60 | 102.77 | 135.33 |
| $30^{\circ}$ | nr | 6.40 | 11.28 | 23.12 | 45.60 | - | - |
| $15^{\circ}$ | nr | 6.40 | 11.28 | 23.12 | 45.60 | - | - |
| $22.5^{\circ}$ | nr | - | - | - | - | 71.96 | 94.74 |
| Single junctions (all zones) | | | | | | | |
| oblique | nr | 9.50 | 14.73 | 34.75 | 71.64 | - | - |
| square | nr | 9.50 | 15.07 | 36.88 | 71.64 | - | - |
| Double junctions (all zones) | nr | - | 30.12 | 67.54 | 143.37 | - | - |
| Saddles (all zones) | | | | | | | |
| oblique | nr | 6.08 | 10.73 | 26.73 | 54.67 | - | - |
| square | nr | 6.64 | 10.73 | 26.73 | 54.67 | - | - |
| Tapers (all zones) | | | | | | | |
| reducers | nr | - | 16.90 | 26.73 | 54.67 | - | - |
| enlargers | nr | - | 16.90 | 26.73 | 54.67 | - | - |

| **Diameter in mm** | | **100** | **150** | **225** | **300** |
|---|---|---|---|---|---|
| **Standard pipe length in m** | | **1.6** | **1.75** | **1.85** | **2.5** |
| | Unit | £ | £ | £ | £ |
| Plain ended vitrified clay pipes and fittings with sleeved joints to BS 65 including 1 Nr coupling per standard pipe length (Supersleve/Hepsleve) | | | | | |
| Straight pipes | | | | | |
| zone 1 | m | 1.78 | 4.04 | 9.46 | 14.44 |
| zone 2 | m | 1.81 | 4.15 | 9.63 | 14.71 |
| zone 3 | m | 1.81 | 4.15 | 9.89 | 15.08 |
| zone 4 | m | 1.81 | 4.15 | 10.17 | 15.37 |
| Additional couplings all zones | nr | 1.04 | 2.55 | 4.69 | 9.60 |
| Bends (all zones) $90^{\circ}$ $45^{\circ}$ $30^{\circ}$ $15^{\circ}$ | nr | 2.09 | 5.38 | 18.24 | 33.92 |

BASIC MATERIAL PRICES

**CLAYWARE PIPES AND FITTINGS** - *continued*

| **Diameter in mm** | | **100** | **150** | **225** | **300** |
|---|---|---|---|---|---|
| **Standard pipe length in m** | | **1.6** | **1.75** | **1.85** | **2.5** |
| | Unit | £ | £ | £ | £ |
| Rest bends | | | | | |
| all zones | nr | 3.46 | 6.90 | 19.85 | 55.30 |
| Saddles (all zones) | | | | | |
| oblique | nr | 4.32 | 7.87 | 25.18 | 42.47 |
| square | nr | 4.17 | 7.54 | 25.18 | 42.47 |

| | Unit | £ |
|---|---|---|
| Taper pipe (all zones) | | |
| 100 - 150 mm | nr | 5.08 |
| 150 - 225 mm | nr | 19.85 |
| 225 - 300 mm | nr | 47.12 |
| Junctions (all zones) | | |
| barrel dai x arm dia | | |
| 100 x 100 mm | nr | 4.40 |
| 150 x 100 mm | nr | 7.21 |
| 150 x 150 mm | nr | 7.89 |
| 225 x 100 mm | nr | 27.22 |
| 225 x 150 mm | nr | 27.22 |
| 225 x 225 mm | nr | 28.54 |
| 300 x 100 mm | nr | 57.38 |
| 300 x 150 mm | nr | 57.38 |
| 300 x 225 mm | nr | 57.38 |
| 300 x 300 mm | nr | 57.38 |

| **Diameter in mm** | | **100** | **150** | **225** | **300** |
|---|---|---|---|---|---|
| **Standard pipe length in m** | | **1.6** | **1.75** | **1.85** | **1.85** |
| | Unit | £ | £ | £ | £ |
| Perforated vitrified clay pipes with joint (Hepline) | | | | | |
| Straight pipes | | | | | |
| zone 1 | m | 3.04 | 5.62 | 10.33 | 19.39 |
| zone 2 | m | 3.15 | 5.77 | 10.60 | 19.85 |
| zone 3 | m | 3.18 | 5.89 | 10.79 | 20.38 |
| zone 4 | m | 3.25 | 6.13 | 11.14 | 20.89 |
| Bends (all zones) | | | | | |
| $90^{o}$ $45^{o}$ $30^{o}$ $15^{o}$ | nr | 2.73 | 5.44 | 12.93 | 21.42 |
| Junctions (all zones) | | | | | |
| oblique and square | nr | 4.93 | 8.36 | 20.19 | 35.83 |

| | Unit | £ |
|---|---|---|
| Road gullies | | |
| 300 mm dia x 600 mm deep (100 mm or 150 mm outlet) | nr | 26.40 |
| 375 mm dia x 750 mm deep (100 mm or 150 mm outlet) | nr | 30.13 |
| 450 mm dia x 750 mm deep (150 mm outlet) | nr | 45.66 |

BASIC MATERIAL PRICES

**CLAYWARE PIPES AND FITTINGS** - *continued*

| | Unit | £ |
|---|---|---|
| Yard gullies with rodding eye & stopper | | |
| 225 mm dia x 610 mm deep | nr | 22.52 |
| 300 mm dia x 585 mm deep | nr | 32.30 |
| 225 mm dia buckets | nr | 12.03 |
| 300 mm dia buckets | nr | 14.81 |
| Grease traps with galvanized tray and lid | | |
| 610 x 457 mm and 610 mm deep | nr | 220.24 |
| 457 x 305 mm and 610 mm deep | nr | 189.48 |

BASIC MATERIAL PRICES

**CONCRETE AND CEMENT**

Ready mixed concrete supplied in full loads delivered to a site within 5 miles (8km) radius of concrete mixing plant.

| | Unit | £ | £ | £ |
|---|---|---|---|---|
| | | Aggregate size | | |
| | | 10 mm | 20 mm | 40 mm |
| Grade C7.5 Prescribed mix | $m^3$ | - | 40.13 | 39.76 |
| Grade C10 Prescribed mix | $m^3$ | - | 42.41 | 41.90 |
| Grade C15 Prescribed mix | $m^3$ | - | 43.27 | 42.85 |
| Grade C20 Prescribed mix | $m^3$ | 48.25 | 44.93 | 43.56 |
| Grade C25 Prescribed mix | $m^3$ | 49.88 | 47.68 | 46.03 |
| Grade C30 Prescribed mix | $m^3$ | 53.18 | 49.61 | 47.68 |
| Grade C30 Designed mix | $m^3$ | 45.76 | 43.56 | - |
| Grade C40 Designed mix | $m^3$ | 49.33 | 47.13 | - |
| Grade C50 Designed mix | $m^3$ | 53.73 | 51.53 | - |
| Grade C60 Designed mix | $m^3$ | 58.14 | 55.94 | - |

| *Add* to the above prices for | Unit | £ |
|---|---|---|
| Rapid-hardening cement to BS12 | $m^3$ | 10.00 to 15.00 |
| Sulphate resisting cement to BS4027 | $m^3$ | 4.5% |
| Fibremesh reinforcement additive | $m^3$ | 8.00 |
| Distance per mile in excess of 5 mile (8km) | $m^3$ | 0.55 |
| Air entrained concrete | $m^3$ | 0.55 |
| Water repellent concrete | $m^3$ | 2.00 |
| Part loads per $m^3$ below full load | $m^3$ | 8.50 |
| Waiting time | hr | 20.00 |

| | | |
|---|---|---|
| Cement Ordinary Portland to BS12 bulk | t | 51.90 |
| bagged | 50kg | 4.04 |
| Cement high Alumina bulk | t | 185.00 |
| bagged | 50kg | 9.65 |
| Cement Admixtures | | |
| Cementone No. 1 Brown | kg | 3.27 |
| Black | kg | 3.43 |
| Cementone No. 2 Waterproofer liquid | 5 L | 6.75 |
| PLA PVA bonding agent | 5 L | 10.41 |
| PLA plasticiser | 5 L | 2.83 |
| Cementone No. 8 Frostproofer | 5 L | 4.44 |

| | | |
|---|---|---|
| Cement Sulphate Resisting | t | 72.07 |
| Cement Rapid-hardening | t | 67.40 |
| Cement White Portland | t | 113.30 |
| Cement Bentonite (2:1) | t | 80.00 |
| Cement Grout | t | 67.40 |

BASIC MATERIAL PRICES

**CONCRETE MANHOLES**

| | Unit | £ |
|---|---|---|
| Shaft and chamber rings (reinforced) | | |
| 675mm diameter | m | 23.35 |
| 900mm diameter | m | 38.61 |
| 1050mm diameter | m | 48.66 |
| 1200mm diameter | m | 65.14 |
| 1350mm diameter | m | 97.20 |
| 1500mm diameter | m | 112.85 |
| 1800mm diameter | m | 142.94 |
| Step irons (built-in) | nr | 2.80 |
| Rung irons (built-in) | nr | 8.40 |
| Reducing slabs (675 mm diameter access) | | |
| 900mm diameter | nr | 49.84 |
| 1050mm diameter | nr | 62.50 |
| 1200mm diameter | nr | 84.85 |
| 1350mm diameter | nr | 118.03 |
| 1500mm diameter | nr | 155.04 |
| 1800mm diameter | nr | 203.77 |
| Landing slabs (900 mm diameter access) | | |
| 1500 mm diameter | nr | 146.70 |
| 1800 mm diameter | nr | 195.39 |
| Cover slabs (heavy duty) (600 mm square access) | | |
| 675mm diameter | nr | 19.95 |
| 900mm diameter | nr | 33.64 |
| 1050mm diameter | nr | 43.30 |
| 1200mm diameter | nr | 59.88 |
| 1350mm diameter | nr | 82.75 |
| 1500mm diameter | nr | 112.85 |
| 1800mm diameter | nr | 150.31 |

BASIC MATERIAL PRICES

**CONCRETE PIPES AND FITTINGS**

Spun/vibrated concrete pipes with flexible joints to BS 5911

| **Diameter in mm** | Unit | **300** £ | **375** £ | **450** £ | **525** £ |
|---|---|---|---|---|---|
| Straight pipes (pipe length 2.5m) | | | | | |
| class H | m | - | 17.00 | 19.92 | 23.29 |
| class M | m | 13.06 | 16.72 | 19.53 | 22.83 |
| class L | m | 12.66 | 16.53 | 19.34 | 22.65 |
| Bends 22.5°/45° | | | | | |
| class H | nr | - | 85.00 | 99.60 | 116.45 |
| class M | nr | 65.30 | 83.60 | 97.65 | 114.15 |
| class L | nr | 63.30 | 82.65 | 96.70 | 113.25 |

Extra over for Sulphate Resisting Concrete = 5%

| **Diameter in mm** | Unit | **750** £ | **900** £ | **1200** £ | **1500** £ |
|---|---|---|---|---|---|
| Straight pipes (pipe length 2.5m) | | | | | |
| class H | m | 41.55 | 54.82 | 97.32 | 144.39 |
| class M | m | 37.78 | 53.19 | 91.51 | 139.22 |
| class L | m | 37.41 | 51.82 | 82.94 | 130.76 |
| Bends 22.5°/45° | | | | | |
| class H | nr | 207.75 | 274.10 | 486.60 | 721.95 |
| class M | nr | 188.90 | 265.95 | 457.55 | 696.10 |
| class L | nr | 187.05 | 259.10 | 414.70 | 653.80 |

| | Unit | £ |
|---|---|---|
| Junctions on concrete pipes (extra over straight pipes), branch diameter | | |
| 150 mm | nr | 26.00 |
| 225 mm | nr | 45.00 |
| 300 mm | nr | 55.00 |
| 375 mm | nr | 65.00 |
| 450 mm | nr | 72.50 |
| 525 mm | nr | 90.00 |
| 600 mm | nr | 112.50 |
| 675 mm | nr | 130.00 |
| 750 mm | nr | 150.00 |
| 900 mm | nr | 235.00 |

| Concrete road gullies to BS 5911 | Unit | £ |
|---|---|---|
| 375mm diameter 750mm deep | nr | 20.07 |
| 450mm diameter 750mm deep | nr | 21.64 |
| 450mm diameter 900mm deep | nr | 22.14 |
| 450mm diameter 1050mm deep | nr | 23.21 |

Extra over for Sulphate Resisting Concrete = (7.5%)

| Porous concrete pipes to BS 1194 (pipe length 1 m) | Unit | £ |
|---|---|---|
| 150mm diameter | m | 2.22 |
| 225mm diameter | m | 3.60 |
| 300mm diameter | m | 5.91 |

## BASIC MATERIAL PRICES

### CONSUMABLE STORES

| | Unit | £ | |
|---|---|---|---|
| Abrasive discs | | | |
| for disc cutters (9") | each | 3.81 | |
| for angle grinders (4.5") | each | 2.02 | |
| for floorsaws (9") | each | 3.81 | |
| Bushing heads | | | |
| for breakers | each | 17.80 | |
| for scabblers | each | 19.50 | |
| Cartridges for Hilti gun (DX 450) (100 Box) | each | 6.90 | |
| Hilti nails NK 32 (100 Box) | each | 8.95 | |
| Hilti nails NK 54 (100 Box) | each | 10.65 | |
| Concrete cutting disc for Stihl saw (12") | each | 3.68 | |
| Diamond blades (wet or dry) | | | |
| asphalt (18") | each | 355.00 | |
| concrete (12") | each | 255.00 | |
| brick (12") | each | 275.00 | |
| Gas | | | |
| Propane 46kg (fuel only) | | 22.40 | |
| Propane 19kg (fuel only) | | 9.70 | |
| Butane 13kg (fuel only) | | 8.05 | |
| Oxygen 6kg small cylinder (gas only) | | 4.05 | |
| Acetylene; ditto | | 4.70 | |
| Compressed Air; ditto | | 2.65 | |
| Lubrication/Oils | | | |
| crankcase | L | 0.76 | |
| transmission | L | | 0.625 |
| hydraulic | L | 0.40 | |
| grease | kg | 0.65 | |
| two stroke | L | 0.38 | |
| airline oil | L | 0.40 | |
| cutting oil | L | 1.62 | |
| Petroleum products | | | |
| petrol (4 star) | L | 0.349 | |
| petrol (unleaded) | L | 0.330 | |
| fuel oil (derv) | L | 0.295 | |
| fuel oil (gas oil) | L | 0.115 | |
| paraffin premium grade | L | 0.22 | |
| Rope | | | |
| Rope (polypropelene) 6mm (draw cord) | 220m | 9.47 | |
| 10mm | 220m | 21.23 | |
| 12mm | 220m | 30.03 | |
| Rope (sisal) 12mm | 220m | 33.45 | |
| Chains 25mm link | ft | 0.80 | |
| 35mm link | ft | 1.00 | |
| Chain brothers | 3m set | 15.00 | |

BASIC MATERIAL PRICES

**CONSUMABLE STORES** - *continued*

| | Unit | £ |
|---|---|---|
| Dewatering disposables | | |
| semi rigid riser pipe | 6m | 2.45 |
| flexible riser pipe | m | 0.35 |
| 65mm x 0.5m cocos | each | 1.75 |
| 65mm x 1.0m cocos | each | 2.30 |
| jointing tape | roll | 1.60 |
| Pneumatic breakers, steels (hex shank) | | |
| MP1 (moil point) | each | 7.75 |
| MP4 | each | 6.35 |
| NC1 (narrow chisel) | each | 7.95 |
| NC4 | each | 6.60 |
| CS1 (clay spade) | each | 24.75 |
| CS4 | each | 25.20 |
| AC1 (asphalt cutter) | each | 16.15 |
| AC3 | each | 17.10 |
| DC1 (digging chisel) | each | 21.15 |
| DC3 | each | 20.70 |
| P1 (Plugs) | each | 17.10 |
| F1 (Feathers) | each | 4.95 |
| ET1/ET2 (combined 7" tampers) | each | 45.65 |
| SM1 (speediburst demolition/rock steel) | each | 14.70 |
| CP9 (point) | each | 4.25 |
| CP9 (chisel) | each | 4.25 |
| FL22 square | each | 7.20 |
| FL22 round | each | 6.40 |
| Kango 900 | each | 3.80 |
| 950 | each | 3.80 |
| 2500 | each | 7.20 |

BASIC MATERIAL PRICES

**CONTRACTORS SITE EQUIPMENT (PURCHASED)**

| | Unit | £ |
|---|---|---|
| Traffic cones and cylinders | | |
| 2 part cone 1m high rubber base | each | 8.95 |
| 2 part cone 0.75m high rubber base | each | 7.10 |
| low index polythene cone 18" including reflective sleeve | each | 3.71 |
| low index polythene cone 21" including reflective sleeve | each | 4.98 |
| low index polythene cone 30" including reflective sleeve | each | 8.31 |
| high visibility motorway cone 38.5" | each | 9.20 |
| traffic cylinder (deliniator) 18" | each | 3.00 |
| traffic cylinder 30" | each | 5.55 |
| base for same | each | 8.50 |
| Fencing pins 4'6" with loop | each | 1.20 |
| Crowd barrier 8'3" galvanized tubular | each | 57.33 |
| Temporary road signs and frames | | |
| triangular x 600 | each | 8.28 |
| rectangular 1050 x 750 | each | 22.64 |
| circular x 600 | each | 10.42 |
| Paraffin road warning lamp | each | 4.04 |
| Battery operated "Trafilite" | each | 8.07 |
| Hydrant stand pipe | each | 77.54 |
| Hydrant key | each | 11.67 |
| Drain stoppers | | |
| 100 (4") steel with plastic cap | each | 4.18 |
| 150 (6") | each | 6.10 |
| 225 (9") | each | 13.85 |
| 300 (12") | each | 23.88 |
| 450 (18") cast iron with metal cap | each | 109.33 |
| 600 (24") | each | 146.67 |
| 750 (30") | each | 368.00 |
| Air bag stopper | | |
| 4" | each | 10.37 |
| 6" | each | 11.95 |
| 9" | each | 15.57 |
| 12" | each | 18.61 |
| Drain testing manometer kit (air gauge) | each | 18.98 |
| Drain testing mandrel | | |
| (PVC excluding draw rope) 4" | each | 7.92 |
| (PVC excluding draw rope) 6" | each | 8.70 |
| (PVC excluding draw rope) 9" | each | 10.45 |
| Bar croppers | | |
| 24" (3/4" jaw) | each | 70.84 |
| high tensile cut 42" (1" jaw) | each | 181.78 |
| Soil pipe cutter (spun pipes or cast) 2" - 24" | each | 271.90 |
| Glazed pipe cutter 6" - 9" | each | 128.24 |

BASIC MATERIAL PRICES

**CONTRACTORS SITE EQUIPMENT (PURCHASED)** - *continued*

| | Unit | £ |
|---|---|---|
| Sandbag polypropelene | each | 0.26 |
| Debris netting (scaffolding net, 50 x 4 m roll, heavyweight) | each | 171.60 |
| Scaffold sheeting (weather sheet, 48 x 2 m roll) | each | 188.77 |
| | | |
| Compressor hose 3/4" x 50' complete with couplings | each | 38.11 |
| Rubber water hose 3/4" bore x 40m | each | 34.76 |
| PVC water hose 1/2" bore x 75m | each | 20.88 |
| | | |
| Post hole borer 6" | each | 42.29 |

BASIC MATERIAL PRICES

**CULVERTS**

| **Diameter in mm** | Unit | **500** £ | **1000** £ | **1500** £ | **2000** £ |
|---|---|---|---|---|---|
| Well-void extra plus steel culverts | | | | | |
| 1.25 mm thick | m | 19.80 | 39.60 | - | - |
| 1.50 mm thick | m | 28.05 | 56.10 | 84.70 | 123.20 |
| 2.00 mm thick | m | - | - | 104.50 | 148.50 |

| **Diameter in mm** | Unit | **1000** £ | **1600** £ | **2000** £ | **2200** £ |
|---|---|---|---|---|---|
| Asset International corrugated steel, galvanised, bitumen coated culverts | | | | | |
| 1.5 mm thick | m | 52.00 | 113.00 | - | - |
| 2.0 mm thick | m | - | - | 172.00 | 185.00 |

| | Unit | £ |
|---|---|---|
| Precast concrete rectangular culverts | | |
| 1000 x 500 mm 2.0 m long | nr | 250.00 |
| 1500 x 1000 mm 2.0 m long | nr | 460.00 |
| 1500 x 1500 mm 2.0 m long | nr | 530.00 |
| 2000 x 1000 mm 2.0 m long | nr | 640.00 |
| 2000 x 1500 mm 2.0 m long | nr | 720.00 |
| 2500 x 1000 mm 2.0 m long | nr | 970.00 |
| 2500 x 1750 mm 1.5 m long | nr | 850.00 |
| 3000 x 2000 mm 1.0 m long | nr | 830.00 |
| 3000 x 2750 mm 1.0 m long | nr | 930.00 |
| 4000 x 2500 mm 1.0 m long | nr | 1100.00 |

BASIC MATERIAL PRICES

**EARTH RETENTION, INCLUDING GABIONS, CRIB WALLING, ETC**

Maccaferri woven wire mesh gabions in mesh size 80 x 100 mm, wire diameter 2.7 mm galvanised to BS 443:1982

PVC coated galvanised wire overall diameter 3.7 mm

| Unit size | £ |
|---|---|
| 1.5 x 1 x 1 m | 27.25 |
| 2 x 1 x 1 m | 33.50 |
| 3 x 1 x 1 m | 46.60 |
| 2 x 1 x 0.5 m | 25.75 |

Galvanised only

| Unit Size | £ |
|---|---|
| 1.5 x 1 x 1 m | 16.40 |
| 2 x 1 x 1 m | 21.65 |
| 3 x 1 x 1 m | 30.65 |
| 2 x 1 x 0.5 m | 16.80 |

Reno woven wire mattresses in mesh size 80 x 100 mm, wire galvanised to BS 443 :1982

PCV coated galvanised wire 2.0 mm diameter, overall diameter 3.0 mm

| Unit size | £ |
|---|---|
| 6 x 2 x 0.17 m | 61.80 |
| 6 x 2 x 0.23 m | 64.20 |

PVC coated galvanised wire 2.4 mm diameter, overall diameter 3.4 mm

| Unit Size | £ |
|---|---|
| 6 x 2 x 0.3 m | 66.15 |

Galvanised wire 2.0 mm diameter

| Unit size | £ |
|---|---|
| 6 x 2 x 0.17 m | 39.00 |
| 6 x 2 x 0.23 m | 40.20 |
| 6 x 2 x 0.30 m | 42.60 |

Tensar GM4 gabions (62 x 62 x 1.75mm)

| Unit size | £ |
|---|---|
| 2 x 1 x 0.5 m | 25.45 |
| 2 x 1 x 1 m | 29.66 |
| 4 x 1 x 1 m | 69.16 |
| 4 x 1 x 0.5 m | 59.30 |

Tensar GM4 mattresses

| Unit size | £ |
|---|---|
| 6 x 2 x 0.17 m | 86.59 |
| 6 x 2 x 0.23 m | 90.62 |
| 6 x 2 x 0.30 m | 95.47 |

BASIC MATERIAL PRICES

**EARTH RETENTION, INCLUDING GABIONS, CRIB WALLING, ETC** - *continued*

| | Unit | £ |
|---|---|---|
| Pre-cast concrete crib units | | |
| Andacrib maxi system | $m^2$ | 65.00 |
| Andacrib mini system | $m^2$ | 54.00 |
| Timber crib walling system | | |
| Permacrib Goliath for retaining walls up to 3.6m high | $m^2$ | 45.00 |
| Permacrib Hercules for retaining walls up to 6.3m high | $m^2$ | 90.00 |
| Precast concrete wall units | | |
| Milton Wall units | | |
| 1050 mm high x 914 mm long | nr | 40.84 |
| 1500 mm high x 914 mm long | nr | 55.41 |
| 1800 mm high x 914 mm long | nr | 90.84 |
| 2400 mm high x 914 mm long | nr | 134.27 |
| 3000 mm high x 914 mm long | nr | 237.09 |
| 3600 mm high x 610 mm long | nr | 199.86 |

| | Unit | Toe £ | Heel £ |
|---|---|---|---|
| Milton Filler Units | | | |
| for 1050 mm high wall unit | nr | 20.03 | 14.35 |
| for 1500 mm high wall unit | nr | 49.33 | 22.05 |
| for 1800 mm high wall unit | nr | 63.11 | 28.69 |
| for 2400 mm high wall unit | nr | 107.11 | 43.43 |
| for 3000 mm high wall unit | nr | 149.46 | 60.64 |
| for 3600 mm high wall unit | nr | 165.83 | 80.83 |
| Mild steel straps and bolts for filler units | | | |
| 1050 mm - 3000 mm high units | nr | 74.48 | 68.91 |
| 3600 mm high units | nr | 111.71 | 103.33 |

Extra over for Sulphate Resisting Cement = 15%

BASIC MATERIAL PRICES

**FENCING (SITE BOUNDARY FENCING)**

| | Unit | £ |
|---|---|---|
| **Chainlink fencing (25 m rolls)** | | |
| galvanised, 1200 mm wide | nr | 65.00 |
| galvanised, 1800 mm wide | nr | 85.33 |
| plastic coated, 1200 mm wide | nr | 74.33 |
| plastic coated, 1800 mm wide | nr | 100.35 |
| **Concrete Posts** | | |
| straight posts for 1200 mm high fencing | nr | 6.15 |
| straight posts for 1800 mm high fencing | nr | 8.00 |
| cranked posts for 1800 mm high fencing | nr | 10.00 |
| struts | nr | 7.25 |
| **Chestnut posts** | | |
| 64 mm diameter posts for 900 mm high fencing | nr | 1.27 |
| 64 mm diameter posts for 1200 mm high fencing | nr | 1.67 |
| 75 x 51 mm struts | nr | 1.60 |
| **Chestnut pale fencing (9 m rolls)** | | |
| 900 mm wide | nr | 23.75 |
| 1200 mm wide | nr | 25.50 |
| **Wire** | | |
| galvanised line wire (25 m rolls) | nr | 6.00 |
| plastic coated wire (25 m rolls) | nr | 6.50 |
| barbed wire (200 m rolls) | nr | 18.00 |
| eye bolts | nr | 0.40 |
| **Softwood fencing** | | |
| 100 x 100 mm posts | m | 2.65 |
| 89 x 19 mm rails | m | 0.64 |
| 100 x 25 mm rails | m | 1.13 |
| 100 x 25 mm feather edge rails | m | 0.55 |
| 200 x 38 mm rails | m | 2.12 |

BASIC MATERIAL PRICES

**GEOTEXTILES**

| | Unit | £ |
|---|---|---|
| Tensar SS1 (50 x 4m roll) for weak soils | nr | 192.00 |
| Tensar SS2 (50 x 4m roll) for very weak soil | nr | 301.50 |
| Tensar SS3 (50 x 4m roll) for trafficked areas | nr | 274.00 |
| Tensar SS35 (30 x 3.5m roll) for over mine workings | nr | 246.00 |
| Tensar ARI (50 x 3.8m roll) for reinforcement of asphalt pavement | nr | 276.50 |
| Tensar GM4 (40 x 4m roll) for gabion constructions | nr | 304.50 |
| Lotrak permeable membrane (100 x 4.5m rolls) | | |
| 10/7 | nr | 130.50 |
| 16/15 | nr | 166.50 |
| 45/45 (100 x 5m rolls) | nr | 306.00 |
| MT22/20 monofilament (100 x 4.5m rolls) | nr | 193.50 |
| Netlon (30 x 2m rolls) | | |
| CE111 | nr | 73.50 |
| CE121 | nr | 121.50 |
| CE131 | nr | 108.50 |
| CE152 (15 x 2m rolls) | nr | 56.50 |
| Terram 1000 (100 x 4.5m roll) permeable membranes | nr | 168.75 |
| Terram 2000 (100 x 4.5m roll) | nr | 300.15 |
| Terram 700 (150 x 4.5m roll) | nr | 224.77 |
| Filtram laminated filter drainage membrane (25 x 1.9m roll) | | |
| Filtram IBI | nr | 166.25 |
| Filtram IBZ | nr | 178.15 |
| Visqueen Polythene (25 x 4m roll) | nr | 29.58 |
| Polythene to PIFA standard 100 mym (4 x 50m) | nr | 28.70 |
| 200 mym (4 x 25m) | nr | 28.70 |
| Hessian (1.83 x 100m roll) | nr | 80.00 |
| Trammel Drainage Fabric (for fin drains) | | |
| Fabric F40 (100 x 3.1m roll) | $m^2$ | 0.85 |
| Core (50 x 2m roll) | $m^2$ | 1.96 |
| Fixing staples (2000 box) | nr | 3.00 |

BASIC MATERIAL PRICES

**GRATES AND FRAMES TO BS 497**

| | Unit | £ |
|---|---|---|
| Grade 'A' class 1, heavy duty double triangular pattern | | |
| 508 x 451 mm; (152 kg); ref GA1-450 | nr | 53.27 |
| 475 x 475 mm; (102 kg); ref GA1-450 | nr | 39.30 |
| Grade 'A' ductile iron, double traingular | | |
| 460 x 460 mm; (60 kg); ref GA1-450 | nr | 50.55 |
| 505 x 455 mm; (62 kg); ref GA1-450 | nr | 52.00 |
| 600 x 600 mm; (124 kg); ref GA1-450 | nr | 84.37 |
| Grade 'A' class 2 heavy duty straight bar, hinged, lift-out pattern | | |
| 400 x 349 x 100 mm depth; (55 kg); ref GA2-325 | nr | 26.97 |
| 502 x 349 x 100 mm depth; (70 kg); ref GA2-325 | nr | 35.49 |
| 400 x 349 x 150 mm depth; (71 kg); ref GA2-325 | nr | 33.14 |
| 502 x 349 x 150 mm depth; (91 kg); ref GA2-325 | nr | 39.95 |
| 457 x 457 x 115 mm depth; (92 kg); ref GA2-450 | nr | 36.80 |
| Grade 'A' ductile iron, end hinged, anti theft | | |
| 385 x 317 mm; (31 kg); ref GA2-325 | nr | 42.64 |
| 510 x 360 mm; (47.5 kg); ref GA2-325 | nr | 44.34 |
| 385 x 440 mm; (40 kg); ref GA2-450 | nr | 47.58 |
| Grade 'B' class 2 medium duty straight bar, hinged, lift-out pattern | | |
| 400 x 349 x 100 mm depth; (55 kg); ref GB-325 | nr | 26.97 |
| 502 x 349 x 100 mm depth; (70 kg); ref GB-325 | nr | 35.49 |
| 457 x 457 x 115 mm depth; (92 kg); ref GB-450 | nr | 36.80 |
| Grade 'B' ductile iron, end hinged, anti-theft | | |
| 385 x 317 mm; (29 kg); ref GB-325 | nr | 34.57 |
| 510 x 360 mm; (40 kg); ref GB-325 | nr | 35.97 |
| 385 x 317 mm pedestrianised; (29.5 kg); ref GB-325 | nr | 35.90 |
| Straight bar, dished top, hinged, lift-out pattern | | |
| 400 x 349 mm; (76 kg) | nr | 33.55 |
| 502 x 349 mm; (140 kg) | nr | 49.10 |
| Kerb type pattern - to suit standard kerb profile | | |
| 483 x 381 mm; (114 kg) | nr | 58.69 |
| road retaining bars for last | nr | 3.45 |
| Hinged and bolted, dished top, straight bar pattern | | |
| 229 x 229 mm; (17 kg) | nr | 14.09 |
| 305 x 305 mm; (29.5 kg) | nr | 24.88 |
| 305 x 279 mm; (64 kg) | nr | 38.00 |

**MANHOLE STEP IRONS**

| | Unit | £ |
|---|---|---|
| Step irons, galvanized malleable iron | | |
| 118mm tail (4 5/8") | nr | 2.41 |
| 230mm tail (9 1/8") | nr | 3.15 |

BASIC MATERIAL PRICES

**INSTRUMENTATION (SOIL INSTRUMENTATION)**

| | Unit | £ |
|---|---|---|
| Tacseal pipe 450mm diameter x 5m long | each | 115.00 |
| x 2.15m long | each | 79.00 |
| x 0.45m long | each | 43.50 |
| x 0.35mm long | each | 38.00 |
| Joints to suit | each | 25.00 |
| 150mm UPVC class C (as protective sleeve) | m | 4.75 |
| 32mm mild steel rods (as settlement rod) | m | 2.20 |
| 150mm diameter steel tube (as protective sleeve) | m | 11.72 |
| Inclinometer tube | m | 9.80 |
| Piezometer tube (nylon) | m | 11.32 |

READOUT EQUIPMENT

Prices vary depending on number of instruments, locations, distance from each instrument. The following is an indicative price for modern portable equipment used for reading up to 30 separate instrument. Item complete and installed.

| | Unit | £ |
|---|---|---|
| Readout equipment for Inclinometer | Item | 13,750.00 |
| Readout equipment for Peizometer | Item | 2,500.00 |

BASIC MATERIAL PRICES

**INTERLOCKING STEEL SHEET PILING AND BEARING PILES**

The following prices are ex works in lots of 20 tonnes and over of one section and one size and in lengths between 7 and 14m for delivery to one destination. Carriage is not included

INTERLOCKING SHEET PILING

| Section | Unit | BS 4360:1986 Grade 43A £ | BS 4360:1986 Grade 50A £ |
|---|---|---|---|
| **Frodingham** | | | |
| 1BXN | t | 388.00 | 421.00 |
| 1N | t | 378.00 | 411.00 |
| 2N | t | 378.00 | 411.00 |
| 3N | t | 378.00 | 411.00 |
| 3NA | t | 378.00 | 411.00 |
| 4N | t | 378.00 | 411.00 |
| 5 | t | 378.00 | 411.00 |
| SW1 and SW1A | t | 417.00 | 450.00 |
| **Larssen** | | | |
| 6W | t | 417.00 | 450.00 |
| 9W | t | 407.00 | 440.00 |
| 12W | t | 392.00 | 425.00 |
| 16W | t | 392.00 | 425.00 |
| 20W | t | 392.00 | 425.00 |
| 25W | t | 392.00 | 425.00 |
| 32W | t | 397.00 | 430.00 |
| 3 | t | 387.00 | 420.00 |
| 4A | t | 387.00 | 420.00 |
| 6 x 122kg per metre | t | 392.00 | 425.00 |
| 6 x 131kg per metre | t | 392.00 | 425.00 |
| 6 x 138kg per metre | t | 392.00 | 425.00 |

Add to the above prices for:

| | | Unit | £ |
|---|---|---|---|
| Quantity | under 5 tonnes | t | 45.00 |
| | under 10 tonnes to 5 tonnes | t | 30.00 |
| | under 20 tonnes to 10 tonnes | t | 15.00 |
| Quality | copper content between 0.20% and 0.35% | t | 10.00 |
| | copper content over 0.35% not exceeding 0.50% | t | 14.00 |

Size the following are £ extra per tonne for various alternative lengths

Frodingham section

| Length | IBXN £ | 1N £ | 2N £ | 3N £ | 3NA £ | 4N £ | 5 £ | SW1 £ | SW1A £ |
|---|---|---|---|---|---|---|---|---|---|
| 2 - 4m | 35.00 | 45.00 | 40.00 | 30.00 | 35.00 | 25.00 | 20.00 | - | - |
| 4 - 7m | 15.00 | 20.00 | 20.00 | 15.00 | 15.00 | 15.00 | 10.00 | 20.00 | 20.00 |
| 14 - 20m | 10.00 | 10.00 | 10.00 | 5.00 | 5.00 | 5.00 | 5.00 | 10.00 | 10.00 |
| 20 - 25m | 25.00 | - | - | 25.00 | 25.00 | 20.00 | 15.00 | - | - |
| 25 - 29m | - | - | - | - | - | 30.00 | 25.00 | - | - |

BASIC MATERIAL PRICES

**STEEL PILING** - *continued*

Larssen Section

| Length | 6W | 9W | 12W | 16W | 20W | 25W | 32W | 3 | 4A | 6 |
|---|---|---|---|---|---|---|---|---|---|---|
| | £ | £ | £ | £ | £ | £ | £ | £ | £ | £ |
| 2 - 4m | 45.00 | 40.00 | 30.00 | 30.00 | 25.00 | 20.00 | 20.00 | 30.00 | 25.00 | 15.00 |
| 4 - 7m | 20.00 | 20.00 | 15.00 | 15.00 | 15.00 | 10.00 | 10.00 | 15.00 | 15.00 | 10.00 |
| 14 - 20m | 10.00 | 10.00 | 10.00 | 5.00 | 5.00 | 5.00 | 5.00 | 5.00 | 5.00 | 5.00 |
| 20 - 25m | - | - | - | 25.00 | 25.00 | 20.00 | 20.00 | 30.00 | 25.00 | 15.00 |
| 25 - 29m | - | - | - | - | 35.00 | 30.00 | 25.00 | - | 35.00 | 25.00 |

UNIVERSAL BEARING PILES

| Size mm | Weight kg per metre | Unit | BS 4360: 1986 Grade 43A £ | BS 4360: 1986 Grade 50A £ |
|---|---|---|---|---|
| 356 x 368 x 174 | | t | 384.00 | 409.00 |
| 356 x 368 x 152 | | t | 380.00 | 405.00 |
| 356 x 368 x 133 | | t | 380.00 | 405.00 |
| 356 x 368 x 109 | | t | 380.00 | 405.00 |
| 305 x 305 x 223 | | t | 362.00 | 387.00 |
| 305 x 305 x 186 | | t | 356.00 | 381.00 |
| 305 x 305 x 149 | | t | 356.00 | 381.00 |
| 305 x 305 x 126 | | t | 356.00 | 381.00 |
| 305 x 305 x 110 | | t | 348.00 | 373.00 |
| 305 x 305 x 95 | | t | 348.00 | 373.00 |
| 305 x 305 x 88 | | t | 348.00 | 373.00 |
| 305 x 305 x 79 | | t | 348.00 | 373.00 |
| 254 x 254 x 85 | | t | 325.00 | 350.00 |
| 254 x 254 x 71 | | t | 316.00 | 341.00 |
| 254 x 254 x 63 | | t | 316.00 | 341.00 |
| 203 x 203 x 54 | | t | 320.00 | 345.00 |
| 203 x 203 x 45 | | t | 320.00 | 345.00 |

| Add to the above prices for: | | Unit | £ |
|---|---|---|---|
| Quantity | under 5 tonnes | t | 20.00 |
| | under 10 tonnes to 5 tonnes | t | 10.00 |
| | under 20 tonnes to 10 tonnes | t | 5.00 |
| Quality | copper content between 0.20% and 0.35% | t | 10.00 |
| | copper content over 0.35% not exceeding 0.50% | t | 14.00 |
| Size | length under 3 m | t | 15.00 |
| | length 3 m to under 5 m | t | 10.00 |
| | length over 14 m to 24 m | t | 5.00 |
| | length over 24 m to 26 m | t | 10.00 |
| Holes | lifting holes | nr | 8.00 |

BASIC MATERIAL PRICES

**JOINT FILLERS AND WATERSTOPS**

| | Unit | £ |
|---|---|---|
| Flexible epoxy resin joint sealant (Epoflex 800) | | |
| 2 litre tin | nr | 24.99 |
| Cold applied waterproofing membrane (Mulseal) | | |
| 5 litre can | nr | 5.36 |
| 25 litre drum | nr | 23.56 |
| 200 litre drum | nr | 153.55 |
| Hot bonded bitumen sheet waterproofing membrane (Famflex) | | |
| 350g/m$^2$ rolls 1 m x 8 m | nr | 34.78 |
| Non absorbent joint filler (Hydrocell) | | |
| sheet size: 10 x 960 x 2250 mm | nr | 21.67 |
| 15 x 940 x 2250 mm | nr | 33.01 |
| 20 x 920 x 2250 mm | nr | 41.82 |
| 25 x 920 x 2250 mm | nr | 53.24 |
| Polysulphide sealant (Thioflex 600) | | |
| Black gun grade | L | 8.60 |
| Grey gun grade | L | 8.60 |
| Mahogany gun grade | L | 9.03 |
| Stone gun grade | L | 9.03 |
| Off white gun grade | L | 9.03 |
| Brick red gun grade | L | 9.03 |
| Grey pouring grade | L | 8.74 |
| Hot applied sealant | | |
| hand applied (Plastijoint) | L | 2.64 |
| poured, low extension grade, Pliastic A2 (25 kg sack) | nr | 25.98 |
| poured, hard grade, Pliastic 77 (25 kg sack) | nr | 24.36 |
| Resin mortars (Expocrete) | | |
| GP | L | 7.56 |
| BC | L | 8.21 |
| UA | L | 9.64 |
| High duty elastomeric pavement sealant (Colpor 200) | L | 6.55 |
| Non extruding expansion joint filler (Flexcell) | | |
| 10 x 2440 x 1200 mm | nr | 8.21 |
| 12.5 x 2440 x 1200 mm | nr | 9.19 |
| 20 x 2440 x 1200 mm | nr | 15.75 |
| 25 x 2440 x 1200 mm | nr | 19.10 |
| Flexible expansion joint strip membrane (Expoband) | | |
| 190mm x 20 m roll | nr | 147.42 |
| 250mm x 20 m roll | nr | 195.41 |
| 1520mm x 20 m roll | nr | 776.62 |
| Bituthene rolls | | |
| 500 grade (20m$^2$) | m$^2$ | 3.92 |
| 1000 grade (20m$^2$) | m$^2$ | 4.36 |
| heavy duty (16m$^2$) | m$^2$ | 4.70 |
| Bitu-shield green (10m$^2$) | m$^2$ | 5.19 |

BASIC MATERIAL PRICES

**JOINT FILLERS AND WATERSTOPS** - *continued*

| | Unit | £ |
|---|---|---|
| Deck flashings (Type DF6) | | |
| 235mm wide (inverted) | m | 14.13 |
| | | |
| Biodegration resistant sealant (Secoseal 12) | | |
| sealant (2 litre tin) | nr | 24.61 |
| accelerator | nr | 4.36 |
| | | |
| Cold applied bituminous sheet waterproofing membrane (Proofex) | | |
| plain (1 m x 15 m roll) | nr | 73.33 |
| reinforced (1 m x 12 m roll) | nr | 65.18 |
| | | |
| Cold applied bitumen/rubber sealant | | |
| gun grade cartridge | nr | 1.78 |
| trowel grade | L | 3.00 |
| | | |
| Servi-dek standard grade | | |
| 18 litre unit | nr | 26.35 |
| | | |
| Parapet flashing 3 m length | | |
| PF10 | length | 5.04 |
| PF20 | length | 10.40 |
| | | |
| Pile cap Serviseal | | |
| 15 metre coil | nr | 173.70 |
| angles | nr | 39.93 |
| | | |
| PVC waterstop (15 m coil) | | |
| 100 mm flat dumbell | nr | 47.55 |
| 170 mm flat dumbell | nr | 66.90 |
| 210 mm flat dumbell | nr | 79.20 |
| 250 mm flat dumbell | nr | 115.65 |
| 160 mm centre bulb | nr | 64.20 |
| 210 mm centre bulb | nr | 92.55 |
| 260 mm centre bulb | nr | 131.55 |
| 325 mm centre bulb | nr | 221.10 |
| | | |
| Rubber waterstop | | |
| 150 mm flat dumbell | m | 12.83 |
| 230 mm flat dumbell | m | 19.40 |
| 150 mm centre bulb | m | 14.76 |
| 230 mm centre bulb | m | 22.05 |
| 305 mm centre bulb | m | 36.36 |
| | | |
| Servicore | | |
| 6 mm dia standard section (30 m coil) | nr | 94.48 |
| 10 mm dia tubular section (23 m coil) | nr | 76.52 |
| 13 mm dia standard section (23 m coil) | nr | 96.39 |
| 19 mm dia standard section (15 m coil) | nr | 76.39 |
| 25 mm dia standard section (12 m coil) | nr | 81.37 |
| 32 mm dia standard section (3 m lengths) | nr | 29.12 |
| 38 mm dia standard section (3 m lengths) | nr | 39.81 |

BASIC MATERIAL PRICES

**LANDSCAPING/SOILING**

(For a more detailed appraisal of landscaping prices, please refer to Spons Landscape and External Works Price Book).

| | Unit | £ |
|---|---|---|
| CULTIVATED TURF (per square yard, 81" x 16") | | |
| Rowlawn select | $yd^2$ | 1.70 |
| Rowlawn R2 olympic | $yd^2$ | 1.45 |
| Rowlawn RB medallion | $yd^2$ | 1.25 |

IMPORTED TOPSOIL

The price for topsoil is so variable depending on site location, quantity available, season, etc. The following prices are a guide for use in the unit cost build ups and reflect reasonable quality soils from a single source delivered to site in 20 tonne tipper trucks, within a 20km radius of source.

| | Unit | 0-1000 £ | 1000-5000 £ | 5000+ £ |
|---|---|---|---|---|
| Subsoil | $m^3$ | 3.75 | 3.15 | 3.15 |
| Reasonable quality Topsoil for seeding and general, sward establishment (1.6t/m3) | $m^3$ | 8.75 | 8.20 | 7.55 |
| Better quality Topsoil for species planting and for use in raised beds, planters, etc (1.6t/m3) | $m^3$ | 10.00 | 9.50 | 8.70 |

Other imported growing mediums/soil improvers

| | Unit | £ |
|---|---|---|
| Peat, as dug | $m^3$ | 12.00 |
| Peat, sphagnum | $m^3$ | 16.50 |
| Coarse bark | $m^3$ | 18.40 |
| Composted bark | $m^3$ | 26.40 |
| Bark nuggets | $m^3$ | 22.40 |
| Graded bark flakes | $m^3$ | 21.40 |
| Ground limestone | $m^3$ | 11.70 |
| Woodfibre mulch | $m^3$ | 12.40 |
| Fisons STC sports turf planting compost | $m^3$ | 45.07 |
| Fisons TPMC tree and shrub planting compost | $m^3$ | 20.19 |
| Standard bark mulch | $m^3$ | 24.40 |

BASIC MATERIAL PRICES

**LANDSCAPING/SOILING** - *continued*

GRASS SEEDS/FERTILIZERS etc

| | Unit | £ |
|---|---|---|
| Pre seeding fertilizers | | |
| Fisons PS5 | 25kg | 5.17 |
| SA1 longlife | 25kg | 8.85 |
| Grass seed BSH ref A15 (Reclamation) | 25kg | 67.55 |
| A16 (Country Park) | 25kg | 81.75 |
| A18 (Road Verges) | 25kg | 55.60 |
| A3 (Landscape) | 25kg | 62.90 |
| A4 (Parkland) | 25kg | 79.10 |
| Fisons "Ficote" fertilizer (per 10kg) | 10kg | 18.24 |
| Growmore granular fertilizer (per 25kg) | 25kg | 8.95 |
| Selective weedkiller | 25L | 80.50 |

Hydroseeding - specialist process; refer to unit cost landscaping section for typical specification and situation.

BASIC MATERIAL PRICES

**MANHOLE COVERS AND FRAMES TO BS497**

Manhole covers and frames (all dimensions are clear opening sizes)

| | Unit | £ |
|---|---|---|
| Grade 'A' ductile iron, double triangular, hinged, lockable | | |
| 600 x 600 mm; (136 kg); ref MA-60 | nr | 115.90 |
| Grade 'A' ductile iron, double triangular | | |
| 600 mm dia; (108 kg); ref MA-60 | nr | 77.83 |
| 610 x 610 mm; (103 kg); ref MA-60 | nr | 77.83 |
| 675 mm dia; (142 kg); ref MA-60 | nr | 146.85 |
| 675 x 675 mm; (134 kg) | nr | 146.85 |
| 600 x 900 mm; (229 kg); ref MA-60 | nr | 180.22 |
| 1220 x 685 mm Super Heavy Duty; (413 kg); ref 'A' Plus | nr | 390.74 |
| Grade 'A' heavy duty, double triangular | | |
| 508 mm dia; (229 kg); ref MA-50 | nr | 74.05 |
| 559 mm dia; (254 kg); ref MA-55 | nr | 75.90 |
| 610 x 610 mm; (229 kg); ref MA-60 | nr | 71.50 |
| Grade 'A' ductile iron, single triangular hinged | | |
| 600 x 600 mm; (105 kg); ref MA-T | nr | 93.66 |
| Grade 'A' heavy duty, 3 point suspension, non rocking, single triangular | | |
| 495 mm; (203 kg); ref MA-T | nr | 72.90 |
| 610 x 610 mm; (216 kg); ref MA-T | nr | 82.45 |
| Grade 'B' class 1, ductile iron, circular | | |
| 610 mm dia; (71 kg); ref MB1-60 | nr | 70.38 |
| 550 mm dia; (67.5 kg); ref MB1-55 | nr | 54.28 |
| Grade 'B' class 1, ductile iron double triangular | | |
| 600 x 600 mm; (83.5 kg); ref MB1-60 | nr | 68.46 |
| Grade 'B' class 2, medium duty, circular | | |
| 508 mm dia; (114 kg); ref MB2-50 | nr | 47.65 |
| 559 mm dia; (127 kg); ref MB2-55 | nr | 48.09 |
| 610 mm dia; (152 kg); ref MB2-60 | nr | 50.15 |
| Grade 'B' class 2, medium duty | | |
| 610 x 457 mm; (143 kg); ref MB2-60/45 | nr | 49.55 |
| 610 x 610 mm; (152 kg); ref MB2-60/60 | nr | 59.80 |
| Grade 'B' class 2, ductile iron, single seal | | |
| 600 x 600 mm; (75 kg); ref MB2-60/60 | nr | 68.46 |
| 450 x 600 mm; (58 kg); ref MB2-60/45 | nr | 53.23 |
| Grade 'C' light duty, double seal, flat type inspection covers & frames | | |
| 457 x 457 mm; (44 kg); ref MC2-45/45 | nr | 44.15 |
| 610 x 457 mm; (52 kg); ref MC2-60/45 | nr | 45.05 |
| 610 x 610 mm; (78 kg); ref MC2-60/60 | nr | 80.65 |

BASIC MATERIAL PRICES

**MANHOLE COVERS AND FRAMES TO BS497** - *continued*

| | Unit | £ |
|---|---|---|
| Grade 'C' light duty, single seal, flat type inspection covers & frames | | |
| 457 x 457 mm; (31 kg); ref MC1-45/45 | nr | 19.93 |
| 610 x 457 mm; (32 kg); ref MC1-60/45 | nr | 18.30 |
| 610 x 610 mm; (51 kg); ref MC1-60/60 | nr | 40.95 |
| Light duty, single seal, inspection covers and frames | | |
| 610 x 457mm; (25kg) | nr | 13.44 |
| 610 x 457mm; (51kg) | nr | 37.54 |

BASIC MATERIAL PRICES

## PAINT/STAINS/PROTECTIVE COATINGS

| | Unit | £ |
|---|---|---|
| Crown trade gloss | | |
| white and magnolia | 5 L | 18.27 |
| brilliant white | 5 L | 18.97 |
| colours | 5 L | 19.72 |
| Crown trade undercoat | | |
| white | 5 L | 18.27 |
| colours | 5 L | 19.72 |
| Crown trade matt emulsion | | |
| white and magnolia | 5 L | 15.12 |
| brilliant white | 5 L | 15.52 |
| colours | 5 L | 16.09 |
| Crown trade silk emulsion | | |
| white and magnolia | 5 L | 16.76 |
| brilliant white | 5 L | 17.06 |
| colours | 5 L | 17.49 |
| Crown trade masonry paint | 5 L | 23.71 |
| Crown woodstain | | |
| matt | 5 L | 23.77 |
| satin | 5 L | 25.60 |
| satin for hardwoods | 2.5 L | 14.63 |
| hardwood base coat | 1 L | 6.70 |
| Cuprinol trade varnishes | | |
| gloss, eggshell and matt varnishes | 2.5 L | 9.47 |
| yacht varnish | 2.5 L | 10.06 |
| Crown Primers | | |
| alkali resisting plaster primer | 5 L | 16.13 |
| oil based non-toxic wood primer | 5 L | 18.76 |
| acrylic wood primer | 5 L | 17.29 |
| calcium plumbate metal primer | 5 L | 27.89 |
| zinc phosphate metal primer | 5 L | 24.78 |
| red lead metal primer | 5 L | 35.54 |
| Crown road marking paint, white/yellow | 5 L | 23.21 |
| Sadolin Holdex | 5 L | 31.64 |
| Classic | 5 L | 26.78 |
| Solignum Architectural | 5 L | 30.75 |
| Clear | 5 L | 14.94 |
| Wykamol 'plus' | 25 L | 26.96 |
| Aluminium paint | 5 L | 21.70 |
| Anti condensation paint | 5 L | 20.79 |
| Red oxide paint | 5 L | 18.82 |
| Concrete floor paint | 5 L | 21.06 |

BASIC MATERIAL PRICES

**PAINT/STAINS/PROTECTIVE COATINGS** - continued

Protective coatings (examples only, used for fabricated steels, lighting columns, parapets and guard railings, etc)

| | Litre £ | Theoretical metre$^2$ £ |
|---|---|---|
| Epoxy zinc rich primer (pre fabrication) | 8.24 | 0.24 |
| Epoxy zinc phosphate primer (post fabrication) | 5.13 | 0.64 |
| Etch primer | 3.79 | 0.18 |
| Passivating T wash (pre galvanising) | 1.71 | 0.07 |
| Lead free primers - zinc phosphate | 4.85 | 0.28 |
| Phenolin based micaceus iron oxide | 4.43 | 0.47 |
| Urethane alkyd undercoat | 3.33 | 0.16 |
| High build alkyd gloss | 3.69 | 0.32 |
| Urethane alkyd gloss | 3.33 | 0.16 |
| High build epoxy micaceous iron oxide undercoat/finish | 5.33 | 0.92 |
| Epoxy enamel finish | 5.95 | 0.30 |
| Epoxy tie coat | 5.00 | 0.25 |
| Ferox urethane zinc primer | 7.69 | 0.46 |
| Ferox urethane intermediate/undercoat | 5.49 | 0.61 |
| Ferox urethane gloss finish | 9.28 | 0.55 |
| Non slip deck coating | 4.61 | 1.29 |

| | Unit | £ | |
|---|---|---|---|
| RIW liquid asphatic composition (two coats) | L | 1.84 | 1.84 |
| RIW 808 fibrous waterproofing composition (two coats) | L | 2.06 | 2.75 |
| RIW 120 waterproofer (two coats) | kg | 1.26 | 3.78 |
| RIW 300 SBR latex | L | 2.23 | 6.03 |
| RIW protection board 2.44 x 1.23m | sheet | 6.22 | - |
| Crown anti graffiti paint systems | | | |
| Ferox urethane undercoat | L | 5.50 | - |
| Crown Ferox concrete sealer | L | 4.90 | - |
| Crown Ferox gloss finish | L | 9.25 | - |
| Crown Cleanwall basecoat | L | 7.04 | - |
| Crown Cleanwall topcoat | L | 7.04 | - |
| Crown Cleanwall graffiti remover | L | 2.23 | - |
| Crown Cleanwall graffiti remover gel | L | 2.23 | - |
| Road lining paint (white or yellow) | L | 4.64 | - |

BASIC MATERIAL PRICES

**POLYMER CHANNELS AND FITTINGS**

| | Unit | £ |
|---|---|---|
| ACO S100 interlocking, stepped channel with locked ductile iron grating | | |
| length 1000 mm | nr | 83.75 |
| length 500 mm | nr | 49.50 |
| step connector | nr | 4.95 |
| ACO Frame interlocking, pre-sloped channel with ductile iron frame, grating and bolts | | |
| length 1000 mm | nr | 59.99 |
| concrete anchors | nr | 1.35 |
| ACO Slot Channel interlocking channel | | |
| length 1000 mm | nr | 21.95 |
| length 500 mm | nr | 15.95 |
| length 1000 mm; with removable inspection panel | nr | 27.50 |
| ACO NW100 interlocking, pre-sloped channel | | |
| length 1000 mm | nr | 15.99 |
| length 500 mm | nr | 12.25 |
| ACO K100S interlocking, pre-sloped channel with galvanised steel frame | | |
| length 1000 mm | nr | 22.95 |
| length 500 mm | nr | 17.50 |
| ACO end caps | | |
| end cap | nr | 2.75 |
| outlet end cap including PVC union | nr | 5.10 |
| inlet end cap including PVC union | nr | 5.10 |
| ACO gullies complete with grating and galvanised bucket | | |
| shallow gully | nr | 195.00 |
| deep gully | nr | 215.00 |
| deep gully with four air trap | nr | 230.00 |
| gully extension unit | nr | 25.00 |
| ACO sump units with galvanised steel bucket | | |
| for S100 channel including ductile iron grating | nr | 117.50 |
| for Frame channel including ductile iron frame and grating | nr | 62.85 |
| for NW100 channel | nr | 40.85 |
| for K100S channel including galvanised steel frame | nr | 44.75 |
| for K100S channel including galvanised steel frame, roddable | nr | 65.00 |
| 100 mm diameter pvc traps | nr | 8.85 |
| ACO channel gratings | | |
| galvanised steel, slotted, light duty, length 1000 mm | nr | 6.50 |
| galvanised steel, slotted, heavy duty, length 1000 mm | nr | 11.85 |
| galvanised steel, mesh (10 x 30 mm), length 1000 mm | nr | 9.25 |
| EVA coated steel, slotted, light duty, length 1000 mm | nr | 13.50 |
| stainless steel, slotted, light duty, length 1000 mm | nr | 22.00 |
| stainless steel, slotted, heavy duty, length 1000 mm | nr | 47.95 |
| galvanised steel, perforated, heavy duty, length 1000 mm | nr | 18.95 |
| stainless steel, perforated, heavy duty, length 1000 mm | nr | 53.50 |
| cast iron, slotted, heavy duty, length 500 mm | nr | 7.35 |
| galvanised steel locking bolt and bar | nr | 1.10 |
| stainless steel locking bolt and bar | nr | 3.85 |

BASIC MATERIAL PRICES

**POLYMER CHANNELS AND FITTINGS** - *continued*

| | Unit | £ |
|---|---|---|
| Bridge gutters | | |
| including cover, length 1000mm | nr | 31.50 |
| with 150mm UPVC base outlet and cover length 1000mm | nr | 42.00 |
| Curved gutters | | |
| including cover length 1000mm | nr | 40.25 |
| with 150mm UPVC base outlet and cover length 1000mm | nr | 53.50 |
| End cap | nr | 4.50 |
| End cap with 100 mm dia upvc pipe | nr | 8.75 |
| Inter-connecting channel with inbuilt fall (Stanton Standrain) | | |
| 1000 mm long | nr | 14.95 |
| 1000 mm long, galvanised steel edging | nr | 22.50 |
| Drain end caps | nr | 2.50 |
| Drain end caps with 100 mm PVC outlet | nr | 4.50 |
| PVC accessories | | |
| 100 mm drain union | nr | 1.10 |
| 150 mm drain union | nr | 3.10 |
| 200 mm drain union | nr | 5.10 |
| 150 mm oval to round union | nr | 3.40 |
| 100 mm foul air trap | nr | 8.25 |
| 150 mm foul air trap | nr | 11.25 |
| Gratings | | |
| Class C ductile iron, slotted, 500 mm long | nr | 6.50 |
| Class E ductile iron, slotted, 500 mm long | nr | 6.75 |
| Class A galvanised steel, slotted, 500 mm long | nr | 3.95 |
| Class A galvanised steel, slotted, 1000 mm long | nr | 5.90 |
| Class A galvanised steel, mesh, 500 mm long | nr | 6.50 |
| Class A galvanised steel, mesh, 1000 mm long | nr | 9.50 |
| Class A stainless steel, slotted, 500 mm long | nr | 16.50 |
| Class A stainless steel, slotted, 1000 mm long | nr | 19.50 |
| Class A stainless steel, mesh, 500 mm long | nr | 38.00 |
| Class A stainless steel, mesh, 1000 mm long | nr | 46.00 |
| Ductile iron locking bar and mild steel bolt | nr | 1.00 |
| Steel locking bar and mild steel bolt | nr | 1.00 |
| Silt boxes | | |
| standard | nr | 38.00 |
| roddable with in-line outlet | nr | 62.00 |
| roddable with side outlet | nr | 62.00 |
| with PVC foul air trap | nr | 42.00 |
| galvanised steel edges | nr | 4.00 |
| Gullies (inclusive of ductile iron grating, galvanised steel silt bucket and locking mechanism) | | |
| shallow | nr | 172.00 |
| deep | nr | 195.00 |
| deep, roddable | nr | 210.00 |
| deep with foul air trap | nr | 230.00 |

BASIC MATERIAL PRICES

**PRECAST CONCRETE KERBS, EDGINGS AND PAVING SLABS**

| | Unit | £ |
|---|---|---|
| Hydraulically Pressed Kerbs (2′ and 3′ lengths) | | |
| 6" x 12" bullnosed | nr | 3.08 |
| 6" x 12" battered | nr | 3.08 |
| 6" x 12" half battered | nr | 3.08 |
| 5" x 10" bullnosed | nr | 1.94 |
| 5" x 10" battered | nr | 1.94 |
| 5" x 10" half battered | nr | 1.94 |
| 5" x 10" square | nr | 1.94 |
| 5" x 6" bullnosed | nr | 1.49 |
| 5" x 6" battered | nr | 1.49 |
| 5" x 6" half battered | nr | 1.49 |
| 5" x 6" square | nr | 1.49 |
| 6" x 4" square | nr | 1.49 |
| Extra Overs | | |
| Dished channel | nr | 0.60 |
| 5" x 10" radius and dropper | nr | 0.95 |
| 5" x 6" radius and dropper | nr | 0.95 |
| 10" x 5" radius and dropper | nr | 0.95 |
| 6" x 5" radius and dropper | nr | 0.95 |
| All other sizes of radius | nr | 1.50 |
| Quadrants | nr | 3.00 |
| Angles | nr | 3.00 |
| 5" x 10" transitions | nr | 3.00 |
| Offlets (weir kerbs) laid flat radius and other transitions | nr | 4.00 |

| | Unit | Natural £ | Coloured £ |
|---|---|---|---|
| Block Kerbs | | | |
| Main kerb 200mm lengths | m | 3.20 | 3.61 |
| External angle | nr | 1.71 | 1.95 |
| Internal angle | nr | 1.71 | 1.95 |
| Radius unit | nr | 1.26 | 1.44 |
| Crossover unit | nr | 1.17 | 1.95 |
| Transition to crossover | nr | 1.71 | 1.95 |
| Transition to half battered kerb | nr | 3.27 | 3.69 |

| | Unit | £ |
|---|---|---|
| Ellis Trief safety kerbs | | |
| Standard (length 914mm) | nr | 28.90 |
| Taper | nr | 28.90 |
| Quadrant | nr | 21.00 |
| Crane off load | load | 25.00 |

BASIC MATERIAL PRICES

**PRECAST CONCRETE KERBS, EDGINGS AND PAVING SLABS** - *continued*

| | Unit | Full Load Delivered Chester £ | Full Load Delivered London £ |
|---|---|---|---|
| Beany Blocks | | | |
| base block, 500 mm long | nr | 8.53 | 8.83 |
| base block, outfall | nr | 12.42 | 12.72 |
| base block, outfall/junction | nr | 12.42 | 12.72 |
| base block, junction | nr | 12.42 | 12.72 |
| base block, bend | nr | 9.58 | 9.74 |
| base block, splay cut one end for radius work | nr | 12.69 | 12.99 |
| top block, 500 mm long | nr | 9.26 | 9.56 |
| top block, splay cut one end for radius work | nr | 13.42 | 13.72 |
| cast iron cover and frame | nr | 74.83 | 75.19 |

| | Unit | £ |
|---|---|---|
| Safeticurb | | |
| Standard capacity slot units (914 mm long) | | |
| 250 x 250 mm, 125 mm bore unit | nr | 10.48 |
| 250 x 250 mm, 125 mm bore unit with cast iron insert | nr | 18.56 |
| Standard capacity grid units (914 mm long) | | |
| 250 x 250 mm, 125 mm bore unit with pressed steel grid insert | nr | 16.78 |
| 250 x 250 mm, 125 mm bore unit with cast iron grid insert | nr | 28.86 |
| Cast iron silt box for standard capacity units | nr | 76.33 |
| Large capacity slot units (914 mm long) | | |
| 305 x 305 mm, 150 mm bore unit | nr | 17.66 |
| 305 x 305 mm, 150 mm bore unit with cast iron insert | nr | 28.36 |
| Large capacity grid units (914 mm long) | | |
| 305 x 305 mm, 150 mm bore unit with pressed steel grid insert | nr | 21.90 |
| 305 x 305 mm, 150 mm bore unit with cast iron grid insert | nr | 34.35 |
| Cast iron silt box for large capacity units | nr | 128.75 |
| Standard capacity kerb units (914 mm long) | | |
| 250 x 320 mm and 250 x 350 mm units, 125 mm bore | nr | 12.99 |
| inspection units | nr | 38.64 |
| transition units | nr | 15.49 |
| cast iron manhole covers (610 mm long) | nr | 142.74 |
| Hydraulically Pressed Edgings (3' lengths) | | |
| 2" x 6" square | nr | 0.86 |
| 2" x 6" bullnosed | nr | 0.86 |
| 2" x 6" chamfered | nr | 0.86 |
| 2" x 8" square | nr | 0.92 |
| 2" x 8" bullnosed | nr | 0.92 |
| 2" x 10" square | nr | 1.10 |
| 2" x 10" bullnosed | nr | 1.10 |

BASIC MATERIAL PRICES

**PRECAST CONCRETE KERBS, EDGINGS AND PAVING SLABS** - *continued*

| | Unit | £ |
|---|---|---|
| Vibrated Path Edgings | | |
| 2" x 6" round top (straight) | nr | 1.09 |
| 2" x 6" radius | nr | 1.43 |
| 2" x 8" round top (straight) | nr | 1.15 |
| 2" x 8" radius | nr | 1.52 |
| 2" x 10" round top (straight) | nr | 1.38 |
| 2" x 10" radius | nr | 1.80 |

| | Unit | Natural £ | Coloured £ |
|---|---|---|---|
| Paving Flags BSS 368 | | | |
| 450 x 600 x 50mm | nr | 1.29 | 1.77 |
| 600 X 600 X 50mm | nr | 1.47 | 2.11 |
| 750 x 600 x 50mm | nr | 1.78 | 2.52 |
| 900 x 600 x 50mm | nr | 1.94 | 2.90 |
| 600 x 300 x 50mm | nr | 0.94 | - |
| 450 x 600 x 63mm | nr | 1.48 | - |
| 600 x 600 x 63mm | nr | 1.70 | - |
| 750 x 600 x 63mm | nr | 1.98 | - |
| 900 x 600 x 63mm | nr | 2.22 | - |
| 900 x 300 x 50mm | nr | 1.32 | - |

| | Unit | Natural £ | Coloured £ | Brindle £ |
|---|---|---|---|---|
| Block Paving BSS 6717 | | | | |
| Rectangular block paving (Charcon Europa) | | | | |
| 200 x 100 x 65mm | $m^2$ | 7.00 | 7.90 | 8.30 |
| 200 x 100 x 80mm | $m^2$ | 7.90 | 8.80 | 9.20 |
| Interlocking block paving (Charcon Elite) | | | | |
| 200 x 150 x 65mm | $m^2$ | 7.30 | 8.25 | - |
| 200 x 150 x 80mm | $m^2$ | 8.05 | 9.20 | - |

| | Unit | Natural £ | Coloured £ |
|---|---|---|---|
| Block Starter Units | | | |
| herringbone 65 mm thick | Pack of 80 | 41.60 | 50.40 |
| herringbone 80 mm thick | Pack of 80 | 45.60 | 54.40 |
| interlocking 65 mm thick | Pack of 98 | 31.36 | 34.30 |
| interlocking 80 mm thick | Pack of 98 | 32.34 | 38.22 |

BASIC MATERIAL PRICES

**PRECAST CONCRETE KERBS, EDGINGS AND PAVING SLABS** - *continued*

Landscape Paving

| | Unit | £ | £ |
|---|---|---|---|
| **Textured** | | | |
| 400 x 400 x 65mm | nr | 1.42 | 1.72 |
| 450 x 450 x 70mm | nr | 1.86 | 2.28 |
| 600 x 600 x 50mm | nr | 2.48 | 3.32 |
| **Ground** | | | |
| 400 x 400 x 65mm | nr | 2.19 | 2.43 |
| 600 x 600 x 50mm | nr | 3.77 | 4.49 |
| **Sett** | | | |
| 400 x 400 x 80mm | nr | 1.94 | - |
| **Tactile** | | | |
| 400 x 400 x 50mm | nr | 1.64 | - |

| | Unit | £ |
|---|---|---|
| **Grassgrid** | | |
| 366 x 274 x 100mm | $m^2$ | 10.30 |

| | Unit | £ |
|---|---|---|
| Granite setts new 100 x 100 (approx) | t | 99.00 |
| Granite setts recelaimed random sized | t | 50.00 |
| Granite setts reclaimed 100 x 100 sorted | t | 60.00 |
| Random size cobbles reclaimed | t | 45.00 |

BASIC MATERIAL PRICES

**QUARRY PRODUCTS, AGGREGATES, ETC**

Cost per tonne delivered to site with a 30 mile radius of quarry or production centre, cost based on most economically available materials, minimum order 500 tonnes (average to nearest 5p - see footnote)

| | Scotland North & Wales | Midland S West | South General |
|---|---|---|---|
| Graded granular material | | | |
| natural gravels (DOT class 1 A, B or C) general fill | 2.50 | 2.55 | 2.75 |
| crushed gravel or rock ditto | 2.95 | 3.00 | 3.55 |
| Reclaimed material (DOT class 2E) | | | |
| blast furnace slag; general fill | 2.00 | 2.00 | - |
| burnt colliery shale; ditto | 2.00 | 2.10 | - |
| Well graded granular material | | | |
| natural gravel or sand (DOT class 6 A) fill below water | 3.80 | 4.00 | 4.15 |
| crushed gravel or rock; ditto (free draining) | 4.20 | 4.40 | 4.60 |
| Selected granular material | | | |
| natural gravel (DOT class 6 F) capping layers | 2.95 | 3.20 | 3.45 |
| crushed rock; ditto | 3.55 | 3.70 | 4.15 |
| Selected uniformly graded material | | | |
| natural gravel or crushed rock (DOT class 6 I, J) fill to reinforced earth - blinding material | 4.60 | 4.70 | 5.00 |
| pea gravel fill on bridges and other structures | 12.75 | 14.00 | 16.00 |
| Selected graded material | | | |
| natural gravel (DOT class 6 N, P) fill to structure | 2.85 | 3.00 | 3.25 |
| natural rock; ditto | 3.60 | 3.75 | 4.00 |
| Selected graded material | | | |
| natural gravel (DOT class 1 C) rock fill | 4.15 | 4.20 | 4.25 |
| crushed gravel; ditto | 4.20 | 4.20 | 4.25 |
| Wet cohesive materials | | | |
| natural boulder clay (DOT class 2 A) general fill | 4.00 | 4.00 | 4.05 |
| oxford clays; ditto | 4.20 | 4.20 | 4.20 |
| Dry cohesive materials | | | |
| natural boulder clay (DOT class 2 B) general fill | 4.00 | 4.05 | 4.05 |
| oxford clays; ditto | 4.20 | 4.20 | 4.25 |
| Chalk fill | | | |
| chalk & associated materials (DOT class 3) general fill | 2.20 | 2.30 | 2.25 |
| Rock fill selected well graded | | | |
| crushed rock, core fill, rock punching | 4.60 | 4.60 | 4.75 |
| Armour stone single size (minimum costs) | | | |
| sea defence shore protection 0.5 tonne | 8.20 | 9.05 | 9.30 |
| 1 tonne | 11.50 | 12.20 | 13.80 |
| 3 tonne | 17.50 | 19.25 | N/A |

BASIC MATERIAL PRICES

**QUARRY PRODUCTS, AGGREGATES, ETC** - continued

| | Scotland North & Wales | Midland S West | South General |
|---|---|---|---|
| Granular material single sized aggregate | | | |
| natural gravel (DOT class 503) pipe bedding/haunching | 4.40 | 4.45 | 4.75 |
| crushed rock; ditto | 5.00 | 5.05 | 5.35 |
| Filter material | | | |
| natural or crushed other than unburnt slag or chalk (DOT class 505 A, B, C) backfill to filter drains | | | |
| Type A | 4.55 | 4.70 | 5.15 |
| Type B | 3.90 | 4.00 | 4.40 |
| Graded granular material | | | |
| crushed rock | 5.30 | 5.45 | 5.90 |
| natural sand or gravels (DOT class 803/804) | | | |
| subbase Type 1/2 | 6.00 | 6.55 | 7.50 |
| crushed concrete (reclaimed) | 2.15 | 2.15 | 2.20 |
| Coated roadstone products | | | |
| DBM roadbases (DOT class 900) carrigeway construction | 20.25 | 20.25 | 20.35 |
| HRA base and wearing; carrigeway construction | 25.65 | 25.65 | 25.75 |
| Bitumen Macadam | | | |
| 40mm aggregate | 17.95 | 17.95 | 18.10 |
| 20mm aggregate | 18.45 | 18.45 | 18.55 |
| 10mm aggregate | 19.10 | 19.15 | 19.30 |
| Concrete aggregates | | | |
| All in aggregate 10mm for concrete production | 9.05 | 9.05 | 9.10 |
| 20mm for concrete production | 8.80 | 8.85 | 8.85 |
| Fine screen sharp sand for concrete production | 7.95 | 7.95 | 7.95 |

**PLEASE NOTE:**

The above prices are based upon information obtained from a number of companies or individual quarries. An average of the above prices is used in Part 4 and 5 (unit cost compilation), and provides a reasonable guide as to comparative costs of common or DOT specification materials.

Transport costs form a high proportion of delivered stone prices: if a 20T tipper costs £20 per hour to keep on the road then it follows that for each hours journey time £1 will be added to the cost of each tonne delivered. The cost shown above are based on a 1.5 hour return journey in 20T tippers with minimum standing time.

Market conditions and commercial considerations could affect these prices to a surprising degree up to double for the more selective materials. We have included such variations only where we feel that there is a sound reason for them, and so have attempted to produce a reasonable guide to market conditions.

If specific requirements are known then local quarries are usually more than helpful with information and guide prices.

BASIC MATERIAL PRICES

**SCAFFOLDING AND PROPS**

| | Unit | £ |
|---|---|---|
| Scaffold tube | | |
| black tube | m | 1.57 |
| galvanised tube | m | 1.97 |
| alloy tube | m | 4.53 |
| Scaffold fittings | | |
| double | nr | 1.70 |
| single | nr | 1.06 |
| swivel | nr | 2.02 |
| sleeve | nr | 1.45 |
| joint pin | nr | 1.18 |
| Scaffold boards | | |
| grade A selected (3.97 m long) | nr | 6.50 |
| Towers | | |
| alloy, 2.5 x 1.35 x 3.73 m | nr | 973.60 |
| alloy, 2.5 x 1.35 x 6.05 m | nr | 1503.25 |
| steel, 4.88 x 1.22 x 0.61 m | nr | 207.60 |
| steel, 4.88 x 1.22 x 0.22 m | nr | 248.75 |
| Adjustable struts | | |
| size 0 | nr | 8.15 |
| size 1 | nr | 9.50 |
| size 2 | nr | 10.75 |
| size 3 | nr | 12.45 |
| Adjustable props | | |
| size 0 | nr | 12.70 |
| size 1 | nr | 15.85 |
| size 2 | nr | 16.45 |
| size 3 | nr | 17.95 |
| size 4 | nr | 21.15 |
| Lightweight stagings | | |
| 3.66 m | nr | 75.70 |
| 4.27 m | nr | 88.60 |
| 4.88 m | nr | 117.15 |
| 5.49 m | nr | 130.00 |
| 6.10 m | nr | 141.45 |
| 6.71 m | nr | 155.70 |
| 7.32 m | nr | 177.15 |
| Timber pole ladders | | |
| 4 m | nr | 36.70 |
| 5 m | nr | 46.70 |
| 6 m | nr | 54.70 |
| 7 m | nr | 69.35 |
| 8 m | nr | 80.00 |
| 9 m | nr | 88.00 |

BASIC MATERIAL PRICES

**SCOUR AND EROSION PROTECTION**

| | Unit | £ |
|---|---|---|
| $150m^2$ Armoflex 140 delivered Derbyshire including geotextile and Duckbills - crane needed to offload and lay | $m^2$ | 14.40 |
| $150m^2$ Armorloc delivered Derbyshire including geotextile and Duckbills - delivered on self offloading vehicle - lay by hand | $m^2$ | 13.20 |
| Enkamat delivered England/Wales and Scotland $1500\text{-}2000m^2$ including steel fixing pins roll sizes 4.8m x 60m | $m^2$ | 4.05 |
| Nicolon geotextiles as reinforcing layer under sub base | | |
| Geolon 15: $1000m^2$-$3000m^2$ delivered in $1000m^2$ rolls | $m^2$ | 0.39 |
| Geolon 40: $1000m^2$-$3000m^2$ delivered in $1000m^2$ rolls | $m^2$ | 0.77 |
| Enkadrain P3 roll size 2m x 30m up to $500m^2$ - delivered | $m^2$ | 4.30 |
| Enkadrain ST roll size 1m x 30m up to $500m^2$ - delivered | $m^2$ | 5.50 |
| "Duckbill" anchors | | |
| standard anchor 28 DB1 tendon 420 x 2mm | nr | 2.39 |
| 68 DB1 tendon 760 x 4mm | nr | 5.02 |
| 88 DB1 tendon 1150 x 6mm | nr | 14.25 |
| 138 DB1 tendon 1525 x 8mm | nr | 21.73 |
| 138 DB1M tendon 2000 x 10mm | nr | 31.34 |
| Spreader plates and other fixings; refer to manufacturer | | |
| Gabions and mattresses; refer to previous lists | | |
| ICI Terram 2000 (100 x 4.5m roll) | nr | 300.15 |
| Tensar GM4 ( 40 x 4.0m roll) | nr | 304.50 |
| Tensar mat ( 30 x 3.0m roll) | nr | 213.00 |
| Tensar mat ( 30 x 4.5m roll) | nr | 319.50 |

BASIC MATERIAL PRICES

**SCEPTIC TANKS, CESSPOOLS AND INTERCEPTORS**

| | Unit | £ |
|---|---|---|
| Klargester GRP septic tanks including cover and frame | | |
| 2720 litre capacity, standard grade | nr | 466.00 |
| 2720 litre capacity, heavy grade | nr | 580.00 |
| 6000 litre capacity, standard grade | nr | 899.00 |
| 6000 litre capacity, heavy grade | nr | 1134.00 |
| 12000 litre capacity, standard grade | nr | 1733.00 |
| 12000 litre capacity, heavy grade | nr | 2194.00 |
| Klargester GRP cesspools including covers and frames | | |
| 9000 litre capacity, standard grade | nr | 1327.00 |
| 9000 litre capacity, heavy grade | nr | 1676.00 |
| 18200 litre capacity, standard grade | nr | 2186.00 |
| Klargester 3 stage GRP petrol interceptors | | |
| 2000 litre capacity | nr | 726.00 |
| 3250 litre capacity | nr | 1055.00 |
| 4000 litre capacity | nr | 1193.00 |

BASIC MATERIAL PRICES

**SHUTTERING, TIMBER, NAILS ETC**

| | Unit | £ |
|---|---|---|
| Birch faced plywood exterior WBP size 1220 x 2440 | | |
| 6.5mm thick | $m^2$ | 4.28 |
| 12mm thick | $m^2$ | 6.72 |
| 18mm thick | $m^2$ | 9.80 |
| Marine plywood gaboon exterior WBP size 1220 x 2500 | | |
| 4.0mm thick | $m^2$ | 2.79 |
| 6.0mm thick | $m^2$ | 3.80 |
| 12.0mm thick | $m^2$ | 6.96 |
| 18.0mm thick | $m^2$ | 10.30 |
| 25.0mm thick | $m^2$ | 15.38 |
| Coated shuttering plywood | | |
| Phenolic film faced birch faced exterior WBP edges sealed | | |
| size 1525 x 3050 x 6.5mm | $m^2$ | 5.95 |
| size 1220 x 2440 x 12.0mm | $m^2$ | 8.58 |
| size 1220 x 2440 x 18.0mm | $m^2$ | 11.85 |
| Phenolic film faced 'Wisaform' multi layer coating | | |
| size 1220 x 2440 x 12.0mm | $m^2$ | 9.26 |
| size 1220 x 2440 x 18.0mm | $m^2$ | 12.78 |
| Resin impregnated overlay size Pourform/Silvaform/Crown 44 size 1220 x 2440 x 17.5mm | $m^2$ | 7.13 |

Good quality timber for formwork, falseworks, temporary works and trench supports, delivered to site

Sawn softwood random lengths per metre (used)

| | £ | £ | £ |
|---|---|---|---|
| Thickness | 25mm | 50mm | 754mm |
| 50mm wide | 0.23 | 0.49 | 0.63 |
| 75mm wide | 0.36 | 0.63 | 1.15 |
| 150mm wide | 0.69 | 1.32 | 2.20 |

Wrought (planed) timber; European softwood per metre (new)

| | £ | £ | £ | £ | £ |
|---|---|---|---|---|---|
| Thickness | 12.5mm | 25mm | 38mm | 50mm | 75mm |
| 25mm wide | 0.30 | - | - | 0.36 | 0.53 |
| 50mm wide | 0.59 | 0.36 | 0.82 | 0.92 | 1.32 |
| 75mm wide | - | 0.53 | 1.18 | 1.32 | 2.46 |
| 100mm wide | - | 0.66 | 1.64 | 1.71 | 3.12 |
| 125mm wide | - | 0.82 | - | - | - |
| 150mm wide | - | 0.99 | 2.30 | 3.28 | 4.92 |
| 175mm wide | - | 1.15 | - | - | - |
| 200mm wide | - | 1.71 | 2.96 | 4.92 | 7.55 |

Sawn European softwood ex-wharf per cubic metre

| | New | Used |
|---|---|---|
| | £ | £ |
| 100 x 225 | 210 | 155 |
| 200 x 200 | 210 | 129 |
| 75 x 175 | 203 | 140 |
| 75 x 125 | 203 | 138 |
| 50 x 225 | 208 | 121 |
| 50 x 175 | 208 | 114 |

BASIC MATERIAL PRICES

**SHUTTERING, TIMBER AND NAILS** - *continued*

| | Unit | £ |
|---|---|---|
| Construction hardwood (for piers, jettys, groynes, dolphins, etc) | | |
| Greenheart (225 and 300 mm x 50 mm, 75 mm and 100 mm) (lengths up to 7.32 m) | $m^3$ | 424.00 |
| Greeheart (150 mm x 150 mm, 225 mm and 300 mm, 200 x 200 mm and 225 x 225 mm) (lengths up to 8.84 m) | $m^3$ | 406.00 |
| Greenheart (250 x 250 mm, 300 x 300 mm, 350 x 350 mm) lengths up to 11.89 m) | $m^3$ | 441.50 |
| 100 x 100 Greenheart (African) | $m^3$ | 497.00 |
| 350 x 350 Greenheart (African) | $m^3$ | 602.00 |
| extra for pre-treatment | $m^2$ | 30.00 |
| Prime Parana pine | | |
| 100 x 100 | $m^3$ | 530.00 |
| 38 x 300 | $m^3$ | 412.00 |
| 38 x 300 (planed) | $m^3$ | 452.00 |
| **Used Timber** | | |
| Beams, baulks, pit props (used as kerbing timbers, temporary barriers, plant support, etc) | $m^3$ | 110-130 |
| Steel nails | | |
| oval lost head 60 mm long | kg | 1.60 |
| 100 mm long | kg | 1.46 |
| pressed head 60 mm long | kg | 1.79 |
| 100 mm long | kg | 1.68 |
| plain head 30 mm long (2.00 mm gauge) | kg | 2.11 |
| 100 mm long (4.00 mm gauge) | kg | 1.23 |
| Galvanised nails | | |
| clout nails 20 mm long (2.65 mm gauge) | kg | 3.11 |
| 75 mm long (3.75 mm gauge) | kg | 2.49 |
| plain head 25 mm long (1.80 mm gauge) | kg | 3.85 |
| 75 mm long (3.75 mm gauge) | kg | 2.33 |
| wire staples 15 mm long (1.60 mm gauge) | kg | 2.58 |
| 50 mm long (5.00 mm gauge) | kg | 2.06 |
| Bonded polystyrene void formers circular in section | | |
| 225mm diameter | m | 1.38 |
| 300mm diameter | m | 2.51 |
| 450mm diameter | m | 5.54 |
| 600mm diameter | m | 10.05 |
| 750mm diameter | m | 15.58 |
| Expanded polystyrene in blocks | $m^3$ | 27.80 |

BASIC MATERIAL PRICES

**SHUTTERING, TIMBER AND NAILS** - *continued*

Disposable column shutters

| | Unit | 2 m length £ | 3 m length £ | 4 m length £ |
|---|---|---|---|---|
| Inform cardboard disposable formwork for columns | | | | |
| 150 mm diameter | nr | - | 22.00 | 30.00 |
| 200 mm diameter | nr | - | 29.00 | 39.00 |
| 300 mm diameter | nr | - | 47.00 | 30.00 |
| 400 mm diameter | nr | - | 72.00 | 39.00 |
| 500 mm diameter | nr | - | 95.00 | 30.00 |
| 600 mm diameter | nr | - | 141.00 | 39.00 |
| 1000 mm diameter | nr | 373.00 | - | - |
| 150 x 200 mm | nr | - | 58.00 | 77.50 |
| 200 x 200 mm | nr | - | 59.00 | 79.00 |
| 200 x 300 mm | nr | - | 73.50 | 98.50 |
| 300 x 300 mm | nr | - | 88.50 | 118.00 |
| 300 x 400 mm | nr | - | 103.50 | 138.00 |
| 400 x 400 mm | nr | - | 118.00 | 157.00 |

BASIC MATERIAL PRICES

**STEEL PIPES AND FITTINGS**

| **Diameter in mm** | Unit | **100** £ | **125** £ | **150** £ | **200** £ | **250** £ | **300** £ |
|---|---|---|---|---|---|---|---|
| Steel pipe to BS 3601 or equivalent | m | 17.83 | 24.20 | 33.17 | 52.96 | 75.12 | 96.34 |
| Fittings to BS 1640, standard strength | | | | | | | |
| 90° bend, long radius | nr | 8.68 | 19.90 | 21.70 | 43.87 | 81.39 | 114.86 |
| 45° bend, long radius | nr | 7.02 | 15.17 | 15.88 | 31.52 | 58.70 | 83.44 |
| single junction, equal | nr | 23.32 | 41.17 | 42.12 | 82.34 | 147.59 | 233.54 |
| single junction, 100mm branch | nr | - | - | 48.55 | 94.72 | - | - |
| single junction, 150mm branch | nr | - | - | - | 94.72 | - | - |
| single junction, 200mm branch | nr | - | - | - | - | 169.55 | - |
| concentric reducers, reducing to 100 mm | nr | - | - | 12.63 | 24.29 | - | - |
| concentric reducers, reducing to 150 mm | nr | - | - | - | 19.43 | 29.13 | - |
| concentric reducers, reducing to 200 mm | nr | - | - | - | - | 29.13 | 45.65 |
| concentric reducers, reducing to 250 mm | nr | - | - | - | - | - | 43.7[illegible] |

| **Diameter in mm** | Unit | **75** £ | **110** £ | **160** £ |
|---|---|---|---|---|
| Stainless steel pipe | m | 13.50 | 17.01 | 36.04 |
| 90° bend | m | 17.39 | 22.31 | 56.30 |
| 45° bend | m | 12.86 | 16.83 | 47.66 |
| Single junction | m | 18.90 | 22.87 | 46.98 |

| **Diameter in mm** | Unit | **100** £ | **150** £ | **200** £ | **250** £ | **300** £ |
|---|---|---|---|---|---|---|
| Fusion bonded epoxy coated steel pipe | m | 300.00 | 502.00 | 576.00 | 738.00 | 955.00 |
| 90° bend | nr | 61.00 | 110.00 | 171.00 | 271.00 | 344.00 |
| 45° bend | nr | 55.00 | 96.00 | 146.00 | 227.00 | 305.00 |
| Single juctions | | | | | | |
| branch size 100 mm | nr | - | 144.00 | 179.00 | 236.00 | 304.00 |
| branch size 150 mm | nr | - | - | 203.00 | 260.00 | 329.00 |
| branch size 200 mm | nr | - | - | 224.00 | 281.00 | 351.00 |
| branch size 250 mm | nr | - | - | - | 314.00 | 384.00 |
| branch size 300 mm | nr | - | - | - | - | 420.00 |

BASIC MATERIAL PRICES

**STRUCTURAL STEELWORK**

Note: The following basic prices are for Basic quantities of grade 43A steel (over 20 tonnes, of one quality, one section, in lengths between 6 m and 15 m, for delivery to one destination). Transport charges are shown in a separate schedule.

Figures in brackets are available weights in kg per metre, at the prices indicated.

| | £ | | £ |
|---|---|---|---|
| Universal beams (£/tonne) | | | |
| 914 x 419 mm (383,343) | 390.00 | 356 x 171 mm (57,67) | 340.00 |
| 914 x 305 mm (201,224,253,289) | 385.00 | 356 x 127 mm (33,39) | 345.00 |
| 838 x 292 mm (176,194,226) | 380.00 | 305 x 165 mm (40) | 315.00 |
| 762 x 267 mm (147,173,197) | 380.00 | 305 x 165 mm (46,54) | 330.00 |
| 686 x 254 mm (125,140,152,170) | 380.00 | 305 x 127 mm (37) | 335.00 |
| 610 x 305 mm (149,179,238) | 360.00 | 305 x 127 mm (42,48) | 340.00 |
| 610 x 229 mm (101,113,125,140) | 360.00 | 305 x 102 mm (25,28,33) | 340.00 |
| 533 x 210 mm (82,92,101) | 345.00 | 254 x 146 mm (31,37) | 305.00 |
| 533 x 210 mm (109,122) | 360.00 | 254 x 146 mm (43) | 310.00 |
| 457 x 191 mm (67,74) | 315.00 | 254 x 102 mm (22,25) | 340.00 |
| 457 x 191 mm (82,89,98) | 325.00 | 254 x 102 mm (28) | 335.00 |
| 457 x 152 mm (52,60,67,74,82) | 350.00 | 203 x 133 mm (25,30) | 295.00 |
| 406 x 178 mm (54) | 325.00 | 203 x 102 mm (23) | 290.00 |
| 406 x 178 mm (60,67,74) | 330.00 | 178 x 102 mm (19) | 285.00 |
| 406 x 140 mm (39,46) | 340.00 | 152 x 89 mm (16) | 325.00 |
| 356 x 171 mm (45,51) | 325.00 | 127 x 76 mm | 345.00 |
| Universal columns (£/tonne) | | | |
| 356 x 406 mm (235,287,340) | 390.00 | 254 x 254 mm (107) | 325.00 |
| 356 x 406 mm (393,467,551,634) | 395.00 | 254 x 254 mm (132) | 335.00 |
| 356 x 368 mm (129,153,177,202) | 390.00 | 254 x 254 mm (167) | 355.00 |
| 305 x 305 mm (97) | 360.00 | 203 x 203 mm (46) | 310.00 |
| 305 x 305 mm (118) | 370.00 | 203 x 203 mm (52) | 320.00 |
| 305 x 305 mm (137,158) | 365.00 | 203 x 203 mm (60) | 335.00 |
| 305 x 305 mm (198) | 350.00 | 203 x 203 mm (71) | 325.00 |
| 305 x 305 mm (240,283) | 370.00 | 203 x 203 mm (86) | 345.00 |
| 254 x 254 mm (73) | 320.00 | 152 x 152 mm (23) | 310.00 |
| 254 x 254 mm (89) | 335.00 | 152 x 152 mm (30,37) | 315.00 |
| Joists (£/tonne) | | | |
| 254 x 203 mm (81.85) | 340.00 | 114 x 114 mm (26.79) | 320.00 |
| 203 x 152 mm (52.09) | 330.00 | 102 x 102 mm (23.06) | 320.00 |
| 152 x 127 mm (37.20) | 330.00 | 89 x 89 mm (19.35) | 320.00 |
| 127 x 114 mm (26.79, 29.76) | 320.00 | 76 x 76 mm (12.65) | 370.00 |
| Channels (£/tonne) | | | |
| 432 x 102 mm (65.54) | 370.00 | 203 x 89 mm (29.78) | 340.00 |
| 381 x 102 mm (55.10) | 370.00 | 203 x 76 mm (23.82) | 290.00 |
| 305 x 102 mm (46.18) | 340.00 | 178 x 89 mm (26.81) | 330.00 |
| 305 x 89 mm (41.69) | 340.00 | 178 x 76 mm (20.84) | 290.00 |
| 254 x 89 mm (35.74) | 340.00 | 152 x 89 mm (23.84) | 330.00 |
| 254 x 76 mm (28.29) | 340.00 | 152 x 76 mm (17.88) | 290.00 |
| 229 x 89 mm (32.76) | 340.00 | 127 x 64 mm (14.90) | 280.00 |
| 229 x 76 mm (26.06) | 340.00 | | |

BASIC MATERIAL PRICES

**STRUCTURAL STEELWORK** - *continued*

Equal angles (£/tonne)

| | £ | | £ | | £ |
|---|---|---|---|---|---|
| 200 x 200 x 16mm | 295.00 | 150 x 150 x 18mm | 310.00 | 100 x 100 x 12mm | 239.0 |
| 200 x 200 x 18mm | 300.00 | 120 x 120 x 8mm | 280.00 | 100 x 100 x 15mm | 265.0 |
| 200 x 200 x 20mm | 300.00 | 120 x 120 x 10mm | 270.00 | 90 x 90 x 6mm | 245.0 |
| 200 x 200 x 24mm | 300.00 | 120 x 120 x 12mm | 265.00 | 90 x 90 x 8mm | 245.0 |
| 150 x 150 x 10mm | 295.00 | 120 x 120 x 15mm | 270.00 | 90 x 90 x 10mm | 245.0 |
| 150 x 150 x 12mm | 295.00 | 100 x 100 x 8mm | 239.00 | 90 x 90 x 12mm | 245.0 |
| 150 x 150 x 15mm | 300.00 | 100 x 100 x 10mm | 239.00 | | |

Unequal angles (£/tonne)

| | £ | | £ | | £ |
|---|---|---|---|---|---|
| 200 x 150 x 12mm | 305.00 | 150 x 90 x 12mm | 300.00 | 125 x 75 x 12mm | 255.0 |
| 200 x 150 x 15mm | 310.00 | 150 x 90 x 15mm | 300.00 | 125 x 75 x 15mm | 290.0 |
| 200 x 150 x 18mm | 310.00 | 150 x 75 x 10mm | 290.00 | 100 x 75 x 8mm | 255.0 |
| 200 x 100 x 10mm | 305.00 | 150 x 75 x 12mm | 290.00 | 100 x 75 x 10mm | 255.0 |
| 200 x 100 x 12mm | 305.00 | 150 x 75 x 15mm | 290.00 | 100 x 75 x 12mm | 255.0 |
| 200 x 100 x 15mm | 325.00 | 125 x 75 x 8mm | 255.00 | 100 x 65 x 7mm | 255.0 |
| 150 x 90 x 10mm | 300.00 | 125 x 75 x 10mm | 255.00 | 100 x 65 x 8mm | 255.0 |
| | | | | 100 x 65 x 10mm | 255.0 |

| Special purlin angles (£/tonne) | £ |
|---|---|
| 137 x 102 x 6.4mm | 305.00 |
| 137 x 102 x 7.9mm | 305.00 |
| 137 x 102 x 9.5mm | 305.00 |

Add to the aforementioned prices for:

| | Unit | £ |
|---|---|---|
| Quantity - Universal beams, columns | | |
| Under 20 tonnes to 10 tonnes | t | 6.00 |
| Under 10 tonnes to 5 tonnes | t | 10.00 |
| Under 5 tonnes to 2 tonnes | t | 20.00 |
| Under 2 tonnes to 1 tonne | t | 30.00 |
| Joists, channels and angles | | |
| Under 20 tonnes to 10 tonnes | t | 6.00 |
| Under 10 tonnes to 5 tonnes | t | 10.00 |
| Under 5 tonnes | t | 25.00 |
| Quality - BS 4360, Grade 43B | t | 5.00 |
| Grade 43C | t | 25.00 |
| Grade 43D | t | 30.00 |
| Grade 43DD | t | 55.00 |
| Grade 50A | t | 25.00 |
| Grade 50B | t | 25.00 |
| Grade 50C | t | 40.00 |
| Grade 50D | t | 75.00 |
| Grade 55C (up to 19 mm flange thickness) | t | 65.00 |
| Grade 55C (over 19 mm flange thickness) | t | 100.00 |
| Size - length under 6 m | t | 15.00 |
| non standard sizes (obtained by raising the rolls of the parent BS4 size) | t | 25.00 |

BASIC MATERIAL PRICES

**STRUCTURAL STEELWORK** - *continued*

TRANSPORT CHARGES BASED ON RADIAL DISTANCES AND QUANTITIES

| RADIAL DISTANCES MILES FROM BASING POINT | 0 - under 5 tonnes £/tonne | 5 - under 10 tonnes £/tonne | 10 - under 20 tonnes £/tonne | 20 tonnes and over £/tonne |
|---|---|---|---|---|
| up to 10 | 7.05 | 4.90 | 3.70 | 3.35 |
| over 10 up to 15 | 7.75 | 5.30 | 4.15 | 3.65 |
| over 15 up to 20 | 8.25 | 5.80 | 4.60 | 4.00 |
| over 20 up to 25 | 8.75 | 6.15 | 4.90 | 4.45 |
| over 25 up to 30 | 9.15 | 6.70 | 5.40 | 4.75 |
| over 30 up to 35 | 9.60 | 7.05 | 5.75 | 5.05 |
| over 35 up to 40 | 10.10 | 7.20 | 6.15 | 5.50 |
| over 40 up to 45 | 10.50 | 7.70 | 6.45 | 5.70 |
| over 45 up to 50 | 10.85 | 7.95 | 6.75 | 6.10 |
| over 50 up to 60 | 11.40 | 8.55 | 7.30 | 6.75 |
| over 60 up to 70 | 11.90 | 9.30 | 8.00 | 7.30 |
| over 70 up to 80 | 12.70 | 10.10 | 8.70 | 8.05 |
| over 80 up to 90 | 13.60 | 10.80 | 9.40 | 8.55 |
| over 90 up to 100 | 14.40 | 11.55 | 10.05 | 9.35 |
| over 100 up to 110 | 15.05 | 12.25 | 10.70 | 10.00 |
| over 110 up to 120 | 15.60 | 13.00 | 11.35 | 10.40 |
| over 120 up to 130 | 16.10 | 13.65 | 12.00 | 11.10 |
| over 130 up to 140 | 16.70 | 14.30 | 12.55 | 11.50 |
| over 140 up to 150 | 16.85 | 15.00 | 13.00 | 12.15 |
| over 150 up to 160 | 17.25 | 15.65 | 13.50 | 12.70 |
| over 160 up to 170 | 17.85 | 16.30 | 14.05 | 13.30 |
| over 170 up to 180 | 18.30 | 16.95 | 14.60 | 13.80 |
| over 180 up to 190 | 18.95 | 17.55 | 15.15 | 14.30 |
| over 190 up to 200 | 19.55 | 18.20 | 15.70 | 14.65 |
| over 200 up to 210 | 20.10 | 18.80 | 16.05 | 15.25 |
| over 210 up to 220 | 20.70 | 19.40 | 16.65 | 15.75 |
| over 220 up to 230 | 21.20 | 20.00 | 17.05 | 16.40 |
| over 230 up to 240 | 22.00 | 20.60 | 17.50 | 16.85 |
| over 240 up to 250 | 22.45 | 21.20 | 18.00 | 17.20 |
| over 250 up to 275 | 23.35 | 22.20 | 18.80 | 18.10 |
| over 275 up to 300 | 24.70 | 23.65 | 19.90 | 19.20 |
| over 300 up to 325 | 25.95 | 25.05 | 21.05 | 20.40 |
| over 325 up to 350 | 27.30 | 26.45 | 22.20 | 21.55 |
| over 350 up to 375 | 28.65 | 27.85 | 23.15 | 22.80 |
| over 375 up to 400 | 30.00 | 29.20 | 24.15 | 23.80 |

Note: - The minimum charge under this schedule will be as for 2 tonnes

Collection by customer from works will incur an extra cost of £3.00/tonne in addition to any additional transport charges between the basing point and the producing works.

BASIC MATERIAL PRICES

**STRUCTURAL STEELWORK** - *continued*

| Structural hollow section to BS4848 Part 2 (£/100m) | BS 4360 Grade 43C £ | BS 4360 Grade 50C £ |
|---|---|---|
| Circular | | |
| 21.3 x 3.2 mm (1.43) | 57.91 | - |
| 26.9 x 3.2 mm (1.87) | 75.73 | - |
| 42.4 x 3.2 mm (3.09) | 124.53 | - |
| 48.3 x 4.0 mm (4.37) | 177.81 | 195.59 |
| 60.3 x 4.0 mm (5.55) | 234.64 | 258.10 |
| 76.1 x 5.0 mm (8.77) | 370.77 | 407.85 |
| 114.3 x 6.3 mm (16.80) | 710.25 | 781.28 |
| 139.7 x 6.3 mm (20.70) | 914.05 | 1005.46 |
| 168.3 x 6.3 mm (25.20) | 1112.76 | 1224.04 |
| 219.1 x 10.0 mm (51.60) | 2855.44 | 3283.77 |
| 273 x 10.0 mm (64.90) | 3591.44 | 4130.17 |
| 323.9 x 12.5 mm (96.00) | 5607.84 | 6448.99 |
| 193.7 x 16 mm (70.10) seamless | - | 5457.36 |
| 406.4 x 20 mm (191.00) seamless | - | 14869.54 |
| Rectangular and square | | |
| 20 x 20 x 2.5 mm (1.35) | 52.56 | - |
| 50 x 50 x 3.2 mm (4.66) | 177.97 | 195.77 |
| 60 x 60 x 4.0 mm (6.97) | 281.94 | 310.14 |
| 80 x 80 x 5.0 mm (11.70) | 501.77 | 551.95 |
| 100 x 50 x 4.0 mm (8.86) | 368.83 | 405.72 |
| 100 x 100 x 6.3 mm (18.40) | 789.10 | 868.02 |
| 100 x 100 x 10.0 mm (27.90) | 1293.00 | 1422.29 |
| 120 x 80 x 6.3 mm (18.40) | 789.10 | 868.02 |
| 150 x 150 x 6.3 mm (28.30) | 1279.87 | 1407.87 |
| 200 x 100 x 8.0 mm (35.40) | 1600.96 | 1761.08 |
| 200 x 200 x 10.0 mm (59.30) | 3361.54 | 3865.77 |
| 300 x 200 x 12.5 mm (92.60) | 5541.00 | 6372.18 |
| 150 x 150 x 16.0 mm (66.40) seamless | 5555.56 | 6111.12 |
| 200 x 100 x 16.0 mm (66.40) seamless | 5555.56 | 6111.12 |

Note: The following basic prices are for orders for 10 tonnes and over in one size, thickness, length, steel grade and surfce finish and include delivery to mainland Great Britain to one destination. Additional costs for variaitons to these factors vary between sections and should be ascertained from the supplier.

BASIC MATERIAL PRICES

**UPVC DRAIN PIPES AND FITTINGS**

| **Diameter in mm** | Unit | **82** £ | **110** £ | **160** £ | **200** £ | **250** £ | **315** £ |
|---|---|---|---|---|---|---|---|
| Plain ended pipes with sleeved joints (OsmaDrain) 3 m or 6 m lengths | m | 2.91 | 3.14 | 6.00 | 11.93 | 18.98 | 29.16 |
| Socketed pipes with ring seal joints 6 m lengths | m | - | - | - | - | 20.93 | 33.14 |
| Couplers | | | | | | | |
| for jointing pipes | nr | 2.58 | 2.46 | 4.92 | 7.92 | 29.81 | - |
| for new branch entry connections | nr | - | 4.25 | 9.40 | 14.86 | 33.52 | 58.11 |
| Reducers | nr | - | 4.11 | 8.36 | 10.87 | 32.09 | 45.39 |
| Junctions, equal | | | | | | | |
| 87.5° | nr | 7.28 | 8.67 | 25.14 | - | - | - |
| 45° | nr | - | 9.06 | 25.20 | 38.18 | 124.49 | - |
| Bends | | | | | | | |
| up to 30° | nr | - | 5.89 | 12.44 | - | - | - |
| 45° to 90° | nr | 4.70 | 6.13 | 13.86 | - | - | - |
| 90° | nr | - | - | - | - | 120.92 | 174.92 |
| 87.5° | nr | - | - | - | 31.57 | - | - |
| 45° | nr | - | - | - | 23.92 | 83.43 | 146.92 |
| 30° | nr | - | - | - | 21.53 | 68.28 | 94.78 |
| 15° | nr | - | - | - | 19.06 | 49.43 | 93.92 |
| Suspended bracketing system adjustable pipe bracket assembly (pack containing threaded brakcet, bracket plate and pipe/socket bracket) | nr | - | 7.46 | 9.44 | - | - | - |
| Adjustable socket bracket and brace assembly (pack containing threaded rod, threaded bracket, bracket plate, two adjustable braces and pipe/socket bracket) | nr | - | 13.95 | 16.01 | - | - | - |

| **Diameter in mm** | Unit | **82** £ | **110** £ | **160** £ |
|---|---|---|---|---|
| Marley underground drainage system | | | | |
| Straight pipes | | | | |
| ring seal socket 6m | m | - | 3.36 | 7.36 |
| ring seal socket 3m | m | 3.74 | 3.95 | 8.12 |
| double spigot 6m | m | - | 3.02 | 6.37 |

BASIC MATERIAL PRICES

**UPVC DRAIN PIPES AND FITTINGS** - *continued*

| **Diameter in mm** | Unit | **82** £ | **110** £ | **160** £ |
|---|---|---|---|---|
| Couplings | | | | |
| double ring seal straight | nr | - | 2.77 | 5.55 |
| double ring seal slip | nr | - | 4.53 | 9.49 |
| loose pipe socket | nr | 2.27 | 2.74 | 5.27 |
| double solvent socket | nr | - | 2.74 | - |
| triple socket | nr | - | 6.25 | - |
| Bends | | | | |
| short radius socket/spigot (22.5°/45° /87.5°) | nr | 4.59 | 6.23 | 14.25 |
| adjustable (11° - 87.5°) | nr | 7.77 | - | - |
| adjustable (21°-90°) | nr | - | 8.57 | - |
| adjustable (15°-90°) | nr | - | - | 13.57 |
| short radius socket/socket (11.25°/ 22.5°/45°/87.5°) | nr | - | 7.75 | - |
| long radius socket/socket 87.5° | nr | - | 12.81 | - |
| long radius socket/spigot 87.5° | nr | - | 19.38 | - |
| Branches | | | | |
| socket/spigot 45° equal | nr | 7.26 | 9.69 | 24.94 |
| socket/spigot 87.5° equal | nr | - | 9.69 | - |
| socket/spigot 45°, 160 x 110 mm | nr | - | - | 18.66 |
| socket/socket 45°/87.5° equal | nr | - | 10.88 | - |
| Access components | | | | |
| socket/spigot pipe | nr | 11.67 | 15.58 | - |
| double solvent socket pipe | nr | - | - | 30.05 |
| socket/spigot bend | nr | - | 14.12 | - |
| socket/spigot branches | | | | |
| 45° branch | nr | - | 24.26 | 33.72 |
| 45°, 160 x 110mm branch | nr | - | - | 27.42 |
| 45° double branch | nr | - | 25.90 | - |
| socket/socket branches | | | | |
| 45° branch | nr | - | 24.26 | - |
| 45° double branch | nr | - | 25.90 | - |
| rodding point 45° socketed | nr | - | 20.08 | - |
| cap and pressure plug | nr | - | 5.17 | 12.65 |
| cap | nr | 4.30 | - | - |
| pressure plug | nr | - | 3.24 | 6.35 |
| socket plug | nr | - | 2.91 | - |
| Standing plug components | | | | |
| socket/spigot boss branch | | | | |
| 110 x 32 mm | nr | - | 4.70 | - |
| 4 upstands | nr | - | - | 8.25 |
| boss connector, ring seal | nr | 1.16 | - | - |

BASIC MATERIAL PRICES

**UPVC DRAIN PIPES AND FITTINGS** - *continued*

| **Diameter in mm** | Unit | **82** £ | **110** £ | **160** £ |
|---|---|---|---|---|
| Open channels | | | | |
| straight double spigot (1500 mm long) | nr | - | 13.70 | 23.52 |
| long radius channel bend | nr | - | 15.63 | - |
| slipper bend | nr | - | 6.72 | 17.29 |
| Reducers | | | | |
| level invert | | | | |
| 110 mm spigot to 110 mm socket | nr | - | 4.55 | - |
| 160 mm spigot to 110 mm socket | nr | - | - | 8.62 |
| eccentric | | | | |
| 82 mm socket to boss upstand | nr | 4.02 | - | - |
| 82 mm socket to 68 mm socket | nr | 1.79 | - | - |
| 110 mm socket to boss upstand | nr | - | 3.27 | - |
| 110 mm socket to 68 mm socket | nr | - | 2.35 | - |
| concentric | | | | |
| 110 mm spigot to boss upstand | nr | - | 2.91 | - |
| Adaptors | | | | |
| upvc spigot to salt glazed/pitch fibre socket | nr | - | 3.28 | 10.56 |
| upvc spigot to cast iron socket | nr | 9.95 | 9.46 | 25.60 |
| upvc spigot to clay plastic coupling | nr | - | 4.70 | 6.73 |
| salt glazed/cast iron spigot to upvc socket | nr | - | 6.39 | 10.98 |
| Special equipment | | | | |
| sealed rodding point (internal) | nr | - | 201.62 | - |
| Access point covers | | | | |
| concrete lid and upvc frame 324mm sq overall | nr | - | 20.16 | - |
| concrete lid and frame 276mm sq | nr | - | 16.71 | - |
| Inspection chambers | | | | |
| 450 mm chamber, excluding lid | | | | |
| 600 mm high | nr | - | 75.85 | - |
| 1000 mm high | nr | - | 100.25 | - |
| 450 mm chamber base 230mm high | nr | - | 41.72 | - |
| 450 mm chamber riser 400 mm high | nr | - | 34.87 | - |
| 450 mm cast iron cover and frame (grade C) | nr | - | 45.17 | - |
| 250 mm straight chamber base | nr | - | 25.06 | - |
| 250 mm branch chamber base | nr | - | 27.50 | - |
| 250 mm double branch chamber base | nr | - | 31.92 | - |
| 250 mm bottom outlet chamber body 4 x 110 mm upstands | nr | - | 27.05 | - |
| 250 mm pressure plug | nr | - | 15.65 | - |
| 250 mm chamber riser 375 mm long | nr | - | 8.23 | - |
| concrete lid, 300 x 300 mm | nr | - | 8.38 | - |
| upvc frame, 300 x 300 mm to 250 mm dia | nr | - | 11.78 | - |

BASIC MATERIAL PRICES

**UPVC DRAIN PIPES AND FITTINGS** - *continued*

| **Diameter in mm** | Unit | **82** £ | **110** £ | **160** £ |
|---|---|---|---|---|
| Gully components | | | | |
| compact gully | nr | - | 19.83 | - |
| gully trap base | nr | - | 7.11 | - |
| bend socket/spigot | nr | - | 14.12 | - |
| hoppers (rectangular) | nr | - | 7.11 | - |
| (square) | nr | - | 5.33 | - |
| inlet raising pieces | | | | |
| 2 x 82 mm upstands | nr | - | 2.58 | - |
| 4 boss upstands | nr | - | 4.79 | - |
| 87.5° branch socket/spigot | nr | - | 9.69 | - |
| grating assembly | nr | - | 3.73 | - |

| **Diameter in mm** | Unit | **200** £ | **250** £ |
|---|---|---|---|
| Marley large diameter underground drainage system | | | |
| Straight pipes | | | |
| ring seal socket 6 m | m | 12.09 | 18.64 |
| double spigot 6 m | m | 10.90 | 17.90 |
| Couplings | | | |
| double ring seal slip | nr | 12.32 | 35.47 |
| Bends | | | |
| socket/spigot 87.5° | nr | 32.22 | - |
| socket/spigot 45° | nr | 24.17 | 60.58 |
| socket/spigot 15° | nr | 19.53 | 48.20 |
| Branches | | | |
| socket/spigot 45° equal | nr | 38.18 | - |
| socket/spigot 45° unequal x 160mm | nr | 30.90 | 77.79 |
| socket/spigot 45° unequal x 200mm | nr | - | 82.19 |
| Access components | | | |
| socket/spigot pipe | nr | 83.00 | 91.58 |
| pressure plug | nr | - | 15.65 |
| socket plug | nr | 6.01 | - |
| Open channels | | | |
| double spigot straight (2000 mm long) | nr | 35.34 | 51.87 |
| Reducers | | | |
| level invert 200mm spigot to 160mm socket | nr | 12.71 | - |
| level invert 250mm spigot to 160mm socket | nr | - | 33.99 |
| level invert 250mm spigot to 200mm socket | nr | - | 28.79 |
| Adaptors | | | |
| UPVC spigot to stoneware/concrete socket | nr | - | 30.77 |

BASIC MATERIAL PRICES

**UPVC DRAINS PIPES AND FITTINGS** - *continued*

| | Unit | £ |
|---|---|---|
| Ancillary Items | | |
| Spare Ring Seals | | |
| 82 mm 'T' or 'O' ring | nr | 0.48 |
| 110 mm 'T' or 'O' ring | nr | 0.53 |
| 160 'T' or 'O' ring | nr | 1.16 |

**UPVC RAINWATER GUTTERS AND PIPES**

| | Unit | £ |
|---|---|---|
| Gutters, Marley Deep Flow | | |
| 110 x 75 mm gutters (3 m length) | m | 2.54 |
| 110 x 75 mm gutters (4 m length) | m | 2.20 |
| union clips | nr | 2.43 |
| fascia brackets | nr | 0.60 |
| angles, 90° | nr | 2.97 |
| angles, 45° | nr | 4.48 |
| running outlets | nr | 2.73 |
| stopend outlets | nr | 2.39 |
| stopends | nr | 1.45 |
| Gutters, Marley Industrial | | |
| 150 mm half round gutters (4 m length) | m | 4.72 |
| union clips | m | 3.76 |
| fascia brackets | nr | 2.05 |
| angles, 90° | nr | 9.11 |
| running outlets | nr | 7.71 |
| stopend, internal | nr | 1.76 |
| stopend, external | nr | 2.48 |

| **Diameter in mm** | | **68** | **110** |
|---|---|---|---|
| | Unit | £ | £ |
| Downpipes, Marley circular | | | |
| ring seal socket 2.5 m | m | 2.16 | - |
| ring seal socket 3 m | m | 2.00 | 4.14 |
| ring seal socket 5.5 m | m | 1.98 | - |
| sockets | nr | 1.44 | 2.91 |
| bends | nr | 2.59 | 7.35 |
| branches | nr | 5.17 | 10.32 |
| shoes | nr | 2.23 | 8.06 |
| hopper heads | nr | 8.30 | 42.14 |
| one piece pipe clip | nr | 0.72 | - |
| socket/pipe clip | nr | 0.60 | 2.53 |
| barrel pipe clip | nr | - | 1.26 |
| backplate | nr | 0.37 | - |

BASIC MATERIAL PRICES

## UPVC SUBSOIL DRAINAGE SYSTEMS

| **Diameter in mm** | Unit | **80** £ | **100** £ |
|---|---|---|---|
| Corrugated, perforated pipes (OsmaDrain) | | | |
| 25 m coils | m | 0.99 | 1.62 |
| end caps | nr | 0.40 | 0.43 |
| couplers | nr | 1.20 | 1.56 |
| reducers | nr | - | 1.07 |
| equal junctions | nr | 2.39 | 2.74 |
| unequal junctions | nr | - | 2.66 |

| **Diameter in mm** | Unit | **82** £ | **110** £ | **160** £ | **200** £ | **250** £ |
|---|---|---|---|---|---|---|
| Marley perforated pipes | | | | | | |
| ring seal socket 6 m | m | - | 3.74 | 9.09 | 13.50 | 19.57 |
| ring seal socket 3 m | m | 3.89 | - | - | - | - |

BASIC MATERIAL PRICES

**MISCELLANEOUS AND MINOR ITEMS**

| | Unit | £ |
|---|---|---|
| Reinforcement spacers: | | |
| spacemen SF20 | 1000 | 24.95 |
| spacemen SF75 | 1000 | 104.20 |
| meshmen MSM | 1000 | 152.80 |
| plasclips 30 NC | 1000 | 30.39 |
| dekchairs (ND 120 - 120 mm high) | 100 | 119.93 |
| hychairs (HYS 200 - 200 mm high) | 100 | 26.27 |
| Lenton reinforcement couplers | | |
| Type A2 25.4 mm dia | nr | 1.86 |
| Type P4 30 mm dia | nr | 8.59 |
| Dowel bars round | | |
| 32 x 600 mm | nr | 2.01 |
| 20 x 600 mm | nr | 0.72 |
| 16 x 600 mm | nr | 0.47 |
| 12 x 600 mm | nr | 0.27 |
| Expansion dowel caps | | |
| 16 x 100 mm | 100 | 11.06 |
| 20 x 100 mm | 100 | 13.68 |
| 32 x 100 mm | 100 | 24.40 |
| Debonding sleeves for dowel bars | | |
| 16 x 200 mm | nr | 0.28 |
| 20 x 300 mm | nr | 0.28 |
| 32 x 375 mm | nr | 0.51 |
| Debonding compound | L | 1.67 |
| Crack inducers (5 m lengths) | | |
| 10 mm wide x 40 mm deep top type (one piece) | m | 2.35 |
| 10 mm wide x 75 mm deep top type (one piece) | m | 3.41 |
| 10 mm wide x 50 mm deep top type (two piece) | m | 1.49 |
| 10 mm wide x 75 mm deep top type (two piece) | m | 1.87 |
| 10 mm wide x 40 mm deep bottom Y type | m | 1.29 |
| 10 mm wide x 75 mm deep bottom Y type | m | 2.00 |
| General purpose mould oil | L | 0.65 |
| Heat resistant mould oil | L | 0.99 |
| Chemical release agent | L | 0.69 |
| Epoxy grout high density | kg | 2.12 |
| Epoxy mortar general purpose | kg | 2.25 |
| 5 star grout | kg | 1.75 |
| Silver sand (bagged) | 25kg | 5.00 |

BASIC MATERIAL PRICES

**MISCELLANEOUS AND MINOR ITEMS**

| | Unit | £ |
|---|---|---|
| Resin bonded anchors | | |
| Resin capsules | | |
| RM8 (10 per box) | 100 | 87.80 |
| RM10 (10 per box) | 100 | 89.45 |
| RM12 (10 per box) | 100 | 93.20 |
| RM16 (10 per box) | 100 | 138.75 |
| RM20 (6 per box) | 100 | 228.35 |
| RM24 (6 per box) | 100 | 245.50 |
| RM30 (6 per box) | 100 | 413.65 |
| Zinc plated steel threaded rod with nut and washer | | |
| 8 x 110 mm (10 per box) | 100 | 35.95 |
| 10 x 130 mm (10 per box) | 100 | 51.95 |
| 12 x 160 mm (10 per box) | 100 | 64.15 |
| 16 x 190 mm (10 per box) | 100 | 107.60 |
| 20 x 260 mm (6 per box) | 100 | 271.95 |
| 24 x 300 mm (6 per box) | 100 | 411.15 |
| 30 x 380 mm (6 per box) | 100 | 849.90 |
| Stainless steel threaded rod with nut and washer | | |
| 8 x 110 mm (10 per box) | 100 | 180.05 |
| 10 x 130 mm (10 per box) | 100 | 217.50 |
| 12 x 160 mm (10 per box) | 100 | 330.30 |
| 16 x 190 mm (10 per box) | 100 | 553.20 |
| 20 x 260 mm (6 per box) | 100 | 1743.30 |
| 24 x 300 mm (6 per box) | 100 | 3008.30 |

BASIC MATERIAL PRICES

**MISCELLANEOUS AND MINOR ITEMS**

| | Unit | £ |
|---|---|---|
| Loose bolt type Wallbolts | | |
| M6 max fixing thickness 25 mm (50 per box) | 100 | 64.30 |
| M8 max fixing thickness 25 mm (50 per box) | 100 | 77.95 |
| M10 max fixing thickness 25 mm (25 per box) | 100 | 103.50 |
| M12 max fixing thickness 25 mm (20 per box) | 100 | 165.05 |
| M16 max fixing thickness 30 mm (10 per box) | 100 | 378.20 |
| M20 max fixing thickness 60 mm (5 per box) | 100 | 605.10 |
| M24 max fixing thickness 100 mm (5 per box) | 100 | 1038.25 |
| M24 max fixing thickness 150 mm (5 per box) | 100 | 1114.55 |
| Bolt projecting type wallbolts | | |
| M6 max fixing thickness 25 mm (50 per box) | 100 | 66.80 |
| M8 max fixing thickness 25 mm (50 per box) | 100 | 81.00 |
| M10 max fixing thickness 30 mm (25 per box) | 100 | 109.90 |
| M12 max fixing thickness 35 mm (20 per box) | 100 | 178.20 |
| M16 max fixing thickness 35 mm (10 per box) | 100 | 420.15 |
| M20 max fixing thickness 30 mm (5 per box) | 100 | 594.00 |
| M20 max fixing thickness 100 mm (5 per box) | 100 | 675.20 |
| M24 max fixing thickness 75 mm (5 per box) | 100 | 1036.35 |
| M24 max fixing thickness 120 mm (5 per box) | 100 | 1111.35 |
| Hyrib permanent shuttering ref 2411 | $m^2$ | 8.45 |
| ref 2611 | $m^2$ | 7.15 |
| Expamet galvanised brickwork reinforcement (100 m rolls) | | |
| 65 mm wide | m | 0.16 |
| 115 mm wide | m | 0.29 |
| 175 mm wide | m | 0.42 |
| 225 mm wide | m | 0.58 |
| Expamet stainless steel brickwork reinforcment (25 m rolls) | | |
| 65 mm wide | m | 0.30 |
| 115 mm wide | m | 0.50 |
| 175 mm wide | m | 0.80 |
| 225 mm wide | m | 1.09 |
| Expamet cast in foundation bolt cylinders | | |
| 75 mm dia x 150 mm long | nr | 0.68 |
| 75 mm dia x 300 mm long | nr | 0.96 |
| 100 mm dia x 375 mm long | nr | 1.53 |
| 100 mm dia x 450 mm long | nr | 1.73 |
| 100 mm dia x 600 mm long | nr | 1.91 |
| Granite setts new 100 x 100 (approx) | t | 99.00 |
| Granite setts reclaimed 100 x 100 (approx) | t | 60.00 |
| Random size cobbles reclaimed | t | 45.00 |

PLANT

**INTRODUCTION**

The information, rates and prices included in this section are calculated examples of actual owning and operating costs of a range of construction plant and equipment (excluding drivers or operators). To an extent they will also serve as a guide to prevailing commercial plant hire rates. But be aware many factors will influence actual hire rates. For example, rates could be lower because of long term hire, use of older or second hand machines, low market demands/loss leaders, easy tasks or sites near to depot. Rates could be higher due to short term hire, high production demands, low utilisation factors, specialist operations, restricted working/accessability problems, high profit or overhead demands, finance fluctuations. The use factors will mean a hired machine may not be used productively for each hour on site, it is therefore up to the estimator to allow for reasonable outputs in his unit rate calculations.

These and other considerations must be borne in mind if using these costs as a comparison to hire rates especially operated plant.

All rates quoted are exclusive of Value Added Tax.

CONTRACTOR OWNED PLANT

Plant owned by a Contractor generally falls under two headings:

1. Small plant and tools which are the subject of a direct charge to contracts and, for estimating purposes, are normally allowed for as a percentage of the labour cost, in site oncosts (see Preliminaries and General Items), although many items are shown in this section for information.

2. Power driven plant and major non-mechanical plant such as steel trestling, scaffolding, gantries, etc. Such plant is normally charged to the contract on a rental basis except in the case of purpose made or special plant bought specifically for a particular operation; this latter is normally charged in full to contracts and allowance made for disposal on completion (often at scrap value).

A very wide range of plant is readily available from plant hire companies; it is not usually economical for Contractors to own plant unless they can ensure at least 75 to 80% utilisation factor based on the Contractor's normal working hours. Where a Contractor does own plant, however, it is essential that he maintains reasonably accurate records of the working hours and detailed costs of maintenance and repairs in order that he may estimate the charges to be made for each item of plant. For this reason, where a Contractor owns a large quantity of plant, it is normal for a separate plant hire department or company to be formed.

This department or company should be financially self supporting and may hire plant to other Contractors when it is not needed. Maintenance of Contractor owned plant can be carried out on site; it is not necessary to maintain a centrally based repair workshop. It is, however, desirable to have some storage facilities available for plant when not in use. The cost of owning plant and hence the rental charges to be made to contracts must take into account:

capital cost
depreciation charges
maintenance and repairs
cost of finance
insurances and licences
administraction, head office, depot and other overhead charges

PLANT

Calculated examples of hourly owning costs of a range of Plant and equipment

| TITLE | A UTILIZATION % FACTOR | B P.A. HOURS | C PLANT YEARS | D LIFE HOURS | E PURCHASE PRICE £ | F RESALE % | G RESALE PRICE £ | H LOSS IN VALUE £ | I DEPREC. £ | J MAINTEN. £ | K FINANCE @18% £ | L INSURANCE £ | M ADMIN. £ | N TOTAL ITEM £ |
|---|---|---|---|---|---|---|---|---|---|---|---|---|---|---|
| **ACCESS PLATFORMS** | | | | | | | | | | | | | | |
| 40′ telescopic | 65 | 1,365 | 10 | 13,650 | 19,000 | 20.00 | 3,800 | 15,200 | 1.11 | 1.39 | 1.00 | 0.28 | 0.76 | 4.54 |
| **AGRO TRACTORS** | | | | | | | | | | | | | | |
| County 774 | 75 | 1,575 | 8 | 12,600 | 26,740 | 23.00 | 6,150 | 20,590 | 1.63 | 1.70 | 1.18 | 0.34 | 0.97 | 5.82 |
| **ASPHALT PAVER** | | | | | | | | | | | | | | |
| BK 171 | 65 | 1,365 | 7 | 9,555 | 110,000 | 27.50 | 30,250 | 79,750 | 8.35 | 8.06 | 5.26 | 1.61 | 4.66 | 27.94 |
| **COMPRESSORS** | | | | | | | | | | | | | | |
| 450 cfm | 85 | 1,785 | 9 | 16,065 | 15,500 | 20.00 | 3,100 | 12,400 | 0.77 | 0.87 | 0.63 | 0.17 | 0.49 | 2.93 |
| **CRANES** | | | | | | | | | | | | | | |
| 15 t mobile | 60 | 1,260 | 12 | 15,120 | 50,000 | 25.00 | 12,500 | 37,500 | 2.48 | 3.97 | 2.68 | 0.79 | 1.98 | 11.90 |
| 40 t | 60 | 1,260 | 15 | 18,900 | 116,000 | 27.50 | 31,900 | 84,100 | 4.45 | 9.21 | 6.01 | 1.84 | 4.30 | 25.81 |
| **GENERATORS** | | | | | | | | | | | | | | |
| 7.5 kVA | 85 | 1,785 | 5 | 8,925 | 3,000 | 15.00 | 450 | 2,550 | 0.29 | 0.30 | 0.13 | 0.03 | 0.15 | 0.90 |
| **EXCAVATORS** | | | | | | | | | | | | | | |
| 16 B/A track | 80 | 1,680 | 9 | 15,120 | 50,000 | 27.50 | 13,750 | 36,250 | 2.40 | 2.98 | 1.94 | 0.60 | 1.58 | 9.50 |
| 60 B/A track | 75 | 1,575 | 9 | 14,175 | 175,000 | 27.50 | 48,125 | 126,875 | 8.95 | 10.05 | 7.25 | 2.22 | 5.69 | 34.16 |
| JCB 3CX | 85 | 1,785 | 6 | 10,710 | 33,000 | 27.50 | 9,075 | 23,925 | 2.23 | 1.85 | 1.21 | 0.37 | 1.13 | £6.79 |
| **ROLLERS** | | | | | | | | | | | | | | |
| Bomag BW90S | 85 | 1,785 | 6 | 10,710 | 7,850 | 27.50 | 2,159 | 5,691 | 0.53 | 0.44 | 0.29 | 0.09 | 0.27 | 1.62 |
| **CRAWLER LOADERS** | | | | | | | | | | | | | | |
| 963 | 80 | 1,680 | 9 | 15,120 | 92,000 | 27.50 | 25,300 | 66,700 | 4.41 | 5.48 | 3.57 | 1.10 | 2.91 | 17.47 |
| **TIPPERS** | | | | | | | | | | | | | | |
| 6 x 4 280 | 90 | 1,890 | 6 | 11,340 | 52,750 | 22.00 | 11,605 | 41,145 | 3.63 | 2.79 | 1.96 | 0.56 | 1.79 | 10.73 |
| **DUMP TRUCK** | | | | | | | | | | | | | | |
| Volvo A35 | 75 | 1,575 | 8 | 12,600 | 81,500 | 25.00 | 20,375 | 61,125 | 4.85 | 5.17 | 3.49 | 1.03 | 2.91 | 17.45 |

PLANT

**Notes**

1. The above costs would be updated annually or bi-annually, the rental rate being revised to ensure complete recovery of the costs associated with the item of plant and return of capital to enable the machine to be replaced at the end of its life with the Contractor. The purchase price must also be adjusted to ensure recovery of the replacement cost and not the original cost of purchase.

2. Driver's wages and costs should be charged direct to site wages.

3. Columns C and D: The period over which the Contractor owns the plant and during which time the items of plant can be maintained at reasonable efficiency will vary according to the Contractor's experience, the type of plant under consideration, and the work on which is it employed. The owning period of an item of plant can vary considerably on this basis. The data we have used in our calculations varies considerably (see table in 4 below).

4. Column A: The utilisation factors used in conjunction with the period of ownership is the percentage of time (per annum) that an item of plant can be expected to be used productively on site and therefore is a very important influence of hourly costs.

   (Note: Utilisation factors are not the same as site utilisation rates - see note 12 below).

   Economical Owing Periods/Use Factors

| | | | |
|---|---|---|---|
| Hydraulic Excavators | large | 9 yrs | 75% |
| Hydraulic Excavators | medium | 7 yrs | 80% |
| Hydraulic Excavators | mini | 6 yrs | 85% |
| Dozers/Scrapers | | 10 yrs | 80% |
| Loaders/Shovels | | 9 yrs | 80% |
| Mobile Cranes | | 12 yrs | 60% |
| Crawler Cranes | | 15 yrs | 60% |
| Dump Trucks | | 8 yrs | 75% |
| Dumpers | | 6 yrs | 90% |
| Rollers/Compaction | | 6 yrs | 80% |
| Compressors/Generators | | 7-9 yrs | 85% |
| Diesel engine road vehicles | | 6-7 yrs | 90% |
| Petrol engine road vehicles | | 5-6 yrs | 90% |

5. Column K: Finance interest charges are taken on the average base rate over the previous 12 months + 3%.

6. Column L: Insurance premiums have been taken as 2% on the replacement value of the plant (purchase price).

7. Column J: Maintenance costs cover major overhauls and replacement costs for wear items only, excluding insurable damage, and will vary with each type of plant. This does not include day to day general servicing on site.

8. Column M: Administration, head office, depot and other overhead charges have been taken at + 20%.

PLANT

9. Consumables should be charged direct to site costs. The costs included with this section are based upon manufacturers data and are used in good faith. (For outline estimates fuel consumption can be taken as approx 0.15 - 0.20 litres per kW per working hour; lube oils, filters, grease etc, can be taken at 2 - 10% of the fuel cost, depending upon the working conditions of the machine and its attachments). See note 10 for details.

10. Please note that the costs in this section have been completely reviewed for this edition, based on latest available advice and data from manufacturers and Plant Hire firms, the basis of calculation, including retention period of the plant, has been changed in a number of cases, providing a realistic overall owning and operating cost of the machine.

11. These examples are for machine costs only, and do not include for operator costs or profit element, or for transport costs to and from site.

12. Fuel consumption is based on the following site utilisation rates. (Site utilization being the percentage of time that the machine is operating at its average fuel consumption during a working day).

| | |
|---|---|
| Excavators | - 75% |
| Dozers/Scrapers | - 80% |
| Loaders/Shovels/Graders | - 75% |
| Mobile Cranes | - 50% |
| Crawler Cranes | - 25% |
| Dump Trucks (not Tippers) | - 75% |
| Dumpers | - 80% |
| Compressors, mixers, generators | - 85% |
| Tractors | - 75% |
| Hoists, etc | - 65% |
| Drills and Saws | - 90% |
| Rollers/Compaction Plant | - 75% |
| Piling/Asphalt Equipment | - 65% |
| All other items | - 100% |

Approx fuel consumption in litres per hour and percentage addition for consumables (examples only) based on the above Site Utilisation rates.

| | | | |
|---|---|---|---|
| Ford 660 Tractor | 9 + 2% | BW 90 roller | 3.5 + 7.5% |
| Blaw Knox BK191 paver | 16 + 5% | 120AD roller | 4 + 7.5% |
| Wacker plate 180kg | 0.7 + 3% | BW6 towed roller | 6 + 7.5% |
| 125 cfm compressor | 12 + 3% | CAT D6 dozer | 21 + 6% |
| 375 cfm compressor | 20 + 3% | CAT D8 dozer | 38 + 3% |
| Tractor/compressor | 9 + 5% | CAT 953 loader | 17 + 4% |
| 5/3.5 mixer | 1.2 + 2% | CAT 621 scraper | 41.5 + 3% |
| 15t mobile crane | 12 + 10% | CAT 657E p/p scraper | 110 + 6% |
| 40t mobile crane | 25 + 10% | Bobcat 543 Skidsteer | 4.5 + 5% |
| 4 kVA generator | 1.6 + 3% | 140G grader | 20 + 5% |
| 25 kVA generator | 4.7 + 5% | 30T dumper | 25 + 5% |
| 11.5 tonne backacter | 9.5 + 7.5% | Volvo A25 dumper | 17 + 3% |
| 19 tonne backacter | 14 + 7.5% | 2T dumper | 3 + 3% |
| 30 tonne backacter | 24 + 7.5% | 3" spate pump | 2 + 2% |
| JCB 3CX 7.3 tonne | 7.5 + 6% | 6" univace pump | 6 + 3% |

PLANT

**PLANT COSTS**

The costs included in this section are based upon the methodology of the worked examples, these have then been compared to average costs, allowances and hire rates in force at the time of writing (Summer 1990). These costs then form the basis for plant costs included in parts 4 and 5 (except for specialist advice). Weekly costs are based upon a 40 hour week, daily costs are based on an 8 hour day, all costs are exclusive of labour for operation. Also given here is a reference to the 1990 FCEC Daywork Schedule (Plant Section).

Rates given for consumables are in line with notes 9 & 12 without loss or wastage, and are based on the following (All plant except those noted are priced on Gas Oil usage):

| | |
|---|---|
| Petrol 4 star | 34.85 pence per litre |
| Petrol Unleaded | 32.95 pence per litre |
| Fuel Oil (taxed for use in licensed vehicles) DERV | 29.50 pence per litre |
| Fuel Oil (untaxed, for use on site) Gas Oil | 11.50 pence per litre |
| Lubrication Oil (15/40) | 76.00 pence per litre |
| Mains Electric power | 6.00 pence per kWh |

Also included here are allowances for Transmission Oils, Hydraulic Oil, Filters, Grease, etc. N/A = information not available.

Imperial units of measurements are used frequently in this section reflecting their continuing usage in this sector of the industry.

| Type of plant | Period | Cost £ | Consumables per working hour £ | 1990 F.C.E.C. Daywork Schedule Reference |
|---|---|---|---|---|
| **Access Platforms** | | | | |
| Access platform, scissor type, towed, height of platform 15 ft | hour | 2.48 | - | 1(8) |
| Access platform, scissor type electric, height of platform 15 ft | hour | 3.05 | 0.40 | 1(1) |
| Access platfrom, scissor type rough terrain, petrol driven height of platform 25 ft | hour | 4.25 | 0.15 | 1(6) |
| Access platfrom telescopic, towed, height of platfrom 35 ft | hour | 3.36 | - | 1(-) |
| Access platfrom telescopic petrol driven, height of platfrom 40 ft | hour | 4.54 | 0.25 | 1(11) |
| Access platform telescopic vehicle mounted height of platfrom 35 ft (Simon Roto Zoom) (DERV) | hour | 8.95 | 1.21 | 1(14) |
| Access platform telescopic vehicle mounted, height of platform 50 ft (Simon Roto Zoom) (DERV) | hour | 10.50 | 1.33 | 1(13) |

PLANT

| Type of plant | Period | Cost £ | Consumables per working hour £ | 1990 F.C.E.C. Daywork Schedule Reference |
|---|---|---|---|---|
| **Access Platform** (Cont'd) | | | | |
| Access platfrom underbridge type (Simon Gala topper) | hour | 18.75 | 1.33 | 1(13) |
| **Agricultural Type Tractors** | | | | |
| County 774 (78 HP) | hour | 5.82 | 1.09 | 32(61) |
| Ford 5000 (68 HP) c/w 1T Hiab lift | hour | 6.70 | 0.91 | 32(*) |
| Ford 5000 (68 HP) with front end loading bucket ($0.33m^3$) | hour | 6.25 | 0.91 | 37(51) |
| Ford 5000 (68 HP) with Hydro seeding equipment | hour | 6.75 | 0.91 | 37(*) |
| International Tractair includes 2 tool compressor and front end loading bucket | hour | 6.95 | 1.08 | - |
| Ford 5000 with Fencing Auger | hour | 6.25 | 0.92 | - |
| **sphalt/Road Construction** | | | | |
| Asphalt paver, extending up to 4.00 m, 81 kW engine | hour | 27.94 | 1.54 | 2(2) |
| Asphalt paver, extending up to 4.75 m, 100 kW engine | hour | 32.76 | 1.90 | 2(4) |
| Asphalt paver, maximum paving width 3.66 m, 19 kW engine | hour | 14.32 | 0.48 | 2(1) |
| Tar sprayer (100 litre) | hour | 1.15 | - | 29(5) |
| Self propelled chip spreader | hour | 4.21 | - | 2(9) |
| Heating iron | hour | 0.75 | - | - |
| Insulated (20 tonne) tipper (DERV) | hour | 13.96 | 4.34 | 14(18) |
| Aveling 3 point roller | hour | 4.90 | 1.07 | 23(4) |
| Surface planer (cold plane) planing width 0.5 m - CAT PR105 | hour | 7.07 | 0.95 | 2(12) |
| Surface planer (cold plane) planing width 3.76 m - CAT PM800 | hour | 25.50 | 2.53 | 2(-) |
| Surface planer (heat plane) planing width up to 1.0 m - wirtgen mini re-mixer | hour | 9.50 | N/A | 2(11) |

PLANT

| Type of plant | Period | Cost £ | Consumables per working hour £ | 1990 F.C.E.C. Daywork Schedule Reference |
|---|---|---|---|---|
| **Asphalt/Road Construction** (Cont'd) | | | | |
| Surface planer (heat plane) planing width up to 4.5 m - wirtgen remixer | hour | 40.00 | N/A | 2(-) |
| Concrete paver; maximum paving width 6.00 m, depth 300 mm, 81 kW engine (SGME MF450) | hour | 35.00 | 1.40 | - |
| Concrete paver, maximum paving width 12.8 m depth 500 mm 340 kW engine (Gomaco 3500 slipform) | hour | 55.70 | 1.40 | - |
| Concrete Slipform paver/trimmer, maximum paving width 5.00 m (Gomaco Commander) | hour | 41.94 | N/A | - |
| Slipformed Concrete joint and bar inserter | hour | 2.50 | N/A | - |
| Concrete Slipform finisher (Gomaco 8500) | hour | 7.50 | N/A | - |
| **Compaction** | | | | |
| Plate compactors | | | | |
| Vibrating compaction plate 60 kg unit weight 350 mm wide (petrol) | week | 28.50 | 0.42 | 22(6) |
| Vibrating compaction plate 140 kg unit weight 400 mm wide (petrol) | week | 30.00 | 0.50 | 22(8) |
| Vibrating compaction plate 180 kg unit weight 450 mm wide | week | 30.75 | 0.10 | 22(9) |
| Vibrating compaction plate 300 kg unit weight 450 mm wide | week | 33.75 | 0.12 | 22(10) |
| Vibrating compaction trench plate 300 mm wide | week | 25.60 | 0.10 | 22(*) |
| Tamper BT58 58 kg (petrol) | week | 26.40 | 0.40 | 22(1) |
| Tamper BT68 68 kg (petrol) | week | 30.75 | 0.40 | 22(2) |
| **Cleaners/Sweepers** | | | | |
| Front Tractor or Truck mounted 7ft wide (excluding vehicle) | week | 44.00 | - | 14(36) |

'LANT

| ype of plant | Period | Cost £ | Consumables per working hour £ | 1990 F.C.E.C. Daywork Schedule Reference |
|---|---|---|---|---|
| **:leaners/Sweepers** (Cont'd) | | | | |
| Towed sweeper diesel engined | week | 65.75 | 0.24 | - |
| Glutton type mobile gully sucker | week | 79.00 | N/A | - |
| **obile Compressors (Silenced or Super Silenced)** | | | | |
| 1.80 cum/65 cfm single tool | week | 48.80 | 1.07 | 4(1) |
| 2.40 cum/85 cfm two tool | week | 60.00 | 1.18 | 4(1) |
| 3.55 cum/125 cfm two tool | week | 70.00 | 1.42 | 4(3) |
| 4.55 cum/160 cfm two tool | week | 78.40 | 1.53 | 4(4) |
| 7.26 cum/250 cfm two tool | week | 87.50 | 2.25 | 4(6) |
| 10.41 cum/375 cfm four tool | week | 96.00 | 2.37 | 4(7) |
| 12.60 cum/450 cfm four tool | week | 117.20 | 2.49 | 4(8) |
| 13.50 cum/475 cfm four tool | week | 122.00 | 2.61 | 4(8) |
| 17.00 cum/600 cfm four tool | week | 138.00 | 2.85 | 4(9) |
| 21.00 cum/750 cfm four tool | week | 160.00 | 3.40 | 4(11) |
| 25.50 cum/900 cfm | week | 190.00 | 3.75 | 4(11) |
| Reciprocating compressor up to 4 cfm for small air tools 240 v | week | 11.20 | 0.28 | 4(16) |
| Reciprocating compressor up to 9 cfm for small air tools 240 v | week | 29.50 | 0.40 | 4(16) |
| Reciprocating compressor up to 13 cfm for small air tools 240 v | week | 34.00 | 0.60 | 4(16) |
| Petrol driven hydraulic power unit c/w breaker (63 kg) | week | 40.60 | 0.52 | - |
| Metreater self propelled hydraulic ashpalt cutter | week | 70.00 | N/A | - |
| hicle compressor Holman 175 CFM on ord Transit B or C series chassis | hour | 8.50 | 2.40 | 4(14) |
| istol IH4 on International Tractor | hour | 6.25 | 1.08 | 4(*) |
| istol IH4 on International Tractor ith loading bucket | hour | 6.95 | 1.08 | 4(*) |

PLANT

| Type of plant | Period | Cost £ | Consumables per working hour £ | 1990 F.C.E.C. Daywork Schedule Reference |
|---|---|---|---|---|
| **Compressor tools (cw 50ft/15m hose)** | | | Air cons cfm | |
| chipping hammer | week | 16.75 | 25 | 30(4) |
| brick hammer/demolition pick | week | 11.20 | 50 | - |
| clay spade | week | 17.25 | 57 | 30(3) |
| road breaker | week | 13.40 | 84 | 30(1) |
| hand hammer drill | week | 15.20 | 32 | 30(5) |
| light rock drill 28 | week | 16.25 | 42 | 30(5) |
| medium rock drill 30 | week | 17.25 | 54 | 30(6) |
| heavy rock drill 33 | week | 18.25 | 84 | 30(7) |
| scabbler 1 head hand | week | 26.50 | 10 min | 30(4) |
| scabbler 3 head hand | week | 32.50 | 25 | - |
| scabbler 5 head trolley | week | 80.00 | 110 | - |
| scabbler 7 head trolley | week | 110.00 | 160 | - |
| needle gun/descaler | week | 26.40 | 10 | 30(4) |
| concrete crack cutter | week | 62.40 | 25 | - |
| rammer | week | 15.20 | 54 | 22(1) |
| air lance with variable valve | week | 6.00 | 9 min | - |
| impact wrench 0.5" drive | week | 15.00 | 9 min | - |
| impact wrench 1" drive | week | 25.00 | 13 min | - |
| trench sheet driver | week | 46.00 | 84 min | - |
| steel post driver (crash barrier) | week | 46.40 | 84 min | - |
| angle grinder 9" | week | 16.50 | 28 | - |
| disc cutter 12" | week | 20.50 | 50 | - |
| poker vibrator P35 (35 mm) | week | 16.80 | 84 | 6(24) |
| poker vibrator P54 (54 mm) | week | 21.60 | 84 | 6(24) |
| poker vibrator P70 (70 mm) | week | 26.40 | 84 | 6(24) |
| rotary drill | week | 18.00 | 10 | - |
| extra 50ft/15 m hose | week | 4.50 | - | - |
| **Concrete Equipment** | | | | |
| Poker vibrator air pokers (see compressor tables) | | | | 6(24) |
| Poker vibrator 2" petrol | week | 27.00 | 0.37 | 6(21) |
| Poker vibrator 2" electric | week | 26.50 | 0.19 | 6(23) |
| Poker vibrator 3" diesel | week | 27.00 | 0.15 | 6(22) |
| Poker vibrator 3" electric | week | 27.50 | 0.36 | 6(23) |
| Vibrating tamper and handles (petrol) | week | 33.20 | 0.33 | 6(32) |
| Razor-back screeder (per metre) | week | 10.30 | air | - |
| Double beam screeder (7.6 m wide) | week | 80.00 | 0.66 | 6(35) |
| Rotary power float (petrol) | | | | |
| 865 diam | week | 33.20 | 0.50 | 6(36) |
| 687 diam | week | 27.50 | 0.42 | 6(36) |
| Power bar cropper | week | 60.50 | 0.22 | 3(6) |
| Power bar bender (electric) | week | 67.50 | 0.22 | 3(3) |
| Stud Roller | week | 10.00 | - | - |

PLANT

| Type of plant | Period | Cost | Consumables per working hour | 1990 F.C.E.C. Daywork Schedule |
|---|---|---|---|---|
| **Mixers** | | | | |
| Concrete mixer 3/2 tip up 240v | week | 10.60 | 0.10 | 5(1) |
| Concrete mixer 4/3 petrol | week | 18.50 | 0.12 | 5(1) |
| Concrete mixer 5/3.5 diesel | week | 17.50 | 0.14 | 5(2) |
| Concrete mixer 8.5/6 diesel skip fed | week | 25.00 | 0.16 | 5(6) |
| Paddle screed mixer 5/3.5 diesel | week | 22.10 | 0.14 | - |
| Truck mixer (Rite mixer) $6m^3$ | hour | 11.00 | 4.15 truck<br>0.66 mixer | 5(40) |
| **Concrete Pumps** | | | | |
| Air operated 'SEM' pump | hour | 4.20 | air | 6(1) |
| Truck mounted 45cm/hr | hour | 26.80 | 1.45 pump | 6(6) |
| Schwing KSP5 slurry pump | hour | 7.07 | 0.97 | 6(1) |
| **Concrete Skips** | | | | |
| 1/2 yd3 rollover type ($0.38m^3$) | week | 16.20 | - | 8(22) |
| 1/2 yd3 dual flow type ($0.38m^3$) | week | 17.80 | - | 8(22) |
| 3/4 yd3 dual flow type ($0.57m^3$) | week | 20.80 | - | 8(23) |
| 1 yd3 dual flow type ($0.76m^3$) | week | 23.50 | - | 8(24) |
| **Cranes** | | | | |
| 7 t mobile (site use) | hour | 10.17 | 1.12 | 7(2) |
| 15 t SWL truck mounted mobile (wheeled) crane | hour | 11.90 | 1.52 | 7(20) |
| 20 t SWL truck mounted mobile (wheeled) crane | hour | 14.28 | 1.90 | 7(20) |
| 25 t SWL truck mounted mobile (wheeled) crane | hour | 17.50 | 2.28 | 7(21) |
| 40 t SWL truck mounted mobile (wheeled) crane | hour | 25.81 | 3.16 | 7(23) |
| 50 t SWL truck mounted mobile (wheeled) crane | hour | 34.50 | 3.94 | 7(24) |
| 70 t SWL truck mounted mobile (wheeled) crane | hour | 46.87 | 4.44 | 7(24) |
| 90 t SWL truck mounted mobile (wheeled) crane | hour | 55.12 | 4.84 | 7(25) |

PLANT

| Type of plant | Period | Cost £ | Consumables per working hour | 1990 F.C.E.C. Daywork Schedule Reference |
|---|---|---|---|---|
| **Cranes** (Cont'd) | | | | |
| 10 t crawler crane (22RB) | hour | 12.50 | 0.76 | 7(12) |
| 20 t crawler crane (30RB) | hour | 15.90 | 1.01 | 7(12) |
| 30 t crawler crane (NCK Pennine) | hour | 18.15 | 1.52 | 7(13) |
| **Diesel generators** | | | | |
| 500 w Petrol 110 or 240V | week | 16.75 | 0.30 | 11(1) |
| 1.5 kVA Petrol 110 or 240V | week | 19.00 | 0.40 | 11(1) |
| 2.5 kVA Petrol 110 or 240V | week | 20.80 | 0.60 | 11(2) |
| 4 kVA diesel dual voltage | week | 27.50 | 0.22 | 11(2) |
| 7 kVA diesel dual voltage | week | 32.00 | 0.25 | 11(3) |
| 7.5 kVA diesel dual voltage | week | 36.00 | 0.29 | 11(3) |
| 7.5 kVA diesel (super silenced) | week | 70.00 | 0.33 | 11(3) |
| 10 kVA diesel dual voltage | week | 41.50 | 0.35 | 11(3) |
| 15 kVA diesel dual voltage | week | 49.00 | 0.39 | 11(4) |
| 25 kVA diesel 3 phase 440V | week | 75.00 | 0.57 | 11(5) |
| 55 kVA diesel 3 phase 440V | week | 94.50 | 1.21 | 11(7) |
| 75 kVA diesel 3 phase 440V | week | 114.00 | 2.01 | 11(8) |
| 125 kVA diesel 3 phase 440V | week | 177.20 | 2.87 | 11(11) |
| 200 kVA diesel 3 phase 440V | week | 276.00 | 3.22 | 11(12) |
| 250 kVA diesel 3 phase 440V | week | 360.00 | 3.95 | 11(13) |
| **Excavators** (Bucket capacity refers to SAE heaped) | | | | |
| Rope operated, crawler mounted dragline equipment | | | | |
| (1.15m$^3$) 22 RB 27 m reach/drop | hour | 15.50 | 1.21 | 10(4) |
| (1.53m$^3$) 30 RB 33 m reach/drop | hour | 25.08 | 1.58 | 10(7) |
| Mini Excavators (tracked) | | | | |
| Kubota KH36, 1180 kg (0.04 m$^3$) | week | 120.00 | 0.21 | 10(15) |
| Kubota KH41, 1340 kg (0.04 m$^3$) | week | 138.00 | 0.30 | 10(15) |
| Kubota KH66, 2820 kg (0.07 m$^3$) | week | 158.00 | 0.46 | 10(16) |
| Smalley 5QT, 2800 kg (0.12 m$^3$) | week | 176.00 | 0.42 | 10(16) |
| Hitachi UE10, 1235 kg (0.05 m$^3$) | week | 120.00 | 0.23 | 10(15) |
| Thwaites all dig 2340 kg (0.25 m$^3$) | week | 150.00 | 0.41 | 10(16) |

PLANT

| Type of plant | | | Period | Cost £ | Consumables per working hour | 1990 F.C.E.C. Daywork Schedule Reference |
|---|---|---|---|---|---|---|
| Mini Excavators (wheeled) | | | | | | |
| Komatsu PW05, 1370 kg | | | week | 158.00 | 0.30 | 10(15) |
| Smalley 5W, 1750 kg | | | week | 200.00 | 0.43 | 10(15) |
| Hydraulic crawler mounted backacter approximate data | | | | | | |
| weight up to | SAE bucket | kW | | | | |
| 3.5 tonne | 0.10 $m^3$ | 25 | hour | 4.40 | 0.62 | 10(17) |
| 6.5 tonne | 0.25 $m^3$ | 40 | hour | 5.70 | 0.80 | 10(19) |
| 8.5 tonne | 0.30 $m^3$ | 50 | hour | 6.60 | 0.99 | 10(19) |
| 11.5 tonne | 0.40 $m^3$ | 63 | hour | 7.10 | 1.17 | 10(20) |
| 14.5 tonne | 0.45 $m^3$ | 70 | hour | 7.90 | 1.36 | 10(20) |
| 16 tonne | 0.55 $m^3$ | 80 | hour | 9.50 | 1.48 | 10(21) |
| 19 tonne | 0.70 $m^3$ | 90 | hour | 12.30 | 1.73 | 10(22) |
| 21 tonne | 1.00 $m^3$ | 100 | hour | 13.75 | 2.10 | 10(22) |
| 23.5 tonne | 0.90 $m^3$ | 115 | hour | 20.80 | 2.60 | 10(23) |
| 28.5 tonne | 1.20 $m^3$ | 130 | hour | 23.23 | 2.97 | 10(24) |
| 35.5 tonne | 1.20 $m^3$ | 160 | hour | 29.75 | 3.34 | 10(25) |
| 40 tonne | 1.60 $m^3$ | 180 | hour | 31.95 | 3.71 | 10(26) |
| 50 tonne | 2.00 $m^3$ | 205 | hour | 34.16 | 3.95 | 10(26) |
| 60 tonne | 2.50 $m^3$ | 235 | hour | 34.16 | 4.04 | 10(26) |
| 20 tonne (0.6$m^3$) Priestman VC15 15m reach, cable operated | | | hour | 14.60 | 1.75 | 10(22) |
| 30 tonne (0.8$m^3$) Priestman VC20 20m reach, cable operated | | | hour | 25.50 | 2.07 | 10(24) |
| Hydraulic wheeled backacter | | | | | | |
| 3.5 tonne 0.10 $m^3$ | | | hour | 4.95 | 0.65 | 10(17) |
| 6.5 tonne 0.27 $m^3$ | | | hour | 6.25 | 0.82 | 10(19) |

PLANT

| Type of plant | Period | Cost £ | Consumables per working hour | 1990 F.C.E.C. Daywork Schedule Reference |
|---|---|---|---|---|
| Hydraulic wheeled backacter (Cont'd) | | | | |
| 8 tonne 0.34 $m^3$ | hour | 7.15 | 0.98 | 10(19) |
| 11 tonne 0.43 $m^3$ | hour | 8.45 | 1.15 | 10(19) |
| 15 tonne 0.47 $m^3$ | hour | 10.05 | 1.45 | 10(21) |
| 19 tonne 1.00 $m^3$ | hour | 12.85 | 1.75 | 10(22) |
| Backhoe loader, wheeled | | | | |
| JCB 3CX 0.9 $m^3$ | hour | 6.79 | 0.95 | 10(37) |
| MF 50H 0.8 $m^3$ | hour | 6.59 | 0.96 | 10(36) |
| Case 580G 0.96 $m^3$ | hour | 6.18 | 0.96 | 10(37) |
| Cat 416 0.76 $m^3$ | hour | 6.15 | 0.93 | 10(36) |
| Cat 428 1.00 $m^3$ | hour | 6.27 | 0.93 | 10(37) |
| Ford 555 0.86 $m^3$ | hour | 6.60 | 0.95 | 10(37) |
| Hymac 180C 0.95 $m^3$ | hour | 6.80 | 0.93 | 10(37) |
| Percussion breaker attachments | | | | |
| BRH90 (100 kg) cw steels | week | 90.00 | - | |
| BRH125 (300 kg) cw steels | week | 127.17 | - | 10(30) |
| HB501 (1000kg) cw steels | week | 190.00 | - | 10(31) |
| Krupp HM 1300 (2200 kg) cw steels | week | 254.00 | - | 10(34) |
| Other hydraulic attachments | | | | |
| Rock backacter bucket (ripper teeth) | week | 70.00 | - | - |
| Ditch bucket (c/w side cutters) | week | 47.00 | - | - |
| Trapezoidal bucket | week | 52.00 | - | - |
| Clamshell Grab (1.0$m^3$) | week | 94.00 | - | - |
| Rock handling Grab (up to 3T) | week | 88.00 | - | - |
| Scrap handling Grab (up to 1T) | week | 88.00 | - | - |
| Scrap shears | week | 117.00 | - | - |
| "Grab John" bucket | week | 52.00 | - | - |
| Demolition ball | week | 17.00 | - | - |
| Large scoop bucket 1.40 m3 | week | 42.00 | - | - |
| Single shank ripper | week | 22.00 | - | - |
| Flat track shoes (highway use) | week | 12.00 | - | - |
| Grouser track shoes (soft ground) | week | 12.00 | - | - |

PLANT

| Type of plant | Period | Cost | Consumables per working hour | 1990 F.C.E.C. Daywork Schedule Reference |
|---|---|---|---|---|
| **Hoists, Lifting and Handling** | | | | |
| Telescopic handler 4WD lifting up to 2.50 tonnes (load all 525) | hour | 6.30 | 0.94 | 15(12) |
| Electric forklift 2.5 tonnes | hour | 2.51 | N/A | |
| Rough terrain forklift (Liner Giraffe) 5.6 tonnes | hour | 7.85 | 0.60 | 15(28) |
| Lorry (8T) with 1T Hiab | hour | 10.50 | 2.73 | 14(10) |
| Ace Comet tower platform hoist | hour | 2.80 | 0.30 | - |
| Tirfor winch TU 32 (3T SWL) | week | 25.00 | - | 36(14) |
| Tirfor winch TU 16 | week | 18.00 | - | 36(12) |
| Chain hoist upto 1000 kg capacity | week | 15.70 | - | - |
| Chain hoist upto 2000 kg capacity | week | 16.90 | - | - |
| Chain hoist upto 3000 kg capacity | week | 18.00 | - | - |
| Electric cable hoist 200 kg SWL with gantry | week | 35.50 | 0.20 | 36(6) |
| Engine crane upto 10 CWT (500 kg SWL) | week | 16.75 | 0.17 | 36(1) |
| **Piling plant** (weight = piston weight) | | | | |
| Piling hammer, double acting (air) | | | | |
| weight up to 2500 kg | week | 514.00 | air | 17(5) |
| weight up to 3500 kg | week | 650.60 | air | 17(6) |
| weight up to 4500 kg | week | 727.00 | air | 17(7) |
| Piling hammer, double acting hydraulic | | | | |
| weight up to 5,000 kg | week | 250.00 | N/A | - |
| weight up to 10,000 kg | week | 270.00 | N/A | - |
| weight up to 50,000 kg | week | 375.00 | N/A | - |
| weight up to 100,000 kg | week | 780.00 | N/A | - |
| Piling hammer, double acting diesel | | | | |
| weight up to 1,500 kg | week | 299.10 | 2.30 | - |
| weight up to 2,500 kg | week | 542.00 | 3.45 | - |
| weight up to 3,500 kg | week | 707.00 | 4.60 | - |
| weight up to 5,000 kg | week | 855.00 | 5.75 | - |
| weight up to 5,700 kg | week | 898.00 | 6.90 | - |
| weight up to 6,200 kg | week | 1705.00 | 8.05 | - |
| weight up to 7,500 kg | week | 1800.00 | 9.20 | - |
| weight up to 8,800 kg | week | 1975.50 | 10.35 | - |

PLANT

| Type of plant | Period | Cost | Consumables per working hour | 1990 F.C.E.C. Daywork Schedule |
|---|---|---|---|---|
| Hanging leaders for use with diesel hammers | | | | |
| length 8.5m | week | 79.05 | - | 17(53) |
| length 15.5m | week | 145.00 | - | 17(54) |
| length 22.5m | week | 205.00 | - | 17(56) |
| Flexible hose for compressed air (per 30ft length) | | | | |
| 1.5 in diameter | week | 15.00 | - | 17(24) |
| 2 in diameter | week | 19.05 | - | 17(25) |
| Piling extractors (including compressor) | | | | |
| BSP HD10 unit weight 3000kg | week | 610.20 | 2.61 | 17(20) |
| BSP HD15 unit rate 4580kg | week | 712.00 | 5.17 | 17(21) |
| Vibratory hammer/extractor hydraulic cent. force - 16 tonne, pulling force - 12.5 tonne | week | 700.00 | 2.30 | 17(39) |
| Vibratory hammer/extractor hydraulic cent. force - 59 tonne, pulling force - 36 tonne | week | 1000.00 | 6.90 | 17(41) |
| Vibratory hammer/extractor hydraulic cent. force - 200 tonne, pulling force - 80 tonne | week | 2000.00 | 10.35 | 17(43) |
| **Rollers** | | | | |
| Pedestrian rollers | | | | |
| Bomag BW35 trench compactor | hour | 1.16 | 0.14 | 23(11) |
| Bomag BW65S double vibratory | hour | 1.30 | 0.18 | 23(11) |
| Bomag BW75S double vibratory | hour | 1.51 | 0.28 | 23(12) |
| Bomag BW90S double vibratory | hour | 1.62 | 0.41 | 23(13) |
| Bomag BW55E single drum | hour | 0.76 | 0.15 | 23(9) |
| Bomag BW75E single drum | hour | 0.84 | 0.27 | 23(9) |
| Self propelled rollers | | | | |
| Bomag 100 AD double vibratory | hour | 3.03 | 0.46 | 23(21) |
| Bomag 120 AD double vibratory | hour | 3.40 | 0.48 | 23(21) |
| Bomag 130 AD double vibratory | hour | 5.26 | 0.48 | 23(22) |
| Bomag BW142, 4.5 tonne | hour | 7.95 | 0.80 | 23(25) |
| Bomag BW172, 5.5 tonne | hour | 8.05 | 0.98 | 23(25) |
| Bomag BW212, 8.90 tonne | hour | 8.15 | 1.58 | 23(26) |
| Bomag BW 217, 17.40 tonne | hour | 8.30 | 1.62 | 23(*) |
| Bomag BC 601 RB Refuse compactor | hour | 9.25 | 2.00 | 23(*) |
| 3 wheel dead weight | | | | |
| Hamm HW90-10 tonne | hour | 4.84 | 0.95 | 23(3) |
| Aveling DC013 - 10 tonne | hour | 4.50 | 0.95 | 23(3) |
| Aveling DC015 - 12 tonne | hour | 4.90 | 1.07 | 23(4) |

PLANT

| Type of plant | Period | Cost £ | Consumables per working hour | 1990 F.C.E.C. Daywork Schedule |
|---|---|---|---|---|
| Rubber tyred rollers | | | | |
| 9 wheel Dynapac CP27 11 tonne | hour | 6.85 | 2.06 | 23(7) |
| 7 wheel Dynapac CP22 7 tonne | hour | 6.80 | 1.50 | 23(6) |
| Towed vibratory rollers | | | | |
| Bomag BW6 (6 tonne) | hour | 3.70 | 1.10 | 23(17) |
| Bomag BW10 (10 tonne) | hour | 4.80 | 1.19 | 23(18) |
| Bomag BW15 (15 tonne) | hour | 8.08 | 1.55 | 23(*) |
| Bomag BW6 Sheepsfoot | hour | 3.80 | 1.10 | 23(30) |
| Bomag BW10 Sheepsfoot | hour | 4.90 | 1.19 | 23(30) |
| Bomag BW15 Sheepsfoot | hour | 8.20 | 1.55 | 23(30) |
| **Dual Purpose Rollers** | | | | |
| Benford MBR71 CW breaker | hour | 1.30 | 0.45 | - |
| Bomag 75 EHB CW breaker | hour | 1.50 | 0.47 | - |
| **Scaffolding and Accessories** | | | | |
| Tube alloy per metre | 4 week | 0.30 | - | 25(2) |
| Tube galvanised per metre | 4 week | 0.25 | - | 25(1) |
| Boards - 225 x 40 mm per metre | 4 week | 0.33 | - | 25(19) |
| Base plates adjustable each | 4 week | 0.08 | - | 25(8) |
| Base plates fixed each | 4 week | 0.09 | - | 25(7) |
| Clips each | 4 week | 0.07 | - | 25(7) |
| Couplers each | 4 week | 0.07 | - | 25(7) |
| Spigots each | 4 week | 0.07 | - | 25(7) |
| Putlog 1.5 m blade end each steel | 4 week | 0.40 | - | 25(3) |
| Putlog 1.8 m blade end each steel | 4 week | 0.48 | - | 25(5) |
| Reveal screws each | 4 week | 0.05 | - | 25(7) |
| Castor wheel nylon tyred | 4 week | 0.15 | - | 25(16) |
| Access Staging and Towers | | | | |
| Alloy towers base size 1.8 x 0.80 height to platform 2.6 m | week | 17.50 | - | - |
| Extra sections per 2 m rise (max 3) | week | 12.00 | - | - |
| Alloy towers base size 1.8 x 1.80 height to platform 2.6 m | week | 24.00 | - | - |
| Extra sections per 2 m rise (max 7) | week | 14.00 | - | - |
| Alloy stairwell tower per 2 m rise | week | 16.00 | - | - |
| Steel towers base size 1.80 x 1.80 height to platform 1.80 m | week | 11.00 | - | - |
| Extra sections per 0.60 rise | week | 2.50 | - | - |
| Steel towers base size 3.80 x 3.80 height to platform 1.80 m | week | 17.00 | - | - |

PLANT

| Type of plant | Period | Cost £ | Consumables per working hour | 1990 F.C.E.C. Daywork Schedule |
|---|---|---|---|---|
| Access Staging and Towers (Cont'd) | | | | |
| Extra sections per 1.20 rise | week | 7.00 | - | - |
| Set of 4 castor wheels | 4 week | 0.75 | - | 25(16) |
| Toe boards per metre | 4 week | 0.12 | - | 25(19) |
| Toe boards hinged per metre | 4 week | 0.14 | - | - |
| Handrail (included with towers) | 4 week | 0.10 | - | - |
| Stair tread complete | 4 week | 0.15 | - | - |
| Brick guard | 4 week | 0.60 | - | - |
| Bridging Unit per metre | week | 1.95 | - | - |
| Fittings etc refer to scaffold details | | | | |
| **Tractors** | | | | |
| Tractor dozers with single equipment | | | | |
| 50 kW Cat D3C | hour | 7.50 | 1.15 | 32(1) |
| 67 kW Cat D4H | hour | 8.82 | 1.51 | 32(1) |
| 104 kW Cat D6D | hour | 12.90 | 2.54 | 32(4) |
| 212 kW Cat D8N | hour | 19.40 | 4.59 | 32(8) |
| 276 kW Cat D9N | hour | 25.50 | 6.28 | 32(9) |
| 67 kW D4H LGP | hour | 9.75 | 1.51 | - |
| 123 kW D6H LGP | hour | 13.60 | 3.02 | - |
| Dozer attachments | | | | |
| Single shank ripper | hour | 2.20 | - | - |
| Triple shank ripper | hour | 4.00 | - | - |
| 'U' Dozer blade | hour | 1.22 | - | - |
| Angle Dozer blade | hour | 1.25 | - | - |
| Straight tiltdozer blade | hour | 1.30 | - | - |
| Skeleton blade | hour | 1.25 | - | - |
| Tractor loaders | | | | |
| 0.8 $m^3$ Cat 931 | hour | 10.27 | 1.21 | 32(11) |
| 1.5 $m^3$ Cat 953 | hour | 13.64 | 2.05 | 32(15) |
| 2.0 $m^3$ Cat 963 | hour | 17.47 | 2.78 | 32(16) |
| 2.8 $m^3$ Cat 973 | hour | 20.80 | 3.86 | 32(18) |
| 2.8 $m^3$ Cat 973 LGP | hour | 22.09 | 3.86 | 32(18) |

PLANT

| Type of plant | Period | Cost £ | Consumables per working hour | 1990 F.C.E.C. Daywork Schedule Reference |
|---|---|---|---|---|
| **Wheeled loaders** | | | | |
| 0.50 $m^3$ 31 kW | hour | 5.35 | 0.72 | 32(28) |
| 0.80 $m^3$ 41 kW | hour | 6.97 | 0.97 | 32(28) |
| 1.40 $m^3$ 70 kW | hour | 7.21 | 1.57 | 32(30) |
| 2.30 $m^3$ 107 kW | hour | 11.90 | 2.05 | 32(35) |
| 2.70 $m^3$ 140 kW | hour | 13.27 | 3.02 | 32(37) |
| 3.50 $m^3$ 177 kW | hour | 14.20 | 3.86 | 32(38) |
| 5.40 $m^3$ 310 kW | hour | 17.08 | 6.03 | 32(41) |
| 10.50 $m^3$ 588 kW | hour | 33.00 | 9.66 | 32(*) |
| Long reach loader; 1.00 $m^3$, 54 kW | hour | 8.75 | 1.21 | - |
| Wheel dozer, blade capacity 7.5 $m^3$ 338 kW | hour | 23.00 | 6.40 | 32(9) |
| **Loader Attachments** | | | | |
| 4 in 1 bucket | hour | 1.70 | - | - |
| Side dumping bucket | hour | 1.50 | - | - |
| Rock bucket | hour | 1.30 | - | - |
| Skeleton bucket | hour | 1.35 | - | - |
| Single shank ripper (D6) | hour | 2.20 | - | - |
| Multi shank ripper (D8) | hour | 4.00 | - | - |
| Rear mounted backacter | hour | 2.70 | - | - |
| **Hydraulic face shovels** | | | | |
| 32 tonne, 1.50 $m^3$, O&K RH9 | hour | 24.25 | 3.34 | 10(27) |
| 42 tonne, 2.60 $m^3$, Komatsu PC400-3 | hour | 33.00 | 4.32 | 10(27) |
| 62 tonne, 1.30-3.40 $m^3$ CAT E650 | hour | 39.00 | 5.68 | 10(28) |
| **Skidsteer Loaders** | | | | |
| Bobcat 443 | hour | 4.05 | 0.36 | 32(48) |
| Bobcat 543 | hour | 4.60 | 0.50 | 32(49) |
| Bobcat 643 | hour | 5.30 | 0.65 | 32(49) |
| **Motor Scraper - single engine** | | | | |
| Cat 621E - 15.30 $m^3$ | hour | 26.90 | 5.01 | 32(69) |
| Cat 651E - 33.60 $m^3$ | hour | N/A | N/A | 32(*) |
| **Motor Scraper - twin engine** | | | | |
| Cat 627E - 15.30 $m^3$ | hour | 32.80 | 7.24 | 32(71) |
| Cat 657E (Push Pull) - 33.60 $m^3$ | hour | 67.60 | 13.28 | 32(73) |
| **Motor scraper - elevating** | | | | |
| Cat 613C - 8.40 $m^3$ | hour | 23.60 | 3.02 | 32(74) |
| Cat 615 - 12.23 $m^3$ | hour | 28.00 | 4.35 | 32(75) |
| Cat 623E - 16.80 $m^3$ | hour | 31.80 | 5.31 | 32(76) |

PLANT

| Type of plant | Period | Cost £ | Consumables per working hour | 1990 F.C.E.C. Daywork Schedule Reference |
|---|---|---|---|---|
| Tractor (tracked) pipe layer c/w counterweight and boom | | | | |
| Upto 120 kW lifting max 28 t (Komatsu D65C-6/CAT 571) | hour | 17.00 | 3.02 | - |
| Upto 135 kW lifting max 41 t (Komatsu D85C - 12/CAT 572) | hour | 20.00 | 3.39 | - |
| Upto 240 kW, lifting max 70 T (Komatsu D155C-1/CAT 583) | hour | 27.00 | 5.28 | - |
| Tractor (tracked) no equipment for Agro/towing use | | | | |
| 75 kW | hour | 7.00 | 1.59 | - |
| 90 kW | hour | 9.00 | 1.85 | - |
| 125 kW | hour | 12.00 | 2.50 | - |
| Motor Grader (6 wheel) | | | | |
| Upto 93 kW blade 3.66 m | hour | 8.50 | 1.91 | 32(55) |
| Upto 113 kW blade 3.66 m | hour | 10.00 | 2.39 | 32(56) |
| Upto 158 kW blade 3.96 m | hour | 13.90 | 2.97 | 32(57) |
| Upto 205 kW blade 4.88 m | hour | 18.71 | 3.34 | 33(*) |
| Trench excavator (Trencher) | | | | |
| Upto 10 kW max width 0.20 m max depth 1.22 m (Bobcat T114) | hour | 5.75 | N/A | 33(1) |
| Upto 100 kW max width 0.60 m max depth 2.40 m (Vermeer T600D) | hour | 13.61 | 2.00 | 33(5) |
| Upto 230 kW max width 1.07 m max depth 3.66 m Vermeer T850) | hour | 37.76 | 4.30 | 33(7) |
| Towed type from Agro Tractor max width 0.13 m max depth 0.55 m | hour | 3.11 | - | 33(*) |
| **Transport (Tippers and Dumpers)** | | | | |
| Tipping lorries | | | | |
| 4 x 2 wheel tippers | | | | |
| Gross weight up to 8.5 tonnes payload up to 5.5 tonnes - side tipping | hour | 6.74 | 13.0 mpg | 14(15/19) |

PLANT

| Type of plant | Period | Cost £ | Consumables per working hour | 1990 F.C.E.C. Daywork Schedule Reference |
|---|---|---|---|---|
| Tipping Lorries (Cont'd) | | | | |
| Gross weight up to 8.5 tonnes payload up to 5.5 tonnes (side tipping) | hour | 8.05 | 12.0 mpg | 14(15) |
| Gross weight up to 11 tonnes payload up to 8 tonnes | hour | 8.20 | 11.5 mpg | 14(15) |
| Gross weight up to 14 tonnes payload up to 10 tonnes | hour | 8.50 | 11.0 mpg | 14(16) |
| Gross weight up to 17 tonnes payload up to 12 tonnes | hour | 9.07 | 10.2 mpg | 14(16) |
| 6 x 4 wheel tippers | | | | |
| Gross weight up to 24 tonnes payload up to 18 tonnes | hour | 10.73 | 9.8 mpg | 14(17) |
| Gross weight up to 26 tonnes payload up to 20 tonnes | hour | 12.00 | 8.5 mpg | 14(18) |
| 8 x 4 wheel tippers | | | | |
| Gross weight up to 27 tonnes payload up to 20 tonnes | hour | 13.21 | 8.3 mpg | 14(18) |
| Gross weight up to 30 tonnes payload up to 22 tonnes | hour | 14.10 | 7.5 mpg | 14(18) |
| Gross weight up to 33 tonnes payload up to 24 tonnes | hour | 15.25 | 6.0 mpg | 14(*) |
| Articulated tipper trailers (Highway type excluding tractor) | | | | |
| Twin axle (Bogie) payload up to 20 tonnes (40' length) | hour | 3.36 | - | - |
| Triple axle (Bogie) payload up to 25 tonnes (50' length) | hour | 4.42 | - | - |
| Dump truck (on site only) | | | | |
| 10.5 $m^3$ heaped capacity, 15.5 tonne payload (FODEN FC17) | hour | 8.75 | 1.60 | 9(13) |
| 14 $m^3$ heaped capacity, 18 tonne payload (KOMATSU HP 180-4) | hour | 12.50 | 1.64 | 9(15) |
| 19 $m^3$ heaped capacity, 30 tonne payload (TEREX 33-05) | hour | 15.00 | 1.85 | 9(18) |
| 24 $m^3$ heaped capacity, 32 tonne payload (CAT 769-C) | hour | 20.00 | 1.90 | 9(18) |

PLANT

| Type of plant | Period | Cost £ | Consumables per working hour | 1990 F.C.E.C. Daywork Schedule Reference |
|---|---|---|---|---|
| 37 $m^3$ heaped capacity, 78 tonne payload (KOMATSU HD 785-2) | hour | 25.00 | 5.10 | 9(*) |
| 46 $m^3$ heaped capacity, 120 tonne payload (CAT 785) | hour | 30.00 | 5.92 | 9(*) |
| Dump truck (articulated type ADT's) | | | | |
| Volvo A20 6x4, 18.5 tonne payload 11 $m^3$ heaped capacity | hour | 12.50 | 1.54 | 9(23) |
| Volvo A25 6x4, 22.5 tonne payload 12 $m^3$ heaped capacity | hour | 13.61 | 1.60 | 9(24) |
| Terex 2364, 23 tonne payload 13 $m^3$ heaped capacity | hour | 13.95 | 1.65 | 9(24) |
| Moxy 6225B, 25 tonne payload 13.60 $m^3$ heaped capacity | hour | 13.95 | 1.72 | 9(25) |
| Caterpillar D250B, 25 tonne payload 14.2 $m^3$ heaped capacity | hour | 13.95 | 1.75 | 9(25) |
| Volvo A35 6x6, 32 tonne payload 19 $m^3$ heaped capacity | hour | 17.45 | 1.84 | 9(26) |
| JCB 712, 12 tonne payload 7.5 $m^3$ heaped capacity | hour | 9.70 | 1.40 | 9(22) |
| Small dumpers | | | | |
| 750 kg payload, heaped capacity 0.55 $m^3$ (Thwaites Sprite 4x2) | hour | 1.00 | 0.20 | 9(3) |
| 1000 kg payload, heaped capacity 0.75 $m^3$ (Thwaites Nimline 4x2) | hour | 1.15 | 0.22 | 9(3) |
| 2000 kg payload, heaped capacity 1.55 $m^3$ (Thwaites All drive 4x4) | hour | 2.18 | 0.36 | 9(5) |
| 2500 kg payload, heaped capacity 1.80 $m^3$ (Thwaites All drive 4x4) | hour | 2.47 | 0.53 | 9(6) |
| 3000 kg payload, heaped capacity 2.00 $m^3$ (Thwaites All drive 4x4) | hour | 2.85 | 0.60 | 9(7) |
| 4000 kg payload, heaped capacity 2.50 $m^3$ (Thwaites All drive 4x4) | hour | 3.60 | 0.83 | 9(8) |
| Power barrow | hour | 1.10 | 0.12 | 9(49) |

PLANT

| Type of plant | Period | Cost £ | Consumables per working hour | 1990 F.C.E.C. Daywork Schedule Reference |
|---|---|---|---|---|
| **Transport (trucks and vans)** | | | | |
| Ford transit 5t chassis | hour | 3.90 | 2.50 | |
| Leyland road runner 8t chassis | hour | 4.50 | 2.47 | |
| Ford cargo 1011 10t chassis | hour | 6.00 | 3.50 | |
| Mercedes 1617 16t chassis | hour | 7.00 | 3.75 | |
| Skip Loader Ford 1713 17t chassis | hour | 10.73 | 4.15 | |
| Escort 55 van (petrol) | hour | 3.65 | 2.90 | 14(20) |
| P100 pick up (Derv) (1.10 tonnes) | hour | 3.90 | 2.15 | 14(21) |
| Personnel carrier (12 seat transit) | hour | 4.00 | 2.50 | 14(30) |
| SWB Land Rover (Derv) | hour | 4.89 | 2.25 | 14(26) |
| LWB Land Rover (Derv) | hour | 5.85 | 2.47 | 14(27) |
| Flat trailer (2 axles) | 4 week | 25.00 | - | - |
| Drop side trailer (2 axles) | 4 week | 30.00 | - | - |
| Plant Trailer up to 10 ton | 4 week | 35.00 | - | - |
| Plant Trailer up to 20 ton | 4 week | 50.00 | - | - |
| **Trench sheets etc (minimum 50)** | | | | |
| Trench struts (each) | | | | |
| No. 0 1ft 0 in - 1ft 6in | week | 0.23 | - | - |
| No. 1 1ft 6 in - 2ft 4in | week | 0.26 | - | 26(3) |
| No. 2 2ft 3 in - 3ft 7in | week | 0.27 | - | 26(4) |
| No. 3 3ft 5 in - 5ft 7in | week | 0.32 | - | 26(5) |
| Standard props (each) | | | | |
| No. 0 3ft 6 in - 6ft 0in | week | 0.23 | - | 26(13) |
| No. 1 5ft 9 in - 10ft 3in | week | 0.28 | - | 26(14) |
| No. 2 6ft 6 in - 11ft 0in | week | 0.36 | - | 26(15) |
| No. 3 8ft 6 in - 13ft 0in | week | 0.40 | - | - |
| No. 4 10ft 6 in - 16ft 0in | week | 0.45 | - | - |
| No. 5 16ft 0 in - 21ft 0in | week | 0.60 | - | - |
| Beam heads (each) | | | | |
| No. 1 | week | 0.30 | - | - |
| No. 2 | week | 0.35 | - | - |
| No. 3 | week | 0.45 | - | - |
| No. 4 | week | 0.50 | - | - |
| Split heads (each) | | | | |
| No. 1 1ft 9in - 2ft 9in | week | 0.40 | - | - |
| No. 2 3ft 9in - 4ft 6in | week | 0.45 | - | - |
| No. 3 4ft 6in - 6ft 3in | week | 0.55 | - | - |
| Trench Sheets | | | | |
| 6'0" | week | 0.70 | - | - |
| 8'0" | week | 0.80 | - | - |
| 10'0" | week | 1.00 | - | - |
| 12'0" | week | 1.25 | - | - |
| 14'0" | week | 1.40 | - | - |
| 16'0" | week | 1.60 | - | - |

PLANT

| Type of plant | Period | Cost £ | Consumables per working hour | 1990 F.C.E.C. Daywork Schedule Reference |
|---|---|---|---|---|
| **Pumping and Dewatering** | | | | |
| Diesel pumps (inc. 25 ft suction and delivery) | | | | |
| 2" Flygt submersible (electric 110v 3kW) | week | 24.75 | 0.18 | 20(22) |
| 3" Spate 3B | week | 38.10 | 0.25 | 20(7) |
| 4" Spate 4C | week | 56.15 | 0.85 | 20(8) |
| 6" Sykes UV5 head pump/high capac | week | 105.90 | 1.47 | 20(13) |
| 3" Sykes VCD3 jetting pump | week | 126.00 | 0.50 | 20(13) |
| 4" Sykes WP100/60 wellpointing pump | week | 116.20 | 0.50 | 20(12) |
| 6" Sykes WP150/60 wellpointing pump | week | 143.50 | 0.75 | 20(13) |
| Extra hose (per lin. m) | | | | |
| 3" Suction | week | 1.00 | - | 21(3) |
| 3" Delivery | week | 0.60 | - | 21(9) |
| 4" Suction | week | 1.25 | - | 21(4) |
| 4" Delivery | week | 0.95 | - | 21(10) |
| 6" Suction | week | 1.90 | - | 21(5) |
| 6" Delivery | week | 1.85 | - | 21(11) |
| 6" Duraline (lay flat fine hose) | week | (included with | | |
| 6" Jetting tube | week | Jetting Pump) | | |
| 6" PVC header pipe & attachments @ 1.50m centres inc. jet wells for wellpointing to 6.5 deep (per wellpoint) | week | 2.30 | - | - |
| **Miscellaneous** | | | | |
| Road form 10m long (each) | | | | |
| 100mm high | week | 2.08 | - | 27(2) |
| 125mm high | week | 2.20 | - | 27(3) |
| 150mm high | week | 2.35 | - | 27(3) |
| 200mm high | week | 2.50 | - | 27(5) |
| 225mm high | week | 2.65 | - | 27(7) |
| Flexible road form 3m long (each) | | | | |
| 150mm high | week | 1.00 | - | - |
| 200mm high | week | 1.05 | - | - |
| 225mm high | week | 1.09 | - | - |
| Road Safety | | | | |
| Road signs on stands (each) | | | | |
| size 2ft 0in | week | 4.80 | - | 37(43) |
| size 2ft 6in | week | 4.80 | - | 37(44) |
| size 3ft 6in | week | 6.00 | - | 37(46) |
| size 4ft 0in | week | 7.00 | - | 37(46) |
| Flashing hazard lamps (Batteries etc) | week | 1.04 | - | 37(22) |
| Paraffin lamps | week | 0.80 | - | - |

PLANT

| Type of plant | Period | Cost £ | Consumables per working hour | 1990 F.C.E.C. Daywork Schedule Reference |
|---|---|---|---|---|
| Standard cone | week | 0.80 | - | 37(39) |
| Road pins | week | 0.40 | - | - |
| UPVC barrier | week | 2.80 | - | - |
| Cone converter | week | 1.00 | - | - |
| Railway sleepers - baulks - each | week | 1.50 | - | - |
| Traffic Light system (two way radar, 110V) | week | 70.00 | 0.36 | 37(9) |
| Traffic light system (two way, mains, timer operated) | week | 60.00 | 0.40 | 37(9) |
| **Transformers/cables** | | | | |
| 2kVA | week | 6.00 | - | 11(16) |
| 4kVA | week | 11.00 | - | 11(17) |
| 6kVA | week | 17.50 | - | 11(18) |
| 15 m extension cable 240V/110V plug and socket | week | 3.20 | - | - |
| 30 m extension cable 240V/110V plug and socket | week | 4.00 | - | - |
| 15 m 4 $mm^2$ 240V/110V | week | 5.52 | - | - |
| 240V junction box | week | 5.76 | - | - |
| 110V junction box | week | 5.76 | - | - |
| **Welding & cutting sets** | | | | |
| ARC Mains 140 amps portable | week | 14.00 | 1.81 | 35(5) |
| ARC Petrol 200 amps mobile | week | 46.40 | 2.02 | - |
| ARC Diesel 300 amps trailer mounted | week | 60.00 | 2.18 | 35(3) |
| Oxy acetylene set portable | week | 30.00 | N/A | 35(1) |
| MIG welder | week | 18.50 | 0.70 | - |
| **Lighting** | | | | |
| Gas floodlight 6′ stand 1000 watt | week | 10.00 | 0.40 | - |
| 17′ floodlight tower (cw 2kVA gen) | week | 19.22 | 0.40 | 11(39) |
| 25′ floodlight tower (cw 6kVA gen) | week | 30.00 | 0.50 | 11(40) |
| Festoon lighting set (34 m) | week | 7.68 | 0.40 | - |
| Tripod floodlight 1.8 m stand 500 watt | week | 12.00 | 0.11 | - |
| Plasters light Flou 60 W | week | 10.00 | 0.06 | - |
| **Drain tools** | | | | |
| Drain rods (30′ set) (12m) including equipment | week | 6.40 | - | 37(24) |
| Drain plug (rubber diaphragm) 4" | week | 2.24 | - | 37(26) |
| Drain plug (rubber diaphragm) 8" | week | 2.88 | - | 37(28) |
| Drain plug (rubber diaphragm) 18" | week | 2.88 | - | 37(29) |
| Mains powered drain cleaner | week | 28.00 | 0.35 | - |
| Drain tester - smoke | week | 4.80 | - | - |
| Drain tester - U bend | week | 4.80 | - | - |
| Drain tester - pressure pump | week | 15.36 | - | - |
| Drain tester - mandrel | week | 3.00 | - | - |
| Stilson wrench 24" | week | 4.16 | - | - |
| Stilson wrench 36" | week | 6.40 | - | - |
| Clay pipe cutter | week | 8.00 | - | - |
| Steel pipe cutter | week | 6.40 | - | - |
| Drain Bag 4" or 6" | week | 2.88 | - | - |

PLANT

| Type of plant | Period | Cost £ | Consumables per working hour | 1990 F.C.E.C. Daywork Schedule Reference |
|---|---|---|---|---|
| Lorry mounted tar boiler | hour | 9.21 | 0.58 (boiler) | 29(*) |
| Tar boiler portable (inc sprayer) (45 litres) | week | 20.00 | 0.46 | 29(1) |
| Shot blasting equipment 150lb | week | 66.00 | 0.62 | - |
| 350lb | week | 91.25 | 0.97 | - |
| Grit blaster ICE160 72kg | week | 73.60 | 0.62 | - |
| Stone splitter 25" x 4" | week | 30.30 | - | - |
| Block splitter | week | 16.00 | - | - |
| Slab splitter | week | 20.00 | - | - |
| Slab lifter | week | 4.20 | - | - |
| Hilti cartridge gun DX650 | week | 12.76 | - | - |
| DX450 | week | 10.80 | - | - |
| Heating/Drying | | | | |
| Radiant heater (12,500 BTU) | week | 8.32 | - | 37(11) |
| Forced air heater (60,000 BTU) | week | 26.00 | - | 37(12) |
| Forced air heater (150,000 BTU) | week | 36.00 | - | 37(14) |
| Forced air heater (325,000 BTU) | week | 46.00 | - | 37(16) |
| Dehumidifier (40 litres/day) | week | 32.00 | - | 37(34) |
| Fume extractor (6" diameter) | week | 35.00 | - | - |
| Drills/Saws/Tools | | | | |
| Electric diamond 6" max diam | week | 52.40 | 0.22 | - |
| Electric masonry & core bits 11mm | week | 16.80 | 0.19 | - |
| Electric masonry & core bits 127mm | week | 32.20 | 0.22 | - |
| Angle head drill | week | 10.00 | 0.22 | |
| Roto broach milling to 34 mm | week | 36.16 | 0.27 | |
| Rotary hammer K950 | week | 16.80 | 0.32 | |
| Rotary hammer K637 | week | 14.80 | 0.32 | |
| Demolition hammer heavy duty | week | 16.00 | 0.48 | |
| Electric grinder 9" | week | 12.00 | 0.26 | |
| Electric grinder 4.5" | week | 8.80 | 0.24 | |
| 16" diam saw bench petrol | week | 28.00 | 0.24 | 24(10) |
| 22" diam saw bench diesel | week | 30.00 | 0.26 | 24(13) |
| 16" chain saw | week | 24.00 | 0.23 | 24(2) |
| 9" electric grinder (angle) | week | 15.20 | 0.26 | 31(10) |
| 18" conc saw 25hp diesel | week | 70.00 | 0.99 | 6(41) |
| 28" conc saw 40hp diesel | week | 83.10 | 1.34 | 6(42) |
| Stihl saw disc 12" petrol | week | 23.20 | 0.22 | 37(18) |
| Stihl saw disc 12" air | week | 20.50 | air | 37(18) |
| 8" floor sander (electric) | week | 27.60 | 0.22 | 37(13) |
| Water/Fuel Supply | | | | |
| Towed (trailer) water/diesel | | | | |
| Bowser 275 gallon | week | 11.00 | - | 34(9) |
| Bowser 500 gallon | week | 27.50 | - | |
| Lorry mounted tanker | | | | |
| 1000 gallon (Derv) 8.5t chassis | hour | 6.74 | 3.03 | 34(10) |
| 1500 gallon (Derv) 12t chassis | hour | 8.20 | 3.75 | - |

PLANT

| Type of plant | Period | Cost £ | Consumables per working hour | 1990 F.C.E.C. Daywork Schedule Reference |
|---|---|---|---|---|
| Storage tanks | | | | |
| up to 250 gallon | 4 week | 21.44 | - | 34(1) |
| up to 500 gallon | 4 week | 28.00 | - | 34(1) |
| up to 1000 gallon | 4 week | 36.00 | - | 34(1) |
| up to 2000 gallon | 4 week | 40.00 | - | 34(2) |
| Landscaping Items and Tools | | | | |
| post hole auger 6"/9" hand | week | 5.44 | - | - |
| Powered auger petrol | week | 40.00 | 0.21 | - |
| post driver for 4" post | week | 4.80 | - | - |
| post driver for 6" posts | week | 5.44 | - | - |
| linemarker 2", 3" or 4" wide | week | 11.20 | - | |
| bolt croppers 24" or 36" | week | 6.72 | - | |
| tipping and loading skip 6 tonne | week | 20.80 | - | 8(13) |
| cultivator self propelled | week | 88.00 | 0.38 | - |
| Boilers and Sprays | | | | |
| bitumen boiler 10 gallon (45 litres) | week | 16.00 | 0.12 | - |
| bitumen boiler 15 gallon (68 litres) | week | 28.48 | 0.14 | - |
| cold tar spray | week | 28.48 | - | - |
| flame gun propane (road burner) | week | 8.80 | 1.15 | 2(15) |
| Air Tool bits | | | | |
| shove holer | week | 5.76 | - | - |
| points and chisels | week | 0.96 | - | - |
| clayspade | week | 1.28 | - | 30(3) |
| tarmac cutter | week | 2.24 | - | - |
| comb holder | week | 0.96 | - | - |
| self-drill anchor holder | week | 1.28 | - | - |
| rammer foot | week | 1.92 | - | - |
| Floor sander | week | 21.60 | 0.22 | |
| Pinking roller | week | 10.00 | - | |
| Floor grinder, electric | week | 44.80 | 0.28 | |
| Floor saw, 450 mm petrol | week | 49.60 | 0.28 | 6(38) |
| Floor plane, petrol | week | 44.48 | 0.30 | |
| Floor scaler petrol | week | 55.00 | 0.29 | |
| Rubbish basket | week | 2.40 | - | - |
| Diesel elevator up to 15 m | week | 37.00 | - | - |
| up to 30 m | week | 52.00 | - | - |

PLANT

| Type of plant | | Period | Cost £ | | 1990 F.C.E.C. Daywork Schedule Reference |
|---|---|---|---|---|---|
| **Formwork Equipment** | | | | | |
| Adjustable props | | | | | |
| No 0 | 1.07 - 1.82 | week | 0.28 | | |
| No 1 | 1.75 - 3.12 | week | 0.34 | | |
| No 2 | 1.98 - 3.35 | week | 0.36 | | |
| No 3 | 2.59 - 3.95 | week | 0.38 | | |
| No 4 | 3.20 - 4.88 | week | 0.42 | | |
| Decking per SM (Kwikstage) (excluding props) | | week | 1.88 - 2.03 | | |
| Column clamps (set) | | | | | |
| No 1 | 0.3 - 0.53 | week | 0.65 | | |
| No 2 | 0.46 - 0.84 | week | 0.75 | | |
| No 3 | 0.66 - 1.222 | week | 0.85 | | |
| Beam Clamps | | | | | |
| 300 mm arms | | week | 0.75 | | |
| 450 mm arms | | week | 0.82 | | |
| 600 mm arms | | week | 0.95 | | |
| Manhole shutters - 1800 mm high | | | | | |
| Internal diameter - 675 mm | | week | 22.00 | | |
| 900 mm | | week | 27.50 | | |
| 1050 mm | | week | 29.00 | | |
| 1200 mm | | week | 31.50 | | |
| 1500 mm | | week | 33.50 | | |
| 1800 mm | | week | 39.00 | | |
| 2100 mm | | week | 44.50 | | |
| 2400 mm | | week | 47.25 | | |
| 2700 mm | | week | 53.00 | | |
| **Portable Accommodation** (including electrical fittings, etc. excluding additional security items, shutters, grilles, etc) | | | up to 26 weeks | over 26 weeks | |
| Jack Leg portakabin | 12'2"x8'8" | week | 8.00 | 7.40 | 16(1) |
| Jack Leg portakabin | 24'2"x10'2" | week | 11.80 | 9.90 | 16(1) |
| Jack Leg portakabin | 48'2"x12'2" | week | 16.20 | 15.50 | 16(1) |
| Wheeled cabin | 12'1"x7'6" | week | 6.40 | 5.80 | 16(6) |
| Wheeled cabin | 22'1"x7'6" | week | 11.24 | 10.70 | 16(7) |
| Toilet unit Jack Leg | 16'2"x8'8" | week | 14.85 | 13.70 | 16(5) |
| Pollution decontam unit | 22'1"x7'6" | week | 30.00 | 27.25 | - |
| Canteen Unit | 32'4"x8'8" | week | 16.20 | 15.50 | 16(2) |

PART 4

# Unit Costs

## (CIVIL ENGINEERING WORKS)

**INTRODUCTORY NOTES**

*The unit costs in this part represent the net cost to the Contractor of executing the work on site; they are not the prices which would be entered in a tender Bill of Quantities. The costs exclude preliminary and general items, head office overheads, finance charges, risk, profit and value added tax. Reference should be made to Part 2 for the cost of Preliminaries and General Items and to Part 6 for the remainder.*

*It must be emphasised that the unit costs are averages calculated on unit outputs for typical site conditions. Costs can vary considerably from contract to contract depending on individual Contractors, site conditions, methods of working and various other factors. Reference should be made to Section 1 for a general discussion on Civil Engineering Estimating.*

*We have structured this unit cost section around The Civil Engineering Standard Method of Measurement 2nd Edition, endeavouring to cover as many aspects of Civil and Highway works as possible. In most cases, prices are broken down into labour, plant and materials, although certain specialist prices are inserted as advised.*

*Reference to Part 3 giving basic costs of labour materials and plant will assist the reader in making adjustments to unit costs.*

*Attention should be paid throughout to explanatory notes both at the start and within each section.*

*The unit costs are broken down into their Labour, Plant and Material elements, as detailed in Section 3:*

*Columns headed L = Labour*
*P = Plant*
*M = Materials*

*The first line is the output in hours taken per unit of work, the second line is the cost result of these outputs, material columns include allowances for waste and/or distribution, the total cost then appears in the* **£ NETT** *column.*

**CLASS B: GROUND INVESTIGATION**

**TRIAL HOLES**

Labour based on Gang A.1

| | | |
|---|---|---|
| Total hourly cost | = £ | 33.71 |

| | | |
|---|---|---|
| PLANT | | |
| 8 Tonne Excavator | | 7.59 |
| 3 Tonne Dumper | | 3.45 |
| plus | | |
| 2 Tool Compressor | | 4.44 |
| 2 No. Breakers | | 0.67 |
| Plate Compactor | | 1.13 |
| Total hourly cost | = £ | 17.28 |

| | | UNIT | L | P | M | £ NETT |
|---|---|---|---|---|---|---|
| The following costs assume the use of mechanical plant and excavating and backfilling on the same day | | | | | | |
| Number in material other than rock (Maximum depth) | | | | | | |
| plan size 1 x 2m | | | | | | |
| ne 1m | hr | nr | 0.24 | 0.24 | | |
| | £ | | 8.09 | 4.15 | - | 12.24 |
| 1 - 2m | hr | nr | 0.47 | 0.47 | | |
| | £ | | 15.84 | 8.12 | - | 23.96 |
| 2 - 3m and over | hr | nr | 0.53 | 0.53 | | |
| | £ | | 17.87 | 9.16 | - | 27.03 |
| Number in material which includes rock (Maximum depth) | | | | | | |
| plan size 1 x 2m | | | | | | |
| ne 1m | hr | nr | 0.28 | 0.28 | | |
| | £ | | 9.44 | 4.84 | - | 14.28 |
| 1 - 2m | hr | nr | 0.51 | 0.51 | | |
| | £ | | 17.19 | 8.81 | - | 26.00 |
| 2 - 3m and over | hr | nr | 0.58 | 0.58 | | |
| | £ | | 19.55 | 10.02 | - | 29.57 |

CLASS B: GROUND INVESTIGATION

TRIAL HOLES - continued

| | | UNIT | L | P | M | £ NETT |
|---|---|---|---|---|---|---|
| Depth in material other than rock (Maximum depth) plan size 1 x 2m | | | | | | |
| ne 1m | hr | m | 0.24 | 0.24 | | |
| | £ | | 8.09 | 4.15 | - | 12.24 |
| 1 - 2m | hr | m | 0.53 | 0.53 | | |
| | £ | | 17.87 | 9.16 | - | 27.03 |
| 2 - 3m | hr | m | 0.59 | 0.59 | | |
| | £ | | 19.89 | 10.20 | - | 30.09 |
| 3 - 5m | hr | m | 0.65 | 0.65 | | |
| | £ | | 21.91 | 11.23 | - | 33.14 |
| Depth in rock (Maximum depth) plan size 1 x 2m | | | | | | |
| ne 1m | hr | m | 2.95 | 2.95 | | |
| | £ | | 99.44 | 50.98 | - | 150.42 |
| 1 - 2m | hr | m | 3.65 | 3.65 | | |
| | £ | | 123.04 | 63.07 | - | 186.11 |
| 2 - 3m | hr | m | 3.95 | 3.95 | | |
| | £ | | 133.15 | 68.26 | - | 201.41 |
| 3 - 5m | hr | m | 4.50 | 4.50 | | |
| | £ | | 151.70 | 77.76 | - | 229.46 |
| Removal of obstructions (Maximum depth) | | | | | | |
| All depths | hr | hr | 1.00 | 1.00 | | |
| | £ | | 33.71 | 17.28 | - | 50.99 |
| Pumping (Maximum depth 4m) minimum 750 litres per hour | hr | hr | 1.00 | 1.00 | | |
| | £ | | 0.40 | 4.10 | - | 4.50 |

## CLASS B: GROUND INVESTIGATION

## ROTARY DRILLED BOREHOLES

| | | UNIT | L | P | M | £ NETT |
|---|---|---|---|---|---|---|
| The following costs are based on using Specialist Contractors and are for guidance only | | | | | | |
| Establishment of standard plant and equipment and removal completion | £ | sum | - | - | - | 1,475.00 |
| | £ | nr | - | - | - | 125.00 |
| Depth without core recovery (Maximum depth) | | | | | | |
| 100mm diameter | | | | | | |
| ne 5m | hr | m | - | - | | |
| | £ | | - | - | - | 19.00 |
| 5 - 10m | hr | m | - | - | | |
| | £ | | - | - | - | 19.00 |
| 10 - 20m | hr | m | - | - | | |
| | £ | | - | - | - | 20.00 |
| 20 - 30m | hr | m | - | - | | |
| | £ | | - | - | - | 21.00 |
| Depth with core recovery (Maximum depth) | | | | | | |
| 75mm diameter | | | | | | |
| ne 5m | hr | m | - | - | | |
| | £ | | - | - | - | 58.00 |
| 5 - 10m | hr | m | - | - | | |
| | £ | | - | - | - | 61.00 |
| 10 - 20m | hr | m | - | - | | |
| | £ | | - | - | - | 65.00 |
| 20 - 30m | hr | m | - | - | | |
| | £ | | - | - | - | 70.00 |
| Extra for use of semi-rigid plastic core barrel liner | hr | m | - | - | | |
| | £ | | - | - | - | 7.00 |

## CLASS B: GROUND INVESTIGATION

### CONSOLIDATION OF ABANDONED MINE WORKS

| | UNIT | SPECIALIST ADVICE £ NETT |
|---|---|---|
| Transport plant Labour and all equipment to and from site (maximum 100 miles) | | 2000.00 |
| **Drilling** | | |
| Erect move and dismantle equipment and labour to each separate bore | nr | 24.50 |
| Drill 50mm diameter boreholes | m | 7.50 |
| Drill 100mm diameter boreholes for pea gravel injection | m | 12.25 |
| Extra over for casing when required | m | 10.00 |
| Standing time for drilling rig and oven | hour | 65.00 |
| **Grouting** | | |
| Connecting grout lines | m | 13.00 |
| Injection of grout | t | 50.00 |
| Add pea gravel at boreholes | t | 60.00 |
| Supply ordinary Portland cement for grouting | t | 51.90 |
| Supply sulphate resistant cement for grouting | t | 72.07 |
| Supply pulverised fuel ash for grouting | t | 12.00 |
| Supply sand for grouting | t | 7.95 |
| Supply pea gravel for grouting | t | 14.00 |
| Supply bentonite for grouting (2:1) | t | 80.00 |
| Supply cement grout for grouting | t | 67.40 |
| Standing time for grouting rig and crew | hour | 65.00 |

## CLASS B: GROUND INVESTIGATION

## CONSOLIDATION OF ABANDONED MINE WORKS

| | UNIT | SPECIALIST ADVICE £ NETT |
|---|---|---|
| In capping to old shafts or similar | | |
| Concrete | | |
| (C20/20 concreted) | | |
| thickness | | |
| ne 150mm | m3 | 66.75 |
| 150 - 300mm | m3 | 64.65 |
| 300 - 500mm | m3 | 63.30 |
| exceeding 500mm | m3 | 62.60 |
| Reinforcement cut and bent | | |
| mild steel bars | | |
| 6mm | t | 919.49 |
| 8mm | t | 817.21 |
| 10mm | t | 785.90 |
| 12mm | t | 760.65 |
| 16mm | t | 702.26 |
| 20mm | t | 617.69 |
| high yield bars | | |
| 6mm | t | 923.54 |
| 8mm | t | 821.25 |
| 10mm | t | 789.94 |
| 12mm | t | 764.69 |
| 16mm | t | 706.30 |
| 20mm | t | 621.73 |

Formwork
Refer to class G

**CLASS C: GEOTECHNICAL AND OTHER SPECIALIST PROCESSES**

The processes referred to in this Class are generally carried out by Specialist Contractors and the costs are therefore an indication of the probable costs based on average site conditions.

**DRILLING FOR GROUT HOLES THROUGH MATERIAL OTHER THAN ROCK OR ARTIFICIAL HARD MATERIAL**

| | | UNIT | L | P | M | £ NETT |
|---|---|---|---|---|---|---|
| The following unit costs are based on drilling 100 grout holes on a clear site with reasonable access | | | | | | |
| Establishment of standard plant and equipment and removal on completion | hr | sum | - | - | | |
| | £ | | - | - | - | 7087.50 |
| Standing time | hr | hr | 1.00 | 1.00 | | |
| | £ | | 34.56 | 84.24 | - | 118.80 |
| Downwards at an angle 0-45 to the vertical (In holes of depth) | | | | | | |
| ne 5m | hr | m | 0.10 | 0.10 | | |
| | £ | | 3.97 | 8.32 | - | 12.29 |
| 5-10m | hr | m | 0.11 | 0.11 | | |
| | £ | | 4.65 | 9.45 | - | 14.10 |
| 10-20m | hr | m | 0.13 | 0.13 | | |
| | £ | | 5.73 | 11.15 | - | 16.88 |
| 20-30m | hr | m | 0.17 | 0.17 | | |
| | £ | | 6.57 | 13.82 | - | 20.39 |
| Upwards at an angle 0-45 to the horizontal (In holes of depth) | | | | | | |
| ne 5m | hr | m | 0.17 | 0.17 | | |
| | £ | | 6.57 | 13.82 | - | 20.39 |
| 5-10m | hr | m | 0.19 | 0.19 | | |
| | £ | | 7.59 | 15.88 | - | 23.47 |
| 10-20m | hr | m | 0.22 | 0.22 | | |
| | £ | | 8.27 | 17.01 | - | 25.28 |
| 20-30m | hr | m | 0.23 | 0.23 | | |
| | £ | | 9.35 | 18.72 | - | 28.07 |

## CLASS C: GEOTECHNICAL AND OTHER SPECIALIST PROCESSES

### DRILLING FOR GROUT HOLES THROUGH ROCK OR ARTIFICIAL HARD MATERIAL - continued

| | | UNIT | L | P | M | £ NETT |
|---|---|---|---|---|---|---|
| The following unit costs are based on drilling 100 grout holes on a clear site with reasonable access | | | | | | |
| Establishment of standard plant and equipment and removal on completion | hr | sum | - | - | | |
| | £ | | - | - | - | 7087.50 |
| Standing time | hr | hr | 1.00 | 1.00 | | |
| | £ | | 33.28 | 84.24 | - | 117.52 |
| Downwards at an angle 0-45 to the vertical (In holes of depth) | | | | | | |
| ne 5m | hr | m | 0.09 | 0.09 | | |
| | £ | | 3.05 | 7.58 | - | 10.63 |
| 5-10m | hr | m | 0.11 | 0.11 | | |
| | £ | | 4.11 | 9.25 | - | 13.36 |
| 10-20m | hr | m | 0.13 | 0.13 | | |
| | £ | | 5.18 | 10.95 | - | 16.13 |
| 20-30m | hr | m | 0.16 | 0.16 | | |
| | £ | | 6.52 | 13.48 | - | 20.00 |
| Upwards at an angle 0-45 to the horizontal (In holes of depth) | | | | | | |
| ne 5m | hr | m | 0.16 | 0.16 | | |
| | £ | | 6.02 | 13.48 | - | 19.50 |
| 5-10m | hr | m | 0.18 | 0.18 | | |
| | £ | | 7.10 | 15.16 | - | 22.26 |
| 10-20m | hr | m | 0.20 | 0.20 | | |
| | £ | | 8.15 | 16.85 | - | 25.00 |
| 20-30m | hr | m | 0.23 | 0.23 | | |
| | £ | | 8.52 | 19.38 | - | 27.90 |
| **GROUT HOLES** | | | | | | |
| Number of holes | hr | nr | - | - | | |
| | £ | | - | - | - | 69.30 |
| Multiple water pressure tests | hr | nr | - | - | | |
| | £ | | - | - | - | 5.03 |

**CLASS C: GEOTECHNICAL AND OTHER SPECIALIST PROCESSES**

**GROUT MATERIALS AND INJECTIONS**

| | | | £ NETT |
|---|---|---|---|
| Establishment of standard plant and removal on completion | hr | sum | |
| | £ | | 6587.50 |
| Standing time | hr | hr | |
| | £ | | 96.90 |
| Materials | | | |
| cement grout | hr | t | |
| | £ | | 70.10 |
| cement P.F.A. | hr | t | |
| | £ | | 40.65 |
| bentonite | hr | t | |
| | £ | | 87.53 |
| Injection | | | |
| number of injections | hr | nr | |
| | £ | | 40.95 |

## CLASS C: GEOTECHNICAL AND OTHER SPECIALIST PROCESSES

### DIAPHRAGM WALLS

Diaphragm walls are the construction of vertical walls, cast in place in a trench excavation. They can be formed in reinforced concrete to provide structural elements for temporary or permanent retaining walls. Wall thicknesses of 500 to 1500m up to 40m deep may be constructed. Special equipment such as the Hydrofraise can construct walls up to 100m deep. Restricted urban sites will significantly increase the costs.

The following costs are based on constructing a diaphragm wall with an excavated volume of 4000m3 using a grab. Typical progress would be up to 500m3 per week. Disposal of excavated material is not included in the unit costs.

| | | UNIT | L | P | M | £ NETT |
|---|---|---|---|---|---|---|
| Establishment of standard plant and equipment and removal on completion | hr | sum | - | - | | |
| | £ | | - | - | - | 86000.00 |
| Standing time | hr | hr | 1.00 | 1.00 | | |
| | £ | | 116.00 | 234.00 | - | 350.00 |
| Excavation in material other than rock or artificial hard material (Maximum depth) | | | | | | |
| 5-10m | hr | m3 | 0.29 | 0.29 | | |
| | £ | | 37.14 | 67.86 | - | 105.00 |
| 10-15m | hr | m3 | 0.31 | 0.31 | | |
| | £ | | 37.46 | 72.54 | - | 110.00 |
| 15-20m | hr | m3 | 0.33 | 0.33 | | |
| | £ | | 40.78 | 77.22 | - | 118.00 |
| Concrete grade C25P; 20mm aggregate (Maximum depth) | hr | m3 | - | - | | |
| | £ | | 17.30 | 18.00 | 45.70 | 81.00 |
| Guide walls (twin) | hr | m | 0.42 | 0.42 | | |
| | £ | | 61.72 | 98.28 | - | 160.00 |

| | | | 12mm | 20mm | 25mm |
|---|---|---|---|---|---|
| Mild steel bar reinforcement | hr | t | | | |
| | £ | | 580.00 | 550.00 | 520.00 |

## CLASS C: GEOTECHNICAL AND OTHER SPECIALIST PROCESSES

### GROUND ANCHORAGES

Ground anchorages consist of the installation of a cable or solid bar tendon fixed in the ground by grouting and tensioned to exceed the working load to be carried. Ground anchors may be of a permanent or temporary nature and can be used in conjunction with diaphragm walling or sheet piling to eliminate the use of strutting etc. The following costs are based on the installation of 50 ground anchors.

| | | UNIT | L | P | M | £ NETT |
|---|---|---|---|---|---|---|
| Establishment of standard plant and equipment and removal on completion | hr | sum | - | - | | |
| | £ | | - | - | - | 8000.00 |
| Standing time | hr | hr | 1.00 | 1.00 | | |
| | £ | | 35.00 | 65.00 | - | 100.00 |
| Number in material other than rock or artificial hard material (Type of ground) (Load in tonnes) | | | | | | |
| temporary or permanent; maximum depth 15m | | | | | | |
| Rock alluvial or clay | | | | | | |
| 0-50 tonnes load | hr | nr | 0.58 | 0.58 | | |
| | £ | | 12.30 | 37.70 | - | 50.00 |
| Rock or alluvial | | | | | | |
| 50-90 tonnes load | hr | nr | 0.58 | 0.58 | | |
| | £ | | 22.30 | 37.70 | - | 60.00 |
| Rock only | | | | | | |
| 90-150 tonnes load | hr | nr | 0.58 | 0.58 | | |
| | £ | | 32.30 | 37.70 | - | 70.00 |
| Total length of tendons in material other than rock or artificial hard material | | | | | | |
| temporary | | | | | | |
| Rock alluvial or clay | | | | | | |
| 0-50 tonnes load | hr | nr | 0.54 | 0.54 | | |
| | £ | | 4.90 | 35.10 | - | 40.00 |
| Rock or alluvial | | | | | | |
| 50-90 tonnes load | hr | nr | 0.58 | 0.58 | | |
| | £ | | 22.30 | 37.70 | - | 60.00 |
| Rock only | | | | | | |
| 90-150 tonnes load | hr | nr | 0.65 | 0.65 | | |
| | £ | | 37.75 | 42.25 | - | 80.00 |
| permanent | | | | | | |
| Rock alluvial or clay | | | | | | |
| 0-50 tonnes load | hr | nr | 0.81 | 0.81 | | |
| | £ | | 7.35 | 52.65 | - | 60.00 |
| Rock or alluvial | | | | | | |
| 50-90 tonnes load | hr | nr | 0.85 | 0.85 | | |
| | £ | | 24.75 | 55.25 | - | 80.00 |
| Rock only | | | | | | |
| 90-150 tonnes load | hr | nr | 0.92 | 0.92 | | |
| | £ | | 40.20 | 59.80 | - | 100.00 |

## CLASS C: GEOTECHNICAL AND OTHER SPECIALIST PROCESSES

### SAND, BAND AND WICK DRAINS

Vertical drains are a technique by which the rate of consolidation of fine grained soils can be considerably increased by the installation of vertical drainage paths in the form of columns of sand or high-quality plastic material consisting of a centre core encased in a filter sleeve. Band drains are generally 100mm wide and 3 - 5mm thick. Water is extracted through the drain from the soft soils when the surface is surcharged. The rate of consolidation is dependent on the drain spacing and the height of surcharge. Drains are usually quickly installed up to depths of 40m by special lances either pulled or vibrated into the ground. Typical drain spacing would be one per 1 - 2.5m with the rate of installation varying between 1500 to 6000m per day depending on ground conditions and depths.

The following costs are based on the installation of 500 vertical band drains.

| | | UNIT | L | P | M | £ NETT |
|---|---|---|---|---|---|---|
| Establishment of standard plant and equipment and removal on completion | hr | sum | - | - | | |
| | £ | | - | - | - | 15000.00 |
| Standing time | hr | hr | 1.00 | 1.00 | | |
| | £ | | 63.64 | 111.36 | - | 175.00 |
| Set up at each drain position | hr | nr | 0.02 | 0.02 | | |
| | £ | | 2.77 | 2.23 | - | 5.00 |
| Install drains maximum depth 10-15m | hr | m | - | - | | |
| | £ | | 0.18 | 0.32 | 0.50 | 1.00 |
| Where the upper strata is hard pre-drilling at each drain position may be necessary | | | | | | |
| Establishment of standard plant and equipment and removal on completion | hr | sum | - | - | | |
| | £ | | - | - | - | 2500.00 |
| Set up at each drain position | hr | nr | 0.02 | 0.02 | | |
| | £ | | 2.77 | 2.23 | - | 5.00 |
| Drilling for vertical band drains up to a maximum depth of 3m | hr | m | 0.03 | 0.03 | | |
| | £ | | 2.66 | 3.34 | - | 6.00 |

**CLASS C: GEOTECHNICAL AND OTHER SPECIALIST PROCESSES**

**GROUND CONSOLIDATION - VIBROREPLACEMENT (dry and wet processes)**

Vibroreplacement is a method of considerably increasing the ground bearing pressure and consists of a specifically designed powerful poker vibrator penetrating vertically into the ground hydraulically. Air and water jets may be used to assist penetration. In cohesive soils a hole is formed into which granular backfill is placed and compacted by the poker, forming a dense stone column. In natural sands and gravels the existing loose deposits may be compacted without the addition of extra material other than making up levels after settlement resulting from compaction.

There are many considerations regarding the soil types to be treated, whether cohesive or non-cohesive, made-up or natural ground, which influence the choice of wet or dry processes, pure densification or stone column techniques with added granular backfill. It is therefore possible only to give indicative costs; a Specialist Contractor should be consulted for more accurate costs for a particular site.

Testing of conditions after consolidation can be static or dynamic penetration tests plate bearing tests or zone bearing tests. A frequently adopted specification calls for plate bearing tests at 1 per 300 and zone loading tests at 1 per 1000 stone columns. Allowable bearing pressures of up to 400kN/m2 can be obtained by densification of suitable granular soils but more typically up to 150kN/m2 by the installation of stone columns made or natural ground.

| | | UNIT | L | P | M | £ NETT |
|---|---|---|---|---|---|---|
| Establishment of standard plant and equipment and removal on completion | hr | sum | - | - | | |
| | £ | | - | - | - | 3500.00 |
| Standing time | hr | h | 1.00 | 1.00 | | |
| | £ | | 70.04 | 116.00 | - | 186.04 |
| Construct stone columns to a depth ne 4m (based on number of columns in original design) | | | | | | |
| dry | hr | m | 0.04 | 0.04 | | |
| | £ | | 2.80 | 4.64 | 4.00 | 11.44 |
| wet | hr | m | 0.07 | 0.07 | | |
| | £ | | 4.90 | 8.12 | 4.00 | 17.02 |
| Plate bearing test ne 11t or 2 hour duration | hr | nr | - | - | | |
| | £ | | - | - | - | 280.00 |
| Zone loading test to specification | hr | nr | - | - | | |
| | £ | | - | - | - | 6035.00 |

## CLASS C: GEOTECHNICAL AND OTHER SPECIALIST PROCESSES

### GROUND CONSOLIDATION - DYNAMIC COMPACTION

Ground consolidation by dynamic compaction is a technique which involves the dropping of a steel or concrete pounder several times in each location on a grid pattern that covers the whole site. For ground compaction up to 10m a 15t pounder from a free fall of 20m would be typical. Several passes over the site are normally required to achieve full compaction. The process is recommended for naturally cohesive soils and is usually uneconomic for areas of less than 4000m2 for sites with granular or mixed granular cohesive soils and 6000m2 for a site with weak cohesive soils. The main considerations to be taken into account when using this method of consolidation are:

sufficient area to be viable

proximity and condition of adjacent property and services

need for blanket layer of granular material for a working surface and as backfill to offset induced settlement

water table level

The final bearing capacity and settlement criteria that can be achieved depends on the nature of the material being compacted. Allowable bearing capacity may be increased by up to twice the pre-treated value for the same settlement. Control testing can be by crater volume measurements, site levelling between passes, penetration tests or plate loading tests.

The following range of costs are averages based on treating an area of about 10,000m2 for a 5 - 6m compaction depth. Typical progress would be 1500 - 2000m2 per week.

| | | UNIT | L | P | M | £ NETT |
|---|---|---|---|---|---|---|
| Establishment of standard plant and equipment and removal on completion | hr | sum | - | - | | |
| | £ | | - | - | - | **30000.00** |
| Ground treatment | hr | m2 | 0.02 | 0.02 | | |
| | £ | | **1.80** | **4.20** | - | **6.00** |
| Free-draining granular blanket layer as working surface and backfill material (300mm thickness required of filter material) | hr | m2 | - | - | | |
| | £ | | **0.65** | **0.55** | **8.05** | **9.25** |
| Control testing including levelling, piezometers and penetrameter testing | hr | m2 | - | - | | |
| | £ | | - | - | - | **2.50** |
| Kentledge load test | hr | nr | - | - | | |
| | £ | | - | - | - | **8000.00** |

## CLASS D: DEMOLITION AND SITE CLEARANCE

### GENERAL CLEARANCE

| | | UNIT | SPECIALIST ADVICE £ NETT |
|---|---|---|---|
| The rates for site clearance include for all sundry items, small trees (i.e. under 500mm diameter), hedges etc., but exclude items that are measured separately under the CESMM2; examples of which are given in this section | | | |
| generally | hr<br>£ | ha | <br>**581.00** |
| wooded areas | hr<br>£ | ha | <br>**1761.00** |
| areas below tidal level | hr<br>£ | ha | <br>**2107.00** |
| **TREES**<br>The following rates are based on removing a minimum 100 trees generally in a group. Cutting down a single tree on a site would be many times these costs | | | |
| Girth | | | |
| 500mm - 1m | hr<br>£ | nr | <br>17.05 |
| 1 - 2m | hr<br>£ | nr | <br>34.20 |
| 2 - 3m | hr<br>£ | nr | <br>126.00 |
| 3 - 5m | hr<br>£ | nr | <br>632.00 |
| 7m | hr<br>£ | nr | <br>945.00 |
| **STUMPS** | | | |
| Diameter | | | |
| 150 - 500mm | hr<br>£ | nr | <br>13.39 |
| 500mm - 1m | hr<br>£ | nr | <br>26.75 |
| 2m | hr<br>£ | nr | <br>53.00 |

**CLASS D: DEMOLITION AND SITE CLEARANCE**

**GENERAL CLEARANCE - continued**

| | | UNIT | SPECIALIST ADVICE £ NETT |
|---|---|---|---|
| **BUILDINGS** | | | |
| The following rates are based on assuming a non urban location where structure does not take up a significant area of the site | | | |
| Demolish building to ground level and dispose off site | | | |
| brickwork with timber floor and roof | hr | m3 | |
| | £ | | 3.35 |
| brickwork with concrete floor | hr | m3 | |
| | £ | | 5.50 |
| masonry with timber floor and roof | hr | m3 | |
| | £ | | 4.30 |
| reinforced concrete frame with brick infill | hr | m3 | |
| | £ | | 5.75 |
| steel frame with brick cladding | hr | m3 | |
| | £ | | 3.20 |
| steel frame with sheet cladding | hr | m3 | |
| | £ | | 3.00 |
| timber | hr | m3 | |
| | £ | | 2.65 |
| Demolish buildings with asbestos linings to ground level and dispose off site | | | |
| brickwork with concrete floor | hr | m3 | |
| | £ | | 12.10 |
| reinforced concrete frame with brick infill | hr | m3 | |
| | £ | | 12.65 |
| steel frame with brick cladding | hr | m3 | |
| | £ | | 7.04 |
| steel frame with sheet cladding | hr | m3 | |
| | £ | | 6.60 |

**CLASS D: DEMOLITION AND SITE CLEARANCE**

**GENERAL CLEARANCE - continued**

| | | UNIT | SPECIALIST ADVICE £ NETT |
|---|---|---|---|
| **PIPELINES** | | | |
| Removal of Redundant Services | | | |
| L.V. Electric Cable | hr | m | |
| | £ | | 1.46 |
| 150 bore L.P. Gas Main | hr | m | |
| | £ | | 2.80 |
| B.T. Ducts Earthenware (4 way) | hr | m | |
| | £ | | 1.96 |
| Sewer or drain 100 bore | hr | m | |
| | £ | | 2.90 |
| Sewer or drain 300 bore | hr | m | |
| | £ | | 5.60 |
| Sewer or drain 750 bore | hr | m | |
| | £ | | 15.79 |

## CLASS E: EARTHWORKS

The cost of earth moving and other associated works is dependent on matching the overall quantities and production rate called for by the programme of works with the most appropriate plant and assessing the most suitable version of that plant that will:

deal with the site conditions (e.g. type of ground, type of excavation, length of haul, prevailing weather, etc.);
comply with the specification requirements (e.g. compaction, separation of materials, surface tolerances, etc.);
complete the work economically (e.g. provide surface tolerances which will avoid undue excess thickness of expensive imported materials);

The following are comparative costs using various types of equipment and include loading into transport. All costs assume 50 minutes productive work in every hour and adequate transport being available to obviate any delays.

| | BUCKET CAPACITY | | UNIT | L | P | M | £ NETT |
|---|---|---|---|---|---|---|---|
| Dragline (For excavations requiring a long reach and long discharge, mainly clearing streams & rivers) | up to 1.15m3 | hr | m3 | 0.06 | 0.06 | | |
| | | £ | | **0.48** | **1.00** | - | **1.48** |
| | up to 1.91m3 | hr | m3 | 0.03 | 0.03 | | |
| | | £ | | **0.24** | **1.01** | - | **1.25** |
| | up to 3.44m3 | hr | m3 | 0.02 | 0.02 | | |
| | | £ | | **0.16** | **0.84** | - | **1.00** |
| Hydraulic backacter (For all types of excavation and loading including trenches, breaking hard ground, etc.) | 0.40m3 | hr | m3 | 0.05 | 0.05 | | |
| | | £ | | **0.40** | **0.42** | - | **0.82** |
| | 0.45 - 1.0m3 | hr | m3 | 0.02 | 0.02 | | |
| | | £ | | **0.16** | **0.47** | - | **0.63** |
| | 1.00 - 1.6m3 | hr | m3 | 0.01 | 0.01 | | |
| | | £ | | **0.08** | **0.36** | - | **0.44** |
| Hydraulic face shovel (predominantly for excavating cuttings and embankments over 2.0m high requiring high output) | up to 1.0m3 | hr | m3 | 0.04 | 0.04 | | |
| | | £ | | **0.32** | **0.55** | - | **0.87** |
| | 1.00 - 1.5m3 | hr | m3 | 0.02 | 0.02 | | |
| | | £ | | **0.16** | **0.46** | - | **0.62** |
| | 1.50 - 2.6m3 | hr | m3 | 0.01 | 0.01 | | |
| | | £ | | **0.08** | **0.37** | - | **0.45** |

**CLASS E: EARTHWORKS**

| | | | UNIT | L | P | M | £ NETT |
|---|---|---|---|---|---|---|---|
| Tractor Loader (For loading, carrying, placing materials, spreading, levelling. Some site clearance operations and reducing levels) | 0.5 - 0.80m3 | hr | m3 | 0.02 | 0.02 | | |
| | | £ | | **0.16** | **0.24** | - | **0.40** |
| | 0.8 - 1.50m3 | hr | m3 | 0.01 | 0.01 | | |
| | | £ | | **0.08** | **0.20** | - | **0.28** |
| | 1.5 - 2.50m3 | hr | m3 | 0.007 | 0.007 | | |
| | | £ | | **0.06** | **0.18** | - | **0.24** |
| Multipurpose loader /backhoe (Versatile machine for small to medium excavations, trenches, loading carrying, backfilling) | 0.3 - 0.50m3 | hr | m3 | 0.08 | 0.08 | | |
| | | £ | | **0.64** | **0.61** | - | **1.25** |
| | 0.5 - 0.75m3 | hr | m3 | 0.06 | 0.06 | | |
| | | £ | | **0.48** | **0.50** | - | **0.98** |

The following are comparative costs using various types of equipment to transport excavated loose material. The capacity of the transport must be suitable for the output of the loading machine. The cost will vary depending on the number of transport units required to meet the output of the loading unit and distance travelled.

| | | | | | | | |
|---|---|---|---|---|---|---|---|
| Loading loose material into transport | | | | | | | |
| wheeled shovel (up to 2.3m3 capacity) | | hr | m3 | 0.02 | 0.02 | | |
| | | £ | | **0.16** | **0.28** | - | **0.44** |
| wheeled shovel (up to 5.4m3 capacity) | | hr | m3 | 0.01 | 0.01 | | |
| | | £ | | **0.08** | **0.23** | - | **0.31** |
| wheeled shovel (up to 10.5m3 capacity) | | hr | m3 | 0.007 | 0.007 | | |
| | | £ | | **0.06** | **0.28** | - | **0.34** |
| Dump truck (rear dump) - on site use only up to 25m3 heaped capacity | | | | | | | |
| | 0 - 0.5Km | hr | m3 | 0.03 | 0.03 | | |
| | | £ | | **0.24** | **0.71** | - | **0.95** |
| | Add per 0.5Km | hr | m3 | 0.015 | 0.02 | | |
| | | £ | | **0.12** | **0.46** | - | **0.58** |
| up to 46m3 heaped capacity | 0 - 0.5Km | hr | m3 | 0.02 | 0.02 | | |
| | | £ | | **0.18** | **0.72** | - | **0.90** |
| | Add per 0.5Km | hr | m3 | 0.01 | 0.01 | | |
| | | £ | | **0.09** | **0.36** | - | **0.45** |

## CLASS E: EARTHWORKS

| | | | UNIT | L | P | M | £ NETT |
|---|---|---|---|---|---|---|---|
| Dump truck (articulated) - on site use only up to 30t payload | 0 - 0.5Km | hr | m3 | 0.04 | 0.04 | | |
| | | £ | | **0.32** | **0.76** | - | **1.08** |
| | Add per 0.5Km | hr | m3 | 0.02 | 0.02 | | |
| | | £ | | **0.16** | **0.38** | - | **0.54** |
| Tipping lorry on or off site use | | | | | | | |
| ne 10t capacity | ne 1Km | hr | m3 | 0.05 | 0.05 | | |
| | | £ | | **0.40** | **0.64** | - | **1.04** |
| | Add per Km | hr | m3 | 0.025 | 0.025 | | |
| | | £ | | **0.20** | **0.32** | - | **0.52** |
| 10-15t capacity | ne 1Km | hr | m3 | 0.04 | 0.04 | | |
| | | £ | | **0.32** | **0.67** | - | **0.99** |
| | Add per Km | hr | m3 | 0.02 | 0.02 | | |
| | | £ | | **0.16** | **0.33** | - | **0.49** |
| 15-25t capacity | ne 1Km | hr | m3 | 0.03 | 0.03 | | |
| | | £ | | **0.24** | **0.71** | - | **0.95** |
| | Add per Km | hr | m3 | 0.015 | 0.015 | | |
| | | £ | | **0.12** | **0.35** | - | **0.47** |

**CLASS E: EARTHWORKS**

The following are comparative costs using various types of equipment to include excavation, transport, spreading and levelling.

| | AVERAGE PULSH ONE WAY | UNIT | | L | P | M | £ NETT |
|---|---|---|---|---|---|---|---|
| Bulldozer up to 70KW (CAT. D4 size machine used for site strip, reducing levels or grading and spreading materials over smaller sites) | 10m | hr | m3 | 0.03 | 0.03 | | |
| | | £ | | **0.24** | **0.36** | - | **0.60** |
| | 30m | hr | m3 | 0.08 | 0.08 | | |
| | | £ | | **0.64** | **0.96** | - | **1.60** |
| | 50m | hr | m3 | 0.14 | 0.14 | | |
| | | £ | | **1.11** | **1.68** | - | **2.79** |
| | 100m | hr | m3 | 0.29 | 0.29 | | |
| | | £ | | **2.31** | **3.48** | - | **5.79** |
| Bull or Angle Dozer up to 115KW (CAT. D6 for reducing levels excavating to greater depths at steeper inclines, grading surfaces small cut and fill operations max. push 100m) | 10m | hr | m3 | 0.03 | 0.03 | | |
| | | £ | | **0.24** | **0.60** | - | **0.84** |
| | 30m | hr | m3 | 0.07 | 0.07 | | |
| | | £ | | **0.56** | **1.22** | - | **1.78** |
| | 50m | hr | m3 | 0.09 | 0.09 | | |
| | | £ | | **0.72** | **1.57** | - | **2.29** |
| | 100m | hr | m3 | 0.19 | 0.19 | | |
| | | £ | | **1.51** | **3.31** | - | **4.82** |
| Bull or Angle Dozer up to 250KW (CAT D8 high output ripping and excavating by reducing levels at steeper inclines harder materal than D6 larger cut and fill operations used in conjunction with towed or S.P. scrapers. Spreading and grading over large areas (max. push still 100m) | 10m | hr | m3 | 0.03 | 0.03 | | |
| | | £ | | **0.27** | **0.80** | - | **1.07** |
| | 30m | hr | m3 | 0.06 | 0.06 | | |
| | | £ | | **0.54** | **1.60** | - | **2.14** |
| | 50m | hr | m3 | 0.07 | 0.07 | | |
| | | £ | | **0.62** | **1.87** | - | **2.49** |
| | 100m | hr | m3 | 0.11 | 0.11 | | |
| | | £ | | **0.98** | **2.94** | - | **3.92** |

**CLASS E: EARTHWORKS**

| | AVERAGE PUSH ONE WAY | | | UNIT | L | P | M | £ NETT |
|---|---|---|---|---|---|---|---|---|
| Motorised scraper capacity (Excavating larger volumes over large haul lengths excavated by reducing levels, also levelling ground, grades large sites, moves and tips material including hard material (used in open cast sites) | 15m3 | 500m | hr | m3 | 0.04 | 0.04 | | |
| | | | £ | | **0.36** | **1.23** | - | **1.59** |
| | | 1000m | hr | m3 | 0.06 | 0.06 | | |
| | | | £ | | **0.54** | **1.85** | - | **2.39** |
| | | 2000m | hr | m3 | 0.08 | 0.08 | | |
| | | | £ | | **0.71** | **2.46** | - | **3.17** |
| | | 3000m | hr | m3 | 0.10 | 0.10 | | |
| | | | £ | | **0.89** | **3.08** | - | **3.97** |
| | 23m3 | 500m | hr | m3 | 0.03 | 0.03 | | |
| | | | £ | | **0.27** | **1.17** | - | **1.44** |
| | | 1000m | hr | m3 | 0.04 | 0.04 | | |
| | | | £ | | **0.36** | **1.55** | - | **1.91** |
| | | 2000m | hr | m3 | 0.05 | 0.05 | | |
| | | | £ | | **0.45** | **1.94** | - | **2.39** |
| | | 3000m | hr | m3 | 0.07 | 0.07 | | |
| | | | £ | | **0.62** | **2.71** | - | **3.33** |
| | 34m3 | 500m | hr | m3 | 0.018 | 0.018 | | |
| | | | £ | | **0.16** | **1.17** | - | **1.33** |
| | | 1000m | hr | m3 | 0.025 | 0.025 | | |
| | | | £ | | **0.22** | **1.95** | - | **2.17** |
| | | 2000m | hr | m3 | 0.03 | 0.03 | | |
| | | | £ | | **0.27** | **2.34** | - | **2.61** |
| | | 3000m | hr | m3 | 0.04 | 0.04 | | |
| | | | £ | | **0.36** | **3.12** | - | **3.48** |

**CLASS E: EARTHWORKS**

The following are comparative costs using various types of equipment for ancillary earthworks and excavation items found in many Civil Engineering/ Highway works. (Maximum depth 1.0m)

| | | UNIT | L | P | M | £ NETT |
|---|---|---|---|---|---|---|
| Excavation by hand (Desirable for work around live services or in areas of highly restricted access) | | | | | | |
| Excavate and load loose material into skip or dumper bucket | hr | m3 | 2.02 | - | | |
| | £ | | 12.52 | - | - | 12.52 |
| Ditto compacted soil/clay | hr | m3 | 3.15 | - | | |
| | £ | | 19.53 | - | - | 19.53 |
| Ditto Mass concrete or sandstone | hr | m3 | 30.00 | - | | |
| | £ | | 186.00 | - | - | 186.00 |
| Ditto broken rock | hr | m3 | 2.95 | - | | |
| | £ | | 18.29 | - | - | 18.29 |
| Ditto existing sub-base/pipe surrounds | hr | m3 | 3.12 | - | | |
| | £ | | 19.34 | - | - | 19.34 |
| Excavate by hand using pneumatic equipment: | | | | | | |
| Excavate rock below ground using compressor and pneumatic breaker and load material into skip or dumper bucket | hr | m3 | 2.12 | 2.12 | | |
| | £ | | 13.14 | 15.05 | - | 28.19 |
| Ditto Brickwork/Mass Concrete | hr | m3 | 2.76 | 2.76 | | |
| | £ | | 17.11 | 19.60 | - | 36.71 |
| Ditto Reinforced concrete | hr | m3 | 3.87 | 3.87 | | |
| | £ | | 23.99 | 27.48 | - | 51.47 |
| Ditto Asphalt in carriageways | hr | m3 | 1.15 | 1.15 | | |
| | £ | | 7.13 | 8.17 | - | 15.30 |

## CLASS E: EARTHWORKS

| | | UNIT | L | P | M | £ NETT |
|---|---|---|---|---|---|---|
| Excavate medium hard rock using equipment attached to backacter | | | | | | |
| 6 tonne machine (to break out) | hr | m3 | 0.43 | 0.43 | | |
| (JCB 3C + Montabert 125 breaker) | £ | | **3.42** | **4.73** | - | **8.15** |
| 12 tonne machine (to load) | hr | m3 | 0.07 | 0.07 | | |
| (CAT 205 + Rock-bucket 0.8m3) | £ | | **0.56** | **0.91** | - | **1.47** |
| Ditto Brickwork/Mass Concrete | | | | | | |
| (JCB/Montabert 125) | hr | m3 | 0.54 | 0.54 | | |
| | £ | | **4.29** | **5.94** | - | **10.23** |
| (CAT 205/Rock-bucket) | hr | m3 | 0.08 | 0.08 | | |
| | £ | | **0.64** | **1.04** | - | **1.68** |
| Ditto Reinforced Concrete | | | | | | |
| (JCB/Montabert 125) | hr | m3 | 0.64 | 0.64 | | |
| | £ | | **5.09** | **7.04** | - | **12.13** |
| (CAT 205/Rock-bucket) | hr | m3 | 0.09 | 0.09 | | |
| | £ | | **0.72** | **1.16** | - | **1.88** |

**CLASS E: EARTHWORKS**

Unless stated the units for excavation in material other than topsoil, rock or artificial hard materials are based on excavation in firm sand and gravel soils Multiply the nett rates by the following factors for differing types of soil.

| | |
|---|---|
| Scrapers | 1.5 for stiff clay<br>2.5 for chalk<br>3.5 for soft rock<br>3.7 for broken rock |
| Backacters (minimum bucket Size 0.5M3) | 1.7 for stiff clay<br>2.0 for chalk<br>2.0 for soft rock<br>1.7 for broken rock |
| Tractor Dozers and Loaders | 2.0 for stiff clay<br>3.0 for chalk<br>2.5 for soft rock<br>2.5 for broken rock |

The other important consideration in excavation of material is the bulkage of material after it is dug and loaded onto transport.

All pricing and estimating is based on the volume of solid material.

| | |
|---|---|
| 1.10 | Sand, bulkage |
| 1.20 | Gravel, bulkage |
| 1.30 | Compacted soil, bulkage |
| 1.30 | Compacted subbase suitable fill, etc., bulkage |
| 1.50 | Stiff clay, bulkage |

To provide an overall cost comparison, rates prices and outputs have been based on a medium sized Civil Engineering project of £5-6 million location neither city centre nor excessively remote, with no abnormal ground conditions that would affect the stated output and consistency of work produced. The rates are optimum rates and assume continuous output with no delays caused by other operations or works.

**CLASS E: EARTHWORKS**

**EXCAVATION FOR FOUNDATIONS**

| | | | |
|---|---|---|---|
| LABOUR | | | |
| Plant Operator | | | 7.95 |
| (Part time Plant Operator) | | | 2.38 |
| Banksman | | | 6.54 |
| Total hourly cost | = | £ | 16.87 |

| | | | |
|---|---|---|---|
| PLANT (AVERAGE COST) | | | |
| Excavator - 20 tonne | | | 14.03 |
| Excavator - 13 tonne with Hydraulic Breaker | | | 12.13 |
| Tractor, Loader - 0.80m3 | | | 12.48 |
| Average cost | = | £ | 12.88 |

**CLASS E: EARTHWORKS**

**EXCAVATION FOR FOUNDATIONS - continued**

| | | UNIT | L | P | M | £ NETT |
|---|---|---|---|---|---|---|
| The following unit costs are based on backacter machine excavations | | | | | | |
| Topsoil | | | | | | |
| maximum depth ne 0.25m | hr | m3 | 0.03 | 0.03 | | |
| | £ | | 0.51 | 0.39 | - | 0.90 |
| Material other than topsoil rock or artificial hard material | | | | | | |
| 0.25 - 0.5m | hr | m3 | 0.05 | 0.05 | | |
| | £ | | 0.84 | 0.64 | - | 1.48 |
| 0.50 - 1.0m | hr | m3 | 0.06 | 0.06 | | |
| | £ | | 1.01 | 0.77 | - | 1.78 |
| 1.00 - 2.0m | hr | m3 | 0.07 | 0.07 | | |
| | £ | | 1.18 | 0.90 | - | 2.08 |
| 2.00 - 5.0m | hr | m3 | 0.123 | 0.123 | | |
| | £ | | 2.08 | 1.58 | - | 3.66 |
| 5.00 - 10.0m | hr | m3 | 0.23 | 0.23 | | |
| | £ | | 3.88 | 2.96 | - | 6.84 |
| Artificial hard material exposed at the commencing surface; maximum depth | | | | | | |
| ne 0.25m | | | | | | |
| Plain Concrete (using hydraulic breaker) | hr | m3 | 0.60 | 0.60 | | |
| | £ | | 10.12 | 7.73 | - | 17.85 |
| Reinforced Concrete (using hydraulic breaker) | hr | m3 | 0.94 | 0.94 | | |
| | £ | | 15.86 | 12.11 | - | 27.97 |
| Tarmacadam (using hydraulic breaker) | hr | m3 | 0.30 | 0.30 | | |
| | £ | | 5.06 | 3.86 | - | 8.92 |
| 0.25 - 0.5m | | | | | | |
| Plain Concrete (using hydraulic breaker) | hr | m3 | 0.66 | 0.66 | | |
| | £ | | 11.13 | 8.50 | - | 19.63 |

**CLASS E: EARTHWORKS**

**EXCAVATION FOR FOUNDATIONS - continued**

| | | UNIT | L | P | M | £ NETT |
|---|---|---|---|---|---|---|
| Reinforced Concrete (using hydraulic breaker) | hr | m3 | 0.94 | 0.94 | | |
| | £ | | 15.86 | 12.11 | - | 27.97 |
| Tarmacadam (using hydraulic breaker) | hr | m3 | 0.30 | 0.30 | | |
| | £ | | 5.06 | 3.86 | - | 8.92 |
| 0.5 - 1.0m | | | | | | |
| Plain Concrete (using hydraulic breaker) | hr | m3 | 0.65 | 0.65 | | |
| | £ | | 10.97 | 8.37 | - | 19.34 |
| Reinforced Concrete (using hydraulic breaker) | hr | m3 | 0.94 | 0.94 | | |
| | £ | | 15.86 | 12.11 | - | 27.97 |
| Tarmacadam (using hydraulic breaker) | hr | m3 | 0.34 | 0.34 | | |
| | £ | | 5.74 | 4.38 | - | 10.12 |
| 1.0 - 2.0m | | | | | | |
| Plain Concrete (using hydraulic breaker) | hr | m3 | 0.80 | 0.80 | | |
| | £ | | 13.50 | 10.30 | - | 23.80 |
| Reinforced Concrete (using hydraulic breaker) | hr | m3 | 1.04 | 1.04 | | |
| | £ | | 17.54 | 13.40 | - | 30.94 |
| Tarmacadam (using hydraulic breaker) | hr | m3 | 0.40 | 0.40 | | |
| | £ | | 6.75 | 5.15 | - | 11.90 |
| Artificial hard material not exposed at the commencing surface; maximum depth below commencing depth; plain concrete | | | | | | |
| ne 0.25m (using hydraulic breaker) | hr | m3 | 0.70 | 0.70 | | |
| | £ | | 11.81 | 9.02 | - | 20.83 |
| 0.25 - 0.5m (using hydraulic breaker) | hr | m3 | 0.85 | 0.85 | | |
| | £ | | 14.34 | 10.95 | - | 25.29 |
| 0.50 - 1.0m (using hydraulic breaker) | hr | m3 | 0.85 | 0.85 | | |
| | £ | | 14.34 | 10.95 | - | 25.29 |
| 1.00 - 2.0m (using hydraulic breaker) | hr | m3 | 1.02 | 1.02 | | |
| | £ | | 17.21 | 13.14 | - | 30.35 |

**CLASS E: EARTHWORKS**

**GENERAL EXCAVATION**

The following unit costs are based on backacter, tractor loader, or face shovel machines.

| | | UNIT | L | P | M | £ NETT |
|---|---|---|---|---|---|---|
| Topsoil | | | | | | |
| maximum depth ne 0.25m | hr | m3 | 0.03 | 0.03 | | |
| | £ | | 0.51 | 0.39 | - | 0.90 |
| Material other than topsoil, rock or artificial hard material; | | | | | | |
| maximum depth | | | | | | |
| ne 0.25m | hr | m3 | 0.03 | 0.03 | | |
| | £ | | 0.51 | 0.39 | - | 0.90 |
| 0.25 - 0.5m | hr | m3 | 0.03 | 0.03 | | |
| | £ | | 0.51 | 0.39 | - | 0.90 |
| 0.50 - 1.0m | hr | m3 | 0.04 | 0.04 | | |
| | £ | | 0.67 | 0.52 | - | 1.19 |
| 1.00 - 2.0m | hr | m3 | 0.06 | 0.06 | | |
| | £ | | 1.01 | 0.77 | - | 1.78 |
| 2.00 - 5.0m | hr | m3 | 0.10 | 0.10 | | |
| | £ | | 1.69 | 1.29 | - | 2.98 |
| 5.00 - 10.0m | hr | m3 | 0.20 | 0.20 | | |
| | £ | | 3.37 | 2.58 | - | 5.95 |
| 10.0 - 15.0m | hr | m3 | 0.29 | 0.29 | | |
| | £ | | 4.89 | 3.74 | - | 8.63 |
| Rock (medium hard) (using hydraulic breaker) | | | | | | |
| maximum depth | | | | | | |
| ne 0.25m | hr | m3 | 0.31 | 0.31 | | |
| | £ | | 5.23 | 3.99 | - | 9.22 |
| 0.25 - 0.5m | hr | m3 | 0.43 | 0.43 | | |
| | £ | | 7.25 | 5.54 | - | 12.79 |
| 0.50 - 1.0m | hr | m3 | 0.58 | 0.58 | | |
| | £ | | 9.78 | 7.47 | - | 17.25 |
| 1.00 - 2.0m | hr | m3 | 0.80 | 0.80 | | |
| | £ | | 13.50 | 10.30 | - | 23.80 |

**CLASS E: EARTHWORKS**

**GENERAL EXCAVATION - continued**

| | | UNIT | L | P | M | £ NETT |
|---|---|---|---|---|---|---|
| Artificial hard material exposed at the commencing surface; maximum depth | | | | | | |
| Mass Concrete (using hydraulic breaker) | | | | | | |
| ne 0.25m | hr | m3 | 0.57 | 0.57 | | |
| | £ | | 9.62 | 7.34 | - | 16.96 |
| 0.25 - 0.5m | hr | m3 | 0.60 | 0.60 | | |
| | £ | | 10.12 | 7.73 | - | 17.85 |
| 0.50 - 1.0m | hr | m3 | 0.67 | 0.67 | | |
| | £ | | 11.30 | 8.63 | - | 19.93 |
| 1.00 - 2.0m | hr | m3 | 0.70 | 0.70 | | |
| | £ | | 11.81 | 9.02 | - | 20.83 |
| Reinforced Concrete (using hydraulic breaker) | | | | | | |
| ne 0.25m | hr | m3 | 0.90 | 0.90 | | |
| | £ | | 15.18 | 11.59 | - | 26.77 |
| 0.25 - 0.5m | hr | m3 | 0.90 | 0.90 | | |
| | £ | | 15.18 | 11.59 | - | 26.77 |
| 0.50 - 1.0m | hr | m3 | 0.95 | 0.95 | | |
| | £ | | 16.03 | 12.24 | - | 28.27 |
| 1.00 - 2.0m | hr | m3 | 1.08 | 1.08 | | |
| | £ | | 18.22 | 13.91 | - | 32.13 |
| Artificial hard material not exposed at the commencing surface; maximum depth (mass concrete) (using hydraulic breaker) | | | | | | |
| ne 0.25m | hr | m3 | 0.65 | 0.65 | | |
| | £ | | 10.97 | 8.37 | - | 19.34 |
| 0.25 - 0.5m (using hydraulic breaker) | hr | m3 | 0.67 | 0.67 | | |
| | £ | | 11.30 | 8.63 | - | 19.93 |
| 0.50 - 1.0m (using hydraulic breaker) | hr | m3 | 0.69 | 0.69 | | |
| | £ | | 11.64 | 8.89 | - | 20.53 |
| 1.00 - 2.0m (using hydraulic breaker) | hr | m3 | 0.72 | 0.72 | | |
| | £ | | 12.15 | 9.27 | - | 21.42 |

**CLASS E: EARTHWORKS**

**EXCAVATION ANCILLARIES**

| | | | |
|---|---|---|---|
| **LABOUR** | | | |
| Plant Operator | | | 7.95 |
| Banksman | | | 6.54 |
| Total hourly cost | = | £ | 14.49 |

| | | |
|---|---|---|
| **PLANT** | | |
| D4 (Bulldozer) or | = | 10.33 |
| D6 (Angle/Bulldozer) or | = | 15.44 |
| 953 (Loader) or | = | 19.52 |
| 12G (Grader) | = | 12.39 |

**CLASS E: EARTHWORKS**

**EXCAVATION ANCILLARIES**

The following unit costs are for various machines appropriate to the work:

| | | UNIT | L | P | M | £ NETT |
|---|---|---|---|---|---|---|
| Trimming of excavated surfaces topsoil (Using D4 Dozer) | | | | | | |
| Horizontal | hr | m2 | 0.02 | 0.02 | | |
| | £ | | **0.29** | **0.21** | - | **0.50** |
| 10 - 45 degrees to horizontal | hr | m2 | 0.025 | 0.025 | | |
| | £ | | **0.36** | **0.26** | - | **0.62** |
| material other than rock or artificial hard material (Using D4 Dozer or Tractor loader or grader) | | | | | | |
| Horizontal | hr | m2 | 0.02 | 0.02 | | |
| | £ | | **0.29** | **0.28** | - | **0.57** |
| 10 - 45 degrees to horizontal | hr | m2 | 0.02 | 0.02 | | |
| | £ | | **0.29** | **0.28** | - | **0.57** |
| 45 - 90 degrees to horizontal | hr | m2 | 0.03 | 0.03 | | |
| | £ | | **0.43** | **0.42** | - | **0.85** |
| Rock (Using D6 Dozer) | | | | | | |
| Horizontal | hr | m2 | 0.40 | 0.40 | | |
| | £ | | **5.80** | **6.18** | - | **11.98** |
| 10 - 45 degrees to horizontal | hr | m2 | 0.40 | 0.40 | | |
| | £ | | **5.80** | **6.18** | - | **11.98** |
| 45 - 90 degrees to horizontal | hr | m2 | 0.53 | 0.53 | | |
| | £ | | **7.68** | **8.18** | - | **15.86** |
| Vertical | hr | m2 | 0.53 | 0.53 | | |
| | £ | | **7.68** | **8.18** | - | **15.86** |

**CLASS E: EARTHWORKS**

**EXCAVATION ANCILLARIES - continued**

| | | UNIT | L | P | M | £ NETT |
|---|---|---|---|---|---|---|
| Preparation of excavated surfaces topsoil (Using D4 Dozer or Tractor loader) | | | | | | |
| Horizontal | hr | m2 | 0.03 | 0.03 | | |
| | £ | | **0.43** | **0.45** | - | **0.88** |
| 10 - 45 degress to horizontal | hr | m2 | 0.03 | 0.03 | | |
| | £ | | **0.43** | **0.45** | - | **0.88** |
| material other than rock or artificial hard material (Using D6 Dozer or Tractor loader or grader) | | | | | | |
| Horizontal | hr | m2 | 0.03 | 0.03 | | |
| | £ | | **0.43** | **0.46** | - | **0.89** |
| 10 - 45 degrees to horizontal | hr | m2 | 0.03 | 0.03 | | |
| | £ | | **0.43** | **0.46** | - | **0.89** |
| 45 - 90 degrees to horizontal | hr | m2 | 0.03 | 0.03 | | |
| | £ | | **0.43** | **0.46** | - | **0.89** |
| Rock (Using D6 Dozer) | | | | | | |
| Horizontal | hr | m2 | 0.40 | 0.40 | | |
| | £ | | **5.80** | **6.18** | - | **11.98** |
| 10 - 45 degrees to horizontal | hr | m2 | 0.30 | 0.30 | | |
| | £ | | **4.35** | **4.63** | - | **8.98** |
| 45 - 90 degrees to horizontal | hr | m2 | 0.30 | 0.30 | | |
| | £ | | **4.35** | **4.63** | - | **8.98** |

**CLASS E: EARTHWORKS**

**EXCAVATION ANCILLARIES - continued**

| | | | |
|---|---|---|---|
| LABOUR | | | |
| Plant Operator/Driver | | | 7.95 |
| Total hourly cost | = | £ | 7.95 |
| PLANT (DISPOSAL) | | | |
| Tipper 8 Wheel | | | 21.10 |
| Volvo A35 | | | 19.17 |
| 37m3 Dumper | | | 30.10 |
| 953 loader | | | 19.52 |
| Excavator (0.5m3) | | | 10.98 |

| | | UNIT | L | P | M | £ NETT |
|---|---|---|---|---|---|---|
| Disposal of excavated material topsoil | | | | | | |
| Stored on site (Volvo) | hr | m3 | 0.05 | 0.05 | | |
| | £ | | **0.40** | **0.96** | - | **1.36** |
| Remove 5km (Tipper) | hr | m3 | 0.11 | 0.11 | | |
| | £ | | **0.87** | **2.32** | - | **3.19** |
| Remove 15km (Tipper) | hr | m3 | 0.21 | 0.21 | | |
| | £ | | **1.67** | **4.43** | - | **6.10** |
| material other than rock or artificial hard material | | | | | | |
| Stored on site(Max 100m)(Volvo) | hr | m3 | 0.05 | 0.05 | | |
| | £ | | **0.40** | **0.96** | - | **1.36** |
| Remove 5km (Tipper) | hr | m3 | 0.12 | 0.12 | | |
| | £ | | **0.95** | **2.53** | - | **3.48** |
| Remove 15km (Tipper) | hr | m3 | 0.22 | 0.22 | | |
| | £ | | **1.75** | **4.64** | - | **6.39** |
| rock | | | | | | |
| Stored on site(Max 100m)(Volvo) | hr | m3 | 0.06 | 0.06 | | |
| | £ | | **0.48** | **1.15** | - | **1.63** |
| Remove 5km (Tipper) | hr | m3 | 0.13 | 0.13 | | |
| | £ | | **1.03** | **2.74** | - | **3.77** |
| Remove 15km (Tipper) | hr | m3 | 0.23 | 0.23 | | |
| | £ | | **1.83** | **4.85** | - | **6.68** |

**CLASS E: EARTHWORKS**

**EXCAVATION ANCILLARIES - continued**

| | | UNIT | L | P | M | £ NETT |
|---|---|---|---|---|---|---|
| Double handling of excavated material | | | | | | |
| topsoil (using tractor loader and Volvo) (average distance moved) | | | | | | |
| 300m | hr | m3 | 0.04 | 0.04 | | |
| | £ | | 0.64 | 1.55 | - | 2.19 |
| 600m | hr | m3 | 0.05 | 0.05 | | |
| | £ | | 0.80 | 1.93 | - | 2.73 |
| 1000m | hr | m3 | 0.06 | 0.06 | | |
| | £ | | 0.95 | 2.32 | - | 3.27 |
| material other than rock or artificial hard material (average distance moved) | | | | | | |
| 300m | hr | m3 | 0.04 | 0.04 | | |
| | £ | | 0.64 | 1.55 | - | 2.19 |
| 600m | hr | m3 | 0.05 | 0.05 | | |
| | £ | | 0.80 | 1.93 | - | 2.73 |
| 1000m | hr | m3 | 0.06 | 0.06 | | |
| | £ | | 0.95 | 2.32 | - | 3.27 |
| rock and artificial hard material (average distance moved) | | | | | | |
| 300m | hr | m3 | 0.08 | 0.08 | | |
| | £ | | 1.27 | 3.10 | - | 4.37 |
| 600m | hr | m3 | 0.09 | 0.09 | | |
| | £ | | 1.43 | 3.48 | - | 4.91 |
| 1000m | hr | m3 | 0.10 | 0.10 | | |
| | £ | | 1.59 | 3.87 | - | 5.46 |
| Excavation of material below the Final Surface and replacement with: | | | | | | |
| granular fill | hr | m3 | 0.60 | 0.60 | | |
| | £ | | 9.54 | 6.59 | 4.50 | 20.63 |
| concrete class E | hr | m3 | 0.30 | 0.30 | | |
| | £ | | 4.77 | 3.29 | 41.75 | 49.81 |
| Timber supports left in | hr | m2 | 0.15 | - | | |
| | £ | | 1.19 | - | 5.00 | 6.19 |
| Metal supports left in | hr | m2 | 0.41 | - | | |
| | £ | | 3.26 | - | 22.00 | 25.26 |

**CLASS E: EARTHWORKS**

**FILLING**

The following unit costs are based mainly on Tractor Loaders

**LABOUR**

| | | | |
|---|---|---|---|
| Plant Operator | | | 7.95 |
| 2 Skilled Operatives | | | 13.08 |
| Total hourly cost | = | £ | 21.03 |

**PLANT**

| | | | |
|---|---|---|---|
| 953 loader | | | 19.52 |
| Bomag BW10 (Pedestrian Roller) | | | 5.99 |
| Bomag BW35 (Towed Roller) | | | 1.20 |
| Total hourly cost | = | £ | 26.71 |

**CLASS E: EARTHWORKS**

**FILLING - continued**

| | | UNIT | L | P | M | £ NETT |
|---|---|---|---|---|---|---|
| To structures | | | | | | |
| Non-selected excavated material other than topsoil or rock | hr | m3 | 0.036 | 0.036 | | |
| | £ | | **0.76** | **0.96** | - | **1.72** |
| Selected excavated material other than topsoil or rock | hr | m3 | 0.041 | 0.041 | | |
| | £ | | **0.86** | **1.10** | - | **1.96** |
| Imported natural material other than topsoil or rock | | | | | | |
| Subsoil (PC = £1.58/tonne) | hr | m3 | 0.036 | 0.036 | | |
| | £ | | **0.76** | **0.96** | 3.00 | **4.72** |
| Granular Fill graded (PC = £2.38/tonne) | hr | m3 | 0.036 | 0.036 | | |
| | £ | | **0.76** | **0.96** | 4.50 | **6.22** |
| Granular Fill selected (PC = £1.90/tonne) | hr | m3 | 0.036 | 0.036 | | |
| | £ | | **0.76** | **0.96** | 3.59 | **5.31** |
| Excavated rock | hr | m3 | 0.04 | 0.04 | | |
| | £ | | **0.84** | **1.07** | - | **1.91** |
| Imported rock (PC = £4.40/tonne) | hr | m3 | 0.04 | 0.04 | | |
| | £ | | **0.84** | **1.07** | 8.32 | **10.23** |
| Embankments | | | | | | |
| Excavated topsoil | hr | m3 | 0.021 | 0.021 | | |
| | £ | | **0.44** | **0.56** | - | **1.00** |
| Imported topsoil (PC £7.75/m3) | hr | m3 | 0.021 | 0.021 | | |
| | £ | | **0.44** | **0.56** | 8.14 | **9.14** |
| Non selected excavated material other than topsoil or rock | hr | m3 | 0.015 | 0.015 | | |
| | £ | | **0.32** | **0.40** | - | **0.72** |
| Selected excavated material other than topsoil or rock | hr | m3 | 0.017 | 0.017 | | |
| | £ | | **0.36** | **0.45** | - | **0.81** |
| Excavated rock | hr | m3 | 0.05 | 0.05 | | |
| | £ | | **1.05** | **1.34** | - | **2.39** |
| Imported rock (PC = £4.40/tonne) | hr | m3 | 0.02 | 0.02 | | |
| | £ | | **0.42** | **0.53** | 8.32 | **9.27** |

**CLASS E: EARTHWORKS**

**FILLING - continued**

| | | UNIT | L | P | M | £ NETT |
|---|---|---|---|---|---|---|
| Imported natural material other than topsoil or rock | | | | | | |
| Subsoil (PC = £1.58/tonne) | hr | m3 | 0.015 | 0.015 | | |
| | £ | | **0.32** | **0.40** | **3.00** | **3.72** |
| Granular Material graded (PC = £2.38/tonne) | hr | m3 | 0.017 | 0.017 | | |
| | £ | | **0.36** | **0.45** | **4.50** | **5.31** |
| Granular Material selected (PC = £1.90/tonne) | hr | m3 | 0.017 | 0.017 | | |
| | £ | | **0.36** | **0.45** | **3.59** | **4.40** |
| General | | | | | | |
| Excavated topsoil | hr | m3 | 0.021 | 0.021 | | |
| | £ | | **0.44** | **0.56** | **-** | **1.00** |
| Imported topsoil (PC £7.75/m3) | hr | m3 | 0.021 | 0.021 | | |
| | £ | | **0.44** | **0.56** | **8.14** | **9.14** |
| Non-selected excavated material other than topsoil or rock | hr | m3 | 0.011 | 0.011 | | |
| | £ | | **0.23** | **0.29** | **-** | **0.52** |
| Selected excavated material other than topsoil or rock | hr | m3 | 0.012 | 0.012 | | |
| | £ | | **0.25** | **0.32** | **-** | **0.57** |
| Imported natural material other than topsoil or rock | | | | | | |
| Subsoil (PC = £1.58/tonne) | hr | m3 | 0.015 | 0.015 | | |
| | £ | | **0.32** | **0.40** | **3.00** | **3.72** |
| Granular Material graded (PC = £2.38/tonne) | hr | m3 | 0.017 | 0.017 | | |
| | £ | | **0.36** | **0.45** | **4.50** | **5.31** |
| Granular Material selected (PC = £1.90/tonne) | hr | m3 | 0.017 | 0.017 | | |
| | £ | | **0.36** | **0.45** | **3.59** | **4.40** |
| Excavated rock | hr | m3 | 0.05 | 0.05 | | |
| | £ | | **1.05** | **1.34** | **-** | **2.39** |
| Imported rock (PC = £4.40/tonne) | hr | m3 | 0.02 | 0.02 | | |
| | £ | | **0.42** | **0.53** | **8.32** | **9.27** |

**CLASS E: EARTHWORKS**

**FILLING - continued**

| | | UNIT | L | P | M | £ NETT |
|---|---|---|---|---|---|---|
| To stated depth or thickness | | | | | | |
| Excavated topsoil (953 loader) | | | | | | |
| 150mm | hr | m2 | 0.007 | 0.007 | | |
| | £ | | **0.06** | **0.14** | - | **0.20** |
| 250mm | hr | m2 | 0.012 | 0.012 | | |
| | £ | | **0.10** | **0.23** | - | **0.33** |
| 400mm | hr | m2 | 0.017 | 0.017 | | |
| | £ | | **0.14** | **0.33** | - | **0.47** |
| 600mm | hr | m2 | 0.024 | 0.024 | | |
| | £ | | **0.19** | **0.47** | - | **0.66** |
| Imported topsoil (PC £7.75/m3) | | | | | | |
| 150mm | hr | m2 | 0.021 | 0.021 | | |
| | £ | | **0.17** | **0.41** | **1.22** | **1.80** |
| 250mm | hr | m2 | 0.03 | 0.03 | | |
| | £ | | **0.24** | **0.59** | **2.03** | **2.86** |
| 400mm | hr | m2 | 0.035 | 0.035 | | |
| | £ | | **0.28** | **0.68** | **3.25** | **4.21** |
| 600mm | hr | m2 | 0.037 | 0.037 | | |
| | £ | | **0.29** | **0.72** | **4.88** | **5.89** |
| Non-selected excavated material other than topsoil or rock | | | | | | |
| 150mm | hr | m2 | 0.004 | 0.004 | | |
| | £ | | **0.03** | **0.08** | - | **0.11** |
| 250mm | hr | m2 | 0.006 | 0.006 | | |
| | £ | | **0.05** | **0.12** | - | **0.17** |
| 400mm | hr | m2 | 0.009 | 0.009 | | |
| | £ | | **0.07** | **0.18** | - | **0.25** |
| 600mm | hr | m2 | 0.012 | 0.012 | | |
| | £ | | **0.10** | **0.23** | - | **0.33** |
| Selected excavated material other than topsoil or rock | | | | | | |
| 150mm | hr | m2 | 0.004 | 0.004 | | |
| | £ | | **0.03** | **0.08** | - | **0.11** |
| 250mm | hr | m2 | 0.007 | 0.007 | | |
| | £ | | **0.06** | **0.14** | - | **0.20** |
| 400mm | hr | m2 | 0.010 | 0.010 | | |
| | £ | | **0.08** | **0.20** | - | **0.28** |
| 600mm | hr | m2 | 0.013 | 0.013 | | |
| | £ | | **0.10** | **0.25** | - | **0.35** |

**CLASS E: EARTHWORKS**

**FILLING - continued**

| | | UNIT | L | P | M | £ NETT |
|---|---|---|---|---|---|---|
| Imported natural material other than topsoil or rock; Subsoil | | | | | | |
| 150mm | hr | m2 | 0.009 | 0.009 | | |
| | £ | | **0.07** | **0.18** | **0.45** | **0.70** |
| 250mm | hr | m2 | 0.013 | 0.013 | | |
| | £ | | **0.09** | **0.26** | **0.75** | **1.10** |
| 400mm | hr | m2 | 0.015 | 0.015 | | |
| | £ | | **0.11** | **0.29** | **1.20** | **1.60** |
| 600mm | hr | m2 | 0.017 | 0.017 | | |
| | £ | | **0.12** | **0.33** | **1.80** | **2.25** |
| Granular Fill selected | | | | | | |
| 150mm | hr | m2 | 0.012 | 0.012 | | |
| | £ | | **0.09** | **0.24** | **0.54** | **0.87** |
| 250mm | hr | m2 | 0.017 | 0.017 | | |
| | £ | | **0.12** | **0.33** | **0.90** | **1.35** |
| 400mm | hr | m2 | 0.020 | 0.020 | | |
| | £ | | **0.15** | **0.39** | **1.44** | **1.98** |
| 600mm | hr | m2 | 0.020 | 0.020 | | |
| | £ | | **0.15** | **0.39** | **2.15** | **2.69** |
| Granular Fill graded | | | | | | |
| 150mm | hr | m2 | 0.012 | 0.012 | | |
| | £ | | **0.09** | **0.24** | **0.68** | **1.01** |
| 250mm | hr | m2 | 0.017 | 0.017 | | |
| | £ | | **0.12** | **0.33** | **1.13** | **1.58** |
| 400mm | hr | m2 | 0.02 | 0.02 | | |
| | £ | | **0.15** | **0.39** | **1.80** | **2.34** |
| 600mm | hr | m2 | 0.02 | 0.02 | | |
| | £ | | **0.15** | **0.39** | **2.70** | **3.24** |
| Excavated rock | | | | | | |
| 150mm | hr | m2 | 0.017 | 0.017 | | |
| | £ | | **0.12** | **0.33** | - | **0.45** |
| 250mm | hr | m2 | 0.026 | 0.026 | | |
| | £ | | **0.19** | **0.51** | - | **0.70** |
| 400mm | hr | m2 | 0.04 | 0.04 | | |
| | £ | | **0.29** | **0.79** | - | **1.08** |
| 600mm | hr | m2 | 0.055 | 0.055 | | |
| | £ | | **0.40** | **1.08** | - | **1.48** |

**CLASS E: EARTHWORKS**

**FILLING - continued**

| | | UNIT | L | P | M | £ NETT |
|---|---|---|---|---|---|---|
| Imported rock | | | | | | |
| 150mm | hr | m2 | 0.01 | 0.01 | | |
| | £ | | **0.07** | **0.20** | 1.23 | 1.50 |
| 250mm | hr | m2 | 0.02 | 0.02 | | |
| | £ | | **0.15** | **0.39** | 2.05 | 2.59 |
| 400mm | hr | m2 | 0.03 | 0.03 | | |
| | £ | | **0.22** | **0.59** | 3.28 | 4.09 |
| 600mm | hr | m2 | 0.03 | 0.03 | | |
| | £ | | **0.22** | **0.59** | 4.92 | 5.73 |

**FILLING ANCILLARIES**

Trimming of filled surfaces

| | | UNIT | L | P | M | £ NETT |
|---|---|---|---|---|---|---|
| topsoil | | | | | | |
| Horizontal | hr | m2 | 0.021 | 0.021 | | |
| | £ | | **0.17** | **0.41** | - | **0.58** |
| 10-45 to horizontal | hr | m2 | 0.021 | 0.021 | | |
| | £ | | **0.17** | **0.41** | - | **0.58** |
| 45-90 to horizontal | hr | m2 | 0.027 | 0.027 | | |
| | £ | | **0.21** | **0.53** | - | **0.74** |
| Material other than rock or artificial hard material | | | | | | |
| Horizontal | hr | m2 | 0.021 | 0.021 | | |
| | £ | | **0.17** | **0.41** | - | **0.58** |
| 10-45 to horizontal | hr | m2 | 0.021 | 0.021 | | |
| | £ | | **0.17** | **0.41** | - | **0.58** |
| 45-90 to horizontal | hr | m2 | 0.027 | 0.027 | | |
| | £ | | **0.21** | **0.53** | - | **0.74** |
| Rock | | | | | | |
| Horizontal | hr | m2 | 0.51 | 0.51 | | |
| | £ | | **4.05** | **9.96** | - | **14.01** |
| 10-45 degrees to horizontal | hr | m2 | 0.52 | 0.52 | | |
| | £ | | **4.13** | **10.15** | - | **14.28** |
| 45-90 degrees to horizontal | hr | m2 | 0.70 | 0.70 | | |
| | £ | | **5.57** | **13.66** | - | **19.23** |

**CLASS E: EARTHWORKS**

**FILLING ANCILLARIES - continued**

| | | UNIT | L | P | M | £ NETT |
|---|---|---|---|---|---|---|
| Preparation of filled surfaces | | | | | | |
| Topsoil | | | | | | |
| Horizontal | hr | m2 | 0.03 | 0.03 | | |
| | £ | | **0.22** | **0.59** | - | **0.81** |
| 10-45 degrees to horizontal | hr | m2 | 0.03 | 0.03 | | |
| | £ | | **0.22** | **0.59** | - | **0.81** |
| 45-90 degrees to horizontal | hr | m2 | 0.037 | 0.037 | | |
| | £ | | **0.27** | **0.73** | - | **1.00** |
| Material other than rock or artificial hard material | | | | | | |
| Horizontal | hr | m2 | 0.03 | 0.03 | | |
| | £ | | **0.22** | **0.59** | - | **0.81** |
| 10-45 degrees to horizontal | hr | m2 | 0.03 | 0.03 | | |
| | £ | | **0.22** | **0.59** | - | **0.81** |
| 45-90 degrees to horizontal | hr | m2 | 0.037 | 0.037 | | |
| | £ | | **0.27** | **0.73** | - | **1.00** |
| Rock | | | | | | |
| Horizontal | hr | m2 | 0.33 | 0.33 | | |
| | £ | | **2.39** | **6.49** | - | **8.88** |
| 10-45 degrees to horizontal | hr | m2 | 0.33 | 0.33 | | |
| | £ | | **2.39** | **6.49** | - | **8.88** |
| 45-90 degrees to horizontal | hr | m2 | 0.52 | 0.52 | | |
| | £ | | **3.77** | **10.22** | - | **13.99** |

**CLASS E: EARTHWORKS**

**DRILLING AND BLASTING IN ROCK**

The cost of blasting is controlled by the number of holes and the length of drilling required to achieve the tolerances and degree of shatter required e.g. line drilling to trenches, depth of drilling to control horizontal overbreak, etc.

| | | UNIT | L | P | M | £ NETT |
|---|---|---|---|---|---|---|
| Drilling with rotary percussion drills (type of rock) 105-110mm diameter | | | | | | |
| Hard rock | hr | m | - | - | | |
| | £ | | 2.05 | 0.98 | 0.70 | 3.73 |
| Sandstone | hr | m | - | - | | |
| | £ | | 2.60 | 1.02 | 0.75 | 4.37 |
| 125mm diameter | | | | | | |
| Hard rock | hr | m | - | - | | |
| | £ | | 2.60 | 1.02 | 0.75 | 4.37 |
| Sandstone | hr | m | - | - | | |
| | £ | | 3.70 | 1.40 | 0.80 | 5.90 |
| 165mm diameter | | | | | | |
| Hard rock | hr | m | - | - | | |
| | £ | | 3.70 | 1.40 | 0.80 | 5.90 |
| Sandstone | hr | m | - | - | | |
| | £ | | 4.60 | 2.05 | 0.85 | 7.50 |
| Drilling and blasting in open cut for bulk excavation excluding cost of excavation or trimming | | | | | | |
| Hard rock | hr | m3 | - | - | | |
| | £ | | 1.10 | 0.41 | 0.42 | 1.93 |
| Sandstone | hr | m3 | - | - | | |
| | £ | | 1.45 | 0.60 | 0.53 | 2.58 |
| Drilling and blasting for quarry operations with face height exceeding 10m | | | | | | |
| Hard rock | hr | m3 | - | - | | |
| | £ | | 0.88 | 0.37 | 0.32 | 1.57 |
| Sandstone | hr | m3 | - | - | | |
| | £ | | 1.05 | 0.40 | 0.37 | 1.82 |

**CLASS E: EARTHWORKS**

**DRILLING AND BLASTING IN ROCK**

| | | UNIT | L | P | M | £ NETT |
|---|---|---|---|---|---|---|
| Drilling and blasting in trenches excluding cost of excavation or trimming | | | | | | |
| trench width ne 1m | hr | m3 | - | - | | |
| | £ | | 7.90 | 3.70 | 2.88 | 14.48 |
| trench width 1.0 - 1.5m | hr | m3 | - | - | | |
| | £ | | 7.15 | 2.85 | 2.60 | 12.60 |
| trench width over 1.5m | hr | m3 | - | - | | |
| | £ | | 6.05 | 2.40 | 2.15 | 10.60 |
| Secondary blasting to boulders | | | | | | |
| pop shooting | hr | nr | - | - | | |
| | £ | | 2.35 | 1.00 | 1.10 | 4.45 |
| plaster shooting | hr | nr | - | - | | |
| | £ | | 1.10 | 0.50 | 0.77 | 2.37 |

**CLASS E: EARTHWORKS**

**DEWATERING**

The following unit costs are for dewatering pervious ground only and are for sets of equipment comprising:

| | | |
|---|---|---|
| 1nr diesel driven pump (WP 150/60 or similar) | | 46.70 |
| 50m of header pipe 150mm diameter<br>35nr well points (removable) | | 15.75 |
| 18m of delivery pipe | | 6.66 |
| 1nr diesel driven standby pump | | 23.24 |
| 1nr jetting pump and hoses | | 37.20 |
| Attendant labour and Plant (2hrs per day) | | 47.80 |
| Total cost per day | = £ | 293.45 |

Costs are based on 24hr operation in 12hr shifts with attendant operators (specialist advice)

| | | UNIT | L | P | M | £ NETT |
|---|---|---|---|---|---|---|
| Bring to site and install equipment | hr | set | - | - | | |
| | £ | | - | - | - | 1375.00 |
| Dismantle, move to new location and re-install | hr | set | - | - | | |
| | £ | | - | - | - | 1550.00 |
| Dismantle equipment and remove from site | hr | set | - | - | | |
| | £ | | - | - | - | 580.00 |
| Operating costs | | | | | | |
| 1 set | hr | day | - | - | | |
| | £ | | - | - | - | 177.35 |
| 2 sets | hr | day | - | - | | |
| | £ | | - | - | - | 331.46 |
| 3 sets | hr | day | - | - | | |
| | £ | | - | - | - | 485.57 |

## CLASS E: EARTHWORKS

### DREDGING

Dredging can be carried out by land based machines or by floating plant. The cost of the former can be assessed by reference to the excavation costs of the various types of plant given at the beginning of this Class, suitably adjusted to take account of the type of material to be excavated, depth of water and method of disposal. The cost of the latter is governed by many factors which affect the rates and leads to wide variations.

For reliable estimates it is advisable to seek the advice of a Specialist Contractor. The prices included here are for some typical dredging situations and are shown for a cost comparison and exclude initial mobilisation charges which can range from a minimum of £3000.

The main factors affecting the cost of floating plant are:

- type of material to be dredged
- depth of water
- depth of cut
- tidal range
- disposal location
- size and type of plant required
- current location of plant
- method of disposal of dredged material

| | UNIT | £ |
|---|---|---|
| (a) Floating pontoon with excavator loading material onto floating barges with bottom dumping facility maximum of 500m offshore, maximum depth of dig 15m non-tidal location, 10m tidal location, for use creating new channels on approaches to quays and similar or within dock systems. Approximate total hire cost daily £2220 realising 480m3 (60m3 per hour output) | m3 | 4.63 |
| (b) Ladder type dredger excavating sand, grit or silt using rope operated clamshell grabs each 1.50m3 capacity optimum depths 20 - 27.5m for use in rivers, estuaries or, deltas etc. for general bed lowering or maintaining shipping channels. Material can be deposited in stated locations or can be reclaimed for commercial use (e.g. Mersey grit used in concrete production and general fills) cost totally dependent on materials excavated cost here shown per tonne (relative densities should be known for conversion to m3) | | |
| Dredger capacity = 3500 tonnes (hold) | | |
| locate, load, deposit, relocate total 7-10 hours (8.5 average for 3500 tonnes.) | | |
| hourly cost (contract hire basis) = £450 | tonne | 1.20 |

**CLASS E: EARTHWORKS**

**DREDGING - continued**

| | UNIT | £ |
|---|---|---|
| (c) Bucket ladder dredger (excluding dumb barge) optimum dredge depth 25m max elevation 20m. Stationary dredger for harbour bed control output average ouput 540 tonnes per hour. | | |
| hourly cost (contract hire basis) = £535 | tonne | **0.99** |
| (d) Floating craft using diesel driven suction method 750mm diameter flexible pipe maximum distance from point of suction to point of discharge 5000m. For use in land reclamation situation by pumping from river/marine environment to point of fill, or for removing material from silted locations close to shore/ within dock systems. Maximum height of lift 10m. Average pumping capacity is 750,000 litres per hour containing maximum 30% solid volume equivalent to 225m3 per hour assuming 10 hours pumping per day = 2250m3 of material moved. | | |
| Daily cost (hire basis) £5900 including all floating equipment and discharge pipes, maintainence and all labour and plant to service but excluding mobilisation/initial set-up costs (min. £10,000) equivalent cost per:- | m3 | **2.62** |
| (e) Floating craft using diesel driven suction method 200mm diameter flexible pipe maximum distance from point of suction to point of discharge 1500m. For use in lakes, canals, rivers, industrial lagoons and from silted locations in dock systems. Maximum dredge depth 5m. Average pumping output = 30m3 per hour approximate average costs excluding mobilisation costs of about £3,000 | m3 | **4.00** |

## CLASS E: EARTHWORKS

## GEOTEXTILES

**(Refer also to MMHW Section 6 page 512)**

| | | UNIT | L | P | M | £ NETT |
|---|---|---|---|---|---|---|
| Geotextiles - proprietory products including wastage | | | | | | |
| "Typar" spunbonded polypropylene Geotextile (waste = 5%) | | | | | | |
| Horizontal | hr | m2 | 0.06 | - | | |
| | £ | | **0.44** | - | **1.16** | **1.60** |
| 10 - 45 degrees to horizontal | hr | m2 | 0.065 | - | | |
| | £ | | **0.47** | - | **1.16** | **1.63** |
| 45 - 90 degrees to horizontal | hr | m2 | 0.069 | - | | |
| | £ | | **0.50** | - | **1.16** | **1.66** |
| Vertical | hr | m2 | 0.10 | - | | |
| | £ | | **0.73** | - | **1.16** | **1.89** |
| "Terram" 2000 permeable membrane | | | | | | |
| 10 - 90 degrees (waste = 7 - 10%) | hr | m2 | 0.03 | - | | |
| | £ | | **0.22** | - | **0.72** | **0.94** |
| "Tensar" SS2 Geogrid | | | | | | |
| 10 - 90 degrees (waste = 5 - 7%) | hr | m2 | 0.06 | - | | |
| | £ | | **0.44** | - | **1.65** | **2.09** |
| "Enkamat" Erosion control matting (including fixing pegs) | | | | | | |
| 10 - 90 degrees (waste = 0%) | hr | m2 | 0.11 | - | | |
| | £ | | **0.80** | - | **3.65** | **4.45** |

Labour cost = £7.25

waste allowance in brackets

**CLASS E: EARTHWORKS**

**LANDSCAPING**

**TURFING AND SEEDING**

| | | UNIT | L | P | M | £ NETT |
|---|---|---|---|---|---|---|
| Turfing; PC £1.07/m2 | | | | | | |
| Horizontal | hr | m2 | 0.12 | - | | |
| | £ | | **1.49** | - | **1.07** | **2.56** |
| 10 - 45 to horizontal | hr | m2 | 0.14 | - | | |
| | £ | | **1.74** | - | **1.07** | **2.81** |
| 45 - 90 to horizontal (including pegging down) | hr | m2 | 0.19 | - | | |
| | £ | | **2.36** | - | **1.41** | **3.77** |
| hydraulic mulch grass seeding | | | | | | |
| Horizontal | hr | m2 | 0.005 | 0.005 | | |
| | £ | | **0.06** | **0.05** | **0.05** | **0.16** |
| 10 - 45 to horizontal | hr | m2 | 0.005 | 0.005 | | |
| | £ | | **0.06** | **0.05** | **0.05** | **0.16** |
| 45 - 90 to horizontal | hr | m2 | 0.007 | 0.005 | | |
| | £ | | **0.09** | **0.05** | **0.05** | **0.19** |
| seeding; PC £2.80/kg @ 50g/m2 supply and sow in two operations | | | | | | |
| Horizontal | hr | m2 | 0.01 | - | | |
| | £ | | **0.12** | - | **0.14** | **0.26** |
| 10 - 45 to horizontal | hr | m2 | 0.015 | - | | |
| | £ | | **0.19** | - | **0.14** | **0.33** |
| 45 - 90 to horizontal | hr | m2 | 0.02 | - | | |
| | £ | | **0.25** | - | **0.14** | **0.39** |

**CLASS E: EARTHWORKS**

**LANDSCAPING - continued**

**PLANTING TREES**

The cost of planting semi-mature trees will depend on the size and species, and on the access to the site for tree handling machines. Prices should be obtained for individual trees and planting. (PC = Prime Cost. Total rate = Labour, Plant and Materials, including PC).

| | PC £ | MATERIAL £ | UNIT | TOTAL RATE £ |
|---|---|---|---|---|
| Break up subsoil in treepit to a depth of 200m | - | - | each | 0.72 |
| Supply and plant Extra Heavy Standard trees in prepared pit and backfill with excavated topsoil, minimum 600 mm deep | 22.50 | 24.35 | each | 43.93 |
| Supply and plant Heavy Standard tree in prepared pit and backfill with excavated topsoil, minimum 600 mm deep | 16.85 | 18.25 | each | 29.42 |
| Supply and plant Selected Standard tree in prepared pit and backfill with excavated topsoil, minimum 600 mm deep | 8.90 | 9.70 | each | 19.72 |
| Supply and plant Standard tree in prepared pit and backfill with excavated topsoil, minimum 600 mm deep | 6.65 | 7.20 | each | 13.38 |
| Supply and plant Light Standard tree in prepared pit and backfill with excavated topsoil, minimum 600 mm deep | 5.15 | 5.52 | each | 9.70 |
| Extra for filling with topsoil from spoil heap not exceeding 100 m travel | - | - | m3 | 12.02 |
| Extra for filling with imported topsoil | 11.70 | - | m3 | 23.22 |
| Extra for incorporating manure or compost into topsoil 1 m3:5 m3 | - -| - - | - | +60% |
| Extra for supply and fix tree stake 3.0m long x 100 mm diam. driven 500 mm into firm ground and trimmed to approved height, including two tree ties to approved pattern | 5.60 | 6.40 | each | 9.88 |
| As above but with tree stakes 2.4m x 100mm diameter | 4.40 | 5.23 | each | 7.73 |
| As above but with two stakes 3.0m x 100mm diameter | 11.20 | 12.13 | each | 19.28 |
| As above but with two stakes 2.4m x 100mm diameter | 8.80 | 9.67 | each | 14.26 |

**CLASS E: EARTHWORKS**

**LANDSCAPING - continued**

**PLANTING TREES - continued**

| | PC £ | MATERIAL £ | UNIT | TOTAL RATE £ |
|---|---|---|---|---|
| Supply and support tree with three collars and wire guys; including pickets: | | | | |
| galvanised steel 50mm x 600mm | - | 19.69 | each | 45.26 |
| hardwood 75mm x 600mm | - | 19.96 | each | 48.15 |
| Extra for land drain to tree pit 100mm diameter clayware agricultural drain including excavating drain trench, laying pipes and backfilling | - | 2.63 | each | 5.90 |
| Supply and fix standard steel tree guard | - | 15.62 | each | 21.25 |

PLANTING HEDGES

| | PC £ | Hand 100 M £ | M £ |
|---|---|---|---|
| Set out, nick out and excavate trench minimum 400mm deep and break up subsoil to minimum depth 300mm. | - | 310.00 | 3.15 |
| Supply the plant hedging plants including backfill with excavated topsoil: | price per 100 | | |
| single row plants at 150 mm centres | 27.00 | 335.95 | 4.02 |
| 230 mm centres | 27.00 | 219.35 | 2.51 |
| 300 mm centres | 27.00 | 168.00 | 2.01 |
| 375 mm centres | 27.00 | 134.80 | 1.51 |
| 450 mm centres | 27.00 | 111.30 | 1.01 |
| double row plants at 230 mm centres | 27.00 | 437.65 | 4.53 |
| 300 mm centres | 27.00 | 336.00 | 4.02 |
| 375 mm centres | 27.00 | 267.50 | 3.02 |
| 450 mm centres | 27.00 | 223.65 | 2.51 |
| Extra for incorporating manure at 1m3 per 30m | m3 2.25 | | 3.83 |

**CLASS E: EARTHWORKS**

**LANDSCAPING - continued**

**PLANTING SHRUBS**

| | PC | MATERIAL | RATE 100 | RATE EACH |
|---|---|---|---|---|
| | £ | £ | £ | £ |
| Form planting hole in cultivated area measured elsewhere, supply and plant specified shrub and backfill with excavated material: | 100 rate each | | | |
| shrub 300mm high | 2.07 | - | 235.40 | 2.35 |
| shrub 600mm high | 1.33 | - | 154.10 | 1.55 |
| shrub 900mm high | 1.64 | - | 188.30 | 1.89 |
| shrub 1.0m + high | 2.20 | - | 246.10 | 2.46 |
| Extra for supply and fix shrub stake 1.5m x 75mm including two ties | each 3.32 | 4.10 | - | 6.37 |
| Extra for incorporating manure | m3 | | m3 | m3 |
| at 1m3/10m2 | 2.25 | | 275.75 | 2.80 |
| at 1m3/15m2 | 2.25 | | 266.06 | 2.69 |
| at 1m3/20m2 | 2.25 | | 252.51 | 2.54 |
| Extra for incorporating fertiliser | kg | | 100 m2 | m2 |
| at 35g/m2 | 1.84 | | 17.59 | 0.19 |
| at 50g/m2 | 1.84 | | 21.25 | 0.22 |
| at 70g/m2 | 1.84 | | 26.82 | 0.28 |
| at 100g/m2 | 1.84 | | 33.39 | 0.35 |

**PLANTING HERBACEOUS AND GROUNDCOVER PLANTS**

| | PC | MATERIAL | RATE 100 | RATE EACH |
|---|---|---|---|---|
| Form planting hole in cultivated area measured elsewhere, supply and plant specified herbaceous plant and backfill with excavated material: | 100 | | 100m2 | m3 |
| 1 - 3 plants per m2 | 129.00 | | 434.90 | 4.42 |
| 4 - 11 plants per m2 | 129.00 | | 1534.90 | 15.47 |
| 12 - 25 plants per m2 | 129.00 | | 3418.50 | 34.34 |
| 26 - 35 plants per m2 | 129.00 | | 4722.30 | 47.70 |
| 36 - 50 plants per m2 | 129.00 | | 6656.80 | 67.20 |

**CLASS E: EARTHWORKS**

**LANDSCAPING - continued**

**PLANTING HERBACEOUS AND GROUNDCOVER PLANTS - continued**

| | PC £ | RATE 100 £ | EACH £ |
|---|---|---|---|
| Extra for incorporating manure | m3 | 100m2 | m2 |
| at 1 m3/10m2 | 2.25 | 275.95 | 2.80 |
| at 1 m3/15m2 | 2.25 | 266.06 | 2.69 |
| at 1 m3/20m2 | 2.25 | 252.51 | 2.59 |
| Extra for incorporating fertilizer | kg | 100m2 | m2 |
| at 35g/m2 | 1.84 | 17.59 | 0.19 |
| at 50g/m2 | 1.84 | 21.25 | 0.22 |
| at 70g/m2 | 1.84 | 26.82 | 0.28 |
| at 100g/m2 | 1.84 | 33.39 | 0.35 |
| Supply and fix Plant Support Netting on 50mm diameter stakes 750mm long driven into ground at 1.5m centres green extruded plastic mesh 125mm sq | roll 300 x 1.5m 37.85 | 1.80 | m2 |

**PLANTING BULBS**

| | PC £ | RATE 100 £ | EACH £ |
|---|---|---|---|
| Form planting hole in cultivated area measured elsewhere, supply and plant bulbs and backfill with excavated material: | 100 | 100 | each |
| small | 11.15 | 30.10 | 0.32 |
| medium | 17.80 | 40.00 | 0.42 |
| large | 21.20 | 49.00 | 0.51 |
| Supply and plant bulbs in grassed area using bulb planter and backfill with screened topsoil or peat and cut turf plug: | | | |
| small | 11.15 | 29.15 | 0.31 |
| medium | 17.80 | 38.80 | 0.42 |
| large | 21.20 | 48.50 | 0.51 |
| Extra to items above for planting in prefabricated or in-situ planters | | | |
| Shrubs | - | - | +20% |
| Herbaceous plants | - | - | +20% |
| Bulbs | - | - | +15% |

**CLASS E: EARTHWORKS**

**LANDSCAPING - continued**

**OPERATIONS AFTER PLANTING (MAINTENANCE)**

| | | HAND | |
|---|---|---|---|
| | PC | 100M | M |
| | £ | £ | £ |
| Initial cut back to shrubs and hedge plants, clear away all cuttings | | 31.00 | 0.32 |
| Protect planted areas with windbreak fencing 1m high fixed to stakes 1.50m x 50mm diameter at 2m centres and clear away on completion | m<br>2.25 | 516.75 | 5.25 |
| Cut out dead and diseased wood, prune trim and cut to shape treat with sealant | | | |
| shrubs up to 1.0m high | each | 0.43 | |
| 1-2m high | each | 0.95 | |
| 2-3m high | each | 1.69 | |
| Cut and trim hedge to specified profile clean away cuttings | | | |
| 2.0m high | m | 0.95 | |
| | 100m | 93.36 | |
| 2.4m high | m | 1.65 | |
| | 100m | 165.05 | |
| Trim field hedge to specified height and shape | | | |
| using flail | 100m | 310.00 | |
| using cutting bar | 100m | 465.00 | |
| Weed and handfork planted areas including removing and dumping weeds and debris on site | m2 | 0.64 | |
| | 100m2 | 63.00 | |
| Supply and apply selective weed killer at: | | | |
| 35g/m2 | m2 | 0.22 | |
| 70g/m2 | m2 | 0.34 | |
| 100g/m2 | m2 | 0.44 | |

**CLASS F: IN-SITU CONCRETE**

The unit rates in this Class are based on nett measurements and appropriate adjustments should be made for unmeasured excess (e.g. additional blinding thickness as a result of ground conditions).

The unit rates for the provision of concrete are based on ready mixed concrete in full loads delivered to site within 5 miles (8km) of the concrete mixing plant and include an allowance for wastage prior to placing.

This section assumes optimum outputs of an efficiently controlled pour with no delays caused by out of sequence working and no abnormal conditions that would affect continuity of work.

LABOUR based on Gang K

| | | | |
|---|---|---|---|
| Total hourly cost | = | £ | 46.93 |

PLANT

| | | | |
|---|---|---|---|
| 22RB Crane. (Part time) | | | 7.11 |
| Audex Skip. (Part time) | | | 0.84 |
| Compressor 10m3/min. | | | 4.83 |
| Air vibrators | | | 2.52 |
| Total hourly cost | = | £ | 15.30 |

Wastage allowances (class F and G)

5% onto delivered readymixed concrete

5 - 7% onto made-up formwork

1% onto bent reinforcement

2.5% onto straight reinforcement cut and bent on site

15% onto fabric reinforcement (for laps and waste)

Bent reinforcement manhours do not include for bending cutting or shaping reinforcement (refer to separate table for this information page 215)

**CLASS F: IN-SITU CONCRETE**

| | | UNIT | L | P | M | £ NETT |
|---|---|---|---|---|---|---|
| The following costs do not reflect in the rates and should be considered separately. | | | | | | |
| Delivery to site for each additional mile (1.6km) from the concrete plant | £ | m3 | - | - | - | 0.55 |
| Mix design | £ | per trial mix | - | - | - | 125.00 |
| Part loads | £ | per m3 below full load | - | - | - | 8.50 |
| Waiting time (after 40 minutes discharge time) | £ | hr | - | - | - | 20.00 |
| Making and testing concrete cube | hr<br>£ | per m3 | 2.75 | - | 1.75 | 4.50 |
| Pumping from ready mix truck to point of placing average 25m3/hour | hr<br>£ | m3 | -<br>0.98 | -<br>1.40 | - | 2.38 |
| 45m3/hour | hr<br>£ | m3 | -<br>0.56 | -<br>1.20 | - | 1.76 |

**PROVISION OF CONCRETE; ORDINARY PRESCRIBED MIX - CEMENT TO BS12**

| | Unit | Aggregate size 10mm £ | 20mm £ | 40mm £ |
|---|---|---|---|---|
| Grade C7.5P | m3 | | 42.24 | 41.75 |
| Grade C10P | m3 | | 44.54 | 44.00 |
| Grade C15P | m3 | | 45.44 | 45.00 |
| Grade C20P | m3 | 50.66 | 47.18 | 45.74 |
| Grade C25P | m3 | 52.37 | 50.06 | 48.33 |
| Grade C30P | m3 | 55.84 | 52.09 | 50.06 |

**CLASS F: IN-SITU CONCRETE**

**PROVISION OF CONCRETE; ORDINARY PRESCRIBED MIX - CEMENT TO BS12 - continued**

| | Unit | £ |
|---|---|---|
| ADD to the above for using sulphate resisting cement to BS4027 | | |
| Grade C7.5P | m3 | 0.94 |
| Grade C10P | m3 | 1.03 |
| Grade C15P | m3 | 1.25 |
| Grade C20P | m3 | 1.68 |
| Grade C25P | m3 | 2.08 |
| Grade C30P | m3 | 2.23 |

**PROVISION OF CONCRETE; DESIGNED MIX - CEMENT TO BS12**

| | Unit | Aggregate size 10mm £ | Aggregate size 20mm £ |
|---|---|---|---|
| Grade C30 | m3 | 48.05 | 45.74 |
| Grade C40 | m3 | 51.80 | 49.49 |
| Grade C50 | m3 | 56.42 | 54.11 |
| Grade C60 | m3 | 61.05 | 58.74 |
| ADD to the above for using sulphate resisting cement to BS4027 | | | |
| Grade C30 | m3 | | 2.10 |
| Grade C40 | m3 | | 2.20 |
| Grade C50 | m3 | | 2.24 |
| Grade C60 | m3 | | 2.35 |

**CLASS F: IN-SITU CONCRETE**

**PLACING OF CONCRETE; MASS**

| | | UNIT | L | P | M | £ NETT |
|---|---|---|---|---|---|---|
| Blinding; thickness | | | | | | |
| ne 150mm | hr | m3 | 0.18 | 0.01 | - | |
| | £ | | 8.45 | 0.15 | - | 8.60 |
| 150 - 300mm | hr | m3 | 0.16 | 0.03 | | |
| | £ | | 7.51 | 0.46 | - | 7.97 |
| 300 - 500mm | hr | m3 | 0.14 | 0.05 | | |
| | £ | | 6.57 | 0.77 | - | 7.34 |
| Bases, footings, pile caps and ground beams; thickness | | | | | | |
| ne 150mm | hr | m3 | 0.20 | 0.20 | | |
| | £ | | 9.39 | 3.06 | - | 12.45 |
| 150 - 300mm | hr | m3 | 0.17 | 0.17 | | |
| | £ | | 7.98 | 2.60 | - | 10.58 |
| 300 - 500mm | hr | m3 | 0.15 | 0.15 | | |
| | £ | | 7.04 | 2.30 | - | 9.34 |
| exceeding 500mm | hr | m3 | 0.14 | 0.14 | | |
| | £ | | 6.57 | 2.14 | - | 8.71 |
| ADD to the above for placing against an excavated surface | hr | m3 | - | - | | |
| | £ | | - | - | - | 1.44 |
| Walls; thickness | | | | | | |
| ne 150mm | hr | m3 | 0.21 | 0.21 | | |
| | £ | | 9.86 | 3.21 | - | 13.07 |
| 150 - 300mm | hr | m3 | 0.15 | 0.15 | | |
| | £ | | 7.04 | 2.30 | - | 9.34 |
| 300 - 500mm | hr | m3 | 0.13 | 0.13 | | |
| | £ | | 6.10 | 1.99 | - | 8.09 |
| exceeding 500mm | hr | m3 | 0.12 | 0.12 | | |
| | £ | | 5.63 | 1.84 | - | 7.47 |
| ADD to the above for placing against an excavated surface | hr | m3 | - | - | | |
| | £ | | - | - | - | 1.65 |

**CLASS F: IN-SITU CONCRETE**

**PLACING OF CONCRETE; MASS - continued**

| | | UNIT | L | P | M | £ NETT |
|---|---|---|---|---|---|---|
| **Other concrete forms** | | | | | | |
| plinth 1000 x 1000 x 600mm | hr | m3 | 0.33 | 0.33 | | |
| | £ | | 15.49 | 5.05 | - | 20.54 |
| plinth 1500 x 1500 x 750mm | hr | m3 | 0.25 | 0.25 | | |
| | £ | | 11.73 | 3.83 | - | 15.56 |
| plinth 2000 x 2000 x 600mm | hr | m3 | 0.20 | 0.20 | | |
| | £ | | 9.39 | 3.06 | - | 12.45 |
| surround to precast concrete manhole chambers 200mm thick | hr | m3 | 0.29 | 0.29 | | |
| | £ | | 13.61 | 4.44 | - | 18.05 |

**PLACING OF CONCRETE; REINFORCED**

| | | UNIT | L | P | M | £ NETT |
|---|---|---|---|---|---|---|
| **Bases, footings, pile caps and ground beams; thickness** | | | | | | |
| ne 150mm | hr | m3 | 0.21 | 0.21 | | |
| | £ | | 9.86 | 3.21 | - | 13.07 |
| 150 - 300mm | hr | m3 | 0.18 | 0.18 | | |
| | £ | | 8.45 | 2.75 | - | 11.20 |
| 300 - 500mm | hr | m3 | 0.16 | 0.16 | | |
| | £ | | 7.51 | 2.45 | - | 9.96 |
| exceeding 500mm | hr | m3 | 0.15 | 0.15 | | |
| | £ | | 7.04 | 2.30 | - | 9.34 |
| **Suspended slabs; thickness** | | | | | | |
| ne 150mm | hr | m3 | 0.27 | 0.27 | | |
| | £ | | 12.67 | 4.13 | - | 16.80 |
| 150 - 300mm | hr | m3 | 0.21 | 0.21 | | |
| | £ | | 9.86 | 3.21 | - | 13.07 |
| 300 - 500mm | hr | m3 | 0.19 | 0.19 | | |
| | £ | | 8.92 | 2.91 | - | 11.83 |
| exceeding 500mm | hr | m3 | 0.19 | 0.19 | | |
| | £ | | 8.92 | 2.91 | - | 11.83 |

**CLASS F: IN-SITU CONCRETE**

**PLACING OF CONCRETE; REINFORCED - continued**

| | | UNIT | L | P | M | £ NETT |
|---|---|---|---|---|---|---|
| Walls; thickness | | | | | | |
| ne 150mm | hr | m3 | 0.29 | 0.29 | | |
| | £ | | 13.61 | 4.44 | - | 18.05 |
| 150 - 300mm | hr | m3 | 0.22 | 0.22 | | |
| | £ | | 10.32 | 3.37 | - | 13.69 |
| 300 - 500mm | hr | m3 | 0.20 | 0.20 | | |
| | £ | | 9.39 | 3.06 | - | 12.45 |
| exceeding 500mm | hr | m3 | 0.20 | 0.20 | | |
| | £ | | 9.39 | 3.06 | - | 12.45 |
| Columns and piers; cross-sectional area | | | | | | |
| ne 0.03m2 | hr | m3 | 0.50 | 0.50 | | |
| | £ | | 23.47 | 7.65 | - | 31.12 |
| 0.03 - 0.10m2 | hr | m3 | 0.40 | 0.40 | | |
| | £ | | 18.77 | 6.12 | - | 24.89 |
| 0.10 - 0.25m2 | hr | m3 | 0.35 | 0.35 | | |
| | £ | | 16.43 | 5.36 | - | 21.79 |
| 0.25 - 1.00m2 | hr | m3 | 0.35 | 0.35 | | |
| | £ | | 16.43 | 5.36 | - | 21.79 |
| exceeding 1m2 | hr | m3 | 0.28 | 0.28 | | |
| | £ | | 13.14 | 4.28 | - | 17.42 |
| Beams; cross-sectional area | | | | | | |
| ne 0.03m2 | hr | m3 | 0.50 | 0.50 | | |
| | £ | | 23.47 | 7.65 | - | 31.12 |
| 0.03 - 0.10m2 | hr | m3 | 0.40 | 0.40 | | |
| | £ | | 18.77 | 6.12 | - | 24.89 |
| 0.10 - 0.25m2 | hr | m3 | 0.35 | 0.35 | | |
| | £ | | 16.43 | 5.36 | - | 21.79 |
| 0.25 - 1.00m2 | hr | m3 | 0.35 | 0.35 | | |
| | £ | | 16.43 | 5.36 | - | 21.79 |
| exceeding 1m2 | hr | m3 | 0.28 | 0.28 | | |
| | £ | | 13.14 | 4.28 | - | 17.42 |

**CLASS F: IN-SITU CONCRETE**

**PLACING OF CONCRETE; REINFORCED - continued**

| | | UNIT | L | P | M | £ NETT |
|---|---|---|---|---|---|---|
| Casing to metal sections; cross-sectional area | | | | | | |
| ne 0.03m2 | hr | m3 | 0.47 | 0.47 | | |
| | £ | | 22.06 | 7.19 | - | 29.25 |
| 0.03 - 0.10m2 | hr | m3 | 0.47 | 0.47 | | |
| | £ | | 22.06 | 7.19 | - | 29.25 |
| 0.10 - 0.25m2 | hr | m3 | 0.40 | 0.40 | | |
| | £ | | 18.77 | 6.12 | - | 24.89 |
| 0.25 - 1.00m2 | hr | m3 | 0.40 | 0.40 | | |
| | £ | | 18.77 | 6.12 | - | 24.89 |
| exceeding 1m2 | hr | m3 | 0.35 | 0.35 | | |
| | £ | | 16.43 | 5.36 | - | 21.79 |

**PLACING OF CONCRETE; PRESTRESSED**

| | | UNIT | L | P | M | £ NETT |
|---|---|---|---|---|---|---|
| Suspended slabs; thickness | | | | | | |
| ne 150mm | hr | m3 | 0.28 | 0.28 | | |
| | £ | | 13.14 | 4.28 | - | 17.42 |
| 150 - 300mm | hr | m3 | 0.22 | 0.22 | | |
| | £ | | 10.32 | 3.37 | - | 13.69 |
| 300 - 500mm | hr | m3 | 0.20 | 0.20 | | |
| | £ | | 9.39 | 3.06 | - | 12.45 |
| exceeding 500mm | hr | m3 | 0.19 | 0.19 | | |
| | £ | | 8.92 | 2.91 | - | 11.83 |
| Beams; cross-sectional area | | | | | | |
| ne 0.03m2 | hr | m3 | 0.50 | 0.50 | | |
| | £ | | 23.47 | 7.65 | - | 31.12 |
| 0.03 - 0.10m2 | hr | m3 | 0.40 | 0.40 | | |
| | £ | | 18.77 | 6.12 | - | 24.89 |
| 0.10 - 0.25m2 | hr | m3 | 0.35 | 0.35 | | |
| | £ | | 16.43 | 5.36 | - | 21.79 |
| 0.25 - 1.00m2 | hr | m3 | 0.35 | 0.35 | | |
| | £ | | 16.43 | 5.36 | - | 21.79 |
| exceeding 1m2 | hr | m3 | 0.28 | 0.28 | | |
| | £ | | 13.14 | 4.28 | - | 17.42 |

**CLASS G: CONCRETE ANCILLARIES**

The following unit costs do not include for formwork outside the payline and are based on an optimum of a minimum 8 uses.
ADD to formwork costs generally for number of uses.

| | | |
|---|---|---|
| 1 | 90 - 170% | allow for waste + 7% |
| 2 | 50 - 80% | allow for waste + 7% |
| 3 | 15 - 30% | allow for waste + 6% |
| 6 | 5 - 10% | allow for waste + 6% |
| 8 | 0 | allow for waste + 5% |
| 10 - | 5 - 7% | allow for waste + 5% |

Formwork materials include for shutter, bracing, ties, support, kentledge and all consumables.

LABOUR based on Gang C

Large Area

| | | | |
|---|---|---|---|
| Total hourly cost | = | £ | 35.21 * |

Small Area

| | | | |
|---|---|---|---|
| Total hourly cost | = | £ | 29.20 |

PLANT

| | | | |
|---|---|---|---|
| 20 tonne Mobile Crane (part time) | | | 4.04 |
| Saw Bench | | | 1.01 |
| Small Power Tools | | | 0.50 |
| Total hourly cost | = | £ | 5.55 |

Note: sloping formwork is 10 - 85 degrees to the vertical
battered formwork is 0 - 10 degress to the vertical

Please refer to tables of multipliers preceeding section F (page 190) which are equally applicable to this section of work.

**CLASS G: CONCRETE ANCILLARIES**

**FORMWORK; ROUGH FINISH**

| | | UNIT | L | P | M | £ NETT |
|---|---|---|---|---|---|---|
| Plane horizontal; width | | | | | | |
| ne 0.1m | hr | m | 0.10 | 0.10 | | |
| | £ | | 2.92 | 0.56 | 0.50 | 3.98 |
| 0.1 - 0.20m | hr | m | 0.18 | 0.13 | | |
| | £ | | 5.26 | 0.72 | 0.54 | 6.52 |
| * 0.2 - 0.40m | hr | m2 | 0.41 | 0.40 | | |
| | £ | | 14.44 | 2.22 | 2.80 | 19.46 |
| * 0.4 - 1.22m | hr | m2 | 0.41 | 0.40 | | |
| | £ | | 14.44 | 2.22 | 2.80 | 19.46 |
| * exceeding 1.22m | hr | m2 | 0.38 | 0.37 | | |
| | £ | | 13.38 | 2.05 | 2.80 | 18.23 |
| Plane sloping; width | | | | | | |
| ne 0.1m | hr | m | 0.11 | 0.11 | | |
| | £ | | 3.21 | 0.61 | 0.50 | 4.32 |
| 0.1 - 0.20m | hr | m | 0.20 | 0.15 | | |
| | £ | | 5.84 | 0.83 | 0.57 | 7.24 |
| * 0.2 - 0.40m | hr | m2 | 0.43 | 0.40 | | |
| | £ | | 15.14 | 2.22 | 2.94 | 20.30 |
| * 0.4 - 1.22m | hr | m2 | 0.43 | 0.40 | | |
| | £ | | 15.14 | 2.22 | 2.94 | 20.30 |
| * exceeding 1.22m | hr | m2 | 0.38 | 0.37 | | |
| | £ | | 13.38 | 2.05 | 2.51 | 17.94 |
| Plane battered; width | | | | | | |
| ne 0.1m | hr | m | 0.12 | 0.12 | | |
| | £ | | 3.50 | 0.67 | 0.70 | 4.87 |
| 0.1 - 0.20m | hr | m | 0.20 | 0.19 | | |
| | £ | | 5.84 | 1.05 | 0.70 | 7.59 |
| * 0.2 - 0.40m | hr | m2 | 0.43 | 0.42 | | |
| | £ | | 15.14 | 2.33 | 4.21 | 21.68 |
| * 0.4 - 1.22m | hr | m2 | 0.43 | 0.42 | | |
| | £ | | 15.14 | 2.33 | 4.21 | 21.68 |
| * exceeding 1.22m | hr | m2 | 0.40 | 0.47 | | |
| | £ | | 14.08 | 2.61 | 4.31 | 21.00 |

**CLASS G: CONCRETE ANCILLARIES**

**FORMWORK; ROUGH FINISH - continued**

| | | UNIT | L | P | M | £ NETT |
|---|---|---|---|---|---|---|
| Plane vertical; width | | | | | | |
| ne 0.1m | hr | m | 0.12 | 0.12 | | |
| | £ | | 3.50 | 0.67 | 0.54 | 4.71 |
| 0.1 - 0.20m | hr | m | 0.19 | 0.19 | | |
| | £ | | 5.55 | 1.05 | 0.67 | 7.27 |
| * 0.2 - 0.40m | hr | m2 | 0.72 | 0.72 | | |
| | £ | | 25.35 | 4.00 | 3.32 | 32.67 |
| * 0.4 - 1.22m | hr | m2 | 0.51 | 0.47 | | |
| | £ | | 17.96 | 2.61 | 3.32 | 23.89 |
| * exceeding 1.22m | hr | m2 | 0.47 | 0.47 | | |
| | £ | | 16.55 | 2.61 | 3.21 | 22.37 |
| Curved to one radius in one plane, 0.5m radius; width | | | | | | |
| ne 0.1m | hr | m | 0.19 | 0.19 | | |
| | £ | | 5.55 | 1.05 | 0.54 | 7.14 |
| 0.1 - 0.20m | hr | m | 0.25 | 0.25 | | |
| | £ | | 7.30 | 1.39 | 0.69 | 9.38 |
| * 0.2 - 0.40m | hr | m2 | 0.95 | 0.90 | | |
| | £ | | 33.45 | 5.00 | 3.63 | 42.08 |
| * 0.4 - 1.22m | hr | m2 | 0.75 | 0.72 | | |
| | £ | | 26.41 | 4.00 | 3.59 | 34.00 |
| * exceeding 1.22m | hr | m2 | 0.65 | 0.60 | | |
| | £ | | 22.89 | 3.33 | 3.52 | 29.74 |
| Curved to one radius in one plane, 1m radius or greater; width | | | | | | |
| ne 0.1m | hr | m | 0.19 | 0.19 | | |
| | £ | | 5.55 | 1.05 | 0.54 | 7.14 |
| 0.1 - 0.20m | hr | m | 0.25 | 0.25 | | |
| | £ | | 7.30 | 1.39 | 0.69 | 9.38 |
| * 0.2 - 0.40m | hr | m2 | 0.95 | 0.93 | | |
| | £ | | 33.45 | 5.16 | 3.63 | 42.24 |
| * 0.4 - 1.22m | hr | m2 | 0.72 | 0.70 | | |
| | £ | | 25.35 | 3.89 | 3.63 | 32.87 |
| * exceeding 1.22m | hr | m2 | 0.60 | 0.57 | | |
| | £ | | 21.13 | 3.16 | 3.57 | 27.86 |

**CLASS G: CONCRETE ANCILLARIES**

**FORMWORK; ROUGH FINISH - continued**

| | | UNIT | L | P | M | £ NETT |
|---|---|---|---|---|---|---|
| For concrete components of constant cross-section | | | | | | |
| beams; 200 x 200mm | hr | m | 0.48 | 0.40 | | |
| | £ | | 14.02 | 2.22 | 1.00 | 17.24 |
| beams; 500 x 500mm | hr | m | 0.55 | 0.45 | | |
| | £ | | 16.06 | 2.50 | 2.94 | 21.50 |
| beams; 500 x 800mm | hr | m | 0.67 | 0.60 | | |
| | £ | | 19.56 | 3.33 | 3.63 | 26.52 |
| columns; 200 x 200mm | hr | m | 0.55 | 0.40 | | |
| | £ | | 16.06 | 2.22 | 1.00 | 19.28 |
| columns; 300 x 300mm | hr | m | 0.55 | 0.55 | | |
| | £ | | 16.06 | 3.05 | 2.80 | 21.91 |
| columns; 300 x 500mm | hr | m | 0.62 | 0.62 | | |
| | £ | | 18.10 | 3.44 | 3.23 | 24.77 |
| * walls; 1.0m high thickness 250mm | hr | m | 1.35 | 1.35 | | |
| | £ | | 47.53 | 7.49 | 4.09 | 59.11 |
| * walls; 1.5m high thickness 300mm | hr | m | 1.10 | 1.10 | | |
| | £ | | 38.73 | 6.11 | 3.23 | 48.07 |
| * box culvert; 2 x 2m internally and wall thickness 300mm | hr | m | 4.50 | 3.95 | | |
| | £ | | 158.45 | 21.92 | 38.45 | 218.82 |
| projections (100mm deep) | hr | m | 0.10 | 0.10 | | |
| | £ | | 2.92 | 0.56 | 2.20 | 5.68 |
| intrusions (100mm deep) | hr | m | 0.10 | 0.10 | | |
| | £ | | 2.92 | 0.56 | 1.00 | 4.48 |
| For voids (void former) | | | | | | |
| small void; depth ne 0.5m | hr | nr | 0.07 | - | | |
| | £ | | 2.04 | - | 2.20 | 4.24 |
| small void; depth 0.5 - 1m | hr | nr | 0.12 | - | | |
| | £ | | 3.50 | - | 4.26 | 7.76 |
| small void; depth 1.0 - 2m | hr | nr | 0.16 | - | | |
| | £ | | 4.67 | - | 8.32 | 12.99 |
| large void; depth ne 0.5m | hr | nr | 0.14 | - | | |
| | £ | | 4.09 | - | 4.68 | 8.77 |

**CLASS G: CONCRETE ANCILLARIES**

**FORMWORK; ROUGH FINISH - continued**

| | | UNIT | L | P | M | £ NETT |
|---|---|---|---|---|---|---|
| large void; depth 0.5 - 1m | hr | nr | 0.33 | - | | |
| | £ | | 9.64 | - | 9.36 | 19.00 |
| large void; depth 1.0 - 2m | hr | nr | 0.58 | - | | |
| | £ | | 16.94 | - | 16.64 | 33.58 |
| ADD for craneage and rub up (where required) | £ | m2 | 1.16 | 1.26 | 0.02 | 2.44 |
| ADD for craneage and rub up (where required) | £ | m | 3.88 | 4.20 | 0.08 | 8.16 |

**CLASS G: CONCRETE ANCILLARIES**

**FORMWORK; FAIR FINISH**

| | | UNIT | L | P | M | £ NETT |
|---|---|---|---|---|---|---|
| Plane horizontal; width | | | | | | |
| ne 0.1m | hr | m | 0.10 | 0.10 | | |
| | £ | | 2.92 | 0.56 | 0.67 | 4.15 |
| 0.1 - 0.20m | hr | m | 0.18 | 0.13 | | |
| | £ | | 5.26 | 0.72 | 0.70 | 6.68 |
| * 0.2 - 0.40m | hr | m2 | 0.41 | 0.40 | | |
| | £ | | 14.44 | 2.22 | 3.95 | 20.61 |
| * 0.4 - 1.22m | hr | m2 | 0.41 | 0.40 | | |
| | £ | | 14.44 | 2.22 | 3.64 | 20.30 |
| * exceeding 1.22m | hr | m2 | 0.41 | 0.37 | | |
| | £ | | 14.44 | 2.05 | 3.64 | 20.13 |
| Plane sloping; width | | | | | | |
| ne 0.1m | hr | m | 0.11 | 0.11 | | |
| | £ | | 3.21 | 0.61 | 0.67 | 4.49 |
| 0.1 - 0.20m | hr | m | 0.20 | 0.15 | | |
| | £ | | 5.84 | 0.83 | 0.75 | 7.42 |
| * 0.2 - 0.40m | hr | m2 | 0.43 | 0.40 | | |
| | £ | | 15.14 | 2.22 | 4.13 | 21.49 |
| * 0.4 - 1.22m | hr | m2 | 0.43 | 0.40 | | |
| | £ | | 15.14 | 2.22 | 3.82 | 21.18 |
| * exceeding 1.22m | hr | m2 | 0.38 | 0.37 | | |
| | £ | | 13.38 | 2.05 | 3.82 | 19.25 |
| Plane battered; width | | | | | | |
| ne 0.1m | hr | m | 0.12 | 0.12 | | |
| | £ | | 3.50 | 0.67 | 0.70 | 4.87 |
| 0.1 - 0.20m | hr | m | 0.20 | 0.19 | | |
| | £ | | 5.84 | 1.05 | 0.81 | 7.70 |
| * 0.2 - 0.40m | hr | m2 | 0.43 | 0.42 | | |
| | £ | | 15.14 | 2.33 | 4.52 | 21.99 |
| * 0.4 - 1.22m | hr | m2 | 0.43 | 0.42 | | |
| | £ | | 15.14 | 2.33 | 4.21 | 21.68 |
| * exceeding 1.22m | hr | m2 | 0.40 | 0.40 | | |
| | £ | | 14.08 | 2.22 | 4.21 | 20.51 |

**CLASS G: CONCRETE ANCILLARIES**

**FORMWORK; FAIR FINISH - continued**

| | | UNIT | L | P | M | £ NETT |
|---|---|---|---|---|---|---|
| Plane vertical; width | | | | | | |
| ne 0.1m | hr | m | 0.12 | 0.12 | | |
| | £ | | 3.50 | 0.67 | 0.70 | 4.87 |
| 0.1 - 0.20m | hr | m | 0.20 | 0.19 | | |
| | £ | | 5.84 | 1.05 | 0.88 | 7.77 |
| * 0.2 - 0.40m | hr | m2 | 0.72 | 0.72 | | |
| | £ | | 25.35 | 4.00 | 4.41 | 33.76 |
| * 0.4 - 1.22m | hr | m2 | 0.53 | 0.47 | | |
| | £ | | 18.66 | 2.61 | 4.31 | 25.58 |
| * exceeding 1.22m | hr | m2 | 0.48 | 0.48 | | |
| | £ | | 16.90 | 2.66 | 4.24 | 23.80 |
| Curved to one radius in one plane, 0.5m radius; width | | | | | | |
| ne 0.1m | hr | m | 0.20 | 0.19 | | |
| | £ | | 5.84 | 1.05 | 0.70 | 7.59 |
| 0.1 - 0.20m | hr | m | 0.26 | 0.25 | | |
| | £ | | 7.59 | 1.39 | 0.90 | 9.88 |
| * 0.2 - 0.40m | hr | m2 | 0.95 | 0.90 | | |
| | £ | | 33.45 | 5.00 | 5.25 | 43.70 |
| * 0.4 - 1.22m | hr | m2 | 0.77 | 0.72 | | |
| | £ | | 27.11 | 4.00 | 4.81 | 35.92 |
| * exceeding 1.22m | hr | m2 | 0.65 | 0.60 | | |
| | £ | | 22.89 | 3.33 | 4.76 | 30.98 |
| Curved to one radius in one plane, 1m radius or greater; width | | | | | | |
| ne 0.1m | hr | m | 0.20 | 0.19 | | |
| | £ | | 5.84 | 1.05 | 0.70 | 7.59 |
| 0.1 - 0.20m | hr | m | 0.26 | 0.25 | | |
| | £ | | 7.59 | 1.39 | 0.90 | 9.88 |
| * 0.2 - 0.40m | hr | m2 | 0.95 | 0.93 | | |
| | £ | | 33.45 | 5.16 | 4.76 | 43.37 |
| * 0.4 - 1.22m | hr | m2 | 0.74 | 0.70 | | |
| | £ | | 26.06 | 3.89 | 4.36 | 34.31 |
| * exceeding 1.22m | hr | m2 | 0.60 | 0.57 | | |
| | £ | | 21.13 | 3.16 | 4.29 | 28.58 |

**CLASS G: CONCRETE ANCILLARIES**

**FORMWORK; FAIR FINISH - continued**

| | | UNIT | L | P | M | £ NETT |
|---|---|---|---|---|---|---|
| For concrete components of constant cross-section | | | | | | |
| beams; 200 x 200mm | hr | m | 0.49 | 0.40 | | |
| | £ | | 14.31 | 2.22 | 1.31 | 17.84 |
| beams; 500 x 500mm | hr | m | 0.57 | 0.45 | | |
| | £ | | 16.64 | 2.50 | 3.82 | 22.96 |
| beams; 500 x 800mm | hr | m | 0.69 | 0.60 | | |
| | £ | | 20.15 | 3.33 | 4.71 | 28.19 |
| columns; 200 x 200mm | hr | m | 0.57 | 0.40 | | |
| | £ | | 16.64 | 2.22 | 1.31 | 20.17 |
| columns; 300 x 300mm | hr | m | 0.57 | 0.55 | | |
| | £ | | 16.64 | 3.05 | 3.64 | 23.33 |
| columns; 300 x 500mm | hr | m | 0.64 | 0.62 | | |
| | £ | | 18.69 | 3.44 | 4.23 | 26.36 |
| * walls; 1.0m high thickness 250mm | hr | m | 1.39 | 1.35 | | |
| | £ | | 48.94 | 7.49 | 5.33 | 61.76 |
| * walls; 1.5m high thickness 300mm | hr | m | 1.13 | 1.10 | | |
| | £ | | 39.79 | 6.11 | 4.23 | 50.13 |
| * box culvert; 2 x 2m internally wall thickness 300mm | hr | m | 4.64 | 3.95 | | |
| | £ | | 163.37 | 21.92 | 49.20 | 234.49 |
| projections (100mm deep) | hr | m | 0.10 | 0.10 | | |
| | £ | | 2.92 | 0.56 | 2.86 | 6.34 |
| intrusions (100mm deep) | hr | m | 0.10 | 0.10 | | |
| | £ | | 2.92 | 0.56 | 1.31 | 4.79 |
| For voids (void former) | | | | | | |
| small void; depth ne 0.5m | hr | nr | 0.07 | - | | |
| | £ | | 2.04 | - | 2.87 | 4.91 |
| small void; depth 0.5 - 1m | hr | nr | 0.12 | - | | |
| | £ | | 3.50 | - | 5.54 | 9.04 |
| small void; depth 1.0 - 2m | hr | nr | 0.16 | - | | |
| | £ | | 4.67 | - | 10.82 | 15.49 |
| large void; depth ne 0.5m | hr | nr | 0.14 | - | | |
| | £ | | 4.09 | - | 6.08 | 10.17 |

CLASS G: CONCRETE ANCILLARIES

FORMWORK; FAIR FINISH - continued

| | | UNIT | L | P | M | £ NETT |
|---|---|---|---|---|---|---|
| large void; depth 0.5 - 1m | hr | nr | 0.34 | - | | |
| | £ | | 9.93 | - | 12.17 | 22.10 |
| large void; depth 1.0 - 2m | hr | nr | 0.60 | - | | |
| | £ | | 17.52 | - | 21.63 | 39.15 |
| ADD for additional craneage and rub up (where required) | £ | m2 | 1.19 | 1.26 | 0.03 | 2.48 |
| ADD for additional craneage and rub up (where required) | £ | m | 4.00 | 4.20 | 0.10 | 8.30 |

**CLASS G: CONCRETE ANCILLARIES**

**FORMWORK; EXTRA SMOOTH FINISH**

| | | UNIT | L | P | M | £ NETT |
|---|---|---|---|---|---|---|
| Plane horizontal; width | | | | | | |
| ne 0.1m | hr | m | 0.10 | 0.10 | | |
| | £ | | 2.92 | 0.56 | 0.83 | 4.31 |
| 0.1 - 0.20m | hr | m | 0.18 | 0.13 | | |
| | £ | | 5.26 | 0.72 | 0.86 | 6.84 |
| * 0.2 - 0.40m | hr | m2 | 0.42 | 0.40 | | |
| | £ | | 14.79 | 2.22 | 4.48 | 21.49 |
| * 0.4 - 1.22m | hr | m2 | 0.42 | 0.40 | | |
| | £ | | 14.79 | 2.22 | 4.48 | 21.49 |
| * exceeding 1.22m | hr | m2 | 0.42 | 0.37 | | |
| | £ | | 14.79 | 2.05 | 4.16 | 21.00 |
| Plane sloping; width | | | | | | |
| ne 0.1m | hr | m | 0.11 | 0.11 | | |
| | £ | | 3.21 | 0.61 | 0.83 | 4.65 |
| 0.1 - 0.20m | hr | m | 0.20 | 0.15 | | |
| | £ | | 5.84 | 0.83 | 0.91 | 7.58 |
| * 0.2 - 0.40m | hr | m2 | 0.44 | 0.40 | | |
| | £ | | 15.49 | 2.22 | 4.70 | 22.41 |
| * 0.4 - 1.22m | hr | m2 | 0.44 | 0.40 | | |
| | £ | | 15.49 | 2.22 | 4.70 | 22.41 |
| * exceeding 1.22m | hr | m2 | 0.38 | 0.37 | | |
| | £ | | 13.38 | 2.05 | 4.13 | 19.56 |
| Plane battered; width | | | | | | |
| ne 0.1m | hr | m | 0.12 | 0.12 | | |
| | £ | | 3.50 | 0.67 | 0.86 | 5.03 |
| 0.1 - 0.20m | hr | m | 0.20 | 0.19 | | |
| | £ | | 5.84 | 1.05 | 0.99 | 7.88 |
| * 0.2 - 0.40m | hr | m2 | 0.44 | 0.42 | | |
| | £ | | 15.49 | 2.33 | 5.17 | 22.99 |
| * 0.4 - 1.22m | hr | m2 | 0.44 | 0.42 | | |
| | £ | | 15.49 | 2.33 | 5.17 | 22.99 |
| * exceeding 1.22m | hr | m2 | 0.40 | 0.40 | | |
| | £ | | 14.08 | 2.22 | 4.86 | 21.16 |

**CLASS G: CONCRETE ANCILLARIES**

**FORMWORK; EXTRA SMOOTH FINISH - continued**

| | | UNIT | L | P | M | £ NETT |
|---|---|---|---|---|---|---|
| Plane vertical; width | | | | | | |
| ne 0.1m | hr | m | 0.12 | 0.12 | | |
| | £ | | 3.50 | 0.67 | 0.86 | 5.03 |
| 0.1 - 0.20m | hr | m | 0.20 | 0.19 | | |
| | £ | | 5.84 | 1.05 | 1.08 | 7.97 |
| * 0.2 - 0.40m | hr | m2 | 0.74 | 0.72 | | |
| | £ | | 26.06 | 4.00 | 5.93 | 35.99 |
| * 0.4 - 1.22m | hr | m2 | 0.53 | 0.47 | | |
| | £ | | 18.66 | 2.61 | 5.93 | 27.20 |
| * exceeding 1.22m | hr | m2 | 0.48 | 0.48 | | |
| | £ | | 16.90 | 2.66 | 5.80 | 25.36 |
| Curved to one radius in one plane, 0.5m radius; width | | | | | | |
| ne 0.1m | hr | m | 0.20 | 0.19 | | |
| | £ | | 5.84 | 1.05 | 0.86 | 7.75 |
| 0.1 - 0.20m | hr | m | 0.28 | 0.25 | | |
| | £ | | 8.18 | 1.39 | 1.11 | 10.68 |
| * 0.2 - 0.40m | hr | m2 | 0.99 | 0.90 | | |
| | £ | | 34.86 | 5.00 | 5.80 | 45.66 |
| * 0.4 - 1.22m | hr | m2 | 0.78 | 0.72 | | |
| | £ | | 27.46 | 4.00 | 5.92 | 37.38 |
| * exceeding 1.22m | hr | m2 | 0.67 | 0.62 | | |
| | £ | | 23.59 | 3.44 | 6.46 | 33.49 |
| Curved to one radius in one plane, 1m radius or greater; width | | | | | | |
| ne 0.1m | hr | m | 0.20 | 0.19 | | |
| | £ | | 5.84 | 1.05 | 0.86 | 7.75 |
| 0.1 - 0.20m | hr | m | 0.28 | 0.25 | | |
| | £ | | 8.18 | 1.39 | 1.11 | 10.68 |
| * 0.2 - 0.40m | hr | m2 | 0.99 | 0.93 | | |
| | £ | | 34.86 | 5.16 | 5.80 | 45.82 |
| * 0.4 - 1.22m | hr | m2 | 0.75 | 0.70 | | |
| | £ | | 26.41 | 3.89 | 6.00 | 36.30 |
| * exceeding 1.22m | hr | m2 | 0.62 | 0.59 | | |
| | £ | | 21.83 | 3.27 | 6.55 | 31.65 |

**CLASS G: CONCRETE ANCILLARIES**

**FORMWORK; EXTRA SMOOTH FINISH** - continued

| | UNIT | | L | P | M | £ NETT |
|---|---|---|---|---|---|---|
| For concrete components of constant cross-section | | | | | | |
| beams; 200 x 200mm | hr | m | 0.50 | 0.40 | | |
| | £ | | **14.60** | **2.22** | **1.65** | **18.47** |
| beams; 500 x 500mm | hr | m | 0.57 | 0.45 | | |
| | £ | | **16.64** | **2.50** | **4.70** | **23.84** |
| beams; 500 x 800mm | hr | m | 0.70 | 0.60 | | |
| | £ | | **20.44** | **3.33** | **5.80** | **29.57** |
| columns; 200 x 200mm | hr | m | 0.57 | 0.40 | | |
| | £ | | **16.64** | **2.22** | **1.65** | **20.51** |
| columns; 300 x 300mm | hr | m | 0.57 | 0.55 | | |
| | £ | | **16.64** | **3.05** | **4.48** | **24.17** |
| columns; 300 x 500mm | hr | m | 0.64 | 0.62 | | |
| | £ | | **18.69** | **3.44** | **5.17** | **27.30** |
| * walls; 1.0m high thickness 250mm | hr | m | 1.40 | 1.35 | | |
| | £ | | **49.29** | **7.49** | **6.55** | **63.33** |
| * walls; 1.5m high thickness 300mm | hr | m | 1.14 | 1.10 | | |
| | £ | | **40.14** | **6.11** | **5.17** | **51.42** |
| * box culvert; 2 x 2m internally wall thickness 300mm | hr | m | 4.68 | 3.95 | | |
| | £ | | **164.78** | **21.92** | **60.57** | **247.27** |
| projections (100mm deep) | hr | m | 0.10 | 0.10 | | |
| | £ | | **2.92** | **0.56** | **3.51** | **6.99** |
| intrusions (100mm deep) | hr | m | 0.10 | 0.10 | | |
| | £ | | **2.92** | **0.56** | **1.65** | **5.13** |
| For voids (void former) | | | | | | |
| small void; depth ne 0.5m | hr | nr | 0.07 | - | | |
| | £ | | **2.04** | - | **3.53** | **5.57** |
| small void; depth 0.5 - 1m | hr | nr | 0.12 | - | | |
| | £ | | **3.50** | - | **6.84** | **10.34** |
| small void; depth 1.0 - 2m | hr | nr | 0.17 | - | | |
| | £ | | **4.96** | - | **13.31** | **18.27** |
| large void; depth ne 0.5m | hr | nr | 0.15 | - | | |
| | £ | | **4.38** | - | **7.49** | **11.87** |

**CLASS G: CONCRETE ANCILLARIES**

**FORMWORK; EXTRA SMOOTH FINISH - continued**

| | | UNIT | L | P | M | £ NETT |
|---|---|---|---|---|---|---|
| large void; depth 0.5 - 1m | hr | nr | 0.34 | - | | |
| | £ | | 9.93 | - | 14.98 | 24.91 |
| large void; depth 1.0 - 2m | hr | nr | 0.60 | - | | |
| | £ | | 17.52 | - | 26.62 | 44.14 |
| ADD for additional craneage and rub up (where required) | £ | m2 | 1.21 | 1.26 | 0.03 | 2.50 |
| ADD for additional craneage and rub up (where required) | £ | m | 4.03 | 4.20 | 0.13 | 8.36 |

**CLASS G: CONCRETE ANCILLARIES**

**FORMWORK; STEEL FORMS - Proprietary Systems**

(For these items the 'M' column shows the value of the modular units)

(eg. of hire cost:- curved formwork 4 days x 85p per M2 per day = 340p)
(0.20 hours per m2 equates to 5m2 per hour gang output)).

| | | |
|---|---|---|
| LABOUR based on Gang I | | |
| Total hourly cost | = £ | 38.51 |
| PLANT | | |
| 20T Crawler Crane(30RB) (Part time) | | 7.92 |
| Temporary props, ties tirfors and kentlidge | | 1.70 |
| Total hourly cost | = £ | 9.62 |

| | | UNIT | L | P | M | £ NETT |
|---|---|---|---|---|---|---|
| Extra Smooth Finish width up to 0.4m | | | | | | |
| Plane Horizontal | hr | m | 0.08 | 0.02 | | |
| | £ | | 3.08 | 0.19 | 1.41 | 4.68 |
| Plane sloping or battered | hr | m | 0.08 | 0.02 | | |
| | £ | | 3.08 | 0.19 | 1.41 | 4.68 |
| Plane Vertical | hr | m | 0.11 | 0.03 | | |
| | £ | | 4.24 | 0.29 | 1.41 | 5.94 |
| Curved to one radius in one plane, greater than 1m radius | hr | m | 0.11 | 0.03 | | |
| | £ | | 4.24 | 0.29 | 1.82 | 6.35 |
| width over 0.4m | | | | | | |
| Plane Horizontal | hr | m2 | 0.20 | 0.20 | | |
| | £ | | 7.70 | 1.92 | 2.60 | 12.22 |
| Plane sloping or battered | hr | m2 | 0.20 | 0.20 | | |
| | £ | | 7.70 | 1.92 | 2.60 | 12.22 |
| Plane Vertical | hr | m2 | 0.16 | 0.16 | | |
| | £ | | 6.16 | 1.54 | 2.60 | 10.30 |
| Curved to one radius in one plane, greater than 1m radius | hr | m2 | 0.16 | 0.16 | | |
| | £ | | 6.16 | 1.54 | 3.40 | 11.10 |

## CLASS G: CONCRETE ANCILLARIES

### REINFORCEMENT

LABOUR

All based on major fix Gang D = £ 41.41 Hourly cost

PLANT (PART TIME)

| | |
|---|---|
| Crane (30T Crawler) | 3.20 |
| Bar Cropper | 0.50 |
| Small Power Tools | 0.20 |
| Support Acrows, Tirfors, kentledge etc. | 0.60 |
| Total hourly cost = £ | 4.50 |

MATERIALS
Reinforcement materials include for bars, tying wire, spacers, couplers and steel (re-bar) supports for bottom layer reinforcement (stools, chairs or risers).
Wastage and consumables are allowed at the following rates:

1.0% for cut and bent bar
2.5% for straight bar
15.0% on to fabric to include for minimum lapping

| | | UNIT | L | P | M | £ NETT |
|---|---|---|---|---|---|---|
| Stainless steel bars bent and cut lengths | | | | | | |
| 8mm | hr | t | 6.74 | 6.74 | | |
| | £ | | 279.10 | 30.33 | 4100.91 | 4410.34 |
| 10mm | hr | t | 6.74 | 6.74 | | |
| | £ | | 279.10 | 30.33 | 3860.00 | 4169.43 |
| 12mm | hr | t | 6.74 | 6.74 | | |
| | £ | | 279.10 | 30.33 | 3457.92 | 3767.35 |
| 16mm | hr | t | 6.15 | 6.15 | | |
| | £ | | 254.67 | 27.68 | 3141.11 | 3423.46 |
| 20mm | hr | t | 4.44 | 4.44 | | |
| | £ | | 183.86 | 19.98 | 3019.14 | 3222.98 |
| 25mm | hr | t | 4.44 | 4.44 | | |
| | £ | | 183.86 | 19.98 | 2992.89 | 3196.73 |

**CLASS G: CONCRETE ANCILLARIES**

**REINFORCEMENT - continued**

| | | UNIT | L | P | M | £ NETT |
|---|---|---|---|---|---|---|
| Straight bars | | | | | | |
| nominal size 6mm | | | | | | |
| Mild steel to BS4449 | hr | t | 8.00 | 8.00 | | |
| | £ | | 331.28 | 36.00 | 473.29 | 840.57 |
| High yield steel to BS4449 or BS4461 | hr | t | 8.00 | 8.00 | | |
| | £ | | 331.28 | 36.00 | 477.39 | 844.67 |
| nominal size 8mm | | | | | | |
| Mild steel to BS4449 | hr | t | 6.74 | 6.74 | | |
| | £ | | 279.10 | 30.33 | 433.32 | 742.75 |
| High yield steel to BS4449 or BS4461 | hr | t | 6.74 | 6.74 | | |
| | £ | | 279.10 | 30.33 | 437.42 | 746.85 |
| nominal size 10mm | | | | | | |
| Mild steel to BS4449 | hr | t | 6.74 | 6.74 | | |
| | £ | | 279.10 | 30.33 | 422.04 | 731.47 |
| High yield steel to BS4449 or BS4461 | hr | t | 6.74 | 6.74 | | |
| | £ | | 279.10 | 30.33 | 427.39 | 736.82 |
| nominal size 12mm | | | | | | |
| Mild steel to BS4449 | hr | t | 6.74 | 6.74 | | |
| | £ | | 279.10 | 30.33 | 411.79 | 721.22 |
| High yield steel to BS4449 or BS4461 | hr | t | 6.74 | 6.74 | | |
| | £ | | 279.10 | 30.33 | 415.89 | 725.32 |
| nominal size 16mm | | | | | | |
| Mild steel to BS4449 | hr | t | 6.15 | 6.15 | | |
| | £ | | 254.67 | 27.68 | 390.27 | 672.62 |
| High yield steel to BS4449 or BS4461 | hr | t | 6.15 | 6.15 | | |
| | £ | | 254.67 | 27.68 | 394.37 | 676.72 |
| nominal size 20mm | | | | | | |
| Mild steel to BS4449 | hr | t | 4.44 | 4.44 | | |
| | £ | | 183.86 | 19.98 | 389.24 | 593.08 |
| High yield steel to BS4449 or BS4461 | hr | t | 4.44 | 4.44 | | |
| | £ | | 183.86 | 19.98 | 393.34 | 597.18 |
| nominal size 25mm | | | | | | |
| Mild steel to BS4449 | hr | t | 4.44 | 4.44 | | |
| | £ | | 183.86 | 19.98 | 392.32 | 596.16 |
| High yield steel to BS4449 or BS4461 | hr | t | 4.44 | 4.44 | | |
| | £ | | 183.86 | 19.98 | 396.42 | 600.26 |
| nominal size 32mm | | | | | | |
| Mild steel to BS4449 | hr | t | 4.44 | 4.44 | | |
| | £ | | 183.86 | 19.98 | 399.49 | 603.33 |
| High yield steel to BS4449 or BS4461 | hr | t | 4.44 | 4.44 | | |
| | £ | | 183.86 | 19.98 | 403.59 | 607.43 |
| nominal size 40mm | | | | | | |
| Mild steel to BS4449 | hr | t | 4.44 | 4.44 | | |
| | £ | | 183.86 | 19.98 | 411.79 | 615.63 |
| High yield steel to BS4449 or BS4461 | hr | t | 4.44 | 4.44 | | |
| | £ | | 183.86 | 19.98 | 415.89 | 619.73 |

**CLASS G: CONCRETE ANCILLARIES**

**REINFORCEMENT - continued**

| | | UNIT | L | P | M | £ NETT |
|---|---|---|---|---|---|---|
| Bent bars | | | | | | |
| nominal size 6mm | | | | | | |
| Mild steel to BS4449 | hr | t | 8.00 | 8.00 | | |
| | £ | | 331.28 | 36.00 | 552.21 | 919.49 |
| High yield steel to BS4449 or BS4461 | hr | t | 8.00 | 8.00 | | |
| | £ | | 331.28 | 36.00 | 556.26 | 923.54 |
| nominal size 8mm | | | | | | |
| Mild steel to BS4449 | hr | t | 6.74 | 6.74 | | |
| | £ | | 279.10 | 30.33 | 507.78 | 817.21 |
| High yield steel to BS4449 or BS4461 | hr | t | 6.74 | 6.74 | | |
| | £ | | 279.10 | 30.33 | 511.82 | 821.25 |
| nominal size 10mm | | | | | | |
| Mild steel to BS4449 | hr | t | 6.74 | 6.74 | | |
| | £ | | 279.10 | 30.33 | 476.47 | 785.90 |
| High yield steel to BS4449 or BS4461 | hr | t | 6.74 | 6.74 | | |
| | £ | | 279.10 | 30.33 | 480.51 | 789.94 |
| nominal size 12mm | | | | | | |
| Mild steel to BS4449 | hr | t | 6.74 | 6.74 | | |
| | £ | | 279.10 | 30.33 | 451.22 | 760.65 |
| High yield steel to BS4449 or BS4461 | hr | t | 6.74 | 6.74 | | |
| | £ | | 279.10 | 30.33 | 455.26 | 764.69 |
| nominal size 16mm | | | | | | |
| Mild steel to BS4449 | hr | t | 6.15 | 6.15 | | |
| | £ | | 254.67 | 27.68 | 419.91 | 702.26 |
| High yield steel to BS4449 or BS4461 | hr | t | 6.15 | 6.15 | | |
| | £ | | 254.67 | 27.68 | 423.95 | 706.30 |
| nominal size 20mm | | | | | | |
| Mild steel to BS4449 | hr | t | 4.44 | 4.44 | | |
| | £ | | 183.86 | 19.98 | 413.85 | 617.69 |
| High yield steel to BS4449 or BS4461 | hr | t | 4.44 | 4.44 | | |
| | £ | | 183.86 | 19.98 | 417.89 | 621.73 |
| nominal size 25mm | | | | | | |
| Mild steel to BS4449 | hr | t | 4.44 | 4.44 | | |
| | £ | | 183.86 | 19.98 | 416.88 | 620.72 |
| High yield steel to BS4449 or BS4461 | hr | t | 4.44 | 4.44 | | |
| | £ | | 183.86 | 19.98 | 420.92 | 624.76 |
| nominal size 32mm | | | | | | |
| Mild steel to BS4449 | hr | t | 4.44 | 4.44 | | |
| | £ | | 183.86 | 19.98 | 418.90 | 622.74 |
| High yield steel to BS4449 or BS4461 | hr | t | 4.44 | 4.44 | | |
| | £ | | 183.86 | 19.98 | 422.93 | 626.77 |
| nominal size 40mm | | | | | | |
| Mild steel to BS4449 | hr | t | 4.44 | 4.44 | | |
| | £ | | 183.86 | 19.98 | 431.02 | 634.86 |
| High yield steel to BS4449 or BS4461 | hr | t | 4.44 | 4.44 | | |
| | £ | | 183.86 | 19.98 | 435.06 | 638.90 |

**CLASS G: CONCRETE ANCILLARIES**

**REINFORCEMENT - continued**

| | | UNIT | L | P | M | £ NETT |
|---|---|---|---|---|---|---|
| ADD to the above for bars 12-15m long | | | | | | |
| Mild steel to BS4449 | hr | t | - | - | | |
| | £ | | - | - | - | 14.00 |
| High yield steel to BS4449 or BS4461 | hr | t | - | - | | |
| | £ | | - | - | - | 14.00 |
| ADD to the above for bars over 15m long, per 500mm increments | | | | | | |
| Mild steel to BS4449 | hr | t | - | - | | |
| | £ | | - | - | - | 3.05 |
| High yield steel to BS4449 or BS4461 | hr | t | - | - | | |
| | £ | | - | - | - | 3.05 |
| High yield steel fabric to BS4483 | | | | | | |
| nominal mass ne 2kg/m2; Ref D49 | hr | m2 | 0.02 | - | | |
| | £ | | 0.83 | - | 0.84 | 1.67 |
| nominal mass ne 2kg/m2; Ref A98 | hr | m2 | 0.03 | - | | |
| | £ | | 1.24 | - | 1.04 | 2.28 |
| nominal mass 2-3 kg/m2; Ref A142 | hr | m2 | 0.03 | - | | |
| | £ | | 1.24 | - | 1.32 | 2.56 |
| nominal mass 2-3 kg/m2; Ref C283 | hr | m2 | 0.03 | - | | |
| | £ | | 1.24 | - | 1.57 | 2.81 |
| nominal mass 3-4 kg/m2; Ref A193 | hr | m2 | 0.04 | - | | |
| | £ | | 1.66 | - | 1.79 | 3.45 |
| nominal mass 3-4 kg/m2; Ref B196 | hr | m2 | 0.04 | - | | |
| | £ | | 1.66 | - | 1.96 | 3.62 |
| nominal mass 3-4 kg/m2; Ref C385 | hr | m2 | 0.04 | - | | |
| | £ | | 1.66 | - | 2.16 | 3.82 |
| nominal mass 3-4 kg/m2; Ref B283 | hr | m2 | 0.04 | - | | |
| | £ | | 1.66 | - | 2.34 | 4.00 |
| nominal mass 3-4 kg/m2; Ref A252 | hr | m2 | 0.04 | - | | |
| | £ | | 1.66 | - | 2.36 | 4.02 |
| nominal mass 4-5 kg/m2; Ref C503 | hr | m2 | 0.05 | - | | |
| | £ | | 2.07 | - | 2.62 | 4.69 |
| nominal mass 4-5 kg/m2; Ref B385 | hr | m2 | 0.05 | - | | |
| | £ | | 2.07 | - | 2.81 | 4.88 |
| nominal mass 5-6 kg/m2; Ref C636 | hr | m2 | 0.05 | - | | |
| | £ | | 2.07 | - | 3.38 | 5.45 |

**CLASS G: CONCRETE ANCILLARIES**

**REINFORCEMENT - continued**

| | | UNIT | L | P | M | £ NETT |
|---|---|---|---|---|---|---|
| nominal mass 5-6 kg/m2; Ref B503 | hr | m2 | 0.05 | - | | |
| | £ | | **2.07** | - | **3.55** | **5.62** |
| nominal mass 6-7 kg/m2; Ref A393 | hr | m2 | 0.07 | - | | |
| | £ | | **2.90** | - | **3.66** | **6.56** |
| nominal mass 6-7 kg/m2; Ref C785 | hr | m2 | 0.07 | - | | |
| | £ | | **2.90** | - | **4.09** | **6.99** |
| nominal mass 8.14 kg/m2; Ref B785 | hr | m2 | 0.08 | - | | |
| | £ | | **3.31** | - | **4.85** | **8.16** |
| nominal mass 10.90 kg/m2; Ref; B1131 | hr | m2 | 0.09 | - | | |
| | £ | | **3.73** | - | **6.56** | **10.29** |

Cutting bending tagging and baling reinforcement on site

| | | | | |
|---|---|---|---|---|
| LABOUR: | skilled | 1 | x | 6.54 |
| | unskilled | 1 | x | 6.20 |
| | | = | £ | 12.74 |
| PLANT: | as previous list | = | £ | 4.50 |

| | | UNIT | L | P | M | £ NETT |
|---|---|---|---|---|---|---|
| nominal size 6mm | hr | t | 4.87 | 4.87 | | |
| | £ | | **62.04** | **21.92** | **1.00** | **84.96** |
| nominal size 8mm | hr | t | 4.58 | 4.58 | | |
| | £ | | **58.35** | **20.61** | **1.00** | **79.96** |
| nominal size 10mm | hr | t | 3.42 | 3.42 | | |
| | £ | | **43.57** | **15.39** | **1.00** | **59.96** |
| nominal size 12mm | hr | t | 2.55 | 2.55 | | |
| | £ | | **32.49** | **11.48** | **1.00** | **44.97** |
| nominal size 16mm | hr | t | 2.03 | 2.03 | | |
| | £ | | **25.86** | **9.14** | **1.00** | **36.00** |
| nominal size 20mm | hr | t | 1.68 | 1.68 | | |
| | £ | | **21.40** | **7.56** | **1.00** | **29.96** |
| nominal size 25mm | hr | t | 1.68 | 1.68 | | |
| | £ | | **21.40** | **7.56** | **1.00** | **29.96** |
| nominal size 32mm | hr | t | 1.39 | 1.39 | | |
| | £ | | **17.71** | **6.26** | **1.00** | **24.97** |
| nominal size 40mm | hr | t | 1.39 | 1.39 | | |
| | £ | | **17.71** | **6.26** | **1.00** | **24.97** |

**CLASS G: CONCRETE ANCILLARIES**

| | | UNIT | L | P | M | £ NETT |
|---|---|---|---|---|---|---|
| **JOINTS** (5% WASTAGE ALLOWED) LABOUR based on Gang D2 | | | | | | |
| Open surface plain; width or depth ne 0.5m; scabbling concrete for subsequent pour | hr | m2 | 0.04 | 0.04 | | |
| | £ | | 1.66 | 0.18 | - | 1.84 |
| 0.5m; scabbling concrete for subsequent pour | hr | m2 | 0.03 | 0.03 | | |
| | £ | | 1.24 | 0.14 | - | 1.38 |
| Open surface with filler; width or depth ne 0.5m; 10mm Flexcell joint filler | hr | m2 | 0.04 | - | | |
| | £ | | 1.66 | - | 3.89 | 5.55 |
| 0.5-1m; 10mm Flexcell joint filler | hr | m2 | 0.04 | - | | |
| | £ | | 1.66 | - | 3.98 | 5.64 |
| Formed surface plain; width or depth (including Formwork) ne 0.5m | hr | m2 | - | - | | |
| | £ | | - | - | - | 3.07 |
| 0.5-1m | hr | m2 | - | - | | |
| | £ | | - | - | - | 3.07 |
| Formed surface with filler; width or depth (including Formwork) ne 0.5m; 10mm Flexcell joint filler | hr | m2 | 0.40 | 0.30 | | |
| | £ | | 16.56 | 1.35 | 7.38 | 25.29 |
| 0.5-1m; 10mm Flexcell joint filler | hr | m2 | 0.41 | 0.30 | | |
| | £ | | 16.98 | 1.35 | 7.38 | 25.71 |
| ne 0.5m; 20mm Flexcell joint filler | hr | m2 | 0.42 | 0.30 | | |
| | £ | | 17.39 | 1.35 | 10.16 | 28.90 |
| 0.5-1m; 20mm Flexcell joint filler | hr | m2 | 0.43 | 0.30 | | |
| | £ | | 17.81 | 1.35 | 10.16 | 29.32 |
| ne 0.5m; 25mm Flexcell joint filler | hr | m2 | 0.45 | 0.30 | | |
| | £ | | 18.63 | 1.35 | 11.38 | 31.36 |
| 0.5-1m; 25mm Flexcell joint filler | hr | m2 | 0.44 | 0.30 | | |
| | £ | | 18.22 | 1.35 | 11.38 | 30.95 |

**CLASS G: CONCRETE ANCILLARIES**

**JOINTS - continued**

| | | UNIT | L | P | M | £ NETT |
|---|---|---|---|---|---|---|
| Waterstops | | | | | | |
| 160mm centre bulb pvc | hr | m | 0.036 | - | | |
| | £ | | 1.49 | - | 4.26 | 5.75 |
| junction piece | hr | nr | 0.04 | - | | |
| | £ | | 1.66 | - | 8.79 | 10.45 |
| 210mm centre bulb pvc | hr | m | 0.04 | - | | |
| | £ | | 1.66 | - | 6.32 | 7.98 |
| junction piece | hr | nr | 0.04 | - | | |
| | £ | | 1.66 | - | 10.23 | 11.89 |
| 260mm centre bulb pvc | hr | m | 0.05 | - | | |
| | £ | | 2.07 | - | 9.00 | 11.07 |
| junction piece | hr | nr | 0.04 | - | | |
| | £ | | 1.66 | - | 12.84 | 14.50 |
| 150mm centre bulb rubber | hr | m | 0.036 | - | | |
| | £ | | 1.49 | - | 15.13 | 16.62 |
| junction piece | hr | nr | 0.05 | - | | |
| | £ | | 2.07 | - | 40.84 | 42.91 |
| 230mm centre bulb rubber | hr | m | 0.045 | | | |
| | £ | | 1.86 | - | 22.60 | 24.46 |
| junction piece | hr | nr | 0.07 | - | | |
| | £ | | 2.90 | - | 46.81 | 49.71 |
| 305mm centre bulb rubber | hr | m | 0.054 | - | | |
| | £ | | 2.24 | - | 37.27 | 39.51 |
| junction piece | hr | nr | 0.09 | - | | |
| | £ | | 3.73 | - | 85.97 | 89.70 |
| Sealed rebates or grooves | | | | | | |
| 10 x 20mm; hot poured rubber bitumen | hr | m | 0.05 | - | | |
| | £ | | 2.07 | - | 0.44 | 2.51 |
| 20 x 20mm; hot poured rubber bitumen | hr | m | 0.08 | - | | |
| | £ | | 3.31 | - | 0.89 | 4.20 |
| 25 x 20mm; hot poured rubber bitumen | hr | m | 0.03 | - | | |
| | £ | | 1.24 | - | 1.12 | 2.36 |
| Dowels, plain or greased | | | | | | |
| 12mm mild steel 500mm long | hr | nr | 0.02 | - | | |
| | £ | | 0.83 | - | 0.28 | 1.11 |

**CLASS G: CONCRETE ANCILLARIES**

**JOINTS - continued**

| | | UNIT | L | P | M | £ NETT |
|---|---|---|---|---|---|---|
| Dowels, plain or greased | | | | | | |
| 16mm mild steel 750mm long | hr | nr | 0.02 | - | | |
| | £ | | 0.25 | - | 0.62 | 0.87 |
| 16mm mild steel 750mm long, debonding agent for 375mm | hr | nr | 0.03 | - | | |
| | £ | | 0.38 | - | 0.74 | 1.12 |
| Dowels, sleeved or capped | | | | | | |
| 12mm mild steel 500mm long, debonding agent for 250mm and capped with pvc dowel cap | hr | nr | 0.03 | - | | |
| | £ | | 0.38 | - | 0.50 | 0.88 |

**POST-TENSIONED PRESTRESSING**

The design of prestressing is based on standard patented systems, each of which has produced its own method of anchoring, joining and stressing the cables or wires. The companies marketing the systems will either supply all the materials and fittings required together with the sale or hire of suitable jacks and equipment for stressing and grouting or they will undertake to complete the work on a Subcontract basis. The rates given below are therefore indicative only of the probable labour and plant costs and do not include for any permanent materials. The advice of Specialist Contractors should be sought for more accurate rates based on the design for a particular contract. Pretensioned prestressing is normally used only in the manufacture of precast units utilising special beds set up in the supplier's factory.

Labour and plant cost in post-tensioning

| | | UNIT | L | P | M | £ NETT |
|---|---|---|---|---|---|---|
| form ducts to profile including supports and fixings | | | | | | |
| 50mm (internal diameter) | hr | m | - | - | | |
| | £ | | 0.70 | 2.30 | - | 3.00 |
| 80mm | hr | m | - | - | | |
| | £ | | 1.10 | 2.70 | - | 3.80 |
| 100mm | hr | m | - | - | | |
| | £ | | 1.50 | 2.85 | - | 4.35 |
| extra for grout vents | | | | | | |
| 50mm | hr | nr | - | - | | |
| | £ | | 3.50 | - | - | 3.50 |
| 80mm | hr | nr | - | - | | |
| | £ | | 3.50 | - | - | 3.50 |
| 100mm | hr | nr | - | - | | |
| | £ | | 3.50 | - | - | 3.50 |

**CLASS G: CONCRETE ANCILLARIES**

**POST TENSIONED PRESTRESSING - continued**

| | | UNIT | L | P | M | £ NETT |
|---|---|---|---|---|---|---|
| grout ducts including provision of equipment | | | | | | |
| 50mm | hr | m | - | - | | |
| | £ | | 1.10 | 0.40 | - | 1.50 |
| 80mm | hr | m | - | - | | |
| | £ | | 1.10 | 0.50 | - | 1.60 |
| 100mm | hr | m | - | - | | |
| | £ | | 1.50 | 0.60 | - | 2.10 |
| form tendons including spacers, etc. and pull through ducts (Number of strands) | | | | | | |
| 7 | hr | m | 0.25 | 0.25 | | |
| | £ | | 3.90 | 2.60 | - | 6.50 |
| 12 | hr | m | 0.45 | 0.45 | | |
| | £ | | 7.02 | 4.68 | - | 11.70 |
| 19 | hr | m | 0.65 | 0.65 | | |
| | £ | | 10.14 | 6.76 | - | 16.90 |
| dead end anchorage | | | | | | |
| 7 | hr | nr | 0.75 | 0.75 | | |
| | £ | | 11.70 | 7.80 | - | 19.50 |
| 12 | hr | nr | 0.95 | 0.95 | | |
| | £ | | 14.82 | 9.88 | - | 24.70 |
| 19 | hr | nr | 1.15 | 1.15 | | |
| | £ | | 17.94 | 11.96 | - | 29.90 |
| looped buried dead end anchorage | | | | | | |
| 7 | hr | nr | 0.50 | 0.50 | | |
| | £ | | 7.80 | 5.20 | - | 13.00 |
| 12 | hr | nr | 0.66 | 0.66 | | |
| | £ | | 10.30 | 6.86 | - | 17.16 |
| 19 | hr | nr | 0.89 | 0.89 | | |
| | £ | | 13.88 | 9.26 | - | 23.14 |
| end anchorage including reinforcement | | | | | | |
| 7 | hr | nr | 1.67 | 1.67 | | |
| | £ | | 26.05 | 17.37 | - | 43.42 |
| 12 | hr | nr | 2.04 | 2.04 | | |
| | £ | | 31.82 | 21.22 | - | 53.04 |
| 19 | hr | nr | 2.56 | 2.56 | | |
| | £ | | 39.94 | 26.62 | - | 66.56 |

**CLASS G: CONCRETE ANCILLARIES**

**POST TENSIONED PRESTRESSING - continued**

| | | UNIT | L | P | M | £ NETT |
|---|---|---|---|---|---|---|
| add to last for anchorage coupling | | | | | | |
| 7 | hr | nr | 0.46 | 0.46 | | |
| | £ | | 7.18 | 4.78 | - | 11.96 |
| 12 | hr | nr | 0.79 | 0.79 | | |
| | £ | | 12.32 | 8.22 | - | 20.54 |
| 19 | hr | nr | 1.20 | 1.20 | | |
| | £ | | 18.72 | 12.48 | - | 31.20 |
| stress and lock of including multimatic jack | | | | | | |
| 7 | hr | nr | 3.21 | 3.21 | | |
| | £ | | 50.08 | 33.38 | - | 83.46 |
| 12 | hr | nr | 4.19 | 4.19 | | |
| | £ | | 65.36 | 43.58 | - | 108.94 |
| 19 | hr | nr | 5.58 | 5.58 | | |
| | £ | | 87.05 | 58.03 | - | 145.08 |
| cut off and seal ends of tendons | | | | | | |
| 7 | hr | nr | 0.20 | 0.20 | | |
| | £ | | 3.12 | 2.08 | - | 5.20 |
| 12 | hr | nr | 0.35 | 0.35 | | |
| | £ | | 5.46 | 3.64 | - | 9.10 |
| 19 | hr | nr | 0.55 | 0.55 | | |
| | £ | | 8.58 | 5.72 | - | 14.30 |

## CLASS G: CONCRETE ANCILLARIES

## CONCRETE ACCESSORIES

**LABOUR**

**Based on Gang H x 1/2**

| | | | |
|---|---|---|---|
| **Total hourly cost** | = | £ | **29.43** |

**PLANT**

| | | | |
|---|---|---|---|
| **Compressor** | | | **4.44** |
| **Tools (Bush Hammer)** | | | **1.32** |
| **Dumper (750Kg)** | | | **1.20** |
| **Total hourly cost** | = | £ | **6.96** |

Included in this section:-

Finishing top surfaces

- float
- trowel
- granolithic

Finishing formed surfaces

- aggregate exposure
- bush hammering

Grouting under plates

**CLASS G: CONCRETE ANCILLARIES**

**CONCRETE ACCESSORIES - continued**

| | | UNIT | L | P | M | £ NETT |
|---|---|---|---|---|---|---|
| Finishing of top surfaces | | | | | | |
| wood float; level | hr | m2 | 0.02 | - | | |
| | £ | | **0.59** | - | - | 0.59 |
| wood float; falls or cross falls | hr | m2 | 0.03 | - | | |
| | £ | | **0.88** | - | - | 0.88 |
| steel trowel; level | hr | m2 | 0.03 | - | | |
| | £ | | **0.88** | - | - | 0.88 |
| steel trowel; falls or cross falls | hr | m2 | 0.05 | - | | |
| | £ | | **1.47** | - | - | 1.47 |
| granolithic finish 20mm thick laid monolithically | hr | m2 | 0.07 | 0.04 | | |
| | £ | | **2.06** | **0.28** | 3.60 | 5.94 |
| Finishing of formed surfaces | | | | | | |
| aggregate exposure using retarder | hr | m2 | 0.05 | - | | |
| | £ | | **1.47** | - | 0.30 | 1.77 |
| bush hammering; kango hammer | hr | m2 | 0.28 | 0.28 | | |
| | £ | | **8.24** | **1.95** | - | 10.19 |
| rubbing down concrete surfaces after striking shutters | hr | m2 | 0.02 | - | | |
| | £ | | **0.59** | - | 0.35 | 0.94 |
| Grouting under plates; 25mm cement and sand (1:3) | | | | | | |
| area ne 0.1m2 | hr | m2 | 0.93 | - | | |
| | £ | | **27.37** | - | 1.58 | 28.95 |
| area 0.1 - 0.5 m2 | hr | m2 | 0.86 | - | | |
| | £ | | **25.31** | - | 1.58 | 26.89 |
| area 0.5 - 1.0 m2 | hr | m2 | 0.78 | - | | |
| | £ | | **22.96** | - | 1.58 | 24.54 |

**CLASS H: PRECAST CONCRETE**

The cost of precast concrete items is very much dependent on the complexity of the moulds, the number of units to be cast from each mould and the size and weight of the unit to be handled. The unit rates below are for standard precast items that are often to be found on a Civil Engineering project. It would be misleading to quote for indicative costs for tailor-made precast concrete units and it is advisable to contact Specialist Manufacturers for guide prices.

**BEAMS**

| | | UNIT | L | P | M | £ NETT |
|---|---|---|---|---|---|---|
| Concrete mix C20 | | | | | | |
| size 100 x 150 x 1050mm long | hr | nr | - | - | | |
| | £ | | 5.45 | - | 7.05 | 12.50 |
| size 225 x 150 x 1200mm long | hr | nr | - | - | | |
| | £ | | 2.30 | 1.10 | 17.30 | 20.70 |
| size 225 x 225 x 1800mm long | hr | nr | - | - | | |
| | £ | | 3.36 | 2.18 | 36.20 | 41.74 |

**PRESTRESSED PRE-TENSIONED BEAMS**

| | | UNIT | L | P | M | £ NETT |
|---|---|---|---|---|---|---|
| Concrete mix C20 | | | | | | |
| size 100 x 65 x 1050mm long | hr | nr | - | - | | |
| | £ | | 2.40 | - | 4.76 | 7.16 |
| size 265 x 65 x 1800mm long | hr | nr | - | - | | |
| | £ | | 2.29 | 2.18 | 19.05 | 23.52 |
| Inverted 'T' beams, flange width 495mm | | | | | | |
| section T4; length 12m and depth 575mm | hr | nr | - | - | | |
| | £ | | 78.00 | 90.00 | 635.00 | 803.00 |
| section T7; length 12m and depth 695mm | hr | nr | - | - | | |
| | £ | | 105.00 | 122.00 | 760.00 | 987.00 |
| section T9; length 16m and depth 775mm | hr | nr | - | - | | |
| | £ | | 138.00 | 160.00 | 1125.00 | 1423.00 |
| section T10; length 18m and depth 815mm | hr | nr | - | - | | |
| | £ | | 162.00 | 189.00 | 1330.00 | 1681.00 |

**CLASS H: PRECAST CONCRETE**

**PRESTRESSED PRE-TENSIONED BEAMS - continued**

| | | UNIT | L | P | M | £ NETT |
|---|---|---|---|---|---|---|
| 'M' beams, flange width 970mm | | | | | | |
| section M3; length 18m and depth 800mm | hr | nr | - | - | | |
| | £ | | 250.00 | 291.00 | 2890.00 | 3431.00 |
| section M6; length 22m and depth 1040mm | hr | nr | - | - | | |
| | £ | | 340.00 | 397.00 | 3920.00 | 4657.00 |
| section M8; length 25m and depth 1200mm | hr | nr | - | - | | |
| | £ | | 444.00 | 518.00 | 4725.00 | 5687.00 |
| 'U' beams, base width 970mm | | | | | | |
| section U3; length 16m and depth 900mm | hr | nr | - | - | | |
| | £ | | 460.00 | 536.00 | 4400.00 | 5396.00 |
| section U5; length 20m and depth 1000mm | hr | nr | - | - | | |
| | £ | | 587.00 | 684.00 | 5860.00 | 7131.00 |
| section U8; length 24m and depth 1200mm | hr | nr | - | - | | |
| | £ | | 786.00 | 918.00 | 8200.00 | 9904.00 |
| section U12; length 30m and depth 1600mm | hr | nr | - | - | | |
| | £ | | 1255.00 | 1460.00 | 13150.00 | 15865.00 |

All four Spon Price Books – *Architects' and Builders', Civil Engineering and Highway Works, Landscape and External Works* and *Mechanical and Electrical Services* – are supported by an updating service. Updates are issued each November, February and May and give details of changes in prices of materials, wages rates and other significant items, with regional price level adjustments for Northern Ireland, Scotland and Wales and regions of England.

As a purchaser of this edition you are entitled to this updating service – *free of charge*. Simply complete this registration card and return it to us. Your completion of the reader survey data section – whilst not obligatory – would help ensure that we continue to provide the ideal reference book to facilitate your work.

---

**REGISTRATION CARD for Spon's Price Book Update** **1991**

*Please print your details clearly*

Name..........................................................................................................

*(Please indicate membership of any professional body eg RICS, RIBA, CIOB etc.)*

Address ......................................................................................................

..................................................................................................................

.................................................................................. Postcode ..................

Signature ....................................................................... Date .........................

---

**READERSHIP SURVEY** – *Your assistance here would be appreciated*

**How many other persons access your copy of the Price Book?**

☐ myself alone ☐ 2 ☐ 3-5 ☐ more than 5

**Profession:**

☐ Architect ☐ Building Contractor ☐ Building Surveyor
☐ Civil Engineer ☐ Chemical Engineer ☐ Civil Engineer
☐ Cost Engineer ☐ Electrical Engineer ☐ Landscape Contractor
☐ Landscape Architect ☐ Mechanical Engineer ☐ Quantity Surveyor

☐ Other – please specify..........................................................................

**Would your employer be best described as:**

☐ Commercial concern ☐ Local Authority
☐ Professional practice ☐ Public Utility

---

*Thank you very much for your assistance.*

AFFIX
STAMP
HERE

**E. & F.N. SPON**
2-6 Boundary Row
London
SE1 8HN

**CLASS H: PRECAST CONCRETE**

**PRESTRESSED FLOOR UNITS**

| | | UNIT | L | P | M | £ NETT |
|---|---|---|---|---|---|---|
| Bison 'Drycast or similar flooring planks | | | | | | |
| 110mm thick x 400mm wide planks | hr | m2 | - | - | | |
| | £ | | 6.55 | 7.00 | 22.00 | 35.55 |
| 110mm thick x 1200mm wide planks | hr | m2 | - | - | | |
| | £ | | 4.40 | 5.00 | 16.00 | 25.40 |
| 150mm thick x 400mm wide planks | hr | m2 | - | - | | |
| | £ | | 6.55 | 8.00 | 24.00 | 38.55 |
| 150mm thick x 1200mm wide planks | hr | m2 | - | - | | |
| | £ | | 4.40 | 5.00 | 16.00 | 25.40 |

**COPINGS, SILLS AND WEIR BLOCKS**

| | | | | | | |
|---|---|---|---|---|---|---|
| Concrete mix C30 | | | | | | |
| coping; size 178 x 64mm, weathered and throated | hr | m | - | - | | |
| | £ | | 4.40 | 3.00 | 12.30 | 19.70 |
| coping; size 305 x 76mm, weathered and throated | hr | m | - | - | | |
| | £ | | 3.30 | 2.00 | 9.00 | 14.30 |

**CLASS I: PIPEWORK - PIPES**

**CLAY PIPES**

LABOUR based on Gang B

| | | |
|---|---|---|
| Total hourly cost | = £ | 32.51 |

PLANT

| | | |
|---|---|---|
| Excavator (average 0.5m3 - 1.0m3) | | 10.90 |
| Plate Compactor (part time) | | 0.90 |
| Trench sheets, shores, props etc. | | 2.30 |
| Part time use of pumps, dumper concrete equipment, power tools (average useage) | | 2.03 |
| Total hourly cost | = £ | 16.13 |

To provide a basic, overall cost comparison, the rates, prices and outputs have been based on a medium sized Civil Engineering project of £5-6 million which is neither city centre nor excessively remote.

All rates for materials are now based on the most economically available materials with a minimum waste allowance - 2-3% and the most efficient items of plant (Excavator). The rates are optimum rates and assume continuous output with no delays caused by other operations or works. Ground conditions are assumed to be good soil with no abnormal conditions that would affect outputs and consistency of work.

Multiplier table for Labour and Plant for various site conditions (Please note previous editions have taken an average of the following conditions)

| | | | |
|---|---|---|---|
| Working out of sequence with other works | 2.75 | - | min. |
| Working in hard clay | 1.75 | - | 2.00 |
| Working in running sand | 3.50 | - | min. |
| Working in broken rock | 2.75 | - | 3.50 |
| Working below water Table levels | 2.00 | - | min. |

## CLASS I: PIPEWORK - PIPES

### CLAY PIPES

All rates for clay pipes are based on Zone 2 prices
Waste Allowance 2 - 3%

Vitrified clay pipes to BS65, plain ends with Push-fit polypropylene flexible couplings (Supersleve/Hepsleve), nominal bore

| | | UNIT | L | P | M | £ NETT |
|---|---|---|---|---|---|---|
| 100mm | | | | | | |
| Not in trenches | hr | m | 0.06 | 0.02 | | |
| | £ | | 1.95 | 0.32 | 1.78 | 4.05 |
| In trenches | | | | | | |
| ne 1.5m | hr | m | 0.15 | 0.15 | | |
| | £ | | 4.88 | 2.42 | 1.78 | 9.08 |
| 1.5 - 2.0m | hr | m | 0.19 | 0.19 | | |
| | £ | | 6.18 | 3.06 | 1.78 | 11.02 |
| 2.0 - 2.5m | hr | m | 0.24 | 0.24 | | |
| | £ | | 7.80 | 3.87 | 1.78 | 13.45 |
| 2.5 - 3.0m | hr | m | 0.29 | 0.29 | | |
| | £ | | 9.43 | 4.68 | 1.78 | 15.89 |
| 3.0 - 3.5m | hr | m | 0.36 | 0.36 | | |
| | £ | | 11.70 | 5.81 | 1.78 | 19.29 |
| 3.5 - 4.0m | hr | m | 0.44 | 0.44 | | |
| | £ | | 14.30 | 7.10 | 1.78 | 23.18 |
| 4.0 - 4.5m | hr | m | 0.55 | 0.55 | | |
| | £ | | 17.88 | 8.87 | 1.78 | 28.53 |
| 4.5 - 5.0m | hr | m | 0.70 | 0.70 | | |
| | £ | | 22.76 | 11.29 | 1.78 | 35.83 |
| 5.0 - 5.5m | hr | m | 0.90 | 0.90 | | |
| | £ | | 29.26 | 14.52 | 1.78 | 45.56 |
| 150mm | | | | | | |
| Not in trenches | hr | m | 0.06 | 0.02 | | |
| | £ | | 1.95 | 0.32 | 4.06 | 6.33 |
| In trenches | | | | | | |
| ne 1.5m | hr | m | 0.17 | 0.17 | | |
| | £ | | 5.53 | 2.74 | 4.06 | 12.33 |
| 1.5 - 2.0m | hr | m | 0.20 | 0.20 | | |
| | £ | | 6.50 | 3.23 | 4.06 | 13.79 |
| 2.0 - 2.5m | hr | m | 0.24 | 0.24 | | |
| | £ | | 7.80 | 3.87 | 4.06 | 15.73 |
| 2.5 - 3.0m | hr | m | 0.29 | 0.29 | | |
| | £ | | 9.43 | 4.68 | 4.06 | 18.17 |

**CLASS I: PIPEWORK - PIPES**

**CLAY PIPES - continued**

| | | UNIT | L | P | M | £ NETT |
|---|---|---|---|---|---|---|
| 150mm | | | | | | |
| In trenches | | | | | | |
| 3.0 - 3.5m | hr | m | 0.39 | 0.39 | | |
| | £ | | 12.68 | 6.29 | 4.06 | 23.03 |
| 3.5 - 4.0m | hr | m | 0.45 | 0.45 | | |
| | £ | | 14.63 | 7.26 | 4.06 | 25.95 |
| 4.0 - 4.5m | hr | m | 0.58 | 0.58 | | |
| | £ | | 18.86 | 9.36 | 4.06 | 32.28 |
| 4.5 - 5.0m | hr | m | 0.75 | 0.75 | | |
| | £ | | 24.38 | 12.10 | 4.06 | 40.54 |
| 5.0 - 5.5m | hr | m | 0.96 | 0.96 | | |
| | £ | | 31.21 | 15.48 | 4.06 | 50.75 |
| 225mm | | | | | | |
| Not in trenches | hr | m | 0.08 | 0.02 | | |
| | £ | | 2.60 | 0.32 | 9.41 | 12.33 |
| In trenches | | | | | | |
| ne 1.5m | hr | m | 0.18 | 0.18 | | |
| | £ | | 5.85 | 2.90 | 9.41 | 18.16 |
| 1.5 - 2.0m | hr | m | 0.22 | 0.22 | | |
| | £ | | 7.15 | 3.55 | 9.41 | 20.11 |
| 2.0 - 2.5m | hr | m | 0.26 | 0.26 | | |
| | £ | | 8.45 | 4.19 | 9.41 | 22.05 |
| 2.5 - 3.0m | hr | m | 0.30 | 0.30 | | |
| | £ | | 9.75 | 4.84 | 9.41 | 24.00 |
| 3.0 - 3.5m | hr | m | 0.41 | 0.41 | | |
| | £ | | 13.33 | 6.61 | 9.41 | 29.35 |
| 3.5 - 4.0m | hr | m | 0.47 | 0.47 | | |
| | £ | | 15.28 | 7.58 | 9.41 | 32.27 |
| 4.0 - 4.5m | hr | m | 0.65 | 0.65 | | |
| | £ | | 21.13 | 10.48 | 9.41 | 41.02 |
| 4.5 - 5.0m | hr | m | 0.80 | 0.80 | | |
| | £ | | 26.01 | 12.90 | 9.41 | 48.32 |
| 5.0 - 5.5m | hr | m | 1.02 | 1.02 | | |
| | £ | | 33.16 | 16.45 | 9.41 | 59.02 |

**CLASS I: PIPEWORK - PIPES**

**CLAY PIPES - continued**

| | UNIT | | L | P | M | £ NETT |
|---|---|---|---|---|---|---|
| 300mm | | | | | | |
| Not in trenches | hr | m | 0.10 | 0.05 | | |
| | £ | | 3.25 | 0.81 | 14.36 | 18.42 |
| In trenches | | | | | | |
| ne 1.5m | hr | m | 0.19 | 0.19 | | |
| | £ | | 6.18 | 3.06 | 14.36 | 23.60 |
| 1.5 - 2.0m | hr | m | 0.24 | 0.24 | | |
| | £ | | 7.80 | 3.87 | 14.36 | 26.03 |
| 2.0 - 2.5m | hr | m | 0.28 | 0.28 | | |
| | £ | | 9.10 | 4.52 | 14.36 | 27.98 |
| 2.5 - 3.0m | hr | m | 0.32 | 0.32 | | |
| | £ | | 10.40 | 5.16 | 14.36 | 29.92 |
| 3.0 - 3.5m | hr | m | 0.42 | 0.42 | | |
| | £ | | 13.65 | 6.77 | 14.36 | 34.78 |
| 3.5 - 4.0m | hr | m | 0.52 | 0.52 | | |
| | £ | | 16.91 | 8.39 | 14.36 | 39.66 |
| 4.0 - 4.5m | hr | m | 0.70 | 0.70 | | |
| | £ | | 22.76 | 11.29 | 14.36 | 48.41 |
| 4.5 - 5.0m | hr | m | 0.90 | 0.90 | | |
| | £ | | 29.26 | 14.52 | 14.36 | 58.14 |
| 5.0 - 5.5m | hr | m | 1.12 | 1.12 | | |
| | £ | | 36.41 | 18.07 | 14.36 | 68.84 |
| Vitrified clay pipes to BS65, plain ends with push-fit polypropylene flexible couplings (Densleeve), nominal bore | | | | | | |
| 100mm | | | | | | |
| Not in trenches | hr | m | 0.06 | 0.02 | | |
| | £ | | 1.95 | 0.32 | 1.83 | 4.10 |
| In trenches | | | | | | |
| ne 1.5m | hr | m | 0.15 | 0.15 | | |
| | £ | | 4.88 | 2.42 | 1.83 | 9.13 |
| 1.5 - 2.0m | hr | m | 0.19 | 0.19 | | |
| | £ | | 6.18 | 3.06 | 1.83 | 11.07 |
| 2.0 - 2.5m | hr | m | 0.24 | 0.24 | | |
| | £ | | 7.80 | 3.87 | 1.83 | 13.50 |
| 2.5 - 3.0m | hr | m | 0.29 | 0.29 | | |
| | £ | | 9.43 | 4.68 | 1.83 | 15.94 |

**CLASS I: PIPEWORK - PIPES**

**CLAY PIPES - continued**

| | | | | | | |
|---|---|---|---|---|---|---|
| 100mm | | | | | | |
| In trenches | | | | | | |
| 3.0 - 3.5m | hr | m | 0.36 | 0.36 | | |
| | £ | | **11.70** | **5.81** | **1.83** | **19.34** |
| 3.5 - 4.0m | hr | m | 0.44 | 0.44 | | |
| | £ | | **14.30** | **7.10** | **1.83** | **23.23** |
| 4.0 - 4.5m | hr | m | 0.55 | 0.55 | | |
| | £ | | **17.88** | **8.87** | **1.83** | **28.58** |
| 4.5 - 5.0m | hr | m | 0.70 | 0.70 | | |
| | £ | | **22.76** | **11.29** | **1.83** | **35.88** |
| 5.0 - 5.5m | hr | m | 0.90 | 0.90 | | |
| | £ | | **29.26** | **14.52** | **1.83** | **45.61** |
| 150mm | | | | | | |
| Not in trenches | hr | m | 0.06 | 0.02 | | |
| | £ | | **1.95** | **0.32** | **4.01** | **6.28** |
| In trenches | | | | | | |
| ne 1.5m | hr | m | 0.17 | 0.17 | | |
| | £ | | **5.53** | **2.74** | **4.01** | **12.28** |
| 1.5 - 2.0m | hr | m | 0.20 | 0.20 | | |
| | £ | | **6.50** | **3.23** | **4.01** | **13.74** |
| 2.0 - 2.5m | hr | m | 0.24 | 0.24 | | |
| | £ | | **7.80** | **3.87** | **4.01** | **15.68** |
| 2.5 - 3.0m | hr | m | 0.29 | 0.29 | | |
| | £ | | **9.43** | **4.68** | **4.01** | **18.12** |
| 3.0 - 3.5m | hr | m | 0.39 | 0.39 | | |
| | £ | | **12.68** | **6.29** | **4.01** | **22.98** |
| 3.5 - 4.0m | hr | m | 0.45 | 0.45 | | |
| | £ | | **14.63** | **7.26** | **4.01** | **25.90** |
| 4.0 - 4.5m | hr | m | 0.58 | 0.58 | | |
| | £ | | **18.86** | **9.36** | **4.01** | **32.23** |
| 4.5 - 5.0m | hr | m | 0.75 | 0.75 | | |
| | £ | | **24.38** | **12.10** | **4.01** | **40.49** |
| 5.0 - 5.5m | hr | m | 0.96 | 0.96 | | |
| | £ | | **31.21** | **15.48** | **4.01** | **50.70** |

**CLASS I: PIPEWORK - PIPES**

**CLAY PIPES - continued**

| | | UNIT | L | P | M | £ NETT |
|---|---|---|---|---|---|---|
| 200mm | | | | | | |
| Not in trenches | hr | m | 0.07 | 0.02 | | |
| | £ | | **2.28** | **0.32** | **8.68** | **11.28** |
| In trenches | | | | | | |
| ne 1.5m | hr | m | 0.18 | 0.18 | | |
| | £ | | **5.85** | **2.90** | **8.68** | **17.43** |
| 1.5 - 2.0m | hr | m | 0.21 | 0.21 | | |
| | £ | | **6.83** | **3.39** | **8.68** | **18.90** |
| 2.0 - 2.5m | hr | m | 0.25 | 0.25 | | |
| | £ | | **8.13** | **4.03** | **8.68** | **20.84** |
| 2.5 - 3.0m | hr | m | 0.30 | 0.30 | | |
| | £ | | **9.75** | **4.84** | **8.68** | **23.27** |
| 3.0 - 3.5m | hr | m | 0.40 | 0.40 | | |
| | £ | | **13.00** | **6.45** | **8.68** | **28.13** |
| 3.5 - 4.0m | hr | m | 0.46 | 0.46 | | |
| | £ | | **14.95** | **7.42** | **8.68** | **31.05** |
| 4.0 - 4.5m | hr | m | 0.63 | 0.63 | | |
| | £ | | **20.48** | **10.16** | **8.68** | **39.32** |
| 4.5 - 5.0m | hr | m | 0.78 | 0.78 | | |
| | £ | | **25.36** | **12.58** | **8.68** | **46.62** |
| 5.0 - 5.5m | hr | m | 1.00 | 1.00 | | |
| | £ | | **32.51** | **16.13** | **8.68** | **57.32** |
| 225mm | | | | | | |
| Not in trenches | hr | m | 0.08 | 0.02 | | |
| | £ | | **2.60** | **0.32** | **8.82** | **11.74** |
| In trenches | | | | | | |
| ne 1.5m | hr | m | 0.18 | 0.18 | | |
| | £ | | **5.85** | **2.90** | **8.82** | **17.57** |
| 1.5 - 2.0m | hr | m | 0.22 | 0.22 | | |
| | £ | | **7.15** | **3.55** | **8.82** | **19.52** |
| 2.0 - 2.5m | hr | m | 0.26 | 0.26 | | |
| | £ | | **8.45** | **4.19** | **8.82** | **21.46** |
| 2.5 - 3.0m | hr | m | 0.30 | 0.30 | | |
| | £ | | **9.75** | **4.84** | **8.82** | **23.41** |

## CLASS I: PIPEWORK - PIPES

### CLAY PIPES - continued

| | | UNIT | L | P | M | £ NETT |
|---|---|---|---|---|---|---|
| 225mm | | | | | | |
| In trenches | | | | | | |
| 3.0 - 3.5m | hr | m | 0.41 | 0.41 | | |
| | £ | | 13.33 | 6.61 | 8.82 | 28.76 |
| 3.5 - 4.0m | hr | m | 0.47 | 0.47 | | |
| | £ | | 15.28 | 7.58 | 8.82 | 31.68 |
| 4.0 - 4.5m | hr | m | 0.65 | 0.65 | | |
| | £ | | 21.13 | 10.48 | 8.82 | 40.43 |
| 4.5 - 5.0m | hr | m | 0.80 | 0.80 | | |
| | £ | | 26.01 | 12.90 | 8.82 | 47.73 |
| 5.0 - 5.5m | hr | m | 1.02 | 1.02 | | |
| | £ | | 33.16 | 16.45 | 8.82 | 58.43 |
| Vitrified clay pipes to BS65, spigot and socket joints with sealing ring (Hepseal), nominal bore | | | | | | |
| 100mm | | | | | | |
| Not in trenches | hr | m | 0.06 | 0.02 | | |
| | £ | | 1.95 | 0.32 | 4.84 | 7.11 |
| In trenches | | | | | | |
| ne 1.5m | hr | m | 0.15 | 0.15 | | |
| | £ | | 4.88 | 2.42 | 4.84 | 12.14 |
| 1.5 - 2.0m | hr | m | 0.19 | 0.19 | | |
| | £ | | 6.18 | 3.06 | 4.84 | 14.08 |
| 2.0 - 2.5m | hr | m | 0.24 | 0.24 | | |
| | £ | | 7.80 | 3.87 | 4.84 | 16.51 |
| 2.5 - 3.0m | hr | m | 0.29 | 0.29 | | |
| | £ | | 9.43 | 4.68 | 4.84 | 18.95 |
| 3.0 - 3.5m | hr | m | 0.36 | 0.36 | | |
| | £ | | 11.70 | 5.81 | 4.84 | 22.35 |
| 3.5 - 4.0m | hr | m | 0.44 | 0.44 | | |
| | £ | | 14.30 | 7.10 | 4.84 | 26.24 |
| 4.0 - 4.5m | hr | m | 0.55 | 0.55 | | |
| | £ | | 17.88 | 8.87 | 4.84 | 31.59 |
| 4.5 - 5.0m | hr | m | 0.70 | 0.70 | | |
| | £ | | 22.76 | 11.29 | 4.84 | 38.89 |
| 5.0 - 5.5m | hr | m | 0.90 | 0.90 | | |
| | £ | | 29.26 | 14.52 | 4.84 | 48.62 |

**CLASS I: PIPEWORK - PIPES**

**CLAY PIPES - continued**

| | UNIT | | L | P | M | £ NETT |
|---|---|---|---|---|---|---|
| 150mm | | | | | | |
| Not in trenches | hr | m | 0.06 | 0.02 | | |
| | £ | | 1.95 | 0.32 | 6.26 | 8.53 |
| In trenches | | | | | | |
| ne 1.5m | hr | m | 0.17 | 0.17 | | |
| | £ | | 5.53 | 2.74 | 6.26 | 14.53 |
| 1.5 - 2.0m | hr | m | 0.20 | 0.20 | | |
| | £ | | 6.50 | 3.23 | 6.26 | 15.99 |
| 2.0 - 2.5m | hr | m | 0.24 | 0.24 | | |
| | £ | | 7.80 | 3.87 | 6.26 | 17.93 |
| 2.5 - 3.0m | hr | m | 0.29 | 0.29 | | |
| | £ | | 9.43 | 4.68 | 6.26 | 20.37 |
| 3.0 - 3.5m | hr | m | 0.39 | 0.39 | | |
| | £ | | 12.68 | 6.29 | 6.26 | 25.23 |
| 3.5 - 4.0m | hr | m | 0.45 | 0.45 | | |
| | £ | | 14.63 | 7.26 | 6.26 | 28.15 |
| 4.0 - 4.5m | hr | m | 0.58 | 0.58 | | |
| | £ | | 18.86 | 9.36 | 6.26 | 34.48 |
| 4.5 - 5.0m | hr | m | 0.75 | 0.75 | | |
| | £ | | 24.38 | 12.10 | 6.26 | 42.74 |
| 5.0 - 5.5m | hr | m | 0.96 | 0.96 | | |
| | £ | | 31.21 | 15.48 | 6.26 | 52.95 |
| 225mm | | | | | | |
| Not in trenches | hr | m | 0.08 | 0.02 | | |
| | £ | | 2.60 | 0.32 | 12.26 | 15.18 |
| In trenches | | | | | | |
| ne 1.5m | hr | m | 0.18 | 0.18 | | |
| | £ | | 5.85 | 2.90 | 12.26 | 21.01 |
| 1.5 - 2.0m | hr | m | 0.22 | 0.22 | | |
| | £ | | 7.15 | 3.55 | 12.26 | 22.96 |
| 2.0 - 2.5m | hr | m | 0.26 | 0.26 | | |
| | £ | | 8.45 | 4.19 | 12.26 | 24.90 |
| 2.5 - 3.0m | hr | m | 0.30 | 0.30 | | |
| | £ | | 9.75 | 4.84 | 12.26 | 26.85 |
| 3.0 - 3.5m | hr | m | 0.41 | 0.41 | | |
| | £ | | 13.33 | 6.61 | 12.26 | 32.20 |

**CLASS I: PIPEWORK - PIPES**

**CLAY PIPES - continued**

| | | UNIT | L | P | M | £ NETT |
|---|---|---|---|---|---|---|
| 225mm | | | | | | |
| In trenches | | | | | | |
| 3.5 - 4.0m | hr | m | 0.47 | 0.47 | | |
| | £ | | 15.28 | 7.58 | 12.26 | 35.12 |
| 4.0 - 4.5m | hr | m | 0.65 | 0.65 | | |
| | £ | | 21.13 | 10.48 | 12.26 | 43.87 |
| 4.5 - 5.0m | hr | m | 0.80 | 0.80 | | |
| | £ | | 26.01 | 12.90 | 12.26 | 51.17 |
| 5.0 - 5.5m | hr | m | 1.02 | 1.02 | | |
| | £ | | 33.16 | 16.45 | 12.26 | 61.87 |
| 300mm | | | | | | |
| Not in trenches | hr | m | 0.10 | 0.05 | | |
| | £ | | 3.25 | 0.81 | 19.19 | 23.25 |
| In trenches | | | | | | |
| ne 1.5m | hr | m | 0.19 | 0.19 | | |
| | £ | | 6.18 | 3.06 | 19.19 | 28.43 |
| 1.5 - 2.0m | hr | m | 0.24 | 0.24 | | |
| | £ | | 7.80 | 3.87 | 19.19 | 30.86 |
| 2.0 - 2.5m | hr | m | 0.28 | 0.28 | | |
| | £ | | 9.10 | 4.52 | 19.19 | 32.81 |
| 2.5 - 3.0m | hr | m | 0.32 | 0.32 | | |
| | £ | | 10.40 | 5.16 | 19.19 | 34.75 |
| 3.0 - 3.5m | hr | m | 0.42 | 0.42 | | |
| | £ | | 13.65 | 6.77 | 19.19 | 39.61 |
| 3.5 - 4.0m | hr | m | 0.52 | 0.52 | | |
| | £ | | 16.91 | 8.39 | 19.19 | 44.49 |
| 4.0 - 4.5m | hr | m | 0.70 | 0.70 | | |
| | £ | | 22.76 | 11.29 | 19.19 | 53.24 |
| 4.5 - 5.0m | hr | m | 0.90 | 0.90 | | |
| | £ | | 29.26 | 14.52 | 19.19 | 62.97 |
| 5.0 - 5.5m | hr | m | 1.12 | 1.12 | | |
| | £ | | 36.41 | 18.07 | 19.19 | 73.67 |
| 400mm | | | | | | |
| Not in trenches | hr | m | 0.10 | 0.08 | | |
| | £ | | 3.25 | 1.29 | 37.48 | 42.02 |
| In trenches | | | | | | |
| ne 1.5m | hr | m | 0.23 | 0.23 | | |
| | £ | | 7.48 | 3.71 | 37.48 | 48.67 |

**CLASS I: PIPEWORK - PIPES**

**CLAY PIPES - continued**

| | | UNIT | L | P | M | £ NETT |
|---|---|---|---|---|---|---|
| 400mm | | | | | | |
| In trenches | | | | | | |
| 1.5 - 2.0m | hr | m | 0.29 | 0.29 | | |
| | £ | | 9.43 | 4.68 | 37.48 | 51.59 |
| 2.0 - 2.5m | hr | m | 0.32 | 0.32 | | |
| | £ | | 10.40 | 5.16 | 37.48 | 53.04 |
| 2.5 - 3.0m | hr | m | 0.38 | 0.38 | | |
| | £ | | 12.35 | 6.13 | 37.48 | 55.96 |
| 3.0 - 3.5m | hr | m | 0.46 | 0.46 | | |
| | £ | | 14.95 | 7.42 | 37.48 | 59.85 |
| 3.5 - 4.0m | hr | m | 0.58 | 0.58 | | |
| | £ | | 18.86 | 9.36 | 37.48 | 65.70 |
| 4.0 - 4.5m | hr | m | 0.75 | 0.75 | | |
| | £ | | 24.38 | 12.10 | 37.48 | 73.96 |
| 4.5 - 5.0m | hr | m | 0.95 | 0.95 | | |
| | £ | | 30.88 | 15.32 | 37.48 | 83.68 |
| 5.0 - 5.5m | hr | m | 1.20 | 1.20 | | |
| | £ | | 39.01 | 19.36 | 37.48 | 95.85 |
| 450mm | | | | | | |
| Not in trenches | hr | m | 0.12 | 0.12 | | |
| | £ | | 3.90 | 1.94 | 48.56 | 54.40 |
| In trenches | | | | | | |
| ne 1.5m | hr | m | 0.23 | 0.23 | | |
| | £ | | 7.48 | 3.71 | 48.56 | 59.75 |
| 1.5 - 2.0m | hr | m | 0.30 | 0.30 | | |
| | £ | | 9.75 | 4.84 | 48.56 | 63.15 |
| 2.0 - 2.5m | hr | m | 0.32 | 0.32 | | |
| | £ | | 10.40 | 5.16 | 48.56 | 64.12 |
| 2.5 - 3.0m | hr | m | 0.38 | 0.38 | | |
| | £ | | 12.35 | 6.13 | 48.56 | 67.04 |
| 3.0 - 3.5m | hr | m | 0.47 | 0.47 | | |
| | £ | | 15.28 | 7.58 | 48.56 | 71.42 |
| 3.5 - 4.0m | hr | m | 0.60 | 0.60 | | |
| | £ | | 19.51 | 9.68 | 48.56 | 77.75 |

**CLASS I: PIPEWORK - PIPES**

**CLAY PIPES - continued**

| | | UNIT | L | P | M | £ NETT |
|---|---|---|---|---|---|---|
| 450mm | | | | | | |
| In trenches | | | | | | |
| 4.0 - 4.5m | hr | m | 0.77 | 0.77 | | |
| | £ | | **25.03** | **12.42** | **48.56** | **86.01** |
| 4.5 - 5.0m | hr | m | 0.97 | 0.97 | | |
| | £ | | **31.53** | **15.65** | **48.56** | **95.74** |
| 5.0 - 5.5m | hr | m | 1.20 | 1.20 | | |
| | £ | | **39.01** | **19.36** | **48.56** | **106.93** |
| Vitrified clay pipes spigot and socket joints with sealing ring (Denseal), nominal bore | | | | | | |
| 100mm | | | | | | |
| Not in trenches | hr | m | 0.06 | 0.02 | | |
| | £ | | **1.95** | **0.32** | **3.14** | **5.41** |
| In trenches | | | | | | |
| ne 1.5m | hr | m | 0.15 | 0.15 | | |
| | £ | | **4.88** | **2.42** | **3.14** | **10.44** |
| 1.5 - 2.0m | hr | m | 0.19 | 0.19 | | |
| | £ | | **6.18** | **3.06** | **3.14** | **12.38** |
| 2.0 - 2.5m | hr | m | 0.24 | 0.24 | | |
| | £ | | **7.80** | **3.87** | **3.14** | **14.81** |
| 2.5 - 3.0m | hr | m | 0.29 | 0.29 | | |
| | £ | | **9.43** | **4.68** | **3.14** | **17.25** |
| 3.0 - 3.5m | hr | m | 0.36 | 0.36 | | |
| | £ | | **11.70** | **5.81** | **3.14** | **20.65** |
| 3.5 - 4.0m | hr | m | 0.44 | 0.44 | | |
| | £ | | **14.30** | **7.10** | **3.14** | **24.54** |
| 4.0 - 4.5m | hr | m | 0.55 | 0.55 | | |
| | £ | | **17.88** | **8.87** | **3.14** | **29.89** |
| 4.5 - 5.0m | hr | m | 0.70 | 0.70 | | |
| | £ | | **22.76** | **11.29** | **3.14** | **37.19** |
| 5.0 - 5.5m | hr | m | 0.90 | 0.90 | | |
| | £ | | **29.26** | **14.52** | **3.14** | **46.92** |
| 150mm | | | | | | |
| Not in trenches | hr | m | 0.06 | 0.02 | | |
| | £ | | **1.95** | **0.32** | **4.48** | **6.75** |
| In trenches | | | | | | |
| ne 1.5m | hr | m | 0.17 | 0.17 | | |
| | £ | | **5.53** | **2.74** | **4.48** | **12.75** |

**CLASS I: PIPEWORK - PIPES**

**CLAY PIPES - continued**

| | | UNIT | L | P | M | £ NETT |
|---|---|---|---|---|---|---|
| 150mm | | | | | | |
| In trenches | | | | | | |
| 1.5 - 2.0m | hr | m | 0.20 | 0.20 | | |
| | £ | | 6.50 | 3.23 | 4.48 | 14.21 |
| 2.0 - 2.5m | hr | m | 0.24 | 0.24 | | |
| | £ | | 7.80 | 3.87 | 4.48 | 16.15 |
| 2.5 - 3.0m | hr | m | 0.29 | 0.29 | | |
| | £ | | 9.43 | 4.68 | 4.48 | 18.59 |
| 3.0 - 3.5m | hr | m | 0.39 | 0.39 | | |
| | £ | | 12.68 | 6.29 | 4.48 | 23.45 |
| 3.5 - 4.0m | hr | m | 0.45 | 0.45 | | |
| | £ | | 14.63 | 7.26 | 4.48 | 26.37 |
| 4.0 - 4.5m | hr | m | 0.58 | 0.58 | | |
| | £ | | 18.86 | 9.36 | 4.48 | 32.70 |
| 4.5 - 5.0m | hr | m | 0.75 | 0.75 | | |
| | £ | | 24.38 | 12.10 | 4.48 | 40.96 |
| 5.0 - 5.5m | hr | m | 0.96 | 0.96 | | |
| | £ | | 31.21 | 15.48 | 4.48 | 51.17 |
| 225mm | | | | | | |
| Not in trenches | hr | m | 0.08 | 0.023 | | |
| | £ | | 2.60 | 0.37 | 8.84 | 11.81 |
| In trenches | | | | | | |
| ne 1.5m | hr | m | 0.18 | 0.18 | | |
| | £ | | 5.85 | 2.90 | 8.84 | 17.59 |
| 1.5 - 2.0m | hr | m | 0.22 | 0.22 | | |
| | £ | | 7.15 | 3.55 | 8.84 | 19.54 |
| 2.0 - 2.5m | hr | m | 0.26 | 0.26 | | |
| | £ | | 8.45 | 4.19 | 8.84 | 21.48 |
| 2.5 - 3.0m | hr | m | 0.30 | 0.30 | | |
| | £ | | 9.75 | 4.84 | 8.84 | 23.43 |
| 3.0 - 3.5m | hr | m | 0.41 | 0.41 | | |
| | £ | | 13.33 | 6.61 | 8.84 | 28.78 |
| 3.5 - 4.0m | hr | m | 0.47 | 0.47 | | |
| | £ | | 15.28 | 7.58 | 8.84 | 31.70 |
| 4.0 - 4.5m | hr | m | 0.65 | 0.65 | | |
| | £ | | 21.13 | 10.48 | 8.84 | 40.45 |
| 4.5 - 5.0m | hr | m | 0.80 | 0.80 | | |
| | £ | | 26.01 | 12.90 | 8.84 | 47.75 |

**CLASS I: PIPEWORK - PIPES**

**CLAY PIPES - continued**

| | | UNIT | L | P | M | £ NETT |
|---|---|---|---|---|---|---|
| 225mm | | | | | | |
| In trenches | | | | | | |
| 5.0 - 5.5m | hr | m | 1.02 | 1.02 | | |
| | £ | | 33.16 | 16.45 | 8.84 | 58.45 |
| 300mm | | | | | | |
| Not in trenches | hr | m | 0.10 | 0.05 | | |
| | £ | | 3.25 | 0.81 | 14.11 | 18.17 |
| In trenches | | | | | | |
| ne 1.5m | hr | m | 0.19 | 0.19 | | |
| | £ | | 6.18 | 3.06 | 14.11 | 23.35 |
| 1.5 - 2.0m | hr | m | 0.24 | 0.24 | | |
| | £ | | 7.80 | 3.87 | 14.11 | 25.78 |
| 2.0 - 2.5m | hr | m | 0.28 | 0.28 | | |
| | £ | | 9.10 | 4.52 | 14.11 | 27.73 |
| 2.5 - 3.0m | hr | m | 0.32 | 0.32 | | |
| | £ | | 10.40 | 5.16 | 14.11 | 29.67 |
| 3.0 - 3.5m | hr | m | 0.42 | 0.42 | | |
| | £ | | 13.65 | 6.77 | 14.11 | 34.53 |
| 3.5 - 4.0m | hr | m | 0.52 | 0.52 | | |
| | £ | | 16.91 | 8.39 | 14.11 | 39.41 |
| 4.0 - 4.5m | hr | m | 0.70 | 0.70 | | |
| | £ | | 22.76 | 11.29 | 14.11 | 48.16 |
| 4.5 - 5.0m | hr | m | 0.90 | 0.90 | | |
| | £ | | 29.26 | 14.52 | 14.11 | 57.89 |
| 5.0 - 5.5m | hr | m | 1.12 | 1.12 | | |
| | £ | | 36.41 | 18.07 | 14.11 | 68.59 |
| 375mm | | | | | | |
| Not in trenches | hr | m | 0.10 | 0.07 | | |
| | £ | | 3.25 | 1.13 | 27.40 | 31.78 |
| In trenches | | | | | | |
| ne 1.5m | hr | m | 0.22 | 0.22 | | |
| | £ | | 7.15 | 3.55 | 27.40 | 38.10 |
| 1.5 - 2.0m | hr | m | 0.28 | 0.28 | | |
| | £ | | 9.10 | 4.52 | 27.40 | 41.02 |
| 2.0 - 2.5m | hr | m | 0.31 | 0.31 | | |
| | £ | | 10.08 | 5.00 | 27.40 | 42.48 |
| 2.5 - 3.0m | hr | m | 0.36 | 0.36 | | |
| | £ | | 11.70 | 5.81 | 27.40 | 44.91 |

**CLASS I: PIPEWORK - PIPES**

**CLAY PIPES - continued**

| | | UNIT | L | P | M | £ NETT |
|---|---|---|---|---|---|---|
| 375mm | | | | | | |
| In trenches | | | | | | |
| 3.0 - 3.5m | hr | m | 0.45 | 0.45 | | |
| | £ | | 14.63 | 7.26 | 27.40 | 49.29 |
| 3.5 - 4.0m | hr | m | 0.56 | 0.56 | | |
| | £ | | 18.21 | 9.03 | 27.40 | 54.64 |
| 4.0 - 4.5m | hr | m | 0.74 | 0.74 | | |
| | £ | | 24.06 | 11.94 | 27.40 | 63.40 |
| 4.5 - 5.0m | hr | m | 0.94 | 0.94 | | |
| | £ | | 30.56 | 15.16 | 27.40 | 73.12 |
| 5.0 - 5.5m | hr | m | 1.18 | 1.18 | | |
| | £ | | 38.36 | 19.03 | 27.40 | 84.79 |
| 400mm | | | | | | |
| Not in trenches | hr | m | 0.10 | 0.08 | | |
| | £ | | 3.25 | 1.29 | 27.57 | 32.11 |
| In trenches | | | | | | |
| ne 1.5m | hr | m | 0.23 | 0.23 | | |
| | £ | | 7.48 | 3.71 | 27.57 | 38.76 |
| 1.5 - 2.0m | hr | m | 0.29 | 0.29 | | |
| | £ | | 9.43 | 4.68 | 27.57 | 41.68 |
| 2.0 - 2.5m | hr | m | 0.32 | 0.32 | | |
| | £ | | 10.40 | 5.16 | 27.57 | 43.13 |
| 2.5 - 3.0m | hr | m | 0.38 | 0.38 | | |
| | £ | | 12.35 | 6.13 | 27.57 | 46.05 |
| 3.0 - 3.5m | hr | m | 0.46 | 0.46 | | |
| | £ | | 14.95 | 7.42 | 27.57 | 49.94 |
| 3.5 - 4.0m | hr | m | 0.58 | 0.58 | | |
| | £ | | 18.86 | 9.36 | 27.57 | 55.79 |
| 4.0 - 4.5m | hr | m | 0.75 | 0.75 | | |
| | £ | | 24.38 | 12.10 | 27.57 | 64.05 |
| 4.5 - 5.0m | hr | m | 0.95 | 0.95 | | |
| | £ | | 30.88 | 15.32 | 27.57 | 73.77 |
| 5.0 - 5.5m | hr | m | 1.20 | 1.20 | | |
| | £ | | 39.01 | 19.36 | 27.57 | 85.94 |
| 450mm | | | | | | |
| Not in trenches | hr | m | 0.12 | 0.12 | | |
| | £ | | 3.90 | 1.94 | 35.71 | 41.55 |

## CLASS I: PIPEWORK - PIPES

### CLAY PIPES - continued

| | | UNIT | L | P | M | £ NETT |
|---|---|---|---|---|---|---|
| 450mm | | | | | | |
| In trenches | | | | | | |
| ne 1.5m | hr | m | 0.23 | 0.23 | | |
| | £ | | 7.48 | 3.71 | 35.71 | 46.90 |
| 1.5 - 2.0m | hr | m | 0.30 | 0.30 | | |
| | £ | | 9.75 | 4.84 | 35.71 | 50.30 |
| 2.0 - 2.5m | hr | m | 0.32 | 0.32 | | |
| | £ | | 10.40 | 5.16 | 35.71 | 51.27 |
| 2.5 - 3.0m | hr | m | 0.38 | 0.38 | | |
| | £ | | 12.35 | 6.13 | 35.71 | 54.19 |
| 3.0 - 3.5m | hr | m | 0.47 | 0.47 | | |
| | £ | | 15.28 | 7.58 | 35.71 | 58.57 |
| 3.5 - 4.0m | hr | m | 0.60 | 0.60 | | |
| | £ | | 19.51 | 9.68 | 35.71 | 64.90 |
| 4.0 - 4.5m | hr | m | 0.77 | 0.77 | | |
| | £ | | 25.03 | 12.42 | 35.71 | 73.16 |
| 4.5 - 5.0m | hr | m | 0.97 | 0.97 | | |
| | £ | | 31.53 | 15.65 | 35.71 | 82.89 |
| 5.0 - 5.5m | hr | m | 1.20 | 1.20 | | |
| | £ | | 39.01 | 19.36 | 35.71 | 94.08 |
| Field drains to BS1196 butt joints, nominal bore | | | | | | |
| 75mm | | | | | | |
| Not in trenches | hr | m | 0.06 | 0.02 | | |
| | £ | | 1.95 | 0.32 | 0.75 | 3.02 |
| In trenches | | | | | | |
| ne 1.5m | hr | m | 0.09 | 0.09 | | |
| | £ | | 2.93 | 1.45 | 0.75 | 5.13 |
| 1.5 - 2.0m | hr | m | 0.13 | 0.13 | | |
| | £ | | 4.23 | 2.10 | 0.75 | 7.08 |
| 2.0 - 2.5m | hr | m | 0.18 | 0.18 | | |
| | £ | | 5.85 | 2.90 | 0.75 | 9.50 |
| 2.5 - 3.0m | hr | m | 0.23 | 0.23 | | |
| | £ | | 7.48 | 3.71 | 0.75 | 11.94 |
| 100mm | | | | | | |
| Not in trenches | hr | m | 0.06 | 0.02 | | |
| | £ | | 1.95 | 0.32 | 1.33 | 3.60 |
| In trenches | | | | | | |
| ne 1.5m | hr | m | 0.10 | 0.10 | | |
| | £ | | 3.25 | 1.61 | 1.33 | 6.19 |

**CLASS I: PIPEWORK - PIPES**

**CLAY PIPES - continued**

| | UNIT | | L | P | M | £ NETT |
|---|---|---|---|---|---|---|
| 100mm | | | | | | |
| In trenches | | | | | | |
| 1.5 - 2.0m | hr | m | 0.14 | 0.14 | | |
| | £ | | 4.55 | 2.26 | 1.33 | 8.14 |
| 2.0 - 2.5m | hr | m | 0.19 | 0.19 | | |
| | £ | | 6.18 | 3.06 | 1.33 | 10.57 |
| 2.5 - 3.0m | hr | m | 0.24 | 0.24 | | |
| | £ | | 7.80 | 3.87 | 1.33 | 13.00 |
| 150mm | | | | | | |
| Not in trenches | hr | m | 0.06 | 0.02 | | |
| | £ | | 1.95 | 0.32 | 2.73 | 5.00 |
| In trenches | | | | | | |
| ne 1.5m | hr | m | 0.11 | 0.11 | | |
| | £ | | 3.58 | 1.77 | 2.73 | 8.08 |
| 1.5 - 2.0m | hr | m | 0.15 | 0.15 | | |
| | £ | | 4.88 | 2.42 | 2.73 | 10.03 |
| 2.0 - 2.5m | hr | m | 0.20 | 0.20 | | |
| | £ | | 6.50 | 3.23 | 2.73 | 12.46 |
| 2.5 - 3.0m | hr | m | 0.25 | 0.25 | | |
| | £ | | 8.13 | 4.03 | 2.73 | 14.89 |
| 225mm | | | | | | |
| Not in trenches | hr | m | 0.06 | 0.02 | | |
| | £ | | 1.95 | 0.32 | 7.21 | 9.48 |
| In trenches | | | | | | |
| ne 1.5m | hr | m | 0.13 | 0.13 | | |
| | £ | | 4.23 | 2.10 | 7.21 | 13.54 |
| 1.5 - 2.0m | hr | m | 0.17 | 0.17 | | |
| | £ | | 5.53 | 2.74 | 7.21 | 15.48 |
| 2.0 - 2.5m | hr | m | 0.22 | 0.22 | | |
| | £ | | 7.15 | 3.55 | 7.21 | 17.91 |
| 2.5 - 3.0m | hr | m | 0.27 | 0.27 | | |
| | £ | | 8.78 | 4.36 | 7.21 | 20.35 |
| Vitrified clay perforated field drains to BS65, sleeved joints (Hepline), nominal bore | | | | | | |
| 100mm | | | | | | |
| Not in trenches | hr | m | 0.06 | 0.02 | | |
| | £ | | 1.95 | 0.32 | 3.06 | 5.33 |
| In trenches | | | | | | |
| ne 1.5m | hr | m | 0.15 | 0.15 | | |
| | £ | | 4.88 | 2.42 | 3.06 | 10.36 |

**CLASS I: PIPEWORK - PIPES**

**CLAY PIPES - continued**

| | | UNIT | L | P | M | £ NETT |
|---|---|---|---|---|---|---|
| 100mm | | | | | | |
| In trenches | | | | | | |
| 1.5 - 2.0m | hr | m | 0.19 | 0.19 | | |
| | £ | | 6.18 | 3.06 | 3.06 | 12.30 |
| 2.0 - 2.5m | hr | m | 0.24 | 0.24 | | |
| | £ | | 7.80 | 3.87 | 3.06 | 14.73 |
| 2.5 - 3.0m | hr | m | 0.29 | 0.29 | | |
| | £ | | 9.43 | 4.68 | 3.06 | 17.17 |
| 150mm | | | | | | |
| Not in trenches | hr | m | 0.06 | 0.02 | | |
| | £ | | 1.95 | 0.32 | 5.62 | 7.89 |
| In trenches | | | | | | |
| ne 1.5m | hr | m | 0.17 | 0.17 | | |
| | £ | | 5.53 | 2.74 | 5.62 | 13.89 |
| 1.5 - 2.0m | hr | m | 0.20 | 0.20 | | |
| | £ | | 6.50 | 3.23 | 5.62 | 15.35 |
| 2.0 - 2.5m | hr | m | 0.24 | 0.24 | | |
| | £ | | 7.80 | 3.87 | 5.62 | 17.29 |
| 2.5 - 3.0m | hr | m | 0.29 | 0.29 | | |
| | £ | | 9.43 | 4.68 | 5.62 | 19.73 |
| 225mm | | | | | | |
| Not in trenches | hr | m | 0.08 | 0.02 | | |
| | £ | | 2.60 | 0.32 | 10.32 | 13.24 |
| In trenches | | | | | | |
| ne 1.5m | hr | m | 0.18 | 0.18 | | |
| | £ | | 5.85 | 2.90 | 10.32 | 19.07 |
| 1.5 - 2.0m | hr | m | 0.22 | 0.22 | | |
| | £ | | 7.15 | 3.55 | 10.32 | 21.02 |
| 2.0 - 2.5m | hr | m | 0.26 | 0.26 | | |
| | £ | | 8.45 | 4.19 | 10.32 | 22.96 |
| 2.5 - 3.0m | hr | m | 0.30 | 0.30 | | |
| | £ | | 9.75 | 4.84 | 10.32 | 24.91 |
| 300mm | | | | | | |
| Not in trenches | hr | m | 0.10 | 0.05 | | |
| | £ | | 3.25 | 0.81 | 19.33 | 23.39 |
| In trenches | | | | | | |
| ne 1.5m | hr | m | 0.19 | 0.19 | | |
| | £ | | 6.18 | 3.06 | 19.33 | 28.57 |
| 1.5 - 2.0m | hr | m | 0.24 | 0.24 | | |
| | £ | | 7.80 | 3.87 | 19.33 | 31.00 |

**CLASS I: PIPEWORK - PIPES**

**CLAY PIPES - continued**

| | | UNIT | L | P | M | £ NETT |
|---|---|---|---|---|---|---|
| 300mm | | | | | | |
| In trenches | | | | | | |
| 2.0 - 2.5m | hr | m | 0.28 | 0.28 | | |
| | £ | | 9.10 | 4.52 | 19.33 | 32.95 |
| 2.5 - 3.0m | hr | m | 0.32 | 0.32 | | |
| | £ | | 10.40 | 5.16 | 19.33 | 34.89 |
| Vitrified clay perforated field drains to BS65, sleeved joints (Denline), nominal bore | | | | | | |
| 100mm | | | | | | |
| Not in trenches | hr | m | 0.06 | 0.02 | | |
| | £ | | 1.95 | 0.32 | 2.34 | 4.61 |
| In trenches | | | | | | |
| ne 1.5m | hr | m | 0.15 | 0.15 | | |
| | £ | | 4.88 | 2.42 | 2.34 | 9.64 |
| 1.5 - 2.0m | hr | m | 0.19 | 0.19 | | |
| | £ | | 6.18 | 3.06 | 2.34 | 11.58 |
| 2.0 - 2.5m | hr | m | 0.24 | 0.24 | | |
| | £ | | 7.80 | 3.87 | 2.34 | 14.01 |
| 2.5 - 3.0m | hr | m | 0.29 | 0.29 | | |
| | £ | | 9.43 | 4.68 | 2.34 | 16.45 |
| 150mm | | | | | | |
| Not in trenches | hr | m | 0.06 | 0.02 | | |
| | £ | | 1.95 | 0.32 | 4.20 | 6.47 |
| In trenches | | | | | | |
| ne 1.5m | hr | m | 0.17 | 0.17 | | |
| | £ | | 5.53 | 2.74 | 4.20 | 12.47 |
| 1.5 - 2.0m | hr | m | 0.20 | 0.20 | | |
| | £ | | 6.50 | 3.23 | 4.20 | 13.93 |
| 2.0 - 2.5m | hr | m | 0.24 | 0.24 | | |
| | £ | | 7.80 | 3.87 | 4.20 | 15.87 |
| 2.5 - 3.0m | hr | m | 0.29 | 0.29 | | |
| | £ | | 9.43 | 4.68 | 4.20 | 18.31 |
| 225mm | | | | | | |
| Not in trenches | hr | m | 0.08 | 0.02 | | |
| | £ | | 2.60 | 0.32 | 7.70 | 10.62 |
| In trenches | | | | | | |
| ne 1.5m | hr | m | 0.18 | 0.18 | | |
| | £ | | 5.85 | 2.90 | 7.70 | 16.45 |

**CLASS I: PIPEWORK - PIPES**

**CLAY PIPES - continued**

| | | UNIT | L | P | M | £ NETT |
|---|---|---|---|---|---|---|
| 225mm | | | | | | |
| In trenches | | | | | | |
| 1.5 - 2.0m | hr | m | 0.22 | 0.22 | | |
| | £ | | 7.15 | 3.55 | 7.70 | 18.40 |
| 2.0 - 2.5m | hr | m | 0.26 | 0.26 | | |
| | £ | | 8.45 | 4.19 | 7.70 | 20.34 |
| 2.5 - 3.0m | hr | m | 0.30 | 0.30 | | |
| | £ | | 9.75 | 4.84 | 7.70 | 22.29 |
| 300mm | | | | | | |
| Not in trenches | hr | m | 0.10 | 0.05 | | |
| | £ | | 3.25 | 0.81 | 14.50 | 18.56 |
| In trenches | | | | | | |
| ne 1.5m | hr | m | 0.19 | 0.19 | | |
| | £ | | 6.18 | 3.06 | 14.50 | 23.74 |
| 1.5 - 2.0m | hr | m | 0.24 | 0.24 | | |
| | £ | | 7.80 | 3.87 | 14.50 | 26.17 |
| 2.0 - 2.5m | hr | m | 0.28 | 0.28 | | |
| | £ | | 9.10 | 4.52 | 14.50 | 28.12 |
| 2.5 - 3.0m | hr | m | 0.32 | 0.32 | | |
| | £ | | 10.40 | 5.16 | 14.50 | 30.06 |

**CLASS I: PIPEWORK - PIPES**

**CONCRETE PIPES**

| | | UNIT | L | P | M | £ NETT |
|---|---|---|---|---|---|---|
| Concrete pipe with rebated flexible joints to BS5911 Class L, nominal bore | | | | | | |
| 300mm | | | | | | |
| Not in trenches | hr | m | 0.12 | 0.06 | | |
| | £ | | 3.90 | 0.97 | 12.93 | 17.80 |
| In trenches | | | | | | |
| ne 1.5m | hr | m | 0.22 | 0.22 | | |
| | £ | | 7.15 | 3.55 | 12.93 | 23.63 |
| 1.5 - 2.0m | hr | m | 0.26 | 0.26 | | |
| | £ | | 8.45 | 4.19 | 12.93 | 25.57 |
| 2.0 - 2.5m | hr | m | 0.30 | 0.30 | | |
| | £ | | 9.75 | 4.84 | 12.93 | 27.52 |
| 2.5 - 3.0m | hr | m | 0.34 | 0.34 | | |
| | £ | | 11.05 | 5.48 | 12.93 | 29.46 |
| 3.0 - 3.5m | hr | m | 0.44 | 0.44 | | |
| | £ | | 14.30 | 7.10 | 12.93 | 34.33 |
| 3.5 - 4.0m | hr | m | 0.56 | 0.56 | | |
| | £ | | 18.21 | 9.03 | 12.93 | 40.17 |
| 4.0 - 4.5m | hr | m | 0.73 | 0.73 | | |
| | £ | | 23.73 | 11.77 | 12.93 | 48.43 |
| 4.5 - 5.0m | hr | m | 0.92 | 0.92 | | |
| | £ | | 29.91 | 14.84 | 12.93 | 57.68 |
| 5.0 - 5.5m | hr | m | 1.19 | 1.19 | | |
| | £ | | 38.69 | 19.19 | 12.93 | 70.81 |
| 5.5 - 6.0m | hr | m | 1.42 | 1.42 | | |
| | £ | | 46.16 | 22.90 | 12.93 | 81.99 |
| 375mm | | | | | | |
| Not in trenches | hr | m | 0.15 | 0.10 | | |
| | £ | | 4.88 | 1.61 | 16.88 | 23.37 |
| In trenches | | | | | | |
| ne 1.5m | hr | m | 0.24 | 0.24 | | |
| | £ | | 7.80 | 3.87 | 16.88 | 28.55 |
| 1.5 - 2.0m | hr | m | 0.29 | 0.29 | | |
| | £ | | 9.43 | 4.68 | 16.88 | 30.99 |
| 2.0 - 2.5m | hr | m | 0.33 | 0.33 | | |
| | £ | | 10.73 | 5.32 | 16.88 | 32.93 |
| 2.5 - 3.0m | hr | m | 0.38 | 0.38 | | |
| | £ | | 12.35 | 6.13 | 16.88 | 35.36 |

**CLASS I: PIPEWORK - PIPES**

**CONCRETE PIPES - continued**

| | | UNIT | L | P | M | £ NETT |
|---|---|---|---|---|---|---|
| 375mm | | | | | | |
| In trenches | | | | | | |
| 3.0 - 3.5m | hr | m | 0.46 | 0.46 | | |
| | £ | | 14.95 | 7.42 | 16.88 | 39.25 |
| 3.5 - 4.0m | hr | m | 0.58 | 0.58 | | |
| | £ | | 18.86 | 9.36 | 16.88 | 45.10 |
| 4.0 - 4.5m | hr | m | 0.75 | 0.75 | | |
| | £ | | 24.38 | 12.10 | 16.88 | 53.36 |
| 4.5 - 5.0m | hr | m | 0.95 | 0.95 | | |
| | £ | | 30.88 | 15.32 | 16.88 | 63.08 |
| 5.0 - 5.5m | hr | m | 1.20 | 1.20 | | |
| | £ | | 39.01 | 19.36 | 16.88 | 75.25 |
| 5.5 - 6.0m | hr | m | 1.45 | 1.45 | | |
| | £ | | 47.14 | 23.39 | 16.88 | 87.41 |
| 450mm | | | | | | |
| Not in trenches | hr | m | 0.17 | 0.12 | | |
| | £ | | 5.53 | 1.94 | 19.76 | 27.23 |
| In trenches | | | | | | |
| ne 1.5m | hr | m | 0.25 | 0.25 | | |
| | £ | | 8.13 | 4.03 | 19.76 | 31.92 |
| 1.5 - 2.0m | hr | m | 0.31 | 0.31 | | |
| | £ | | 10.08 | 5.00 | 19.76 | 34.84 |
| 2.0 - 2.5m | hr | m | 0.35 | 0.35 | | |
| | £ | | 11.38 | 5.65 | 19.76 | 36.79 |
| 2.5 - 3.0m | hr | m | 0.40 | 0.40 | | |
| | £ | | 13.00 | 6.45 | 19.76 | 39.21 |
| 3.0 - 3.5m | hr | m | 0.48 | 0.48 | | |
| | £ | | 15.60 | 7.74 | 19.76 | 43.10 |
| 3.5 - 4.0m | hr | m | 0.63 | 0.63 | | |
| | £ | | 20.48 | 10.16 | 19.76 | 50.40 |
| 4.0 - 4.5m | hr | m | 0.80 | 0.80 | | |
| | £ | | 26.01 | 12.90 | 19.76 | 58.67 |
| 4.5 - 5.0m | hr | m | 1.00 | 1.00 | | |
| | £ | | 32.51 | 16.13 | 19.76 | 68.40 |
| 5.0 - 5.5m | hr | m | 1.25 | 1.25 | | |
| | £ | | 40.64 | 20.16 | 19.76 | 80.56 |
| 5.5 - 6.0m | hr | m | 1.51 | 1.51 | | |
| | £ | | 49.09 | 24.36 | 19.76 | 93.21 |

**CLASS I: PIPEWORK - PIPES**

**CONCRETE PIPES - continued**

| | | UNIT | L | P | M | £ NETT |
|---|---|---|---|---|---|---|
| 525mm | | | | | | |
| Not in trenches | hr | m | 0.20 | 0.15 | | |
| | £ | | 6.50 | 2.42 | 23.13 | 32.05 |
| In trenches | | | | | | |
| ne 1.5m | hr | m | 0.27 | 0.27 | | |
| | £ | | 8.78 | 4.36 | 23.13 | 36.27 |
| 1.5 - 2.0m | hr | m | 0.33 | 0.33 | | |
| | £ | | 10.73 | 5.32 | 23.13 | 39.18 |
| 2.0 - 2.5m | hr | m | 0.37 | 0.37 | | |
| | £ | | 12.03 | 5.97 | 23.13 | 41.13 |
| 2.5 - 3.0m | hr | m | 0.42 | 0.42 | | |
| | £ | | 13.65 | 6.77 | 23.13 | 43.55 |
| 3.0 - 3.5m | hr | m | 0.49 | 0.49 | | |
| | £ | | 15.93 | 7.90 | 23.13 | 46.96 |
| 3.5 - 4.0m | hr | m | 0.65 | 0.65 | | |
| | £ | | 21.13 | 10.48 | 23.13 | 54.74 |
| 4.0 - 4.5m | hr | m | 0.83 | 0.83 | | |
| | £ | | 26.98 | 13.39 | 23.13 | 63.50 |
| 4.5 - 5.0m | hr | m | 1.03 | 1.03 | | |
| | £ | | 33.49 | 16.61 | 23.13 | 73.23 |
| 5.0 - 5.5m | hr | m | 1.28 | 1.28 | | |
| | £ | | 41.61 | 20.65 | 23.13 | 85.39 |
| 5.5 - 6.0m | hr | m | 1.55 | 1.55 | | |
| | £ | | 50.39 | 25.00 | 23.13 | 98.52 |

Large Bore Pipes will require additional plant to assist in lowering pipes into trenches.
(Part time use of 20T Crane)

Labour = 33.48 Plant = 20.05

| | | UNIT | L | P | M | £ NETT |
|---|---|---|---|---|---|---|
| 750mm | | | | | | |
| Not in trenches | hr | m | 0.22 | 0.17 | | |
| | £ | | 7.37 | 3.41 | 38.19 | 48.97 |
| In trenches | | | | | | |
| ne 1.5m | hr | m | 0.30 | 0.30 | | |
| | £ | | 10.04 | 6.02 | 38.19 | 54.25 |
| 1.5 - 2.0m | hr | m | 0.36 | 0.36 | | |
| | £ | | 12.05 | 7.22 | 38.19 | 57.46 |

**CLASS I: PIPEWORK - PIPES**

**CONCRETE PIPES - continued**

| | | UNIT | L | P | M | £ NETT |
|---|---|---|---|---|---|---|
| 750mm | | | | | | |
| In trenches | | | | | | |
| 2.0 - 2.5m | hr | m | 0.41 | 0.41 | | |
| | £ | | 13.73 | 8.22 | 38.19 | 60.14 |
| 2.5 - 3.0m | hr | m | 0.47 | 0.47 | | |
| | £ | | 15.74 | 9.42 | 38.19 | 63.35 |
| 3.0 - 3.5m | hr | m | 0.58 | 0.58 | | |
| | £ | | 19.42 | 11.63 | 38.19 | 69.24 |
| 3.5 - 4.0m | hr | m | 0.80 | 0.80 | | |
| | £ | | 26.78 | 16.04 | 38.19 | 81.01 |
| 4.0 - 4.5m | hr | m | 1.05 | 1.05 | | |
| | £ | | 35.15 | 21.05 | 38.19 | 94.39 |
| 4.5 - 5.0m | hr | m | 1.30 | 1.30 | | |
| | £ | | 43.52 | 26.07 | 38.19 | 107.78 |
| 5.0 - 5.5m | hr | m | 1.55 | 1.55 | | |
| | £ | | 51.89 | 31.08 | 38.19 | 121.16 |
| 5.5 - 6.0m | hr | m | 1.82 | 1.82 | | |
| | £ | | 60.93 | 36.49 | 38.19 | 135.61 |
| 900mm | | | | | | |
| Not in trenches | hr | m | 0.25 | 0.20 | | |
| | £ | | 8.37 | 4.01 | 52.90 | 65.28 |
| In trenches | | | | | | |
| ne 1.5m | hr | m | 0.33 | 0.33 | | |
| | £ | | 11.05 | 6.62 | 52.90 | 70.57 |
| 1.5 - 2.0m | hr | m | 0.40 | 0.40 | | |
| | £ | | 13.39 | 8.02 | 52.90 | 74.31 |
| 2.0 - 2.5m | hr | m | 0.46 | 0.46 | | |
| | £ | | 15.40 | 9.22 | 52.90 | 77.52 |
| 2.5 - 3.0m | hr | m | 0.53 | 0.53 | | |
| | £ | | 17.74 | 10.63 | 52.90 | 81.27 |

**CLASS I: PIPEWORK - PIPES**

**CONCRETE PIPES - continued**

| | | UNIT | L | P | M | £ NETT |
|---|---|---|---|---|---|---|
| 900mm | | | | | | |
| In trenches | | | | | | |
| 3.0 - 3.5m | hr | m | 0.70 | 0.70 | | |
| | £ | | 23.44 | 14.04 | 52.90 | 90.38 |
| 3.5 - 4.0m | hr | m | 0.92 | 0.92 | | |
| | £ | | 30.80 | 18.45 | 52.90 | 102.15 |
| 4.0 - 4.5m | hr | m | 1.20 | 1.20 | | |
| | £ | | 40.18 | 24.06 | 52.90 | 117.14 |
| 4.5 - 5.0m | hr | m | 1.50 | 1.50 | | |
| | £ | | 50.22 | 30.08 | 52.90 | 133.20 |
| 5.0 - 5.5m | hr | m | 1.80 | 1.80 | | |
| | £ | | 60.26 | 36.09 | 52.90 | 149.25 |
| 5.5 - 6.0m | hr | m | 2.10 | 2.10 | | |
| | £ | | 70.31 | 42.11 | 52.90 | 165.32 |
| 1200mm | | | | | | |
| Not in trenches | hr | m | 0.30 | 0.25 | | |
| | £ | | 10.04 | 5.01 | 84.65 | 99.70 |
| In trenches | | | | | | |
| ne 1.5m | hr | m | 0.46 | 0.46 | | |
| | £ | | 15.40 | 9.22 | 84.65 | 109.27 |
| 1.5 - 2.0m | hr | m | 0.53 | 0.53 | | |
| | £ | | 17.74 | 10.63 | 84.65 | 113.02 |
| 2.0 - 2.5m | hr | m | 0.60 | 0.60 | | |
| | £ | | 20.09 | 12.03 | 84.65 | 116.77 |
| 2.5 - 3.0m | hr | m | 0.70 | 0.70 | | |
| | £ | | 23.44 | 14.04 | 84.65 | 122.13 |
| 3.0 - 3.5m | hr | m | 0.85 | 0.85 | | |
| | £ | | 28.46 | 17.04 | 84.65 | 130.15 |
| 3.5 - 4.0m | hr | m | 1.12 | 1.12 | | |
| | £ | | 37.50 | 22.46 | 84.65 | 144.61 |
| 4.0 - 4.5m | hr | m | 1.45 | 1.45 | | |
| | £ | | 48.55 | 29.07 | 84.65 | 162.27 |
| 4.5 - 5.0m | hr | m | 1.75 | 1.75 | | |
| | £ | | 58.59 | 35.09 | 84.65 | 178.33 |
| 5.0 - 5.5m | hr | m | 2.05 | 2.05 | | |
| | £ | | 68.63 | 41.10 | 84.65 | 194.38 |
| 5.5 - 6.0m | hr | m | 2.36 | 2.36 | | |
| | £ | | 79.01 | 47.32 | 84.65 | 210.98 |

**CLASS I: PIPEWORK - PIPES**

**CONCRETE PIPES - continued**

| | | UNIT | L | P | M | £ NETT |
|---|---|---|---|---|---|---|
| 1500mm | | | | | | |
| Not in trenches | hr | m | 0.35 | 0.30 | | |
| | £ | | 11.72 | 6.02 | 133.44 | 151.18 |
| In trenches | | | | | | |
| ne 1.5m | hr | m | 0.60 | 0.60 | | |
| | £ | | 20.09 | 12.03 | 133.44 | 165.56 |
| 1.5 - 2.0m | hr | m | 0.70 | 0.70 | | |
| | £ | | 23.44 | 14.04 | 133.44 | 170.92 |
| 2.0 - 2.5m | hr | m | 0.81 | 0.81 | | |
| | £ | | 27.12 | 16.24 | 133.44 | 176.80 |
| 2.5 - 3.0m | hr | m | 0.92 | 0.92 | | |
| | £ | | 30.80 | 18.45 | 133.44 | 182.69 |
| 3.0 - 3.5m | hr | m | 1.05 | 1.05 | | |
| | £ | | 35.15 | 21.05 | 133.44 | 189.64 |
| 3.5 - 4.0m | hr | m | 1.27 | 1.27 | | |
| | £ | | 42.52 | 25.46 | 133.44 | 201.42 |
| 4.0 - 4.5m | hr | m | 1.70 | 1.70 | | |
| | £ | | 56.92 | 34.09 | 133.44 | 224.45 |
| 4.5 - 5.0m | hr | m | 2.05 | 2.05 | | |
| | £ | | 68.63 | 41.10 | 133.44 | 243.17 |
| 5.0 - 5.5m | hr | m | 2.40 | 2.40 | | |
| | £ | | 80.35 | 48.12 | 133.44 | 261.91 |
| 5.5 - 6.0m | hr | m | 2.75 | 2.75 | | |
| | £ | | 92.07 | 55.14 | 133.44 | 280.65 |
| Concrete pipe with rebated flexible joints to BS5911 Class M, nominal bore | | | | | | |
| 300mm | | | | | | |
| Not in trenches | hr | m | 0.12 | 0.06 | | |
| | £ | | 3.90 | 0.97 | 13.34 | 18.21 |
| In trenches | | | | | | |
| ne 1.5m | hr | m | 0.22 | 0.22 | | |
| | £ | | 7.15 | 3.55 | 13.34 | 24.04 |
| 1.5 - 2.0m | hr | m | 0.26 | 0.26 | | |
| | £ | | 8.45 | 4.19 | 13.34 | 25.98 |
| 2.0 - 2.5m | hr | m | 0.30 | 0.30 | | |
| | £ | | 9.75 | 4.84 | 13.34 | 27.93 |
| 2.5 - 3.0m | hr | m | 0.34 | 0.34 | | |
| | £ | | 11.05 | 5.48 | 13.34 | 29.87 |

**CLASS I: PIPEWORK - PIPES**

**CONCRETE PIPES - continued**

| | | UNIT | L | P | M | £ NETT |
|---|---|---|---|---|---|---|
| 300mm | | | | | | |
| In trenches | | | | | | |
| 3.0 - 3.5m | hr | m | 0.44 | 0.44 | | |
| | £ | | 14.30 | 7.10 | 13.34 | 34.74 |
| 3.5 - 4.0m | hr | m | 0.56 | 0.56 | | |
| | £ | | 18.21 | 9.03 | 13.34 | 40.58 |
| 4.0 - 4.5m | hr | m | 0.73 | 0.73 | | |
| | £ | | 23.73 | 11.77 | 13.34 | 48.84 |
| 4.5 - 5.0m | hr | m | 0.92 | 0.92 | | |
| | £ | | 29.91 | 14.84 | 13.34 | 58.09 |
| 5.0 - 5.5m | hr | m | 1.19 | 1.19 | | |
| | £ | | 38.69 | 19.19 | 13.34 | 71.22 |
| 5.5 - 6.0m | hr | m | 1.42 | 1.42 | | |
| | £ | | 46.16 | 22.90 | 13.34 | 82.40 |
| 375mm | | | | | | |
| Not in trenches | hr | m | 0.15 | 0.10 | | |
| | £ | | 4.88 | 1.61 | 17.07 | 23.56 |
| In trenches | | | | | | |
| ne 1.5m | hr | m | 0.24 | 0.24 | | |
| | £ | | 7.80 | 3.87 | 17.07 | 28.74 |
| 1.5 - 2.0m | hr | m | 0.29 | 0.29 | | |
| | £ | | 9.43 | 4.68 | 17.07 | 31.18 |
| 2.0 - 2.5m | hr | m | 0.33 | 0.33 | | |
| | £ | | 10.73 | 5.32 | 17.07 | 33.12 |
| 2.5 - 3.0m | hr | m | 0.38 | 0.38 | | |
| | £ | | 12.35 | 6.13 | 17.07 | 35.55 |
| 3.0 - 3.5m | hr | m | 0.46 | 0.46 | | |
| | £ | | 14.95 | 7.42 | 17.07 | 39.44 |
| 3.5 - 4.0m | hr | m | 0.58 | 0.58 | | |
| | £ | | 18.86 | 9.36 | 17.07 | 45.29 |
| 4.0 - 4.5m | hr | m | 0.75 | 0.75 | | |
| | £ | | 24.38 | 12.10 | 17.07 | 53.55 |
| 4.5 - 5.0m | hr | m | 0.95 | 0.95 | | |
| | £ | | 30.88 | 15.32 | 17.07 | 63.27 |
| 5.0 - 5.5m | hr | m | 1.20 | 1.20 | | |
| | £ | | 39.01 | 19.36 | 17.07 | 75.44 |
| 5.5 - 6.0m | hr | m | 1.45 | 1.45 | | |
| | £ | | 47.14 | 23.39 | 17.07 | 87.60 |

**CLASS I: PIPEWORK - PIPES**

**CONCRETE PIPES - continued**

| | | UNIT | L | P | M | £ NETT |
|---|---|---|---|---|---|---|
| 450mm | | | | | | |
| Not in trenches | hr | m | 0.17 | 0.12 | | |
| | £ | | 5.53 | 1.94 | 19.95 | 27.42 |
| In trenches | | | | | | |
| ne 1.5m | hr | m | 0.25 | 0.25 | | |
| | £ | | 8.13 | 4.03 | 19.95 | 32.11 |
| 1.5 - 2.0m | hr | m | 0.31 | 0.31 | | |
| | £ | | 10.08 | 5.00 | 19.95 | 35.03 |
| 2.0 - 2.5m | hr | m | 0.35 | 0.35 | | |
| | £ | | 11.38 | 5.65 | 19.95 | 36.98 |
| 2.5 - 3.0m | hr | m | 0.40 | 0.40 | | |
| | £ | | 13.00 | 6.45 | 19.95 | 39.40 |
| 3.0 - 3.5m | hr | m | 0.48 | 0.48 | | |
| | £ | | 15.60 | 7.74 | 19.95 | 43.29 |
| 3.5 - 4.0m | hr | m | 0.63 | 0.63 | | |
| | £ | | 20.48 | 10.16 | 19.95 | 50.59 |
| 4.0 - 4.5m | hr | m | 0.80 | 0.80 | | |
| | £ | | 26.01 | 12.90 | 19.95 | 58.86 |
| 4.5 - 5.0m | hr | m | 1.00 | 1.00 | | |
| | £ | | 32.51 | 16.13 | 19.95 | 68.59 |
| 5.0 - 5.5m | hr | m | 1.25 | 1.25 | | |
| | £ | | 40.64 | 20.16 | 19.95 | 80.75 |
| 5.5 - 6.0m | hr | m | 1.51 | 1.51 | | |
| | £ | | 49.09 | 24.36 | 19.95 | 93.40 |
| 525mm | | | | | | |
| Not in trenches | hr | m | 0.20 | 0.15 | | |
| | £ | | 6.50 | 2.42 | 23.32 | 32.24 |
| In trenches | | | | | | |
| ne 1.5m | hr | m | 0.27 | 0.27 | | |
| | £ | | 8.78 | 4.36 | 23.32 | 36.46 |
| 1.5 - 2.0m | hr | m | 0.33 | 0.33 | | |
| | £ | | 10.73 | 5.32 | 23.32 | 39.37 |
| 2.0 - 2.5m | hr | m | 0.37 | 0.37 | | |
| | £ | | 12.03 | 5.97 | 23.32 | 41.32 |
| 2.5 - 3.0m | hr | m | 0.42 | 0.42 | | |
| | £ | | 13.65 | 6.77 | 23.32 | 43.74 |

**CLASS I: PIPEWORK - PIPES**

**CONCRETE PIPES - continued**

| | | UNIT | L | P | M | £ NETT |
|---|---|---|---|---|---|---|
| 525mm | | | | | | |
| In trenches | | | | | | |
| 3.0 - 3.5m | hr | m | 0.49 | 0.49 | | |
| | £ | | **15.93** | **7.90** | **23.32** | **47.15** |
| 3.5 - 4.0m | hr | m | 0.65 | 0.65 | | |
| | £ | | **21.13** | **10.48** | **23.32** | **54.93** |
| 4.0 - 4.5m | hr | m | 0.83 | 0.83 | | |
| | £ | | **26.98** | **13.39** | **23.32** | **63.69** |
| 4.5 - 5.0m | hr | m | 1.03 | 1.03 | | |
| | £ | | **33.49** | **16.61** | **23.32** | **73.42** |
| 5.0 - 5.5m | hr | m | 1.28 | 1.28 | | |
| | £ | | **41.61** | **20.65** | **23.32** | **85.58** |
| 5.5 - 6.0m | hr | m | 1.55 | 1.55 | | |
| | £ | | **50.39** | **25.00** | **23.32** | **98.71** |

Large Bore Pipes will require additional plant to assist in lowering pipes into trenches
(Part time use of 20T Crane)

Labour = 33.48 Plant = 20.05

| | | UNIT | L | P | M | £ NETT |
|---|---|---|---|---|---|---|
| 750mm | | | | | | |
| Not in trenches | hr | m | 0.22 | 0.17 | | |
| | £ | | **7.37** | **3.41** | **38.57** | **49.35** |
| In trenches | | | | | | |
| ne 1.5m | hr | m | 0.30 | 0.30 | | |
| | £ | | **10.04** | **6.02** | **38.57** | **54.63** |
| 1.5 - 2.0m | hr | m | 0.36 | 0.36 | | |
| | £ | | **12.05** | **7.22** | **38.57** | **57.84** |
| 2.0 - 2.5m | hr | m | 0.41 | 0.41 | | |
| | £ | | **13.73** | **8.22** | **38.57** | **60.52** |
| 2.5 - 3.0m | hr | m | 0.47 | 0.47 | | |
| | £ | | **15.74** | **9.42** | **38.57** | **63.73** |

**CLASS I: PIPEWORK - PIPES**

**CONCRETE PIPES - continued**

| | | UNIT | L | P | M | £ NETT |
|---|---|---|---|---|---|---|
| 750mm | | | | | | |
| In trenches | | | | | | |
| 3.0 - 3.5m | hr | m | 0.58 | 0.58 | | |
| | £ | | 19.42 | 11.63 | 38.57 | 69.62 |
| 3.5 - 4.0m | hr | m | 0.80 | 0.80 | | |
| | £ | | 26.78 | 16.04 | 38.57 | 81.39 |
| 4.0 - 4.5m | hr | m | 1.05 | 1.05 | | |
| | £ | | 35.15 | 21.05 | 38.57 | 94.77 |
| 4.5 - 5.0m | hr | m | 1.30 | 1.30 | | |
| | £ | | 43.52 | 26.07 | 38.57 | 108.16 |
| 5.0 - 5.5m | hr | m | 1.55 | 1.55 | | |
| | £ | | 51.89 | 31.08 | 38.57 | 121.54 |
| 5.5 - 6.0m | hr | m | 1.82 | 1.82 | | |
| | £ | | 60.93 | 36.49 | 38.57 | 135.99 |
| 900mm | | | | | | |
| Not in trenches | hr | m | 0.25 | 0.20 | | |
| | £ | | 8.37 | 4.01 | 54.29 | 66.67 |
| In trenches | | | | | | |
| ne 1.5m | hr | m | 0.33 | 0.33 | | |
| | £ | | 11.05 | 6.62 | 54.29 | 71.96 |
| 1.5 - 2.0m | hr | m | 0.40 | 0.40 | | |
| | £ | | 13.39 | 8.02 | 54.29 | 75.70 |
| 2.0 - 2.5m | hr | m | 0.46 | 0.46 | | |
| | £ | | 15.40 | 9.22 | 54.29 | 78.91 |
| 2.5 - 3.0m | hr | m | 0.53 | 0.53 | | |
| | £ | | 17.74 | 10.63 | 54.29 | 82.66 |
| 3.0 - 3.5m | hr | m | 0.70 | 0.70 | | |
| | £ | | 23.44 | 14.04 | 54.29 | 91.77 |
| 3.5 - 4.0m | hr | m | 0.92 | 0.92 | | |
| | £ | | 30.80 | 18.45 | 54.29 | 103.54 |
| 4.0 - 4.5m | hr | m | 1.20 | 1.20 | | |
| | £ | | 40.18 | 24.06 | 54.29 | 118.53 |
| 4.5 - 5.0m | hr | m | 1.50 | 1.50 | | |
| | £ | | 50.22 | 30.08 | 54.29 | 134.59 |
| 5.0 - 5.5m | hr | m | 1.80 | 1.80 | | |
| | £ | | 60.26 | 36.09 | 54.29 | 150.64 |
| 5.5 - 6.0m | hr | m | 2.10 | 2.10 | | |
| | £ | | 70.31 | 42.11 | 54.29 | 166.71 |

**CLASS I: PIPEWORK - PIPES**

**CONCRETE PIPES - continued**

| | | UNIT | L | P | M | £ NETT |
|---|---|---|---|---|---|---|
| 1200mm | | | | | | |
| Not in trenches | hr | m | 0.30 | 0.25 | | |
| | £ | | 10.04 | 5.01 | 93.39 | 108.44 |
| In trenches | | | | | | |
| ne 1.5m | hr | m | 0.46 | 0.46 | | |
| | £ | | 15.40 | 9.22 | 93.39 | 118.01 |
| 1.5 - 2.0m | hr | m | 0.53 | 0.53 | | |
| | £ | | 17.74 | 10.63 | 93.39 | 121.76 |
| 2.0 - 2.5m | hr | m | 0.60 | 0.60 | | |
| | £ | | 20.09 | 12.03 | 93.39 | 125.51 |
| 2.5 - 3.0m | hr | m | 0.70 | 0.70 | | |
| | £ | | 23.44 | 14.04 | 93.39 | 130.87 |
| 3.0 - 3.5m | hr | m | 0.85 | 0.85 | | |
| | £ | | 28.46 | 17.04 | 93.39 | 138.89 |
| 3.5 - 4.0m | hr | m | 1.12 | 1.12 | | |
| | £ | | 37.50 | 22.46 | 93.39 | 153.35 |
| 4.0 - 4.5m | hr | m | 1.45 | 1.45 | | |
| | £ | | 48.55 | 29.07 | 93.39 | 171.01 |
| 4.5 - 5.0m | hr | m | 1.75 | 1.75 | | |
| | £ | | 58.59 | 35.09 | 93.39 | 187.07 |
| 5.0 - 5.5m | hr | m | 2.05 | 2.05 | | |
| | £ | | 68.63 | 41.10 | 93.39 | 203.12 |
| 5.5 - 6.0m | hr | m | 2.36 | 2.36 | | |
| | £ | | 79.01 | 47.32 | 93.39 | 219.72 |
| 1500mm | | | | | | |
| Not in trenches | hr | m | 0.35 | 0.30 | | |
| | £ | | 11.72 | 6.02 | 142.06 | 159.80 |
| In trenches | | | | | | |
| ne 1.5m | hr | m | 0.60 | 0.60 | | |
| | £ | | 20.09 | 12.03 | 142.06 | 174.18 |
| 1.5 - 2.0m | hr | m | 0.70 | 0.70 | | |
| | £ | | 23.44 | 14.04 | 142.06 | 179.54 |
| 2.0 - 2.5m | hr | m | 0.81 | 0.81 | | |
| | £ | | 27.12 | 16.24 | 142.06 | 185.42 |
| 2.5 - 3.0m | hr | m | 0.92 | 0.92 | | |
| | £ | | 30.80 | 18.45 | 142.06 | 191.31 |

**CLASS I : PIPEWORK - PIPES**

**CONCRETE PIPES - continued**

| | | UNIT | L | P | M | £ NETT |
|---|---|---|---|---|---|---|
| 1500mm | | | | | | |
| In trenches | | | | | | |
| 3.0 - 3.5m | hr | m | 1.05 | 1.05 | | |
| | £ | | 35.15 | 21.05 | 142.06 | 198.26 |
| 3.5 - 4.0m | hr | m | 1.27 | 1.27 | | |
| | £ | | 42.52 | 25.46 | 142.06 | 210.04 |
| 4.0 - 4.5m | hr | m | 1.70 | 1.70 | | |
| | £ | | 56.92 | 34.09 | 142.06 | 233.07 |
| 4.5 - 5.0m | hr | m | 2.05 | 2.05 | | |
| | £ | | 68.63 | 41.10 | 142.06 | 251.79 |
| 5.0 - 5.5m | hr | m | 2.40 | 2.40 | | |
| | £ | | 80.35 | 48.12 | 142.06 | 270.53 |
| 5.5 - 6.0m | hr | m | 2.75 | 2.75 | | |
| | £ | | 92.07 | 55.14 | 142.06 | 289.27 |
| Concrete pipe with rebated flexible joints to BS5911 Class H, nominal bore | | | | | | |
| 375mm | | | | | | |
| Not in trenches | hr | m | 0.15 | 0.10 | | |
| | £ | | 4.88 | 1.61 | 17.36 | 23.85 |
| In trenches | | | | | | |
| ne 1.5m | hr | m | 0.24 | 0.24 | | |
| | £ | | 7.80 | 3.87 | 17.36 | 29.03 |
| 1.5 - 2.0m | hr | m | 0.29 | 0.29 | | |
| | £ | | 9.43 | 4.68 | 17.36 | 31.47 |
| 2.0 - 2.5m | hr | m | 0.33 | 0.33 | | |
| | £ | | 10.73 | 5.32 | 17.36 | 33.41 |
| 2.5 - 3.0m | hr | m | 0.38 | 0.38 | | |
| | £ | | 12.35 | 6.13 | 17.36 | 35.84 |
| 3.0 - 3.5m | hr | m | 0.46 | 0.46 | | |
| | £ | | 14.95 | 7.42 | 17.36 | 39.73 |
| 3.5 - 4.0m | hr | m | 0.58 | 0.58 | | |
| | £ | | 18.86 | 9.36 | 17.36 | 45.58 |

**CLASS I : PIPEWORK - PIPES**

**CONCRETE PIPES - continued**

| | | UNIT | L | P | M | £ NETT |
|---|---|---|---|---|---|---|
| 375mm | | | | | | |
| In trenches | | | | | | |
| 4.0 - 4.5m | hr | m | 0.75 | 0.75 | | |
| | £ | | 24.38 | 12.10 | 17.36 | 53.84 |
| 4.5 - 5.0m | hr | m | 0.95 | 0.95 | | |
| | £ | | 30.88 | 15.32 | 17.36 | 63.56 |
| 5.0 - 5.5m | hr | m | 1.20 | 1.20 | | |
| | £ | | 39.01 | 19.36 | 17.36 | 75.73 |
| 5.5 - 6.0m | hr | m | 1.45 | 1.45 | | |
| | £ | | 47.14 | 23.39 | 17.36 | 87.89 |
| 450mm | | | | | | |
| Not in trenches | hr | m | 0.17 | 0.12 | | |
| | £ | | 5.53 | 1.94 | 20.35 | 27.82 |
| In trenches | | | | | | |
| ne 1.5m | hr | m | 0.25 | 0.25 | | |
| | £ | | 8.13 | 4.03 | 20.35 | 32.51 |
| 1.5 - 2.0m | hr | m | 0.31 | 0.31 | | |
| | £ | | 10.08 | 5.00 | 20.35 | 35.43 |
| 2.0 - 2.5m | hr | m | 0.35 | 0.35 | | |
| | £ | | 11.38 | 5.65 | 20.35 | 37.38 |
| 2.5 - 3.0m | hr | m | 0.40 | 0.40 | | |
| | £ | | 13.00 | 6.45 | 20.35 | 39.80 |
| 3.0 - 3.5m | hr | m | 0.48 | 0.48 | | |
| | £ | | 15.60 | 7.74 | 20.35 | 43.69 |
| 3.5 - 4.0m | hr | m | 0.63 | 0.63 | | |
| | £ | | 20.48 | 10.16 | 20.35 | 50.99 |
| 4.0 - 4.5m | hr | m | 0.80 | 0.80 | | |
| | £ | | 26.01 | 12.90 | 20.35 | 59.26 |
| 4.5 - 5.0m | hr | m | 1.00 | 1.00 | | |
| | £ | | 32.51 | 16.13 | 20.35 | 68.99 |
| 5.0 - 5.5m | hr | m | 1.25 | 1.25 | | |
| | £ | | 40.64 | 20.16 | 20.35 | 81.15 |
| 5.5 - 6.0m | hr | m | 1.51 | 1.51 | | |
| | £ | | 49.09 | 24.36 | 20.35 | 93.80 |

**CLASS I : PIPEWORK - PIPES**

**CONCRETE PIPES - continued**

| | | UNIT | L | P | M | £ NETT |
|---|---|---|---|---|---|---|
| 525mm | | | | | | |
| Not in trenches | hr | m | 0.20 | 0.15 | | |
| | £ | | 6.50 | 2.42 | 23.79 | 32.71 |
| In trenches | | | | | | |
| ne 1.5m | hr | m | 0.27 | 0.27 | | |
| | £ | | 8.78 | 4.36 | 23.79 | 36.93 |
| 1.5 - 2.0m | hr | m | 0.33 | 0.33 | | |
| | £ | | 10.73 | 5.32 | 23.79 | 39.84 |
| 2.0 - 2.5m | hr | m | 0.37 | 0.37 | | |
| | £ | | 12.03 | 5.97 | 23.79 | 41.79 |
| 2.5 - 3.0m | hr | m | 0.42 | 0.42 | | |
| | £ | | 13.65 | 6.77 | 23.79 | 44.21 |
| 3.0 - 3.5m | hr | m | 0.49 | 0.49 | | |
| | £ | | 15.93 | 7.90 | 23.79 | 47.62 |
| 3.5 - 4.0m | hr | m | 0.65 | 0.65 | | |
| | £ | | 21.13 | 10.48 | 23.79 | 55.40 |
| 4.0 - 4.5m | hr | m | 0.83 | 0.83 | | |
| | £ | | 26.98 | 13.39 | 23.79 | 64.16 |
| 4.5 - 5.0m | hr | m | 1.03 | 1.03 | | |
| | £ | | 33.49 | 16.61 | 23.79 | 73.89 |
| 5.0 - 5.5m | hr | m | 1.28 | 1.28 | | |
| | £ | | 41.61 | 20.65 | 23.79 | 86.05 |
| 5.5 - 6.0m | hr | m | 1.55 | 1.55 | | |
| | £ | | 50.39 | 25.00 | 23.79 | 99.18 |

Large Bore Pipes will require additional plant to assist in lowering pipes into trenches.
(Part time use of 20T Crane)

Labour = 33.48 Plant = 20.05

| | | UNIT | L | P | M | £ NETT |
|---|---|---|---|---|---|---|
| 750mm | | | | | | |
| Not in trenches | hr | m | 0.22 | 0.17 | | |
| | £ | | 7.37 | 3.41 | 42.41 | 53.19 |
| In trenches | | | | | | |
| ne 1.5m | hr | m | 0.30 | 0.30 | | |
| | £ | | 10.04 | 6.02 | 42.41 | 58.47 |
| 1.5 - 2.0m | hr | m | 0.36 | 0.36 | | |
| | £ | | 12.05 | 7.22 | 42.41 | 61.68 |

**CLASS I : PIPEWORK - PIPES**

**CONCRETE PIPES - continued**

| | | UNIT | L | P | M | £ NETT |
|---|---|---|---|---|---|---|
| 750mm | | | | | | |
| In trenches | | | | | | |
| 2.0 - 2.5m | hr | m | 0.41 | 0.41 | | |
| | £ | | 13.73 | 8.22 | 42.41 | 64.36 |
| 2.5 - 3.0m | hr | m | 0.47 | 0.47 | | |
| | £ | | 15.74 | 9.42 | 42.41 | 67.57 |
| 3.0 - 3.5m | hr | m | 0.58 | 0.58 | | |
| | £ | | 19.42 | 11.63 | 42.41 | 73.46 |
| 3.5 - 4.0m | hr | m | 0.80 | 0.80 | | |
| | £ | | 26.78 | 16.04 | 42.41 | 85.23 |
| 4.0 - 4.5m | hr | m | 1.05 | 1.05 | | |
| | £ | | 35.15 | 21.05 | 42.41 | 98.61 |
| 4.5 - 5.0m | hr | m | 1.30 | 1.30 | | |
| | £ | | 43.52 | 26.07 | 42.41 | 112.00 |
| 5.0 - 5.5m | hr | m | 1.55 | 1.55 | | |
| | £ | | 51.89 | 31.08 | 42.41 | 125.38 |
| 5.5 - 6.0m | hr | m | 1.82 | 1.82 | | |
| | £ | | 60.93 | 36.49 | 42.41 | 139.83 |
| 900mm | | | | | | |
| Not in trenches | hr | m | 0.25 | 0.20 | | |
| | £ | | 8.37 | 4.01 | 55.96 | 68.34 |
| In trenches | | | | | | |
| ne 1.5m | hr | m | 0.33 | 0.33 | | |
| | £ | | 11.05 | 6.62 | 55.96 | 73.63 |
| 1.5 - 2.0m | hr | m | 0.40 | 0.40 | | |
| | £ | | 13.39 | 8.02 | 55.96 | 77.37 |
| 2.0 - 2.5m | hr | m | 0.46 | 0.46 | | |
| | £ | | 15.40 | 9.22 | 55.96 | 80.58 |
| 2.5 - 3.0m | hr | m | 0.53 | 0.53 | | |
| | £ | | 17.74 | 10.63 | 55.96 | 84.33 |
| 3.0 - 3.5m | hr | m | 0.70 | 0.70 | | |
| | £ | | 23.44 | 14.04 | 55.96 | 93.44 |
| 3.5 - 4.0m | hr | m | 0.92 | 0.92 | | |
| | £ | | 30.80 | 18.45 | 55.96 | 105.21 |
| 4.0 - 4.5m | hr | m | 1.20 | 1.20 | | |
| | £ | | 40.18 | 24.06 | 55.96 | 120.20 |

**CLASS I : PIPEWORK - PIPES**

**CONCRETE PIPES - continued**

| | | UNIT | L | P | M | £ NETT |
|---|---|---|---|---|---|---|
| 900mm | | | | | | |
| In trenches | | | | | | |
| 4.5 - 5.0m | hr | m | 1.50 | 1.50 | | |
| | £ | | 50.22 | 30.08 | 55.96 | 136.26 |
| 5.0 - 5.5m | hr | m | 1.80 | 1.80 | | |
| | £ | | 60.26 | 36.09 | 55.96 | 152.31 |
| 5.5 - 6.0m | hr | m | 2.10 | 2.10 | | |
| | £ | | 70.31 | 42.11 | 55.96 | 168.38 |
| 1200mm | | | | | | |
| Not in trenches | hr | m | 0.30 | 0.25 | | |
| | £ | | 10.04 | 5.01 | 99.32 | 114.37 |
| In trenches | | | | | | |
| ne 1.5m | hr | m | 0.46 | 0.46 | | |
| | £ | | 15.40 | 9.22 | 99.32 | 123.94 |
| 1.5 - 2.0m | hr | m | 0.53 | 0.53 | | |
| | £ | | 17.74 | 10.63 | 99.32 | 127.69 |
| 2.0 - 2.5m | hr | m | 0.60 | 0.60 | | |
| | £ | | 20.09 | 12.03 | 99.32 | 131.44 |
| 2.5 - 3.0m | hr | m | 0.70 | 0.70 | | |
| | £ | | 23.44 | 14.04 | 99.32 | 136.80 |
| 3.0 - 3.5m | hr | m | 0.85 | 0.85 | | |
| | £ | | 28.46 | 17.04 | 99.32 | 144.82 |
| 3.5 - 4.0m | hr | m | 1.12 | 1.12 | | |
| | £ | | 37.50 | 22.46 | 99.32 | 159.28 |
| 4.0 - 4.5m | hr | m | 1.45 | 1.45 | | |
| | £ | | 48.55 | 29.07 | 99.32 | 176.94 |
| 4.5 - 5.0m | hr | m | 1.75 | 1.75 | | |
| | £ | | 58.59 | 35.09 | 99.32 | 193.00 |
| 5.0 - 5.5m | hr | m | 2.05 | 2.05 | | |
| | £ | | 68.63 | 41.10 | 99.32 | 209.05 |
| 5.5 - 6.0m | hr | m | 2.36 | 2.36 | | |
| | £ | | 79.01 | 47.32 | 99.32 | 225.65 |

**CLASS I : PIPEWORK - PIPES**

**CONCRETE PIPES - continued**

| | | UNIT | L | P | M | £ NETT |
|---|---|---|---|---|---|---|
| 1500mm | | | | | | |
| Not in trenches | hr | m | 0.35 | 0.30 | | |
| | £ | | 11.72 | 6.02 | 147.34 | 165.08 |
| In trenches | | | | | | |
| ne 1.5m | hr | m | 0.60 | 0.60 | | |
| | £ | | 20.09 | 12.03 | 147.34 | 179.46 |
| 1.5 - 2.0m | hr | m | 0.70 | 0.70 | | |
| | £ | | 23.44 | 14.04 | 147.34 | 184.82 |
| 2.0 - 2.5m | hr | m | 0.81 | 0.81 | | |
| | £ | | 27.12 | 16.24 | 147.34 | 190.70 |
| 2.5 - 3.0m | hr | m | 0.92 | 0.92 | | |
| | £ | | 30.80 | 18.45 | 147.34 | 196.59 |
| 3.0 - 3.5m | hr | m | 1.05 | 1.05 | | |
| | £ | | 35.15 | 21.05 | 147.34 | 203.54 |
| 3.5 - 4.0m | hr | m | 1.27 | 1.27 | | |
| | £ | | 42.52 | 25.46 | 147.34 | 215.32 |
| 4.0 - 4.5m | hr | m | 1.70 | 1.70 | | |
| | £ | | 56.92 | 34.09 | 147.34 | 238.35 |
| 4.5 - 5.0m | hr | m | 2.05 | 2.05 | | |
| | £ | | 68.63 | 41.10 | 147.34 | 257.07 |
| 5.0 - 5.5m | hr | m | 2.40 | 2.40 | | |
| | £ | | 80.35 | 48.12 | 147.34 | 275.81 |
| 5.5 - 6.0m | hr | m | 2.75 | 2.75 | | |
| | £ | | 92.07 | 55.14 | 147.34 | 294.55 |

**CLASS I: PIPEWORK - PIPES**

**CAST OR SPUN IRON PIPES**

| | UNIT | | L | P | M | £ NETT |
|---|---|---|---|---|---|---|
| Cast iron pipe to BS437 plain ended pipe with "Timesaver" mechanical coupling joints, nominal bore | | | | | | |
| 75mm | | | | | | |
| Not in trenches | hr | m | 0.12 | 0.05 | | |
| | £ | | 4.02 | 1.00 | 19.87 | 24.89 |
| In trenches | | | | | | |
| ne 1.5m | hr | m | 0.19 | 0.19 | | |
| | £ | | 6.36 | 3.81 | 19.87 | 30.04 |
| 1.5 - 2.0m | hr | m | 0.21 | 0.21 | | |
| | £ | | 7.03 | 4.21 | 19.87 | 31.11 |
| 2.0 - 2.5m | hr | m | 0.28 | 0.28 | | |
| | £ | | 9.37 | 5.61 | 19.87 | 34.85 |
| 2.5 - 3.0m | hr | m | 0.34 | 0.34 | | |
| | £ | | 11.38 | 6.82 | 19.87 | 38.07 |
| 3.0 - 3.5m | hr | m | 0.43 | 0.43 | | |
| | £ | | 14.40 | 8.62 | 19.87 | 42.89 |
| 3.5 - 4.0m | hr | m | 0.55 | 0.55 | | |
| | £ | | 18.41 | 11.03 | 19.87 | 49.31 |
| 4.0 - 4.5m | hr | m | 0.71 | 0.71 | | |
| | £ | | 23.77 | 14.24 | 19.87 | 57.88 |
| 4.5 - 5.0m | hr | m | 0.89 | 0.89 | | |
| | £ | | 29.80 | 17.84 | 19.87 | 67.51 |
| 5.0 - 5.5m | hr | m | 1.15 | 1.15 | | |
| | £ | | 38.50 | 23.06 | 19.87 | 81.43 |
| 100mm | | | | | | |
| Not in trenches | hr | m | 0.13 | 0.05 | | |
| | £ | | 4.35 | 1.00 | 19.84 | 25.19 |
| In trenches | | | | | | |
| ne 1.5m | hr | m | 0.21 | 0.21 | | |
| | £ | | 7.03 | 4.21 | 19.84 | 31.08 |
| 1.5 - 2.0m | hr | m | 0.23 | 0.23 | | |
| | £ | | 7.70 | 4.61 | 19.84 | 32.15 |
| 2.0 - 2.5m | hr | m | 0.30 | 0.30 | | |
| | £ | | 10.04 | 6.02 | 19.84 | 35.90 |
| 2.5 - 3.0m | hr | m | 0.37 | 0.37 | | |
| | £ | | 12.39 | 7.42 | 19.84 | 39.65 |

**CLASS I: PIPEWORK - PIPES**

**CAST OR SPUN IRON PIPES - continued**

| | UNIT | | L | P | M | £ NETT |
|---|---|---|---|---|---|---|
| 100mm | | | | | | |
| In trenches | | | | | | |
| 3.0 - 3.5m | hr | m | 0.48 | 0.48 | | |
| | £ | | 16.07 | 9.62 | 19.84 | 45.53 |
| 3.5 - 4.0m | hr | m | 0.60 | 0.60 | | |
| | £ | | 20.09 | 12.03 | 19.84 | 51.96 |
| 4.0 - 4.5m | hr | m | 0.75 | 0.75 | | |
| | £ | | 25.11 | 15.04 | 19.84 | 59.99 |
| 4.5 - 5.0m | hr | m | 0.95 | 0.95 | | |
| | £ | | 31.81 | 19.05 | 19.84 | 70.70 |
| 5.0 - 5.5m | hr | m | 1.22 | 1.22 | | |
| | £ | | 40.85 | 24.46 | 19.84 | 85.15 |
| 150mm | | | | | | |
| Not in trenches | hr | m | 0.14 | 0.05 | | |
| | £ | | 4.69 | 1.00 | 35.02 | 40.71 |
| In trenches | | | | | | |
| ne 1.5m | hr | m | 0.24 | 0.24 | | |
| | £ | | 8.04 | 4.81 | 35.02 | 47.87 |
| 1.5 - 2.0m | hr | m | 0.28 | 0.28 | | |
| | £ | | 9.37 | 5.61 | 35.02 | 50.00 |
| 2.0 - 2.5m | hr | m | 0.34 | 0.34 | | |
| | £ | | 11.38 | 6.82 | 35.02 | 53.22 |
| 2.5 - 3.0m | hr | m | 0.40 | 0.40 | | |
| | £ | | 13.39 | 8.02 | 35.02 | 56.43 |
| 3.0 - 3.5m | hr | m | 0.54 | 0.54 | | |
| | £ | | 18.08 | 10.83 | 35.02 | 63.93 |
| 3.5 - 4.0m | hr | m | 0.61 | 0.61 | | |
| | £ | | 20.42 | 12.23 | 35.02 | 67.67 |
| 4.0 - 4.5m | hr | m | 0.79 | 0.79 | | |
| | £ | | 26.45 | 15.84 | 35.02 | 77.31 |
| 4.5 - 5.0m | hr | m | 1.05 | 1.05 | | |
| | £ | | 35.15 | 21.05 | 35.02 | 91.22 |
| 5.0 - 5.5m | hr | m | 1.32 | 1.32 | | |
| | £ | | 44.19 | 26.47 | 35.02 | 105.68 |

**CLASS I: PIPEWORK - PIPES**

**CAST OR SPUN IRON PIPES**

| | UNIT | | L | P | M | £ NETT |
|---|---|---|---|---|---|---|
| 225mm | | | | | | |
| Not in trenches | hr | m | 0.15 | 0.05 | | |
| | £ | | **5.02** | **1.00** | **83.27** | **89.29** |
| In trenches | | | | | | |
| ne 1.5m | hr | m | 0.25 | 0.25 | | |
| | £ | | **8.37** | **5.01** | **83.27** | **96.65** |
| 1.5 - 2.0m | hr | m | 0.30 | 0.30 | | |
| | £ | | **10.04** | **6.02** | **83.27** | **99.33** |
| 2.0 - 2.5m | hr | m | 0.35 | 0.35 | | |
| | £ | | **11.72** | **7.02** | **83.27** | **102.01** |
| 2.5 - 3.0m | hr | m | 0.41 | 0.41 | | |
| | £ | | **13.73** | **8.22** | **83.27** | **105.22** |
| 3.0 - 3.5m | hr | m | 0.54 | 0.54 | | |
| | £ | | **18.08** | **10.83** | **83.27** | **112.18** |
| 3.5 - 4.0m | hr | m | 0.61 | 0.61 | | |
| | £ | | **20.42** | **12.23** | **83.27** | **115.92** |
| 4.0 - 4.5m | hr | m | 0.77 | 0.77 | | |
| | £ | | **25.78** | **15.44** | **83.27** | **124.49** |
| 4.5 - 5.0m | hr | m | 1.02 | 1.02 | | |
| | £ | | **34.15** | **20.45** | **83.27** | **137.87** |
| 5.0 - 5.5m | hr | m | 1.30 | 1.30 | | |
| | £ | | **43.52** | **26.07** | **83.27** | **152.86** |
| Ductile spun iron pipes to BS4772, "Tyton" joints, nominal bore | | | | | | |
| 100mm | | | | | | |
| Not in trenches | hr | m | 0.10 | 0.05 | | |
| | £ | | **3.35** | **1.00** | **7.53** | **11.88** |
| In trenches | | | | | | |
| ne 1.5m | hr | m | 0.18 | 0.18 | | |
| | £ | | **6.03** | **3.61** | **7.53** | **17.17** |
| 1.5 - 2.0m | hr | m | 0.20 | 0.20 | | |
| | £ | | **6.70** | **4.01** | **7.53** | **18.24** |
| 2.0 - 2.5m | hr | m | 0.26 | 0.26 | | |
| | £ | | **8.70** | **5.21** | **7.53** | **21.44** |
| 2.5 - 3.0m | hr | m | 0.32 | 0.32 | | |
| | £ | | **10.71** | **6.42** | **7.53** | **24.66** |

**CLASS I: PIPEWORK - PIPES**

**CAST OR SPUN IRON PIPES - continued**

| | UNIT | | L | P | M | £ NETT |
|---|---|---|---|---|---|---|
| 100mm | | | | | | |
| In trenches | | | | | | |
| 3.0 - 3.5m | hr | m | 0.42 | 0.42 | | |
| | £ | | 14.06 | 8.42 | 7.53 | 30.01 |
| 3.5 - 4.0m | hr | m | 0.53 | 0.53 | | |
| | £ | | 17.74 | 10.63 | 7.53 | 35.90 |
| 4.0 - 4.5m | hr | m | 0.66 | 0.66 | | |
| | £ | | 22.10 | 13.23 | 7.53 | 42.86 |
| 4.5 - 5.0m | hr | m | 0.84 | 0.84 | | |
| | £ | | 28.12 | 16.84 | 7.53 | 52.49 |
| 5.0 - 5.5m | hr | m | 1.08 | 1.08 | | |
| | £ | | 36.16 | 21.65 | 7.53 | 65.34 |
| 150mm | | | | | | |
| Not in trenches | hr | m | 0.11 | 0.05 | | |
| | £ | | 3.68 | 1.00 | 11.14 | 15.82 |
| In trenches | | | | | | |
| ne 1.5m | hr | m | 0.21 | 0.21 | | |
| | £ | | 7.03 | 4.21 | 11.14 | 22.38 |
| 1.5 - 2.0m | hr | m | 0.25 | 0.25 | | |
| | £ | | 8.37 | 5.01 | 11.14 | 24.52 |
| 2.0 - 2.5m | hr | m | 0.30 | 0.30 | | |
| | £ | | 10.04 | 6.02 | 11.14 | 27.20 |
| 2.5 - 3.0m | hr | m | 0.37 | 0.37 | | |
| | £ | | 12.39 | 7.42 | 11.14 | 30.95 |
| 3.0 - 3.5m | hr | m | 0.49 | 0.49 | | |
| | £ | | 16.41 | 9.82 | 11.14 | 37.37 |
| 3.5 - 4.0m | hr | m | 0.55 | 0.55 | | |
| | £ | | 18.41 | 11.03 | 11.14 | 40.58 |
| 4.0 - 4.5m | hr | m | 0.72 | 0.72 | | |
| | £ | | 24.11 | 14.44 | 11.14 | 49.69 |
| 4.5 - 5.0m | hr | m | 0.94 | 0.94 | | |
| | £ | | 31.47 | 18.85 | 11.14 | 61.46 |
| 5.0 - 5.5m | hr | m | 1.19 | 1.19 | | |
| | £ | | 39.84 | 23.86 | 11.14 | 74.84 |

CLASS I: PIPEWORK - PIPES

CAST OR SPUN IRON PIPES - continued

| | UNIT | | L | P | M | £ NETT |
|---|---|---|---|---|---|---|
| 250mm | | | | | | |
| Not in trenches | hr | m | 0.16 | 0.05 | | |
| | £ | | 5.36 | 1.00 | 20.11 | 26.47 |
| In trenches | | | | | | |
| ne 1.5m | hr | m | 0.24 | 0.24 | | |
| | £ | | 8.04 | 4.81 | 20.11 | 32.96 |
| 1.5 - 2.0m | hr | m | 0.30 | 0.30 | | |
| | £ | | 10.04 | 6.02 | 20.11 | 36.17 |
| 2.0 - 2.5m | hr | m | 0.35 | 0.35 | | |
| | £ | | 11.72 | 7.02 | 20.11 | 38.85 |
| 2.5 - 3.0m | hr | m | 0.42 | 0.42 | | |
| | £ | | 14.06 | 8.42 | 20.11 | 42.59 |
| 3.0 - 3.5m | hr | m | 0.56 | 0.56 | | |
| | £ | | 18.75 | 11.23 | 20.11 | 50.09 |
| 3.5 - 4.0m | hr | m | 0.64 | 0.64 | | |
| | £ | | 21.43 | 12.83 | 20.11 | 54.37 |
| 4.0 - 4.5m | hr | m | 0.81 | 0.81 | | |
| | £ | | 27.12 | 16.24 | 20.11 | 63.47 |
| 4.5 - 5.0m | hr | m | 1.02 | 1.02 | | |
| | £ | | 34.15 | 20.45 | 20.11 | 74.71 |
| 5.0 - 5.5m | hr | m | 1.27 | 1.27 | | |
| | £ | | 42.52 | 25.46 | 20.11 | 88.09 |
| 350mm | | | | | | |
| Not in trenches | hr | m | 0.21 | 0.10 | | |
| | £ | | 7.03 | 2.01 | 31.95 | 40.99 |
| In trenches | | | | | | |
| ne 1.5m | hr | m | 0.30 | 0.30 | | |
| | £ | | 10.04 | 6.02 | 31.95 | 48.01 |
| 1.5 - 2.0m | hr | m | 0.39 | 0.39 | | |
| | £ | | 13.06 | 7.82 | 31.95 | 52.83 |
| 2.0 - 2.5m | hr | m | 0.43 | 0.43 | | |
| | £ | | 14.40 | 8.62 | 31.95 | 54.97 |
| 2.5 - 3.0m | hr | m | 0.52 | 0.52 | | |
| | £ | | 17.41 | 10.43 | 31.95 | 59.79 |

**CLASS I: PIPEWORK - PIPES**

**CAST OR SPUN IRON PIPES - continued**

| | UNIT | | L | P | M | £ NETT |
|---|---|---|---|---|---|---|
| 350mm | | | | | | |
| 3.0 - 3.5m | hr | m | 0.66 | 0.66 | | |
| | £ | | 22.10 | 13.23 | 31.95 | 67.28 |
| 3.5 - 4.0m | hr | m | 0.85 | 0.85 | | |
| | £ | | 28.46 | 17.04 | 31.95 | 77.45 |
| 4.0 - 4.5m | hr | m | 1.09 | 1.09 | | |
| | £ | | 36.49 | 21.85 | 31.95 | 90.29 |
| 4.5 - 5.0m | hr | m | 1.38 | 1.38 | | |
| | £ | | 46.20 | 27.67 | 31.95 | 105.82 |
| 5.0 - 5.5m | hr | m | 1.68 | 1.68 | | |
| | £ | | 56.25 | 33.68 | 31.95 | 121.88 |
| 450mm | | | | | | |
| Not in trenches | hr | m | 0.26 | 0.12 | | |
| | £ | | 8.70 | 2.41 | 44.86 | 55.97 |
| In trenches | | | | | | |
| ne 1.5m | hr | m | 0.36 | 0.36 | | |
| | £ | | 12.05 | 7.22 | 44.86 | 64.13 |
| 1.5 - 2.0m | hr | m | 0.47 | 0.47 | | |
| | £ | | 15.74 | 9.42 | 44.86 | 70.02 |
| 2.0 - 2.5m | hr | m | 0.53 | 0.53 | | |
| | £ | | 17.74 | 10.63 | 44.86 | 73.23 |
| 2.5 - 3.0m | hr | m | 0.62 | 0.62 | | |
| | £ | | 20.76 | 12.43 | 44.86 | 78.05 |
| 3.0 - 3.5m | hr | m | 0.77 | 0.77 | | |
| | £ | | 25.78 | 15.44 | 44.86 | 86.08 |
| 3.5 - 4.0m | hr | m | 0.98 | 0.98 | | |
| | £ | | 32.81 | 19.65 | 44.86 | 97.32 |
| 4.0 - 4.5m | hr | m | 1.23 | 1.23 | | |
| | £ | | 41.18 | 24.66 | 44.86 | 110.70 |
| 4.5 - 5.0m | hr | m | 1.55 | 1.55 | | |
| | £ | | 51.89 | 31.08 | 44.86 | 127.83 |
| 5.0 - 5.5m | hr | m | 1.86 | 1.86 | | |
| | £ | | 62.27 | 37.29 | 44.86 | 144.42 |

**CLASS I: PIPEWORK - PIPES**

**CAST OR SPUN IRON PIPES - continued**

| | UNIT | | L | P | M | £ NETT |
|---|---|---|---|---|---|---|
| 600mm | | | | | | |
| Not in trenches | hr | m | 0.34 | 0.15 | | |
| | £ | | 11.38 | 3.01 | 68.65 | 83.04 |
| In trenches | | | | | | |
| ne 1.5m | hr | m | 0.47 | 0.47 | | |
| | £ | | 15.74 | 9.42 | 68.65 | 93.81 |
| 1.5 - 2.0m | hr | m | 0.55 | 0.55 | | |
| | £ | | 18.41 | 11.03 | 68.65 | 98.09 |
| 2.0 - 2.5m | hr | m | 0.66 | 0.66 | | |
| | £ | | 22.10 | 13.23 | 68.65 | 103.98 |
| 2.5 - 3.0m | hr | m | 0.78 | 0.78 | | |
| | £ | | 26.11 | 15.64 | 68.65 | 110.40 |
| 3.0 - 3.5m | hr | m | 0.89 | 0.89 | | |
| | £ | | 29.80 | 17.84 | 68.65 | 116.29 |
| 3.5 - 4.0m | hr | m | 1.09 | 1.09 | | |
| | £ | | 36.49 | 21.85 | 68.65 | 126.99 |
| 4.0 - 4.5m | hr | m | 1.37 | 1.37 | | |
| | £ | | 45.87 | 27.47 | 68.65 | 141.99 |
| 4.5 - 5.0m | hr | m | 1.70 | 1.70 | | |
| | £ | | 56.92 | 34.09 | 68.65 | 159.66 |
| 5.0 - 5.5m | hr | m | 2.03 | 2.03 | | |
| | £ | | 67.96 | 40.70 | 68.65 | 177.31 |
| Ductile spun iron pipes to BS4772, Stanlock joints, nominal bore | | | | | | |
| 100mm | | | | | | |
| Not in trenches | hr | m | 0.10 | 0.05 | | |
| | £ | | 3.35 | 1.00 | 12.11 | 16.46 |
| In trenches | | | | | | |
| ne 1.5m | hr | m | 0.18 | 0.18 | | |
| | £ | | 6.03 | 3.61 | 12.11 | 21.75 |
| 1.5 - 2.0m | hr | m | 0.20 | 0.20 | | |
| | £ | | 6.70 | 4.01 | 12.11 | 22.82 |
| 2.0 - 2.5m | hr | m | 0.26 | 0.26 | | |
| | £ | | 8.70 | 5.21 | 12.11 | 26.02 |
| 2.5 - 3.0m | hr | m | 0.32 | 0.32 | | |
| | £ | | 10.71 | 6.42 | 12.11 | 29.24 |

**CLASS I: PIPEWORK - PIPES**

**CAST OR SPUN IRON PIPES - continued**

| | UNIT | | L | P | M | £ NETT |
|---|---|---|---|---|---|---|
| 100mm | | | | | | |
| In trenches | | | | | | |
| 3.0 - 3.5m | hr | m | 0.42 | 0.42 | | |
| | £ | | 14.06 | 8.42 | 12.11 | 34.59 |
| 3.5 - 4.0m | hr | m | 0.53 | 0.53 | | |
| | £ | | 17.74 | 10.63 | 12.11 | 40.48 |
| 4.0 - 4.5m | hr | m | 0.66 | 0.66 | | |
| | £ | | 22.10 | 13.23 | 12.11 | 47.44 |
| 4.5 - 5.0m | hr | m | 0.84 | 0.84 | | |
| | £ | | 28.12 | 16.84 | 12.11 | 57.07 |
| 5.0 - 5.5m | hr | m | 1.08 | 1.08 | | |
| | £ | | 36.16 | 21.65 | 12.11 | 69.92 |
| 150mm | | | | | | |
| Not in trenches | hr | m | 0.11 | 0.05 | | |
| | £ | | 3.68 | 1.00 | 18.08 | 22.76 |
| In trenches | | | | | | |
| ne 1.5m | hr | m | 0.21 | 0.21 | | |
| | £ | | 7.03 | 4.21 | 18.08 | 29.32 |
| 1.5 - 2.0m | hr | m | 0.25 | 0.25 | | |
| | £ | | 8.37 | 5.01 | 18.08 | 31.46 |
| 2.0 - 2.5m | hr | m | 0.30 | 0.30 | | |
| | £ | | 10.04 | 6.02 | 18.08 | 34.14 |
| 2.5 - 3.0m | hr | m | 0.37 | 0.37 | | |
| | £ | | 12.39 | 7.42 | 18.08 | 37.89 |
| 3.0 - 3.5m | hr | m | 0.49 | 0.49 | | |
| | £ | | 16.41 | 9.82 | 18.08 | 44.31 |
| 3.5 - 4.0m | hr | m | 0.55 | 0.55 | | |
| | £ | | 18.41 | 11.03 | 18.08 | 47.52 |
| 4.0 - 4.5m | hr | m | 0.72 | 0.72 | | |
| | £ | | 24.11 | 14.44 | 18.08 | 56.63 |
| 4.5 - 5.0m | hr | m | 0.94 | 0.94 | | |
| | £ | | 31.47 | 18.85 | 18.08 | 68.40 |
| 5.0 - 5.5m | hr | m | 1.19 | 1.19 | | |
| | £ | | 39.84 | 23.86 | 18.08 | 81.78 |

**CLASS I: PIPEWORK - PIPES**

**CAST OR SPUN IRON PIPES - continued**

| | UNIT | | L | P | M | £ NETT |
|---|---|---|---|---|---|---|
| 250mm | | | | | | |
| Not in trenches | hr | m | 0.16 | 0.05 | | |
| | £ | | 5.36 | 1.00 | 34.13 | 40.49 |
| In trenches | | | | | | |
| ne 1.5m | hr | m | 0.24 | 0.24 | | |
| | £ | | 8.04 | 4.81 | 34.13 | 46.98 |
| 1.5 - 2.0m | hr | m | 0.30 | 0.30 | | |
| | £ | | 10.04 | 6.02 | 34.13 | 50.19 |
| 2.0 - 2.5m | hr | m | 0.35 | 0.35 | | |
| | £ | | 11.72 | 7.02 | 34.13 | 52.87 |
| 2.5 - 3.0m | hr | m | 0.42 | 0.42 | | |
| | £ | | 14.06 | 8.42 | 34.13 | 56.61 |
| 3.0 - 3.5m | hr | m | 0.56 | 0.56 | | |
| | £ | | 18.75 | 11.23 | 34.13 | 64.11 |
| 3.5 - 4.0m | hr | m | 0.64 | 0.64 | | |
| | £ | | 21.43 | 12.83 | 34.13 | 68.39 |
| 4.0 - 4.5m | hr | m | 0.81 | 0.81 | | |
| | £ | | 27.12 | 16.24 | 34.13 | 77.49 |
| 4.5 - 5.0m | hr | m | 1.02 | 1.02 | | |
| | £ | | 34.15 | 20.45 | 34.13 | 88.73 |
| 5.0 - 5.5m | hr | m | 1.27 | 1.27 | | |
| | £ | | 42.52 | 25.46 | 34.13 | 102.11 |
| 400mm | | | | | | |
| Not in trenches | hr | m | 0.24 | 0.11 | | |
| | £ | | 8.04 | 2.21 | 66.01 | 76.26 |
| In trenches | | | | | | |
| ne 1.5m | hr | m | 0.33 | 0.33 | | |
| | £ | | 11.05 | 6.62 | 66.01 | 83.68 |
| 1.5 - 2.0m | hr | m | 0.43 | 0.43 | | |
| | £ | | 14.40 | 8.62 | 66.01 | 89.03 |
| 2.0 - 2.5m | hr | m | 0.48 | 0.48 | | |
| | £ | | 16.07 | 9.62 | 66.01 | 91.70 |
| 2.5 - 3.0m | hr | m | 0.57 | 0.57 | | |
| | £ | | 19.08 | 11.43 | 66.01 | 96.52 |
| 3.0 - 3.5m | hr | m | 0.71 | 0.71 | | |
| | £ | | 23.77 | 14.24 | 66.01 | 104.02 |
| 3.5 - 4.0m | hr | m | 0.91 | 0.91 | | |
| | £ | | 30.47 | 18.25 | 66.01 | 114.73 |

**CLASS I: PIPEWORK - PIPES**

**CAST OR SPUN IRON PIPES - continued**

| | UNIT | | L | P | M | £ NETT |
|---|---|---|---|---|---|---|
| 400mm | | | | | | |
| In trenches | | | | | | |
| 4.0 - 4.5m | hr | m | 1.16 | 1.16 | | |
| | £ | | 38.84 | 23.26 | 66.01 | 128.11 |
| 4.5 - 5.0m | hr | m | 1.47 | 1.47 | | |
| | £ | | 49.22 | 29.47 | 66.01 | 144.70 |
| 5.0 - 5.5m | hr | m | 1.77 | 1.77 | | |
| | £ | | 59.26 | 35.49 | 66.01 | 160.76 |
| 450mm | | | | | | |
| Not in trenches | hr | m | 0.26 | 0.12 | | |
| | £ | | 8.70 | 2.41 | 74.50 | 85.61 |
| In trenches | | | | | | |
| ne 1.5m | hr | m | 0.36 | 0.36 | | |
| | £ | | 12.05 | 7.22 | 74.50 | 93.77 |
| 1.5 - 2.0m | hr | m | 0.47 | 0.47 | | |
| | £ | | 15.74 | 9.42 | 74.50 | 99.66 |
| 2.0 - 2.5m | hr | m | 0.53 | 0.53 | | |
| | £ | | 17.74 | 10.63 | 74.50 | 102.87 |
| 2.5 - 3.0m | hr | m | 0.62 | 0.62 | | |
| | £ | | 20.76 | 12.43 | 74.50 | 107.69 |
| 3.0 - 3.5m | hr | m | 0.77 | 0.77 | | |
| | £ | | 25.78 | 15.44 | 74.50 | 115.72 |
| 3.5 - 4.0m | hr | m | 0.98 | 0.98 | | |
| | £ | | 32.81 | 19.65 | 74.50 | 126.96 |
| 4.0 - 4.5m | hr | m | 1.23 | 1.23 | | |
| | £ | | 41.18 | 24.66 | 74.50 | 140.34 |
| 4.5 - 5.0m | hr | m | 1.55 | 1.55 | | |
| | £ | | 51.89 | 31.08 | 74.50 | 157.47 |
| 5.0 - 5.5m | hr | m | 1.86 | 1.86 | | |
| | £ | | 62.27 | 37.29 | 74.50 | 174.06 |
| 600mm | | | | | | |
| Not in trenches | hr | m | 0.34 | 0.15 | | |
| | £ | | 11.38 | 3.01 | 114.28 | 128.67 |
| In trenches | | | | | | |
| ne 1.5m | hr | m | 0.47 | 0.47 | | |
| | £ | | 15.74 | 9.42 | 114.28 | 139.44 |
| 1.5 - 2.0m | hr | m | 0.55 | 0.55 | | |
| | £ | | 18.41 | 11.03 | 114.28 | 143.72 |

**CLASS I: PIPEWORK - PIPES**

**CAST OR SPUN IRON PIPES - continued**

| | | UNIT | L | P | M | £ NETT |
|---|---|---|---|---|---|---|
| 600mm | | | | | | |
| In trenches | | | | | | |
| 2.0 - 2.5m | hr | m | 0.66 | 0.66 | | |
| | £ | | 22.10 | 13.23 | 114.28 | 149.61 |
| 2.5 - 3.0m | hr | m | 0.78 | 0.78 | | |
| | £ | | 26.11 | 15.64 | 114.28 | 156.03 |
| 3.0 - 3.5m | hr | m | 0.89 | 0.89 | | |
| | £ | | 29.80 | 17.84 | 114.28 | 161.92 |
| 3.5 - 4.0m | hr | m | 1.09 | 1.09 | | |
| | £ | | 36.49 | 21.85 | 114.28 | 172.62 |
| 4.0 - 4.5m | hr | m | 1.37 | 1.37 | | |
| | £ | | 45.87 | 27.47 | 114.28 | 187.62 |
| 4.5 - 5.0m | hr | m | 1.70 | 1.70 | | |
| | £ | | 56.92 | 34.09 | 114.28 | 205.29 |
| 5.0 - 5.5m | hr | m | 2.03 | 2.03 | | |
| | £ | | 67.96 | 40.70 | 114.28 | 222.94 |

**CLASS I: PIPEWORK - PIPES**

**STEEL PIPES**

| | UNIT | | L | P | M | £ NETT |
|---|---|---|---|---|---|---|
| Carbon steel pipes to BS3601 including welded joints to pipes (For protection and lining refer to manufacturer) nominal bore | | | | | | |
| 100mm | | | | | | |
| Not in trenches | hr | m | 0.07 | 0.05 | | |
| | £ | | 2.34 | 1.00 | 19.28 | 22.62 |
| In trenches | | | | | | |
| ne 1.5m | hr | m | 0.15 | 0.15 | | |
| | £ | | 5.02 | 3.01 | 19.28 | 27.31 |
| 1.5 - 2.0m | hr | m | 0.17 | 0.17 | | |
| | £ | | 5.69 | 3.41 | 19.28 | 28.38 |
| 2.0 - 2.5m | hr | m | 0.22 | 0.22 | | |
| | £ | | 7.37 | 4.41 | 19.28 | 31.06 |
| 2.5 - 3.0m | hr | m | 0.27 | 0.27 | | |
| | £ | | 9.04 | 5.41 | 19.28 | 33.73 |
| 3.0 - 3.5m | hr | m | 0.35 | 0.35 | | |
| | £ | | 11.72 | 7.02 | 19.28 | 38.02 |
| 3.5 - 4.0m | hr | m | 0.44 | 0.44 | | |
| | £ | | 14.73 | 8.82 | 19.28 | 42.83 |
| 4.0 - 4.5m | hr | m | 0.55 | 0.55 | | |
| | £ | | 18.41 | 11.03 | 19.28 | 48.72 |
| 4.5 - 5.0m | hr | m | 0.70 | 0.70 | | |
| | £ | | 23.44 | 14.04 | 19.28 | 56.76 |
| 5.0 - 5.5m | hr | m | 0.87 | 0.87 | | |
| | £ | | 29.13 | 17.44 | 19.28 | 65.85 |
| 150mm | | | | | | |
| Not in trenches | hr | m | 0.07 | 0.05 | | |
| | £ | | 2.34 | 1.00 | 35.88 | 39.22 |
| In trenches | | | | | | |
| ne 1.5m | hr | m | 0.17 | 0.17 | | |
| | £ | | 5.69 | 3.41 | 35.88 | 44.98 |
| 1.5 - 2.0m | hr | m | 0.20 | 0.20 | | |
| | £ | | 6.70 | 4.01 | 35.88 | 46.59 |
| 2.0 - 2.5m | hr | m | 0.24 | 0.24 | | |
| | £ | | 8.04 | 4.81 | 35.88 | 48.73 |
| 2.5 - 3.0m | hr | m | 0.29 | 0.29 | | |
| | £ | | 9.71 | 5.81 | 35.88 | 51.40 |

**CLASS I: PIPEWORK - PIPES**

**STEEL PIPES - continued**

| | UNIT | | L | P | M | £ NETT |
|---|---|---|---|---|---|---|
| 150mm | | | | | | |
| In trenches | | | | | | |
| 3.0 - 3.5m | hr | m | 0.39 | 0.39 | | |
| | £ | | 13.06 | 7.82 | 35.88 | 56.76 |
| 3.5 - 4.0m | hr | m | 0.44 | 0.44 | | |
| | £ | | 14.73 | 8.82 | 35.88 | 59.43 |
| 4.0 - 4.5m | hr | m | 0.57 | 0.57 | | |
| | £ | | 19.08 | 11.43 | 35.88 | 66.39 |
| 4.5 - 5.0m | hr | m | 0.72 | 0.72 | | |
| | £ | | 24.11 | 14.44 | 35.88 | 74.43 |
| 5.0 - 5.5m | hr | m | 0.90 | 0.90 | | |
| | £ | | 30.13 | 18.05 | 35.88 | 84.06 |
| 200mm | | | | | | |
| Not in trenches | hr | m | 0.09 | 0.05 | | |
| | £ | | 3.01 | 1.00 | 57.28 | 61.29 |
| In trenches | | | | | | |
| ne 1.5m | hr | m | 0.18 | 0.18 | | |
| | £ | | 6.03 | 3.61 | 57.28 | 66.92 |
| 1.5 - 2.0m | hr | m | 0.21 | 0.21 | | |
| | £ | | 7.03 | 4.21 | 57.28 | 68.52 |
| 2.0 - 2.5m | hr | m | 0.25 | 0.25 | | |
| | £ | | 8.37 | 5.01 | 57.28 | 70.66 |
| 2.5 - 3.0m | hr | m | 0.30 | 0.30 | | |
| | £ | | 10.04 | 6.02 | 57.28 | 73.34 |
| 3.0 - 3.5m | hr | m | 0.40 | 0.40 | | |
| | £ | | 13.39 | 8.02 | 57.28 | 78.69 |
| 3.5 - 4.0m | hr | m | 0.46 | 0.46 | | |
| | £ | | 15.40 | 9.22 | 57.28 | 81.90 |
| 4.0 - 4.5m | hr | m | 0.59 | 0.59 | | |
| | £ | | 19.75 | 11.83 | 57.28 | 88.86 |
| 4.5 - 5.0m | hr | m | 0.74 | 0.74 | | |
| | £ | | 24.78 | 14.84 | 57.28 | 96.90 |
| 5.0 - 5.5m | hr | m | 0.92 | 0.92 | | |
| | £ | | 30.80 | 18.45 | 57.28 | 106.53 |

**CLASS I: PIPEWORK - PIPES**

**STEEL PIPES - continued**

| | UNIT | | L | P | M | £ NETT |
|---|---|---|---|---|---|---|
| 250mm | | | | | | |
| Not in trenches | hr | m | 0.10 | 0.05 | | |
| | £ | | **3.35** | **1.00** | **81.24** | **85.59** |
| In trenches | | | | | | |
| ne 1.5m | hr | m | 0.18 | 0.18 | | |
| | £ | | **6.03** | **3.61** | **81.24** | **90.88** |
| 1.5 - 2.0m | hr | m | 0.22 | 0.22 | | |
| | £ | | **7.37** | **4.41** | **81.24** | **93.02** |
| 2.0 - 2.5m | hr | m | 0.26 | 0.26 | | |
| | £ | | **8.70** | **5.21** | **81.24** | **95.15** |
| 2.5 - 3.0m | hr | m | 0.31 | 0.31 | | |
| | £ | | **10.38** | **6.22** | **81.24** | **97.84** |
| 3.0 - 3.5m | hr | m | 0.41 | 0.41 | | |
| | £ | | **13.73** | **8.22** | **81.24** | **103.19** |
| 3.5 - 4.0m | hr | m | 0.47 | 0.47 | | |
| | £ | | **15.74** | **9.42** | **81.24** | **106.40** |
| 4.0 - 4.5m | hr | m | 0.60 | 0.60 | | |
| | £ | | **20.09** | **12.03** | **81.24** | **113.36** |
| 4.5 - 5.0m | hr | m | 0.75 | 0.75 | | |
| | £ | | **25.11** | **15.04** | **81.24** | **121.39** |
| 5.0 - 5.5m | hr | m | 0.94 | 0.94 | | |
| | £ | | **31.47** | **18.85** | **81.24** | **131.56** |
| 300mm | | | | | | |
| Not in trenches | hr | m | 0.11 | 0.08 | | |
| | £ | | **3.68** | **1.60** | **104.19** | **109.47** |
| In trenches | | | | | | |
| ne 1.5m | hr | m | 0.20 | 0.20 | | |
| | £ | | **6.70** | **4.01** | **104.19** | **114.90** |
| 1.5 - 2.0m | hr | m | 0.25 | 0.25 | | |
| | £ | | **8.37** | **5.01** | **104.19** | **117.57** |
| 2.0 - 2.5m | hr | m | 0.28 | 0.28 | | |
| | £ | | **9.37** | **5.61** | **104.19** | **119.17** |
| 2.5 - 3.0m | hr | m | 0.34 | 0.34 | | |
| | £ | | **11.38** | **6.82** | **104.19** | **122.39** |

**CLASS I: PIPEWORK - PIPES**

**STEEL PIPES - continued**

| | UNIT | | L | P | M | £ NETT |
|---|---|---|---|---|---|---|
| 450mm | | | | | | |
| In trenches | | | | | | |
| 3.0 - 3.5m | hr | m | 0.44 | 0.44 | | |
| | £ | | 14.73 | 8.82 | 104.19 | 127.74 |
| 3.5 - 4.0m | hr | m | 0.53 | 0.53 | | |
| | £ | | 17.74 | 10.63 | 104.19 | 132.56 |
| 4.0 - 4.5m | hr | m | 0.68 | 0.68 | | |
| | £ | | 22.77 | 13.63 | 104.19 | 140.59 |
| 4.5 - 5.0m | hr | m | 0.86 | 0.86 | | |
| | £ | | 28.79 | 17.24 | 104.19 | 150.22 |
| 5.0 - 5.5m | hr | m | 1.06 | 1.06 | | |
| | £ | | 35.49 | 21.25 | 104.19 | 160.93 |

## CLASS I: PIPEWORK - PIPES

### PLASTIC PIPES

| | UNIT | | L | P | M | £ NETT |
|---|---|---|---|---|---|---|
| Unplasticised pvc pipes with ring seal sockets (Marley) nominal bore | | | | | | |
| 82mm | | | | | | |
| Not in trenches | hr | m | 0.06 | 0.03 | | |
| | £ | | 1.95 | 0.48 | 3.65 | 6.08 |
| In trenches | | | | | | |
| ne 1.5m | hr | m | 0.10 | 0.10 | | |
| | £ | | 3.25 | 1.61 | 3.65 | 8.51 |
| 1.5 - 2.0m | hr | m | 0.13 | 0.13 | | |
| | £ | | 4.23 | 2.10 | 3.65 | 9.98 |
| 2.0 - 2.5m | hr | m | 0.16 | 0.16 | | |
| | £ | | 5.20 | 2.58 | 3.65 | 11.43 |
| 2.5 - 3.0m | hr | m | 0.19 | 0.19 | | |
| | £ | | 6.18 | 3.06 | 3.65 | 12.89 |
| 3.0 - 3.5m | hr | m | 0.22 | 0.22 | | |
| | £ | | 7.15 | 3.55 | 3.65 | 14.35 |
| 3.5 - 4.0m | hr | m | 0.25 | 0.25 | | |
| | £ | | 8.13 | 4.03 | 3.65 | 15.81 |
| 4.0 - 4.5m | hr | m | 0.28 | 0.28 | | |
| | £ | | 9.10 | 4.52 | 3.65 | 17.27 |
| 4.5 - 5.0m | hr | m | 0.32 | 0.32 | | |
| | £ | | 10.40 | 5.16 | 3.65 | 19.21 |
| 5.0 - 5.5m | hr | m | 0.35 | 0.35 | | |
| | £ | | 11.38 | 5.65 | 3.65 | 20.68 |
| 110mm | | | | | | |
| Not in trenches | hr | m | 0.06 | 0.03 | | |
| | £ | | 1.95 | 0.48 | 3.28 | 5.71 |
| In trenches | | | | | | |
| ne 1.5m | hr | m | 0.10 | 0.10 | | |
| | £ | | 3.25 | 1.61 | 3.28 | 8.14 |
| 1.5 - 2.0m | hr | m | 0.13 | 0.13 | | |
| | £ | | 4.23 | 2.10 | 3.28 | 9.61 |
| 2.0 - 2.5m | hr | m | 0.16 | 0.16 | | |
| | £ | | 5.20 | 2.58 | 3.28 | 11.06 |
| 2.5 - 3.0m | hr | m | 0.19 | 0.19 | | |
| | £ | | 6.18 | 3.06 | 3.28 | 12.52 |

**CLASS I: PIPEWORK - PIPES**

**PLASTIC PIPES - continued**

| | UNIT | | L | P | M | £ NETT |
|---|---|---|---|---|---|---|
| 110mm | | | | | | |
| In trenches | | | | | | |
| 3.0 - 3.5m | hr | m | 0.22 | 0.22 | | |
| | £ | | 7.15 | 3.55 | 3.28 | 13.98 |
| 3.5 - 4.0m | hr | m | 0.25 | 0.25 | | |
| | £ | | 8.13 | 4.03 | 3.28 | 15.44 |
| 4.0 - 4.5m | hr | m | 0.28 | 0.28 | | |
| | £ | | 9.10 | 4.52 | 3.28 | 16.90 |
| 4.5 - 5.0m | hr | m | 0.32 | 0.32 | | |
| | £ | | 10.40 | 5.16 | 3.28 | 18.84 |
| 5.0 - 5.5m | hr | m | 0.36 | 0.36 | | |
| | £ | | 11.70 | 5.81 | 3.28 | 20.79 |
| 160mm | | | | | | |
| Not in trenches | hr | m | 0.06 | 0.03 | | |
| | £ | | 1.95 | 0.48 | 7.18 | 9.61 |
| In trenches | | | | | | |
| ne 1.5m | hr | m | 0.11 | 0.11 | | |
| | £ | | 3.58 | 1.77 | 7.18 | 12.53 |
| 1.5 - 2.0m | hr | m | 0.14 | 0.14 | | |
| | £ | | 4.55 | 2.26 | 7.18 | 13.99 |
| 2.0 - 2.5m | hr | m | 0.18 | 0.18 | | |
| | £ | | 5.85 | 2.90 | 7.18 | 15.93 |
| 2.5 - 3.0m | hr | m | 0.22 | 0.22 | | |
| | £ | | 7.15 | 3.55 | 7.18 | 17.88 |
| 3.0 - 3.5m | hr | m | 0.23 | 0.23 | | |
| | £ | | 7.48 | 3.71 | 7.18 | 18.37 |
| 3.5 - 4.0m | hr | m | 0.26 | 0.26 | | |
| | £ | | 8.45 | 4.19 | 7.18 | 19.82 |
| 4.0 - 4.5m | hr | m | 0.30 | 0.30 | | |
| | £ | | 9.75 | 4.84 | 7.18 | 21.77 |
| 4.5 - 5.0m | hr | m | 0.35 | 0.35 | | |
| | £ | | 11.38 | 5.65 | 7.18 | 24.21 |
| 5.0 - 5.5m | hr | m | 0.40 | 0.40 | | |
| | £ | | 13.00 | 6.45 | 7.18 | 26.63 |

**CLASS I: PIPEWORK - PIPES**

**PLASTIC PIPES - continued**

| | UNIT | | L | P | M | £ NETT |
|---|---|---|---|---|---|---|
| 200mm | | | | | | |
| Not in trenches | hr | m | 0.07 | 0.03 | | |
| | £ | | 2.28 | 0.48 | 11.81 | 14.57 |
| In trenches | | | | | | |
| ne 1.5m | hr | m | 0.12 | 0.12 | | |
| | £ | | 3.90 | 1.94 | 11.81 | 17.65 |
| 1.5 - 2.0m | hr | m | 0.15 | 0.15 | | |
| | £ | | 4.88 | 2.42 | 11.81 | 19.11 |
| 2.0 - 2.5m | hr | m | 0.20 | 0.20 | | |
| | £ | | 6.50 | 3.23 | 11.81 | 21.54 |
| 2.5 - 3.0m | hr | m | 0.23 | 0.23 | | |
| | £ | | 7.48 | 3.71 | 11.81 | 23.00 |
| 3.0 - 3.5m | hr | m | 0.24 | 0.24 | | |
| | £ | | 7.80 | 3.87 | 11.81 | 23.48 |
| 3.5 - 4.0m | hr | m | 0.27 | 0.27 | | |
| | £ | | 8.78 | 4.36 | 11.81 | 24.95 |
| 4.0 - 4.5m | hr | m | 0.32 | 0.32 | | |
| | £ | | 10.40 | 5.16 | 11.81 | 27.37 |
| 4.5 - 5.0m | hr | m | 0.36 | 0.36 | | |
| | £ | | 11.70 | 5.81 | 11.81 | 29.32 |
| 5.0 - 5.5m | hr | m | 0.45 | 0.45 | | |
| | £ | | 14.63 | 7.26 | 11.81 | 33.70 |
| 250mm | | | | | | |
| Not in trenches | hr | m | 0.08 | 0.03 | | |
| | £ | | 2.60 | 0.48 | 18.19 | 21.27 |
| In trenches | | | | | | |
| ne 1.5m | hr | m | 0.13 | 0.13 | | |
| | £ | | 4.23 | 2.10 | 18.19 | 24.52 |
| 1.5 - 2.0m | hr | m | 0.16 | 0.16 | | |
| | £ | | 5.20 | 2.58 | 18.19 | 25.97 |
| 2.0 - 2.5m | hr | m | 0.21 | 0.21 | | |
| | £ | | 6.83 | 3.39 | 18.19 | 28.41 |
| 2.5 - 3.0m | hr | m | 0.23 | 0.23 | | |
| | £ | | 7.48 | 3.71 | 18.19 | 29.38 |
| 3.0 - 3.5m | hr | m | 0.25 | 0.25 | | |
| | £ | | 8.13 | 4.03 | 18.19 | 30.35 |

**CLASS I: PIPEWORK - PIPES**

**PLASTIC PIPES - continued**

| | UNIT | | L | P | M | £ NETT |
|---|---|---|---|---|---|---|
| 250mm | | | | | | |
| In trenches | | | | | | |
| 3.5 - 4.0m | hr | m | 0.28 | 0.28 | | |
| | £ | | **9.10** | **4.52** | **18.19** | **31.81** |
| 4.0 - 4.5m | hr | m | 0.34 | 0.34 | | |
| | £ | | **11.05** | **5.48** | **18.19** | **34.72** |
| 4.5 - 5.0m | hr | m | 0.38 | 0.38 | | |
| | £ | | **12.35** | **6.13** | **18.19** | **36.67** |
| 5.0 - 5.5m | hr | m | 0.47 | 0.47 | | |
| | £ | | **15.28** | **7.58** | **18.19** | **41.05** |
| 315mm | | | | | | |
| Not in trenches | hr | m | 0.08 | 0.05 | | |
| | £ | | **2.60** | **0.81** | **32.32** | **35.73** |
| In trenches | | | | | | |
| ne 1.5m | hr | m | 0.18 | 0.18 | | |
| | £ | | **5.85** | **2.90** | **32.32** | **41.07** |
| 1.5 - 2.0m | hr | m | 0.22 | 0.22 | | |
| | £ | | **7.15** | **3.55** | **32.32** | **43.02** |
| 2.0 - 2.5m | hr | m | 0.26 | 0.26 | | |
| | £ | | **8.45** | **4.19** | **32.32** | **44.96** |
| 2.5 - 3.0m | hr | m | 0.28 | 0.28 | | |
| | £ | | **9.10** | **4.52** | **32.32** | **45.94** |
| 3.0 - 3.5m | hr | m | 0.38 | 0.38 | | |
| | £ | | **12.35** | **6.13** | **32.32** | **50.80** |
| 3.5 - 4.0m | hr | m | 0.43 | 0.43 | | |
| | £ | | **13.98** | **6.94** | **32.32** | **53.24** |
| 4.0 - 4.5m | hr | m | 0.47 | 0.47 | | |
| | £ | | **15.28** | **7.58** | **32.32** | **55.18** |
| 4.5 - 5.0m | hr | m | 0.51 | 0.51 | | |
| | £ | | **16.58** | **8.23** | **32.32** | **57.13** |
| 5.0 - 5.5m | hr | m | 0.55 | 0.55 | | |
| | £ | | **17.88** | **8.87** | **32.32** | **59.07** |

**CLASS I: PIPEWORK - PIPES**

**PLASTIC PIPES - continued**

| | UNIT | | L | P | M | £ NETT |
|---|---|---|---|---|---|---|
| Unplasticised pvc pipes with ring seal sockets (OsmaDrain) nominal bore | | | | | | |
| 250mm | | | | | | |
| Not in trenches | hr | m | 0.08 | 0.03 | | |
| | £ | | 2.60 | 0.48 | 20.42 | 23.50 |
| In trenches | | | | | | |
| ne 1.5m | hr | m | 0.13 | 0.13 | | |
| | £ | | 4.23 | 2.10 | 20.42 | 26.75 |
| 1.5 - 2.0m | hr | m | 0.16 | 0.16 | | |
| | £ | | 5.20 | 2.58 | 20.42 | 28.20 |
| 2.0 - 2.5m | hr | m | 0.21 | 0.21 | | |
| | £ | | 6.83 | 3.39 | 20.42 | 30.64 |
| 2.5 - 3.0m | hr | m | 0.23 | 0.23 | | |
| | £ | | 7.48 | 3.71 | 20.42 | 31.61 |
| 3.0 - 3.5m | hr | m | 0.25 | 0.25 | | |
| | £ | | 8.13 | 4.03 | 20.42 | 32.58 |
| 3.5 - 4.0m | hr | m | 0.28 | 0.28 | | |
| | £ | | 9.10 | 4.52 | 20.42 | 34.04 |
| 4.0 - 4.5m | hr | m | 0.34 | 0.34 | | |
| | £ | | 11.05 | 5.48 | 20.42 | 36.95 |
| 4.5 - 5.0m | hr | m | 0.38 | 0.38 | | |
| | £ | | 12.35 | 6.13 | 20.42 | 38.90 |
| 5.0 - 5.5m | hr | m | 0.47 | 0.47 | | |
| | £ | | 15.28 | 7.58 | 20.42 | 43.28 |
| Unplasticised pvc pipes with polypropylene couplings (Marley) nominal bore | | | | | | |
| 110mm | | | | | | |
| Not in trenches | hr | m | 0.06 | 0.03 | | |
| | £ | | 1.95 | 0.48 | 3.76 | 6.19 |
| In trenches | | | | | | |
| ne 1.5m | hr | m | 0.10 | 0.10 | | |
| | £ | | 3.25 | 1.61 | 3.76 | 8.62 |
| 1.5 - 2.0m | hr | m | 0.13 | 0.13 | | |
| | £ | | 4.23 | 2.10 | 3.76 | 10.09 |
| 2.0 - 2.5m | hr | m | 0.16 | 0.16 | | |
| | £ | | 5.20 | 2.58 | 3.76 | 11.54 |
| 2.5 - 3.0m | hr | m | 0.19 | 0.19 | | |
| | £ | | 6.18 | 3.06 | 3.76 | 13.00 |

**CLASS I: PIPEWORK - PIPES**

**PLASTIC PIPES - continued**

| | UNIT | | L | P | M | £ NETT |
|---|---|---|---|---|---|---|
| 110mm | | | | | | |
| In trenches | | | | | | |
| 3.0 - 3.5m | hr | m | 0.22 | 0.22 | | |
| | £ | | 7.15 | 3.55 | 3.76 | 14.46 |
| 3.5 - 4.0m | hr | m | 0.25 | 0.25 | | |
| | £ | | 8.13 | 4.03 | 3.76 | 15.92 |
| 4.0 - 4.5m | hr | m | 0.28 | 0.28 | | |
| | £ | | 9.10 | 4.52 | 3.76 | 17.38 |
| 4.5 - 5.0m | hr | m | 0.32 | 0.32 | | |
| | £ | | 10.40 | 5.16 | 3.76 | 19.32 |
| 5.0 - 5.5m | hr | m | 0.36 | 0.36 | | |
| | £ | | 11.70 | 5.81 | 3.76 | 21.27 |
| 160mm | | | | | | |
| Not in trenches | hr | m | 0.06 | 0.03 | | |
| | £ | | 1.95 | 0.48 | 7.13 | 9.56 |
| In trenches | | | | | | |
| ne 1.5m | hr | m | 0.11 | 0.11 | | |
| | £ | | 3.58 | 1.77 | 7.13 | 12.48 |
| 1.5 - 2.0m | hr | m | 0.14 | 0.14 | | |
| | £ | | 4.55 | 2.26 | 7.13 | 13.94 |
| 2.0 - 2.5m | hr | m | 0.18 | 0.18 | | |
| | £ | | 5.85 | 2.90 | 7.13 | 15.88 |
| 2.5 - 3.0m | hr | m | 0.22 | 0.22 | | |
| | £ | | 7.15 | 3.55 | 7.13 | 17.83 |
| 3.0 - 3.5m | hr | m | 0.23 | 0.23 | | |
| | £ | | 7.48 | 3.71 | 7.13 | 18.32 |
| 3.5 - 4.0m | hr | m | 0.26 | 0.26 | | |
| | £ | | 8.45 | 4.19 | 7.13 | 19.77 |
| 4.0 - 4.5m | hr | m | 0.30 | 0.30 | | |
| | £ | | 9.75 | 4.84 | 7.13 | 21.72 |
| 4.5 - 5.0m | hr | m | 0.35 | 0.35 | | |
| | £ | | 11.38 | 5.65 | 7.13 | 24.16 |
| 5.0 - 5.5m | hr | m | 0.40 | 0.40 | | |
| | £ | | 13.00 | 6.45 | 7.13 | 26.58 |
| 200mm | | | | | | |
| Not in trenches | hr | m | 0.07 | 0.03 | | |
| | £ | | 2.28 | 0.48 | 12.65 | 15.41 |
| In trenches | | | | | | |
| ne 1.5m | hr | m | 0.12 | 0.12 | | |
| | £ | | 3.90 | 1.94 | 12.65 | 18.49 |

**CLASS I: PIPEWORK - PIPES**

**PLASTIC PIPES - continued**

| | UNIT | | L | P | M | £ NETT |
|---|---|---|---|---|---|---|
| 200mm | | | | | | |
| In trenches | | | | | | |
| 1.5 - 2.0m | hr | m | 0.15 | 0.15 | | |
| | £ | | 4.88 | 2.42 | 12.65 | 19.95 |
| 2.0 - 2.5m | hr | m | 0.20 | 0.20 | | |
| | £ | | 6.50 | 3.23 | 12.65 | 22.38 |
| 2.5 - 3.0m | hr | m | 0.23 | 0.23 | | |
| | £ | | 7.48 | 3.71 | 12.65 | 23.84 |
| 3.0 - 3.5m | hr | m | 0.24 | 0.24 | | |
| | £ | | 7.80 | 3.87 | 12.65 | 24.32 |
| 3.5 - 4.0m | hr | m | 0.27 | 0.27 | | |
| | £ | | 8.78 | 4.36 | 12.65 | 25.79 |
| 4.0 - 4.5m | hr | m | 0.32 | 0.32 | | |
| | £ | | 10.40 | 5.16 | 12.65 | 28.21 |
| 4.5 - 5.0m | hr | m | 0.36 | 0.36 | | |
| | £ | | 11.70 | 5.81 | 12.65 | 30.16 |
| 5.0 - 5.5m | hr | m | 0.45 | 0.45 | | |
| | £ | | 14.63 | 7.26 | 12.65 | 34.54 |
| 250mm | | | | | | |
| Not in trenches | hr | m | 0.08 | 0.03 | | |
| | £ | | 2.60 | 0.48 | 24.24 | 27.32 |
| In trenches | | | | | | |
| ne 1.5m | hr | m | 0.13 | 0.13 | | |
| | £ | | 4.23 | 2.10 | 24.24 | 30.57 |
| 1.5 - 2.0m | hr | m | 0.16 | 0.16 | | |
| | £ | | 5.20 | 2.58 | 24.24 | 32.02 |
| 2.0 - 2.5m | hr | m | 0.21 | 0.21 | | |
| | £ | | 6.83 | 3.39 | 24.24 | 34.46 |
| 2.5 - 3.0m | hr | m | 0.23 | 0.23 | | |
| | £ | | 7.48 | 3.71 | 24.24 | 35.43 |
| 3.0 - 3.5m | hr | m | 0.25 | 0.25 | | |
| | £ | | 8.13 | 4.03 | 24.24 | 36.40 |
| 3.5 - 4.0m | hr | m | 0.28 | 0.28 | | |
| | £ | | 9.10 | 4.52 | 24.24 | 37.86 |
| 4.0 - 4.5m | hr | m | 0.34 | 0.34 | | |
| | £ | | 11.05 | 5.48 | 24.24 | 40.77 |
| 4.5 - 5.0m | hr | m | 0.38 | 0.38 | | |
| | £ | | 12.35 | 6.13 | 24.24 | 42.72 |

**CLASS I: PIPEWORK - PIPES**

**PLASTIC PIPES - continued**

| | UNIT | | L | P | M | £ NETT |
|---|---|---|---|---|---|---|
| **250mm** | | | | | | |
| In trenches | | | | | | |
| 5.0 - 5.5m | hr | m | 0.47 | 0.47 | | |
| | £ | | 15.28 | 7.58 | 24.24 | 47.10 |
| Unplasticised pvc pipes with polypropylene couplings (OsmaDrain) nominal bore | | | | | | |
| **82mm** | | | | | | |
| Not in trenches | hr | m | 0.06 | 0.03 | | |
| | £ | | 1.95 | 0.48 | 3.27 | 5.70 |
| In trenches | | | | | | |
| ne 1.5m | hr | m | 0.10 | 0.10 | | |
| | £ | | 3.25 | 1.61 | 3.27 | 8.13 |
| 1.5 - 2.0m | hr | m | 0.13 | 0.13 | | |
| | £ | | 4.23 | 2.10 | 3.27 | 9.60 |
| 2.0 - 2.5m | hr | m | 0.16 | 0.16 | | |
| | £ | | 5.20 | 2.58 | 3.27 | 11.05 |
| 2.5 - 3.0m | hr | m | 0.19 | 0.19 | | |
| | £ | | 6.18 | 3.06 | 3.27 | 12.51 |
| 3.0 - 3.5m | hr | m | 0.22 | 0.22 | | |
| | £ | | 7.15 | 3.55 | 3.27 | 13.97 |
| 3.5 - 4.0m | hr | m | 0.25 | 0.25 | | |
| | £ | | 8.13 | 4.03 | 3.27 | 15.43 |
| 4.0 - 4.5m | hr | m | 0.28 | 0.28 | | |
| | £ | | 9.10 | 4.52 | 3.27 | 16.89 |
| 4.5 - 5.0m | hr | m | 0.32 | 0.32 | | |
| | £ | | 10.40 | 5.16 | 3.27 | 18.83 |
| 5.0 - 5.5m | hr | m | 0.35 | 0.35 | | |
| | £ | | 11.38 | 5.65 | 3.27 | 20.30 |
| **110mm** | | | | | | |
| Not in trenches | hr | m | 0.06 | 0.03 | | |
| | £ | | 1.95 | 0.48 | 3.43 | 5.86 |
| In trenches | | | | | | |
| ne 1.5m | hr | m | 0.10 | 0.10 | | |
| | £ | | 3.25 | 1.61 | 3.43 | 8.29 |
| 1.5 - 2.0m | hr | m | 0.13 | 0.13 | | |
| | £ | | 4.23 | 2.10 | 3.43 | 9.76 |
| 2.0 - 2.5m | hr | m | 0.16 | 0.16 | | |
| | £ | | 5.20 | 2.58 | 3.43 | 11.21 |

**CLASS I: PIPEWORK - PIPES**

**PLASTIC PIPES - continued**

| | UNIT | L | P | M | £ NETT |
|---|---|---|---|---|---|
| 110mm | | | | | |
| In trenches | | | | | |
| 2.5 - 3.0m | hr m | 0.19 | 0.19 | | |
| | £ | 6.18 | 3.06 | 3.43 | 12.67 |
| 3.0 - 3.5m | hr m | 0.22 | 0.22 | | |
| | £ | 7.15 | 3.55 | 3.43 | 14.13 |
| 3.5 - 4.0m | hr m | 0.25 | 0.25 | | |
| | £ | 8.13 | 4.03 | 3.43 | 15.59 |
| 4.0 - 4.5m | hr m | 0.28 | 0.28 | | |
| | £ | 9.10 | 4.52 | 3.43 | 17.05 |
| 4.5 - 5.0m | hr m | 0.32 | 0.32 | | |
| | £ | 10.40 | 5.16 | 3.43 | 18.99 |
| 5.0 - 5.5m | hr m | 0.36 | 0.36 | | |
| | £ | 11.70 | 5.81 | 3.43 | 20.94 |
| 160mm | | | | | |
| Not in trenches | hr m | 0.06 | 0.03 | | |
| | £ | 1.95 | 0.48 | 6.67 | 9.10 |
| In trenches | | | | | |
| ne 1.5m | hr m | 0.11 | 0.11 | | |
| | £ | 3.58 | 1.77 | 6.67 | 12.02 |
| 1.5 - 2.0m | hr m | 0.14 | 0.14 | | |
| | £ | 4.55 | 2.26 | 6.67 | 13.48 |
| 2.0 - 2.5m | hr m | 0.18 | 0.18 | | |
| | £ | 5.85 | 2.90 | 6.67 | 15.42 |
| 2.5 - 3.0m | hr m | 0.22 | 0.22 | | |
| | £ | 7.15 | 3.55 | 6.67 | 17.37 |
| 3.0 - 3.5m | hr m | 0.23 | 0.23 | | |
| | £ | 7.48 | 3.71 | 6.67 | 17.86 |
| 3.5 - 4.0m | hr m | 0.26 | 0.26 | | |
| | £ | 8.45 | 4.19 | 6.67 | 19.31 |
| 4.0 - 4.5m | hr m | 0.30 | 0.30 | | |
| | £ | 9.75 | 4.84 | 6.67 | 21.26 |
| 4.5 - 5.0m | hr m | 0.35 | 0.35 | | |
| | £ | 11.38 | 5.65 | 6.67 | 23.70 |
| 5.0 - 5.5m | hr m | 0.40 | 0.40 | | |
| | £ | 13.00 | 6.45 | 6.67 | 26.12 |

**CLASS I: PIPEWORK - PIPES**

**PLASTIC PIPES - continued**

| | | UNIT | L | P | M | £ NETT |
|---|---|---|---|---|---|---|
| 200mm | | | | | | |
| Not in trenches | hr | m | 0.07 | 0.03 | | |
| | £ | | 2.28 | 0.48 | 12.94 | 15.70 |
| In trenches | | | | | | |
| ne 1.5m | hr | m | 0.12 | 0.12 | | |
| | £ | | 3.90 | 1.94 | 12.94 | 18.78 |
| 1.5 - 2.0m | hr | m | 0.15 | 0.15 | | |
| | £ | | 4.88 | 2.42 | 12.94 | 20.24 |
| 2.0 - 2.5m | hr | m | 0.20 | 0.20 | | |
| | £ | | 6.50 | 3.23 | 12.94 | 22.67 |
| 2.5 - 3.0m | hr | m | 0.23 | 0.23 | | |
| | £ | | 7.48 | 3.71 | 12.94 | 24.13 |
| 3.0 - 3.5m | hr | m | 0.24 | 0.24 | | |
| | £ | | 7.80 | 3.87 | 12.94 | 24.61 |
| 3.5 - 4.0m | hr | m | 0.27 | 0.27 | | |
| | £ | | 8.78 | 4.36 | 12.94 | 26.08 |
| 4.0 - 4.5m | hr | m | 0.32 | 0.32 | | |
| | £ | | 10.40 | 5.16 | 12.94 | 28.50 |
| 4.5 - 5.0m | hr | m | 0.36 | 0.36 | | |
| | £ | | 11.70 | 5.81 | 12.94 | 30.45 |
| 5.0 - 5.5m | hr | m | 0.45 | 0.45 | | |
| | £ | | 14.63 | 7.26 | 12.94 | 34.83 |
| 250mm | | | | | | |
| Not in trenches | hr | m | 0.08 | 0.03 | | |
| | £ | | 2.60 | 0.48 | 21.75 | 24.83 |
| In trenches | | | | | | |
| ne 1.5m | hr | m | 0.13 | 0.13 | | |
| | £ | | 4.23 | 2.10 | 21.75 | 28.08 |
| 1.5 - 2.0m | hr | m | 0.16 | 0.16 | | |
| | £ | | 5.20 | 2.58 | 21.75 | 29.53 |
| 2.0 - 2.5m | hr | m | 0.21 | 0.21 | | |
| | £ | | 6.83 | 3.39 | 21.75 | 31.97 |
| 2.5 - 3.0m | hr | m | 0.23 | 0.23 | | |
| | £ | | 7.48 | 3.71 | 21.75 | 32.94 |
| 3.0 - 3.5m | hr | m | 0.25 | 0.25 | | |
| | £ | | 8.13 | 4.03 | 21.75 | 33.91 |
| 3.5 - 4.0m | hr | m | 0.28 | 0.28 | | |
| | £ | | 9.10 | 4.52 | 21.75 | 35.37 |

**CLASS I: PIPEWORK - PIPES**

**PLASTIC PIPES - continued**

| | UNIT | | L | P | M | £ NETT |
|---|---|---|---|---|---|---|
| 250mm | | | | | | |
| In trenches | | | | | | |
| 4.0 - 4.5m | hr | m | 0.34 | 0.34 | | |
| | £ | | 11.05 | 5.48 | 21.75 | 38.28 |
| 4.5 - 5.0m | hr | m | 0.38 | 0.38 | | |
| | £ | | 12.35 | 6.13 | 21.75 | 40.23 |
| 5.0 - 5.5m | hr | m | 0.47 | 0.47 | | |
| | £ | | 15.28 | 7.58 | 21.75 | 44.61 |
| 315mm | | | | | | |
| Not in trenches | hr | m | 0.08 | 0.05 | | |
| | £ | | 2.60 | 0.81 | 37.90 | 41.31 |
| In trenches | | | | | | |
| ne 1.5m | hr | m | 0.18 | 0.18 | | |
| | £ | | 5.85 | 2.90 | 37.90 | 46.65 |
| 1.5 - 2.0m | hr | m | 0.22 | 0.22 | | |
| | £ | | 7.15 | 3.55 | 37.90 | 48.60 |
| 2.0 - 2.5m | hr | m | 0.26 | 0.26 | | |
| | £ | | 8.45 | 4.19 | 37.90 | 50.54 |
| 2.5 - 3.0m | hr | m | 0.28 | 0.28 | | |
| | £ | | 9.10 | 4.52 | 37.90 | 51.52 |
| 3.0 - 3.5m | hr | m | 0.38 | 0.38 | | |
| | £ | | 12.35 | 6.13 | 37.90 | 56.38 |
| 3.5 - 4.0m | hr | m | 0.43 | 0.43 | | |
| | £ | | 13.98 | 6.94 | 37.90 | 58.82 |
| 4.0 - 4.5m | hr | m | 0.47 | 0.47 | | |
| | £ | | 15.28 | 7.58 | 37.90 | 60.76 |
| 4.5 - 5.0m | hr | m | 0.51 | 0.51 | | |
| | £ | | 16.58 | 8.23 | 37.90 | 62.71 |
| 5.0 - 5.5m | hr | m | 0.55 | 0.55 | | |
| | £ | | 17.88 | 8.87 | 37.90 | 64.65 |

**CLASS I: PIPEWORK - PIPES**

**PLASTIC PIPES - continued**

| | UNIT | | L | P | M | £ NETT |
|---|---|---|---|---|---|---|
| Unplasticised pvc perforated pipes with ring seal sockets (Marley) nominal bore | | | | | | |
| 82mm | | | | | | |
| Not in trenches | hr | m | 0.06 | 0.03 | | |
| | £ | | 1.95 | 0.48 | 3.80 | 6.23 |
| In trenches | | | | | | |
| ne 1.5m | hr | m | 0.10 | 0.10 | | |
| | £ | | 3.25 | 1.61 | 3.80 | 8.66 |
| 1.5 - 2.0m | hr | m | 0.13 | 0.13 | | |
| | £ | | 4.23 | 2.10 | 3.80 | 10.13 |
| 2.0 - 2.5m | hr | m | 0.16 | 0.16 | | |
| | £ | | 5.20 | 2.58 | 3.80 | 11.58 |
| 2.5 - 3.0m | hr | m | 0.19 | 0.19 | | |
| | £ | | 6.18 | 3.06 | 3.80 | 13.04 |
| 3.0 - 3.5m | hr | m | 0.22 | 0.22 | | |
| | £ | | 7.15 | 3.55 | 3.80 | 14.50 |
| 3.5 - 4.0m | hr | m | 0.25 | 0.25 | | |
| | £ | | 8.13 | 4.03 | 3.80 | 15.96 |
| 4.0 - 4.5m | hr | m | 0.28 | 0.28 | | |
| | £ | | 9.10 | 4.52 | 3.80 | 17.42 |
| 4.5 - 5.0m | hr | m | 0.32 | 0.32 | | |
| | £ | | 10.40 | 5.16 | 3.80 | 19.36 |
| 5.0 - 5.5m | hr | m | 0.35 | 0.35 | | |
| | £ | | 11.38 | 5.65 | 3.80 | 20.83 |
| 110mm | | | | | | |
| Not in trenches | hr | m | 0.06 | 0.03 | | |
| | £ | | 1.95 | 0.48 | 3.65 | 6.08 |
| In trenches | | | | | | |
| ne 1.5m | hr | m | 0.10 | 0.10 | | |
| | £ | | 3.25 | 1.61 | 3.65 | 8.51 |
| 1.5 - 2.0m | hr | m | 0.13 | 0.13 | | |
| | £ | | 4.23 | 2.10 | 3.65 | 9.98 |
| 2.0 - 2.5m | hr | m | 0.16 | 0.16 | | |
| | £ | | 5.20 | 2.58 | 3.65 | 11.43 |
| 2.5 - 3.0m | hr | m | 0.19 | 0.19 | | |
| | £ | | 6.18 | 3.06 | 3.65 | 12.89 |
| 3.0 - 3.5m | hr | m | 0.22 | 0.22 | | |
| | £ | | 7.15 | 3.55 | 3.65 | 14.35 |

**CLASS I: PIPEWORK - PIPES**

**PLASTIC PIPES - continued**

| | UNIT | | L | P | M | £ NETT |
|---|---|---|---|---|---|---|
| 110mm | | | | | | |
| In trenches | | | | | | |
| 3.5 - 4.0m | hr | m | 0.25 | 0.25 | | |
| | £ | | 8.13 | 4.03 | 3.65 | 15.81 |
| 4.0 - 4.5m | hr | m | 0.28 | 0.28 | | |
| | £ | | 9.10 | 4.52 | 3.65 | 17.27 |
| 4.5 - 5.0m | hr | m | 0.32 | 0.32 | | |
| | £ | | 10.40 | 5.16 | 3.65 | 19.21 |
| 5.0 - 5.5m | hr | m | 0.36 | 0.36 | | |
| | £ | | 11.70 | 5.81 | 3.65 | 21.16 |
| 160mm | | | | | | |
| Not in trenches | hr | m | 0.06 | 0.03 | | |
| | £ | | 1.95 | 0.48 | 8.88 | 11.31 |
| In trenches | | | | | | |
| ne 1.5m | hr | m | 0.11 | 0.11 | | |
| | £ | | 3.58 | 1.77 | 8.88 | 14.23 |
| 1.5 - 2.0m | hr | m | 0.14 | 0.14 | | |
| | £ | | 4.55 | 2.26 | 8.88 | 15.69 |
| 2.0 - 2.5m | hr | m | 0.18 | 0.18 | | |
| | £ | | 5.85 | 2.90 | 8.88 | 17.63 |
| 2.5 - 3.0m | hr | m | 0.22 | 0.22 | | |
| | £ | | 7.15 | 3.55 | 8.88 | 19.58 |
| 3.0 - 3.5m | hr | m | 0.23 | 0.23 | | |
| | £ | | 7.48 | 3.71 | 8.88 | 20.07 |
| 3.5 - 4.0m | hr | m | 0.26 | 0.26 | | |
| | £ | | 8.45 | 4.19 | 8.88 | 21.52 |
| 4.0 - 4.5m | hr | m | 0.30 | 0.30 | | |
| | £ | | 9.75 | 4.84 | 8.88 | 23.47 |
| 4.5 - 5.0m | hr | m | 0.35 | 0.35 | | |
| | £ | | 11.38 | 5.65 | 8.88 | 25.91 |
| 5.0 - 5.5m | hr | m | 0.40 | 0.40 | | |
| | £ | | 13.00 | 6.45 | 8.88 | 28.33 |

**CLASS I: PIPEWORK - PIPES**

**CONCRETE POROUS PIPES**

| | UNIT | | L | P | M | £ NETT |
|---|---|---|---|---|---|---|
| Concrete porous pipes BS1194 nominal bore | | | | | | |
| 150mm | | | | | | |
| Not in trenches | hr | m | 0.06 | 0.02 | | |
| | £ | | 1.95 | 0.32 | 2.33 | 4.60 |
| In trenches | | | | | | |
| ne 1.5m | hr | m | 0.17 | 0.17 | | |
| | £ | | 5.53 | 2.74 | 2.33 | 10.60 |
| 1.5 - 2.0m | hr | m | 0.20 | 0.20 | | |
| | £ | | 6.50 | 3.23 | 2.33 | 12.06 |
| 2.0 - 2.5m | hr | m | 0.24 | 0.24 | | |
| | £ | | 7.80 | 3.87 | 2.33 | 14.00 |
| 2.5 - 3.0m | hr | m | 0.29 | 0.29 | | |
| | £ | | 9.43 | 4.68 | 2.33 | 16.44 |
| 3.0 - 3.5m | hr | m | 0.39 | 0.39 | | |
| | £ | | 12.68 | 6.29 | 2.33 | 21.30 |
| 3.5 - 4.0m | hr | m | 0.45 | 0.45 | | |
| | £ | | 14.63 | 7.26 | 2.33 | 24.22 |
| 4.0 - 4.5m | hr | m | 0.58 | 0.58 | | |
| | £ | | 18.86 | 9.36 | 2.33 | 30.55 |
| 4.5 - 5.0m | hr | m | 0.75 | 0.75 | | |
| | £ | | 24.38 | 12.10 | 2.33 | 38.81 |
| 5.0 - 5.5m | hr | m | 0.96 | 0.96 | | |
| | £ | | 31.21 | 15.48 | 2.33 | 49.02 |
| 225mm | | | | | | |
| Not in trenches | hr | m | 0.08 | 0.02 | | |
| | £ | | 2.60 | 0.32 | 3.78 | 6.70 |
| In trenches | | | | | | |
| ne 1.5m | hr | m | 0.18 | 0.18 | | |
| | £ | | 5.85 | 2.90 | 3.78 | 12.53 |
| 1.5 - 2.0m | hr | m | 0.22 | 0.22 | | |
| | £ | | 7.15 | 3.55 | 3.78 | 14.48 |
| 2.0 - 2.5m | hr | m | 0.26 | 0.26 | | |
| | £ | | 8.45 | 4.19 | 3.78 | 16.42 |
| 2.5 - 3.0m | hr | m | 0.30 | 0.30 | | |
| | £ | | 9.75 | 4.84 | 3.78 | 18.37 |

**CLASS I: PIPEWORK - PIPES**

**CONCRETE POROUS PIPES - continued**

| | UNIT | | L | P | M | £ NETT |
|---|---|---|---|---|---|---|
| 225mm | | | | | | |
| In trenches | | | | | | |
| 3.0 - 3.5m | hr | m | 0.41 | 0.41 | | |
| | £ | | **13.33** | **6.61** | **3.78** | **23.72** |
| 3.5 - 4.0m | hr | m | 0.47 | 0.47 | | |
| | £ | | **15.28** | **7.58** | **3.78** | **26.64** |
| 4.0 - 4.5m | hr | m | 0.65 | 0.65 | | |
| | £ | | **21.13** | **10.48** | **3.78** | **35.39** |
| 4.5 - 5.0m | hr | m | 0.80 | 0.80 | | |
| | £ | | **26.01** | **12.90** | **3.78** | **42.69** |
| 5.0 - 5.5m | hr | m | 1.02 | 1.02 | | |
| | £ | | **33.16** | **16.45** | **3.78** | **53.39** |
| 300mm | | | | | | |
| Not in trenches | hr | m | 0.10 | 0.05 | | |
| | £ | | **3.25** | **0.81** | **6.21** | **10.27** |
| In trenches | | | | | | |
| ne 1.5m | hr | m | 0.19 | 0.19 | | |
| | £ | | **6.18** | **3.06** | **6.21** | **15.45** |
| 1.5 - 2.0m | hr | m | 0.24 | 0.24 | | |
| | £ | | **7.80** | **3.87** | **6.21** | **17.88** |
| 2.0 - 2.5m | hr | m | 0.28 | 0.28 | | |
| | £ | | **9.10** | **4.52** | **6.21** | **19.83** |
| 2.5 - 3.0m | hr | m | 0.32 | 0.32 | | |
| | £ | | **10.40** | **5.16** | **6.21** | **21.77** |
| 3.0 - 3.5m | hr | m | 0.42 | 0.42 | | |
| | £ | | **13.65** | **6.77** | **6.21** | **26.63** |
| 3.5 - 4.0m | hr | m | 0.52 | 0.52 | | |
| | £ | | **16.91** | **8.39** | **6.21** | **31.51** |
| 4.0 - 4.5m | hr | m | 0.70 | 0.70 | | |
| | £ | | **22.76** | **11.29** | **6.21** | **40.26** |
| 4.5 - 5.0m | hr | m | 0.90 | 0.90 | | |
| | £ | | **29.26** | **14.52** | **6.21** | **49.99** |
| 5.0 - 5.5m | hr | m | 1.12 | 1.12 | | |
| | £ | | **36.41** | **18.07** | **6.21** | **60.69** |

**CLASS J: PIPEWORK - FITTINGS AND VALVES**

LABOUR based on Gang B.2

| | | |
|---|---|---|
| Total hourly cost | = £ | 32.51 |

**PLANT**

| | | |
|---|---|---|
| Excavator (average 0.5m - 1.0m3) | | 10.90 |
| Plate Compactor (part time) | | 0.90 |
| Trench sheets, shores, props, etc. | | 2.30 |
| Part time use of pumps, dumper concrete equipment, power tools. (average useage) | | 2.03 |
| Total hourly cost | = £ | 16.13 |

This section of unit costs has been completely reviewed for this edition.
To provide a basic, overall cost comparison, the rates, prices and outputs have been based on a medium sized Civil Engineering project of £5-6 million which is neither city centre nor excessively remote.

All rates for materials are now based on the most economically available materials with a minimum waste allowance - 2-3% and the most efficient items of plant (Excavator). The rates in this section are the extra cost of installing the fittings in sequence with the pipe runs in open trenches.

Multiplier table for Labour and Plant for various site conditions. (Please note previous editions have taken an average of the following conditions)

| | | | | |
|---|---|---|---|---|
| Working out of sequence with other works | | 2.75 | - | min. |
| Working in depths up to | 1.5m | - | | 1.00 |
| | 1.5m | 3.0m | - | 1.30 |
| | 3.0m | 4.5m | - | 1.50 |
| | over 4.5m | - | | 1.80 |
| Working below water Table levels | | 2.00 | - | min. |

**CLASS J: PIPEWORK - FITTINGS AND VALVES**
(No allowance for waste or loss)

**CLAY PIPE FITTINGS**

| | | UNIT | L | P | M | £ NETT |
|---|---|---|---|---|---|---|
| Vitrified clay, polypropylene couplings (Supersleve/Hepsleve) | | | | | | |
| bends | | | | | | |
| 100mm | hr | nr | 0.05 | - | | |
| | £ | | 1.63 | - | 2.21 | 3.84 |
| 150mm | hr | nr | 0.08 | - | | |
| | £ | | 2.60 | - | 10.52 | 13.12 |
| 225mm | hr | nr | 0.10 | - | | |
| | £ | | 3.25 | - | 27.66 | 30.91 |
| 300mm | hr | nr | 0.15 | - | | |
| | £ | | 4.88 | - | 53.16 | 58.04 |
| bends; rest | | | | | | |
| 100mm | hr | nr | 0.06 | - | | |
| | £ | | 1.95 | - | 5.58 | 7.53 |
| 150mm | hr | nr | 0.09 | - | | |
| | £ | | 2.93 | - | 12.04 | 14.97 |
| 225mm | hr | nr | 0.11 | - | | |
| | £ | | 3.58 | - | 29.27 | 32.85 |
| 300mm | hr | nr | 0.16 | - | | |
| | £ | | 5.20 | - | 74.54 | 79.74 |
| single junctions; equal | | | | | | |
| 100mm | hr | nr | 0.09 | - | | |
| | £ | | 2.93 | - | 7.58 | 10.51 |
| 150mm | hr | nr | 0.11 | - | | |
| | £ | | 3.58 | - | 15.60 | 19.18 |
| 225mm | hr | nr | 0.16 | - | | |
| | £ | | 5.20 | - | 42.67 | 47.87 |
| 300mm | hr | nr | 0.19 | - | | |
| | £ | | 6.18 | - | 86.24 | 92.42 |
| tapers (maximum diameter) | | | | | | |
| 150mm | hr | nr | 0.07 | - | | |
| | £ | | 2.28 | - | 8.71 | 10.99 |
| 225mm | hr | nr | 0.12 | - | | |
| | £ | | 3.90 | - | 27.13 | 31.03 |
| 300mm | hr | nr | 0.15 | - | | |
| | £ | | 4.88 | - | 61.45 | 66.33 |

**CLASS J: PIPEWORK - FITTINGS AND VALVES**

**CLAY PIPE FITTINGS - continued**

| | | UNIT | L | P | M | £ NETT |
|---|---|---|---|---|---|---|
| saddles | | | | | | |
| 100mm | hr | nr | 0.23 | - | | |
| | £ | | 7.48 | - | 5.31 | 12.79 |
| 150mm | hr | nr | 0.29 | - | | |
| | £ | | 9.43 | - | 10.28 | 19.71 |
| 225mm | hr | nr | 0.36 | - | | |
| | £ | | 11.70 | - | 29.89 | 41.59 |
| 300mm | hr | nr | 0.44 | - | | |
| | £ | | 14.30 | - | 52.09 | 66.39 |
| Vitrified clay, ring seal joints (Hepseal) | | | | | | |
| bends; 90 degrees | | | | | | |
| 100mm | hr | nr | 0.05 | - | | |
| | £ | | 1.63 | - | 6.85 | 8.48 |
| 150mm | hr | nr | 0.08 | - | | |
| | £ | | 2.60 | - | 11.30 | 13.90 |
| 225mm | hr | nr | 0.10 | - | | |
| | £ | | 3.25 | - | 23.14 | 26.39 |
| 300mm | hr | nr | 0.15 | - | | |
| | £ | | 4.88 | - | 45.62 | 50.50 |
| 400mm | hr | nr | 0.24 | - | | |
| | £ | | 7.80 | - | 143.93 | 151.73 |
| 450mm | hr | nr | 0.29 | - | | |
| | £ | | 9.43 | - | 189.50 | 198.93 |
| bends; 45 degrees | | | | | | |
| 100mm | hr | nr | 0.05 | - | | |
| | £ | | 1.63 | - | 6.85 | 8.48 |
| 150mm | hr | nr | 0.08 | - | | |
| | £ | | 2.60 | - | 11.30 | 13.90 |
| 225mm | hr | nr | 0.10 | - | | |
| | £ | | 3.25 | - | 23.14 | 26.39 |
| 300mm | hr | nr | 0.15 | - | | |
| | £ | | 4.88 | - | 45.62 | 50.50 |
| 400mm | hr | nr | 0.24 | - | | |
| | £ | | 7.80 | - | 102.81 | 110.61 |
| 450mm | hr | nr | 0.29 | - | | |
| | £ | | 9.43 | - | 135.37 | 144.80 |

**CLASS J: PIPEWORK - FITTINGS AND VALVES**

**CLAY PIPE FITTINGS - continued**

| | | UNIT | L | P | M | £ NETT |
|---|---|---|---|---|---|---|
| bends; 30 degrees and 15 degrees | | | | | | |
| 100mm | hr | nr | 0.05 | - | | |
| | £ | | 1.63 | - | 6.42 | 8.05 |
| 150mm | hr | nr | 0.08 | - | | |
| | £ | | 2.60 | - | 11.30 | 13.90 |
| 225mm | hr | nr | 0.10 | - | | |
| | £ | | 3.25 | - | 23.14 | 26.39 |
| 300mm | hr | nr | 0.15 | - | | |
| | £ | | 4.88 | - | 45.62 | 50.50 |
| bends; 22.5 degrees | | | | | | |
| 400mm | hr | nr | 0.24 | - | | |
| | £ | | 7.80 | - | 72.00 | 79.80 |
| 450mm | hr | nr | 0.29 | - | | |
| | £ | | 9.43 | - | 94.78 | 104.21 |
| bends; rest | | | | | | |
| 100mm | hr | nr | 0.06 | - | | |
| | £ | | 1.95 | - | 8.14 | 10.09 |
| 150mm | hr | nr | 0.09 | - | | |
| | £ | | 2.93 | - | 13.47 | 16.40 |
| 225mm | hr | nr | 0.11 | - | | |
| | £ | | 3.58 | - | 32.22 | 35.80 |
| 300mm | hr | nr | 0.16 | - | | |
| | £ | | 5.20 | - | 68.32 | 73.52 |
| junctions; single | | | | | | |
| 100mm | hr | nr | 0.09 | - | | |
| | £ | | 2.93 | - | 9.54 | 12.47 |
| 150mm | hr | nr | 0.11 | - | | |
| | £ | | 3.58 | - | 14.94 | 18.52 |
| 225mm | hr | nr | 0.16 | - | | |
| | £ | | 5.20 | - | 35.86 | 41.06 |
| 300mm | hr | nr | 0.19 | - | | |
| | £ | | 6.18 | - | 71.68 | 77.86 |
| junctions; double | | | | | | |
| 150mm | hr | nr | 0.13 | - | | |
| | £ | | 4.23 | - | 30.14 | 34.37 |
| 225mm | hr | nr | 0.18 | - | | |
| | £ | | 5.85 | - | 67.56 | 73.41 |
| 300mm | hr | nr | 0.21 | - | | |
| | £ | | 6.83 | - | 143.39 | 150.22 |

**CLASS J: PIPEWORK - FITTINGS AND VALVES**

**CLAY PIPE FITTINGS - continued**

| | | UNIT | L | P | M | £ NETT |
|---|---|---|---|---|---|---|
| **tapers (maximum diameter)** | | | | | | |
| **150mm** | hr | nr | 0.07 | - | | |
| | £ | | 2.28 | - | 16.92 | 19.20 |
| **225mm** | hr | nr | 0.12 | - | | |
| | £ | | 3.90 | - | 26.75 | 30.65 |
| **300mm** | hr | nr | 0.15 | - | | |
| | £ | | 4.88 | - | 54.69 | 59.57 |
| **saddles** | | | | | | |
| **100mm** | hr | nr | 0.23 | - | | |
| | £ | | 7.48 | - | 6.38 | 13.86 |
| **150mm** | hr | nr | 0.29 | - | | |
| | £ | | 9.43 | - | 10.75 | 20.18 |
| **225mm** | hr | nr | 0.36 | - | | |
| | £ | | 11.70 | - | 26.75 | 38.45 |
| **300mm** | hr | nr | 0.44 | - | | |
| | £ | | 14.30 | - | 54.69 | 68.99 |
| Vitrified clay, polypropylene couplings (Densleeve) | | | | | | |
| **bends** | | | | | | |
| **100mm** | hr | nr | 0.05 | - | | |
| | £ | | 1.63 | - | 4.06 | 5.69 |
| **150mm** | hr | nr | 0.08 | - | | |
| | £ | | 2.60 | - | 9.71 | 12.31 |
| **200mm** | hr | nr | 0.10 | - | | |
| | £ | | 3.25 | - | 23.75 | 27.00 |
| **225mm** | hr | nr | 0.11 | - | | |
| | £ | | 3.58 | - | 23.75 | 27.33 |
| **bends; rest** | | | | | | |
| **100mm** | hr | nr | 0.06 | - | | |
| | £ | | 1.95 | - | 5.12 | 7.07 |
| **150mm** | hr | nr | 0.09 | - | | |
| | £ | | 2.93 | - | 10.89 | 13.82 |
| **225mm** | hr | nr | 0.11 | - | | |
| | £ | | 3.58 | - | 24.93 | 28.51 |

**CLASS J: PIPEWORK - FITTINGS AND VALVES**

**CLAY PIPE FITTINGS - continued**

| | | UNIT | L | P | M | £ NETT |
|---|---|---|---|---|---|---|
| junctions; equal | | | | | | |
| 100mm | hr | nr | 0.09 | - | | |
| | £ | | 2.93 | - | 6.67 | 9.60 |
| 150mm | hr | nr | 0.11 | - | | |
| | £ | | 3.58 | - | 13.51 | 17.09 |
| 225mm | hr | nr | 0.16 | - | | |
| | £ | | 5.20 | - | 33.79 | 38.99 |
| tapers (maximum diameter) | | | | | | |
| 150mm | hr | nr | 0.07 | - | | |
| | £ | | 2.28 | - | 8.43 | 10.71 |
| 200mm | hr | nr | 0.12 | - | | |
| | £ | | 3.90 | - | 24.81 | 28.71 |
| 225mm | hr | nr | 0.15 | - | | |
| | £ | | 4.88 | - | 24.81 | 29.69 |
| saddles | | | | | | |
| 100mm | hr | nr | 0.23 | - | | |
| | £ | | 7.48 | - | 4.85 | 12.33 |
| 150mm | hr | nr | 0.29 | - | | |
| | £ | | 9.43 | - | 9.48 | 18.91 |
| 225mm | hr | nr | 0.36 | - | | |
| | £ | | 11.70 | - | 25.41 | 37.11 |

CLASS J: PIPEWORK - FITTINGS AND VALVES

CLAY PIPE FITTINGS - continued

| | | UNIT | L | P | M | £ NETT |
|---|---|---|---|---|---|---|
| Vitrified clay, ring seal joints (Denseal) | | | | | | |
| bends | | | | | | |
| 100mm | hr | nr | 0.05 | - | | |
| | £ | | 1.63 | - | 4.68 | 6.31 |
| 150mm | hr | nr | 0.08 | - | | |
| | £ | | 2.60 | - | 8.30 | 10.90 |
| 225mm | hr | nr | 0.10 | - | | |
| | £ | | 3.25 | - | 17.04 | 20.29 |
| 300mm | hr | nr | 0.15 | 0.07 | | |
| | £ | | 4.88 | 1.13 | 33.55 | 39.56 |
| 375mm | hr | nr | 0.19 | 0.09 | | |
| | £ | | 6.18 | 1.45 | 63.71 | 71.34 |
| 400mm | hr | nr | 0.24 | 0.12 | | |
| | £ | | 7.80 | 1.94 | 75.60 | 85.34 |
| 450mm | hr | nr | 0.29 | 0.14 | | |
| | £ | | 9.43 | 2.26 | 99.52 | 111.21 |
| bends; rest | | | | | | |
| 100mm | hr | nr | 0.06 | - | | |
| | £ | | 1.95 | - | 5.82 | 7.77 |
| 150mm | hr | nr | 0.09 | - | | |
| | £ | | 2.93 | - | 9.66 | 12.59 |
| 225mm | hr | nr | 0.11 | - | | |
| | £ | | 3.58 | - | 23.69 | 27.27 |
| single junctions | | | | | | |
| 100mm | hr | nr | 0.09 | - | | |
| | £ | | 2.93 | - | 6.18 | 9.11 |
| 150mm | hr | nr | 0.11 | - | | |
| | £ | | 3.58 | - | 10.09 | 13.67 |
| 225mm | hr | nr | 0.16 | - | | |
| | £ | | 5.20 | - | 23.69 | 28.89 |
| 300mm | hr | nr | 0.19 | 0.08 | | |
| | £ | | 6.18 | 1.29 | 50.23 | 57.70 |
| 375mm | hr | nr | 0.21 | 0.10 | | |
| | £ | | 6.83 | 1.61 | 89.47 | 97.91 |
| 400mm | hr | nr | 0.23 | 0.11 | | |
| | £ | | 7.48 | 1.77 | 99.14 | 108.39 |
| 450mm | hr | nr | 0.29 | 0.14 | | |
| | £ | | 9.43 | 2.26 | 118.59 | 130.28 |

**CLASS J: PIPEWORK - FITTINGS AND VALVES**

**CLAY PIPE FITTINGS - continued**

| | | UNIT | L | P | M | £ NETT |
|---|---|---|---|---|---|---|
| taper/increasers (maximum diameter) | | | | | | |
| 150mm | hr | nr | 0.07 | | | |
| | £ | | 2.28 | - | 12.49 | 14.77 |
| 225mm | hr | nr | 0.12 | | | |
| | £ | | 3.90 | - | 19.66 | 23.56 |
| 300mm | hr | nr | 0.15 | 0.07 | | |
| | £ | | 4.88 | 1.13 | 40.23 | 46.24 |
| 375mm | hr | nr | 0.18 | 0.09 | | |
| | £ | | 5.85 | 1.45 | 59.62 | 66.92 |
| tumbling bay/drop backs | | | | | | |
| 100mm | hr | nr | 0.09 | - | | |
| | £ | | 2.93 | - | 6.18 | 9.11 |
| 150mm | hr | nr | 0.11 | - | | |
| | £ | | 3.58 | - | 10.25 | 13.83 |
| 225mm | hr | nr | 0.16 | - | | |
| | £ | | 5.20 | - | 26.36 | 31.56 |
| 300mm | hr | nr | 0.19 | 0.08 | | |
| | £ | | 6.18 | 1.29 | 56.76 | 64.23 |
| 375mm | hr | nr | 0.21 | 0.10 | | |
| | £ | | 6.83 | 1.61 | 113.24 | 121.68 |
| 400mm | hr | nr | 0.23 | 0.11 | | |
| | £ | | 7.48 | 1.77 | 134.42 | 143.67 |
| 450mm | hr | nr | 0.29 | 0.14 | | |
| | £ | | 9.43 | 2.26 | 155.92 | 167.61 |
| Vitrified clay subsoil perforated drainage (Hepline) | | | | | | |
| bends | | | | | | |
| 100mm | hr | nr | 0.05 | - | | |
| | £ | | 1.63 | - | 2.73 | 4.36 |
| 150mm | hr | nr | 0.08 | - | | |
| | £ | | 2.60 | - | 5.44 | 8.04 |
| 225mm | hr | nr | 0.10 | - | | |
| | £ | | 3.25 | - | 12.93 | 16.18 |
| 300mm | hr | nr | 0.15 | - | | |
| | £ | | 4.88 | - | 21.42 | 26.30 |

**CLASS J: PIPEWORK - FITTINGS AND VALVES**

**CLAY PIPE FITTINGS - continued**

| | | UNIT | L | P | M | £ NETT |
|---|---|---|---|---|---|---|
| junctions | | | | | | |
| 100mm | hr | nr | 0.09 | - | | |
| | £ | | 2.93 | - | 4.93 | 7.86 |
| 150mm | hr | nr | 0.11 | - | | |
| | £ | | 3.58 | - | 8.36 | 11.94 |
| 225mm | hr | nr | 0.16 | - | | |
| | £ | | 5.20 | - | 20.19 | 25.39 |
| 300mm | hr | nr | 0.19 | - | | |
| | £ | | 6.18 | - | 35.83 | 42.01 |

Vitrified clay subsoil perforated drainage (Denline)

| | | UNIT | L | P | M | £ NETT |
|---|---|---|---|---|---|---|
| bends | | | | | | |
| 100mm | hr | nr | 0.05 | - | | |
| | £ | | 1.63 | - | 2.10 | 3.73 |
| 150mm | hr | nr | 0.08 | - | | |
| | £ | | 2.60 | - | 4.07 | 6.67 |
| 225mm | hr | nr | 0.10 | - | | |
| | £ | | 3.25 | - | 9.69 | 12.94 |
| 300mm | hr | nr | 0.15 | - | | |
| | £ | | 4.88 | - | 16.01 | 20.89 |
| junctions | | | | | | |
| 100mm | hr | nr | 0.09 | - | | |
| | £ | | 2.93 | - | 3.81 | 6.74 |
| 150mm | hr | nr | 0.11 | - | | |
| | £ | | 3.58 | - | 6.27 | 9.85 |
| 225mm | hr | nr | 0.16 | - | | |
| | £ | | 5.20 | - | 15.11 | 20.31 |
| 300mm | hr | nr | 0.19 | - | | |
| | £ | | 6.18 | - | 26.83 | 33.01 |

**CLASS J: PIPEWORK - FITTINGS AND VALVES**

**CONCRETE PIPE FITTINGS**

| | | UNIT | L | P | M | £ NETT |
|---|---|---|---|---|---|---|
| Concrete pipes to BS5911 Class L | | | | | | |
| bends | | | | | | |
| 300mm | hr | nr | 0.08 | 0.05 | | |
| | £ | | 2.60 | 0.81 | 63.30 | 66.71 |
| 375mm | hr | nr | 0.10 | 0.07 | | |
| | £ | | 3.25 | 1.13 | 82.65 | 87.03 |
| 450mm | hr | nr | 0.13 | 0.10 | | |
| | £ | | 4.23 | 1.61 | 96.70 | 102.54 |
| 525mm | hr | nr | 0.16 | 0.13 | | |
| | £ | | 5.20 | 2.10 | 113.25 | 120.55 |
| 750mm | hr | nr | 0.24 | 0.24 | | |
| | £ | | 7.80 | 3.87 | 187.05 | 198.72 |
| 900mm | hr | nr | 0.39 | 0.39 | | |
| | £ | | 12.68 | 6.29 | 259.10 | 278.07 |
| 1200mm | hr | nr | 0.51 | 0.51 | | |
| | £ | | 16.58 | 8.23 | 414.70 | 439.51 |
| 1500mm | hr | nr | 0.63 | 0.63 | | |
| | £ | | 20.48 | 10.16 | 653.80 | 684.44 |
| Concrete pipes to BS5911 Class M | | | | | | |
| bends | | | | | | |
| 300mm | hr | nr | 0.08 | 0.05 | | |
| | £ | | 2.60 | 0.81 | 65.30 | 68.71 |
| 375mm | hr | nr | 0.10 | 0.07 | | |
| | £ | | 3.25 | 1.13 | 83.60 | 87.98 |
| 450mm | hr | nr | 0.13 | 0.10 | | |
| | £ | | 4.23 | 1.61 | 97.65 | 103.49 |
| 525mm | hr | nr | 0.16 | 0.13 | | |
| | £ | | 5.20 | 2.10 | 114.15 | 121.45 |

**CLASS J: PIPEWORK - FITTINGS AND VALVES**

**CONCRETE PIPE FITTING - continued**

| | | UNIT | L | P | M | £ NETT |
|---|---|---|---|---|---|---|
| **bends** | | | | | | |
| **750mm** | hr | nr | 0.24 | 0.24 | | |
| | £ | | 7.80 | 3.87 | 188.90 | 200.57 |
| **900mm** | hr | nr | 0.40 | 0.40 | | |
| | £ | | 13.00 | 6.45 | 265.95 | 285.40 |
| **1200mm** | hr | nr | 0.52 | 0.52 | | |
| | £ | | 16.91 | 8.39 | 457.55 | 482.85 |
| **1500mm** | hr | nr | 0.64 | 0.64 | | |
| | £ | | 20.81 | 10.32 | 696.10 | 727.23 |

Concrete pipes to BS5911 Class H

| | | | | | | |
|---|---|---|---|---|---|---|
| **bends** | | | | | | |
| **375mm** | hr | nr | 0.10 | 0.07 | | |
| | £ | | 3.25 | 1.13 | 85.00 | 89.38 |
| **450mm** | hr | nr | 0.13 | 0.10 | | |
| | £ | | 4.23 | 1.61 | 99.60 | 105.44 |
| **525mm** | hr | nr | 0.16 | 0.13 | | |
| | £ | | 5.20 | 2.10 | 116.45 | 123.75 |
| **750mm** | hr | nr | 0.24 | 0.24 | | |
| | £ | | 7.80 | 3.87 | 207.75 | 219.42 |
| **900mm** | hr | nr | 0.40 | 0.40 | | |
| | £ | | 13.00 | 6.45 | 274.10 | 293.55 |
| **1200mm** | hr | nr | 0.52 | 0.52 | | |
| | £ | | 16.91 | 8.39 | 486.60 | 511.90 |
| **1500mm** | hr | nr | 0.64 | 0.64 | | |
| | £ | | 20.81 | 10.32 | 721.95 | 753.08 |

**CLASS J: PIPEWORK - FITTINGS AND VALVES**

**CAST OR SPUN IRON PIPE FITTINGS**

| | | UNIT | L | P | M | £ NETT |
|---|---|---|---|---|---|---|
| Cast and ductile iron to BS437 | | | | | | |
| bends; 87.5 degrees | | | | | | |
| 75mm | hr | nr | 0.31 | 0.02 | | |
| | £ | | 10.08 | 0.32 | 32.90 | 43.30 |
| 100mm | hr | nr | 0.38 | 0.03 | | |
| | £ | | 12.35 | 0.48 | 39.66 | 52.49 |
| 150mm | hr | nr | 0.56 | 0.04 | | |
| | £ | | 18.21 | 0.65 | 63.52 | 82.38 |
| 225mm | hr | nr | 0.77 | 0.04 | | |
| | £ | | 25.03 | 0.65 | 190.73 | 216.41 |
| bends 45 degrees | | | | | | |
| 75mm | hr | nr | 0.31 | 0.02 | | |
| | £ | | 10.08 | 0.32 | 32.00 | 42.40 |
| 100mm | hr | nr | 0.38 | 0.03 | | |
| | £ | | 12.35 | 0.48 | 36.80 | 49.63 |
| 150mm | hr | nr | 0.56 | 0.04 | | |
| | £ | | 18.21 | 0.65 | 57.90 | 76.76 |
| 225mm | hr | nr | 0.77 | 0.04 | | |
| | £ | | 25.03 | 0.65 | 190.73 | 216.41 |
| bends, long radius | | | | | | |
| 100mm | hr | nr | 0.38 | 0.03 | | |
| | £ | | 12.35 | 0.48 | 43.99 | 56.82 |
| 150mm | hr | nr | 0.56 | 0.04 | | |
| | £ | | 18.21 | 0.65 | 75.20 | 94.06 |
| single junctions | | | | | | |
| 75mm | hr | nr | 0.48 | 0.04 | | |
| | £ | | 15.60 | 0.65 | 47.64 | 63.89 |
| 100mm | hr | nr | 0.59 | 0.05 | | |
| | £ | | 19.18 | 0.81 | 56.20 | 76.19 |
| 150mm | hr | nr | 0.88 | 0.08 | | |
| | £ | | 28.61 | 1.29 | 92.32 | 122.22 |
| 225mm | hr | nr | 1.21 | 0.09 | | |
| | £ | | 39.34 | 1.45 | 278.03 | 318.82 |
| double junctions | | | | | | |
| 75mm | hr | nr | - | - | | |
| | £ | | - | - | - | - |
| 100mm | hr | nr | 0.80 | 0.08 | | |
| | £ | | 26.01 | 1.29 | 79.32 | 106.62 |

**CLASS J: PIPEWORK - FITTINGS AND VALVES**

**CAST OR SPUN IRON PIPE FITTINGS - continued**

| | | UNIT | L | P | M | £ NETT |
|---|---|---|---|---|---|---|
| double junctions | | | | | | |
| 150mm | hr | nr | - | - | | |
| | £ | | - | - | - | - |
| 225mm | hr | nr | 1.59 | 0.11 | | |
| | £ | | 51.69 | 1.77 | 392.73 | 446.19 |
| taper | | | | | | |
| 75mm | hr | nr | 0.27 | 0.02 | | |
| | £ | | 8.78 | 0.32 | 26.66 | 35.76 |
| 100mm | hr | nr | 0.40 | 0.05 | | |
| | £ | | 13.00 | 0.81 | 35.22 | 49.03 |
| 150mm | hr | nr | 0.56 | 0.08 | | |
| | £ | | 18.21 | 1.29 | 47.66 | 67.16 |
| 225mm | hr | nr | 0.77 | 0.09 | | |
| | £ | | 25.03 | 1.45 | 104.93 | 131.41 |
| Ductile spun iron pipes to BS4772, Tyton joints | | | | | | |
| bends 90 degrees | | | | | | |
| 100mm | hr | nr | 0.38 | 0.03 | | |
| | £ | | 12.35 | 0.48 | 25.11 | 37.94 |
| 150mm | hr | nr | 0.57 | 0.04 | | |
| | £ | | 18.53 | 0.65 | 44.57 | 63.75 |
| 250mm | hr | nr | 0.96 | 0.06 | | |
| | £ | | 31.21 | 0.97 | 142.21 | 174.39 |
| duckfoot bends | | | | | | |
| 100mm | hr | nr | 0.41 | 0.04 | | |
| | £ | | 13.33 | 0.65 | 58.30 | 72.28 |
| 150mm | hr | nr | 0.60 | 0.06 | | |
| | £ | | 19.51 | 0.97 | 95.66 | 116.14 |
| 250mm | hr | nr | 0.90 | 0.08 | | |
| | £ | | 29.26 | 1.29 | 273.99 | 304.54 |
| single junctions (equal) | | | | | | |
| 100mm | hr | nr | 0.60 | 0.06 | | |
| | £ | | 19.51 | 0.97 | 44.99 | 65.47 |
| 150mm | hr | nr | 0.88 | 0.08 | | |
| | £ | | 28.61 | 1.29 | 67.99 | 97.89 |
| 250mm | hr | nr | 1.32 | 0.10 | | |
| | £ | | 42.91 | 1.61 | 135.02 | 179.54 |

**CLASS J: PIPEWORK - FITTINGS AND VALVES**

**CAST OR SPUN IRON PIPE FITTINGS - continued**

| | | UNIT | L | P | M | £ NETT |
|---|---|---|---|---|---|---|
| Ductile spun iron pipes to BS4772, Stanlock joints | | | | | | |
| bends 90 degrees | | | | | | |
| 100mm | hr | nr | 0.38 | 0.03 | | |
| | £ | | 12.35 | 0.48 | 40.49 | 53.32 |
| 150mm | hr | nr | 0.57 | 0.04 | | |
| | £ | | 18.53 | 0.65 | 62.60 | 81.78 |
| 250mm | hr | nr | 0.96 | 0.06 | | |
| | £ | | 31.21 | 0.97 | 185.31 | 217.49 |
| single junctions (equal) | | | | | | |
| 100mm | hr | nr | 0.60 | 0.06 | | |
| | £ | | 19.51 | 0.97 | 61.88 | 82.36 |
| 150mm | hr | nr | 0.88 | 0.08 | | |
| | £ | | 28.61 | 1.29 | 90.98 | 120.88 |
| 250mm | hr | nr | 1.32 | 0.10 | | |
| | £ | | 42.91 | 1.61 | 229.18 | 273.70 |

STEEL PIPE FITTINGS

| | | UNIT | L | P | M | £ NETT |
|---|---|---|---|---|---|---|
| Steel pipes to BS3601 including welded joints to pipes (for protection and lining refer to manufacturer) | | | | | | |
| bends 45 degrees | | | | | | |
| 100mm | hr | nr | 0.07 | 0.03 | | |
| | £ | | 2.28 | 0.48 | 8.48 | 11.24 |
| 150mm | hr | nr | 0.09 | 0.04 | | |
| | £ | | 2.93 | 0.65 | 18.03 | 21.61 |
| bends 90 degrees | | | | | | |
| 100mm | hr | nr | 0.07 | 0.03 | | |
| | £ | | 2.28 | 0.48 | 10.23 | 12.99 |
| 150mm | hr | nr | 0.09 | 0.04 | | |
| | £ | | 2.93 | 0.65 | 23.00 | 26.58 |
| single junctions | | | | | | |
| 100mm | hr | nr | 0.11 | 0.05 | | |
| | £ | | 3.58 | 0.81 | 26.16 | 30.55 |
| 150mm | hr | nr | 0.16 | 0.08 | | |
| | £ | | 5.20 | 1.29 | 46.38 | 52.87 |

**CLASS J: PIPEWORK - FITTINGS AND VALVES**

**STEEL PIPE FITTINGS - continued**

| | | UNIT | L | P | M | £ NETT |
|---|---|---|---|---|---|---|
| **STEEL PIPE FITTINGS** | | | | | | |
| Steel pipe to BS3601 with welded joints (pipe only) (For protection and lining refer to manufacturer) | | | | | | |
| bends 45 degrees | | | | | | |
| 100mm | hr | nr | 0.07 | 0.03 | | |
| | £ | | 2.28 | 0.48 | 9.90 | 12.66 |
| 150mm | hr | nr | 0.09 | 0.04 | | |
| | £ | | 2.93 | 0.65 | 29.80 | 33.38 |
| bends 90 degrees | | | | | | |
| 100mm | hr | nr | 0.07 | 0.03 | | |
| | £ | | 2.28 | 0.48 | 11.50 | 14.26 |
| 150mm | hr | nr | 0.09 | 0.04 | | |
| | £ | | 2.93 | 0.65 | 35.05 | 38.63 |
| single junctions | | | | | | |
| 100mm | hr | nr | 0.11 | 0.05 | | |
| | £ | | 3.58 | 0.81 | 33.70 | 38.09 |
| 150mm | hr | nr | 0.16 | 0.08 | | |
| | £ | | 5.20 | 1.29 | 101.50 | 107.99 |

**CLASS J: PIPEWORK - FITTINGS AND VALVES**

**PLASTIC PIPE FITTINGS**

| | | UNIT | L | P | M | £ NETT |
|---|---|---|---|---|---|---|
| Unplasticised PVC pipes with ring seal sockets (Marley) | | | | | | |
| bends, short radius (socket/spigot) | | | | | | |
| 82mm | hr | nr | 0.05 | - | | |
| | £ | | 1.63 | - | 6.88 | 8.51 |
| 110mm | hr | nr | 0.05 | - | | |
| | £ | | 1.63 | - | 9.02 | 10.65 |
| 160mm | hr | nr | 0.05 | - | | |
| | £ | | 1.63 | - | 19.82 | 21.45 |
| 200mm (87.5 degrees) | hr | nr | 0.07 | - | | |
| | £ | | 2.28 | - | 44.56 | 46.84 |
| 200mm (45.0 degrees) | hr | nr | 0.07 | - | | |
| | £ | | 2.28 | - | 36.51 | 38.79 |
| 200mm (15.0 degrees) | hr | nr | 0.07 | - | | |
| | £ | | 2.28 | - | 31.87 | 34.15 |
| 250mm (45.0 degrees) | hr | nr | 0.07 | - | | |
| | £ | | 2.28 | - | 96.07 | 98.35 |
| 250mm (15.0 degrees) | hr | nr | 0.07 | - | | |
| | £ | | 2.28 | - | 83.69 | 85.97 |
| bends, short radius (socket/socket) | | | | | | |
| 110mm | hr | nr | 0.05 | - | | |
| | £ | | 1.63 | - | 7.79 | 9.42 |
| bends, long radius (socket/socket) | | | | | | |
| 110mm | hr | nr | 0.05 | - | | |
| | £ | | 1.63 | - | 12.85 | 14.48 |
| branches, equal (socket/spigot) | | | | | | |
| 82mm | hr | nr | 0.07 | - | | |
| | £ | | 2.28 | - | 9.55 | 11.83 |
| 110mm | hr | nr | 0.07 | - | | |
| | £ | | 2.28 | - | 12.48 | 14.76 |
| 160mm | hr | nr | 0.07 | - | | |
| | £ | | 2.28 | - | 30.51 | 32.79 |
| 200mm | hr | nr | 0.09 | - | | |
| | £ | | 2.93 | - | 50.52 | 53.45 |
| 250mm (200mm arm) | hr | nr | 0.09 | - | | |
| | £ | | 2.93 | - | 117.68 | 120.61 |

## CLASS J: PIPEWORK - FITTINGS AND VALVES

### PLASTIC PIPE FITTINGS

| | | UNIT | L | P | M | £ NETT |
|---|---|---|---|---|---|---|
| Unplasticised PVC pipes with ring seal sockets (OsmaDrain) bends, short radius (socket/spigot) | | | | | | |
| 82mm | hr | nr | 0.05 | - | | |
| | £ | | 1.63 | - | 7.30 | 8.93 |
| 110mm (to 30.0 degrees) | hr | nr | 0.05 | - | | |
| | £ | | 1.63 | - | 8.37 | 10.00 |
| 110mm (45 degrees - 87.5 degrees) | hr | nr | 0.05 | - | | |
| | £ | | 1.63 | - | 8.61 | 10.24 |
| 160mm (to 30.0 degrees) | hr | nr | 0.05 | - | | |
| | £ | | 1.63 | - | 17.38 | 19.01 |
| 160mm (45 degrees - 87.5 degrees) | hr | nr | 0.05 | - | | |
| | £ | | 1.63 | - | 18.80 | 20.43 |
| 200mm (87.5 degrees) | hr | nr | 0.07 | - | | |
| | £ | | 2.28 | - | 39.51 | 41.79 |
| 200mm (15.0 degrees) | hr | nr | 0.07 | - | | |
| | £ | | 2.28 | - | 27.00 | 29.28 |
| 250mm (90.0 degrees) | hr | nr | 0.07 | - | | |
| | £ | | 2.28 | - | 150.77 | 153.05 |
| 250mm (15.0 degrees) | hr | nr | 0.07 | - | | |
| | £ | | 2.28 | - | 79.28 | 81.56 |
| 315mm (90.0 degrees) | hr | nr | 0.09 | - | | |
| | £ | | 2.93 | - | 233.09 | 236.02 |
| 315mm (15.0 degrees) | hr | nr | 0.09 | - | | |
| | £ | | 2.93 | - | 152.09 | 155.02 |

**CLASS J: PIPEWORK - FITTINGS AND VALVES**

**PLASTIC PIPE FITTINGS - continued**

| | | UNIT | L | P | M | £ NETT |
|---|---|---|---|---|---|---|
| **bends, short radius (socket/socket)** | | | | | | |
| **110mm (to 30 degrees)** | **hr** | **nr** | **0.05** | **-** | | |
| | **£** | | **1.63** | **-** | **7.09** | **8.72** |
| **110mm (45 degrees - 87.5 degrees)** | **hr** | **nr** | **0.05** | **-** | | |
| | **£** | | **1.63** | **-** | **7.33** | **8.96** |
| **160mm (to 30 degrees)** | **hr** | **nr** | **0.05** | **-** | | |
| | **£** | | **1.63** | **-** | **14.16** | **15.79** |
| **160mm (45 degrees - 87.5 degrees)** | **hr** | **nr** | **0.05** | **-** | | |
| | **£** | | **1.63** | **-** | **15.58** | **17.21** |
| **200mm (45.0 degrees)** | **hr** | **nr** | **0.07** | **-** | | |
| | **£** | | **2.28** | **-** | **24.84** | **27.12** |
| **200mm (87.5 degrees)** | **hr** | **nr** | **0.07** | **-** | | |
| | **£** | | **2.28** | **-** | **32.49** | **34.77** |
| **branches, equal** | | | | | | |
| **82mm** | **hr** | **nr** | **0.07** | **-** | | |
| | **£** | | **2.28** | **-** | **9.88** | **12.16** |
| **110mm** | **hr** | **nr** | **0.07** | **-** | | |
| | **£** | | **2.28** | **-** | **11.35** | **13.63** |
| **160mm** | **hr** | **nr** | **0.07** | **-** | | |
| | **£** | | **2.28** | **-** | **30.11** | **32.39** |
| **200mm** | **hr** | **nr** | **0.09** | **-** | | |
| | **£** | | **2.93** | **-** | **38.18** | **41.11** |
| **250mm** | **hr** | **nr** | **0.09** | **-** | | |
| | **£** | | **2.93** | **-** | **154.34** | **157.27** |
| **315mm (160mm arm)** | **hr** | **nr** | **0.11** | **-** | | |
| | **£** | | **3.58** | **-** | **206.28** | **209.86** |

**CLASS J: PIPEWORK - FITTINGS AND VALVES**

**VALVES AND PENSTOCKS**

| | | UNIT | L | P | M | £ NETT |
|---|---|---|---|---|---|---|
| Non-return valves<br>cast iron single door<br>tidal flap | | | | | | |
| 250mm | hr | nr | 0.40 | 0.20 | | |
| | £ | | 13.00 | 3.23 | 240.00 | 256.23 |
| 350mm | hr | nr | 0.55 | 0.20 | | |
| | £ | | 17.88 | 3.23 | 375.00 | 396.11 |
| 450mm | hr | nr | 0.70 | 0.20 | | |
| | £ | | 22.76 | 3.23 | 445.00 | 470.99 |
| 600mm | hr | nr | 1.10 | 0.30 | | |
| | £ | | 35.76 | 4.84 | 500.00 | 540.60 |
| 800mm | hr | nr | 1.50 | 0.35 | | |
| | £ | | 48.77 | 5.65 | 950.00 | 1004.42 |
| Penstocks<br>cast iron wall mounted<br>hand operated | | | | | | |
| 250mm | hr | nr | 1.66 | 0.35 | | |
| | £ | | 53.97 | 5.65 | 320.00 | 379.62 |
| 350mm | hr | nr | 2.30 | 0.40 | | |
| | £ | | 74.77 | 6.45 | 370.00 | 451.22 |
| 450mm | hr | nr | 2.90 | 0.45 | | |
| | £ | | 94.28 | 7.26 | 400.00 | 501.54 |
| 600 x 600mm | hr | nr | 5.40 | 0.80 | | |
| | £ | | 175.55 | 12.90 | 625.00 | 813.45 |
| 1000 x 1000mm | hr | nr | 9.00 | 1.50 | | |
| | £ | | 292.59 | 24.20 | 1600.00 | 1916.79 |

**CLASS K: PIPEWORK - MANHOLES AND PIPEWORK ANCILLARIES**

**MANHOLES**

| | | | |
|---|---|---|---|
| LABOUR based on Gang B | | | |
| Total hourly cost | = | £ | 30.45 |
| PLANT | | | |
| 0.4m3 Excavator | | | 8.27 |
| Part time small tools,<br>Dumper 2T<br>Disc saw<br>Compaction plate<br>2 Inch vibrator unit<br>Small pump | | | 2.03 |
| Total hourly cost | = | £ | 10.30 |

To provide a basic, overall cost comparison, the rates and outputs have been based on a medium sized Civil Engineering project of £5 - 6 million which is neither city centre nor excessively remote.

All rates for materials are based on the most economically available materials with a minumum waste allowance - 2-3% and the most efficient items of plant (Excavator). The rates are optimum rates and assume continuous output with no delays caused by other operations or works. Ground conditions are assumed to be good soil with no abnormal conditions that would affect outputs and consistency of work.

Multiplier Table for Labour and Plant for various site conditions.

| | | | |
|---|---|---|---|
| Working out of sequence with other works | 2.75 | - | min. |
| Working in hard clay | 1.75 | - | 2.00 |
| Working in running sand | 3.50 | - | min. |
| Working in broken rock | 2.75 | - | 3.50 |
| Working in below water Table levels | 2.00 | - | min. |

**CLASS K: PIPEWORK - MANHOLES AND PIPEWORK ANCILLARIES**

**MANHOLES**

In the CESSM2 manholes are enumerated. To enable a rate to be established the various constituent elements have been broken down and unit rates given against each. (waste allowance, concrete = 10%, manufactured goods = 2%)
(Refer to Approximate Estimates for a list of enumerated manholes)

| | UNIT | | L | P | M | £ NETT |
|---|---|---|---|---|---|---|
| Excavation by machine | | | | | | |
| ne 2m deep | hr | m3 | 0.22 | 0.22 | | |
| | £ | | 6.70 | 2.27 | - | 8.97 |
| 2.0 - 3.0m deep | hr | m3 | 0.24 | 0.24 | | |
| | £ | | 7.31 | 2.47 | - | 9.78 |
| 3.0 - 4.0m deep | hr | m3 | 0.26 | 0.26 | | |
| | £ | | 7.92 | 2.68 | - | 10.60 |
| Backfill | | | | | | |
| ne 2m deep | hr | m3 | 0.11 | 0.11 | | |
| | £ | | 3.35 | 1.13 | - | 4.48 |
| 2.0 - 3.0m deep | hr | m3 | 0.11 | 0.11 | | |
| | £ | | 3.35 | 1.13 | - | 4.48 |
| 3.0 - 4.0m deep | hr | m3 | 0.11 | 0.11 | | |
| | £ | | 3.35 | 1.13 | - | 4.48 |
| Earthwork support | | | | | | |
| ne 2m deep | hr | m2 | - | - | | |
| | £ | | 0.20 | 0.10 | 0.45 | 0.75 |
| 2.0 - 3.0m deep | hr | m2 | - | - | | |
| | £ | | 0.30 | 0.10 | 0.85 | 1.25 |
| 3.0 - 4.0m deep | hr | m2 | - | - | | |
| | £ | | 0.40 | 0.15 | 1.45 | 2.00 |
| Disposal of excavated material | | | | | | |
| off site | hr | m3 | - | - | | |
| | £ | | 1.40 | 3.30 | - | 4.70 |
| to spoil heap average 50m | hr | m3 | - | - | | |
| | £ | | 0.51 | 1.15 | - | 1.66 |

**CLASS K: PIPEWORK - MANHOLES AND PIPEWORK ANCILLARIES**

**MANHOLES - continued**

| | UNIT | | L | P | M | £ NETT |
|---|---|---|---|---|---|---|
| Mass concrete grade C15P blinding | | | | | | |
| thickness ne 150mm | hr | m3 | 0.21 | 0.21 | | |
| | £ | | 6.39 | 2.16 | 47.61 | 56.16 |
| Mass concrete grade C15P surrounds to manholes | | | | | | |
| thickness ne 150mm | hr | m3 | 0.46 | 0.46 | | |
| | £ | | 14.01 | 4.74 | 47.61 | 66.36 |
| thickness 150 - 500mm | hr | m3 | 0.34 | 0.34 | | |
| | £ | | 10.35 | 3.50 | 47.61 | 61.46 |
| Reinforced concrete grade C20P bases | | | | | | |
| thickness 150 - 300mm | hr | m3 | 0.46 | 0.46 | | |
| | £ | | 14.01 | 4.74 | 49.42 | 68.17 |
| thickness 300 - 500mm | hr | m3 | 0.50 | 0.50 | | |
| | £ | | 15.23 | 5.15 | 49.42 | 69.80 |
| Formwork curved vertical to base and benching | hr | m2 | - | - | | |
| | £ | | 3.25 | 1.00 | 6.79 | 11.04 |
| curved to surround (based on GRP forms) | hr | m2 | - | - | | |
| | £ | | 2.00 | 1.00 | 4.84 | 7.84 |

**CLASS K: PIPEWORK - MANHOLES AND PIPEWORK ANCILLARIES**

**MANHOLES - continued**

| | UNIT | | L | P | M | £ NETT |
|---|---|---|---|---|---|---|
| Precast concrete circular manhole rings; BS5911 Part 1 | | | | | | |
| shaft ring; | | | | | | |
| 675mm (plain/reinforced) | hr | m | 0.80 | 0.80 | | |
| | £ | | 24.36 | 8.24 | 23.35 | 55.95 |
| 900mm (plain/reinforced) | hr | m | 0.95 | 0.95 | | |
| | £ | | 28.93 | 9.79 | 38.61 | 77.33 |
| 1050mm (plain/reinforced) | hr | m | 1.05 | 1.05 | | |
| | £ | | 31.97 | 10.82 | 48.66 | 91.45 |
| 1200mm (plain/reinforced) | hr | m | 1.17 | 1.17 | | |
| | £ | | 35.63 | 12.05 | 65.14 | 112.82 |
| 1500mm (reinforced) | hr | m | 1.32 | 1.32 | | |
| | £ | | 40.19 | 13.60 | 112.85 | 166.64 |
| reducing slabs (675mm diameter access) | | | | | | |
| 900mm | hr | nr | 0.57 | 0.57 | | |
| | £ | | 17.36 | 5.87 | 49.84 | 73.07 |
| 1050mm | hr | nr | 0.63 | 0.63 | | |
| | £ | | 19.18 | 6.49 | 62.50 | 88.17 |
| 1200mm | hr | nr | 0.70 | 0.70 | | |
| | £ | | 21.32 | 7.21 | 84.85 | 113.38 |
| 1500mm | hr | nr | 0.79 | 0.79 | | |
| | £ | | 24.06 | 8.14 | 155.04 | 187.24 |
| landing slabs (900mm diameter access) | | | | | | |
| 1200mm | hr | nr | 0.70 | 0.70 | | |
| | £ | | 21.32 | 7.21 | 146.70 | 175.23 |
| 1500mm | hr | nr | 0.79 | 0.79 | | |
| | £ | | 24.06 | 8.14 | 195.39 | 227.59 |

**CLASS K: PIPEWORK - MANHOLES AND PIPEWORK ANCILLARIES**

**MANHOLES - continued**

| | | UNIT | L | P | M | £ NETT |
|---|---|---|---|---|---|---|
| cover slab | | | | | | |
| 675mm | hr | nr | 0.43 | 0.43 | | |
| | £ | | 13.09 | 4.43 | 19.95 | 37.47 |
| 900mm | hr | nr | 0.54 | 0.54 | | |
| | £ | | 16.44 | 5.56 | 33.64 | 55.64 |
| 1050mm | hr | nr | 0.60 | 0.60 | | |
| | £ | | 18.27 | 6.18 | 43.30 | 67.75 |
| 1200mm | hr | nr | 0.70 | 0.70 | | |
| | £ | | 21.32 | 7.21 | 59.88 | 88.41 |
| 1500mm | hr | nr | 0.80 | 0.80 | | |
| | £ | | 24.36 | 8.24 | 112.85 | 145.45 |
| Mass concrete grade C15P benching to bottom of manhole | | | | | | |
| 675mm | hr | nr | 0.20 | 0.20 | | |
| | £ | | 6.09 | 2.06 | 2.73 | 10.88 |
| 900mm | hr | nr | 0.40 | 0.40 | | |
| | £ | | 12.18 | 4.12 | 4.54 | 20.84 |
| 1050mm | hr | nr | 0.72 | 0.72 | | |
| | £ | | 21.92 | 7.42 | 6.18 | 35.52 |
| 1200mm | hr | nr | 0.95 | 0.95 | | |
| | £ | | 28.93 | 9.79 | 8.08 | 46.80 |
| 1500mm | hr | nr | 1.15 | 1.15 | | |
| | £ | | 35.02 | 11.85 | 12.62 | 59.49 |

## CLASS K: PIPEWORK - MANHOLES AND PIPEWORK ANCILLARIES

### MANHOLES (BRICKWORK TO MANHOLES)

| | | |
|---|---|---|
| LABOUR based on Gang J | | |
| Total hourly cost | = £ | 28.35 |
| PLANT | | |
| Mixer | | 0.83 |
| Part time small tools<br>Disc grinder/saw<br>Dumper 2T<br>Access equipment | | 0.53 |
| Total hourly cost | = £ | 1.36 |

| | UNIT | | L | P | M | £ NETT |
|---|---|---|---|---|---|---|
| Class B Engineering brickwork; PC £200.00/1000 | | | | | | |
| one brick thick | | | | | | |
| depths up to 1m deep | hr | m2 | 0.75 | 0.75 | | |
| | £ | | 21.26 | 1.02 | 27.14 | 49.42 |
| 1.0 - 2.0m | hr | m2 | 0.80 | 0.80 | | |
| | £ | | 22.68 | 1.09 | 27.14 | 50.91 |
| 2.0 - 3.0m | hr | m2 | 0.87 | 0.87 | | |
| | £ | | 24.66 | 1.18 | 27.14 | 52.98 |
| 3.0 - 4.0m | hr | m2 | 0.94 | 0.94 | | |
| | £ | | 26.65 | 1.28 | 27.14 | 55.07 |
| 4.0 - 5.0m | hr | m2 | 1.05 | 1.05 | | |
| | £ | | 29.77 | 1.43 | 27.14 | 58.34 |
| one and one half brick thick | | | | | | |
| depths up to 1m deep | hr | m2 | 1.03 | 1.03 | | |
| | £ | | 29.20 | 1.40 | 41.11 | 71.71 |
| 1.0 - 2.0m | hr | m2 | 1.15 | 1.15 | | |
| | £ | | 32.60 | 1.56 | 41.11 | 75.27 |
| 2.0 - 3.0m | hr | m2 | 1.26 | 1.26 | | |
| | £ | | 35.72 | 1.71 | 41.11 | 78.54 |
| 3.0 - 4.0m | hr | m2 | 1.37 | 1.37 | | |
| | £ | | 38.84 | 1.86 | 41.11 | 81.81 |
| 4.0 - 5.0m | hr | m2 | 1.48 | 1.48 | | |
| | £ | | 41.96 | 2.01 | 41.11 | 85.08 |

## CLASS K: PIPEWORK - MANHOLES AND PIPEWORK ANCILLARIES

**MANHOLES (BRICKWORK TO MANHOLES) - continued**

| | UNIT | | L | P | M | £ NETT |
|---|---|---|---|---|---|---|
| Step irons; galvanised mild steel built in brickwork | | | | | | |
| depths up to 1m deep | hr | nr | 0.03 | 0.03 | | |
| | £ | | 0.85 | 0.04 | 4.00 | 4.89 |
| 1.0 - 2.0m | hr | nr | 0.03 | 0.03 | | |
| | £ | | 0.85 | 0.04 | 4.00 | 4.89 |
| 2.0 - 3.0m | hr | nr | 0.04 | 0.04 | | |
| | £ | | 1.13 | 0.05 | 4.00 | 5.18 |
| 3.0 - 4.0m | hr | nr | 0.05 | 0.05 | | |
| | £ | | 1.42 | 0.07 | 4.00 | 5.49 |
| 4.0 - 5.0m | hr | nr | 0.06 | 0.06 | | |
| | £ | | 1.70 | 0.08 | 4.00 | 5.78 |
| cast in precast concrete | | | | | | |
| depths up to 1m deep | hr | nr | 0.03 | 0.03 | | |
| | £ | | 0.85 | 0.04 | 3.20 | 4.09 |
| 1.0 - 2.0m | hr | nr | 0.03 | 0.03 | | |
| | £ | | 0.85 | 0.04 | 3.20 | 4.09 |
| 2.0 - 3.0m | hr | nr | 0.04 | 0.04 | | |
| | £ | | 1.13 | 0.05 | 3.20 | 4.38 |
| 3.0 - 4.0m | hr | nr | 0.05 | 0.05 | | |
| | £ | | 1.42 | 0.07 | 3.20 | 4.69 |
| 4.0 - 5.0m | hr | nr | 0.06 | 0.06 | | |
| | £ | | 1.70 | 0.08 | 3.20 | 4.98 |
| Vitrified clayware channels in base of manhole (Channel diameter) | | | | | | |
| straight | | | | | | |
| 100mm | hr | m | 0.18 | 0.18 | | |
| | £ | | 5.10 | 0.24 | 1.73 | 7.07 |
| 150mm | hr | m | 0.23 | 0.23 | | |
| | £ | | 6.52 | 0.31 | 2.87 | 9.70 |
| 225mm | hr | m | 0.32 | 0.32 | | |
| | £ | | 9.07 | 0.44 | 6.39 | 15.90 |
| 300mm | hr | m | 0.43 | 0.43 | | |
| | £ | | 12.19 | 0.58 | 12.64 | 25.41 |

**CLASS K: PIPEWORK - MANHOLES AND PIPEWORK ANCILLARIES**

**MANHOLES (BRICKWORK TO MANHOLES) - continued**

| | | UNIT | L | P | M | £ NETT |
|---|---|---|---|---|---|---|
| channel bend | | | | | | |
| 100mm | hr | nr | 0.14 | 0.14 | | |
| | £ | | 3.97 | 0.19 | 1.59 | 5.75 |
| 150mm | hr | nr | 0.19 | 0.19 | | |
| | £ | | 5.39 | 0.26 | 2.63 | 8.28 |
| 225mm | hr | nr | 0.28 | 0.28 | | |
| | £ | | 7.94 | 0.38 | 8.78 | 17.10 |
| 300mm | hr | nr | 0.40 | 0.40 | | |
| | £ | | 11.34 | 0.54 | 17.43 | 29.31 |
| channel branch | | | | | | |
| 100mm | hr | nr | 0.16 | 0.16 | | |
| | £ | | 4.54 | 0.22 | 3.03 | 7.79 |
| 150mm | hr | nr | 0.26 | 0.26 | | |
| | £ | | 7.37 | 0.35 | 5.02 | 12.74 |
| 225mm | hr | nr | 0.67 | 0.67 | | |
| | £ | | 18.99 | 0.91 | 14.39 | 34.29 |
| 300mm | hr | nr | 1.07 | 1.07 | | |
| | £ | | 30.33 | 1.46 | 34.24 | 66.03 |
| half section straight taper | | | | | | |
| 150mm | hr | nr | 0.22 | 0.22 | | |
| | £ | | 6.24 | 0.30 | 6.66 | 13.20 |
| 225mm | hr | nr | 0.33 | 0.33 | | |
| | £ | | 9.36 | 0.45 | 14.72 | 24.53 |
| 300mm | hr | nr | 0.48 | 0.48 | | |
| | £ | | 13.61 | 0.65 | 29.12 | 43.38 |
| half section branch channel bends | | | | | | |
| 100mm | hr | nr | 0.14 | 0.14 | | |
| | £ | | 3.97 | 0.19 | 3.26 | 7.42 |
| 150mm | hr | nr | 0.19 | 0.19 | | |
| | £ | | 5.39 | 0.26 | 5.35 | 11.00 |
| 225mm | hr | nr | 0.28 | 0.28 | | |
| | £ | | 7.94 | 0.38 | 17.67 | 25.99 |
| 300mm | hr | nr | 0.40 | 0.40 | | |
| | £ | | 11.34 | 0.54 | 34.98 | 46.86 |

**CLASS K: PIPEWORK - MANHOLES AND PIPEWORK ANCILLARIES**

**MANHOLES (BRICKWORK TO MANHOLES) - continued**

| | UNIT | | L | P | M | £ NETT |
|---|---|---|---|---|---|---|
| three quarter section branch channel bends | | | | | | |
| 100mm | hr | nr | 0.14 | 0.14 | | |
| | £ | | 3.97 | 0.19 | 3.60 | 7.76 |
| 150mm | hr | nr | 0.19 | 0.19 | | |
| | £ | | 5.39 | 0.26 | 6.08 | 11.73 |
| 225mm | hr | nr | 0.28 | 0.28 | | |
| | £ | | 7.94 | 0.38 | 22.03 | 30.35 |
| 300mm | hr | nr | 0.40 | 0.40 | | |
| | £ | | 11.34 | 0.54 | 43.69 | 55.57 |
| Spun concrete channels in base of manhole (Channel diameter) | | | | | | |
| straight | | | | | | |
| 300mm | hr | nr | 0.38 | 0.38 | | |
| | £ | | 10.77 | 0.52 | 5.46 | 16.75 |
| 450mm | hr | nr | 0.75 | 0.75 | | |
| | £ | | 21.26 | 1.02 | 9.40 | 31.68 |
| 600mm | hr | nr | 1.05 | 1.05 | | |
| | £ | | 29.77 | 1.43 | 17.23 | 48.43 |
| 900mm | hr | nr | 1.97 | 1.97 | | |
| | £ | | 55.85 | 2.68 | 33.08 | 91.61 |
| 1200mm | hr | nr | 3.14 | 3.14 | | |
| | £ | | 89.02 | 4.27 | 58.51 | 151.80 |
| Build in pipes | | | | | | |
| to one brick wall | | | | | | |
| 300mm | hr | nr | 0.08 | 0.08 | | |
| | £ | | 2.27 | 0.11 | 0.71 | 3.09 |
| 450mm | hr | nr | 0.11 | 0.11 | | |
| | £ | | 3.12 | 0.15 | 1.01 | 4.28 |
| 600mm | hr | nr | 0.15 | 0.15 | | |
| | £ | | 4.25 | 0.20 | 1.36 | 5.81 |
| 900mm | hr | nr | 0.20 | 0.20 | | |
| | £ | | 5.67 | 0.27 | 2.50 | 8.44 |
| 1200mm | hr | nr | 0.22 | 0.22 | | |
| | £ | | 6.24 | 0.30 | 5.00 | 11.54 |

**CLASS K: PIPEWORK - MANHOLES AND PIPEWORK ANCILLARIES**

**MANHOLES (BRICKWORK TO MANHOLES) - continued**

| | UNIT | | L | P | M | £ NETT |
|---|---|---|---|---|---|---|
| to precast concrete walls | | | | | | |
| 300mm | hr | nr | 0.45 | 0.45 | | |
| | £ | | **12.76** | **0.61** | **0.94** | **14.31** |
| 450mm | hr | nr | 0.67 | 0.67 | | |
| | £ | | **18.99** | **0.91** | **1.37** | **21.27** |
| 600mm | hr | nr | 0.87 | 0.87 | | |
| | £ | | **24.66** | **1.18** | **1.95** | **27.79** |
| 900mm | hr | nr | 1.28 | 1.28 | | |
| | £ | | **36.29** | **1.74** | **4.32** | **42.35** |
| 1200mm | hr | nr | 1.61 | 1.61 | | |
| | £ | | **45.64** | **2.19** | **7.72** | **55.55** |

For material costs of access covers and frames refer to resources section

| | UNIT | | L | P | M | £ NETT |
|---|---|---|---|---|---|---|
| Fix only access covers and frames | | | | | | |
| grade A double triangular solid top | hr | nr | 0.60 | 0.60 | | |
| | £ | | **17.01** | **0.82** | **1.75** | **19.58** |
| grade A single triangular solid top | hr | nr | 0.48 | 0.48 | | |
| | £ | | **13.61** | **0.65** | **1.75** | **16.01** |
| grade B medium duty | hr | nr | 0.39 | 0.39 | | |
| | £ | | **11.06** | **0.53** | **1.75** | **13.34** |
| grade C light duty | hr | nr | 0.29 | 0.29 | | |
| | £ | | **8.22** | **0.39** | **1.75** | **10.36** |

**MANHOLE ANCILLARIES**

| | UNIT | | L | P | M | £ NETT |
|---|---|---|---|---|---|---|
| manhole interceptors | | | | | | |
| 150mm diameter | hr | nr | 0.20 | 0.20 | | |
| | £ | | **5.67** | **0.27** | **26.69** | **32.63** |
| 255mm diameter | hr | nr | 0.30 | 0.30 | | |
| | £ | | **8.51** | **0.41** | **71.97** | **80.89** |
| 300mm diameter | hr | nr | 0.40 | 0.40 | | |
| | £ | | **11.34** | **0.54** | **164.70** | **176.58** |
| Rodding eye points (excluding pipework) | | | | | | |
| 100mm diameter | hr | nr | 0.40 | 0.40 | | |
| | £ | | **11.34** | **0.54** | **17.16** | **29.04** |
| 150mm diameter | hr | nr | 0.45 | 0.45 | | |
| | £ | | **12.76** | **0.61** | **27.49** | **40.86** |

**CLASS K: PIPEWORK - MANHOLES AND PIPEWORK ANCILLARIES**

**GULLIES**

| | | UNIT | L | P | M | £ NETT |
|---|---|---|---|---|---|---|
| **Clay** | | | | | | |
| vitrified clay road gully 450mm diameter x 900mm deep, including excavation and removal, 150mm concrete grade C20P surround, class B semi engineering brick seating, cast iron road gully and frame 475 x 475mm grade A | hr | nr | 0.50 | 0.50 | | |
| | £ | | 14.18 | 5.15 | 141.22 | 160.55 |
| **Clay trapped** | | | | | | |
| vitrified clay yard gully 225 diameter (Deans) with bucket and rodding eye, including excavation and removal, 150mm concrete grade C20P surround, cast iron gully and grating and frame grade B | hr | nr | 0.30 | 0.30 | | |
| | £ | | 8.51 | 3.09 | 67.35 | 78.95 |
| vitrified clay grease trap 600 x 450mm with internal access and bucket, including excavation and removal, 150mm concrete grade C20P surround, cast iron lid and frame grade B class 2 | hr | nr | 0.35 | 0.35 | | |
| | £ | | 9.92 | 3.61 | 291.73 | 305.26 |
| **Precast concrete trapped** | | | | | | |
| precast concrete road gully 450mm diameter with bucket and rodding eye, including excavation and removal, 150mm concrete grade C20P surround, class B semi-engineering brick seating, cast iron road gully grating and frame 475 x 475mm grade A | hr | nr | 0.50 | 0.50 | | |
| | £ | | 14.18 | 5.15 | 124.07 | 143.40 |

**FRENCH DRAINS, RUBBLE DRAINS, DITCHES AND TRENCHES**

| | | UNIT | L | P | M | £ NETT |
|---|---|---|---|---|---|---|
| **Filling French and rubble drains with graded material** | | | | | | |
| 20mm stone aggregate (PC £4.10/t) | hr | m3 | 0.29 | 0.29 | | |
| | £ | | 8.83 | 2.99 | 8.10 | 19.92 |
| **Filling French and rubble drains with rubble** | | | | | | |
| broken brick/concrete rubble (PC £3.10/t) | hr | m3 | 0.29 | 0.29 | | |
| | £ | | 8.83 | 2.99 | 6.15 | 17.97 |

**CLASS K: PIPEWORK - MANHOLES AND PIPEWORK ANCILLARIES**

**FRENCH DRAINS, RUBBLE DRAINS, DITCHES AND TRENCHES - continued**

| | UNIT | | L | P | M | £ NETT |
|---|---|---|---|---|---|---|
| Trenches for unpiped rubble drains (excluding trench support) (Cross-sectional area) | | | | | | |
| 0.25 - 0.50m2 | hr | m | 0.10 | 0.10 | | |
| | £ | | 3.05 | 1.03 | - | 4.08 |
| 0.50 - 0.75m2 | hr | m | 0.12 | 0.12 | | |
| | £ | | 3.65 | 1.24 | - | 4.89 |
| 0.75 - 1.00m2 | hr | m | 0.14 | 0.14 | | |
| | £ | | 4.26 | 1.44 | - | 5.70 |
| 1.00 - 1.50m2 | hr | m | 0.17 | 0.17 | | |
| | £ | | 5.18 | 1.75 | - | 6.93 |
| 1.50 - 2.00m2 | hr | m | 0.20 | 0.20 | | |
| | £ | | 6.09 | 2.06 | - | 8.15 |
| Rectangular section ditches unlined (Cross-sectional area) | | | | | | |
| 0.25 - 0.50m2 | hr | m | 0.11 | 0.11 | | |
| | £ | | 3.35 | 1.13 | - | 4.48 |
| 0.50 - 0.75m2 | hr | m | 0.13 | 0.13 | | |
| | £ | | 3.96 | 1.34 | - | 5.30 |
| 0.75 - 1.00m2 | hr | m | 0.16 | 0.16 | | |
| | £ | | 4.87 | 1.65 | - | 6.52 |
| 1.00 - 1.50m2 | hr | m | 0.20 | 0.20 | | |
| | £ | | 6.09 | 2.06 | - | 8.15 |
| 1.50 - 2.00m2 | hr | m | 0.25 | 0.25 | | |
| | £ | | 7.61 | 2.58 | - | 10.19 |
| Rectangular section ditches lined (Cross-sectional area) | | | | | | |
| 0.25 - 0.50m2 | hr | m | 0.15 | 0.15 | | |
| | £ | | 4.57 | 1.55 | 6.56 | 12.68 |
| 0.50 - 0.75m2 | hr | m | 0.25 | 0.25 | | |
| | £ | | 7.61 | 2.58 | 10.94 | 21.13 |
| 0.75 - 1.00m2 | hr | m | 0.36 | 0.36 | | |
| | £ | | 10.96 | 3.71 | 15.75 | 30.42 |
| 1.00 - 1.50m2 | hr | m | 0.40 | 0.40 | | |
| | £ | | 12.18 | 4.12 | 22.50 | 38.80 |
| 1.50 - 2.00m2 | hr | m | 0.45 | 0.45 | | |
| | £ | | 13.70 | 4.64 | 31.50 | 49.84 |

**CLASS K: PIPEWORK - MANHOLES AND PIPEWORK ANCILLARIES**

**FRENCH DRAINS, RUBBLE DRAINS, DITCHES AND TRENCHES - continued**

| | | UNIT | L | P | M | £ NETT |
|---|---|---|---|---|---|---|
| Vee section ditches unlined (Cross-sectional area) | | | | | | |
| 0.25 - 0.50m2 | hr | m | 0.10 | 0.10 | | |
| | £ | | 3.05 | 1.03 | - | 4.08 |
| 0.50 - 0.75m2 | hr | m | 0.12 | 0.12 | | |
| | £ | | 3.65 | 1.24 | - | 4.89 |
| 0.75 - 1.00m2 | hr | m | 0.14 | 0.14 | | |
| | £ | | 4.26 | 1.44 | - | 5.70 |
| 1.00 - 1.50m2 | hr | m | 0.18 | 0.18 | | |
| | £ | | 5.48 | 1.85 | - | 7.33 |
| 1.50 - 2.00m2 | hr | m | 0.22 | 0.22 | | |
| | £ | | 6.70 | 2.27 | - | 8.97 |
| Trenches for pipes or cables not to be laid by the Contractor (Cross-sectional area) | | | | | | |
| 0.25 - 0.50m2 | hr | m | 0.10 | 0.10 | | |
| | £ | | 3.05 | 1.03 | - | 4.08 |
| 0.50 - 0.75m2 | hr | m | 0.13 | 0.13 | | |
| | £ | | 3.96 | 1.34 | - | 5.30 |
| 0.75 - 1.00m2 | hr | m | 0.17 | 0.17 | | |
| | £ | | 5.18 | 1.75 | - | 6.93 |
| 1.00 - 1.50m2 | hr | m | 0.20 | 0.20 | | |
| | £ | | 6.09 | 2.06 | - | 8.15 |
| 1.50 - 2.00m2 | hr | m | 0.24 | 0.24 | | |
| | £ | | 7.31 | 2.47 | - | 9.78 |

**DUCTS AND METAL CULVERTS**

| | | UNIT | L | P | M | £ NETT |
|---|---|---|---|---|---|---|
| Cable ducts,150mm diameter one way; Hepduct clay ducts | | | | | | |
| Not in trenches | hr | m | 0.08 | 0.08 | | |
| | £ | | 2.44 | 0.82 | 5.46 | 8.72 |
| In trenches | | | | | | |
| ne 1.50m | hr | m | 0.17 | 0.17 | | |
| | £ | | 5.18 | 1.75 | 5.46 | 12.39 |
| 1.50 - 2.00m | hr | m | 0.24 | 0.24 | | |
| | £ | | 7.31 | 2.47 | 5.46 | 15.24 |
| 2.00 - 2.50m | hr | m | 0.30 | 0.30 | | |
| | £ | | 9.14 | 3.09 | 5.46 | 17.69 |
| 2.50 - 3.00m | hr | m | 0.37 | 0.37 | | |
| | £ | | 11.27 | 3.81 | 5.46 | 20.54 |

**CLASS K: PIPEWORK - MANHOLES AND PIPEWORK ANCILLARIES**

**DUCTS AND METAL CULVERTS - continued**

| | UNIT | | L | P | M | £ NETT |
|---|---|---|---|---|---|---|
| two way; Hepduct clay ducts | | | | | | |
| Not in trenches | hr | m | 0.15 | 0.15 | | |
| | £ | | 4.57 | 1.55 | 10.92 | 17.04 |
| In trenches | | | | | | |
| ne 1.50m | hr | m | 0.32 | 0.32 | | |
| | £ | | 9.74 | 3.30 | 10.92 | 23.96 |
| 1.50 - 2.00m | hr | m | 0.44 | 0.44 | | |
| | £ | | 13.40 | 4.53 | 10.92 | 28.85 |
| 2.00 - 2.50m | hr | m | 0.55 | 0.55 | | |
| | £ | | 16.75 | 5.67 | 10.92 | 33.34 |
| 2.50 - 3.00m | hr | m | 0.70 | 0.70 | | |
| | £ | | 21.32 | 7.21 | 10.92 | 39.45 |
| Sectional corrugated metal culverts. Bitumen coated galvanised steel. | | | | | | |
| nominal internal diameter 1000mm | | | | | | |
| Not in trenches | hr | m | 0.15 | 0.15 | | |
| | £ | | 4.57 | 1.55 | 53.30 | 59.42 |
| In trenches | | | | | | |
| ne 1.50m | hr | m | 0.31 | 0.31 | | |
| | £ | | 9.44 | 3.19 | 53.30 | 65.93 |
| 1.50 - 2.00m | hr | m | 0.43 | 0.43 | | |
| | £ | | 13.09 | 4.43 | 53.30 | 70.82 |
| 2.00 - 2.50m | hr | m | 0.51 | 0.51 | | |
| | £ | | 15.53 | 5.25 | 53.30 | 74.08 |
| 2.50 - 3.00m | hr | m | 0.60 | 0.60 | | |
| | £ | | 18.27 | 6.18 | 53.30 | 77.75 |
| nominal internal diameter 1600mm | | | | | | |
| Not in trenches | hr | m | 0.21 | 0.21 | | |
| | £ | | 6.39 | 2.16 | 113.00 | 121.55 |
| In trenches | | | | | | |
| ne 1.50m | hr | m | 0.44 | 0.44 | | |
| | £ | | 13.40 | 4.53 | 113.00 | 130.93 |
| 1.50 - 2.00m | hr | m | 0.56 | 0.56 | | |
| | £ | | 17.05 | 5.77 | 113.00 | 135.82 |
| 2.00 - 2.50m | hr | m | 0.68 | 0.68 | | |
| | £ | | 20.71 | 7.00 | 113.00 | 140.71 |
| 2.50 - 3.00m | hr | m | 0.82 | 0.82 | | |
| | £ | | 24.97 | 8.45 | 113.00 | 146.42 |

**CLASS K: PIPEWORK - MANHOLES AND PIPEWORK ANCILLARIES**

**DUCTS AND METAL CULVERTS - continued**

| | UNIT | | L | P | M | £ NETT |
|---|---|---|---|---|---|---|
| nominal internal diameter 2000mm | | | | | | |
| Not in trenches | hr | m | 0.26 | 0.26 | | |
| | £ | | 7.92 | 2.68 | 172.00 | 182.60 |
| In trenches | | | | | | |
| ne 1.50m | hr | m | 0.46 | 0.46 | | |
| | £ | | 14.01 | 4.74 | 172.00 | 190.75 |
| 1.50 - 2.00m | hr | m | 0.60 | 0.60 | | |
| | £ | | 18.27 | 6.18 | 172.00 | 196.45 |
| 2.00 - 2.50m | hr | m | 0.75 | 0.75 | | |
| | £ | | 22.84 | 7.73 | 172.00 | 202.57 |
| 2.50 - 3.00m | hr | m | 0.93 | 0.93 | | |
| | £ | | 28.32 | 9.58 | 172.00 | 209.90 |
| nominal internal diameter 2200mm | | | | | | |
| Not in trenches | hr | m | 0.33 | 0.33 | | |
| | £ | | 10.05 | 3.40 | 185.00 | 198.45 |
| In trenches | | | | | | |
| ne 1.50m | hr | m | 0.64 | 0.64 | | |
| | £ | | 19.49 | 6.59 | 185.00 | 211.08 |
| 1.50 - 2.00m | hr | m | 0.77 | 0.77 | | |
| | £ | | 23.45 | 7.93 | 185.00 | 216.38 |
| 2.00 - 2.50m | hr | m | 1.02 | 1.02 | | |
| | £ | | 31.06 | 10.51 | 185.00 | 226.57 |
| 2.50 - 3.00m | hr | m | 1.32 | 1.32 | | |
| | £ | | 40.19 | 13.60 | 185.00 | 238.79 |

**CLASS K: PIPEWORK - MANHOLES AND PIPEWORK ANCILLARIES**

**REINSTATEMENT**

| | UNIT | | L | P | M | £ NETT |
|---|---|---|---|---|---|---|
| Temporary and permanent reinstatment 150mm bitumacadam road construction and base under | | | | | | |
| ne 300mm (pipe bore) | hr | m | 0.38 | 0.38 | | |
| | £ | | 11.57 | 3.91 | 4.80 | 20.28 |
| 450mm | hr | m | 0.50 | 0.50 | | |
| | £ | | 15.23 | 5.15 | 8.30 | 28.68 |
| 600mm | hr | m | 0.62 | 0.62 | | |
| | £ | | 18.88 | 6.39 | 9.90 | 35.17 |
| 750mm | hr | m | 0.77 | 0.77 | | |
| | £ | | 23.45 | 7.93 | 11.60 | 42.98 |
| 900mm | hr | m | 0.94 | 0.94 | | |
| | £ | | 28.62 | 9.68 | 13.20 | 51.50 |
| 50mm bitumacadam pavement construction and base under | | | | | | |
| ne 300mm (pipe bore) | hr | m | 0.25 | 0.25 | | |
| | £ | | 7.61 | 2.58 | 1.60 | 11.79 |
| 450mm | hr | m | 0.35 | 0.35 | | |
| | £ | | 10.66 | 3.61 | 2.77 | 17.04 |
| 600mm | hr | m | 0.48 | 0.48 | | |
| | £ | | 14.62 | 4.94 | 3.30 | 22.86 |
| 750mm | hr | m | 0.62 | 0.62 | | |
| | £ | | 18.88 | 6.39 | 3.87 | 29.14 |
| 900mm | hr | m | 0.80 | 0.80 | | |
| | £ | | 24.36 | 8.24 | 4.40 | 37.00 |
| Strip top soil from easement and reinstate per m width | | | | | | |
| depth 250mm (6 x by hand) | hr | m | 0.01 | 0.01 | | |
| | £ | | 0.30 | 0.10 | - | 0.40 |

**OTHER PIPEWORK ANCILLARIES**

| | UNIT | | L | P | M | £ NETT |
|---|---|---|---|---|---|---|
| Connections of new pipes to existing pipes | | | | | | |
| 150 diameter to existing 150, 225 or 300mm diameter | hr | nr | 0.42 | 0.30 | | |
| | £ | | 12.79 | 3.09 | 9.52 | 25.40 |
| 225 diameter to existing 150, 225 or 300mm diameter | hr | nr | 0.60 | 0.50 | | |
| | £ | | 18.27 | 5.15 | 22.14 | 45.56 |
| 300 diameter to existing 150, 225 or 300mm diameter | hr | nr | 1.15 | 1.00 | | |
| | £ | | 35.02 | 10.30 | 43.21 | 88.53 |

**CLASS K: PIPEWORK - MANHOLES AND PIPEWORK ANCILLARIES**

**OTHER PIPEWORK ANCILLARIES**

| | UNIT | | L | P | M | £ NETT |
|---|---|---|---|---|---|---|
| Headwall and outfall works | | | | | | |
| Refer to sections F and G (Concrete and concrete ancillaries) for cost relevant to Construction of Headwall Structure | | | | | | |
| Pipes built into headwall as work proceeds (Note pipe included with pipe run). Clay or concrete | | | | | | |
| 150mm diameter | hr | nr | 0.22 | 0.22 | | |
| | £ | | 6.70 | 2.27 | - | 8.97 |
| 225mm diameter | hr | nr | 0.25 | 0.25 | | |
| | £ | | 7.61 | 2.58 | - | 10.19 |
| 300mm diameter | hr | nr | 0.29 | 0.29 | | |
| | £ | | 8.83 | 2.99 | - | 11.82 |
| 375mm diameter | hr | nr | 0.34 | 0.34 | | |
| | £ | | 10.35 | 3.50 | - | 13.85 |
| 450mm diameter | hr | nr | 0.40 | 0.40 | | |
| | £ | | 12.18 | 4.12 | - | 16.30 |
| 600mm diameter | hr | nr | 0.47 | 0.47 | | |
| | £ | | 14.31 | 4.84 | - | 19.15 |
| 750mm diameter | hr | nr | 0.55 | 0.55 | | |
| | £ | | 16.75 | 5.67 | - | 22.42 |
| 900mm diameter | hr | nr | 0.65 | 0.65 | | |
| | £ | | 19.79 | 6.70 | - | 26.49 |
| 1200mm diameter | hr | nr | 0.75 | 0.75 | | |
| | £ | | 22.84 | 7.73 | - | 30.57 |
| Pipes built into existing headwalls training walls or Brickwork etc. | | | | | | |
| 150mm diameter | hr | nr | 0.60 | 0.60 | | |
| | £ | | 18.27 | 6.18 | 7.20 | 31.65 |
| 225mm diameter | hr | nr | 0.95 | 0.95 | | |
| | £ | | 28.93 | 9.79 | 13.50 | 52.22 |
| 300mm diameter | hr | nr | 1.25 | 1.25 | | |
| | £ | | 38.06 | 12.88 | 20.25 | 71.19 |
| 375mm diameter | hr | nr | 1.45 | 1.45 | | |
| | £ | | 44.15 | 14.94 | 27.00 | 86.09 |
| 450mm diameter | hr | nr | 1.75 | 1.75 | | |
| | £ | | 53.29 | 18.03 | 33.75 | 105.07 |

**CLASS K: PIPEWORK - MANHOLES AND PIPEWORK ANCILLARIES**

**OTHER PIPEWORK ANCILLARIES - continued**

| | UNIT | | L | P | M | £ NETT |
|---|---|---|---|---|---|---|
| Connections to existing manholes and other chambers | | | | | | |
| one brick sides 2.5m deep to invert and reform benching | | | | | | |
| 300mm diameter | hr | m | 1.57 | 1.57 | | |
| | £ | | 47.81 | 16.17 | 3.80 | 67.78 |
| 450mm diameter | hr | m | 2.45 | 2.45 | | |
| | £ | | 74.60 | 25.24 | 4.80 | 104.64 |
| 600mm diameter | hr | m | 3.51 | 3.51 | | |
| | £ | | 106.88 | 36.15 | 7.80 | 150.83 |
| precast concrete sides 2.5m deep invert and reform benching | | | | | | |
| 300mm diameter | hr | m | 0.89 | 0.89 | | |
| | £ | | 27.10 | 9.17 | 3.06 | 39.33 |
| 450mm diameter | hr | m | 1.14 | 1.14 | | |
| | £ | | 34.71 | 11.74 | 3.44 | 49.89 |
| 600mm diameter | hr | m | 1.37 | 1.37 | | |
| | £ | | 41.72 | 14.11 | 3.80 | 59.63 |
| Connections of existing pipes, ducts and culverts | | | | | | |
| connection to new manhole | | | | | | |
| 300mm diameter | hr | m | 1.03 | 1.03 | | |
| | £ | | 31.36 | 10.61 | 3.40 | 45.37 |
| 450mm diameter | hr | m | 1.34 | 1.34 | | |
| | £ | | 40.80 | 13.80 | 3.80 | 58.40 |
| 600mm diameter | hr | m | 1.69 | 1.69 | | |
| | £ | | 51.46 | 17.41 | 4.00 | 72.87 |
| Marker posts | | | | | | |
| 75 x 75mm precast concrete marker posts 600mm long surrounded in concrete grade C10P | hr | m | 0.54 | - | | |
| | £ | | 16.44 | - | 4.30 | 20.74 |
| Timber supports left in excavations | hr | m2 | - | - | | |
| | £ | | - | - | - | 3.51 |
| Metal supports left in excavations | hr | m2 | - | - | | |
| | £ | | - | - | - | 5.68 |

**CLASS L: PIPEWORK - SUPPORTS & PROTECTION, ANCILLARIES TO LAYING & EXCAVATION**

**EXTRAS TO EXCAVATION AND BACKFILLING**

| | | |
|---|---|---|
| **LABOUR** | | |
| **2 LABOURERS** | | 12.40 |
| Total hourly cost | = £ | 12.40 |
| **PLANT** | | |
| 2 tool compressor and tools | | 4.88 |
| Part time<br>Excavator plus<br>Extra cost of rock bucket | | 3.82 |
| Total hourly cost | = £ | 8.70 |

To provide a basic, overall cost comparison, the rates, prices and outputs have been based on a medium sized Civil Engineering project of £5 - 6 million which is neither city centre nor excessively remote.

All rates for materials are now based on the most economically available materials with minimum waste allowance - 2 - 3% and the most efficient items of plant (Excavator). The rates in this section are the extra cost of installing the fittings in sequence with the pipe runs in open trenches.

Multiplier table for Labour and Plant for various site conditions.

| | | | | | |
|---|---|---|---|---|---|
| Working out of sequence with other works | | | 2.75 | - | min. |
| Working in depths up to 1.5m | | | - | | 1.00 |
| | 1.5m | - | 3.0m | - | 1.30 |
| | 3.0m | - | 4.5m | - | 1.50 |
| | over 4.5m | | - | | 1.80 |
| Working below water Table levels | | | 2.00 | 2.00 | min. |

**CLASS L: PIPEWORK - SUPPORTS & PROTECTION, ANCILLARIES TO LAYING & EXCAVATION**

**EXTRAS TO EXCAVATION AND BACKFILLING**

| | | UNIT | L | P | M | £ NETT |
|---|---|---|---|---|---|---|
| In pipe trenches | | | | | | |
| excavation of rock | hr | m3 | 1.30 | 1.30 | | |
| | £ | | 16.12 | 11.31 | - | 27.43 |
| excavation of mass concrete | hr | m3 | 1.69 | 1.69 | | |
| | £ | | 20.96 | 14.70 | - | 35.66 |
| excavation of reinforced concrete | hr | m3 | 2.37 | 2.37 | | |
| | £ | | 29.39 | 20.62 | - | 50.01 |
| excavation of Asphalt and similar | hr | m2 | 0.70 | 0.70 | | |
| | £ | | 8.68 | 6.09 | - | 14.77 |
| backfilling above the final surface with concrete; grade C15P | hr | m3 | 0.20 | 0.20 | | |
| | £ | | 2.48 | 1.74 | 49.77 | 53.99 |
| excavation of natural material below the final surface and backfilling with concrete grade C15P | hr | m3 | 0.60 | 0.60 | | |
| | £ | | 7.44 | 5.22 | 49.77 | 62.43 |
| In manholes and other chambers | | | | | | |
| excavation of rock | hr | m3 | 1.30 | 1.30 | | |
| | £ | | 16.12 | 11.31 | - | 27.43 |
| excavation of mass concrete | hr | m3 | 1.69 | 1.69 | | |
| | £ | | 20.96 | 14.70 | - | 35.66 |
| excavation of reinforced concrete | hr | m3 | 2.37 | 2.37 | | |
| | £ | | 29.39 | 20.62 | - | 50.01 |
| backfilling above the final surface with concrete; grade C15P | hr | m3 | 0.20 | 0.20 | | |
| | £ | | 2.48 | 1.74 | 49.77 | 53.99 |
| excavation of natural material below the final surface and backfilling with concrete grade C15P | hr | m3 | 0.60 | 0.60 | | |
| | £ | | 7.44 | 5.22 | 49.77 | 62.43 |

**CLASS L: PIPEWORK - SUPPORTS & PROTECTION, ANCILLARIES TO LAYING & EXCAVATION**

**EXTRAS TO EXCAVATION AND BACKFILLING**

| | | UNIT | L | P | M | £ NETT |
|---|---|---|---|---|---|---|
| In headwalls, outfalls and headings. (excluding specialist tunnelling procedures) | | | | | | |
| excavation of rock | hr | m3 | 4.00 | 4.00 | | |
| | £ | | 49.60 | 34.80 | - | 84.40 |
| excavation of mass concrete | hr | m3 | 5.15 | 5.15 | | |
| | £ | | 63.86 | 44.81 | - | 108.67 |
| excavation of reinforced concrete | hr | m3 | 7.20 | 7.20 | | |
| | £ | | 89.28 | 62.64 | - | 151.92 |
| backfilling above the final surface with concrete; grade C15P | hr | m3 | 2.10 | 2.10 | | |
| | £ | | 26.04 | 18.27 | 49.77 | 94.08 |
| excavation of natural material below the final surface and backfilling with concrete grade C15P | hr | m3 | 1.90 | 1.90 | | |
| | £ | | 23.56 | 16.53 | 49.77 | 89.86 |
| In pipe jacking | | | | | | |
| excavation of rock | hr | m3 | 8.50 | 8.50 | | |
| | £ | | 105.40 | 73.95 | - | 179.35 |
| excavation of mass concrete | hr | m3 | 11.05 | 11.05 | | |
| | £ | | 137.02 | 96.14 | - | 233.16 |
| excavation of reinforced concrete | hr | m3 | 15.45 | 15.45 | | |
| | £ | | 191.58 | 134.42 | - | 326.00 |
| backfilling above the final surface with concrete; grade C15P | hr | m3 | 4.60 | 4.60 | | |
| | £ | | 57.04 | 40.02 | 49.77 | 146.83 |
| excavation of natural material below the final surface and backfilling with concrete grade C15P | hr | m3 | 1.30 | 1.30 | | |
| | £ | | 16.12 | 11.31 | 49.77 | 77.20 |

**CLASS L: PIPEWORK - SUPPORTS & PROTECTION, ANCILLARIES TO LAYING & EXCAVATION**

**SPECIAL PIPE LAYING METHODS**

There are many factors, apart from design consideration, which influence the cost of pipe jacking, that it is only possible to give guide prices for a sample of the work involved. For more reliable estimates it is advisable to seek the advice of a Specialist Contractor.

The main cost considerations are:-

- the nature of the ground
- length of drive
- location
- presence of water
- depth below surface

| | | UNIT | L | P | M | SPECIALIST ADVICE £ NETT |
|---|---|---|---|---|---|---|
| Pipe jacking, excluding the cost of non drainage materials and shafts (Nominal bore) | | | | | | |
| concrete pipes BS5911 Part 1 Class H with rebated joints, steel reinforcing band. | | | | | | |
| 900mm | hr | m | - | - | | |
| | £ | | 155.00 | 55.00 | 134.65 | 344.65 |
| 1200mm | hr | m | - | - | | |
| | £ | | 225.00 | 80.00 | 203.04 | 508.04 |

**CLASS L: PIPEWORK - SUPPORTS & PROTECTION, ANCILLARIES TO LAYING & EXCAVATION**

**LABOUR**

| | | |
|---|---|---|
| 1 Plant Operator | | 7.95 |
| 2 Labourers | | 12.40 |
| Total hourly cost | = £ | 20.35 |

**PLANT**

| | | |
|---|---|---|
| 0.4m3 Excavator | | 8.27 |
| Bomag BW 65 | | 1.48 |
| Total hourly cost | = £ | 9.75 |

This sub-section includes costs for beds haunches and surrounds in the most commonly specified materials.

The outputs shown are based on the same considerations detailed on page 329.

**CLASS L: PIPEWORK - SUPPORTS & PROTECTION, ANCILLARIES TO LAYING & EXCAVATION**

**BEDS**

| | | UNIT | L | P | M | £ NETT |
|---|---|---|---|---|---|---|
| **Sand; depth 100mm (nominal bore)** | | | | | | |
| 100mm | hr | m | 0.02 | 0.02 | | |
| | £ | | 0.41 | 0.20 | 0.59 | 1.20 |
| 150mm | hr | m | 0.02 | 0.02 | | |
| | £ | | 0.41 | 0.20 | 0.66 | 1.27 |
| 225mm | hr | m | 0.03 | 0.03 | | |
| | £ | | 0.61 | 0.29 | 0.78 | 1.68 |
| 300mm | hr | m | 0.04 | 0.04 | | |
| | £ | | 0.81 | 0.39 | 0.88 | 2.08 |
| **Sand; depth 150mm (nominal bore)** | | | | | | |
| 150mm | hr | m | 0.06 | 0.06 | | |
| | £ | | 1.22 | 0.59 | 0.99 | 2.80 |
| 225mm | hr | m | 0.07 | 0.07 | | |
| | £ | | 1.42 | 0.68 | 1.16 | 3.26 |
| 300mm | hr | m | 0.09 | 0.09 | | |
| | £ | | 1.83 | 0.88 | 1.32 | 4.03 |
| 400mm | hr | m | 0.12 | 0.12 | | |
| | £ | | 2.44 | 1.17 | 1.54 | 5.15 |
| 450mm | hr | m | 0.14 | 0.14 | | |
| | £ | | 2.85 | 1.37 | 1.65 | 5.87 |
| 600mm | hr | m | 0.17 | 0.17 | | |
| | £ | | 3.46 | 1.66 | 1.99 | 7.11 |
| 750mm | hr | m | 0.19 | 0.19 | | |
| | £ | | 3.87 | 1.85 | 2.31 | 8.03 |
| 900mm | hr | m | 0.21 | 0.21 | | |
| | £ | | 4.27 | 2.05 | 2.65 | 8.97 |
| 1200mm | hr | m | 0.25 | 0.25 | | |
| | £ | | 5.09 | 2.44 | 3.31 | 10.84 |

**CLASS L: PIPEWORK - SUPPORTS & PROTECTION, ANCILLARIES TO LAYING & EXCAVATION**

**BEDS - continued**

| | | UNIT | L | P | M | £ NETT |
|---|---|---|---|---|---|---|
| Imported granular material; (nominal bore) depth 100mm | | | | | | |
| 100mm | hr | m | 0.02 | 0.02 | | |
| | £ | | 0.41 | 0.20 | 0.32 | 0.93 |
| 150mm | hr | m | 0.03 | 0.03 | | |
| | £ | | 0.61 | 0.29 | 0.36 | 1.26 |
| 225mm | hr | m | 0.04 | 0.04 | | |
| | £ | | 0.81 | 0.39 | 0.41 | 1.61 |
| 300mm | hr | m | 0.05 | 0.05 | | |
| | £ | | 1.02 | 0.49 | 0.47 | 1.98 |
| depth 150mm | | | | | | |
| 150mm | hr | m | 0.06 | 0.06 | | |
| | £ | | 1.22 | 0.59 | 0.53 | 2.34 |
| 225mm | hr | m | 0.08 | 0.08 | | |
| | £ | | 1.63 | 0.78 | 0.62 | 3.03 |
| 300mm | hr | m | 0.10 | 0.10 | | |
| | £ | | 2.04 | 0.98 | 0.70 | 3.72 |
| 400mm | hr | m | 0.13 | 0.13 | | |
| | £ | | 2.65 | 1.27 | 0.83 | 4.75 |
| 450mm | hr | m | 0.15 | 0.15 | | |
| | £ | | 3.05 | 1.46 | 0.88 | 5.39 |
| 600mm | hr | m | 0.18 | 0.18 | | |
| | £ | | 3.66 | 1.76 | 1.06 | 6.48 |
| 750mm | hr | m | 0.20 | 0.20 | | |
| | £ | | 4.07 | 1.95 | 1.24 | 7.26 |
| 900mm | hr | m | 0.22 | 0.22 | | |
| | £ | | 4.48 | 2.15 | 1.42 | 8.05 |
| 1200mm | hr | m | 0.26 | 0.26 | | |
| | £ | | 5.29 | 2.54 | 1.76 | 9.59 |

**CLASS L: PIPEWORK - SUPPORTS & PROTECTION, ANCILLARIES TO LAYING & EXCAVATION**

**BEDS - continued**

| | | UNIT | L | P | M | £ NETT |
|---|---|---|---|---|---|---|
| Mass concrete; depth 100mm grade C15P, 20mm aggregate (nominal bore) | | | | | | |
| 100mm | hr | m | 0.07 | 0.07 | | |
| | £ | | 1.42 | 0.68 | 2.10 | 4.20 |
| 150mm | hr | m | 0.08 | 0.08 | | |
| | £ | | 1.63 | 0.78 | 2.36 | 4.77 |
| 225mm | hr | m | 0.09 | 0.09 | | |
| | £ | | 1.83 | 0.88 | 2.76 | 5.47 |
| 300mm | hr | m | 0.11 | 0.11 | | |
| | £ | | 2.24 | 1.07 | 3.15 | 6.46 |
| Mass concrete; depth 150mm grade C15P, 20mm aggregate (nominal bore) | | | | | | |
| 150mm | hr | m | 0.12 | 0.12 | | |
| | £ | | 2.44 | 1.17 | 3.55 | 7.16 |
| 225mm | hr | m | 0.14 | 0.14 | | |
| | £ | | 2.85 | 1.37 | 4.14 | 8.36 |
| 300mm | hr | m | 0.16 | 0.16 | | |
| | £ | | 3.26 | 1.56 | 4.73 | 9.55 |
| 400mm | hr | m | 0.19 | 0.19 | | |
| | £ | | 3.87 | 1.85 | 5.51 | 11.23 |
| 450mm | hr | m | 0.21 | 0.21 | | |
| | £ | | 4.27 | 2.05 | 5.91 | 12.23 |
| 600mm | hr | m | 0.24 | 0.24 | | |
| | £ | | 4.88 | 2.34 | 7.09 | 14.31 |
| 750mm | hr | m | 0.26 | 0.26 | | |
| | £ | | 5.29 | 2.54 | 8.27 | 16.10 |
| 900mm | hr | m | 0.28 | 0.28 | | |
| | £ | | 5.70 | 2.73 | 9.45 | 17.88 |
| 1200mm | hr | m | 0.32 | 0.32 | | |
| | £ | | 6.51 | 3.12 | 11.81 | 21.44 |

**CLASS L: PIPEWORK - SUPPORTS & PROTECTION, ANCILLARIES TO LAYING & EXCAVATION**

**HAUNCHES**

| | | UNIT | L | P | M | £ NETT |
|---|---|---|---|---|---|---|
| Mass concrete; depth 100mm grade C15P, 20mm aggregate (nominal bore) | | | | | | |
| 100mm | hr | m | 0.14 | 0.14 | | |
| | £ | | 2.85 | 1.37 | 3.65 | 7.87 |
| 150mm | hr | m | 0.16 | 0.16 | | |
| | £ | | 3.26 | 1.56 | 4.00 | 8.82 |
| 225mm | hr | m | 0.18 | 0.18 | | |
| | £ | | 3.66 | 1.76 | 4.50 | 9.92 |
| 300mm | hr | m | 0.22 | 0.22 | | |
| | £ | | 4.48 | 2.15 | 5.00 | 11.63 |
| Mass concrete; depth 150mm grade C15P, 20mm aggregate hand lay (nominal bore) | | | | | | |
| 150mm | hr | m | 0.24 | - | | |
| | £ | | 4.88 | - | 4.00 | 8.88 |
| 225mm | hr | m | 0.29 | - | | |
| | £ | | 5.90 | - | 5.00 | 10.90 |
| 300mm | hr | m | 0.36 | - | | |
| | £ | | 7.33 | - | 7.00 | 14.33 |
| 400mm | hr | m | 0.43 | - | | |
| | £ | | 8.75 | - | 9.00 | 17.75 |
| 450mm | hr | m | 0.50 | - | | |
| | £ | | 10.18 | - | 10.00 | 20.18 |
| 600mm | hr | m | 0.56 | - | | |
| | £ | | 11.40 | - | 13.50 | 24.90 |
| 750mm | hr | m | 0.62 | - | | |
| | £ | | 12.62 | - | 17.00 | 29.62 |
| 900mm | hr | m | 0.69 | - | | |
| | £ | | 14.04 | - | 20.50 | 34.54 |
| 1200mm | hr | m | 0.75 | - | | |
| | £ | | 15.26 | - | 27.00 | 42.26 |

**CLASS L: PIPEWORK - SUPPORTS & PROTECTION, ANCILLARIES TO LAYING & EXCAVATION**

**HAUNCHES - continued**

| | | UNIT | L | P | M | £ NETT |
|---|---|---|---|---|---|---|
| Reinforced concrete; depth 150mm grade C25P, 20m aggregate, 1 layer type A142 mesh reinforcement hand lay (nominal bore) | | | | | | |
| 450mm | hr | m | 0.60 | - | | |
| | £ | | 12.21 | - | 11.50 | 23.71 |
| 600mm | hr | m | 0.67 | - | | |
| | £ | | 13.63 | - | 16.00 | 29.63 |
| 750mm | hr | m | 0.73 | - | | |
| | £ | | 14.86 | - | 20.50 | 35.36 |
| 900mm | hr | m | 0.80 | - | | |
| | £ | | 16.28 | - | 25.50 | 41.78 |
| 1200mm | hr | m | 0.97 | - | | |
| | £ | | 19.74 | - | 34.00 | 53.74 |

**CLASS L: PIPEWORK - SUPPORTS & PROTECTION, ANCILLARIES TO LAYING & EXCAVATION**

**SURROUNDS**

| | | UNIT | L | P | M | £ NETT |
|---|---|---|---|---|---|---|
| Sand; thickness 100mm (nominal bore) | | | | | | |
| 100mm | hr | m | 0.04 | 0.04 | | |
| | £ | | 0.81 | 0.39 | 1.47 | 2.67 |
| 150mm | hr | m | 0.05 | 0.05 | | |
| | £ | | 1.02 | 0.49 | 2.06 | 3.57 |
| 225mm | hr | m | 0.06 | 0.06 | | |
| | £ | | 1.22 | 0.59 | 2.65 | 4.46 |
| 300mm | hr | m | 0.08 | 0.08 | | |
| | £ | | 1.63 | 0.78 | 3.53 | 5.94 |
| Sand; thickness 150mm (nominal bore) | | | | | | |
| 150mm | hr | m | 0.12 | 0.12 | | |
| | £ | | 2.44 | 1.17 | 2.94 | 6.55 |
| 225mm | hr | m | 0.14 | 0.14 | | |
| | £ | | 2.85 | 1.37 | 3.68 | 7.90 |
| 300mm | hr | m | 0.18 | 0.18 | | |
| | £ | | 3.66 | 1.76 | 4.41 | 9.83 |
| 400mm | hr | m | 0.24 | 0.24 | | |
| | £ | | 4.88 | 2.34 | 5.15 | 12.37 |
| 450mm | hr | m | 0.28 | 0.28 | | |
| | £ | | 5.70 | 2.73 | 5.88 | 14.31 |
| 600mm | hr | m | 0.34 | 0.34 | | |
| | £ | | 6.92 | 3.32 | 7.35 | 17.59 |
| 750mm | hr | m | 0.38 | 0.38 | | |
| | £ | | 7.73 | 3.71 | 8.81 | 20.25 |
| 900mm | hr | m | 0.42 | 0.42 | | |
| | £ | | 8.55 | 4.10 | 11.02 | 23.67 |
| 1200mm | hr | m | 0.50 | 0.50 | | |
| | £ | | 10.18 | 4.88 | 18.36 | 33.42 |

**CLASS L: PIPEWORK - SUPPORTS & PROTECTION, ANCILLARIES TO LAYING & EXCAVATION**

**SURROUNDS - continued**

| | | UNIT | L | P | M | £ NETT |
|---|---|---|---|---|---|---|
| **Imported granular material; thickness 100mm (nominal bore)** | | | | | | |
| 100mm | hr | m | 0.07 | 0.07 | | |
| | £ | | 1.42 | 0.68 | 0.79 | 2.89 |
| 150mm | hr | m | 0.10 | 0.10 | | |
| | £ | | 2.04 | 0.98 | 1.10 | 4.12 |
| 225mm | hr | m | 0.13 | 0.13 | | |
| | £ | | 2.65 | 1.27 | 1.42 | 5.34 |
| 300mm | hr | m | 0.16 | 0.16 | | |
| | £ | | 3.26 | 1.56 | 1.89 | 6.71 |
| **Imported granular material; thickness 150mm (nominal bore)** | | | | | | |
| 150mm | hr | m | 0.12 | 0.12 | | |
| | £ | | 2.44 | 1.17 | 1.58 | 5.19 |
| 225mm | hr | m | 0.14 | 0.14 | | |
| | £ | | 2.85 | 1.37 | 1.96 | 6.18 |
| 300mm | hr | m | 0.18 | 0.18 | | |
| | £ | | 3.66 | 1.76 | 2.35 | 7.77 |
| 400mm | hr | m | 0.24 | 0.24 | | |
| | £ | | 4.88 | 2.34 | 2.75 | 9.97 |
| 450mm | hr | m | 0.28 | 0.28 | | |
| | £ | | 5.70 | 2.73 | 3.14 | 11.57 |
| 600mm | hr | m | 0.34 | 0.34 | | |
| | £ | | 6.92 | 3.32 | 3.93 | 14.17 |
| 750mm | hr | m | 0.38 | 0.38 | | |
| | £ | | 7.73 | 3.71 | 4.71 | 16.15 |
| 900mm | hr | m | 0.42 | 0.42 | | |
| | £ | | 8.55 | 4.10 | 5.89 | 18.54 |
| 1200mm | hr | m | 0.50 | 0.50 | | |
| | £ | | 10.18 | 4.88 | 9.82 | 24.88 |

**CLASS L: PIPEWORK - SUPPORTS & PROTECTION, ANCILLARIES TO LAYING & EXCAVATION**

**SURROUNDS - continued**

| | | UNIT | L | P | M | £ NETT |
|---|---|---|---|---|---|---|
| Mass concrete; thickness 100mm grade C15P, 20mm aggregate (nominal bore) | | | | | | |
| 100mm | hr | m | 0.14 | 0.14 | | |
| | £ | | 2.85 | 1.37 | 6.00 | 10.22 |
| 150mm | hr | m | 0.16 | 0.16 | | |
| | £ | | 3.26 | 1.56 | 7.00 | 11.82 |
| 225mm | hr | m | 0.18 | 0.18 | | |
| | £ | | 3.66 | 1.76 | 9.00 | 14.42 |
| 300mm | hr | m | 0.22 | 0.22 | | |
| | £ | | 4.48 | 2.15 | 12.20 | 18.83 |
| Mass concrete; thickness 150mm grade C15P, 20mm aggregate (nominal bore) | | | | | | |
| 150mm | hr | m | 0.23 | 0.23 | | |
| | £ | | 4.68 | 2.24 | 10.00 | 16.92 |
| 225mm | hr | m | 0.26 | 0.26 | | |
| | £ | | 5.29 | 2.54 | 12.50 | 20.33 |
| 300mm | hr | m | 0.30 | 0.30 | | |
| | £ | | 6.11 | 2.93 | 15.00 | 24.04 |
| 400mm | hr | m | 0.36 | 0.36 | | |
| | £ | | 7.33 | 3.51 | 17.50 | 28.34 |
| 450mm | hr | m | 0.40 | 0.40 | | |
| | £ | | 8.14 | 3.90 | 20.00 | 32.04 |
| 600mm | hr | m | 0.45 | 0.45 | | |
| | £ | | 9.16 | 4.39 | 25.00 | 38.55 |
| 750mm | hr | m | 0.50 | 0.50 | | |
| | £ | | 10.18 | 4.88 | 30.00 | 45.06 |
| 900mm | hr | m | 0.55 | 0.55 | | |
| | £ | | 11.19 | 5.36 | 37.50 | 54.05 |
| 1200mm | hr | m | 0.61 | 0.61 | | |
| | £ | | 12.41 | 5.95 | 62.50 | 80.86 |

**CLASS L: PIPEWORK - SUPPORTS & PROTECTION, ANCILLARIES TO LAYING & EXCAVATION**

**CONCRETE STOOLS AND THRUST BLOCKS**

| | | UNIT | L | P | M | £ NETT |
|---|---|---|---|---|---|---|
| Mass Concrete grade C25P, 20mm aggregate including pipe fittings (Volume *) | | | | | | |
| 0.1m3 | hr | nr | 0.18 | 0.18 | | |
| | £ | | 3.66 | 1.76 | 5.00 | 10.42 |
| 0.1 - 0.2m3 | hr | nr | 0.32 | 0.32 | | |
| | £ | | 6.51 | 3.12 | 10.00 | 19.63 |
| 0.2 - 0.5m3 | hr | nr | 0.62 | 0.62 | | |
| | £ | | 12.62 | 6.05 | 25.00 | 43.67 |
| 0.5 - 1.0m3 | hr | nr | 0.91 | 0.91 | | |
| | £ | | 18.52 | 8.87 | 50.00 | 77.39 |
| 1.0 - 2.0m3 | hr | nr | 1.29 | 1.29 | | |
| | £ | | 26.25 | 12.58 | 100.00 | 138.83 |
| 2.0 - 4.0m3 | hr | nr | 3.15 | 3.15 | | |
| | £ | | 64.10 | 30.71 | 200.00 | 294.81 |

* excluding the volume occupied by pipes

**CLASS M: STRUCTURAL METALWORK**

The following are guide prices for various structural members commonly found in a Civil Engineering contract. It is by no means exhaustive and does not include prices for bridge members as the costs are dependent on the particular design and will vary greatly according to specific requirements.

For more detailed prices reference should be made to Specialist Contractors.

**FABRICATION OF MEMBERS FOR FRAMES - STEEL GRADE 43A**

| | UNIT | SPECIALIST ADVICE £ NETT |
|---|---|---|
| Columns | t | 850.00 |
| Beams | | |
| universal beams; straight on plan | t | 700.00 |
| universal beams; curved on plan | t | 1,250.00 |
| channels; straight on plan | t | 800.00 |
| channels; curved on plan | t | 1,350.00 |
| castellated beams; straight on plan | t | 1,050.00 |
| Portal frames | | |
| straight on plan | t | 750.00 |
| Trusses and built up girders | | |
| straight on plan | t | 950.00 |
| curved on plan | t | 1,500.00 |
| Bracings | | |
| angles; straight on plan | t | 750.00 |
| CHS; straight on plan | t | 1,250.00 |
| Purlins and cladding rails | | |
| straight on plan | t | 700.00 |
| Anchorages and holding down bolt assemblies | t | 2,850.00 |

**CLASS M: STRUCTURAL METALWORK**

| | UNIT | SPECIALIST ADVICE £ NETT |
|---|---|---|
| **ERECTION OF MEMBERS FOR FRAMES** | | |
| Trial erection | t | 200.00 |
| Permanent erection | t | 85.00 to 180.00 |
| Site bolts | | |
| black | t | 1,950.00 |
| HSFG general grade | t | 1,950.00 |
| HSFG higher grade | t | 2,376.00 |
| HSGF load indicating or load limit types, general grade | t | 2,600.00 |
| **OFF SITE SURFACE TREATMENT** | | |
| Blast cleaning | t | 55.00 |
| Galvanising | t | 210.00 |
| Painting | | |
| one coat zinc chromate primer | t | 55.00 |
| one coat Epilux 66 Blast; two coats micaceous oxide high build epoxy finish | t | 200.00 |

## CLASS N: MISCELLANEOUS METALWORK

| | UNIT | SPECIALIST ADVICE £ NETT |
|---|---|---|
| Stairways and landings | t | 2,600.00 |
| Walkways and platforms | t | 2,250.00 |
| Ladders | | |
| cat ladder; 64 x 13mm bar strings; 19mm rungs at 250mm centres 460mm wide with hoops | m | 200.00 |
| Handrails | | |
| galvanised tubular metal; 76mm diameter handrail, 48mm diameter standards at 750mm centres, 48mm diameter middle rail; 1070mm high overall | m | 92.00 |
| Miscellaneous framing | | |
| angle section; 200 x 200 x 16mm | m | 25.00 |
| angle section; 150 x 150 x 12mm | m | 16.20 |
| angle section; 100 x 100 x 10mm | m | 8.20 |
| angle section; 200 x 150 x 15mm | m | 23.70 |
| angle section; 150 x 75 x 10mm | m | 11.05 |
| channel section; 381 x 102mm | m | 33.00 |
| channel section; 254 x 76mm | m | 16.92 |
| channel section; 152 x 76mm | m | 10.60 |
| tubular section; 100 x 100 x 10mm | m | 23.72 |
| tubular section; 200 x 200 x 15mm | m | 67.00 |
| tubular section; 76.1 x 5.0mm | m | 7.45 |
| tubular section; 139.7 x 6.3mm | m | 17.60 |
| Plate flooring | | |
| 8mm (on plain) galvanised mild steel "Durbar" pattern floor plates maximum weight each panel 50kg | m2 | 80.00 |
| Open grid flooring | | |
| 50mm galvanised open grid flooring | m2 | 163.00 |

**CLASS O: TIMBER**

| | | | |
|---|---|---|---|
| LABOUR based on Gang C | | | |
| Total hourly cost | = | £ | 29.20 |
| **PLANT** | | | |
| Part time use of Portable tools and equipment (average useage) | | | 2.09 |
| Tractor/trailer | | | 7.07 |
| Part time: | | | |
| 22RB crane | | | 3.32 |
| Forklift (R.T.) | | | 2.11 |
| Total hourly cost | = | £ | 14.59 |

**CLASS O: TIMBER**

**HARDWOOD COMPONENTS**
**(Greenheart)**

| | | UNIT | L | P | M | £ NETT |
|---|---|---|---|---|---|---|
| Generally in marine works (In length) | | | | | | |
| 100 x 75mm | | | | | | |
| up to 3m | hr | m | 0.13 | 0.13 | | |
| | £ | | 3.80 | 1.90 | 3.60 | 9.30 |
| 3 - 5m | hr | m | 0.12 | 0.12 | | |
| | £ | | 3.50 | 1.75 | 3.60 | 8.85 |
| 5 - 8m | hr | m | - | - | | |
| | £ | | - | - | - | - |
| 150 x 75mm | | | | | | |
| up to 3m | hr | m | 0.15 | 0.15 | | |
| | £ | | 4.38 | 2.19 | 5.20 | 11.77 |
| 3 - 5m | hr | m | 0.135 | 0.135 | | |
| | £ | | 3.94 | 1.97 | 5.20 | 11.11 |
| 5 - 8m | hr | m | - | - | | |
| | £ | | - | - | - | - |
| 200 x 100mm | | | | | | |
| up to 3m | hr | m | 0.21 | 0.21 | | |
| | £ | | 6.13 | 3.06 | 8.40 | 17.59 |
| 3 - 5m | hr | m | 0.195 | 0.195 | | |
| | £ | | 5.69 | 2.85 | 8.40 | 16.94 |
| 5 - 8m | hr | m | - | - | | |
| | £ | | - | - | - | - |
| 200 x 200mm | | | | | | |
| up to 3m | hr | m | 0.40 | 0.40 | | |
| | £ | | 11.68 | 5.84 | 18.40 | 35.92 |
| 3 - 5m | hr | m | 0.38 | 0.38 | | |
| | £ | | 11.10 | 5.54 | 18.40 | 35.04 |
| 5 - 8m | hr | m | 0.34 | 0.34 | | |
| | £ | | 9.93 | 4.96 | 18.40 | 33.29 |

**CLASS O: TIMBER**

**HARDWOOD COMPONENTS - continued**
**(Greenheart)**

| | | UNIT | L | P | M | £ NETT |
|---|---|---|---|---|---|---|
| 300 x 200mm | | | | | | |
| up to 3m | hr | m | 0.45 | 0.45 | | |
| | £ | | 13.14 | 6.57 | 36.55 | 56.26 |
| 3 - 5m | hr | m | 0.40 | 0.40 | | |
| | £ | | 11.68 | 5.84 | 36.55 | 54.07 |
| 5 - 8m | hr | m | 0.35 | 0.35 | | |
| | £ | | 10.22 | 5.11 | 36.55 | 51.88 |
| 300 x 300mm | | | | | | |
| up to 3m | hr | m | 0.475 | 0.475 | | |
| | £ | | 13.87 | 6.93 | 39.75 | 60.55 |
| 3 - 5m | hr | m | 0.435 | 0.435 | | |
| | £ | | 12.70 | 6.35 | 39.75 | 58.80 |
| 5 - 8m | hr | m | 0.40 | 0.40 | | |
| | £ | | 11.68 | 5.84 | 39.75 | 57.27 |
| 450 x 450mm | | | | | | |
| up to 3m | hr | m | 0.90 | 0.90 | | |
| | £ | | 26.28 | 13.13 | 90.60 | 130.01 |
| 3 - 5m | hr | m | 0.83 | 0.83 | | |
| | £ | | 24.24 | 12.11 | 90.60 | 126.95 |
| 5 - 8m | hr | m | 0.75 | 0.75 | | |
| | £ | | 21.90 | 10.94 | 90.60 | 123.44 |
| 600 x 600mm | | | | | | |
| up to 3m | hr | m | - | - | | |
| | £ | | - | - | - | - |
| 3 - 5m | hr | m | 1.15 | 1.15 | | |
| | £ | | 33.58 | 16.78 | 160.80 | 211.16 |
| 5 - 8m | hr | m | 1.00 | 1.00 | | |
| | £ | | 29.20 | 14.59 | 160.80 | 204.59 |

## CLASS O: TIMBER

## SOFTWOOD COMPONENTS

| | | UNIT | L | P | M | £ NETT |
|---|---|---|---|---|---|---|
| Generally in marine works | | | | | | |
| 100 x 75mm | | | | | | |
| up to 3m | hr | m | 0.10 | 0.10 | | |
| | £ | | 2.92 | 1.46 | 2.25 | 6.63 |
| 3 - 5m | hr | m | 0.10 | 0.10 | | |
| | £ | | 2.92 | 1.46 | 2.25 | 6.63 |
| 5 - 8m | hr | m | - | - | | |
| | £ | | - | - | - | - |
| 150 x 75mm | | | | | | |
| up to 3m | hr | m | 0.125 | 0.125 | | |
| | £ | | 3.65 | 1.82 | 3.38 | 8.85 |
| 3 - 5m | hr | m | 0.11 | 0.11 | | |
| | £ | | 3.21 | 1.60 | 3.38 | 8.19 |
| 5 - 8m | hr | m | - | - | | |
| | £ | | - | - | - | - |
| 200 x 100mm | | | | | | |
| up to 3m | hr | m | 0.16 | 0.16 | | |
| | £ | | 4.67 | 2.33 | 6.00 | 13.00 |
| 3 - 5m | hr | m | 0.14 | 0.14 | | |
| | £ | | 4.09 | 2.04 | 6.00 | 12.13 |
| 5 - 8m | hr | m | - | - | | |
| | £ | | - | - | - | - |
| 200 x 200mm | | | | | | |
| up to 3m | hr | m | 0.25 | 0.25 | | |
| | £ | | 7.30 | 3.65 | 12.00 | 22.95 |
| 3 - 5m | hr | m | 0.225 | 0.225 | | |
| | £ | | 6.57 | 3.28 | 12.00 | 21.85 |
| 5 - 8m | hr | m | 0.20 | 0.20 | | |
| | £ | | 5.84 | 2.92 | 12.00 | 20.76 |
| 300 x 200mm | | | | | | |
| up to 3m | hr | m | 0.27 | 0.27 | | |
| | £ | | 7.88 | 3.94 | 18.00 | 29.82 |
| 3 - 5m | hr | m | 0.25 | 0.25 | | |
| | £ | | 7.30 | 3.65 | 18.00 | 28.95 |
| 5 - 8m | hr | m | 0.225 | 0.225 | | |
| | £ | | 6.57 | 3.28 | 18.00 | 27.85 |

**CLASS O: TIMBER**

**SOFTWOOD COMPONENTS - continued**

| | | UNIT | L | P | M | £ NETT |
|---|---|---|---|---|---|---|
| 300 x 300mm | | | | | | |
| up to 3m | hr | m | 0.30 | 0.30 | | |
| | £ | | 8.76 | 4.38 | 27.00 | 40.14 |
| 3 - 5m | hr | m | 0.27 | 0.27 | | |
| | £ | | 7.88 | 3.94 | 27.00 | 38.82 |
| 5 - 8m | hr | m | 0.25 | 0.25 | | |
| | £ | | 7.30 | 3.65 | 27.00 | 37.95 |
| 450 x 450mm | | | | | | |
| up to 3m | hr | m | 0.35 | 0.35 | | |
| | £ | | 10.22 | 5.11 | 60.75 | 76.08 |
| 3 - 5m | hr | m | 0.31 | 0.31 | | |
| | £ | | 9.05 | 4.52 | 60.75 | 74.32 |
| 5 - 8m | hr | m | 0.28 | 0.28 | | |
| | £ | | 8.18 | 4.09 | 60.75 | 73.02 |
| 600 x 600mm | | | | | | |
| up to 3m | hr | m | - | - | | |
| | £ | | - | - | - | - |
| 3 - 5m | hr | m | 0.39 | 0.39 | | |
| | £ | | 11.39 | 5.69 | 108.00 | 125.08 |
| 5 - 8m | hr | m | 0.37 | 0.37 | | |
| | £ | | 10.80 | 5.40 | 108.00 | 124.20 |
| ADD to above prices for vacuum/ pressure impregnating (minimum salt retention 5.30kg/m3) | hr | m3 | - | - | | |
| | £ | | - | - | - | 31.50 |

**CLASS O: TIMBER**

**HARDWOOD DECKING**

| | | UNIT | L | P | M | £ NETT |
|---|---|---|---|---|---|---|
| Thickness | | | | | | |
| 50mm | hr | m2 | 0.575 | 0.575 | | |
| | £ | | 16.79 | 8.39 | 30.60 | 55.78 |
| 75mm | hr | m2 | 0.75 | 0.75 | | |
| | £ | | 21.90 | 10.94 | 39.00 | 71.84 |
| 100mm | hr | m2 | 0.95 | 0.95 | | |
| | £ | | 27.74 | 13.86 | 50.30 | 91.90 |
| **SOFTWOOD DECKING** | | | | | | |
| Thickness | | | | | | |
| 50mm | hr | m2 | 0.39 | 0.39 | | |
| | £ | | 11.39 | 5.69 | 14.30 | 31.38 |
| 75mm | hr | m2 | 0.50 | 0.50 | | |
| | £ | | 14.60 | 7.30 | 18.90 | 40.80 |
| 100mm | hr | m2 | 0.65 | 0.65 | | |
| | £ | | 18.98 | 9.48 | 24.00 | 52.46 |
| **FITTINGS AND FASTENINGS** | | | | | | |
| Straps; galvanised mild steel | | | | | | |
| 30 x 2.5 x 800mm girth | hr | nr | 0.13 | - | | |
| | £ | | 3.80 | - | 0.76 | 4.56 |
| 30 x 2.5 x 1000mm girth | hr | nr | 0.13 | - | | |
| | £ | | 3.80 | - | 0.93 | 4.73 |
| 30 x 2.5 x 1200mm girth | hr | nr | 0.15 | - | | |
| | £ | | 4.38 | - | 1.10 | 5.48 |
| Spikes | | | | | | |
| mild steel material rosehead; | | | | | | |
| 14 x 14 x 275mm long | hr | nr | 0.13 | - | | |
| | £ | | 3.80 | - | 0.75 | 4.55 |
| Steel coach screws, including forming holes | | | | | | |
| 6.0 x 75mm long | hr | nr | 0.04 | - | | |
| | £ | | 1.17 | - | 0.18 | 1.35 |
| 6.0 x 100mm long | hr | nr | 0.05 | - | | |
| | £ | | 1.46 | - | 0.23 | 1.69 |
| 8.0 x 100mm long | hr | nr | 0.05 | - | | |
| | £ | | 1.46 | - | 0.40 | 1.86 |
| 8.0 x 125mm long | hr | nr | 0.06 | - | | |
| | £ | | 1.75 | - | 0.48 | 2.23 |
| 12.5 x 100mm long | hr | nr | 0.06 | - | | |
| | £ | | 1.75 | - | 0.63 | 2.38 |
| 12.5 x 150mm long | hr | nr | 0.07 | - | | |
| | £ | | 2.04 | - | 0.79 | 2.83 |

CLASS O: TIMBER

BOLTS

| | | UNIT | L | P | M | £ NETT |
|---|---|---|---|---|---|---|
| Blackbolts, nuts and washers; including forming holes | | | | | | |
| M 6 x 50mm long | hr | nr | 0.05 | - | | |
| | £ | | 1.46 | - | 0.08 | 1.54 |
| M 6 x 100mm long | hr | nr | 0.07 | - | | |
| | £ | | 2.04 | - | 0.14 | 2.18 |
| M 8 x 50mm long | hr | nr | 0.07 | - | | |
| | £ | | 2.04 | - | 0.12 | 2.16 |
| M 8 x 100mm long | hr | nr | 0.075 | - | | |
| | £ | | 2.19 | - | 0.21 | 2.40 |
| M12 x 100mm long | hr | nr | 0.08 | - | | |
| | £ | | 2.34 | - | 0.37 | 2.71 |
| M12 x 140mm long | hr | nr | 0.09 | - | | |
| | £ | | 2.63 | - | 0.56 | 3.19 |
| M12 x 180mm long | hr | nr | 0.10 | - | | |
| | £ | | 2.92 | - | 1.05 | 3.97 |

PLATES

| | | UNIT | L | P | M | £ NETT |
|---|---|---|---|---|---|---|
| Galvanised round toothplate double sided for 10mm or 12mm bolts | | | | | | |
| 38mm diameter | hr | nr | 0.005 | - | | |
| | £ | | 0.15 | - | 0.12 | 0.27 |
| 50mm diameter | hr | nr | 0.005 | - | | |
| | £ | | 0.15 | - | 0.14 | 0.29 |
| 63mm diameter | hr | nr | 0.008 | - | | |
| | £ | | 0.23 | - | 0.15 | 0.38 |
| 75mm diameter | hr | nr | 0.008 | - | | |
| | £ | | 0.23 | - | 0.35 | 0.58 |
| Galvanised mild steel shear plate connector | | | | | | |
| 102mm diameter | hr | nr | 0.060 | - | | |
| | £ | | 1.75 | - | 8.75 | 10.50 |
| Galvanised mild steel flitch plates | | | | | | |
| 200 x 75 x 10mm | hr | nr | 0.07 | - | | |
| | £ | | 2.04 | - | 1.30 | 3.34 |
| 300 x 100 x 10mm | hr | nr | 0.09 | - | | |
| | £ | | 2.63 | - | 3.30 | 5.93 |
| 450 x 150 x 12mm | hr | nr | 0.15 | - | | |
| | £ | | 4.38 | - | 8.43 | 12.81 |

**CLASS P: PILES**
**CLASS Q: PILING ANCILLARIES**

There are a number of different types of piling which are available for use in differing situations. Selection of the most suitable type of piling for a particular site will depend on a number of factors including the physical conditions likely to be encountered during driving, the loads to be carried, the design of superstructure, etc. The most commonly used systems are included in this section.
It is essential that a thorough and adequate site investigation is carried out to ascertain details of the ground strata and bearing capacities to enable a a proper assessment to be made of the most suitable and economical type of piling to be adopted.
There are so many factors, apart from design considerations, which influence the cost of piling that it is not possible to give more than an approximate indication of costs. To obtain reliable costs for a particular contract advice should be sought from a company specialising in the particular type of piling proposed. Some Specialists Contractors will also provide a design service if required.

The section comprises:
bored in-situ reinforced concrete piling (tripod rig)
bored in-situ reinforced concrete piling (mobile rig)
concrete injected piles
large diameter bored in-situ piles (general note only)
driven in situ reinforced concrete piling
driven timber piles
Frodingham steel sheet piling
Larssen steel sheet piling
steel bearing piles
steel tubular piles

This section has not been annotated to CESMM2 measurement headings approximate proportions of Labour, Plant and Materials are noted.

**BORED IN-SITU REINFORCED CONCRETE PILING (TRIPOD RIG)**

The following unit costs cover the construction of small diameter bored piling using light and compact tripod rigs requiring no expensive site levelling or levelling or access ways. Piling can be constructed in very restricted headroom or on confined and difficult sites. Standard diameters are between 400 and 600mm with a normal maximum depth of 30m.

The costs are based on installing 100 piles of 500mm nominal diameter, a concrete strength of 20N/mm2 with nominal reinforcement, on a clear site with reasonable access. Disposal of excavated material is not included in the unit costs.

| | | UNIT | L | P | M | £ NETT |
|---|---|---|---|---|---|---|
| Provision of all plant, equipment and labour including transporting to and from the site and establishing and dismantling | hr | sum | - | - | | |
| | £ | | 1600.00 | 2800.00 | - | 4400.00 |
| Set up at each pile position | hr | nr | - | - | | |
| | £ | | 15.00 | 15.00 | - | 30.00 |

**CLASS P: PILES**
**CLASS Q: PILING ANCILLARIES**

**BORED IN-SITU REINFORCED CONCRETE PILING (TRIPOD RIG) - continued**

| | | UNIT | L | P | M | £ NETT |
|---|---|---|---|---|---|---|
| Boring to a depth of 10m and placing concrete and reinforcement | hr | m | - | - | | |
| | £ | | 210.00 | 450.00 | - | 660.00 |
| Add for additional length of pile up to a depth of 15m | hr | m | - | - | | |
| | £ | | - | - | 48.00 | 48.00 |
| Deduct for reduction in length of pile below 10m | hr | m | - | - | | |
| | £ | | 9.00 | 15.00 | - | 24.00 |
| Add for boring through obstructions | hr | hr | - | - | | |
| | £ | | 30.00 | 45.00 | - | 75.00 |
| Standing time | hr | hr | - | - | | |
| | £ | | 30.00 | 45.00 | - | 75.00 |
| Carry out pile test for maximum test load of 600kN on a working pile as reaction | | | | | | |
| first pile | hr | nr | - | - | | |
| | £ | | 1300.00 | 2300.00 | - | 3600.00 |
| subsequent pile | hr | nr | - | - | | |
| | £ | | 1050.00 | 1650.00 | - | 2700.00 |
| Take and test undisturbed soil samples | hr | nr | - | - | | |
| | £ | | 50.00 | 65.00 | - | 115.00 |
| Make, cure and test concrete cubes | hr | nr | - | - | | |
| | £ | | - | - | - | 6.00 |
| Trim down heads of piles | hr | nr | - | - | | |
| | £ | | 16.00 | 24.00 | - | 40.00 |

**CLASS P: PILES**
**CLASS Q: PILING ANCILLARIES**

**BORED IN-SITU REINFORCED CONCRETE PILING (MOBILE RIG)**

The following unit costs cover the construction of small diameter bored piles using lorry or crawler mounted rotary boring rigs. This type of plant is more mobile and faster in operation than the tripod rigs and is ideal for large contracts in cohesive ground. Construction of piles under bentonite suspension can be carried out to obviate the use of liners. Standard diameters of 450 to 900mm diameter can be constructed to depths of 30m.

The costs are based on installing 100 piles of 500mm nominal diameter, a concrete strength of 20N/mm2 with nominal reinforcement, on a clear site with reasonable access. Disposal of excavated material is not included in the unit costs.

| | | UNIT | L | P | M | £ NETT |
|---|---|---|---|---|---|---|
| Provision of all plant, equipment and labour including transport to and from site and establishing and dismantling | hr | sum | - | - | | |
| | £ | | 2750.00 | 4250.00 | - | 7000.00 |
| Set up at each pile position | hr | nr | - | - | | |
| | £ | | 7.00 | 28.00 | - | 35.00 |
| Boring to a depth of 10m, placing concrete and reinforcement | hr | nr | - | - | | |
| | £ | | - | - | - | 300.00 |
| Add for additional length of pile up to a depth of 15m | hr | m | - | - | | |
| | £ | | - | - | - | 28.00 |
| Deduct for reduction in length of pile below 10m | hr | m | - | - | | |
| | £ | | - | - | - | 15.00 |
| Add for boring through obstructions | hr | hr | - | - | | |
| | £ | | 35.00 | 105.00 | - | 140.00 |
| Standing time | hr | nr | - | - | | |
| | £ | | 50.00 | 90.00 | - | 140.00 |

**CLASS P: PILES**
**CLASS Q: PILING ANCILLARIES**

**BORED IN-SITU REINFORCED CONCRETE PILING (MOBILE RIG) - continued**

| | | UNIT | L | P | M | £ NETT |
|---|---|---|---|---|---|---|
| Carry out pile test for maximum test load of 600kN on a working pile as reaction | | | | | | |
| first pile | hr | nr | - | - | | |
| | £ | | - | - | - | 3500.00 |
| subsequent piles | hr | nr | - | - | | |
| | £ | | - | - | - | 2500.00 |
| Take and test undisturbed soil samples | hr | nr | - | - | | |
| | £ | | - | - | - | 120.00 |
| Make, cure and test concrete cubes | hr | nr | - | - | | |
| | £ | | - | - | - | 6.00 |
| Trim down heads of piles | hr | nr | - | - | | |
| | £ | | 16.00 | 24.00 | - | 40.00 |

**CLASS P: PILES**
**CLASS Q: PILING ANCILLARIES**

**CONCRETE INJECTED PILES**

The following unit costs cover the construction of piles by screwing a continuous flight auger into the ground to the required design depth. Concrete is then pumped through the hollow stem of the auger to the bottom and the pile formed as the auger is withdrawn. Spoil is removed by the auger as it is withdrawn. This is a fast method of construction without causing disturbance or vibration to adjacent ground. No casing is required even in unsuitable soils. Reinforcement can be placed after grouting is complete. The costs are based on installing 100 piles on a clear site with reasonable access. Disposal of excavated material is not included in the unit costs.

| | | UNIT | L | P | M | £ NETT |
|---|---|---|---|---|---|---|
| Provision of all plant, equipment and labour including transport to and from the site and establishing and dismantling | | | | | | |
| 450mm/ 650kN | hr | sum | - | - | | |
| | £ | | 2300.00 | 4200.00 | - | 6500.00 |
| 600mm/1400kN | hr | sum | - | - | | |
| | £ | | 2300.00 | 4200.00 | - | 6500.00 |
| 750mm/2200kN | hr | sum | - | - | | |
| | £ | | 2300.00 | 4200.00 | - | 6500.00 |
| Set up at each pile position | | | | | | |
| 450mm/ 650kN | hr | nr | - | - | | |
| | £ | | 7.00 | 28.00 | - | 35.00 |
| 600mm/1400kN | hr | nr | - | - | | |
| | £ | | 7.00 | 28.00 | - | 35.00 |
| 750mm/2200kN | hr | nr | - | - | | |
| | £ | | 7.00 | 28.00 | - | 35.00 |
| Construct pile with 20N/mm2 concrete and reinforce with starter bars | | | | | | |
| 450mm/ 650kN | hr | m | - | - | | |
| | £ | | 5.00 | 7.00 | 18.00 | 30.00 |
| 600mm/1400kN | hr | m | - | - | | |
| | £ | | 8.00 | 10.50 | 23.50 | 42.00 |
| 750mm/2200kN | hr | m | - | - | | |
| | £ | | 14.00 | 20.25 | 32.75 | 67.00 |

**CLASS P: PILES**
**CLASS Q: PILING ANCILLARIES**

**CONCRETE INJECTED PILES - continued**

| | | UNIT | L | P | M | £ NETT |
|---|---|---|---|---|---|---|
| Add for additional length of pile over design depth up to a maximum depth of 22m | | | | | | |
| 450mm/ 650kN | hr | m | - | - | | |
| | £ | | 10.00 | 7.00 | 18.00 | 35.00 |
| 600mm/1400kN | hr | m | - | - | | |
| | £ | | 10.00 | 12.50 | 24.50 | 47.00 |
| 750mm/2200kN | hr | m | - | - | | |
| | £ | | 17.00 | 23.25 | 34.75 | 75.00 |
| Deduct for reduction in length of pile under design depth | | | | | | |
| 450mm/ 650kN | hr | m | - | - | | |
| | £ | | 3.50 | 3.50 | 9.00 | 16.00 |
| 600mm/1400kN | hr | m | - | - | | |
| | £ | | 5.00 | 6.00 | 10.00 | 21.00 |
| 750mm/2200kN | hr | m | - | - | | |
| | £ | | 8.00 | 10.00 | 16.00 | 34.00 |
| Standing time | | | | | | |
| 450mm/ 650kN | hr | hr | - | - | | |
| | £ | | 49.00 | 91.00 | - | 140.00 |
| 600mm/1400kN | hr | hr | - | - | | |
| | £ | | 49.00 | 91.00 | - | 140.00 |
| 750mm/2200kN | hr | hr | - | - | | |
| | £ | | 49.00 | 91.00 | - | 140.00 |
| Carry out pile test for maximum load first pile | | | | | | |
| 450mm/ 650kN | hr | nr | - | - | | |
| | £ | | 1300.00 | 2300.00 | - | 3600.00 |
| 600mm/1400kN | hr | nr | - | - | | |
| | £ | | 1630.00 | 2870.00 | - | 4500.00 |
| 750mm/2200kN | hr | nr | - | - | | |
| | £ | | 2000.00 | 3500.00 | - | 5500.00 |

**CLASS P: PILES**
**CLASS Q: PILING ANCILLARIES**

**CONCRETE INJECTED PILES - continued**

| | UNIT | | L | P | M | £ NETT |
|---|---|---|---|---|---|---|
| subsequent piles | | | | | | |
| 450mm/ 650kN | hr | nr | - | - | | |
| | £ | | 920.00 | 1680.00 | - | 2600.00 |
| 600mm/1400kN | hr | nr | - | - | | |
| | £ | | 1260.00 | 2240.00 | - | 3500.00 |
| 750mm/2200kN | hr | nr | - | - | | |
| | £ | | 1630.00 | 2870.00 | - | 4500.00 |
| Make, cure and test concrete cubes | | | | | | |
| 450mm/ 650kN | hr | nr | - | - | | |
| | £ | | - | - | - | 6.00 |
| 600mm/1400kN | hr | nr | - | - | | |
| | £ | | - | - | - | 6.00 |
| 750mm/2200kN | hr | nr | - | - | | |
| | £ | | - | - | - | 6.00 |
| Trim down heads of piles | | | | | | |
| 450mm/ 650kN | hr | nr | - | - | | |
| | £ | | 15.50 | 24.50 | - | 40.00 |
| 600mm/1400kN | hr | nr | - | - | | |
| | £ | | 28.00 | 42.00 | - | 70.00 |
| 750mm/2200kN | hr | nr | - | - | | |
| | £ | | 47.00 | 63.00 | - | 110.00 |

**CLASS P: PILES**
**CLASS Q: PILING ANCILLARIES**

**LARGE DIAMETER BORED IN-SITU PILES**

Large diameter bored in-situ piles may be constructed using rotary driven augers mounted on lorry or crawler cranes. Piles may be straight shafted or belled at the bottom to give greater bearing area. In utilising this type of pile it is of great importance to carry out a very thorough site investigation of the strata to be encountered before design proceeds.
These piles may be constructed in nominal diameters from 750 to 2100mm increments of 150mm and underreamed bells may be formed in suitable ground up to three times the diameter of the shaft. Loads of up to 30,000kN can be carried and piles can be constructed to depths up to 33m. Visual inspection of the shaft and bell can be made using special man-riding cages.
Shafts may be lined where necessary or bentonite suspension may be used in straight shafted piles in poor ground.
It is not practical to give average rates for this type of piling and the advice of a Specialist Contractor should be sought.

**DRIVEN IN-SITU REINFORCED CONCRETE PILING**

The following unit costs cover the construction of piles by driving a tube into the ground either by using an internal hammer acting on a gravel or concrete plug or by using an external hammer on a driving helmet at the top of the tube. After driving to the required depth an enlarged base is formed by hammering out successive charges of concrete down the tube. The tube is then filled with concrete which is compacted as the tube is vibrated and withdrawn piles of 350 to 500mm diameter can be constructed with rakes up to 1 in 4 to carry working loads up to 120t per pile.
The costs are based on installing 100 piles of 430mm nominal diameter, a concrete strength of 20N/mm2 suitably reinforced for a working load not exceeding 750kN, on a clear site with reasonable access.

| | | UNIT | L | P | M | £ NETT |
|---|---|---|---|---|---|---|
| Provision of all plant, equipment and labour including transporting to and from the site and establishing and dismantling | | | | | | |
| Bottom driven | hr | sum | - | - | | |
| | £ | | 1710.00 | 3990.00 | - | 5700.00 |
| Top driven | hr | sum | - | - | | |
| | £ | | 1710.00 | 3990.00 | - | 5700.00 |
| Set up at each pile position | | | | | | |
| Bottom driven | hr | nr | - | - | | |
| | £ | | 25.50 | 59.50 | - | 85.00 |
| Top driven | hr | nr | - | - | | |
| | £ | | 19.50 | 45.50 | - | 65.00 |

**CLASS P: PILES**
**CLASS Q: PILING ANCILLARIES**

**DRIVEN IN-SITU REINFORCED CONCRETE PILING - continued**

| | UNIT | | L | P | M | £ NETT |
|---|---|---|---|---|---|---|
| Drive shell and form pile shaft including reinforcement to a depth of 10m | | | | | | |
| Bottom driven | hr | m | - | - | | |
| | £ | | - | - | - | 22.00 |
| Top driven | hr | m | - | - | | |
| | £ | | - | - | - | 20.00 |
| Add for additional length exceeding 10m but ne 15m | | | | | | |
| Bottom driven | hr | m | - | - | | |
| | £ | | - | - | - | 23.50 |
| Top driven | hr | m | - | - | | |
| | £ | | - | - | - | 21.25 |
| Add for driving through obstructions where within the capabilities of the normal plant | | | | | | |
| Bottom driven | hr | nr | - | - | | |
| | £ | | - | - | - | 140.00 |
| Top driven | hr | nr | - | - | | |
| | £ | | - | - | - | 140.00 |
| Standing time | | | | | | |
| Bottom driven | hr | nr | - | - | | |
| | £ | | - | - | - | 130.00 |
| Top driven | hr | nr | - | - | | |
| | £ | | - | - | - | 130.00 |
| Carry out test on 430mm diameter pile for maximum test load of 1125kN using non working tension piles as reaction | | | | | | |
| first piles | | | | | | |
| Bottom driven | hr | m | - | - | | |
| | £ | | 1005.00 | 2345.00 | - | 3350.00 |
| Top driven | hr | m | - | - | | |
| | £ | | 855.00 | 1995.00 | - | 2850.00 |

**CLASS P: PILES**
**CLASS Q: PILING ANCILLARIES**

**DRIVEN IN-SITU REINFORCED CONCRETE PILING - continued**

| | | UNIT | L | P | M | £ NETT |
|---|---|---|---|---|---|---|
| subsequent piles | | | | | | |
| Bottom driven | hr | nr | - | - | | |
| | £ | | 690.00 | 1610.00 | - | 2300.00 |
| Top driven | hr | nr | - | - | | |
| | £ | | 690.00 | 1610.00 | - | 2300.00 |
| Make, cure and test concrete cubes | | | | | | |
| Bottom driven | hr | nr | - | - | | |
| | £ | | - | - | - | 5.75 |
| Top driven | hr | nr | - | - | | |
| | £ | | - | - | - | 5.75 |
| Trim down heads of piles | | | | | | |
| Bottom driven | hr | nr | - | - | | |
| | £ | | 7.80 | 18.20 | - | 26.00 |
| Top driven | hr | nr | - | - | | |
| | £ | | 7.80 | 18.20 | - | 26.00 |

**CLASS P: PILES**
**CLASS Q: PILING ANCILLARIES**

**DRIVEN TIMBER PILES**

The following unit costs are based on installing 100 piles on a clear site with reasonable access

| | | UNIT | L | P | M | £ NETT |
|---|---|---|---|---|---|---|
| Provision of all plant, equipment and labour including transporting to and from the site and establishing and dismantling | hr | sum | - | - | | |
| | £ | | 1500.00 | 3500.00 | - | 5000.00 |
| Standing time | hr | hr | - | - | | |
| | £ | | 60.00 | 140.00 | - | 200.00 |
| Provide timber piles (Pile size) | | | | | | |
| Douglas fir | | | | | | |
| 225 x 225mm | hr | m | - | - | | |
| | £ | | - | - | 16.05 | 16.05 |
| 300 x 300mm | hr | m | - | - | | |
| | £ | | - | - | 24.85 | 24.85 |
| 350 x 350mm | hr | m | - | - | | |
| | £ | | - | - | 31.50 | 31.50 |
| 450 x 450mm | hr | m | - | - | | |
| | £ | | - | - | 50.15 | 50.15 |
| Greenheart (hewn to mean size) | | | | | | |
| 225 x 225mm | hr | m | - | - | | |
| | £ | | - | - | 31.95 | 31.95 |
| 300 x 300mm | hr | m | - | - | | |
| | £ | | - | - | 53.45 | 53.45 |
| 350 x 350mm | hr | m | - | - | | |
| | £ | | - | - | 70.40 | 70.40 |
| 450 x 450mm | hr | m | - | - | | |
| | £ | | - | - | 113.85 | 113.85 |

**CLASS P: PILES**
**CLASS Q: PILING ANCILLARIES**

**DRIVEN TIMBER PILES - continued**

| | UNIT | | L | P | M | £ NETT |
|---|---|---|---|---|---|---|
| Set top at each pile position | | | | | | |
| 225 x 225mm | hr | nr | - | - | | |
| | £ | | - | - | - | 55.00 |
| 300 x 300mm | hr | nr | - | - | | |
| | £ | | - | - | - | 55.00 |
| 350 x 350mm | hr | nr | - | - | | |
| | £ | | - | - | - | 55.00 |
| 450 x 450mm | hr | nr | - | - | | |
| | £ | | - | - | - | 55.00 |
| Drive piles | | | | | | |
| 225 x 225mm | hr | m | - | - | | |
| | £ | | 6.05 | - | - | 6.05 |
| 300 x 300mm | hr | m | - | - | | |
| | £ | | 7.50 | - | - | 7.50 |
| 350 x 350mm | hr | m | - | - | | |
| | £ | | 8.25 | - | - | 8.25 |
| 450 x 450mm | hr | m | - | - | | |
| | £ | | 9.10 | - | - | 9.10 |
| Cut heads of pile to level | | | | | | |
| Douglas fir | | | | | | |
| 225 x 225mm | hr | nr | - | - | | |
| | £ | | 1.85 | - | - | 1.85 |
| 300 x 300mm | hr | nr | - | - | | |
| | £ | | 3.30 | - | - | 3.30 |
| 350 x 350mm | hr | nr | - | - | | |
| | £ | | 4.24 | - | - | 4.24 |
| 450 x 450mm | hr | nr | - | - | | |
| | £ | | 8.11 | - | - | 8.11 |
| Greenheart | | | | | | |
| 225 x 225mm | hr | nr | - | - | | |
| | £ | | 3.65 | - | - | 3.65 |
| 300 x 300mm | hr | nr | - | - | | |
| | £ | | 10.60 | - | - | 10.60 |
| 350 x 350mm | hr | nr | - | - | | |
| | £ | | 8.50 | - | - | 8.50 |
| 450 x 450mm | hr | nr | - | - | | |
| | £ | | 15.75 | - | - | 15.75 |

**CLASS P: PILES**
**CLASS Q: PILING ANCILLARIES**

**FRODINGHAM STEEL SHEET PILING**

The following unit costs are based on driving/extracting 1500m2 of sheet piling on a clear site with reasonable access

| | | UNIT | L | P | M | £ NETT |
|---|---|---|---|---|---|---|
| Provision of all plant, equipment and labour including transport to and from the site and establishing and dismantling for the driving of sheet piling | hr | sum | - | - | | |
| | £ | | 1300.00 | 3000.00 | - | 4300.00 |
| Standing time | | | | | | |
| | hr | hr | - | - | | |
| | £ | | 60.00 | 140.00 | - | 200.00 |
| Provision of all plant, equipment and labour including transport to and from the site and establishing and dismantling for the extraction of sheet piling | hr | sum | - | - | | |
| | £ | | 1140.00 | 2660.00 | - | 3800.00 |
| Standing time | | | | | | |
| | hr | hr | - | - | | |
| | £ | | 60.00 | 140.00 | - | 200.00 |

**CLASS P: PILES**
**CLASS Q: PILING ANCILLARIES**

**FRODINGHAM STEEL SHEET PILING - continued**

| | UNIT | | L | P | M | £ NETT |
|---|---|---|---|---|---|---|
| Provide new steel sheet piling to BS4360 Grade 43A in lengths 7 - 14m (Piling section reference) (Weight in kg/m2) | | | | | | |
| 1N 99.10 | hr | m2 | - | - | | |
| | £ | | - | - | 37.46 | 37.46 |
| 2N 112.30 | hr | m2 | - | - | | |
| | £ | | - | - | 42.45 | 42.45 |
| 3NA 137.10 | hr | m2 | - | - | | |
| | £ | | - | - | 51.82 | 51.82 |
| 4N 170.80 | hr | m2 | - | - | | |
| | £ | | - | - | 64.56 | 64.56 |
| 5 236.90 | hr | m2 | - | - | | |
| | £ | | - | - | 89.55 | 89.55 |
| Add to last item for | | | | | | |
| copper content ne 0.35% (Piling section reference) (Weight in kg/m2) | | | | | | |
| 1N 99.10 | hr | m2 | - | - | | |
| | £ | | - | - | 0.99 | 0.99 |
| 2N 112.30 | hr | m2 | - | - | | |
| | £ | | - | - | 1.12 | 1.12 |
| 3NA 137.10 | hr | m2 | - | - | | |
| | £ | | - | - | 1.37 | 1.37 |
| 4N 170.80 | hr | m2 | - | - | | |
| | £ | | - | - | 1.70 | 1.70 |
| 5 236.90 | hr | m2 | - | - | | |
| | £ | | - | - | 2.37 | 2.37 |

**CLASS P: PILES**
**CLASS Q: PILING ANCILLARIES**

**FRODINGHAM STEEL SHEET PILING - continued**

| | UNIT | | L | P | M | £ NETT |
|---|---|---|---|---|---|---|
| copper content 0.35 - 0.5% (Piling section reference) (Weight in kg/m2) | | | | | | |
| 1N 99.10 | hr | m2 | - | - | | |
| | £ | | - | - | 1.39 | 1.39 |
| 2N 112.30 | hr | m2 | - | - | | |
| | £ | | - | - | 1.57 | 1.57 |
| 3NA 137.10 | hr | m2 | - | - | | |
| | £ | | - | - | 1.92 | 1.92 |
| 4N 170.80 | hr | m2 | - | - | | |
| | £ | | - | - | 2.39 | 2.39 |
| 5 236.90 | hr | m2 | - | - | | |
| | £ | | - | - | 3.32 | 3.32 |
| one coat black tar varnish (Piling section reference) (Weight in kg/m2) | | | | | | |
| 1N 99.10 | hr | m2 | - | - | | |
| | £ | | 2.80 | - | 4.20 | 7.00 |
| 2N 112.30 | hr | m2 | - | - | | |
| | £ | | 2.80 | - | 4.20 | 7.00 |
| 3NA 137.10 | hr | m2 | - | - | | |
| | £ | | 2.80 | - | 4.20 | 7.00 |
| 4N 170.80 | hr | m2 | - | - | | |
| | £ | | 2.80 | - | 4.20 | 7.00 |
| 5 236.90 | hr | m2 | - | - | | |
| | £ | | 2.80 | - | 4.20 | 7.00 |

**CLASS P: PILES**
**CLASS Q: PILING ANCILLARIES**

**FRODINGHAM STEEL SHEET PILING - continued**

| | | UNIT | L | P | M | £ NETT |
|---|---|---|---|---|---|---|
| fabricated piles (Piling section reference) (Weight in kg/m2) | | | | | | |
| 1N 99.10 | hr | t | - | - | | |
| | £ | | 201.00 | 469.00 | 342.78 | 1012.78 |
| 2N 112.30 | hr | t | - | - | | |
| | £ | | 240.00 | 560.00 | 352.50 | 1152.50 |
| 3NA 137.10 | hr | t | - | - | | |
| | £ | | 300.00 | 700.00 | 352.50 | 1352.50 |
| 4N 170.80 | hr | t | - | - | | |
| | £ | | 360.00 | 840.00 | 352.50 | 1552.50 |
| 5 236.90 | hr | t | - | - | | |
| | £ | | 420.00 | 980.00 | 352.50 | 1752.50 |
| Extra for piles under 4 m long (Piling section reference) (Weight in kg/m2) | | | | | | |
| 1N 99.10 | hr | m2 | - | - | | |
| | £ | | 0.20 | 0.50 | 4.46 | 5.16 |
| 2N 112.30 | hr | m2 | - | - | | |
| | £ | | 0.20 | 0.50 | 4.50 | 5.20 |
| 3NA 137.10 | hr | m2 | - | - | | |
| | £ | | 0.20 | 0.50 | 4.80 | 5.50 |
| 4N 170.80 | hr | m2 | - | - | | |
| | £ | | 0.20 | 0.50 | 4.27 | 4.97 |
| 5 236.90 | hr | m2 | - | - | | |
| | £ | | 0.20 | 0.50 | 4.74 | 5.44 |

**CLASS P: PILES**
**CLASS Q: PILING ANCILLARIES**

**FRODINGHAM STEEL SHEET PILING - continued**

| | | UNIT | | L | P | M | £ NETT |
|---|---|---|---|---|---|---|---|
| Extra for piles 4 - 7m long (Piling section reference) (Weight in kg/m2) | | | | | | | |
| 1N | 99.10 | hr | m2 | - | - | | |
| | | £ | | 0.09 | 0.23 | 1.98 | 2.30 |
| 2N | 112.30 | hr | m2 | - | - | | |
| | | £ | | 0.10 | 0.25 | 2.25 | 2.60 |
| 3NA | 137.10 | hr | m2 | - | - | | |
| | | £ | | 0.09 | 0.22 | 2.06 | 2.37 |
| 4N | 170.80 | hr | m2 | - | - | | |
| | | £ | | 0.12 | 0.28 | 2.56 | 2.96 |
| 5 | 236.90 | hr | m2 | - | - | | |
| | | £ | | 0.11 | 0.27 | 2.37 | 2.75 |
| Extra for piles 14 - 20m long (Piling section reference) (Weight in kg/m2) | | | | | | | |
| 1N | 99.10 | hr | m2 | - | - | | |
| | | £ | | 0.04 | 0.10 | 1.00 | 1.14 |
| 2N | 112.30 | hr | m2 | - | - | | |
| | | £ | | 0.05 | 0.18 | 1.12 | 1.35 |
| 3NA | 137.10 | hr | m2 | - | - | | |
| | | £ | | 0.03 | 0.06 | 0.70 | 0.79 |
| 4N | 170.80 | hr | m2 | - | - | | |
| | | £ | | 0.04 | 0.14 | 0.85 | 1.03 |
| 5 | 236.90 | hr | m2 | - | - | | |
| | | £ | | 0.06 | 0.13 | 1.18 | 1.37 |

**CLASS P: PILES**
**CLASS Q: PILING ANCILLARIES**

**FRODINGHAM STEEL SHEET PILING - continued**

| | | UNIT | L | P | M | £ NETT |
|---|---|---|---|---|---|---|
| Extra for piles 20 - 25m long (Piling section reference) (Weight in kg/m2) | | | | | | |
| 1N 99.10 | hr | m2 | - | - | | |
| | £ | | - | - | - | - |
| 2N 112.30 | hr | m2 | - | - | | |
| | £ | | - | - | - | - |
| 3NA 137.10 | hr | m2 | - | - | | |
| | £ | | 0.16 | 0.37 | 3.43 | 3.96 |
| 4N 170.80 | hr | m2 | - | - | | |
| | £ | | 0.16 | 0.37 | 3.42 | 3.95 |
| 5 236.90 | hr | m2 | - | - | | |
| | £ | | 0.17 | 0.38 | 3.55 | 4.10 |
| Extra for piles 25 - 29m long (Piling section reference) (Weight in kg/m2) | | | | | | |
| 1N 99.10 | hr | m2 | - | - | | |
| | £ | | - | - | - | - |
| 2N 112.30 | hr | m2 | - | - | | |
| | £ | | - | - | - | - |
| 3NA 137.10 | hr | m2 | - | - | | |
| | £ | | - | - | - | - |
| 4N 170.80 | hr | m2 | - | - | | |
| | £ | | 0.24 | 0.55 | 5.13 | 5.92 |
| 5 236.90 | hr | m2 | - | - | | |
| | £ | | 0.27 | 0.62 | 5.95 | 6.84 |

**CLASS P: PILES**
**CLASS Q: PILING ANCILLARIES**

**FRODINGHAM STEEL SHEET PILING - continued**

| | | UNIT | L | P | M | £ NETT |
|---|---|---|---|---|---|---|
| Handle pitch and drive piles (Piling section reference) (Weight in kg/m2) | | | | | | |
| 1N 99.10 | hr | m2 | - | - | | |
| | £ | | 8.24 | 17.52 | - | 25.76 |
| 2N 112.30 | hr | m2 | - | - | | |
| | £ | | 7.40 | 16.80 | - | 24.20 |
| 3NA 137.10 | hr | m2 | - | - | | |
| | £ | | 6.24 | 15.60 | - | 21.84 |
| 4N 170.80 | hr | m2 | - | - | | |
| | £ | | 5.26 | 14.22 | - | 19.48 |
| 5 236.90 | hr | m2 | - | - | | |
| | £ | | 5.78 | 15.62 | - | 21.40 |
| Burn off tops of piles to level (Piling section reference) (Weight in kg/m2) | | | | | | |
| 1N 99.10 | hr | m | - | - | | |
| | £ | | 1.70 | 5.30 | - | 7.00 |
| 2N 112.30 | hr | m | - | - | | |
| | £ | | 1.70 | 5.30 | - | 7.00 |
| 3NA 137.10 | hr | m | - | - | | |
| | £ | | 1.70 | 5.30 | - | 7.00 |
| 4N 170.80 | hr | m | - | - | | |
| | £ | | 1.88 | 5.62 | - | 7.50 |
| 5 236.90 | hr | m | - | - | | |
| | £ | | 2.25 | 6.75 | - | 9.00 |

**CLASS P: PILES**
**CLASS Q: PILING ANCILLARIES**

**FRODINGHAM STEEL SHEET PILING - continued**

| | UNIT | | L | P | M | £ NETT |
|---|---|---|---|---|---|---|
| Extract piling and stack on site (Piling section reference) (Weight in kg/m2) | | | | | | |
| 1N 99.10 | hr | m2 | - | - | | |
| | £ | | 5.13 | 12.00 | - | 17.13 |
| 2N 112.30 | hr | m2 | - | - | | |
| | £ | | 3.95 | 11.25 | - | 15.20 |
| 3NA 137.10 | hr | m2 | - | - | | |
| | £ | | 2.92 | 9.24 | - | 12.16 |
| 4N 170.80 | hr | m2 | - | - | | |
| | £ | | 2.81 | 8.89 | - | 11.70 |
| 5 236.90 | hr | m2 | - | - | | |
| | £ | | 3.11 | 9.81 | - | 12.92 |

**CLASS P: PILES**
**CLASS Q: PILING ANCILLARIES**

**LARSSEN STEEL SHEET PILING**

The following unit costs are based on driving/extracting 1500m2 of sheet piling on a clear site with reasonable access.

| | | UNIT | L | P | M | £ NETT |
|---|---|---|---|---|---|---|
| Provision of all plant, equipment and labour including transport to and from the site and establishing and dismantling for the driving of sheet piling | hr | sum | - | - | | |
| | £ | | 1300.00 | 3000.00 | - | 4300.00 |
| Standing time | hr | hr | - | - | | |
| | £ | | 60.00 | 140.00 | - | 200.00 |
| Provision of all plant, equipment and labour including transport to and from the site and establishing and dismantling for the extraction of sheet piling | hr | sum | - | - | | |
| | £ | | 1140.00 | 2660.00 | - | 3800.00 |
| Standing time | hr | hr | - | - | | |
| | £ | | 60.00 | 140.00 | - | 200.00 |
| Provide new steel sheet piling to BS4360 Grade 43A in lengths 7 - 14m (Piling section reference) (Weight in kg/m2) | | | | | | |
| 6W 85.10 | hr | m2 | - | - | | |
| | £ | | - | - | 35.50 | 35.50 |
| 9W 97.10 | hr | m2 | - | - | | |
| | £ | | - | - | 39.50 | 39.50 |
| 12W 115.10 | hr | m2 | - | - | | |
| | £ | | - | - | 45.15 | 45.15 |
| 16W 130.10 | hr | m2 | - | - | | |
| | £ | | - | - | 51.00 | 51.00 |
| 20W 147.20 | hr | m2 | - | - | | |
| | £ | | - | - | 57.70 | 57.70 |

**CLASS P: PILES**
**CLASS Q: PILING ANCILLARIES**

**LARSSEN STEEL SHEET PILING - continued**

| | UNIT | | L | P | M | £ NETT |
|---|---|---|---|---|---|---|
| Add to last item for copper content ne 0.35% (Piling section reference) (Weight in kg/m2) | | | | | | |
| 6W 85.10 | hr | m2 | - | - | | |
| | £ | | - | - | 0.85 | 0.85 |
| 9W 97.10 | hr | m2 | - | - | | |
| | £ | | - | - | 0.97 | 0.97 |
| 12W 115.10 | hr | m2 | - | - | | |
| | £ | | - | - | 1.15 | 1.15 |
| 16W 130.10 | hr | m2 | - | - | | |
| | £ | | - | - | 1.30 | 1.30 |
| 20W 147.20 | hr | m2 | - | - | | |
| | £ | | - | - | 1.47 | 1.47 |
| copper content 0.35 - 0.5% (Piling section reference) (Weight in kg/m2) | | | | | | |
| 6W 85.10 | hr | m2 | - | - | | |
| | £ | | - | - | 1.19 | 1.19 |
| 9W 97.10 | hr | m2 | - | - | | |
| | £ | | - | - | 1.36 | 1.36 |
| 12W 115.10 | hr | m2 | - | - | | |
| | £ | | - | - | 1.61 | 1.61 |
| 16W 130.10 | hr | m2 | - | - | | |
| | £ | | - | - | 1.82 | 1.82 |
| 20W 147.20 | hr | m2 | - | - | | |
| | £ | | - | - | 2.06 | 2.06 |

**CLASS P: PILES**
**CLASS Q: PILING ANCILLARIES**

**LARSSEN STEEL SHEET PILING - continued**

| | | UNIT | L | P | M | £ NETT |
|---|---|---|---|---|---|---|
| one coat black tar varnish (Piling section reference) (Weight in kg/m2) | | | | | | |
| 6W 85.10 | hr | m2 | - | - | | |
| | £ | | 2.80 | - | 4.20 | 7.00 |
| 9W 97.10 | hr | m2 | - | - | | |
| | £ | | 2.80 | - | 4.20 | 7.00 |
| 12W 115.10 | hr | m2 | - | - | | |
| | £ | | 2.80 | - | 4.20 | 7.00 |
| 16W 130.10 | hr | m2 | - | - | | |
| | £ | | 2.80 | - | 4.20 | 7.00 |
| 20W 147.20 | hr | m2 | - | - | | |
| | £ | | 2.80 | - | 4.20 | 7.00 |
| fabricated piles (Piling section reference) (Weight in kg/m2) | | | | | | |
| 6W 85.10 | hr | t | - | - | | |
| | £ | | 72.00 | 168.00 | 417.15 | 657.15 |
| 9W 97.10 | hr | t | - | - | | |
| | £ | | 120.00 | 280.00 | 406.85 | 806.85 |
| 12W 115.10 | hr | t | - | - | | |
| | £ | | 180.00 | 420.00 | 391.40 | 991.40 |
| 16W 130.10 | hr | t | - | - | | |
| | £ | | 240.00 | 560.00 | 391.40 | 1191.40 |
| 20W 147.20 | hr | t | - | - | | |
| | £ | | 300.00 | 700.00 | 391.06 | 1391.06 |

**CLASS P: PILES**
**CLASS Q: PILING ANCILLARIES**

**LARSSEN STEEL SHEET PILING - continued**

| | UNIT | | L | P | M | £ NETT |
|---|---|---|---|---|---|---|
| Extra for piles under 4m long (Piling section reference) (Weight in kg/m2) | | | | | | |
| 6W 85.10 | hr | m2 | - | - | | |
| | £ | | 0.18 | 0.41 | 3.83 | 4.42 |
| 9W 97.10 | hr | m2 | - | - | | |
| | £ | | 0.18 | 0.42 | 3.88 | 4.48 |
| 12W 115.10 | hr | m2 | - | - | | |
| | £ | | 0.17 | 0.38 | 3.45 | 4.00 |
| 16W 130.10 | hr | m2 | - | - | | |
| | £ | | 0.18 | 0.42 | 3.90 | 4.50 |
| 20W 147.20 | hr | m2 | - | - | | |
| | £ | | 0.17 | 0.40 | 3.68 | 4.25 |
| Extra for piles 4 - 7m long (Piling section reference) (Weight in kg/m2) | | | | | | |
| 6W 85.10 | hr | m2 | - | - | | |
| | £ | | 0.08 | 0.18 | 1.70 | 1.96 |
| 9W 97.10 | hr | m2 | - | - | | |
| | £ | | 0.09 | 0.22 | 1.94 | 2.25 |
| 12W 115.10 | hr | m2 | - | - | | |
| | £ | | 0.08 | 0.19 | 1.73 | 2.00 |
| 16W 130.10 | hr | m2 | - | - | | |
| | £ | | 0.09 | 0.22 | 1.95 | 2.26 |
| 20W 147.20 | hr | m2 | - | - | | |
| | £ | | 0.10 | 0.24 | 2.21 | 2.55 |

**CLASS P: PILES**
**CLASS Q: PILING ANCILLARIES**

**LARSSEN STEEL SHEET PILING - continued**

| | UNIT | L | P | M | £ NETT |
|---|---|---|---|---|---|
| piles 14-20m long (Piling section reference) (Weight in kg/m2) | | | | | |
| 6W 85.10 | hr m2 | - | - | | |
| | £ | 0.04 | 0.10 | 0.85 | 0.99 |
| 9W 97.10 | hr m2 | - | - | | |
| | £ | 0.04 | 0.14 | 0.97 | 1.15 |
| 12W 115.10 | hr m2 | - | - | | |
| | £ | 0.05 | 0.18 | 1.15 | 1.38 |
| 16W 130.10 | hr m2 | - | - | | |
| | £ | 0.03 | 0.06 | 0.65 | 0.74 |
| 20W 147.20 | hr m2 | - | - | | |
| | £ | 0.03 | 0.06 | 0.74 | 0.83 |
| piles 20-25m long (Piling section reference) (Weight in kg/m2) | | | | | |
| 6W 85.10 | hr m2 | - | - | | |
| | £ | - | - | - | - |
| 9W 97.10 | hr m2 | - | - | | |
| | £ | - | - | - | - |
| 12W 115.10 | hr m2 | - | - | | |
| | £ | - | - | - | - |
| 16W 130.10 | hr m2 | - | - | | |
| | £ | 0.15 | 0.36 | 3.25 | 3.76 |
| 20W 147.20 | hr m2 | - | - | | |
| | £ | 0.17 | 0.40 | 3.68 | 4.25 |

**CLASS P: PILES**
**CLASS Q: PILING ANCILLARIES**

**LARSSEN STEEL SHEET PILING - continued**

| | UNIT | | L | P | M | £ NETT |
|---|---|---|---|---|---|---|
| piles over 25m long (Piling section reference) (Weight in kg/m2) | | | | | | |
| 6W 85.10 | hr | m2 | - | - | | |
| | £ | | - | - | - | - |
| 9W 97.10 | hr | m2 | - | - | | |
| | £ | | - | - | - | - |
| 12W 115.10 | hr | m2 | - | - | | |
| | £ | | - | - | - | - |
| 16W 130.10 | hr | m2 | - | - | | |
| | £ | | - | - | - | - |
| 20W 147.20 | hr | m2 | - | - | | |
| | £ | | 0.24 | 0.56 | 5.15 | 5.95 |
| Handle pitch and drive piles (Piling section reference) (Weight in kg/m2) | | | | | | |
| 6W 85.10 | hr | m2 | - | - | | |
| | £ | | 8.46 | 17.96 | - | 26.42 |
| 9W 97.10 | hr | m2 | - | - | | |
| | £ | | 6.76 | 15.78 | - | 22.54 |
| 12W 115.10 | hr | m2 | - | - | | |
| | £ | | 5.78 | 14.18 | - | 19.96 |
| 16W 130.10 | hr | m2 | - | - | | |
| | £ | | 4.84 | 13.06 | - | 17.90 |
| 20W 147.20 | hr | m2 | - | - | | |
| | £ | | 3.98 | 12.58 | - | 16.56 |

**CLASS P: PILES**
**CLASS Q: PILING ANCILLARIES**

**LARSSEN STEEL SHEET PILING - continued**

| | | UNIT | L | P | M | £ NETT |
|---|---|---|---|---|---|---|
| Burn off tops of piles to level (Piling section reference) (Weight in kg/m2) | | | | | | |
| 6W | 85.10 | hr m | - | - | | |
| | | £ | 1.50 | 4.50 | - | 6.00 |
| 9W | 97.10 | hr m | - | - | | |
| | | £ | 1.63 | 4.87 | - | 6.50 |
| 12W | 115.10 | hr m | - | - | | |
| | | £ | 1.68 | 5.32 | - | 7.00 |
| 16W | 130.10 | hr m | - | - | | |
| | | £ | 1.68 | 5.32 | - | 7.00 |
| 20W | 147.20 | hr m | - | - | | |
| | | £ | 1.80 | 5.70 | - | 7.50 |
| Extract piling and stack on site (Piling section reference) (Weight in kg/m2) | | | | | | |
| 6W | 85.10 | hr m2 | - | - | | |
| | | £ | 5.03 | 11.74 | - | 16.77 |
| 9W | 97.10 | hr m2 | - | - | | |
| | | £ | 4.48 | 10.47 | - | 14.95 |
| 12W | 115.10 | hr m2 | - | - | | |
| | | £ | 3.15 | 9.45 | - | 12.60 |
| 16W | 130.10 | hr m2 | - | - | | |
| | | £ | 2.82 | 8.94 | - | 11.76 |
| 20W | 147.20 | hr m2 | - | - | | |
| | | £ | 2.62 | 8.30 | - | 10.92 |

**CLASS P: PILES**
**CLASS Q: PILING ANCILLARIES**

**LARSSEN STEEL SHEET PILING - continued**

| | | UNIT | | L | P | M | £ NETT |
|---|---|---|---|---|---|---|---|
| Provide new steel sheet piling to BS4360 Grade 43A (Piling section reference) (Weight in kg/m2) | | | | | | | |
| 25W | 167.40 | hr | m2 | - | - | | |
| | | £ | | - | - | 65.62 | 65.62 |
| 32W | 197.40 | hr | m2 | - | - | | |
| | | £ | | - | - | 78.37 | 78.37 |
| 3 | 155.50 | hr | m2 | - | - | | |
| | | £ | | - | - | 60.18 | 60.18 |
| 4A | 185.10 | hr | m2 | - | - | | |
| | | £ | | - | - | 71.63 | 71.63 |
| 6/131kg/m | 311.80 | hr | m2 | - | - | | |
| | | £ | | - | - | 122.23 | 122.23 |
| Add to last item for copper content ne 0.35% (Piling section reference) (Weight in kg/m2) | | | | | | | |
| 25W | 167.40 | hr | m2 | - | - | | |
| | | £ | | - | - | 1.68 | 1.68 |
| 32W | 197.40 | hr | m2 | - | - | | |
| | | £ | | - | - | 1.97 | 1.97 |
| 3 | 155.50 | hr | m2 | - | - | | |
| | | £ | | - | - | 1.56 | 1.56 |
| 4A | 185.10 | hr | m2 | - | - | | |
| | | £ | | - | - | 1.85 | 1.85 |
| 6/131kg/m | 311.80 | hr | m2 | - | - | | |
| | | £ | | - | - | 3.12 | 3.12 |

**CLASS P: PILES**
**CLASS Q: PILING ANCILLARIES**

**LARSSEN STEEL SHEET PILING - continued**

| | | | UNIT | L | P | M | £ NETT |
|---|---|---|---|---|---|---|---|
| copper content 0.35-0.5% (Piling section reference) (Weight in kg/m2) | | | | | | | |
| 25W | 167.40 | hr | m2 | - | - | | |
| | | £ | | - | - | 2.35 | 2.35 |
| 32W | 197.40 | hr | m2 | - | - | | |
| | | £ | | - | - | 2.76 | 2.76 |
| 3 | 155.50 | hr | m2 | - | - | | |
| | | £ | | - | - | 2.18 | 2.18 |
| 4A | 185.10 | hr | m2 | - | - | | |
| | | £ | | - | - | 2.60 | 2.60 |
| 6/131kg/m | 311.80 | hr | m2 | - | - | | |
| | | £ | | - | - | 4.36 | 4.36 |
| one coat black tar varnish (Piling section reference) (Weight in kg/m2) | | | | | | | |
| 25W | 167.40 | hr | m2 | - | - | | |
| | | £ | | 2.80 | - | 4.20 | 7.00 |
| 32W | 197.40 | hr | m2 | - | - | | |
| | | £ | | 2.80 | - | 4.20 | 7.00 |
| 3 | 155.50 | hr | m2 | - | - | | |
| | | £ | | 2.80 | - | 4.20 | 7.00 |
| 4A | 185.10 | hr | m2 | - | - | | |
| | | £ | | 2.80 | - | 4.20 | 7.00 |
| 6/131kg/m | 311.80 | hr | m2 | - | - | | |
| | | £ | | 2.80 | - | 4.20 | 7.00 |

**CLASS P: PILES**
**CLASS Q: PILING ANCILLARIES**

**LARSSEN STEEL SHEET PILING - continued**

| | | | UNIT | L | P | M | £ NETT |
|---|---|---|---|---|---|---|---|
| fabricated piles (Piling section reference) (Weight in kg/m2) | | | | | | | |
| 25W | 167.40 | hr | t | - | - | | |
| | | £ | | 90.00 | 210.00 | 398.76 | 698.76 |
| 32W | 197.40 | hr | t | - | - | | |
| | | £ | | 120.00 | 280.00 | 404.61 | 804.61 |
| 3 | 155.50 | hr | t | - | - | | |
| | | £ | | 180.00 | 120.00 | 394.73 | 694.73 |
| 4A | 185.10 | hr | t | - | - | | |
| | | £ | | 240.00 | 560.00 | 394.58 | 1194.58 |
| 6/131kg/m | 311.80 | hr | t | - | - | | |
| | | £ | | 300.00 | 700.00 | 399.36 | 1399.36 |
| piles under 4m long (Piling section reference) (Weight in kg/m2) | | | | | | | |
| 25W | 167.40 | hr | m2 | - | - | | |
| | | £ | | 0.15 | 0.36 | 3.35 | 3.86 |
| 32W | 197.40 | hr | m2 | - | - | | |
| | | £ | | 0.18 | 0.43 | 3.95 | 4.56 |
| 3 | 155.50 | hr | m2 | - | - | | |
| | | £ | | 0.22 | 0.50 | 4.67 | 5.39 |
| 4A | 185.10 | hr | m2 | - | - | | |
| | | £ | | 0.21 | 0.50 | 4.63 | 5.34 |
| 6/131kg/m | 311.80 | hr | m2 | - | - | | |
| | | £ | | 0.22 | 0.51 | 4.67 | 5.40 |

**CLASS P: PILES**
**CLASS Q: PILING ANCILLARIES**

**LARSSEN STEEL SHEET PILING - continued**

piles 4 - 7m long
(Piling section reference)
(Weight in kg/m2)

| | | | UNIT | L | P | M | £ NETT |
|---|---|---|---|---|---|---|---|
| 25W | 167.40 | hr | m2 | - | - | | |
| | | £ | | 0.08 | 0.17 | 1.68 | 1.93 |
| 32W | 197.40 | hr | m2 | - | - | | |
| | | £ | | 0.09 | 0.22 | 1.97 | 2.28 |
| 3 | 155.50 | hr | m2 | - | - | | |
| | | £ | | 0.11 | 0.26 | 2.33 | 2.70 |
| 4A | 185.10 | hr | m2 | - | - | | |
| | | £ | | 0.13 | 0.29 | 2.78 | 3.20 |
| 6/131kg/m | 311.80 | hr | m2 | - | - | | |
| | | £ | | 0.14 | 0.34 | 3.12 | 3.60 |

piles 14 - 20m long
(Piling section reference)
(Weight in kg/m2)

| | | | UNIT | L | P | M | £ NETT |
|---|---|---|---|---|---|---|---|
| 25W | 167.40 | hr | m2 | - | - | | |
| | | £ | | 0.04 | 0.09 | 0.84 | 0.97 |
| 32W | 197.40 | hr | m2 | - | - | | |
| | | £ | | 0.05 | 0.10 | 0.99 | 1.14 |
| 3 | 155.50 | hr | m2 | - | - | | |
| | | £ | | 0.04 | 0.08 | 0.78 | 0.90 |
| 4A | 185.10 | hr | m2 | - | - | | |
| | | £ | | 0.04 | 0.10 | 0.93 | 1.07 |
| 6/131kg/m | 311.80 | hr | m2 | - | - | | |
| | | £ | | 0.07 | 0.17 | 1.56 | 1.80 |

**CLASS P: PILES**
**CLASS Q: PILING ANCILLARIES**

**LARSSEN STEEL SHEET PILING - continued**

| | | UNIT | | L | P | M | £ NETT |
|---|---|---|---|---|---|---|---|
| piles 20 - 25m long (Piling section reference) (Weight in kg/m2) | | | | | | | |
| 25W | 167.40 | hr | m2 | - | - | | |
| | | £ | | 0.15 | 0.36 | 3.35 | 3.86 |
| 32W | 197.40 | hr | m2 | - | - | | |
| | | £ | | 0.18 | 0.42 | 3.94 | 4.54 |
| 3 | 155.50 | hr | m2 | - | - | | |
| | | £ | | 0.22 | 0.50 | 4.67 | 5.39 |
| 4A | 185.10 | hr | m2 | - | - | | |
| | | £ | | 0.21 | 0.50 | 4.63 | 5.34 |
| 6/131kg/m | 311.80 | hr | m2 | - | - | | |
| | | £ | | 0.22 | 0.51 | 4.67 | 5.40 |
| piles over 25m long (Piling section reference) (Weight in kg/m2) | | | | | | | |
| 25W | 167.40 | hr | m2 | - | - | | |
| | | £ | | 0.23 | 0.54 | 5.03 | 5.80 |
| 32W | 197.40 | hr | m2 | - | - | | |
| | | £ | | 0.23 | 0.54 | 4.93 | 5.70 |
| 3 | 155.50 | hr | m2 | - | - | | |
| | | £ | | - | - | - | - |
| 4A | 185.10 | hr | m2 | - | - | | |
| | | £ | | 0.30 | 0.71 | 6.48 | 7.49 |
| 6/131kg/m | 311.80 | hr | m2 | - | - | | |
| | | £ | | 0.36 | 0.85 | 7.79 | 9.00 |

**CLASS P: PILES**
**CLASS Q: PILING ANCILLARIES**

**LARSSEN STEEL SHEET PILING - continued**

| | | UNIT | | L | P | M | £ NETT |
|---|---|---|---|---|---|---|---|
| Handle pitch and drive piles (Piling section reference) (Weight in kg/m2) | | | | | | | |
| 25W | 167.40 | hr | m2 | - | - | | |
| | | £ | | 3.84 | 10.36 | - | 14.20 |
| 32W | 197.40 | hr | m2 | - | - | | |
| | | £ | | 4.26 | 11.54 | - | 15.80 |
| 3 | 155.50 | hr | m2 | - | - | | |
| | | £ | | 6.30 | 17.02 | - | 23.32 |
| 4A | 185.10 | hr | m2 | - | - | | |
| | | £ | | 6.10 | 16.48 | - | 22.58 |
| 6/131kg/m | 311.80 | hr | m2 | - | - | | |
| | | £ | | 6.96 | 18.82 | - | 25.78 |
| Burn off tops of piles to level (Piling section reference) (Weight in kg/m2) | | | | | | | |
| 25W | 167.40 | hr | m | - | - | | |
| | | £ | | 2.11 | 5.69 | - | 7.80 |
| 32W | 197.40 | hr | m | - | - | | |
| | | £ | | 2.22 | 5.98 | - | 8.20 |
| 3 | 155.50 | hr | m | - | - | | |
| | | £ | | 2.32 | 6.28 | - | 8.60 |
| 4A | 185.10 | hr | m | - | - | | |
| | | £ | | 2.10 | 6.65 | - | 8.75 |
| 6/131kg/m | 311.80 | hr | m | - | - | | |
| | | £ | | 3.06 | 9.19 | - | 12.25 |

**CLASS P: PILES**
**CLASS Q: PILING ANCILLARIES**

**LARSSEN STEEL SHEET PILING - continued**

| | | UNIT | | L | P | M | £ NETT |
|---|---|---|---|---|---|---|---|
| **Extract piling and stack on site (Piling section reference) (Weight in kg/m2)** | | | | | | | |
| 25W | 167.40 | hr | m2 | - | - | | |
| | | £ | | 2.62 | 8.30 | - | 10.92 |
| 32W | 197.40 | hr | m2 | - | - | | |
| | | £ | | 2.62 | 8.30 | - | 10.92 |
| 3 | 155.50 | hr | m2 | - | - | | |
| | | £ | | 4.74 | 11.06 | - | 15.80 |
| 4A | 185.10 | hr | m2 | - | - | | |
| | | £ | | 4.74 | 11.06 | - | 15.80 |
| 6/131kg/m | 311.80 | hr | m2 | - | - | | |
| | | £ | | 5.69 | 13.26 | - | 18.95 |

**CLASS P: PILES**
**CLASS Q: PILING ANCILLARIES**

**STEEL BEARING PILES**

Isolated steel bearing piles.
All 356 x 368 x 109 kg/m
Section Grade 50B Steel

| | | UNIT | L | P | M | £ NETT |
|---|---|---|---|---|---|---|
| Establishment of piling plant for steel bearing piles in main piling based on a max 100 mile radius of base | hr | sum | - | - | | |
| | £ | | - | - | - | 6000.00 |
| Establishment of piling plant for steel bearing piles in main piling based on a max 250 mile radius of base | hr | sum | - | - | | |
| | £ | | - | - | - | 15000.00 |
| Moving piling plant for steel bearing piles in main piling | hr | nr | - | - | | |
| | £ | | - | - | - | 180.00 |
| Steel bearing pile exceeding 15 metres in length but not exceeding 20 metres in main piling | hr | m | - | - | | |
| | £ | | - | - | 48.90 | 48.90 |
| Lengthening piece for steel bearing pile not exceeding 5 metres in length in main piling | hr | m | - | - | | |
| | £ | | - | - | 48.90 | 48.90 |
| Driving vertical steel bearing pile exceeding 15 metres in length but not exceeding 20 metres in main piling | hr | m | - | - | | |
| | £ | | 2.40 | 5.60 | - | 8.00 |
| Driving raking steel bearing pile exceeding 15 metres in length but not exceeding 20 metres in main piling | hr | m | - | - | | |
| | £ | | 2.80 | 7.20 | - | 10.00 |
| Driving lengthened vertical steel bearing pile not exceeding 5 metres in length in main piling | hr | m | - | - | | |
| | £ | | 7.50 | 10.50 | - | 18.00 |
| Driving lengthened raking steel bearing pile not exceeding 5 metres in length in main piling | hr | m | - | - | | |
| | £ | | 5.10 | 15.40 | - | 20.50 |

CLASS P: PILES
CLASS Q: PILING ANCILLARIES

STEEL BEARING PILES - continued

| | | UNIT | L | P | M | £ NETT |
|---|---|---|---|---|---|---|
| Welding on lengthening piece to vertical steel bearing pile in main piling | hr | nr | - | - | | |
| | £ | | 86.30 | 50.70 | 3.00 | 140.00 |
| Welding on lengthening piece to raking steel bearing pile in main piling | hr | nr | - | - | | |
| | £ | | 86.30 | 50.70 | 3.00 | 140.00 |
| Cutting or burning off surplus length of vertical steel bearing pipe in main piling | hr | nr | - | - | | |
| | £ | | 4.40 | 3.00 | 0.60 | 8.00 |
| Cutting or burning off surplus length of raking steel bearing pipe in main piling | hr | nr | - | - | | |
| | £ | | 4.40 | 3.00 | 0.60 | 8.00 |

PILING TESTING

| | | UNIT | L | P | M | £ NETT |
|---|---|---|---|---|---|---|
| Establishment of pile testing equipment for steel bearing pile in trial piling as a separate operation in advance of the main piling. | hr | sum | - | - | | |
| | £ | | - | - | - | 16500.00 |
| Vertical load testing of steel bearing pile in trial piling as a separate operation in advance of the main piling to 108 tonnes | hr | nr | - | - | | |
| | £ | | - | - | - | 6200.00 |

**CLASS P: PILES**
**CLASS Q: PILING ANCILLARIES**

**STEEL TUBULAR PILES**

Isolated steel tubular piles. All Grade 43A steel, 12.5mm thickness

| | | UNIT | L | P | M | £ NETT |
|---|---|---|---|---|---|---|
| Establishment of piling plant for steel tubular piles in main piling based on a maximum 100 mile radius of base | hr | sum | - | - | | |
| | £ | | - | - | - | 6000.00 |
| Establishment of piling plant for steel tubular piles in main piling based on a maximum 250 mile radius of base | hr | sum | - | - | | |
| | £ | | - | - | - | 15000.00 |
| Moving piling plant for steel tubular piles in main piling | hr | nr | - | - | | |
| | £ | | - | - | - | 200.00 |
| Steel tubular pile exceeding 15 metres in length but not exceeding 20 metres in length in main piling 508mm diameter | hr | m | - | - | | |
| | £ | | - | - | 61.20 | 61.20 |
| Steel tubular pile exceeding 15 metres in length but not exceeding 20 metres in length in main piling 762mm diameter | hr | m | - | - | | |
| | £ | | - | - | 107.65 | 107.65 |
| Lengthening piece of steel tubular pile not exceeding 5 metres in length in main piling 508mm diameter | hr | m | - | - | | |
| | £ | | - | - | 61.20 | 61.20 |
| Lengthening piece of steel tubular pile not exceeding 5 metres in length in main piling 762mm diameter | hr | m | - | - | | |
| | £ | | - | - | 107.65 | 107.65 |

**CLASS P: PILES**
**CLASS Q: PILING ANCILLARIES**

**STEEL TUBULAR PILES - continued**

| | | UNIT | L | P | M | £ NETT |
|---|---|---|---|---|---|---|
| Driving vertical steel tubular pile exceeding 15 metres in length but not exceeding 20 metres in main piling 508mm diameter | hr | m | - | - | | |
| | £ | | 1.95 | 4.55 | - | 6.50 |
| Driving vertical steel tubular pile exceeding 15 metres in length but not exceeding 20 metres in main piling 762mm diameter | hr | m | - | - | | |
| | £ | | 2.04 | 4.76 | - | 6.80 |
| Driving lengthened vertical steel tubular pile not exceeding 5 metres in length in main piling 508mm diameter | hr | m | - | - | | |
| | £ | | 3.58 | 5.02 | - | 8.60 |
| Driving lengthened vertical steel tubular pile not exceeding 5 metres in length in main piling 762mm diameter | hr | m | - | - | | |
| | £ | | 3.71 | 5.19 | - | 8.90 |
| Welding on lengthening piece to vertical steel tubular pile in main piling 508mm diameter | hr | nr | - | - | | |
| | £ | | 86.30 | 50.70 | 3.00 | 140.00 |
| Welding on lengthening piece to vertical steel tubular pile in main piling 762mm diameter | hr | nr | - | - | | |
| | £ | | 98.00 | 77.00 | 5.00 | 180.00 |
| Cutting or burning off surplus length of vertical steel tubular pipe in main piling 508mm diameter | hr | nr | - | - | | |
| | £ | | 4.84 | 3.30 | 0.66 | 8.80 |
| Cutting or burning off surplus length of vertical steel tubular pipe in main piling 762mm diameter | hr | nr | - | - | | |
| | £ | | 4.98 | 3.39 | 0.68 | 9.05 |

**CLASS P: PILES**
**CLASS Q: PILING ANCILLARIES**

**STEEL TUBULAR PILES - continued**

| | | UNIT | L | P | M | £ NETT |
|---|---|---|---|---|---|---|
| **PILING TESTING** | | | | | | |
| Establishment of pile testing equipment for steel tubular pile in trial piling as a separate operation in advance of the main piling 508mm diameter | hr | sum | - | - | | |
| | £ | | - | - | - | 16500.00 |
| Establishment of pile testing equipment for steel tubular pile in trial piling as a separate operation in advance of the main piling 762mm diameter | hr | sum | - | - | | |
| | £ | | - | - | - | 16500.00 |
| Vertical load testing of steel tubular pile in trial piling as a separate operation in advance of the main piling to 108 tonnes 508mm diameter | hr | nr | - | - | | |
| | £ | | - | - | - | 6000.00 |
| Vertical load testing of steel tubular pile in trial piling as a separate operation in advance of the main piling to 108 tonnes 762mm diameter | hr | nr | - | - | | |
| | £ | | - | - | - | 6200.00 |

## CLASS R: ROADS AND PAVINGS

### SUB BASES

The following unit costs are generally based on the Department of Transport (DTp) Specification for Highway Works and reference is made throughout this section to clauses within that specification.

LABOUR based on Gang A

| | | | |
|---|---|---|---|
| Total hourly cost | = | £ | 33.71 |

PLANT

| | | | |
|---|---|---|---|
| Grader 130G | | | 22.05 |
| Tractor loader 931 | | | 11.48 |
| Towed roller BW10 | | | 5.79 |
| Total hourly cost | = | £ | 39.32 |

Included in this section are:-

- granular sub bases
- lean concrete sub bases
- hardcore
- wet mix macadam

For information on coverage of sub bases and voids in materials, refer to page 793, material prices are on pages 86 - 87.

CLASS R: ROADS AND PAVINGS

SUB BASES - continued

| | | UNIT | L | P | M | £ NETT |
|---|---|---|---|---|---|---|
| Granular material DTp specified type 1 (PC = £6.70/tonne) | | | | | | |
| 75mm depth | hr | m3 | 0.035 | 0.035 | | |
| | £ | | 1.18 | 1.38 | 12.25 | 14.81 |
| 100mm depth | hr | m3 | 0.04 | 0.04 | | |
| | £ | | 1.35 | 1.57 | 12.25 | 15.17 |
| 150mm depth | hr | m3 | 0.045 | 0.045 | | |
| | £ | | 1.52 | 1.77 | 12.25 | 15.54 |
| 200mm depth | hr | m3 | 0.05 | 0.05 | | |
| | £ | | 1.69 | 1.97 | 12.25 | 15.91 |
| Granular material DTp specified type 2 (PC = £6.25/tonne) | | | | | | |
| 75mm depth | hr | m3 | 0.035 | 0.035 | | |
| | £ | | 1.18 | 1.38 | 11.42 | 13.98 |
| 100mm depth | hr | m3 | 0.04 | 0.04 | | |
| | £ | | 1.35 | 1.57 | 11.42 | 14.34 |
| 150mm depth | hr | m3 | 0.045 | 0.045 | | |
| | £ | | 1.52 | 1.77 | 11.42 | 14.71 |
| 200mm depth | hr | m3 | 0.05 | 0.05 | | |
| | £ | | 1.69 | 1.97 | 11.42 | 15.08 |
| Lean concrete DTp specified strength mix (1:3:6 - 20mm aggregate) | | | | | | |
| 100mm depth | hr | m2 | 0.016 | 0.016 | | |
| | £ | | 0.54 | 0.63 | 4.22 | 5.39 |
| 200mm depth | hr | m2 | 0.017 | 0.017 | | |
| | £ | | 0.57 | 0.67 | 8.45 | 9.69 |

**CLASS R: ROADS AND PAVINGS**

**SUB BASES - continued**

| | | UNIT | L | P | M | £ NETT |
|---|---|---|---|---|---|---|
| Hardcore | | | | | | |
| 100mm depth | hr | m3 | 0.04 | 0.04 | | |
| | £ | | 1.35 | 1.57 | 2.60 | 5.52 |
| 150mm depth | hr | m3 | 0.045 | 0.045 | | |
| | £ | | 1.52 | 1.77 | 2.60 | 5.89 |
| 200mm depth | hr | m3 | 0.05 | 0.05 | | |
| | £ | | 1.69 | 1.97 | 2.60 | 6.26 |
| Wet mix macadam; DTp clause 808 | | | | | | |
| 75mm depth | hr | m2 | 0.053 | 0.053 | | |
| | £ | | 1.79 | 2.08 | 1.10 | 4.97 |
| 100mm depth | hr | m2 | 0.055 | 0.055 | | |
| | £ | | 1.85 | 2.16 | 1.47 | 5.48 |
| 200mm depth | hr | m2 | 0.058 | 0.058 | | |
| | £ | | 1.96 | 2.28 | 2.93 | 7.17 |

**CLASS R: ROADS AND PAVINGS**

**FLEXIBLE ROAD BASES AND SURFACING**

| | | | |
|---|---|---|---|
| LABOUR based on Gang G | | | |
| Total hourly cost | = | £ | 61.57 |
| PLANT | | | |
| 2 Asphalt pavers BK171 | | | 58.96 |
| 2 107 Rollers (3 point) | | | 11.58 |
| Tractair + Tools | | | 8.85 |
| (Part time)<br>Tar Sprayer | | | 1.15 |
| S.P. Chip spreader | | | 4.21 |
| Heat Iron | | | 0.75 |
| Total hourly cost | = | £ | 85.50 |

Notes: Material prices include delivery via insulated road tippers and include an allowance of 1% for waste and 1% for sundry materials and extra consumables (diesel/propane/fuels etc.) All outputs are based on clear runs without undue delay to two pavers with a 75% utilisation.

The outputs can be adjusted as follows to take account of space or time influences on the utilisation.

| | | | |
|---|---|---|---|
| 1 paver @ 75% utilisation | = | x | 2.00 |
| 1 paver @ 100% utilisation | = | x | 1.50 |
| 2 paver @ 100% utilisation | = | x | 0.75 |

**CLASS R: ROADS AND PAVINGS**

**FLEXIBLE ROAD BASES AND SURFACING - continued**
**(All references to 1986 DTp Spec)**

| | | UNIT | L | P | M | £ NETT |
|---|---|---|---|---|---|---|
| Dense Bitumen Macadam | | | | | | |
| Road Base to DTp clause 903 | | | | | | |
| 100mm depth | hr | m2 | 0.007 | 0.007 | | |
| | £ | | 0.43 | 0.60 | 3.85 | 4.88 |
| 150mm depth | hr | m2 | 0.009 | 0.009 | | |
| | £ | | 0.55 | 0.77 | 5.70 | 7.02 |
| 200mm depth | hr | m2 | 0.011 | 0.011 | | |
| | £ | | 0.68 | 0.94 | 7.65 | 9.27 |
| Base Course to DTp clause 906 | | | | | | |
| 50mm depth | hr | m2 | 0.010 | 0.010 | | |
| | £ | | 0.62 | 0.86 | 1.70 | 3.18 |
| 100mm depth | hr | m2 | 0.013 | 0.013 | | |
| | £ | | 0.80 | 1.11 | 2.50 | 4.41 |
| Wearing Course to DTp clause 912 | | | | | | |
| 30mm depth | hr | m2 | 0.004 | 0.004 | | |
| | £ | | 0.25 | 0.34 | 1.50 | 2.09 |
| 50mm depth | hr | m2 | 0.007 | 0.007 | | |
| | £ | | 0.43 | 0.60 | 2.50 | 3.53 |
| Bitumen Macadam | | | | | | |
| Base Course to DTp clause 908 | | | | | | |
| 35mm depth | hr | m2 | 0.004 | 0.004 | | |
| | £ | | 0.25 | 0.34 | 1.50 | 2.09 |
| 70mm depth | hr | m2 | 0.006 | 0.006 | | |
| | £ | | 0.37 | 0.51 | 3.00 | 3.88 |
| Dense Tarmacadam | | | | | | |
| Road Base to DTp clause 902 | | | | | | |
| 50mm depth | hr | m2 | 0.004 | 0.004 | | |
| | £ | | 0.25 | 0.34 | 2.07 | 2.66 |
| 100mm depth | hr | m2 | 0.007 | 0.007 | | |
| | £ | | 0.43 | 0.60 | 4.10 | 5.13 |
| Base Course to DTp clause 907 | | | | | | |
| 60mm depth | hr | m2 | 0.004 | 0.004 | | |
| | £ | | 0.25 | 0.34 | 2.65 | 3.24 |
| 80mm depth | hr | m2 | 0.005 | 0.005 | | |
| | £ | | 0.31 | 0.43 | 3.50 | 4.24 |

**CLASS R: ROADS AND PAVINGS**

**FLEXIBLE ROAD BASES AND SURFACING - continued**

| | | UNIT | L | P | M | £ NETT |
|---|---|---|---|---|---|---|
| Dense Tar Surfacing | | | | | | |
| Wearing Course to DTp clause 913 | | | | | | |
| 30mm depth | hr | m2 | 0.004 | 0.004 | | |
| | £ | | **0.25** | **0.34** | **1.50** | **2.09** |
| 50mm depth | hr | m2 | 0.005 | 0.005 | | |
| | £ | | **0.31** | **0.43** | **2.50** | **3.24** |
| Cold Asphalt | | | | | | |
| Wearing course to DTp clause 914 | | | | | | |
| 15mm depth | hr | m2 | 0.004 | 0.004 | | |
| | £ | | **0.25** | **0.34** | **0.77** | **1.36** |
| 30mm depth | hr | m2 | 0.005 | 0.005 | | |
| | £ | | **0.31** | **0.43** | **1.52** | **2.26** |
| Rolled Asphalt | | | | | | |
| Base Course to DTp clause 905 | | | | | | |
| 60mm depth | hr | m2 | 0.005 | 0.005 | | |
| | £ | | **0.31** | **0.43** | **2.60** | **3.34** |
| 80mm depth | hr | m2 | 0.006 | 0.006 | | |
| | £ | | **0.37** | **0.51** | **3.50** | **4.38** |
| Wearing Course to DTp clause 911 | | | | | | |
| 40mm depth | hr | m2 | 0.006 | 0.006 | | |
| | £ | | **0.37** | **0.51** | **2.40** | **3.28** |
| 60mm depth | hr | m2 | 0.007 | 0.007 | | |
| | £ | | **0.43** | **0.60** | **3.60** | **4.63** |
| Slurry sealing | | | | | | |
| BS434 class K3 clause 918 | | | | | | |
| 3mm depth | hr | m2 | 0.003 | 0.003 | | |
| | £ | | **0.18** | **0.26** | **0.75** | **1.19** |
| 3mm depth | hr | m2 | 0.004 | 0.004 | | |
| | £ | | **0.25** | **0.34** | **1.00** | **1.59** |

**CLASS R: ROADS AND PAVINGS**

**FLEXIBLE ROAD BASES AND SURFACING - continued**

| | | UNIT | L | P | M | £ NETT |
|---|---|---|---|---|---|---|
| Surface dressing coated chippings, 9 - 11 kg/m2 - clause 915 | | | | | | |
| 6mm nominal size | hr | m2 | 0.002 | 0.002 | | |
| | £ | | **0.12** | **0.17** | **0.14** | **0.43** |
| 8mm nominal size | hr | m2 | 0.002 | 0.002 | | |
| | £ | | **0.12** | **0.17** | **0.15** | **0.44** |
| 10mm nominal size | hr | m2 | 0.002 | 0.002 | | |
| | £ | | **0.12** | **0.17** | **0.16** | **0.45** |
| 12mm nominal size | hr | m2 | 0.003 | 0.003 | | |
| | £ | | **0.18** | **0.26** | **0.18** | **0.62** |
| Bituminous spray tack coat: BS434 K1 - 40 clause 920 | | | | | | |
| large area over 20m2 | hr | m2 | - | - | | |
| | £ | | - | - | **0.08** | **0.08** |
| small area below 20m2 | hr | m2 | - | - | | |
| | £ | | - | - | **0.08** | **0.08** |
| Cutting and sawing existing slabs | ( Labour = | | 6.54 | Plant = | 2.75 ) | |
| Trimming edges only of existing slabs, floors or similar surfaces (wet or dry) Cutting width 6mm | | | | | | |
| Cutting depth 50mm | hr | m | 0.02 | 0.02 | | |
| | £ | | **0.13** | **0.06** | **0.60** | **0.79** |
| depth 100mm | hr | m | 0.03 | 0.03 | | |
| | £ | | **0.20** | **0.08** | **0.80** | **1.08** |
| Cutting existing slabs, floors or similar surfaces (wet or dry) Cutting width 8mm | | | | | | |
| Cutting depth 50mm | hr | m | 0.025 | 0.025 | | |
| | £ | | **0.16** | **0.07** | **0.60** | **0.83** |
| depth 100mm | hr | m | 0.06 | 0.06 | | |
| | £ | | **0.39** | **0.17** | **0.80** | **1.36** |
| depth 150mm | hr | m | 0.08 | 0.08 | | |
| | £ | | **0.52** | **0.22** | **1.50** | **2.24** |

N.B. The materials column for cutting and sawing shows an allowance for the cost of using diamond tipped cutting discs, at £80 per mm used, using 1mm per 100m per 100mm cut or carborundum discs at £15 each using one per 20m cut. (i.e. similar costs)

**CLASS R: ROADS AND PAVINGS**

**FLEXIBLE ROAD BASES AND SURFACING - continued**

**LABOUR based on Gang G**

| | | |
|---|---|---|
| Total hourly cost | = £ | 61.57 |

**PLANT**

| | | | |
|---|---|---|---|
| Cold planer | 28.03 | Heat planer | 44.00 |
| Wheeled loader (2.30m3) | 13.95 | Wheeled loader | 13.95 |
| Total hourly cost = £ | 41.98 | = £ | 57.95 |

| | | UNIT | L | P | M | £ NETT |
|---|---|---|---|---|---|---|
| Removal of flexible surfaces (assumes disposal on site or re-use as fill but excludes transport if required) | | | | | | |
| depth 75mm | hr | m2 | 0.027 | 0.027 | | |
| | £ | | **1.66** | **1.13** | - | **2.79** |
| depth 100mm | hr | m2 | 0.036 | 0.036 | | |
| | £ | | **2.22** | **1.51** | - | **3.73** |
| depth 50mm; scarifying surface | hr | m2 | 0.022 | 0.022 | | |
| | £ | | **1.35** | **0.92** | - | **2.27** |
| depth 75mm; scarifying surface | hr | m2 | 0.037 | 0.037 | | |
| | £ | | **2.28** | **1.55** | - | **3.83** |
| depth 25mm; heat planing for re-use | hr | m2 | 0.032 | 0.032 | | |
| | £ | | **1.97** | **1.85** | - | **3.82** |
| depth 50mm; heat planing for re-use | hr | m2 | 0.056 | 0.056 | | |
| | £ | | **3.45** | **3.25** | - | **6.70** |
| raking over scarified or heat planed material up to 50mm depth and compacting with 10t roller | hr | m2 | 0.005 | 0.005 | | |
| | £ | | **0.31** | **0.29** | - | **0.60** |

## CLASS R: ROADS AND PAVINGS

### CONCRETE SLABS

Note: The following unit costs are for jointed reinforced concrete slabs, laid in reasonable areas (over 200m2) for slabs laid by paver train/ slipformer see section 6 MMHW page 531.

LABOUR based on Gang G

Total hourly cost = £ 61.57

PLANT

Loading shovel
Screeding equipment
Compaction and ancillary equipment

Total hourly cost = £ 47.50

| | | UNIT | L | P | M | £ NETT |
|---|---|---|---|---|---|---|
| Slab, Runway, access roads or similar in P.Q. concrete grade C30, 20mm aggregate | | | | | | |
| depth 180mm | hr | m2 | 0.015 | 0.015 | | |
| | £ | | 0.92 | 0.71 | 8.91 | 10.54 |
| depth 220mm | hr | m2 | 0.018 | 0.018 | | |
| | £ | | 1.11 | 0.86 | 10.89 | 12.86 |
| depth 260mm | hr | m2 | 0.022 | 0.022 | | |
| | £ | | 1.35 | 1.05 | 12.87 | 15.27 |
| depth 300mm | hr | m2 | 0.025 | 0.025 | | |
| | £ | | 1.54 | 1.19 | 14.85 | 17.58 |
| Steel fabric reinforcement to BS4483 | | | | | | |
| nominal mass 2.22kg Ref A142 | hr | m2 | 0.03 | - | | |
| | £ | | 1.85 | - | 1.32 | 3.17 |
| nominal mass 3.95kg Ref A252 | hr | m2 | 0.04 | - | | |
| | £ | | 2.46 | - | 2.36 | 4.82 |
| nominal mass 4.53kg Ref B385 | hr | m2 | 0.04 | - | | |
| | £ | | 2.46 | - | 2.81 | 5.27 |
| nominal mass 5.55kg Ref C636 | hr | m2 | 0.05 | - | | |
| | £ | | 3.08 | - | 3.38 | 6.46 |
| nominal mass 5.93kg Ref B503 | hr | m2 | 0.05 | - | | |
| | £ | | 3.08 | - | 3.55 | 6.63 |

**CLASS R: ROADS AND PAVINGS**

**CONCRETE SLABS - continued**

| | | UNIT | L | P | M | £ NETT |
|---|---|---|---|---|---|---|
| Mild Steel bar reinforcement to BS4449 cut and bent (nominal size) | | | | | | |
| 6mm | hr | t | 8.00 | - | | |
| | £ | | 492.56 | - | 552.21 | 1044.77 |
| 8mm | hr | t | 6.74 | - | | |
| | £ | | 414.98 | - | 507.78 | 922.76 |
| 10mm | hr | t | 6.74 | - | | |
| | £ | | 414.98 | - | 476.47 | 891.45 |
| 12mm | hr | t | 6.74 | - | | |
| | £ | | 414.98 | - | 451.22 | 866.20 |
| 16mm | hr | t | 6.15 | - | | |
| | £ | | 378.66 | - | 419.91 | 798.57 |
| High yield steel bar reinforcement to BS4449 cut and bent (nominal size) | | | | | | |
| 6mm | hr | t | 8.00 | - | | |
| | £ | | 492.56 | - | 556.26 | 1048.82 |
| 8mm | hr | t | 6.74 | - | | |
| | £ | | 414.98 | - | 511.82 | 926.80 |
| 10mm | hr | t | 6.74 | - | | |
| | £ | | 414.98 | - | 448.51 | 863.49 |
| 12mm | hr | t | 6.74 | - | | |
| | £ | | 414.98 | - | 455.26 | 870.24 |
| 16mm | hr | t | 6.15 | - | | |
| | £ | | 378.66 | - | 423.95 | 802.61 |
| Waterproof membranes below concrete pavements (nominal size) plastic sheeting; 1200 grade | | | | | | |
| 1000 gauge | hr | m2 | 0.01 | - | | |
| | £ | | 0.62 | - | 0.47 | 1.09 |
| 2000 gauge | hr | m2 | 0.01 | | | |
| | £ | | 0.62 | - | 0.83 | 1.45 |

**CLASS R: ROADS AND PAVINGS**

**JOINTS IN CONCRETE SLABS**

| | | UNIT | L | P | M | £ NETT |
|---|---|---|---|---|---|---|
| Longitudinal joints (joint depth) | | | | | | |
| 180mm | hr | m | 0.156 | 0.012 | | |
| | £ | | 9.60 | 0.57 | 12.82 | 22.99 |
| 220mm | hr | m | 0.156 | 0.012 | | |
| | £ | | 9.60 | 0.57 | 13.67 | 23.84 |
| 260mm | hr | m | 0.168 | 0.012 | | |
| | £ | | 10.34 | 0.57 | 16.25 | 27.16 |
| 300mm | hr | m | 0.168 | 0.012 | | |
| | £ | | 10.34 | 0.57 | 17.24 | 28.15 |
| Expansion joints (joint depth) | | | | | | |
| 180mm | hr | m | 0.216 | 0.012 | | |
| | £ | | 13.30 | 0.57 | 19.81 | 33.68 |
| 220mm | hr | m | 0.228 | 0.012 | | |
| | £ | | 14.04 | 0.57 | 23.24 | 37.85 |
| 260mm | hr | m | 0.24 | 0.012 | | |
| | £ | | 14.78 | 0.57 | 26.84 | 42.19 |
| 300mm | hr | m | 0.252 | 0.012 | | |
| | £ | | 15.52 | 0.57 | 27.54 | 43.63 |
| Contraction joints (joint depth) | | | | | | |
| 180mm | hr | m | 0.144 | 0.012 | | |
| | £ | | 8.87 | 0.57 | 10.61 | 20.05 |
| 220mm | hr | m | 0.156 | 0.012 | | |
| | £ | | 9.60 | 0.57 | 11.47 | 21.64 |
| 260mm | hr | m | 0.168 | 0.012 | | |
| | £ | | 10.34 | 0.57 | 15.04 | 25.95 |
| 300mm | hr | m | 0.168 | 0.012 | | |
| | £ | | 10.34 | 0.57 | 15.04 | 25.95 |

**CLASS R: ROADS AND PAVINGS**

**JOINTS IN CONCRETE SLABS - continued**

| | | UNIT | L | P | M | £ NETT |
|---|---|---|---|---|---|---|
| Construction joints (joint depth) | | | | | | |
| 180mm | hr | m | 0.132 | 0.012 | | |
| | £ | | 8.13 | 0.57 | 6.65 | 15.35 |
| 220mm | hr | m | 0.132 | 0.012 | | |
| | £ | | 8.13 | 0.57 | 7.32 | 16.02 |
| 260mm | hr | m | 0.144 | 0.012 | | |
| | £ | | 8.87 | 0.57 | 8.06 | 17.50 |
| 300mm | hr | m | 0.144 | 0.012 | | |
| | £ | | 8.87 | 0.57 | 8.86 | 18.30 |
| Open joints with filler ne 0.5m; 10mm flexcell joint filler | hr | m2 | 0.04 | - | | |
| | £ | | 1.66 | - | 3.89 | 5.55 |
| 0.5 - 1m; 10mm flexcell joint filler | hr | m2 | 0.04 | - | | |
| | £ | | 1.66 | - | 3.98 | 5.64 |
| Hot bitumen sealant 10 x 20mm | hr | m | 0.05 | - | | |
| | £ | | 2.07 | - | 0.44 | 2.51 |
| Cold polysulphide sealant 20 x 20mm | hr | m | 0.065 | - | | |
| | £ | | 1.27 | - | 3.20 | 4.47 |

CLASS R: ROADS AND PAVINGS

KERBS, CHANNELS AND EDGINGS

| | | | |
|---|---|---|---|
| LABOUR based on Gang E | | | |
| Total hourly cost | = | £ | 27.81 |
| PLANT | | | |
| (Part time) JCB 3CX | | | 2.58 |
| Power tools, Roadforms | | | 2.09 |
| Total hourly cost | = | £ | 4.67 |

| | | UNIT | L | P | M | £ NETT |
|---|---|---|---|---|---|---|
| Precast concrete kerbs to BS340 (Bullnosed splayed or half battered) Straight or curved to radius exceeding 12m | | | | | | |
| 150 x 305mm | hr | m | 0.07 | 0.07 | | |
| | £ | | **1.95** | **0.33** | **5.98** | **8.26** |
| 125 x 255mm | hr | m | 0.07 | 0.07 | | |
| | £ | | **1.95** | **0.33** | **4.19** | **6.47** |
| 125 x 150mm | hr | m | 0.06 | 0.06 | | |
| | £ | | **1.67** | **0.28** | **2.99** | **4.94** |
| Curved to radius not exceeding 12m | | | | | | |
| 150 x 305mm | hr | m | 0.075 | 0.075 | | |
| | £ | | **2.09** | **0.35** | **6.90** | **9.34** |
| 125 x 255mm | hr | m | 0.075 | 0.075 | | |
| | £ | | **2.09** | **0.35** | **4.79** | **7.23** |
| 125 x 150mm | hr | m | 0.065 | 0.065 | | |
| | £ | | **1.81** | **0.30** | **3.94** | **6.05** |
| Precast concrete kerbs to BS340 | | | | | | |
| quadrants; 455 x 455 x 255mm | hr | nr | 0.10 | 0.10 | | |
| | £ | | **2.78** | **0.47** | **5.10** | **8.35** |
| quadrants; 305 x 305 x 150mm | hr | nr | 0.08 | 0.08 | | |
| | £ | | **2.22** | **0.37** | **4.33** | **6.92** |
| drop kerbs; 125 x 255mm | hr | m | 0.07 | 0.07 | | |
| | £ | | **1.95** | **0.33** | **4.20** | **6.48** |

**CLASS R: ROADS AND PAVINGS**

**KERBS, CHANNELS AND EDGINGS - continued**

| | | UNIT | L | P | M | £ NETT |
|---|---|---|---|---|---|---|
| drop kerbs; 150 x 305mm | hr | m | 0.07 | 0.07 | | |
| | £ | | 1.95 | 0.33 | 6.90 | 9.18 |
| Precast concrete channels | | | | | | |
| Straight or curved to radius exceeding 12m; 255 x 125mm | hr | m | 0.07 | 0.07 | | |
| | £ | | 1.95 | 0.33 | 4.79 | 7.07 |
| Curved to radius not exceeding 12m; 255 x 125mm | hr | m | 0.07 | 0.07 | | |
| | £ | | 1.95 | 0.33 | 5.69 | 7.97 |
| | | | | SPECIALIST ADVICE | | |
| In-situ kerbing | | | | | | |
| In-situ Asphalt kerbing | | | | | | |
| Straight or curved to radius exceeding 12m (BS5931) | | | | | | |
| Kerb height 75mm | hr | m | - | - | | |
| | £ | | - | - | - | 3.12 |
| Kerb height 100mm | hr | m | - | - | | |
| | £ | | - | - | - | 4.40 |
| Kerb height 125mm | hr | m | - | - | | |
| | £ | | - | - | - | 5.75 |
| Curved to radius not exceeding 12m (BS5931) | | | | | | |
| Kerb height 75mm | hr | m | - | - | | |
| | £ | | - | - | - | 3.30 |
| Kerb height 100mm | hr | m | - | - | | |
| | £ | | - | - | - | 4.55 |
| Kerb height 125mm | hr | m | - | - | | |
| | £ | | - | - | - | 5.92 |
| In-situ Asphalt channelling | | | | | | |
| Straight or curved to radius exceeding 12m (40mm thick) | | | | | | |
| 300mm wide | hr | m | - | - | | |
| | £ | | - | - | - | 8.55 |
| 250mm wide | hr | m | - | - | | |
| | £ | | - | - | - | 6.50 |
| Curved to radius not exceeding 12m | | | | | | |
| 300mm wide | hr | m | - | - | | |
| | £ | | - | - | - | 8.72 |
| 250mm wide | hr | m | - | - | | |
| | £ | | - | - | - | 6.66 |

**CLASS R: ROADS AND PAVINGS**

**KERBS, CHANNELS AND EDGINGS - continued**

| | | | SPECIALIST ADVICE | | | |
|---|---|---|---|---|---|---|
| In-situ concrete kerbing | | | | | | |
| Straight or curved to radius exceeding 12m (BS5931) | | | | | | |
| Kerb height 100mm | hr | m | - | - | | |
| | £ | | - | - | - | 5.20 |
| Kerb height 125mm | hr | m | - | - | | |
| | £ | | - | - | - | 6.12 |
| Curved to radius not exceeding 12m (BS5931) | | | | | | |
| Kerb height 100mm | hr | m | - | - | | |
| | £ | | - | - | - | 5.20 |
| Kerb height 125mm | hr | m | - | - | | |
| | £ | | - | - | - | 5.30 |

| | | UNIT | L | P | M | £ NETT |
|---|---|---|---|---|---|---|
| Precast concrete edgings 150 x 50mm | | | | | | |
| Straight or curved to radius exceeding 12m | hr | m | 0.04 | 0.04 | | |
| | £ | | 1.11 | 0.19 | 1.06 | 2.36 |
| Curved to radius not exceeding 12m | hr | m | 0.045 | 0.045 | | |
| | £ | | 1.25 | 0.21 | 1.46 | 2.92 |
| Precast concrete 'Safticurb' | | | | | | |
| 250 x 254 medium duty | hr | m | 0.075 | 0.075 | | |
| | £ | | 2.09 | 0.35 | 11.98 | 14.42 |
| 305 x 305 heavy duty | hr | m | 0.095 | 0.095 | | |
| | £ | | 2.64 | 0.44 | 19.16 | 22.24 |
| Precast concrete 'Ellis Trief' safety kerb laid | | | | | | |
| Straight or curved to radius exceeding 12m (415 x 380mm) | hr | m | 0.225 | 0.225 | | |
| | £ | | 6.26 | 1.05 | 33.40 | 40.71 |
| Curved to radius not exceeding 12m | hr | m | 0.25 | 0.25 | | |
| | £ | | 6.95 | 1.17 | 33.40 | 41.52 |

**CLASS R: ROADS AND PAVINGS**

**KERBS, CHANNELS AND EDGINGS - continued**

| | | UNIT | L | P | M | £ NETT |
|---|---|---|---|---|---|---|
| Precast concrete combined kerb and drainage block (Beany Block System) | hr | m | 0.06 | 0.06 | | |
| | £ | | 1.67 | 0.28 | 38.30 | 40.25 |
| Precast concrete or pressed clay. Block kerbing straight or curved to radius exceeding 12m (natural shades) | hr | m | 0.05 | 0.05 | | |
| | £ | | 1.39 | 0.23 | 17.75 | 19.37 |
| Extra over for additional concrete for kerbing | hr | m3 | 0.02 | 0.02 | | |
| | £ | | 0.56 | 0.09 | 38.50 | 39.15 |

**FOOTWAYS**

| | | UNIT | L | P | M | £ NETT |
|---|---|---|---|---|---|---|
| Flexible paved footway excluding sub base Bitumen macadam surfacing 40 base course | hr | m2 | 0.04 | 0.04 | | |
| | £ | | 1.11 | 0.19 | 2.47 | 3.77 |
| 20 wearing course | hr | m2 | 0.045 | 0.045 | | |
| | £ | | 1.25 | 0.21 | 2.73 | 4.19 |
| Extra over flexible footways for colour tinting additives (including alterations to designs) Red | hr | m2 | - | - | | |
| | £ | | - | - | 1.10 | 1.10 |
| Green | hr | m2 | - | - | | |
| | £ | | - | - | 1.85 | 1.85 |

Central Reserve Treatments

| | | UNIT | L | P | M | £ NETT |
|---|---|---|---|---|---|---|
| Provide and compact 10mm down limestone 100mm depth, spray twice with cold bituminous emulsion and blind with 6mm Quartzite fine gravels | hr | m2 | - | - | | |
| | £ | | 0.55 | 0.20 | 2.12 | 2.87 |

CLASS R: ROADS AND PAVINGS

KERBS, CHANNELS AND EDGINGS - continued

| | | UNIT | L | P | M | £ NETT |
|---|---|---|---|---|---|---|
| 50mm thick 'Gold Top' gravels in central reserves (onto sub bases) (Johnstones hot bitumen spray and hot chip process) | hr | m2 | - | - | | |
| | £ | | 0.70 | 0.30 | 3.30 | 4.30 |
| Surface dressing, double layer to clause 919 | hr | m2 | - | - | | |
| | £ | | 0.20 | 0.20 | 1.25 | 1.65 |
| 600 x 900 x 63mm thick precast concrete paved footway on granular subbase material type 2, 100mm thick to surface sloping at 10 degrees or less to the horizontal | hr | m2 | 0.27 | 0.27 | | |
| | £ | | 7.51 | 1.26 | 5.00 | 13.77 |
| 600 x 900 x 63mm thick precast concrete paved footway on 20mm mortar bed on class E concrete 100mm thick to surface sloping at 10 degrees or less to the horizontal | hr | m2 | 0.30 | 0.30 | | |
| | £ | | 8.34 | 1.40 | 8.05 | 17.79 |
| 600 x 600 x 50mm thick precast concrete paved footway on granular sub base material type 2, 100mm thick to surface sloping at 10 degrees or less to the horizontal | hr | m2 | 0.30 | 0.30 | | |
| | £ | | 8.34 | 1.40 | 5.52 | 15.26 |
| 450 x 600 x 50mm thick precast concrete paved footway on granular sub base material type 2, 100mm thick to surface sloping at 10 degrees or less to the horizontal | hr | m2 | 0.31 | 0.31 | | |
| | £ | | 8.62 | 1.45 | 6.05 | 16.12 |
| 80mm thick concrete block paving on type 2 sub base, 10mm thick to surfaces sloping at 10 degrees or less to horizontal | hr | m2 | 0.30 | 0.30 | | |
| | £ | | 8.34 | 1.40 | 8.90 | 18.64 |
| 80mm stock flex paver bricks on type 2 sub base 100mm thick to surfaces sloping at 10 degrees or less to horizontal (PC 418/1000) | hr | m2 | 0.30 | 0.30 | | |
| | £ | | 8.34 | 1.40 | 15.25 | 24.99 |

**CLASS R: ROADS AND PAVINGS**

**KERBS, CHANNELS AND EDGINGS - continued**

| | | UNIT | L | P | M | £ NETT |
|---|---|---|---|---|---|---|
| Extra over for class E concrete bases to block paving or paver bricks (50mm thickness) | hr | m2 | 0.03 | 0.03 | | |
| | £ | | 0.83 | 0.14 | 2.50 | 3.47 |
| Remove from store and relay random sized granite setts laid to random pattern on 25mm cement mortar on 100mm granular sub base material type 2 | hr | m2 | 0.20 | 0.20 | | |
| | £ | | 5.56 | 0.93 | 2.60 | 9.09 |
| Remove from store and relay 100mm x 100mm granite setts laid to specified pattern on 25mm cement mortar on 100mm granular sub base material type 2 | hr | m2 | 0.30 | 0.30 | | |
| | £ | | 8.34 | 1.40 | 2.60 | 12.34 |
| Extra over cost for new granite setts (200Kg/m2) | hr | m2 | - | - | | |
| | £ | | - | - | 18.00 | 18.00 |
| Extra over any item of kerbing or paving for excavation of rock, masonry, mass concrete and the like | hr | m3 | 0.71 | 0.71 | | |
| | £ | | 19.75 | 3.32 | - | 23.07 |
| Extra for any item of kerbing or paving for excavation for reinforced concrete | hr | m3 | 1.08 | 1.08 | | |
| | £ | | 30.03 | 5.04 | - | 35.07 |

## CLASS S: RAIL TRACK

For prices for the supplying and laying of rail track reference should be to Specialist Contractors. The following unit costs are for guidance only.

| | UNIT | SPECIALIST ADVICE £ NETT |
|---|---|---|
| **SUPPLYING** | | |
| DECAUVILLE TRACK | | |
| Dock and crane rails | | |
| section 56 crane rail | t | 634.00 |
| section 101 crane rail | t | 550.00 |
| Fittings | | |
| 20mm mild steel sole plate 400mm wide | m | 35.00 |
| **LAYING** | | |
| Crane rails | | |
| plain track: continuous welded section 56 crane rail | m | 25.00 |
| plain track: continuous welded section 101 crane rail | m | 30.00 |
| thermit weld section 56 crane rail | nr | 92.00 |
| thermit weld section 101 crane rail | nr | 112.00 |
| **SUNDRIES** | | |
| Wheel stops | nr | 45.00 |

## CLASS S: RAIL TRACK

| | UNIT | SPECIALIST ADVICE £ NETT |
|---|---|---|
| **SUPPLYING** | | |
| STANDARD GAUGE TRACK | | |
| Standard gauge 1.44m gauge in permanent way construction | | |
| Weighing 55kg/metre | | |
| Flat bottomed rails for jointed track | | |
| Rails for jointed track, delivered in standard 60 foot lengths to BS11 | t | 544.00 |
| Rails for welded track, delivered in standard 60 foot lengths to BS11 | t | 537.00 |
| Additional cost for curved rails form super elevation radius greater than 600m | | 15 percent |
| Pre-stressed pre-cast concrete sleepers 8'6" long, Dowmac or similar | nr | 27.75 |
| Bitumen saturated Swedish pitch pine (softwood) sleeper - new, unused | nr | 20.20 |
| Cast iron chairs | nr | 14.15 |
| Cast iron sole plates | nr | 20.50 |
| Mild steel fishplates c/w nuts and washers | nr | 8.00 |

**CLASS S: RAIL TRACK**

| | | UNIT | L | P | | SPECIALIST ADVICE £ NETT |
|---|---|---|---|---|---|---|
| **LAYING** | | | | | | |
| Jointed track BS11 jointed with fishplates onto | hr | m | - | - | | |
| concrete sleepers | £ | | 15.00 | 6.50 | - | 21.50 |
| Jointed track BS11 jointed with fishplates onto | hr | m | - | - | | |
| timber sleepers | £ | | 12.70 | 5.50 | - | 18.20 |
| Extra over cost for curved | hr | m | - | - | | |
| rail | £ | | - | - | - | 5.00 |
| Welded track BS11 onto | hr | m | - | - | | |
| concrete sleepers | £ | | 17.80 | 7.25 | - | 25.05 |
| Welded track BS11 onto | hr | m | - | - | | |
| timber sleepers | £ | | - | - | - | 21.75 |
| Extra over cost for curved | hr | m | - | - | | |
| rail | £ | | - | - | - | 5.00 |
| Alumino-thermic welding (used for site joints | hr | nr | - | - | | |
| including refractory mould) | £ | | - | - | - | 204.23 |

**CLASS T: TUNNELS**

There are so many factors, apart from design considerations, which influence the cost of tunnelling that it is only possible to give guide prices for a sample of the work involved. For more reliable estimates it is advisable to seek the advice of a Specialist Contractor.

The main cost considerations are:

- the nature of the ground
- size of tunnel
- length of drive
- depth below surface
- anticipated overbreak
- support of face and roof of tunnel (rockbolting etc.)
- necessity for pregrouting
- ventilation
- presence of water
- use of compressed air working

| | UNIT | SPECIALIST ADVICE £ NETT |
|---|---|---|
| **EXCAVATION** | | |
| Tunnels in rock | | |
| 1.5m diameter | m3 | **118.00** |
| 3.0m diameter | m3 | **84.00** |
| Tunnels in soft material | | |
| 1.5m diameter | m3 | **90.00** |
| 3.0m diameter | m3 | **50.00** |
| Shafts in rock | | |
| 1.5m diameter | m3 | **146.00** |
| 3.0m diameter | m3 | **106.00** |
| Shafts in soft material | | |
| 1.5m diameter | m3 | **101.00** |
| 3.0m diameter | m3 | **78.50** |

## CLASS T: TUNNELS

| | UNIT | SPECIALIST ADVICE £ NETT |
|---|---|---|
| **IN-SITU LINING TO TUNNELS** | | |
| Cast concrete primary; grade C30, 20mm aggregate | | |
| 1.5m diameter | m3 | 114.00 |
| 3.0m diameter | m3 | 101.00 |
| Cast concrete secondary; grade C30, 20mm aggregate | | |
| 1.5m diameter | m3 | 114.00 |
| 3.0m diameter | m3 | 101.00 |
| Formwork; rough finish | | |
| 1.5m diameter | m3 | 22.50 |
| 3.0m diameter | m3 | 19.00 |
| **IN-SITU LINING TO SHAFTS** | | |
| Cast concrete primary; grade C30, 20mm aggregate | | |
| 1.5m diameter | m3 | 121.00 |
| 3.0m diameter | m3 | 106.50 |
| Cast concrete secondary; grade C30, 20mm aggregate | | |
| 1.5m diameter | m3 | 121.00 |
| 3.0m diameter | m3 | 106.50 |
| Formwork; rough finish | | |
| 1.5m diameter | m3 | 33.00 |
| 3.0m diameter | m3 | 28.00 |

**CLASS T: TUNNELS**

In-situ lining to tunnels

Based on 5m shaft 15m head.
Multipliers for depth/length
Labour and Plant only

| depth of shaft | 15 | 30 | 60 | 90 |
|---|---|---|---|---|
| 5 | 1.00 | 1.15 | 1.40 | 1.65 |
| 10 | 1.10 | 1.20 | 1.55 | 1.70 |
| 15 | 1.15 | 1.30 | 1.65 | 1.80 |
| 20 | 1.25 | 1.40 | 1.75 | 1.85 |

| | UNIT | SPECIALIST ADVICE £ NETT |
|---|---|---|
| **PREFORMED SEGMENTAL LININGS TO TUNNELS** | | |
| Precast concrete bolted rings; including packing | | |
| 1.5m diameter | m3 | 304.75 |
| 3.0m diameter | m3 | 730.25 |
| **SUPPORT AND STABILIZATION** | | |
| Rock bolts | | |
| mechanical | m | 8.10 |
| chemically filled | m | 13.25 |
| Internal Support | | |
| steel arches; supply | t | 610.40 |
| steel arches; erection | t | 263.00 |
| Pressure grouting | | |
| sets of drilling and grouting plant | nr | 1550.00 |
| drilling and flushing | m | 12.60 |
| injection of grout materials; chemical grout | t | 182.00 |

**CLASS U: BRICKWORK BLOCKWORK AND MASONRY**

**BRICKWORK**

The following unit costs are for brickwork built in cement mortar and include for all cutting, forming openings, building in ties, etc. and pointing as the work proceeds.

| | | |
|---|---|---|
| LABOUR based on Gang J | | |
| Total hourly cost | = £ | 50.51 |
| PLANT | | |
| (Part time) | | |
| Mixer | | 0.83 |
| Dumper | | 1.15 |
| Power tools | | 2.09 |
| Scaffold etc. | | 0.90 |
| Total hourly cost | = £ | 4.97 |

| | | UNIT | L | P | M | £ NETT |
|---|---|---|---|---|---|---|
| Half brick thick | | | | | | |
| vertical straight walls | | | | | | |
| Commons PC £120/1000 | hr | m2 | 0.23 | 0.23 | | |
| | £ | | 11.62 | 1.14 | 8.56 | 21.32 |
| Facings PC £350/1000 | hr | m2 | 0.34 | 0.34 | | |
| | £ | | 17.17 | 1.69 | 22.72 | 41.58 |
| Engineering PC £300/1000 | hr | m2 | 0.29 | 0.29 | | |
| | £ | | 14.65 | 1.44 | 19.49 | 35.58 |
| vertical curved walls | | | | | | |
| Commons | hr | m2 | 0.304 | 0.304 | | |
| | £ | | 15.36 | 1.51 | 8.56 | 25.43 |
| Facings | hr | m2 | 0.45 | 0.45 | | |
| | £ | | 22.73 | 2.24 | 22.72 | 47.69 |
| Engineering | hr | m2 | 0.416 | 0.416 | | |
| | £ | | 21.01 | 2.07 | 19.49 | 42.57 |

**CLASS U: BRICKWORK BLOCKWORK AND MASONRY**

**BRICKWORK - continued**

| | | UNIT | L | P | M | £ NETT |
|---|---|---|---|---|---|---|
| vertical facing to concrete | | | | | | |
| Commons | hr | m2 | 0.262 | 0.262 | | |
| | £ | | **13.23** | **1.30** | **8.56** | **23.09** |
| Facings | hr | m2 | 0.368 | 0.368 | | |
| | £ | | **18.59** | **1.83** | **22.72** | **43.14** |
| Engineering | hr | m2 | 0.318 | 0.318 | | |
| | £ | | **16.06** | **1.58** | **19.49** | **37.13** |
| casing to metal sections | | | | | | |
| Commons | hr | m2 | 0.363 | 0.363 | | |
| | £ | | **18.34** | **1.80** | **8.56** | **28.70** |
| Facings | hr | m2 | 0.49 | 0.49 | | |
| | £ | | **24.75** | **2.44** | **22.72** | **49.91** |
| Engineering | hr | m2 | 0.44 | 0.318 | | |
| | £ | | **22.22** | **1.58** | **19.49** | **43.29** |
| One brick thick | | | | | | |
| vertical straight walls | | | | | | |
| Commons | hr | m2 | 0.44 | 0.44 | | |
| | £ | | **22.22** | **2.19** | **17.17** | **41.58** |
| Facings | hr | m2 | 0.574 | 0.574 | | |
| | £ | | **28.99** | **2.85** | **45.53** | **77.37** |
| Engineering | hr | m2 | 0.485 | 0.485 | | |
| | £ | | **24.50** | **2.41** | **39.03** | **65.94** |
| vertical curved walls | | | | | | |
| Commons | hr | m2 | 0.59 | 0.59 | | |
| | £ | | **29.80** | **2.93** | **17.17** | **49.90** |
| Facings | hr | m2 | 0.84 | 0.84 | | |
| | £ | | **42.43** | **4.17** | **45.53** | **92.13** |
| Engineering | hr | m2 | 0.78 | 0.78 | | |
| | £ | | **39.40** | **3.88** | **39.03** | **82.31** |

**CLASS U: BRICKWORK BLOCKWORK AND MASONRY**

**BRICKWORK - continued**

| | | UNIT | L | P | M | £ NETT |
|---|---|---|---|---|---|---|
| vertical facing to concrete | | | | | | |
| Commons | hr | m2 | 0.51 | 0.51 | | |
| | £ | | 25.76 | 2.53 | 17.17 | 45.46 |
| Facings | hr | m2 | 0.66 | 0.66 | | |
| | £ | | 33.34 | 3.28 | 45.53 | 82.15 |
| Engineering | hr | m2 | 0.56 | 0.56 | | |
| | £ | | 28.29 | 2.78 | 39.03 | 70.10 |
| casing to metal sections | | | | | | |
| Commons | hr | m2 | 0.65 | 0.65 | | |
| | £ | | 32.83 | 3.23 | 17.17 | 53.23 |
| Facings | hr | m2 | 0.82 | 0.82 | | |
| | £ | | 41.42 | 4.08 | 45.53 | 91.03 |
| Engineering | hr | m2 | 0.71 | 0.71 | | |
| | £ | | 35.86 | 3.53 | 39.03 | 78.42 |
| Two brick thick | | | | | | |
| vertical straight walls | | | | | | |
| Commons | hr | m2 | 0.78 | 0.78 | | |
| | £ | | 39.40 | 3.88 | 34.85 | 78.13 |
| Facings (commons in centre) | hr | m2 | 1.08 | 1.08 | | |
| | £ | | 54.55 | 5.37 | 70.04 | 129.96 |
| Engineering | hr | m2 | 0.905 | 0.905 | | |
| | £ | | 45.71 | 4.50 | 78.68 | 128.89 |
| vertical curved walls | | | | | | |
| Commons | hr | m2 | 1.04 | 1.04 | | |
| | £ | | 52.53 | 5.17 | 34.85 | 92.55 |
| Facings (commons in centre) | hr | m2 | 1.39 | 1.39 | | |
| | £ | | 70.21 | 6.91 | 70.04 | 147.16 |
| Engineering | hr | m2 | 1.30 | 1.30 | | |
| | £ | | 65.66 | 6.46 | 78.68 | 150.80 |

**CLASS U: BRICKWORK BLOCKWORK AND MASONRY**

**BRICKWORK - continued**

| | | UNIT | L | P | M | £ NETT |
|---|---|---|---|---|---|---|
| Columns and piers | | | | | | |
| 215 x 215mm | | | | | | |
| Commons | hr | m | 0.13 | 0.13 | | |
| | £ | | 6.57 | 0.65 | 4.06 | 11.28 |
| Facings | hr | m | 0.18 | 0.18 | | |
| | £ | | 9.09 | 0.89 | 10.73 | 20.71 |
| Engineering | hr | m | 0.15 | 0.15 | | |
| | £ | | 7.58 | 0.75 | 9.23 | 17.56 |
| 440 x 215mm | | | | | | |
| Commons | hr | m | 0.24 | 0.24 | | |
| | £ | | 12.12 | 1.19 | 8.31 | 21.62 |
| Facings | hr | m | 0.32 | 0.32 | | |
| | £ | | 16.16 | 1.59 | 22.03 | 39.78 |
| Engineering | hr | m | 0.26 | 0.26 | | |
| | £ | | 13.13 | 1.29 | 18.89 | 33.31 |
| 665 x 327.5mm | | | | | | |
| Commons | hr | m | 0.44 | 0.44 | | |
| | £ | | 22.22 | 2.19 | 17.82 | 42.23 |
| Facings | hr | m | 0.63 | 0.63 | | |
| | £ | | 31.82 | 3.13 | 47.18 | 82.13 |
| Engineering | hr | m | 0.52 | 0.52 | | |
| | £ | | 26.27 | 2.58 | 40.45 | 69.30 |
| Surface features | | | | | | |
| copings; brick-on-edge 215mm wide | | | | | | |
| Commons | hr | m | 0.105 | 0.105 | | |
| | £ | | 5.30 | 0.52 | 1.97 | 7.79 |
| Facings | hr | m | 0.134 | 0.134 | | |
| | £ | | 6.77 | 0.67 | 5.22 | 12.66 |
| Engineering | hr | m | 0.134 | 0.134 | | |
| | £ | | 6.77 | 0.67 | 4.48 | 11.92 |

**CLASS U: BRICKWORK BLOCKWORK AND MASONRY**

**BRICKWORK - continued**

| | | UNIT | L | P | M | £ NETT |
|---|---|---|---|---|---|---|
| rebates and chases; 100 x 50mm deep | | | | | | |
| Commons | hr | m | 0.12 | 0.12 | | |
| | £ | | **6.06** | **0.60** | - | **6.66** |
| Facings | hr | m | 0.34 | 0.34 | | |
| | £ | | **17.17** | **1.69** | - | **18.86** |
| Engineering | hr | m | 0.34 | 0.34 | | |
| | £ | | **17.17** | **1.69** | - | **18.86** |
| fair facings | | | | | | |
| Commons | hr | m2 | 0.035 | 0.035 | | |
| | £ | | **1.77** | **0.17** | - | **1.94** |
| Facings | hr | m2 | 0.057 | 0.057 | | |
| | £ | | **2.88** | **0.28** | - | **3.16** |
| Engineering | hr | m2 | 0.057 | 0.057 | | |
| | £ | | **2.88** | **0.28** | - | **3.16** |
| Expamet joint reinforcement 65mm wide | | | | | | |
| Commons | hr | m | 0.006 | 0.006 | | |
| | £ | | **0.30** | **0.03** | **0.17** | **0.50** |
| Facings | hr | m | 0.006 | 0.006 | | |
| | £ | | **0.30** | **0.03** | **0.17** | **0.50** |
| Engineering | hr | m | 0.006 | 0.006 | | |
| | £ | | **0.30** | **0.03** | **0.17** | **0.50** |
| 115mm wide | | | | | | |
| Commons | hr | m | 0.008 | 0.008 | | |
| | £ | | **0.40** | **0.04** | **0.30** | **0.74** |
| Facings | hr | m | 0.008 | 0.008 | | |
| | £ | | **0.40** | **0.04** | **0.30** | **0.74** |
| Engineering | hr | m | 0.008 | 0.008 | | |
| | £ | | **0.40** | **0.04** | **0.30** | **0.74** |

**CLASS U: BRICKWORK BLOCKWORK AND MASONRY**

**BRICKWORK - continued**

| | | UNIT | L | P | M | £ NETT |
|---|---|---|---|---|---|---|
| 175mm wide | | | | | | |
| Commons | hr | m | 0.011 | 0.011 | | |
| | £ | | **0.56** | **0.05** | **0.44** | **1.05** |
| Facings | hr | m | 0.011 | 0.011 | | |
| | £ | | **0.56** | **0.05** | **0.44** | **1.05** |
| Engineering | hr | m | 0.011 | 0.011 | | |
| | £ | | **0.56** | **0.05** | **0.44** | **1.05** |
| 225mm wide | | | | | | |
| Commons | hr | m | 0.014 | 0.014 | | |
| | £ | | **0.71** | **0.07** | **0.61** | **1.39** |
| Facings | hr | m | 0.014 | 0.014 | | |
| | £ | | **0.71** | **0.07** | **0.61** | **1.39** |
| Engineering | hr | m | 0.014 | 0.014 | | |
| | £ | | **0.71** | **0.07** | **0.61** | **1.39** |
| Hyload damp-proof course half brick wide | | | | | | |
| Commons | hr | m | 0.011 | 0.011 | | |
| | £ | | **0.56** | **0.05** | **0.50** | **1.11** |
| Facings | hr | m | 0.011 | 0.011 | | |
| | £ | | **0.56** | **0.05** | **0.50** | **1.11** |
| Engineering | hr | m | 0.011 | 0.011 | | |
| | £ | | **0.56** | **0.05** | **0.50** | **1.11** |
| one brick wide | | | | | | |
| Commons | hr | m | 0.026 | 0.026 | | |
| | £ | | **1.31** | **0.13** | **0.95** | **2.39** |
| Facings | hr | m | 0.026 | 0.026 | | |
| | £ | | **1.31** | **0.13** | **0.95** | **2.39** |
| Engineering | hr | m | 0.026 | 0.026 | | |
| | £ | | **1.31** | **0.13** | **0.95** | **2.39** |

**CLASS U: BRICKWORK BLOCKWORK AND MASONRY**

**BRICKWORK - continued**

| | | UNIT | L | P | M | £ NETT |
|---|---|---|---|---|---|---|
| Built-in pipes and ducts cross-sectional area ne 0.05m2 | | | | | | |
| half brick wall | | | | | | |
| Commons | hr | nr | 0.06 | 0.06 | | |
| | £ | | 3.03 | 0.30 | 0.40 | 3.73 |
| Facings | hr | nr | 0.095 | 0.095 | | |
| | £ | | 4.80 | 0.47 | 0.53 | 5.80 |
| Engineering | hr | nr | 0.095 | 0.095 | | |
| | £ | | 4.80 | 0.47 | 0.46 | 5.73 |
| one brick wall | | | | | | |
| Commons | hr | nr | 0.115 | 0.115 | | |
| | £ | | 5.81 | 0.57 | 0.60 | 6.98 |
| Facings | hr | nr | 0.15 | 0.15 | | |
| | £ | | 7.58 | 0.75 | 0.82 | 9.15 |
| Engineering | hr | nr | 0.15 | 0.15 | | |
| | £ | | 7.58 | 0.75 | 0.64 | 8.97 |
| Built-in pipes and ducts cross-sectional area 0.05 - 0.25m2 | | | | | | |
| half brick wall | | | | | | |
| Commons | hr | nr | 0.15 | 0.15 | | |
| | £ | | 7.58 | 0.75 | 0.75 | 9.08 |
| Facings | hr | nr | 0.19 | 0.19 | | |
| | £ | | 9.60 | 0.94 | 0.90 | 11.44 |
| Engineering | hr | nr | 0.19 | 0.19 | | |
| | £ | | 9.60 | 0.94 | 0.86 | 11.40 |
| one brick wall | | | | | | |
| Commons | hr | nr | 0.29 | 0.29 | | |
| | £ | | 14.65 | 1.44 | 1.15 | 17.24 |
| Facings | hr | nr | 0.33 | 0.33 | | |
| | £ | | 16.67 | 1.64 | 1.53 | 19.84 |
| Engineering | hr | nr | 0.33 | 0.33 | | |
| | £ | | 16.67 | 1.64 | 1.32 | 19.63 |

**CLASS U: BRICKWORK BLOCKWORK AND MASONRY**

**BRICKWORK - continued**

| | | UNIT | L | P | M | £ NETT |
|---|---|---|---|---|---|---|
| Built-in pipes and ducts cross-sectional area 0.25 - 0.5m2 | | | | | | |
| half brick wall | | | | | | |
| Commons | hr | nr | 0.175 | 0.175 | | |
| | £ | | **8.84** | **0.87** | **0.90** | **10.61** |
| Facings | hr | nr | 0.23 | 0.23 | | |
| | £ | | **11.62** | **1.14** | **1.10** | **13.86** |
| Engineering | hr | nr | 0.23 | 0.23 | | |
| | £ | | **11.62** | **1.14** | **1.04** | **13.80** |
| one brick wall | | | | | | |
| Commons | hr | nr | 0.34 | 0.34 | | |
| | £ | | **17.17** | **1.69** | **1.39** | **20.25** |
| Facings | hr | nr | 0.395 | 0.395 | | |
| | £ | | **19.95** | **1.96** | **1.85** | **23.76** |
| Engineering | hr | nr | 0.395 | 0.395 | | |
| | £ | | **19.95** | **1.96** | **1.60** | **23.51** |

**CLASS U: BRICKWORK BLOCKWORK AND MASONRY**

**BLOCKWORK AND MASONRY**

The following unit costs are for blockwork built in cement-lime mortar and include for all cutting, forming openings, building in ties, etc. and pointing as the work proceeds.

| | | UNIT | L | P | M | £ NETT |
|---|---|---|---|---|---|---|
| 50mm thick | | | | | | |
| vertical facing to concrete | | | | | | |
| Lightweight blocks 3.5N/mm2 | hr | m2 | - | - | | |
| | £ | | - | - | - | - |
| Dense blocks 3.5 or 7N/mm2 | hr | m2 | - | - | | |
| | £ | | - | - | - | - |
| Ashlar Masonry | | | | | | |
| Portland Whitbed limestone | hr | m2 | 0.84 | 0.84 | | |
| | £ | | 42.43 | 4.17 | 113.23 | 159.83 |
| 100mm thick | | | | | | |
| vertical straight walls | | | | | | |
| Lightweight blocks 3.5N/mm2 | hr | m2 | 0.174 | 0.174 | | |
| | £ | | 8.79 | 0.86 | 5.50 | 15.15 |
| Dense blocks 3.5 or 7N/mm2 | hr | m2 | 0.17 | 0.17 | | |
| | £ | | 8.59 | 0.84 | 4.70 | 14.13 |
| Ashlar Masonry | | | | | | |
| Portland Whitbed limestone | hr | m2 | 0.23 | 0.23 | | |
| | £ | | 11.62 | 1.14 | 163.76 | 176.52 |
| casing to metal sections | | | | | | |
| Lightweight blocks 3.5N/mm2 | hr | m2 | 0.21 | 0.21 | | |
| | £ | | 10.61 | 1.04 | 5.50 | 17.15 |
| Dense blocks 3.5 or 7N/mm2 | hr | m2 | 0.21 | 0.21 | | |
| | £ | | 10.61 | 1.04 | 4.70 | 16.35 |

**CLASS U: BRICKWORK BLOCKWORK AND MASONRY**

**BLOCKWORK AND MASONRY - continued**

| | | UNIT | L | P | M | £ NETT |
|---|---|---|---|---|---|---|
| 140mm thick | | | | | | |
| vertical straight walls | | | | | | |
| Lightweight blocks 3.5N/mm2 | hr | m2 | 0.225 | 0.225 | | |
| | £ | | 11.36 | 1.12 | 8.21 | 20.69 |
| Dense blocks 3.5 or 7N/mm2 | hr | m2 | 0.20 | 0.20 | | |
| | £ | | 10.10 | 0.99 | 7.66 | 18.75 |
| casing to metal sections | | | | | | |
| Lightweight blocks 3.5N/mm2 | hr | m2 | 0.27 | 0.27 | | |
| | £ | | 13.64 | 1.34 | 8.21 | 23.19 |
| Dense blocks 3.5 or 7N/mm2 | hr | m2 | 0.24 | 0.24 | | |
| | £ | | 12.12 | 1.19 | 7.66 | 20.97 |
| 215mm thick | | | | | | |
| vertical straight walls | | | | | | |
| Lightweight blocks 3.5N/mm2 | hr | m2 | 0.275 | 0.275 | | |
| | £ | | 13.89 | 1.37 | 15.21 | 30.47 |
| Dense blocks 3.5 or 7N/mm2 | hr | m2 | 0.24 | 0.24 | | |
| | £ | | 12.12 | 1.19 | 15.42 | 28.73 |
| Surface features | | | | | | |
| copings; 350 x 75mm, weathered and twice throated | | | | | | |
| Ashlar Masonry | | | | | | |
| Portland Whitbed limestone | hr | m | 0.23 | 0.23 | | |
| | £ | | 11.62 | 1.14 | 56.09 | 68.85 |
| surrounds to openings 200 x 75mm, sunken splayed | | | | | | |
| Portland Whitbed limestone | hr | m | 0.325 | 0.325 | | |
| | £ | | 16.42 | 1.62 | 43.43 | 61.47 |
| Built-in pipes and ducts cross-sectional area ne 0.05m2 | | | | | | |
| 100mm block wall | | | | | | |
| Lightweight blocks 3.5N/mm2 | hr | nr | 0.04 | 0.04 | | |
| | £ | | 2.02 | 0.20 | 0.21 | 2.43 |
| Dense blocks 3.5 or 7N/mm2 | hr | nr | 0.045 | 0.045 | | |
| | £ | | 2.27 | 0.22 | 0.20 | 2.69 |

**CLASS U: BRICKWORK BLOCKWORK AND MASONRY**

**BLOCKWORK AND MASONRY - continued**

| | | UNIT | L | P | M | £ NETT |
|---|---|---|---|---|---|---|
| 140mm block wall | | | | | | |
| Lightweight blocks 3.5N/mm2 | hr | nr | 0.09 | 0.09 | | |
| | £ | | **4.55** | **0.45** | **0.47** | **5.47** |
| Dense blocks 3.5 or 7N/mm2 | hr | nr | 0.10 | 0.10 | | |
| | £ | | **5.05** | **0.50** | **0.45** | **6.00** |
| 215mm block wall | | | | | | |
| Lightweight blocks 3.5N/mm2 | hr | nr | 0.126 | 0.126 | | |
| | £ | | **6.36** | **0.63** | **0.62** | **7.61** |
| Dense blocks 3.5 or 7N/mm2 | hr | nr | 0.135 | 0.135 | | |
| | £ | | **6.82** | **0.67** | **0.60** | **8.09** |
| Built-in pipes and ducts cross-sectional area 0.05 - 0.25m2 | | | | | | |
| 100mm block wall | | | | | | |
| Lightweight blocks 3.5N/mm2 | hr | nr | 0.126 | 0.126 | | |
| | £ | | **6.36** | **0.63** | **0.62** | **7.61** |
| Dense blocks 3.5 or 7N/mm2 | hr | nr | 0.135 | 0.135 | | |
| | £ | | **6.82** | **0.67** | **0.60** | **8.09** |
| 140mm block wall | | | | | | |
| Lightweight blocks 3.5N/mm2 | hr | nr | 0.176 | 0.176 | | |
| | £ | | **8.89** | **0.87** | **0.92** | **10.68** |
| Dense blocks 3.5 or 7N/mm2 | hr | nr | 0.19 | 0.19 | | |
| | £ | | **9.60** | **0.94** | **0.89** | **11.43** |
| 215mm block wall | | | | | | |
| Lightweight blocks 3.5N/mm2 | hr | nr | 0.25 | 0.25 | | |
| | £ | | **12.63** | **1.24** | **1.23** | **15.10** |
| Dense blocks 3.5 or 7N/mm2 | hr | nr | 0.27 | 0.27 | | |
| | £ | | **13.64** | **1.34** | **1.20** | **16.18** |

**CLASS U: BRICKWORK BLOCKWORK AND MASONRY**

**BLOCKWORK AND MASONRY - continued**

| | | UNIT | L | P | M | £ NETT |
|---|---|---|---|---|---|---|
| Built-in pipes and ducts cross-sectional area 0.25 - 0.50m2 | | | | | | |
| 100mm block wall | | | | | | |
| Lightweight blocks 3.5N/mm2 | hr | nr | 0.15 | 0.15 | | |
| | £ | | **7.58** | **0.75** | **0.78** | **9.11** |
| Dense blocks 3.5 or 7N/mm2 | hr | nr | 0.163 | 0.163 | | |
| | £ | | **8.23** | **0.81** | **0.75** | **9.79** |
| 140mm block wall | | | | | | |
| Lightweight blocks 3.5N/mm2 | hr | nr | 0.21 | 0.21 | | |
| | £ | | **10.61** | **1.04** | **0.94** | **12.59** |
| Dense blocks 3.5 or 7N/mm2 | hr | nr | 0.23 | 0.23 | | |
| | £ | | **11.62** | **1.14** | **0.91** | **13.67** |
| 215mm block wall | | | | | | |
| Lightweight blocks 3.5N/mm2 | hr | nr | 0.295 | 0.295 | | |
| | £ | | **14.90** | **1.47** | **1.48** | **17.85** |
| Dense blocks 3.5 or 7N/mm2 | hr | nr | 0.32 | 0.32 | | |
| | £ | | **16.16** | **1.59** | **1.45** | **19.20** |
| Rubble masonry (based on 2.0 tonne of stone per m3 of wall) | | | | | | |
| Random Rubble walling (PC £38.40 per tonne) (based on thickness of wall) | | | | | | |
| 300mm | hr | m2 | 0.40 | 0.40 | | |
| | £ | | **20.20** | **1.99** | **25.48** | **47.67** |
| 450mm | hr | m2 | 0.55 | 0.55 | | |
| | £ | | **27.78** | **2.73** | **38.23** | **68.74** |
| 600mm | hr | m2 | 0.75 | 0.75 | | |
| | £ | | **37.88** | **3.73** | **50.98** | **92.59** |

**CLASS U: BRICKWORK BLOCKWORK AND MASONRY**

**BLOCKWORK AND MASONRY - continued**

| | | UNIT | L | P | M | £ NETT |
|---|---|---|---|---|---|---|
| Squared Rubble walling (as a facing)(PC £48.00 per tonne) thickness of facing | | | | | | |
| 150mm | hr | m2 | 0.35 | 0.35 | | |
| | £ | | 17.68 | 1.74 | 15.93 | 35.35 |
| 225mm | hr | m2 | 0.40 | 0.40 | | |
| | £ | | 20.20 | 1.99 | 23.89 | 46.08 |
| 300mm | hr | m2 | 0.45 | 0.45 | | |
| | £ | | 22.73 | 2.24 | 31.85 | 56.82 |
| Drystone walling (PC £38.40 per tonne) | | | | | | |
| 300mm | hr | m2 | 0.27 | 0.27 | | |
| | £ | | 13.64 | 1.34 | 23.04 | 38.02 |
| 450mm | hr | m2 | 0.35 | 0.35 | | |
| | £ | | 17.68 | 1.74 | 34.60 | 54.02 |
| 600mm | hr | m2 | 0.70 | 0.70 | | |
| | £ | | 35.36 | 3.48 | 46.08 | 84.92 |

Apply the following multipliers to stone walls

| | |
|---|---|
| Height up to 2m | 1.00 |
| Height 2m to 5m | 1.21 |
| Height 5m to 10m | 1.37 |
| Wall to radius small | 1.75 |
| Wall to radius large | 1.50 |
| Wall to rake or batter | 1.15 |
| Wall in piers or stanchion | 1.50 |
| Wall in buttresses | 1.15 |

**CLASS V: PAINTING**

| | | | |
|---|---|---|---|
| LABOUR based on Gang I | | | |
| Total hourly cost | = | £ | 38.51 |
| **PLANT** | | | |
| Power tools | | | 2.09 |
| Ladders etc. | | | 0.88 |
| Part time transport (van) | | | 1.64 |
| Total hourly cost | = | £ | 4.61 |

The following unit prices include for preparing surfaces. Work on timber includes for knotting and stopping. (waste allowance 5%)

REFER ALSO TO MMHW SECTION 19

**CLASS V: PAINTING**

Costs based on Ferox urethane paints as a protective coating, for alternative paints refer to materials section for details of cost per m2.

| | | UNIT | L | P | M | £ NETT |
|---|---|---|---|---|---|---|
| One coat primer | | | | | | |
| metal | | | | | | |
| ne 300mm | hr | m | 0.01 | 0.01 | | |
| | £ | | 0.39 | 0.05 | 0.16 | 0.60 |
| 300mm - 1m | hr | m | 0.025 | 0.025 | | |
| | £ | | 0.96 | 0.12 | 0.32 | 1.40 |
| Soffits generally | hr | m2 | 0.03 | 0.03 | | |
| | £ | | 1.16 | 0.14 | 0.48 | 1.78 |
| Vertical faces generally | hr | m2 | 0.025 | 0.025 | | |
| | £ | | 0.96 | 0.12 | 0.48 | 1.56 |
| timber | | | | | | |
| ne 300mm | hr | m | 0.01 | 0.01 | | |
| | £ | | 0.39 | 0.05 | 0.17 | 0.61 |
| 300mm - 1m | hr | m | 0.02 | 0.02 | | |
| | £ | | 0.77 | 0.09 | 0.35 | 1.21 |
| Soffits generally | hr | m2 | 0.025 | 0.025 | | |
| | £ | | 0.96 | 0.12 | 0.53 | 1.61 |
| Vertical faces generally | hr | m2 | 0.02 | 0.02 | | |
| | £ | | 0.77 | 0.09 | 0.53 | 1.39 |
| One coat primer, one undercoat and one top coat gloss paint | | | | | | |
| metal | | | | | | |
| ne 300mm | hr | m | 0.025 | 0.025 | | |
| | £ | | 0.96 | 0.12 | 0.52 | 1.60 |
| 300mm - 1m | hr | m | 0.05 | 0.05 | | |
| | £ | | 1.93 | 0.23 | 1.11 | 3.27 |
| Soffits generally | hr | m2 | 0.08 | 0.08 | | |
| | £ | | 3.08 | 0.37 | 1.70 | 5.15 |
| Vertical faces generally | hr | m2 | 0.075 | 0.075 | | |
| | £ | | 2.89 | 0.35 | 1.70 | 4.94 |

**CLASS V: PAINTING**

| | | UNIT | L | P | M | £ NETT |
|---|---|---|---|---|---|---|
| timber | | | | | | |
| ne 300mm | hr | m | 0.025 | 0.025 | | |
| | £ | | **0.96** | **0.12** | **0.53** | **1.61** |
| 300mm - 1m | hr | m | 0.05 | 0.05 | | |
| | £ | | **1.93** | **0.23** | **1.13** | **3.29** |
| Soffits generally | hr | m2 | 0.08 | 0.08 | | |
| | £ | | **3.08** | **0.37** | **1.74** | **5.19** |
| Vertical faces generally | hr | m2 | 0.075 | 0.075 | | |
| | £ | | **2.89** | **0.35** | **1.74** | **4.98** |
| One coat primer, two undercoat and one top gloss paint | | | | | | |
| metal | | | | | | |
| ne 300mm | hr | m | 0.03 | 0.03 | | |
| | £ | | **1.16** | **0.14** | **0.70** | **2.00** |
| 300mm - 1m | hr | m | 0.06 | 0.06 | | |
| | £ | | **2.31** | **0.28** | **1.52** | **4.11** |
| Soffits generally | hr | m2 | 0.075 | 0.075 | | |
| | £ | | **2.89** | **0.35** | **2.34** | **5.58** |
| Vertical faces generally | hr | m2 | 0.07 | 0.07 | | |
| | £ | | **2.70** | **0.32** | **2.34** | **5.36** |
| timber | | | | | | |
| ne 300mm | hr | m | 0.03 | 0.07 | | |
| | £ | | **1.16** | **0.32** | **0.71** | **2.19** |
| 300mm - 1m | hr | m | 0.06 | 0.06 | | |
| | £ | | **2.31** | **0.28** | **1.55** | **4.14** |
| Soffits generally | hr | m2 | 0.08 | 0.08 | | |
| | £ | | **3.08** | **0.37** | **2.38** | **5.83** |
| Vertical faces generally | hr | m2 | 0.075 | 0.075 | | |
| | £ | | **2.89** | **0.35** | **2.38** | **5.62** |

**CLASS V: PAINTING**

| | | UNIT | L | P | M | £ NETT |
|---|---|---|---|---|---|---|
| Two coats matt emulsion paint (Crown Trade) | | | | | | |
| smooth concrete | | | | | | |
| ne 300mm | hr | m | 0.015 | 0.015 | | |
| | £ | | 0.58 | 0.07 | 0.13 | 0.78 |
| 300mm - 1m | hr | m | 0.03 | 0.03 | | |
| | £ | | 1.16 | 0.14 | 0.29 | 1.59 |
| Soffits generally | hr | m2 | 0.04 | 0.04 | | |
| | £ | | 1.54 | 0.18 | 0.44 | 2.16 |
| Vertical faces generally | hr | m2 | 0.045 | 0.045 | | |
| | £ | | 1.73 | 0.21 | 0.44 | 2.38 |
| brickwork and blockwork | | | | | | |
| ne 300mm | hr | m | 0.02 | 0.02 | | |
| | £ | | 0.77 | 0.09 | 0.15 | 1.01 |
| 300mm - 1m | hr | m | 0.04 | 0.04 | | |
| | £ | | 1.54 | 0.18 | 0.33 | 2.05 |
| Soffits generally | hr | m2 | - | - | | |
| | £ | | - | - | - | - |
| Vertical faces generally | hr | m2 | 0.05 | 0.05 | | |
| | £ | | 1.93 | 0.23 | 0.50 | 2.66 |
| Two coats masonry paint, (Crown Stronghold) | | | | | | |
| smooth concrete | | | | | | |
| ne 300mm | hr | m | 0.03 | 0.03 | | |
| | £ | | 1.16 | 0.14 | 0.20 | 1.50 |
| 300mm - 1m | hr | m | 0.07 | 0.07 | | |
| | £ | | 2.70 | 0.32 | 0.38 | 3.40 |
| Soffits generally | hr | m2 | 0.08 | 0.08 | | |
| | £ | | 3.08 | 0.37 | 0.59 | 4.04 |
| Vertical faces generally | hr | m2 | 0.075 | 0.075 | | |
| | £ | | 2.89 | 0.35 | 0.59 | 3.83 |

**CLASS V: PAINTING**

| | | UNIT | L | P | M | £ NETT |
|---|---|---|---|---|---|---|
| brickwork and blockwork | | | | | | |
| ne 300mm | hr | m | 0.035 | 0.035 | | |
| | £ | | 1.35 | 0.16 | 0.20 | 1.71 |
| 300mm - 1m | hr | m | 0.075 | 0.075 | | |
| | £ | | 2.89 | 0.35 | 0.38 | 3.62 |
| Soffits generally | hr | m2 | - | - | | |
| | £ | | - | - | - | - |
| Vertical faces generally | hr | m2 | 0.008 | 0.008 | | |
| | £ | | 0.31 | 0.04 | 0.59 | 0.94 |
| Two coats polyurethane varnish timber | | | | | | |
| ne 300mm | hr | m | 0.025 | 0.025 | | |
| | £ | | 0.96 | 0.12 | 0.15 | 1.23 |
| 300mm - 1m | hr | m | 0.05 | 0.05 | | |
| | £ | | 1.93 | 0.23 | 0.30 | 2.46 |
| Soffits generally | hr | m2 | 0.06 | 0.06 | | |
| | £ | | 2.31 | 0.28 | 0.50 | 3.09 |
| Vertical faces generally | hr | m2 | 0.05 | 0.05 | | |
| | £ | | 1.93 | 0.23 | 0.50 | 2.66 |
| Two coats coloured stained polyurethane varnish timber | | | | | | |
| ne 300mm | hr | m | 0.025 | 0.025 | | |
| | £ | | 0.96 | 0.12 | 0.27 | 1.35 |
| 300mm - 1m | hr | m | 0.05 | 0.05 | | |
| | £ | | 1.93 | 0.23 | 0.59 | 2.75 |
| Soffits generally | hr | m2 | 0.06 | 0.06 | | |
| | £ | | 2.31 | 0.28 | 0.90 | 3.49 |
| Vertical faces generally | hr | m2 | 0.05 | 0.05 | | |
| | £ | | 1.93 | 0.23 | 0.90 | 3.06 |

**CLASS V: PAINTING**

| | | UNIT | L | P | M | £ NETT |
|---|---|---|---|---|---|---|
| One coat primer, one undercoat and one top coat alkyd gloss | | | | | | |
| Pipework | hr | m2 | 0.09 | 0.09 | | |
| | £ | | 3.47 | 0.41 | 1.00 | 4.88 |
| Metal Sections | hr | m2 | 0.08 | 0.08 | | |
| | £ | | 3.08 | 0.37 | 1.00 | 4.45 |
| One coat primer, two undercoat and one top coat alkyd gloss | | | | | | |
| Pipework | hr | m2 | 0.10 | 0.10 | | |
| | £ | | 3.85 | 0.46 | 1.28 | 5.59 |
| Metal Sections | hr | m2 | 0.085 | 0.085 | | |
| | £ | | 3.27 | 0.39 | 1.28 | 4.94 |
| Two coats bituminous paint | | | | | | |
| Pipework | hr | m2 | 0.09 | 0.09 | | |
| | £ | | 3.47 | 0.41 | 0.35 | 4.23 |
| Metal Sections | hr | m2 | 0.08 | 0.08 | | |
| | £ | | 3.08 | 0.37 | 0.35 | 3.80 |

**CLASS W: WATERPROOFING, ROOFING AND CLADDING**

**PROTECTIVE LAYERS**

| LABOUR | | | |
|---|---|---|---|
| Ganger | | | 6.82 |
| Skilled labour | | | 6.54 |
| Unskilled labour | | | 6.20 |
| Total hourly cost | = | £ | 19.56 |

| PLANT | | | |
|---|---|---|---|
| Tar boilers | | | 0.96 |
| Power tools | | | 2.09 |
| Part time dumper/ compressor | | | 1.00 |
| Total hourly cost | = | £ | 4.05 |

**CLASS W: WATERPROOFING, ROOFING AND CLADDING**

**PROTECTIVE LAYERS - continued**

| | | UNIT | L | P | M | £ NETT |
|---|---|---|---|---|---|---|
| Flexible Sheeting | | | | | | |
| 0.3mm polythene sheet | | | | | | |
| ne 300mm | hr | m | 0.009 | 0.009 | | |
| | £ | | 0.18 | 0.04 | 0.15 | 0.37 |
| 300mm - 1m | hr | m | 0.01 | 0.01 | | |
| | £ | | 0.20 | 0.04 | 0.33 | 0.57 |
| Horizontal/inclined surfaces generally | hr | m2 | 0.012 | 0.012 | | |
| | £ | | 0.23 | 0.05 | 0.50 | 0.78 |
| Vertical surfaces generally | hr | m2 | 0.01 | 0.01 | | |
| | £ | | 0.20 | 0.04 | 0.50 | 0.74 |
| 1000 grade bituthene | | | | | | |
| ne 300mm | hr | m | 0.025 | 0.025 | | |
| | £ | | 0.49 | 0.10 | 1.50 | 2.09 |
| 300mm - 1m | hr | m | 0.04 | 0.04 | | |
| | £ | | 0.78 | 0.16 | 3.18 | 4.12 |
| Horizontal/inclined surfaces generally | hr | m2 | 0.05 | 0.05 | | |
| | £ | | 0.98 | 0.20 | 4.89 | 6.07 |
| Vertical surfaces generally | hr | m2 | 0.06 | 0.06 | | |
| | £ | | 1.17 | 0.24 | 5.03 | 6.44 |
| Sand | | | | | | |
| 25mm thick sand | | | | | | |
| ne 300mm | hr | m | 0.005 | 0.005 | | |
| | £ | | 0.10 | 0.02 | 0.12 | 0.24 |
| 300mm - 1m | hr | m | 0.007 | 0.007 | | |
| | £ | | 0.14 | 0.03 | 0.24 | 0.41 |
| Horizontal/inclined surfaces generally | hr | m2 | 0.01 | 0.01 | | |
| | £ | | 0.20 | 0.04 | 0.37 | 0.61 |
| Vertical surfaces generally | hr | m2 | - | - | | |
| | £ | | - | - | - | - |

**CLASS W: WATERPROOFING, ROOFING AND CLADDING**

**PROTECTIVE LAYERS - continued**

| | | UNIT | L | P | M | £ NETT |
|---|---|---|---|---|---|---|
| 50mm thick pea gravel | | | | | | |
| ne 300mm | hr | m | 0.005 | 0.005 | | |
| | £ | | **0.10** | **0.02** | **0.42** | **0.54** |
| 300mm - 1m | hr | m | 0.007 | 0.007 | | |
| | £ | | **0.14** | **0.03** | **1.05** | **1.22** |
| Horizontal/inclined surfaces generally | hr | m2 | 0.01 | 0.01 | | |
| | £ | | **0.20** | **0.04** | **1.25** | **1.49** |
| Vertical surfaces generally | hr | m2 | - | - | | |
| | £ | | - | - | - | - |
| Tiles | | | | | | |
| 50mm hydraulically pressed paving slabs set on support system (support system included elsewhere) | | | | | | |
| ne 300mm | hr | m | 0.45 | 0.45 | | |
| | £ | | **8.80** | **1.82** | **1.64** | **12.26** |
| 300mm - 1m | hr | m | 0.55 | 0.55 | | |
| | £ | | **10.76** | **2.23** | **2.86** | **15.85** |
| Horizontal/inclined surfaces generally | hr | m2 | 0.60 | 0.60 | | |
| | £ | | **11.74** | **2.43** | **4.28** | **18.45** |
| Vertical surfaces generally | hr | m2 | - | - | | |
| | £ | | - | - | - | - |

**SPRAYED OR BRUSHED WATERPROOFING**

| | | UNIT | L | P | M | £ NETT |
|---|---|---|---|---|---|---|
| Synthaprufe in three coats | hr | m2 | 0.08 | 0.08 | | |
| | £ | | **1.56** | **0.32** | **3.25** | **5.13** |

**DAMP-PROOFING**

| | | UNIT | L | P | M | £ NETT |
|---|---|---|---|---|---|---|
| Waterproof sheeting | | | | | | |
| 0.3mm polythene sheet | | | | | | |
| ne 300mm | hr | m | 0.009 | 0.009 | | |
| | £ | | **0.18** | **0.04** | **0.15** | **0.37** |
| 300mm - 1m | hr | m | 0.01 | 0.01 | | |
| | £ | | **0.20** | **0.04** | **0.33** | **0.57** |
| Horizontal/inclined surfaces generally | hr | m2 | 0.012 | 0.012 | | |
| | £ | | **0.23** | **0.05** | **0.50** | **0.78** |
| Vertical surfaces generally | hr | m2 | 0.01 | 0.01 | | |
| | £ | | **0.20** | **0.04** | **0.50** | **0.74** |

CLASS W: WATERPROOFING, ROOFING AND CLADDING

DAMP-PROOFING - continued

| | | UNIT | L | P | M | £ NETT |
|---|---|---|---|---|---|---|
| Render in waterproof cement mortar (1:3) | | | | | | |
| 19mm two coat work | | | | | | |
| ne 300mm | hr | m | 0.11 | 0.11 | | |
| | £ | | 2.15 | 0.45 | 1.70 | 4.30 |
| 300mm - 1m | hr | m | 0.13 | 0.13 | | |
| | £ | | 2.54 | 0.53 | 3.20 | 6.27 |
| Horizontal/inclined surfaces generally | hr | m2 | - | - | | |
| | £ | | - | - | - | - |
| Vertical surfaces generally | hr | m2 | 0.11 | 0.11 | | |
| | £ | | 2.15 | 0.45 | 4.75 | 7.35 |
| 32mm one coat work | | | | | | |
| ne 300mm | hr | m | 0.07 | 0.07 | | |
| | £ | | 1.37 | 0.28 | 1.95 | 3.60 |
| 300mm - 1m | hr | m | 0.11 | 0.11 | | |
| | £ | | 2.15 | 0.45 | 2.21 | 4.81 |
| Horizontal/inclined surfaces generally | hr | m2 | 0.08 | 0.08 | | |
| | £ | | 1.56 | 0.32 | 5.33 | 7.21 |
| Vertical surfaces generally | hr | m2 | - | - | | |
| | £ | | - | - | - | - |
| Waterproof Felting | | | | | | |
| three layer built-up felt type 2B | | | | | | |
| ne 300mm | hr | m | 0.14 | 0.14 | | |
| | £ | | 2.74 | 0.57 | 2.90 | 6.21 |
| 300mm - 1m | hr | m | 0.32 | 0.32 | | |
| | £ | | 6.26 | 1.30 | 5.33 | 12.89 |
| Horizontal/inclined surfaces generally | hr | m2 | 0.29 | 0.29 | | |
| | £ | | 5.67 | 1.17 | 7.75 | 14.59 |
| Vertical surfaces generally | hr | m2 | - | - | | |
| | £ | | - | - | - | - |

**CLASS W: WATERPROOFING, ROOFING AND CLADDING**

**ROOFING**

**ROOF CLADDING**

| | UNIT | SPECIALIST ADVICE £ NETT |
|---|---|---|
| **SUPPLY, DELIVER AND FIX** | | |
| Plastisol coated steel sheeting 0.70mm thick, profile 1000R, fixing to steel purlins at 1800mm maximum centres with plastic headed self-tapping screws (1200m2) | m2 | 15.60 |
| Ridge/hip capping (50m) | m | 10.60 |
| Eaves gutter and flashing (100m) | m | 6.39 |
| Verge capping (50m) | m | 9.60 |
| Plastisol coated steel sheeting 0.70mm thick, profile 1000R, 80mm mineral wool insulation and profile 1000LP enamelled lining panels, fixing to steel purlins at 1800mm max. centres with plastic headed self-tapping screws (1200m2) | m2 | 20.35 |
| Ridge/hip capping including lining (50m) | m | 13.34 |
| Eaves gutter and flashing, including lining (100m) | m | 8.39 |
| Verge capping, including lining (50m) | m | 12.20 |
| Plastisol coated steel Warmclad 1000R bonded panels, comprising 0.70mm thick steel and 40mm thick polyisocyanurate insulation, fixing to steel purlins at 1800mm maximum centres with plastic headed self-tapping screws (1200m2) | m2 | 25.26 |
| Ridge/hip capping including lining (50m) | m | 13.34 |
| Eaves gutter and flashing, including lining (100m) | m | 8.39 |
| Verge capping, including lining (50m) | m | 12.20 |
| Single skin translucent GRP rooflighting (120m2) | m2 | 11.72 |
| Double skin translucent GRP rooflighting (120m2) | m2 | 19.72 |
| Rigidal aluminium building sheeting, Duralcote finish, Type A7, 0.70mm thick; fixing to steel purlins at 1800mm maximum centres with plastic headed self-tapping screws (1200m2) | m2 | 21.60 |
| Ridge/hip capping (50m) | m | 11.75 |
| Eaves gutter and flashing (100m) | m | 7.84 |
| Verge capping (50m) | m | 11.60 |

## CLASS W: WATERPROOFING, ROOFING AND CLADDING

### ROOFING - continued

### ROOF CLADDING - continued

| | UNIT | SPECIALIST ADVICE £ NETT |
|---|---|---|
| **SUPPLY, DELIVER AND FIX - continued** | | |
| Rigidal aluminium insulated sheet, Duralcote finish, Type A7, 0.70mm thick sheeting, 40mm thick polyisocyanurate insulation with white finish; fixing to steel purlins at 1800mm maximum centres with plastic headed self-tapping screws (1200m2) | m2 | 32.30 |
| Ridge/hip capping including lining (50m) | m | 15.30 |
| Eaves gutter and flashing, including lining (100m) | m | 10.40 |
| Verge capping, including lining (50m) | m | 12.90 |
| Rigidal composite panels, Duralcote finish Type WA6, polyisocyanurate insulation; fixing to steel purlins at 1800mm maximum centres with plastic headed self-tapping screws (1200m2) | m2 | 36.80 |
| Ridge/hip capping including lining (50m) | m | 15.30 |
| Eaves gutter and flashing, including lining (100m) | m | 10.40 |
| Verge capping, including lining (50m) | m | 12.90 |
| Single skin translucent rooflighting (120m2) | m2 | 11.72 |
| Double skin translucent rooflighting (120m2) | m2 | 19.72 |

CLASS W: WATERPROOFING, ROOFING AND CLADDING

ROOFING - continued

WALL CLADDING

| | UNIT | SPECIALIST ADVICE £ NETT |
|---|---|---|
| **SUPPLY, DELIVER AND FIX - continued** | | |
| Plastisol coated steel sheeting 0.55mm thick, profile 1000W, fixing to steel rails with plastic headed self-tapping screws (1500m2) | m2 | 16.80 |
| Plastisol coated steel sheeting 0.70mm thick, profile 1000W, fixing to steel rails with plastic headed self-tapping screws (1500m2) | m2 | 17.40 |
| External corners (50m) | m | 7.90 |
| Internal corners (20m) | m | 7.90 |
| Drip (150m) | m | 6.93 |
| Jamb (20m) | m | 9.50 |
| Roller shutter head (20m) | m | 9.50 |
| Plastisol coated steel sheeting 0.55mm thick, profile 1000W, 80mm mineral wool insulation and profile 1000LP enamelled lining panels, fixing to steel rails with plastic headed self-tapping screws (1500m2) | m2 | 22.60 |
| External corners including lining (50m) | m | 10.72 |
| Internal corners including lining (20m) | m | 10.72 |
| Drip (150m) | m | 6.93 |
| Jamb (20m) | m | 9.50 |
| Roller shutter head (20m) | m | 9.50 |
| Plastisol coated steel Warmclad 1000R bonded panels, comprising 0.55mm thick steel and 40mm thick polyisocyanurate insulation, fixing to steel rails with plastic headed self-tapping screws (1500m2) | m2 | 28.70 |
| External corners including lining (50m) | m | 10.72 |
| Internal corners including lining (20m) | m | 10.72 |
| Drip (150m) | m | 6.93 |
| Jamb (20m) | m | 9.50 |
| Roller shutter head (20m) | m | 9.50 |
| Rigidal aluminium building sheeting, Duralcote finish, Type A7, 0.55mm thick; fixing to steel rails with plastic headed self-tapping screws (1500m2) | m2 | 23.60 |
| Rigidal aluminium building sheeting, Duralcote finish, Type WA6M, 0.55mm thick; fixing to steel rails with plastic headed self-tapping screws (1500m2) | m2 | 33.60 |

**CLASS W: WATERPROOFING, ROOFING AND CLADDING**

**ROOFING - continued**

**WALL CLADDING - continued**

| | UNIT | SPECIALIST ADVICE £ NETT |
|---|---|---|
| **SUPPLY, DELIVER AND FIX - continued** | | |
| Rigidal aluminium building sheeting, Duralcote finish, Type WA6M, 0.70mm thick; fixing to steel rails with plastic headed self-tapping screws (1500m2) | m2 | 35.80 |
| External corners (50m) | m | 12.90 |
| Internal corners (20m) | m | 12.90 |
| Drip (150m) | m | 8.30 |
| Jamb (20m) | m | 11.60 |
| Roller shutter head (20m) | m | 11.60 |
| Rigidal aluminium insulated sheeting, Duralcote finish, Type A7, 0.55mm thick sheeting, 40mm thick polyisocyanurate insulation with white finish; fixing to steel rails with plastic headed self-tapping screws (1500m2) | m2 | 28.80 |
| External corners including lining (50m) | m | 15.50 |
| Internal corners including lining (20m) | m | 15.50 |
| Drip (150m) | m | 8.30 |
| Jamb (20m) | m | 11.60 |
| Roller shutter head (20m) | m | 11.60 |
| Rigidal composite panels, Duralcote finish Type WA6, polyisocyanurate insulation; fixing to steel rails with plastic headed self-tapping screws (1500m2) | m2 | 39.60 |
| Rigidal composite panels, Duralcote finish Type WA6M, polyisocyanurate insulation; fixing to steel rails with plastic headed self-tapping screws (1500m2) | m2 | 42.50 |
| External corners including lining (50m) | m | 15.50 |
| Internal corners including lining (20m) | m | 15.50 |
| Drip (150m) | m | 8.30 |
| Jamb (20m) | m | 11.60 |
| Roller shutter head (20m) | m | 11.60 |

**CLASS X: MISCELLANEOUS WORK**

**ROCK FILLED GABIONS**

**LABOUR**

| | | | |
|---|---|---|---|
| Total hourly cost | = | £ | 6.00 |

**PLANT**

| | | | |
|---|---|---|---|
| PC 180 Excavator | | | 13.50 |
| Total hourly cost | = | £ | 13.50 |

| | | UNIT | L | P | M | £ NETT |
|---|---|---|---|---|---|---|
| Gabions | | | | | | |
| PVC coated galvanised wire mesh | | | | | | |
| In module sizes 1m x 1.0m | hr | m3 | 1.50 | 1.50 | | |
| | £ | | 9.00 | 20.25 | 27.50 | 56.75 |
| 1m x 0.5m | hr | m3 | 2.00 | 2.00 | | |
| | £ | | 12.00 | 27.00 | 36.60 | 75.60 |
| Heavily galvanised woven wire mesh | | | | | | |
| In module sizes 1m x 1.0m | hr | m3 | 1.30 | 1.30 | | |
| | £ | | 7.80 | 17.55 | 24.84 | 50.19 |
| 1m x 0.5m | hr | m3 | 1.60 | 1.60 | | |
| | £ | | 9.60 | 21.60 | 27.60 | 58.80 |
| Reno Matresses | | | | | | |
| PVC coated, woven wire mesh, 300mm deep | hr | m2 | 1.30 | 1.30 | | |
| | £ | | 7.80 | 17.55 | 29.18 | 54.53 |
| Heavily galvanised, woven wire mesh 300mm deep | hr | m2 | 0.30 | 0.30 | | |
| | £ | | 1.80 | 4.05 | 22.63 | 28.48 |

**CLASS X: MISCELLANEOUS WORK**

**DRAINAGE TO STRUCTURES ABOVE GROUND**

| | | UNIT | L | P | M | £ NETT |
|---|---|---|---|---|---|---|
| Marley DeepFlow uPVC gutters | | | | | | |
| 110 x 75mm | hr | m | 0.16 | 0.16 | | |
| | £ | | 2.04 | 0.14 | 3.58 | 5.76 |
| stop end | hr | nr | 0.024 | 0.024 | | |
| | £ | | 0.31 | 0.02 | 2.15 | 2.48 |
| running outlet | hr | nr | 0.032 | 0.032 | | |
| | £ | | 0.41 | 0.03 | 3.50 | 3.94 |
| angle | hr | nr | 0.032 | 0.032 | | |
| | £ | | 0.41 | 0.03 | 3.75 | 4.19 |
| Marley Industrial uPVC gutter | | | | | | |
| 150mm diameter half round | hr | m | 0.20 | 0.20 | | |
| | £ | | 2.55 | 0.18 | 8.10 | 10.83 |
| stop end | hr | nr | 0.024 | 0.024 | | |
| | £ | | 0.31 | 0.02 | 4.76 | 5.09 |
| running outlet | hr | nr | 0.032 | 0.032 | | |
| | £ | | 0.41 | 0.03 | 10.25 | 10.69 |
| angle | hr | nr | 0.032 | 0.032 | | |
| | £ | | 0.41 | 0.03 | 11.72 | 12.16 |
| Marley uPVC downpipes | | | | | | |
| 68mm diameter | hr | m | 0.15 | 0.15 | | |
| | £ | | 1.91 | 0.14 | 2.86 | 4.91 |
| bend | hr | nr | 0.005 | 0.005 | | |
| | £ | | 0.06 | 0.01 | 3.48 | 3.55 |
| offset bend | hr | nr | 0.007 | 0.007 | | |
| | £ | | 0.09 | 0.01 | 3.78 | 3.88 |
| shoe | hr | nr | 0.007 | 0.007 | | |
| | £ | | 0.09 | 0.01 | 3.10 | 3.20 |

**CLASS X: MISCELLANEOUS WORK**

**DRAINAGE TO STRUCTURES ABOVE GROUND - continued**

| | | UNIT | L | P | M | £ NETT |
|---|---|---|---|---|---|---|
| Marley uPVC downpipes | | | | | | |
| 110mm diameter | hr | m | 0.16 | 0.16 | | |
| | £ | | **2.04** | **0.14** | **6.56** | **8.74** |
| bend | hr | nr | 0.007 | 0.007 | | |
| | £ | | **0.09** | **0.01** | **10.37** | **10.47** |
| offset bend | hr | nr | 0.007 | 0.007 | | |
| | £ | | **0.09** | **0.01** | **7.84** | **7.94** |
| shoe | hr | nr | 0.007 | 0.007 | | |
| | £ | | **0.09** | **0.01** | **11.12** | **11.22** |

**Outputs are based on heights up to 3m above ground and exclude time spent on erecting access equipment, but include marking, cutting, drilling to wood, brick or concrete and all fixings. Testing of finished work is not included.**

Output multipliers for heights over 3m.

| | | |
|---|---|---|
| 3m - 6m | = | 1.25 |
| 6m - 9m | = | 1.50 |
| 9m - 12m | = | 1.75 |
| 12m - 15m | = | 2.00 |

**CLASS Y: SEWER RENOVATION AND ANCILLARY WORKS**

**PREPARATION OF EXISTING SEWERS**

| | | UNIT | L | P | M | £ NETT |
|---|---|---|---|---|---|---|
| Cleaning | | | | | | |
| eggshape sewer 1300mm high | hr | m | - | - | | |
| | £ | | 6.20 | 1.90 | - | 8.10 |
| Removing intrusions | | | | | | |
| brickwork | hr | m3 | - | - | | |
| | £ | | 31.75 | 9.70 | - | 41.45 |
| concrete | hr | m3 | - | - | | |
| | £ | | 54.75 | 16.75 | - | 71.50 |
| reinforced concrete | hr | m3 | - | - | | |
| | £ | | 69.50 | 21.25 | - | 90.75 |
| Plugging laterals | | | | | | |
| bore not exceeding 300mm; | | | | | | |
| concrete plug | hr | nr | - | - | | |
| | £ | | 29.45 | 9.10 | 1.50 | 40.05 |
| bore 450mm; concrete plug | hr | nr | - | - | | |
| | £ | | 44.75 | 13.70 | 3.50 | 61.95 |
| bore 750mm; brickwork plug | hr | nr | - | - | | |
| | £ | | 136.85 | 41.75 | 7.75 | 186.35 |
| Local internal repairs | | | | | | |
| area not exceeding 0.1m2 | hr | nr | - | - | | |
| | £ | | 5.70 | 1.70 | 1.00 | 8.40 |
| area 0.1 - 0.25m2 | hr | nr | - | - | | |
| | £ | | 14.65 | 4.45 | 2.75 | 21.85 |
| area 1m2 | hr | nr | - | - | | |
| | £ | | 30.10 | 9.20 | 5.50 | 44.80 |
| area 10m2 | hr | nr | - | - | | |
| | £ | | 167.35 | 51.00 | 25.50 | 243.85 |

**CLASS Y: SEWER RENOVATION AND ANCILLARY WORKS**

**STABILIZATION OF EXISTING SEWERS**

| | | UNIT | L | P | M | £ NETT |
|---|---|---|---|---|---|---|
| Pointing cement mortar (1:3) | hr | m2 | - | - | | |
| | £ | | 12.40 | 3.50 | 2.00 | 17.90 |

**RENOVATION OF EXISTING SEWERS**

Sliplining

| | | UNIT | L | P | M | £ NETT |
|---|---|---|---|---|---|---|
| GRP one piece unit; eggshape sewer 1300mm high | hr | m | - | - | | |
| | £ | | - | - | - | 200.00 |

**LATERALS TO RENOVATED SEWERS**

Jointing

| | | UNIT | L | P | M | £ NETT |
|---|---|---|---|---|---|---|
| bore not exceeding 150mm | hr | nr | - | - | | |
| | £ | | 29.45 | - | - | 29.45 |
| bore 150 - 300mm | hr | nr | - | - | | |
| | £ | | 52.35 | - | - | 52.35 |
| bore 450mm | hr | nr | - | - | | |
| | £ | | 73.10 | - | - | 73.10 |

**INTERRUPTIONS**

| | | UNIT | L | P | M | £ NETT |
|---|---|---|---|---|---|---|
| Preparation of existing sewers cleaning | hr | h | - | - | | |
| | £ | | 181.25 | 8.75 | - | 190.00 |
| Stabilisation of existing sewers pointing | hr | h | - | - | | |
| | £ | | 29.75 | 1.45 | 3.00 | 34.20 |
| Renovation of existing sewers sliplining | hr | h | - | - | | |
| | £ | | - | - | - | 48.50 |

PART 5

# Unit Costs

## (HIGHWAY WORKS)

**INTRODUCTORY NOTES**

*The unit costs in this part represent the net cost to the Contractor of executing the work on site; they are not the prices which would be entered in a tender Bill of Quantities. The costs exclude preliminary and general items, head office overheads, finance charges, risk, profit and value added tax. Reference should be made to Part 2 for the cost of Preliminaries and General Items and to Part 6 for the remainder.*

*It must be emphasised that the unit costs are averages calculated on unit outputs for typical site conditions. Costs can vary considerably from contract to contract depending on individual Contractors, site conditions, methods of working and various other factors. Reference should be made to Section 1 for a general discussion on Civil Engineering Estimating.*

*We have structured this unit cost section around The Method of Measurement for Highway Works, endeavouring to cover as many aspects of Civil and Highway works as possible. In most cases, prices are broken down into labour, plant and materials, although certain specialist prices are inserted as advised.*

*Reference to Part 3 giving basic costs of labour materials and plant will assist the reader in making adjustments to unit costs.*

*Attention should be paid throughout to explanatory notes both at the start and within each section.*

*The unit costs are broken down into their Labour, Plant and Material elements, as detailed in Section 3:*

*Columns headed L = Labour*
*P = Plant*
*M = Materials*

*The first line is the output in hours taken per unit of work, the second line is the cost result of these outputs, material columns include allowances for waste and/or distribution, the total cost then appears in the £ **NETT** column.*

**SECTION 2 SITE CLEARANCE**

The prices in this section are to include for the removal of superficial obstructions down to existing ground level. Taking up or down of specified items are measured separately. Examples are included here:

| | | | |
|---|---|---|---|
| LABOUR based on Gang A | | | |
| Total hourly cost | = | £ | 33.71 |
| **PLANT** | | | |
| (Part time) | | | |
| Dumper | | | 1.73 |
| Tractor loader (931) | | | 5.74 |
| 4 Tool Compressor and Tools | | | 3.99 |
| 20T Mobile Crane | | | 4.05 |
| Lorry with Hiab. | | | 4.37 |
| Total hourly cost | = | £ | 19.88 |

## SECTION 2 SITE CLEARANCE

| | | UNIT | L | P | M | £ NETT |
|---|---|---|---|---|---|---|
| **General site clearance** | | | | | | |
| Open field site | hr | ha | 9.91 | 9.91 | | |
| | £ | | 334.07 | 197.01 | - | 531.08 |
| Medium density wooded | hr | ha | 20.61 | 20.61 | | |
| | £ | | 694.76 | 409.73 | - | 1104.49 |
| Heavy density wooded | hr | ha | 32.09 | 32.09 | | |
| | £ | | 1081.75 | 637.95 | - | 1719.70 |
| Urban areas(town centre) | hr | ha | 30.60 | 30.60 | | |
| | £ | | 1031.53 | 608.33 | - | 1639.86 |
| Live dual carriageway | hr | ha | 30.60 | 30.60 | | |
| | £ | | 1031.53 | 608.33 | - | 1639.86 |
| **Take up or down and disposal off site** | | | | | | |
| Existing road sign | hr | nr | 0.38 | 0.38 | | |
| | £ | | 12.81 | 7.55 | - | 20.36 |
| 5m lighting column | hr | nr | 0.82 | 0.82 | | |
| | £ | | 27.64 | 16.30 | - | 43.94 |
| 10m lighting column | hr | nr | 0.85 | 0.85 | | |
| | £ | | 28.65 | 16.90 | - | 45.55 |
| Tensioned corrugated safety fence | hr | m | 0.166 | 0.166 | | |
| | £ | | 5.60 | 3.30 | - | 8.90 |
| Timber post and 4 rail fence | hr | m | 0.09 | 0.09 | | |
| | £ | | 3.03 | 1.79 | - | 4.82 |
| **Take up or down and set aside for re-use or remove to store** | | | | | | |
| Road sign | hr | nr | 0.26 | 0.26 | | |
| | £ | | 8.76 | 5.17 | - | 13.93 |
| 5m lighting column | hr | nr | 0.58 | 0.58 | | |
| | £ | | 19.55 | 11.53 | - | 31.08 |
| 10m lighting column | hr | nr | 0.607 | 0.607 | | |
| | £ | | 20.46 | 12.07 | - | 32.53 |
| Tensioned corrugated safety fence | hr | m | 0.166 | 0.166 | | |
| | £ | | 5.60 | 3.30 | - | 8.90 |
| Timber post and 4 rail fence | hr | m | 0.08 | 0.08 | | |
| | £ | | 2.70 | 1.59 | - | 4.29 |

**SECTION 2 SITE CLEARANCE**

| | | SPECIALIST ADVICE UNIT | £ NETT |
|---|---|---|---|
| Demolition of individual buildings or structure | | | |
| MMHW states that individual structures should be itemised. The following rates are based on per m3. Therefore it will be necessary to obtain the volume of the structure. (Refer also to CESMM2 Class D) | | | |
| Demolish to ground level and disposal off site | | | |
| Reinforced concrete bridge deck or super structure | hr | m3 | |
| | £ | | 4.90 |
| Reinforced concrete bridge abutment or bank seat | hr | m3 | |
| | £ | | 3.52 |
| Reinforced concrete retaining wall | hr | m3 | |
| | £ | | 3.35 |
| Brick or masonry retaining wall | hr | m3 | |
| | £ | | 3.00 |
| Brick or masonry boundary wall | hr | m3 | |
| | £ | | 2.43 |
| Dry stone boundary wall | hr | m3 | |
| | £ | | 2.10 |

Removal of disused sewers, drains or redundant services. With 1 metre or less of cover to existing ground level.

| | | UNIT | L | P | M | £ NETT |
|---|---|---|---|---|---|---|
| Low pressure gas mains up to 150mm diameter | hr | m | 0.044 | 0.044 | | |
| | £ | | 1.48 | 0.87 | - | 2.35 |
| Low pressure water mains up to 75mm diameter | hr | m | 0.03 | 0.03 | | |
| | £ | | 1.01 | 0.60 | - | 1.61 |
| Low voltage electricity cable | hr | m | 0.026 | 0.026 | | |
| | £ | | 0.88 | 0.52 | - | 1.40 |

## SECTION 2 SITE CLEARANCE

| | | UNIT | L | P | M | £ NETT |
|---|---|---|---|---|---|---|
| 100mm diameter earthenware ducts | | | | | | |
| 1 Way | hr | m | 0.034 | 0.034 | | |
| | £ | | 1.15 | 0.68 | - | 1.83 |
| 2 Way | hr | m | 0.044 | 0.044 | | |
| | £ | | 1.48 | 0.87 | - | 2.35 |
| 4 Way | hr | m | 0.053 | 0.053 | | |
| | £ | | 1.79 | 1.05 | - | 2.84 |
| Removal of disused sewers or drains | | | | | | |
| diameter 150mm | hr | m | 0.034 | 0.034 | | |
| | £ | | 1.15 | 0.68 | - | 1.83 |
| diameter 225mm | hr | m | 0.044 | 0.044 | | |
| | £ | | 1.48 | 0.87 | - | 2.35 |
| diameter 300mm | hr | m | 0.06 | 0.06 | | |
| | £ | | 2.02 | 1.19 | - | 3.21 |
| diameter 375mm | hr | m | 0.103 | 0.103 | | |
| | £ | | 3.47 | 2.05 | - | 5.52 |
| diameter 450mm | hr | m | 0.12 | 0.12 | | |
| | £ | | 4.05 | 2.39 | - | 6.44 |
| diameter 600mm | hr | m | 0.18 | 0.18 | | |
| | £ | | 6.07 | 3.58 | - | 9.65 |
| diameter 750mm | hr | m | 0.24 | 0.24 | | |
| | £ | | 8.09 | 4.77 | - | 12.86 |
| Add for concrete surround | hr | m | 0.039 | 0.039 | | |
| | £ | | 1.31 | 0.78 | - | 2.09 |
| Add for reinforced concrete surround | hr | m | 0.17 | 0.17 | | |
| | £ | | 5.73 | 3.38 | - | 9.11 |

**SECTION 3 HEDGES, FENCING AND NOISE BARRIERS**

This section is restricted to those hedges, fences etc. that are most commonly found on Highway Works.

LABOUR based on 1/2 Gang I

| | | | |
|---|---|---|---|
| Total hourly cost | = | £ | 19.26 |

PLANT

| | | | |
|---|---|---|---|
| Tractor Trailer with Auger | | | 7.17 |
| (Part time) Power Tools and concrete equipment | | | 2.03 |
| Total hourly cost | = | £ | 9.20 |

## SECTION 3 HEDGES, FENCING AND NOISE BARRIERS

### PLANTING HEDGES

| | PC £ | Hand 100 M £ | M £ |
|---|---|---|---|
| Set out, nick out and excavate trench minimum 400mm deep and break up subsoil to minimum depth 300mm. | - | 310.00 | 3.15 |
| Supply the plant hedging plants including backfill with excavated topsoil: | price per 100 | | |
| single row plants at 150 mm centres | 27.00 | 335.95 | 4.02 |
| 230 mm centres | 27.00 | 219.35 | 2.51 |
| 300 mm centres | 27.00 | 168.00 | 2.01 |
| 375 mm centres | 27.00 | 134.80 | 1.51 |
| 450 mm centres | 27.00 | 111.30 | 1.01 |
| double row plants at 230 mm centres | 27.00 | 437.65 | 4.53 |
| 300 mm centres | 27.00 | 336.00 | 4.02 |
| 375 mm centres | 27.00 | 267.50 | 3.02 |
| 450 mm centres | 27.00 | 223.65 | 2.51 |
| Extra for incorporating manure at 1m3 per 30m | m3 2.25 | | 3.83 |

### TEMPORARY FENCING

| | | UNIT | L | P | M | £ NETT |
|---|---|---|---|---|---|---|
| Type 2 Wooden post 6 wire | hr | m | 0.06 | 0.06 | | |
| | £ | | 1.16 | 0.55 | 2.30 | 4.01 |
| Type 1 Wooden post 4 wire (2 barbed) | hr | m | 0.06 | 0.06 | | |
| | £ | | 1.16 | 0.55 | 2.50 | 4.21 |
| Wooden post woven wire (1.23m high) | hr | m | 0.06 | 0.06 | | |
| | £ | | 1.16 | 0.55 | 3.50 | 5.21 |
| Extra for sheep netting on post and wire | hr | m | 0.04 | - | | |
| | £ | | 0.77 | - | 0.26 | 1.03 |
| Extra for pig netting on post and wire | hr | m | 0.04 | - | | |
| | £ | | 0.77 | - | 0.29 | 1.06 |
| Extra for straining posts including struts | hr | nr | - | - | | |
| | £ | | - | - | - | 3.20 |

**SECTION 3 HEDGES, FENCING AND NOISE BARRIERS**

| | | UNIT | L | P | M | £ NETT |
|---|---|---|---|---|---|---|
| **TEMPORARY FENCING - continued** | | | | | | |
| Type 3 | | | | | | |
| Cleft chestnut pale fencing 1200mm high | hr | m | 0.07 | 0.07 | | |
| | £ | | 1.35 | 0.64 | 2.55 | 4.54 |
| Type 4 | | | | | | |
| 4 wire and 2 barb | hr | m | 0.13 | 0.13 | | |
| | £ | | 2.50 | 1.20 | 2.30 | 6.00 |
| Extra over for sheep netting | hr | m | 0.04 | - | | |
| | £ | | 0.77 | - | 0.26 | 1.03 |
| Extra over for pig netting | hr | m | 0.04 | - | | |
| | £ | | 0.77 | - | 0.29 | 1.06 |
| **PERMANENT FENCING** | | | | | | |
| Post and 4 rail fencing 1.4m high | hr | m | 0.13 | 0.13 | | |
| | £ | | 2.50 | 1.20 | 8.10 | 11.80 |
| Close boarded fencing concrete posts 1.8m high | hr | m | 0.30 | 0.30 | | |
| | £ | | 5.78 | 2.76 | 22.01 | 30.55 |
| Steel single field gate | hr | m | 0.30 | 0.30 | | |
| | £ | | 5.78 | 2.76 | 22.01 | 30.55 |
| Timber single field gate | | | | | | |
| 3.50m wide | hr | m | 1.90 | 1.90 | | |
| | £ | | 36.59 | 17.48 | 110.00 | 164.07 |
| 4.10m wide | hr | m | 2.00 | 2.00 | | |
| | £ | | 38.52 | 18.40 | 133.00 | 189.92 |
| 4.71m wide | hr | m | 2.00 | 2.00 | | |
| | £ | | 38.52 | 18.40 | 137.00 | 193.92 |
| Timber stile type 1 | hr | m | 1.50 | 1.50 | | |
| | £ | | 28.89 | 13.80 | 78.50 | 121.19 |
| Timber stile type 2 | hr | m | 1.40 | 1.40 | | |
| | £ | | 26.96 | 12.88 | 59.00 | 98.84 |
| Double steel gate 1.8m high 3.00m wide | hr | m | 1.90 | 1.90 | | |
| | £ | | 36.59 | 17.48 | 108.00 | 162.07 |

**SECTION 3 HEDGES, FENCING AND NOISE BARRIERS**

| | | UNIT | L | P | M | £ NETT |
|---|---|---|---|---|---|---|
| Concrete footings to intermediate posts | hr | nr | 0.10 | - | | |
| | £ | | 1.93 | - | 2.20 | 4.13 |
| Plastic coated chain mesh fence 1.40m high | hr | m | 0.05 | 0.05 | | |
| | £ | | 0.96 | 0.46 | 7.30 | 8.72 |
| Plastic coated chain mesh fence 1.80m high | hr | m | 0.06 | 0.06 | | |
| | £ | | 1.16 | 0.55 | 9.61 | 11.32 |
| High tensile strained wire fence 1.35m high | hr | m | 0.16 | 0.16 | | |
| | £ | | 3.08 | 1.47 | 7.75 | 12.30 |
| High tensile strained wire fence 1.80m high | hr | m | 0.20 | 0.20 | | |
| | £ | | 3.85 | 1.84 | 10.33 | 16.02 |
| High tensile strained wire fence 2.10m high | hr | m | 0.22 | 0.22 | | |
| | £ | | 4.24 | 2.02 | 12.05 | 18.31 |
| Turning posts for strained wire fence | hr | m | 0.08 | 0.08 | | |
| | £ | | 1.54 | 0.74 | 3.20 | 5.48 |

## SECTION 3 HEDGES, FENCING AND NOISE BARRIERS

| | UNIT | | L | P | M | £ NETT |
|---|---|---|---|---|---|---|
| **PEDESTRIAN GUARD RAILING** | | | | | | |
| Supply and erect 1.0m high Tubular galvanised mild steel pedestrian guard railing to BS3049 | | | | | | |
| Mesh infill (105wg 50mm Sq.) | | | | | | |
| mounted on posts with concrete footing | hr | m | 0.28 | 0.28 | | |
| | £ | | 5.39 | 2.58 | 92.00 | 99.97 |
| mounted on posts bolted to structure or ground beam | hr | m | 0.26 | 0.26 | | |
| | £ | | 5.01 | 2.39 | 91.00 | 98.40 |
| Supply and erect 1.0m high galvanised steel pedestrian guard railing with vertical rails (group P4 parapet) | | | | | | |
| mounted on posts with concrete footing | hr | m | 0.33 | 0.33 | | |
| | £ | | 6.36 | 3.04 | 75.00 | 84.40 |
| mounted on posts bolted to structure or ground beam | hr | m | 0.31 | 0.31 | | |
| | £ | | 5.97 | 2.85 | 76.00 | 84.82 |
| Supply and erect: | | | | | | |
| Tubular double ball galvanised steel handrails 50mm diameter 1.20m high posts | hr | m | 0.25 | 0.25 | | |
| | £ | | 4.82 | 2.30 | 72.00 | 79.12 |
| Tubular double ball aluminium handrails 63mm diameter 1.20m high posts | hr | m | 0.25 | 0.25 | | |
| | £ | | 4.82 | 2.30 | 107.00 | 114.12 |
| concrete footings for handrails | hr | m | 0.10 | - | | |
| | £ | | 1.93 | - | 2.20 | 4.13 |

**SECTION 4 SAFETY FENCES**

LABOUR based on Gang I

| | | | |
|---|---|---|---|
| Total hourly cost | = | £ | 38.51 |

PLANT

| | | | |
|---|---|---|---|
| Tractor with pneumatic post driver | | | 8.66 |
| Trailer | | | 0.19 |
| Small tools (part time) | | | 1.04 |
| Total hourly cost | = | £ | 9.89 |

**SECTION 4 SAFETY FENCES**

| | | UNIT | L | P | M | £ NETT |
|---|---|---|---|---|---|---|
| **SAFETY BARRIER** | | | | | | |
| Tensioned single sided corrugated beam | hr | m | 0.10 | 0.10 | | |
| | £ | | 3.85 | 0.99 | 11.94 | 16.78 |
| Driven steel post for single sided corrugated beam | hr | nr | 0.055 | 0.055 | | |
| | £ | | 2.12 | 0.54 | 15.65 | 18.31 |
| Steel post for setting in concrete for single sided corrugated beam | hr | nr | 0.06 | 0.06 | | |
| | £ | | 2.31 | 0.59 | 8.90 | 11.80 |
| Terminal for tensioned single sided corrugated beam (including concrete anchorage block) | hr | nr | 0.96 | 0.96 | | |
| | £ | | 36.97 | 9.49 | 240.00 | 286.46 |
| Concrete footing for steel post for corrugated beam | hr | nr | 0.226 | 0.226 | | |
| | £ | | 8.70 | 2.24 | 4.00 | 14.94 |
| Non-standard concrete footing Type 1 spanning French drain for steel post for corrugated beam | hr | nr | 0.37 | 0.37 | | |
| | £ | | 14.25 | 3.66 | 6.00 | 23.91 |
| Untensioned single sided corrugated beam | hr | m | 0.07 | 0.07 | | |
| | £ | | 2.70 | 0.69 | 11.78 | 15.17 |
| Driven wooden post with standard spacers for single sided corrugated beam | hr | nr | 0.35 | 0.35 | | |
| | £ | | 13.48 | 3.46 | 10.90 | 27.84 |
| Terminal for untensioned single sided corrugated beam | hr | nr | 0.71 | 0.71 | | |
| | £ | | 27.34 | 7.02 | 119.00 | 153.36 |
| Tensioned double sided corrugated beam | hr | m | 0.12 | 0.12 | | |
| | £ | | 4.62 | 1.19 | 24.62 | 30.43 |
| Driven steel post for double sided corrugated beam | hr | nr | 0.05 | 0.05 | | |
| | £ | | 1.93 | 0.49 | 15.84 | 18.26 |

**SECTION 4 SAFETY FENCES**

| | | UNIT | L | P | M | £ NETT |
|---|---|---|---|---|---|---|
| Untensioned single sided open box beam | hr | m | 0.07 | 0.07 | | |
| | £ | | 2.70 | 0.69 | 23.04 | 26.43 |
| Driven steel post for open box beam | hr | nr | 0.055 | 0.055 | | |
| | £ | | 2.12 | 0.54 | 22.25 | 24.91 |
| Steel post for setting in concrete for open box beam | hr | nr | 0.06 | 0.06 | | |
| | £ | | 2.31 | 0.59 | 12.62 | 15.52 |
| Terminal for untensioned single sided open box beam (including anchorage block) | hr | nr | 1.01 | 1.01 | | |
| | £ | | 38.90 | 9.99 | 277.50 | 326.39 |
| Concrete footing for steel post for open box beam | hr | nr | 0.23 | 0.23 | | |
| | £ | | 8.86 | 2.27 | 4.00 | 15.13 |
| Mounting bracket fixed to structure for single sided open box beam | hr | nr | 0.16 | 0.16 | | |
| | £ | | 6.16 | 1.58 | 18.80 | 26.54 |
| Standard full height anchorage for single sided open box beam | hr | nr | 4.05 | 4.05 | | |
| | £ | | 155.97 | 40.05 | 460.00 | 656.02 |
| Standard expansion joint anchorage for single sided open box beam | hr | nr | 4.52 | 4.52 | | |
| | £ | | 174.07 | 44.70 | 579.98 | 798.75 |
| Standard connection of single sided open box beam to bridge parapet (Type 048) | hr | nr | 0.70 | 0.70 | | |
| | £ | | 26.96 | 6.92 | 81.74 | 115.62 |
| Transition piece for open box beam to single sided corrugated beam | hr | nr | 0.78 | 0.78 | | |
| | £ | | 30.04 | 7.71 | 86.10 | 123.85 |
| Untensioned double sided open box beam | hr | m | 0.22 | 0.22 | | |
| | £ | | 8.47 | 2.18 | 46.06 | 56.71 |
| Steel post for setting in concrete for double height open box beam | hr | nr | 0.07 | 0.07 | | |
| | £ | | 2.70 | 0.69 | 15.05 | 18.44 |

**SECTION 4 SAFETY FENCES**

| | | UNIT | L | P | M | £ NETT |
|---|---|---|---|---|---|---|
| Standard connection of double height open box beam to bridge parapet (Type 048) | hr | nr | 0.70 | 0.70 | | |
| | £ | | 26.96 | 6.92 | 163.48 | 197.36 |
| Concrete footing for steel post for double height open box beam | hr | nr | 0.225 | 0.225 | | |
| | £ | | 8.66 | 2.23 | 5.00 | 15.89 |
| Driven steel post for untensioned double sided open box beam | hr | nr | 0.05 | 0.05 | | |
| | £ | | 1.93 | 0.49 | 24.68 | 27.10 |
| Steel post for setting in concrete for open box beam | hr | nr | 0.06 | 0.06 | | |
| | £ | | 2.31 | 0.59 | 15.05 | 17.95 |
| Concrete footing for steel post for open box beam | hr | nr | 0.23 | 0.23 | | |
| | £ | | 8.86 | 2.27 | 5.00 | 16.13 |
| Terminal for untensioned double sided open box beam (including concrete anchorage block) | hr | nr | 1.78 | 1.78 | | |
| | £ | | 68.55 | 17.60 | 340.00 | 426.15 |
| Untensioned double height open box beam | hr | m | 0.24 | 0.24 | | |
| | £ | | 9.24 | 2.37 | 47.44 | 59.05 |
| Driven steel post for untensioned double height open box beam | hr | nr | 0.11 | 0.11 | | |
| | £ | | 4.24 | 1.09 | 52.52 | 57.85 |

**WIRE ROPE SAFETY FENCE**

specialist advice (based on 600m lengths)

| | | | | | | |
|---|---|---|---|---|---|---|
| Wire rope safety fence to BS5750 with driven posts 4 rope system (Per Carriageway protected) | hr | m | - | - | | |
| | £ | | - | - | - | 20.00 |
| Wire rope safety fence to BS5750 with posts cast into concrete footings 4 rope system (Per Carriageway protected) | hr | m | - | - | | |
| | £ | | - | - | - | 24.00 |

approximate installation time on a metre by metre basis is one third that required to install Tensioned beam barrier

**SECTION 5 DRAINAGE AND SERVICE DUCTS**

**SEWERS AND DRAINS**

(Refer also to CESMM class I)

LABOUR based on Gang B

| | | | |
|---|---|---|---|
| Total hourly cost | = | £ | 32.51 |

PLANT

| | | | |
|---|---|---|---|
| Excavator (average 0.5m3 - 1.0m3) | | | 10.90 |
| Trench sheets, shores, props etc. (part time) | | | 2.30 |
| Plate Compactor | | | 0.90 |
| Use of power tools, pumps dumper etc. (average use) | | | 2.03 |
| Total hourly cost | = | £ | 16.13 |

## SECTION 5 DRAINAGE AND SERVICE DUCTS

### SEWERS AND DRAINS - continued

| | | UNIT | L | P | M | £ NETT |
|---|---|---|---|---|---|---|
| 150mm clayware to BS65, sleeved joints | | | | | | |
| 150mm granular bed and surround | | | | | | |
| depth to invert of 2.0m or less | hr | m | 0.22 | 0.22 | | |
| | £ | | 7.15 | 3.55 | 6.17 | 16.87 |
| depth to invert of over 2.0m average depth 2.5m | hr | m | 0.245 | 0.245 | | |
| | £ | | 7.96 | 3.95 | 6.17 | 18.08 |
| average depth 3.0m | hr | m | 0.26 | 0.26 | | |
| | £ | | 8.45 | 4.19 | 6.17 | 18.81 |
| average depth 3.5m | hr | m | 0.27 | 0.27 | | |
| | £ | | 8.78 | 4.36 | 6.17 | 19.31 |
| 150mm sand bed and surround | | | | | | |
| depth to invert of 2.0m or less | hr | m | 0.215 | 0.215 | | |
| | £ | | 6.99 | 3.47 | 7.99 | 18.45 |
| depth to invert of over 2.0m average depth 2.5m | hr | m | 0.24 | 0.24 | | |
| | £ | | 7.80 | 3.87 | 7.99 | 19.66 |
| average depth 3.0m | hr | m | 0.25 | 0.25 | | |
| | £ | | 8.13 | 4.03 | 7.99 | 20.15 |
| 150mm class E concrete bed and haunch | | | | | | |
| depth to invert of 2.0m or less | hr | m | 0.24 | 0.24 | | |
| | £ | | 7.80 | 3.87 | 11.61 | 23.28 |
| depth to invert of over 2.0m average depth 2.5m | hr | m | 0.27 | 0.27 | | |
| | £ | | 8.78 | 4.36 | 11.61 | 24.75 |

**SECTION 5 DRAINAGE AND SERVICE DUCTS**

**SEWERS AND DRAINS - continued**

| | | UNIT | L | P | M | £ NETT |
|---|---|---|---|---|---|---|
| 225mm clayware to BS65, sleeved joints | | | | | | |
| 150mm granular bed and surround | | | | | | |
| depth to invert of 2.0m or less | hr | m | 0.20 | 0.20 | | |
| | £ | | 6.50 | 3.23 | 11.99 | 21.72 |
| depth to invert of over 2.0m average depth 2.5m | hr | m | 0.23 | 0.23 | | |
| | £ | | 7.48 | 3.71 | 11.99 | 23.18 |
| average depth 3.0m | hr | m | 0.255 | 0.255 | | |
| | £ | | 8.29 | 4.11 | 11.99 | 24.39 |
| average depth 4.0m | hr | m | 0.295 | 0.295 | | |
| | £ | | 9.59 | 4.76 | 11.99 | 26.34 |
| 150mm sand bed and surround | | | | | | |
| depth to invert of 2.0m or less | hr | m | 0.20 | 0.20 | | |
| | £ | | 6.50 | 3.23 | 14.25 | 23.98 |
| depth to invert of over 2.0m average depth 2.5m | hr | m | 0.23 | 0.23 | | |
| | £ | | 7.48 | 3.71 | 14.25 | 25.44 |
| average depth 3.0m | hr | m | 0.254 | 0.254 | | |
| | £ | | 8.26 | 4.10 | 14.25 | 26.61 |
| average depth 4.0m | hr | m | 0.294 | 0.294 | | |
| | £ | | 9.56 | 4.74 | 14.25 | 28.55 |
| 150mm class E concrete bed and haunch | | | | | | |
| depth to invert of 2.0m or less | hr | m | 0.21 | 0.21 | | |
| | £ | | 6.83 | 3.39 | 18.55 | 28.77 |
| depth to invert of over 2.0m average depth 2.5m | hr | m | 0.23 | 0.23 | | |
| | £ | | 7.48 | 3.71 | 18.55 | 29.74 |
| average depth 3.0m | hr | m | 0.28 | 0.28 | | |
| | £ | | 9.10 | 4.52 | 18.55 | 32.17 |
| average depth 4.0m | hr | m | 0.37 | 0.37 | | |
| | £ | | 12.03 | 5.97 | 18.55 | 36.55 |

**SECTION 5 DRAINAGE AND SERVICE DUCTS**

**SEWERS AND DRAINS - continued**

| | UNIT | | L | P | M | £ NETT |
|---|---|---|---|---|---|---|
| 300 clayware to BS65, ringseal joints | | | | | | |
| 150mm granular bed and surround | | | | | | |
| depth to invert of 2.0m or less | hr | m | 0.20 | 0.20 | | |
| | £ | | 6.50 | 3.23 | 17.41 | 27.14 |
| depth to invert of over 2.0m | | | | | | |
| average depth 2.5m | hr | m | 0.24 | 0.24 | | |
| | £ | | 7.80 | 3.87 | 17.41 | 29.08 |
| average depth 3.0m | hr | m | 0.30 | 0.30 | | |
| | £ | | 9.75 | 4.84 | 17.41 | 32.00 |
| average depth 4m | hr | m | 0.34 | 0.34 | | |
| | £ | | 11.05 | 5.48 | 17.41 | 33.94 |
| average depth 5.0m | hr | m | 0.365 | 0.365 | | |
| | £ | | 11.87 | 5.89 | 17.41 | 35.17 |
| 150mm sand bed and surround | | | | | | |
| depth to invert of 2.0m or less | hr | m | 0.20 | 0.20 | | |
| | £ | | 6.50 | 3.23 | 20.09 | 29.82 |
| depth to invert of over 2.0m | | | | | | |
| average depth 2.5m | hr | m | 0.24 | 0.24 | | |
| | £ | | 7.80 | 3.87 | 20.09 | 31.76 |
| average depth 3.0m | hr | m | 0.30 | 0.30 | | |
| | £ | | 9.75 | 4.84 | 20.09 | 34.68 |
| average depth 4.0m | hr | m | 0.34 | 0.34 | | |
| | £ | | 11.05 | 5.48 | 20.09 | 36.62 |
| average depth 5.0m | hr | m | 0.365 | 0.365 | | |
| | £ | | 11.87 | 5.89 | 20.09 | 37.85 |
| 150mm class E concrete bed and haunch | | | | | | |
| depth to invert of 2.0m or less | hr | m | 0.14 | 0.14 | | |
| | £ | | 4.55 | 2.26 | 26.09 | 32.90 |
| depth to invert of over 2.0m | | | | | | |
| average depth 2.5m | hr | m | 0.185 | 0.185 | | |
| | £ | | 6.01 | 2.98 | 26.09 | 35.08 |
| average depth 3.0m | hr | m | 0.24 | 0.24 | | |
| | £ | | 7.80 | 3.87 | 26.09 | 37.76 |
| average depth 4.0m | hr | m | 0.26 | 0.26 | | |
| | £ | | 8.45 | 4.19 | 26.09 | 38.73 |
| average depth 5.0m | hr | m | 0.32 | 0.32 | | |
| | £ | | 10.40 | 5.16 | 26.09 | 41.65 |

**SECTION 5 DRAINAGE AND SERVICE DUCTS**

**SEWERS AND DRAINS - continued**

| | | UNIT | L | P | M | £ NETT |
|---|---|---|---|---|---|---|
| 375mm clayware to BS65, ringseal joints | | | | | | |
| 150mm granular bed and surround | | | | | | |
| depth to invert of 2.0m or less | hr | m | 0.23 | 0.23 | | |
| | £ | | 7.48 | 3.71 | 30.87 | 42.06 |
| depth to invert of over 2.0m average depth 2.5m | hr | m | 0.275 | 0.275 | | |
| | £ | | 8.94 | 4.44 | 30.87 | 44.25 |
| average depth 3.0m | hr | m | 0.31 | 0.34 | | |
| | £ | | 10.08 | 5.48 | 30.87 | 46.43 |
| average depth 4.0m | hr | m | 0.35 | 0.35 | | |
| | £ | | 11.38 | 5.65 | 30.87 | 47.90 |
| 150mm sand bed and surround | | | | | | |
| depth to invert of 2.0m or less | hr | m | 0.22 | 0.22 | | |
| | £ | | 7.15 | 3.55 | 33.89 | 44.59 |
| depth to invert of over 2.0m average depth 2.5m | hr | m | 0.26 | 0.26 | | |
| | £ | | 8.45 | 4.19 | 33.89 | 46.53 |
| average depth 3.0m | hr | m | 0.31 | 0.31 | | |
| | £ | | 10.08 | 5.00 | 33.89 | 48.97 |
| average depth 4.0m | hr | m | 0.35 | 0.35 | | |
| | £ | | 11.38 | 5.65 | 33.89 | 50.92 |
| 150mm class E concrete bed and haunch | | | | | | |
| depth to invert of 2.0m or less | hr | m | 0.25 | 0.25 | | |
| | £ | | 8.13 | 4.03 | 40.96 | 53.12 |
| depth to invert of over 2.0m | | | | | | |
| average depth 2.5m | hr | m | 0.30 | 0.30 | | |
| | £ | | 9.75 | 4.84 | 40.96 | 55.55 |
| average depth 3.0m | hr | m | 0.36 | 0.36 | | |
| | £ | | 11.70 | 5.81 | 40.96 | 58.47 |
| average depth 4.0m | hr | m | 0.39 | 0.39 | | |
| | £ | | 12.68 | 6.29 | 40.96 | 59.93 |

**SECTION 5 DRAINAGE AND SERVICE DUCTS**

**PIPED CULVERTS**

LABOUR based on Gang B

| | | |
|---|---|---|
| Total hourly cost | = £ | 32.51 |

PLANT

| | | |
|---|---|---|
| 1.0m3 Excavator | | 16.48 |
| Sheets, shores and props | | 3.70 |
| (part time) | | |
| Ped. Roller | | 1.10 |
| 20T Mobile Crane | | 3.38 |
| Tools, pumps etc. | | 2.03 |
| Total hourly cost | = £ | 26.69 |

**SECTION 5 DRAINAGE AND SERVICE DUCTS**

**PIPED CULVERTS - continued**

| | | UNIT | L | P | M | £ NETT |
|---|---|---|---|---|---|---|
| 450mm concrete class H | | | | | | |
| Trench type A (concrete bed) | | | | | | |
| depth to invert of 2.0m or less | hr | m | 0.37 | 0.37 | | |
| | £ | | 12.03 | 9.88 | 26.46 | 48.37 |
| depth to invert of over 2.0m | | | | | | |
| average depth 3.0m | hr | m | 0.434 | 0.434 | | |
| | £ | | 14.11 | 11.58 | 26.46 | 52.15 |
| average depth 4.0m | hr | m | 0.48 | 0.48 | | |
| | £ | | 15.60 | 12.81 | 26.46 | 54.87 |
| average depth 6.0m | hr | m | 0.63 | 0.63 | | |
| | £ | | 20.48 | 16.81 | 26.46 | 63.75 |
| 600mm concrete class H | | | | | | |
| Trench type A (concrete bed) | | | | | | |
| depth to invert of 2.0m or less | hr | m | 0.40 | 0.40 | | |
| | £ | | 13.00 | 10.68 | 39.15 | 62.83 |
| depth to invert of over 2.0m | | | | | | |
| average depth 3.0m | hr | m | 0.46 | 0.46 | | |
| | £ | | 14.95 | 12.28 | 39.15 | 66.38 |
| average depth 4.0m | hr | m | 0.51 | 0.51 | | |
| | £ | | 16.58 | 13.61 | 39.15 | 69.34 |
| average depth 6.0m | hr | m | 0.66 | 0.66 | | |
| | £ | | 21.46 | 17.62 | 39.15 | 78.23 |

**SECTION 5 DRAINAGE AND SERVICE DUCTS**

**PIPED CULVERTS - continued**

| | | UNIT | L | P | M | £ NETT |
|---|---|---|---|---|---|---|
| 900mm concrete class H | | | | | | |
| Trench type A (concrete bed) | | | | | | |
| depth to invert of 2.0m or less | hr | m | 0.55 | 0.55 | | |
| | £ | | 17.88 | 14.68 | 76.59 | 109.15 |
| depth to invert of over 2.0m | | | | | | |
| average depth 3.0m | hr | m | 0.74 | 0.74 | | |
| | £ | | 24.06 | 19.75 | 76.59 | 120.40 |
| average depth 4.0m | hr | m | 1.145 | 1.145 | | |
| | £ | | 37.22 | 30.56 | 76.59 | 144.37 |
| average depth 6.0m | hr | m | 1.77 | 1.77 | | |
| | £ | | 57.54 | 47.24 | 76.59 | 181.37 |
| Corrugated pipe galvanised and hot dip.bitumen coated. (Armco type) Bed types granular bed and surround | | | | | | |
| 1000mm diameter | | | | | | |
| depth to invert of 2.0m or less | hr | m | 0.99 | 0.99 | | |
| | £ | | 32.18 | 26.42 | 73.79 | 132.39 |
| 1600mm diameter | | | | | | |
| depth to invert of 2.0m or less | hr | m | 2.134 | 2.134 | | |
| | £ | | 69.38 | 56.96 | 135.22 | 261.56 |
| 2000mm diameter | | | | | | |
| depth to invert of 2.0m or less | hr | m | 2.97 | 2.97 | | |
| | £ | | 96.55 | 79.27 | 200.41 | 376.23 |
| 2200mm diameter | | | | | | |
| depth to invert of 2.0m or less | hr | m | 3.29 | 3.29 | | |
| | £ | | 106.96 | 87.81 | 216.72 | 411.49 |

**SECTION 5 DRAINAGE AND SERVICE DUCTS**

**FILTER DRAINS (FRENCH DRAINS)**

LABOUR based on Gang B

| | | | |
|---|---|---|---|
| Total hourly cost | = | £ | 32.51 |

| | | | |
|---|---|---|---|
| PLANT | | | |
| Excavator (average 0.5m3 - 1.0m3) | | | 10.90 |
| Trench sheets, shores, props. etc. (part time) | | | 2.30 |
| Plate Compactor | | | 0.90 |
| Use of power tools, pumps dumper etc. (average use) | | | 2.03 |
| Total hourly cost | = | £ | 16.13 |

**SECTION 5 DRAINAGE AND SERVICE DUCTS**

**FILTER DRAINS (FRENCH DRAINS) - continued**

| | | UNIT | L | P | M | £ NETT |
|---|---|---|---|---|---|---|
| 150mm perforated clayware pipe | | | | | | |
| Filter material type A bed<br>Filter material type A Fill | | | | | | |
| depth to invert 2.0m or less<br>average depth 1.0m | hr | m | 0.295 | 0.295 | | |
| | £ | | 9.59 | 4.76 | 9.23 | 23.58 |
| average depth 2.0m | hr | m | 0.30 | 0.30 | | |
| | £ | | 9.75 | 4.84 | 14.41 | 29.00 |
| depth to invert over 2.0m<br>average depth 3.0m | hr | m | 0.30 | 0.30 | | |
| | £ | | 9.75 | 4.84 | 19.58 | 34.17 |
| Filter material type A bed<br>Filter material type B Fill | | | | | | |
| depth to invert 2.0m or less<br>average depth 1.0m | hr | m | 0.295 | 0.295 | | |
| | £ | | 9.59 | 4.76 | 8.53 | 22.88 |
| average depth 2.0m | hr | m | 0.30 | 0.30 | | |
| | £ | | 9.75 | 4.84 | 12.81 | 27.40 |
| depth to invert over 2.0m<br>average depth 3.0m | hr | m | 0.30 | 0.30 | | |
| | £ | | 9.75 | 4.84 | 17.08 | 31.67 |
| Additional cost of concrete bed in place of type A Filter material | hr | m | 0.025 | 0.025 | | |
| | £ | | 0.81 | 0.40 | 3.63 | 4.84 |
| Deduct for porous concrete pipe | hr | m | - | - | | |
| | £ | | - | - | 3.38 | 3.38 |

**SECTION 5 DRAINAGE AND SERVICE DUCTS**

**FILTER DRAINS (FRENCH DRAINS) - continued**

| | | UNIT | L | P | M | £ NETT |
|---|---|---|---|---|---|---|
| 225mm perforated clayware pipe | | | | | | |
| Filter material type A bed<br>Filter material type A Fill | | | | | | |
| depth to invert 2.0m or less<br>average depth 1.0m | hr | m | 0.27 | 0.27 | | |
| | £ | | 8.78 | 4.36 | 13.18 | 26.32 |
| average depth 2.0m | hr | m | 0.30 | 0.30 | | |
| | £ | | 9.75 | 4.84 | 19.01 | 33.60 |
| depth to invert over 2.0m<br>average depth 3.0m | hr | m | 0.315 | 0.315 | | |
| | £ | | 10.24 | 5.08 | 24.84 | 40.16 |
| Filter material type A bed<br>Filter material type B Fill | | | | | | |
| depth to invert 2.0m or less<br>average depth 1.0m | hr | m | 0.27 | 0.27 | | |
| | £ | | 8.78 | 4.36 | 12.47 | 25.61 |
| average depth 2.0m | hr | m | 0.30 | 0.30 | | |
| | £ | | 9.75 | 4.84 | 17.27 | 31.86 |
| depth to invert over 2.0m<br>average depth 3.0m | hr | m | 0.315 | 0.315 | | |
| | £ | | 10.24 | 5.08 | 22.08 | 37.40 |
| Additional cost of concrete bed in place of type A Filter material | hr | m | 0.025 | 0.025 | | |
| | £ | | 0.81 | 0.40 | 5.34 | 6.55 |
| Deduct for porous concrete pipe | hr | m | - | - | | |
| | £ | | - | - | 5.54 | 5.54 |

**SECTION 5 DRAINAGE AND SERVICE DUCTS**

**FILTER DRAINS (FRENCH DRAINS) - continued**

| | | UNIT | L | P | M | £ NETT |
|---|---|---|---|---|---|---|
| 300mm perforated clayware pipe | | | | | | |
| Filter material type A bed<br>Filter material type A Fill | | | | | | |
| depth to invert 2.0m or less<br>average depth 1.0m | hr | m | 0.28 | 0.28 | | |
| | £ | | 9.10 | 4.52 | 20.36 | 33.98 |
| average depth 2.0m | hr | m | 0.27 | 0.27 | | |
| | £ | | 8.78 | 4.36 | 26.84 | 39.98 |
| depth to invert over 2.0m<br>average depth 3.0m | hr | m | 0.306 | 0.306 | | |
| | £ | | 9.95 | 4.94 | 33.40 | 48.29 |
| Filter material type A bed<br>Filter material type B Fill | | | | | | |
| depth to invert 2.0m or less<br>average depth 1.0m | hr | m | 0.28 | 0.28 | | |
| | £ | | 9.10 | 4.52 | 19.63 | 33.25 |
| average depth 2.0m | hr | m | 0.27 | 0.27 | | |
| | £ | | 8.78 | 4.36 | 24.97 | 38.11 |
| depth to invert over 2.0m<br>average depth 3.0m | hr | m | 0.306 | 0.306 | | |
| | £ | | 9.95 | 4.94 | 30.30 | 45.19 |
| Additional cost of concrete bed in place of type A Filter material | hr | m | 0.025 | 0.025 | | |
| | £ | | 0.81 | 0.40 | 7.22 | 8.43 |
| Type A Filter material contiguous with Filter drain | hr | m3 | 0.07 | 0.07 | | |
| | £ | | 2.28 | 1.13 | 8.63 | 12.04 |
| Type B Filter material contiguous with Filter drain | hr | m3 | 0.07 | 0.07 | | |
| | £ | | 2.28 | 1.13 | 7.12 | 10.53 |

## SECTION 5 DRAINAGE AND SERVICE DUCTS

## FIN DRAINS AND NARROW FILTER DRAINS

| | | UNIT | L | P | M | £ NETT |
|---|---|---|---|---|---|---|
| Edge of carriageway drainage. | | | | | | |
| Fin Drain D.O.T. type 6 using 'Trammel' drainage fabrics. pipe surround: sand and granular fill backfill: selected suitable material | | | | | | |
| 100mm clay perforated pipe | | | | | | |
| depth to invert of 2.0m or less | | | | | | |
| average depth 1.0m | hr | m | 0.17 | 0.17 | | |
| | £ | | 5.53 | 2.74 | 9.02 | 17.29 |
| average depth 2.0m | hr | m | 0.23 | 0.23 | | |
| | £ | | 7.48 | 3.71 | 9.02 | 20.21 |
| Fin Drain D.O.T. type 7 using 'Trammel' drainage fabrics. pipe surround: sand and granular fill backfill: selected suitable material | | | | | | |
| 100mm clay slotted pipe. | | | | | | |
| depth to invert of 2.0m or less | | | | | | |
| average depth 1.0m | hr | m | 0.17 | 0.17 | | |
| | £ | | 5.53 | 2.74 | 8.81 | 17.08 |
| average depth 2.0m | hr | m | 0.23 | 0.23 | | |
| | £ | | 7.48 | 3.71 | 8.81 | 20.00 |
| Narrow Filter Drain D.O.T. type 8 using 'Trammel' drainage fabrics. pipe surround: sand and granular fill backfill: selected suitable material | | | | | | |
| 100mm clay perforated pipe | | | | | | |
| depth to invert of 2.0m or less | | | | | | |
| average depth 1.0m | hr | m | 0.165 | 0.165 | | |
| | £ | | 5.36 | 2.66 | 6.36 | 14.38 |
| average depth 2.0m | hr | m | 0.22 | 0.22 | | |
| | £ | | 7.15 | 3.55 | 6.36 | 17.06 |

**SECTION 5 DRAINAGE AND SERVICE DUCTS**

**FIN DRAINS AND NARROW FILTER DRAINS - continued**

| | | UNIT | L | P | M | £ NETT |
|---|---|---|---|---|---|---|
| Narrow Filter Drain D.O.T.type 9 using 'Trammel' drainage fabrics. pipe surround: sand and granular fill backfill: selected suitable material 100mm clay perforated pipe | | | | | | |
| depth to invert of 2.0m or less average depth 1.0m | hr | m | 0.165 | 0.165 | | |
| | £ | | 5.36 | 2.66 | 5.99 | 14.01 |
| average depth 2.0m | hr | m | 0.23 | 0.23 | | |
| | £ | | 7.48 | 3.71 | 5.99 | 17.18 |

**SERVICE DUCTS**

| | | UNIT | L | P | M | £ NETT |
|---|---|---|---|---|---|---|
| One number 100mm diameter service ducts granular bed and surround trench depth to invert of 2.0m or less | hr | m | 0.14 | 0.14 | | |
| | £ | | 4.55 | 2.26 | 2.30 | 9.11 |
| Two number 100mm diameter service ducts granular bed and surround trench depth to invert of 2.0m or less | hr | m | 0.24 | 0.24 | | |
| | £ | | 7.80 | 3.87 | 4.60 | 16.27 |
| Three number 100mm diameter service ducts granular bed and surround trench depth to invert of 2.0m or less | hr | m | 0.32 | 0.32 | | |
| | £ | | 10.40 | 5.16 | 6.82 | 22.38 |
| Four number 100mm diameter service ducts granular bed and surround trench depth to invert of 2.0m or less | hr | m | 0.40 | 0.40 | | |
| | £ | | 13.00 | 6.45 | 8.97 | 28.42 |
| Six number 100mm diameter service ducts granular bed and surround trench depth to invert of 2.0m or less | hr | m | 0.60 | 0.60 | | |
| | £ | | 19.51 | 9.68 | 13.49 | 42.68 |
| Extra over for class E Concrete bed and surround | hr | m | 0.08 | 0.08 | | |
| | £ | | 2.60 | 1.29 | 8.00 | 11.89 |
| Extra cost for 150 diameter conduit | hr | m | 0.005 | 0.005 | | |
| | £ | | 0.16 | 0.08 | 2.70 | 2.94 |
| Extra cost for 225 diameter conduit | hr | m | 0.012 | 0.012 | | |
| | £ | | 0.39 | 0.19 | 7.20 | 7.78 |

**SECTION 5 DRAINAGE AND SERVICE DUCTS**

**MANHOLES & CHAMBERS**

In the MMHW manholes are enumerated. To enable a rate to be established the various constituent elements have been broken down and unit rates given against each. (waste allowance, concrete = 10%, manufactured goods = 2%)
(Refer to Approximate Estimates for a list of enumerated manholes)
See also page 311

| | | UNIT | L | P | M | £ NETT |
|---|---|---|---|---|---|---|
| Excavation by machine | | | | | | |
| ne 2m deep | hr | m3 | 0.22 | 0.22 | | |
| | £ | | 6.70 | 2.27 | - | 8.97 |
| 2.0 - 3.0m deep | hr | m3 | 0.24 | 0.24 | | |
| | £ | | 7.31 | 2.47 | - | 9.78 |
| 3.0 - 4.0m deep | hr | m3 | 0.26 | 0.26 | | |
| | £ | | 7.92 | 2.68 | - | 10.60 |
| Backfill | | | | | | |
| ne 2m deep | hr | m3 | 0.11 | 0.11 | | |
| | £ | | 3.35 | 1.13 | - | 4.48 |
| 2.0 - 3.0m deep | hr | m3 | 0.11 | 0.11 | | |
| | £ | | 3.35 | 1.13 | - | 4.48 |
| 3.0 - 4.0m deep | hr | m3 | 0.11 | 0.11 | | |
| | £ | | 3.35 | 1.13 | - | 4.48 |
| Earthwork support | | | | | | |
| ne 2m deep | hr | m2 | - | - | | |
| | £ | | 0.20 | 0.10 | 0.45 | 0.75 |
| 2.0 - 3.0m deep | hr | m2 | - | - | | |
| | £ | | 0.30 | 0.10 | 0.85 | 1.25 |
| 3.0 - 4.0m deep | hr | m2 | - | - | | |
| | £ | | 0.40 | 0.15 | 1.45 | 2.00 |
| Disposal of excavated material | | | | | | |
| off site | hr | m3 | - | - | | |
| | £ | | 1.40 | 3.30 | - | 4.70 |
| to spoil heap average 50m | hr | m3 | - | - | | |
| | £ | | 0.51 | 1.15 | - | 1.66 |

**SECTION 5 DRAINAGE AND SERVICE DUCTS**

**MANHOLES - continued**

| | | UNIT | L | P | M | £ NETT |
|---|---|---|---|---|---|---|
| Mass concrete grade C15P blinding thickness ne 150mm | hr | m3 | 0.21 | 0.21 | | |
| | £ | | 6.39 | 2.16 | 47.61 | 56.16 |
| Mass concrete grade C15P surrounds to manholes thickness ne 150mm | hr | m3 | 0.46 | 0.46 | | |
| | £ | | 14.01 | 4.74 | 47.61 | 66.36 |
| thickness 150 - 500mm | hr | m3 | 0.34 | 0.34 | | |
| | £ | | 10.35 | 3.50 | 47.61 | 61.46 |
| Reinforced concrete grade C20P bases thickness 150 - 300mm | hr | m3 | 0.46 | 0.46 | | |
| | £ | | 14.01 | 4.74 | 49.42 | 68.17 |
| thickness 300 - 500mm | hr | m3 | 0.50 | 0.50 | | |
| | £ | | 15.23 | 5.15 | 49.42 | 69.80 |
| Formwork curved vertical to base and benching | hr | m2 | - | - | | |
| | £ | | 3.25 | 1.00 | 6.79 | 11.04 |
| curved to surround (based on GRP forms) | hr | m2 | - | - | | |
| | £ | | 2.00 | 1.00 | 4.84 | 7.84 |

**SECTION 5 DRAINAGE AND SERVICE DUCTS**

**MANHOLES - continued**

| | | UNIT | L | P | M | £ NETT |
|---|---|---|---|---|---|---|
| Precast concrete circular manhole rings; BS5911 Part 1 | | | | | | |
| shaft ring; | | | | | | |
| 675mm (plain/reinforced) | hr | m | 0.80 | 0.80 | | |
| | £ | | 24.36 | 8.24 | 23.35 | 55.95 |
| 900mm (plain/reinforced) | hr | m | 0.95 | 0.95 | | |
| | £ | | 28.93 | 9.79 | 38.61 | 77.33 |
| 1050mm (plain/reinforced) | hr | m | 1.05 | 1.05 | | |
| | £ | | 31.97 | 10.82 | 48.66 | 91.45 |
| 1200mm (plain/reinforced) | hr | m | 1.17 | 1.17 | | |
| | £ | | 35.63 | 12.05 | 65.14 | 112.82 |
| 1500mm (reinforced) | hr | m | 1.32 | 1.32 | | |
| | £ | | 40.19 | 13.60 | 112.85 | 166.64 |
| reducing slabs (675mm diameter access) | | | | | | |
| 900mm | hr | nr | 0.57 | 0.57 | | |
| | £ | | 17.36 | 5.87 | 49.84 | 73.07 |
| 1050mm | hr | nr | 0.63 | 0.63 | | |
| | £ | | 19.18 | 6.49 | 62.50 | 88.17 |
| 1200mm | hr | nr | 0.70 | 0.70 | | |
| | £ | | 21.32 | 7.21 | 84.85 | 113.38 |
| 1500mm | hr | nr | 0.79 | 0.79 | | |
| | £ | | 24.06 | 8.14 | 155.04 | 187.24 |
| landing slabs (900mm diameter access) | | | | | | |
| 1200mm | hr | nr | 0.70 | 0.70 | | |
| | £ | | 21.32 | 7.21 | 146.70 | 175.23 |
| 1500mm | hr | nr | 0.79 | 0.79 | | |
| | £ | | 24.06 | 8.14 | 195.39 | 227.59 |

**SECTION 5 DRAINAGE AND SERVICE DUCTS**

**MANHOLES - continued**

| | | UNIT | L | P | M | £ NETT |
|---|---|---|---|---|---|---|
| cover slab | | | | | | |
| 675mm | hr | nr | 0.43 | 0.43 | | |
| | £ | | 13.09 | 4.43 | 19.95 | 37.47 |
| 900mm | hr | nr | 0.54 | 0.54 | | |
| | £ | | 16.44 | 5.56 | 33.64 | 55.64 |
| 1050mm | hr | nr | 0.60 | 0.60 | | |
| | £ | | 18.27 | 6.18 | 43.30 | 67.75 |
| 1200mm | hr | nr | 0.70 | 0.70 | | |
| | £ | | 21.32 | 7.21 | 59.88 | 88.41 |
| 1500mm | hr | nr | 0.80 | 0.80 | | |
| | £ | | 24.36 | 8.24 | 112.85 | 145.45 |
| Mass concrete grade C15P benching to bottom of manhole | | | | | | |
| 675mm | hr | nr | 0.20 | 0.20 | | |
| | £ | | 6.09 | 2.06 | 2.73 | 10.88 |
| 900mm | hr | nr | 0.40 | 0.40 | | |
| | £ | | 12.18 | 4.12 | 4.54 | 20.84 |
| 1050mm | hr | nr | 0.72 | 0.72 | | |
| | £ | | 21.92 | 7.42 | 6.18 | 35.52 |
| 1200mm | hr | nr | 0.95 | 0.95 | | |
| | £ | | 28.93 | 9.79 | 8.08 | 46.80 |
| 1500mm | hr | nr | 1.15 | 1.15 | | |
| | £ | | 35.02 | 11.85 | 12.62 | 59.49 |

**SECTION 5 DRAINAGE AND SERVICE DUCTS**

**MANHOLES (BRICKWORK TO MANHOLES)**

| | | | |
|---|---|---|---|
| LABOUR based on Gang J | | | |
| Total hourly cost | = | £ | 28.35 |
| PLANT | | | |
| Mixer | | | 0.83 |
| Part time small tools<br>Disc grinder/saw<br>Dumper 2T<br>Access equipment | | | 0.53 |
| Total hourly cost | = | £ | 1.36 |

| | | UNIT | L | P | M | £ NETT |
|---|---|---|---|---|---|---|
| Class B Engineering brickwork; PC £200.00/1000 | | | | | | |
| one brick thick | | | | | | |
| depths up to 1m deep | hr | m2 | 0.75 | 0.75 | | |
| | £ | | 21.26 | 1.02 | 27.14 | 49.42 |
| 1.0 - 2.0m | hr | m2 | 0.80 | 0.80 | | |
| | £ | | 22.68 | 1.09 | 27.14 | 50.91 |
| 2.0 - 3.0m | hr | m2 | 0.87 | 0.87 | | |
| | £ | | 24.66 | 1.18 | 27.14 | 52.98 |
| 3.0 - 4.0m | hr | m2 | 0.94 | 0.94 | | |
| | £ | | 26.65 | 1.28 | 27.14 | 55.07 |
| 4.0 - 5.0m | hr | m2 | 1.05 | 1.05 | | |
| | £ | | 29.77 | 1.43 | 27.14 | 58.34 |
| one and one half brick thick | | | | | | |
| depths up to 1m deep | hr | m2 | 1.03 | 1.03 | | |
| | £ | | 29.20 | 1.40 | 41.11 | 71.71 |
| 1.0 - 2.0m | hr | m2 | 1.15 | 1.15 | | |
| | £ | | 32.60 | 1.56 | 41.11 | 75.27 |
| 2.0 - 3.0m | hr | m2 | 1.26 | 1.26 | | |
| | £ | | 35.72 | 1.71 | 41.11 | 78.54 |
| 3.0 - 4.0m | hr | m2 | 1.37 | 1.37 | | |
| | £ | | 38.84 | 1.86 | 41.11 | 81.81 |
| 4.0 - 5.0m | hr | m2 | 1.48 | 1.48 | | |
| | £ | | 41.96 | 2.01 | 41.11 | 85.08 |

**SECTION 5 DRAINAGE AND SERVICE DUCTS**

**MANHOLES (BRICKWORK TO MANHOLES) - continued**

| | | UNIT | L | P | M | £ NETT |
|---|---|---|---|---|---|---|
| Step irons; galvanised mild steel | | | | | | |
| built in brickwork | | | | | | |
| depths up to 1m deep | hr | nr | 0.03 | 0.03 | | |
| | £ | | 0.85 | 0.04 | 4.00 | 4.89 |
| 1.0 - 2.0m | hr | nr | 0.03 | 0.03 | | |
| | £ | | 0.85 | 0.04 | 4.00 | 4.89 |
| 2.0 - 3.0m | hr | nr | 0.04 | 0.04 | | |
| | £ | | 1.13 | 0.05 | 4.00 | 5.18 |
| 3.0 - 4.0m | hr | nr | 0.05 | 0.05 | | |
| | £ | | 1.42 | 0.07 | 4.00 | 5.49 |
| 4.0 - 5.0m | hr | nr | 0.06 | 0.06 | | |
| | £ | | 1.70 | 0.08 | 4.00 | 5.78 |
| cast in precast concrete | | | | | | |
| depths up to 1m deep | hr | nr | 0.03 | 0.03 | | |
| | £ | | 0.85 | 0.04 | 3.20 | 4.09 |
| 1.0 - 2.0m | hr | nr | 0.03 | 0.03 | | |
| | £ | | 0.85 | 0.04 | 3.20 | 4.09 |
| 2.0 - 3.0m | hr | nr | 0.04 | 0.04 | | |
| | £ | | 1.13 | 0.05 | 3.20 | 4.38 |
| 3.0 - 4.0m | hr | nr | 0.05 | 0.05 | | |
| | £ | | 1.42 | 0.07 | 3.20 | 4.69 |
| 4.0 - 5.0m | hr | nr | 0.06 | 0.06 | | |
| | £ | | 1.70 | 0.08 | 3.20 | 4.98 |
| Vitrified clayware channels in base of manhole (Channel diameter) | | | | | | |
| straight | | | | | | |
| 100mm | hr | m | 0.18 | 0.18 | | |
| | £ | | 5.10 | 0.24 | 1.73 | 7.07 |
| 150mm | hr | m | 0.23 | 0.23 | | |
| | £ | | 6.52 | 0.31 | 2.87 | 9.70 |
| 225mm | hr | m | 0.32 | 0.32 | | |
| | £ | | 9.07 | 0.44 | 6.39 | 15.90 |
| 300mm | hr | m | 0.43 | 0.43 | | |
| | £ | | 12.19 | 0.58 | 12.64 | 25.41 |

**SECTION 5 DRAINAGE AND SERVICE DUCTS**

**MANHOLES (BRICKWORK TO MANHOLES) - continued**

| | | UNIT | L | P | M | £ NETT |
|---|---|---|---|---|---|---|
| channel bend | | | | | | |
| 100mm | hr | nr | 0.14 | 0.14 | | |
| | £ | | 3.97 | 0.19 | 1.59 | 5.75 |
| 150mm | hr | nr | 0.19 | 0.19 | | |
| | £ | | 5.39 | 0.26 | 2.63 | 8.28 |
| 225mm | hr | nr | 0.28 | 0.28 | | |
| | £ | | 7.94 | 0.38 | 8.78 | 17.10 |
| 300mm | hr | nr | 0.40 | 0.40 | | |
| | £ | | 11.34 | 0.54 | 17.43 | 29.31 |
| channel branch | | | | | | |
| 100mm | hr | nr | 0.16 | 0.16 | | |
| | £ | | 4.54 | 0.22 | 3.03 | 7.79 |
| 150mm | hr | nr | 0.26 | 0.26 | | |
| | £ | | 7.37 | 0.35 | 5.02 | 12.74 |
| 225mm | hr | nr | 0.67 | 0.67 | | |
| | £ | | 18.99 | 0.91 | 14.39 | 34.29 |
| 300mm | hr | nr | 1.07 | 1.07 | | |
| | £ | | 30.33 | 1.46 | 34.24 | 66.03 |
| half section straight taper | | | | | | |
| 150mm | hr | nr | 0.22 | 0.22 | | |
| | £ | | 6.24 | 0.30 | 6.66 | 13.20 |
| 225mm | hr | nr | 0.33 | 0.33 | | |
| | £ | | 9.36 | 0.45 | 14.72 | 24.53 |
| 300mm | hr | nr | 0.48 | 0.48 | | |
| | £ | | 13.61 | 0.65 | 29.12 | 43.38 |
| half section branch channel bends | | | | | | |
| 100mm | hr | nr | 0.14 | 0.14 | | |
| | £ | | 3.97 | 0.19 | 3.26 | 7.42 |
| 150mm | hr | nr | 0.19 | 0.19 | | |
| | £ | | 5.39 | 0.26 | 5.35 | 11.00 |
| 225mm | hr | nr | 0.28 | 0.28 | | |
| | £ | | 7.94 | 0.38 | 17.67 | 25.99 |
| 300mm | hr | nr | 0.40 | 0.40 | | |
| | £ | | 11.34 | 0.54 | 34.98 | 46.86 |

**SECTION 5 DRAINAGE AND SERVICE DUCTS**

**MANHOLES (BRICKWORK TO MANHOLES) - continued**

| | | UNIT | L | P | M | £ NETT |
|---|---|---|---|---|---|---|
| three quarter section branch channel bends | | | | | | |
| 100mm | hr | nr | 0.14 | 0.14 | | |
| | £ | | 3.97 | 0.19 | 3.60 | 7.76 |
| 150mm | hr | nr | 0.19 | 0.19 | | |
| | £ | | 5.39 | 0.26 | 6.08 | 11.73 |
| 225mm | hr | nr | 0.28 | 0.28 | | |
| | £ | | 7.94 | 0.38 | 22.03 | 30.35 |
| 300mm | hr | nr | 0.40 | 0.40 | | |
| | £ | | 11.34 | 0.54 | 43.69 | 55.57 |
| Spun concrete channels in base of manhole (Channel diameter) | | | | | | |
| straight | | | | | | |
| 300mm | hr | nr | 0.38 | 0.38 | | |
| | £ | | 10.77 | 0.52 | 5.46 | 16.75 |
| 450mm | hr | nr | 0.75 | 0.75 | | |
| | £ | | 21.26 | 1.02 | 9.40 | 31.68 |
| 600mm | hr | nr | 1.05 | 1.05 | | |
| | £ | | 29.77 | 1.43 | 17.23 | 48.43 |
| 900mm | hr | nr | 1.97 | 1.97 | | |
| | £ | | 55.85 | 2.68 | 33.08 | 91.61 |
| 1200mm | hr | nr | 3.14 | 3.14 | | |
| | £ | | 89.02 | 4.27 | 58.51 | 151.80 |
| Build in pipes | | | | | | |
| to one brick wall | | | | | | |
| 300mm | hr | nr | 0.08 | 0.08 | | |
| | £ | | 2.27 | 0.11 | 0.71 | 3.09 |
| 450mm | hr | nr | 0.11 | 0.11 | | |
| | £ | | 3.12 | 0.15 | 1.01 | 4.28 |
| 600mm | hr | nr | 0.15 | 0.15 | | |
| | £ | | 4.25 | 0.20 | 1.36 | 5.81 |
| 900mm | hr | nr | 0.20 | 0.20 | | |
| | £ | | 5.67 | 0.27 | 2.50 | 8.44 |
| 1200mm | hr | nr | 0.22 | 0.22 | | |
| | £ | | 6.24 | 0.30 | 5.00 | 11.54 |

**SECTION 5 DRAINAGE AND SERVICE DUCTS**

**MANHOLES (BRICKWORK TO MANHOLES) - continued**

| | | UNIT | L | P | M | £ NETT |
|---|---|---|---|---|---|---|
| to precast concrete walls | | | | | | |
| 300mm | hr | nr | 0.45 | 0.45 | | |
| | £ | | 12.76 | 0.61 | 0.94 | 14.31 |
| 450mm | hr | nr | 0.67 | 0.67 | | |
| | £ | | 18.99 | 0.91 | 1.37 | 21.27 |
| 600mm | hr | nr | 0.87 | 0.87 | | |
| | £ | | 24.66 | 1.18 | 1.95 | 27.79 |
| 900mm | hr | nr | 1.28 | 1.28 | | |
| | £ | | 36.29 | 1.74 | 4.32 | 42.35 |
| 1200mm | hr | nr | 1.61 | 1.61 | | |
| | £ | | 45.64 | 2.19 | 7.72 | 55.55 |

For material costs of access covers and frames refer to resources section

| | | | | | | |
|---|---|---|---|---|---|---|
| Fix only access covers and frames | | | | | | |
| grade A double triangular solid top | hr | nr | 0.60 | 0.60 | | |
| | £ | | 17.01 | 0.82 | - | 17.83 |
| grade A single triangular solid top | hr | nr | 0.48 | 0.48 | | |
| | £ | | 13.61 | 0.65 | - | 14.26 |
| grade B medium duty | hr | nr | 0.39 | 0.39 | | |
| | £ | | 11.06 | 0.53 | - | 11.59 |
| grade C light duty | hr | nr | 0.29 | 0.29 | | |
| | £ | | 8.22 | 0.39 | - | 8.61 |

**MANHOLE ANCILLARIES**

| | | | | | | |
|---|---|---|---|---|---|---|
| manhole interceptors | | | | | | |
| 150mm diameter | hr | nr | 0.20 | 0.20 | | |
| | £ | | 5.67 | 0.27 | 26.69 | 32.63 |
| 255mm diameter | hr | nr | 0.30 | 0.30 | | |
| | £ | | 8.51 | 0.41 | 71.97 | 80.89 |
| 300mm diameter | hr | nr | 0.40 | 0.40 | | |
| | £ | | 11.34 | 0.54 | 164.70 | 176.58 |
| Rodding eye points (excluding pipework) | | | | | | |
| 100mm diameter | hr | nr | 0.40 | 0.40 | | |
| | £ | | 11.34 | 0.54 | 17.16 | 29.04 |
| 150mm diameter | hr | nr | 0.45 | 0.45 | | |
| | £ | | 12.76 | 0.61 | 27.49 | 40.86 |

## SECTION 5 DRAINAGE AND SERVICE DUCTS

### GULLIES

| | | UNIT | L | P | M | £ NETT |
|---|---|---|---|---|---|---|
| **Clay** | | | | | | |
| vitrified clay road gully 450mm diameter x 900mm deep, including excavation and removal, 150mm concrete grade C20P surround, class B semi engineering brick seating, cast iron road gully and frame 475 x 475mm grade A | hr | nr | 0.50 | 0.50 | | |
| | £ | | 14.18 | 5.15 | 141.22 | 160.55 |
| **Clay trapped** | | | | | | |
| vitrified clay yard gully 225 diameter (Deans) with bucket and rodding eye, including excavation and removal, 150mm concrete grade C20P surround, cast iron gully and grating and frame grade B | hr | nr | 0.30 | 0.30 | | |
| | £ | | 8.51 | 3.09 | 67.35 | 78.95 |
| vitrified clay grease trap 600 x 450mm with internal access and bucket, including excavation and removal, 150mm concrete grade C20P surround, cast iron lid and frame grade B class 2 | hr | nr | 0.35 | 0.35 | | |
| | £ | | 9.92 | 3.61 | 291.73 | 305.26 |
| **Precast concrete trapped** | | | | | | |
| precast concrete road gully 450mm diameter with bucket and rodding eye, including excavation and removal, 150mm concrete grade C20P surround, class B semi-engineering brick seating, cast iron road gully grating and frame 475 x 475mm grade A | hr | nr | 0.50 | 0.50 | | |
| | £ | | 14.18 | 5.15 | 124.07 | 143.40 |

## SECTION 5 DRAINAGE AND SERVICE DUCTS

| | | UNIT | L | P | M | £ NETT |
|---|---|---|---|---|---|---|
| **Excavation in rock, pavement, brickwork, concrete, reinforced concrete, masonry and the like** | | | | | | |
| Extra over any item of drainage for excavation in existing pavement, brickwork, concrete, masonry and the like | hr | m3 | 0.145 | 0.145 | | |
| | £ | | 4.71 | 2.34 | - | 7.05 |
| Extra over any item of drainage for excavation of rock | hr | m3 | 0.35 | 0.35 | | |
| | £ | | 11.38 | 5.65 | - | 17.03 |
| Extra over any item of drainage for excavation in reinforced concrete | hr | m3 | 0.60 | 0.60 | | |
| | £ | | 19.51 | 9.68 | - | 29.19 |
| **Reinstatement of pavement** | | | | | | |
| Extra over one number 100mm diameter service duct for reinstating pavement and foundation under | hr | m | 0.09 | 0.09 | | |
| | £ | | 2.93 | 1.45 | 10.50 | 14.88 |
| Extra over 150mm diameter sewer, drain or service duct for reinstating existing pavement and foundation under | hr | m | 0.10 | 0.10 | | |
| | £ | | 3.25 | 1.61 | 11.22 | 16.08 |
| Extra over 225mm diameter sewer, drain or service duct for reinstating existing pavement and foundation under | hr | m | 0.10 | 0.10 | | |
| | £ | | 3.25 | 1.61 | 12.25 | 17.11 |
| Extra over 300mm diameter sewer, drain or service duct for reinstating existing pavement and foundation under | hr | m | 0.10 | 0.10 | | |
| | £ | | 3.25 | 1.61 | 12.60 | 17.46 |
| Extra over 375mm diameter sewer, drain or service duct for reinstating existing pavement and foundation under | hr | m | 0.10 | 0.10 | | |
| | £ | | 3.25 | 1.61 | 12.95 | 17.81 |

## SECTION 5 DRAINAGE AND SERVICE DUCTS

| | | UNIT | L | P | M | £ NETT |
|---|---|---|---|---|---|---|
| Extra over 450mm diameter sewer, drain or service duct for reinstating existing pavement and foundation under | hr | m | 0.11 | 0.11 | | |
| | £ | | 3.58 | 1.77 | 13.68 | 19.03 |
| Extra over two number 100mm diameter group I service ducts for reinstating pavement and foundation under | hr | m | 0.10 | 0.10 | | |
| | £ | | 3.25 | 1.61 | 10.85 | 15.71 |
| Extra over three number 100mm diameter group I service ducts for reinstating pavement and foundation | hr | m | 0.10 | 0.10 | | |
| | £ | | 3.25 | 1.61 | 11.22 | 16.08 |
| Extra over four number 100mm diameter group I service ducts for reinstating pavement and foundation | hr | m | 0.10 | 0.10 | | |
| | £ | | 3.25 | 1.61 | 12.25 | 17.11 |
| **Soft spots and other voids** | | | | | | |
| Excavation of soft spots in bottom of trenches, manholes, catchpits, drawpits and gullies | hr | m3 | 0.07 | 0.07 | | |
| | £ | | 2.28 | 1.13 | 5.40 | 8.81 |
| Filling of soft spots and other voids in bottom of trenches, manholes, catchpits, drawpits and gullies with pipe bedding material | hr | m3 | 0.112 | 0.112 | | |
| | £ | | 3.64 | 1.81 | 13.25 | 18.70 |
| Raising the level of 100mm x 100mm water stop tap boxes on 100mm x 100mm brick chambers by 150mm or less | hr | nr | 0.06 | 0.06 | | |
| | £ | | 1.95 | 0.97 | 7.35 | 10.27 |
| Lowering the level of 100mm x 100mm water stop tap boxes on 100mm x 100mm brick chambers by 150mm or less | hr | nr | 0.04 | 0.04 | | |
| | £ | | 1.30 | 0.65 | 4.40 | 6.35 |

## SECTION 5 DRAINAGE AND SERVICE DUCTS

| | | UNIT | L | P | M | £ NETT |
|---|---|---|---|---|---|---|
| Raising the level of 420mm x 420mm cover and frame on 420mm x 420mm in-situ concrete chamber by 150mm or less | hr | nr | 0.10 | 0.10 | | |
| | £ | | 3.25 | 1.61 | 12.75 | 17.61 |
| Lowering the level of 420mm x 420mm G.P.O. cover and frame on 420mm x 420mm in-situ concrete chamber by 150mm or less | hr | nr | 0.06 | 0.06 | | |
| | £ | | 1.95 | 0.97 | 7.90 | 10.82 |
| Raising the level of 700mm x 500mm cover and frame on 700mm x 500mm in-situ concrete chamber by 150mm or less | hr | nr | 0.17 | 0.17 | | |
| | £ | | 5.53 | 2.74 | 18.90 | 27.17 |
| Raising the level of 600mm x 600mm grade "A" heavy duty manhole cover and frame on 600mm x 600mm brick chamber by 150mm or less | hr | nr | 0.17 | 0.17 | | |
| | £ | | 5.53 | 2.74 | 18.90 | 27.17 |
| Raising the level of 600mm x 600mm grade "A" heavy duty manhole cover and frame on 600mm x 600mm brick chamber by exceeding 150mm but not exceeding 300mm | hr | nr | 0.21 | 0.21 | | |
| | £ | | 6.83 | 3.39 | 23.80 | 34.02 |
| Lowering the level of 600mm x 600mm grade "A" heavy duty manhole cover and frame on 600mm x 600mm brick chamber by 150mm or less | hr | nr | 0.10 | 0.10 | | |
| | £ | | 3.25 | 1.61 | 11.25 | 16.11 |
| Extra over for back filling trenches with selected sand fill including compaction | hr | m3 | 0.085 | 0.085 | | |
| | £ | | 2.76 | 1.37 | 12.90 | 17.03 |
| Sealing redundant road gullies with class E concrete | hr | nr | 0.023 | 0.023 | | |
| | £ | | 0.75 | 0.37 | 6.50 | 7.62 |

**SECTION 5 DRAINAGE AND SERVICE DUCTS**

| | | UNIT | L | P | M | £ NETT |
|---|---|---|---|---|---|---|
| Sealing Redundant Chambers | | | | | | |
| depth to invert 1m or less | hr | nr | 0.09 | 0.09 | | |
| | £ | | 2.93 | 1.45 | 26.00 | 30.38 |
| depth to invert 1 - 2m | hr | nr | 0.21 | 0.21 | | |
| | £ | | 6.83 | 3.39 | 41.60 | 51.82 |
| depth to invert 2 - 3m | hr | nr | 0.55 | 0.55 | | |
| | £ | | 17.88 | 8.87 | 68.50 | 95.25 |
| Grouting Redundant Drains and Sewers | | | | | | |
| 100mm diameter | hr | m | 0.03 | 0.03 | | |
| | £ | | 0.98 | 0.48 | 2.02 | 3.48 |
| 300mm diameter | hr | m | 0.125 | 0.125 | | |
| | £ | | 4.06 | 2.02 | 8.20 | 14.28 |
| 450mm diameter | hr | m | 0.26 | 0.26 | | |
| | £ | | 8.45 | 4.19 | 17.75 | 30.39 |
| 600mm diameter | hr | m | 0.50 | 0.50 | | |
| | £ | | 16.26 | 8.07 | 36.05 | 60.38 |
| 1200mm diameter | hr | m | 1.70 | 1.70 | | |
| | £ | | 55.27 | 27.42 | 100.80 | 183.49 |

## SECTION 6 EARTHWORKS

The cost of earth moving and other associated works is dependent on matching the overall quantities and production rates called for by the programme of works with the most appropriate plant and assessing the most suitable version of that plant that will:

deal with the site conditions (e.g. type of ground, type of excavation, length of haul, prevailing weather, etc.);

comply with the specification requirements (e.g. compaction, separation of materials, surface tolerances, etc.);

complete the work economically (e.g. provide surface tolerances which will avoid undue excess thickness of expensive imported materials);

Refer to CESMM Class E for details of comparative costs for various types of equipment.

Unless stated the units for excavation in material other than topsoil rock or artificial hard materials, are based on excavation in firm gravel soils. Multiply the rates by factors for differing types of soil.

| | | | |
|---|---|---|---|
| Scrapers | 1.5 for stiff clay<br>2.5 for chalk<br>3.5 for soft rock<br>3.7 for broken rock | Tractor Dozers and loaders | 2.0 for stiff clay<br>3.0 for chalk<br>2.5 for soft rock<br>2.5 for broken rock |
| Backacters (minimum bucket size 0.5m3) | 1.7 for stiff clay<br>2.0 for chalk<br>2.0 for soft rock<br>1.7 for broken rock | | |

The other important consideration in excavation of material is the bulkage of material after it is dug and loaded onto transport.

All pricing and estimating is based on the volume of solid material.

| | |
|---|---|
| 1.10 | Sand, bulkage |
| 1.20 | Gravel, bulkage |
| 1.30 | Compacted soil, bulkage |
| 1.30 | Compacted subbase, suitable fill, etc., bulkage |
| 1.20 | Stiff clay, bulkage |

See also page 769

**SECTION 6 EARTHWORKS**
**PLANT AND LABOUR COSTS**

| | SCRAPER EXCAVATION | GENERAL EXCAVATION |
|---|---|---|
| **LABOUR** | | |
| Plant Operator | 8.92 | 7.95 |
| Plant Operator (part time) | - | 2.37 |
| Banksman | - | 6.54 |
| Total hourly cost | = £ 8.92 | = £ 16.86 |
| **PLANT** | | |
| 0.40 m3 Excavator | - | 8.27 |
| Motor Scraper 623 E | = £ 37.11 | - |
| (Part time) Hydraulic Breaker | - | 1.20 |
| D8 Dozer with Ripper | = £ 28.00 | - |
| Tractor, Loader/Ripper (part time) | - | 3.90 |
| | | = £ 13.37 |

| | GENERAL FILLS, CAPPING LAYERS, COMPACTIONS ETC. | SHORE DEFENCES (ARMOUR STONES) |
|---|---|---|
| **LABOUR** | | |
| Plant Operator | 7.95 | 7.95 |
| Banksman | 6.54 | 6.54 |
| Total hourly cost | = £ 14.49 | = £ 14.49 |
| **PLANT** | | |
| Backacter (28.5T) | - | 26.20 |
| Tractor, Loader | 19.52 | - |
| Special Equipment | - | 2.20 |
| Towed Roller, 10T | 5.99 | - |
| Total hourly cost | = £ 25.51 | = £ 28.40 |

## SECTION 6 EARTHWORKS

### EXCAVATION AND ANCILLARY WORKS

### EXCAVATION FOR CUTTINGS IN BULK IN THE OPEN ABOVE EARTHWORKS OUTLINE

| | | UNIT | L | P | M | £ NETT |
|---|---|---|---|---|---|---|
| The following unit costs are for a typical motorway cutting generally using motorised scrapers and dozers on a average haul of 2000m (one way) | | | | | | |
| Topsoil | hr | m3 | 0.008 | 0.008 | | |
| | £ | | **0.07** | **0.26** | - | **0.33** |
| Material other than topsoil, rock or artificial hard material | hr | m3 | 0.10 | 0.10 | | |
| | £ | | **0.89** | **3.19** | - | **4.08** |
| Artificial hard material exposed at the commencing surface (scraper and Ripper Dozer) | | | | | | |
| Mass Concrete/Medium Hard Rock | hr | m3 | 0.09 | 0.09 | | |
| | £ | | **1.61** | **5.86** | - | **7.47** |
| Reinforced Concrete/Hard Rock | hr | m3 | 0.15 | 0.15 | | |
| | £ | | **2.68** | **9.77** | - | **12.45** |
| Tarmacadam | hr | m3 | 0.016 | 0.016 | | |
| | £ | | **0.29** | **1.04** | - | **1.33** |
| (The following are for general excavation using Backacters and Tractor Loaders) | | | | | | |
| Excavate topsoil | hr | m3 | 0.03 | 0.03 | | |
| | £ | | **0.27** | **0.96** | - | **1.23** |
| Excavate suitable material depths up to 1.0m | hr | m3 | 0.03 | 0.03 | | |
| | £ | | **0.27** | **0.96** | - | **1.23** |
| 1.0 - 2.0m | hr | m3 | 0.04 | 0.04 | | |
| | £ | | **0.36** | **1.28** | - | **1.64** |
| 2.0 - 4.0m | hr | m3 | 0.123 | 0.123 | | |
| | £ | | **1.10** | **3.92** | - | **5.02** |
| 4.0 - 6.0m | hr | m3 | 0.23 | 0.23 | | |
| | £ | | **2.05** | **7.34** | - | **9.39** |

**SECTION 6 EARTHWORKS**

**EXCAVATION AND ANCILLARY WORKS - continued**

| | | UNIT | L | P | M | £ NETT |
|---|---|---|---|---|---|---|
| Excavate unsuitable material depths up to 1.0m | hr | m3 | 0.03 | 0.03 | | |
| | £ | | 0.51 | 0.40 | - | 0.91 |
| 1.0 - 2.0m | hr | m3 | 0.043 | 0.043 | | |
| | £ | | 0.72 | 0.57 | - | 1.29 |
| 2.0 - 4.0m | hr | m3 | 0.13 | 0.13 | | |
| | £ | | 2.19 | 1.74 | - | 3.93 |
| 4.0 - 6.0m | hr | m3 | 0.235 | 0.235 | | |
| | £ | | 3.96 | 3.14 | - | 7.10 |
| Excavate rock up to 1.0m | hr | m3 | 0.36 | 0.36 | | |
| | £ | | 6.07 | 4.81 | - | 10.88 |
| 1.0 - 2.0m | hr | m3 | 0.75 | 0.75 | | |
| | £ | | 12.65 | 10.03 | - | 22.68 |
| 2.0 - 4.0m | hr | m3 | 1.16 | 1.16 | | |
| | £ | | 19.56 | 15.51 | - | 35.07 |
| 4.0 - 6.0m | hr | m3 | 2.14 | 2.14 | | |
| | £ | | 36.08 | 28.61 | - | 64.69 |
| Excavate pavements brickwork concrete masonry up to 1.0m | hr | m3 | 0.46 | 0.46 | | |
| | £ | | 7.76 | 6.15 | - | 13.91 |
| 1.0 - 2.0m | hr | m3 | 0.70 | 0.70 | | |
| | £ | | 11.80 | 9.36 | - | 21.16 |
| 2.0 - 4.0m | hr | m3 | 1.15 | 1.15 | | |
| | £ | | 19.39 | 15.38 | - | 34.77 |
| 4.0 - 6.0m | hr | m3 | 2.12 | 2.12 | | |
| | £ | | 35.74 | 28.34 | - | 64.08 |
| Excavate reinforced concrete up to 1.0m | hr | m3 | 0.86 | 0.86 | | |
| | £ | | 14.50 | 11.50 | - | 26.00 |
| 1.0 - 2.0m | hr | m3 | 1.08 | 1.08 | | |
| | £ | | 18.21 | 14.44 | - | 32.65 |
| 2.0 - 4.0m | hr | m3 | 1.93 | 1.93 | | |
| | £ | | 32.54 | 25.80 | - | 58.34 |

## SECTION 6 EARTHWORKS

## EXCAVATION FOR STRUCTURES

| | | UNIT | L | P | M | £ NETT |
|---|---|---|---|---|---|---|
| (The following are for general excavation using Backacters and Tractor Loaders) | | | | | | |
| Excavate topsoil | hr | m3 | 0.035 | 0.035 | | |
| | £ | | 0.59 | 0.47 | - | 1.06 |
| Excavate suitable material depths up to 1.0m | hr | m3 | 0.04 | 0.04 | | |
| | £ | | 0.67 | 0.53 | - | 1.20 |
| 1.0 - 2.0m | hr | m3 | 0.06 | 0.06 | | |
| | £ | | 1.01 | 0.80 | - | 1.81 |
| 2.0 - 4.0m | hr | m3 | 0.185 | 0.185 | | |
| | £ | | 3.12 | 2.47 | - | 5.59 |
| 4.0 - 6.0m | hr | m3 | 0.34 | 0.34 | | |
| | £ | | 5.73 | 4.55 | - | 10.28 |
| Excavate unsuitable material depths up to 1.0m | hr | m3 | 0.04 | 0.04 | | |
| | £ | | 0.67 | 0.53 | - | 1.20 |
| 1.0 - 2.0m | hr | m3 | 0.065 | 0.065 | | |
| | £ | | 1.10 | 0.87 | - | 1.97 |
| 2.0 - 4.0m | hr | m3 | 0.22 | 0.22 | | |
| | £ | | 3.71 | 2.94 | - | 6.65 |
| 4.0 - 6.0m | hr | m3 | 0.35 | 0.35 | | |
| | £ | | 5.90 | 4.68 | - | 10.58 |
| Excavate rock up to 1.0m | hr | m3 | 0.385 | 0.385 | | |
| | £ | | 6.49 | 5.15 | - | 11.64 |
| 1.0 - 2.0m | hr | m3 | 0.80 | 0.80 | | |
| | £ | | 13.49 | 10.70 | - | 24.19 |
| 2.0 - 4.0m | hr | m3 | 1.75 | 1.75 | | |
| | £ | | 29.51 | 23.40 | - | 52.91 |
| 4.0 - 6.0m | hr | m3 | 3.22 | 3.22 | | |
| | £ | | 54.29 | 43.05 | - | 97.34 |

**SECTION 6 EARTHWORKS**

**EXCAVATION FOR STRUCTURES** - continued

| | | UNIT | L | P | M | £ NETT |
|---|---|---|---|---|---|---|
| Excavate pavements brickwork concrete masonry | | | | | | |
| up to 1.0m | hr | m3 | 0.50 | 0.50 | | |
| | £ | | 8.43 | 6.69 | - | 15.12 |
| 1.0 - 2.0m | hr | m3 | 0.75 | 0.75 | | |
| | £ | | 12.65 | 10.03 | - | 22.68 |
| 2.0 - 4.0m | hr | m3 | 1.19 | 1.19 | | |
| | £ | | 20.06 | 15.91 | - | 35.97 |
| 4.0 - 6.0m | hr | m3 | 2.15 | 2.15 | | |
| | £ | | 36.25 | 28.75 | - | 65.00 |
| Excavate reinforced concrete | | | | | | |
| up to 1.0m | hr | m3 | 0.90 | 0.90 | | |
| | £ | | 15.17 | 12.03 | - | 27.20 |
| 1.0 - 2.0m | hr | m3 | 1.12 | 1.12 | | |
| | £ | | 18.88 | 14.97 | - | 33.85 |
| 2.0 - 4.0m | hr | m3 | 2.00 | 2.00 | | |
| | £ | | 33.72 | 26.74 | - | 60.46 |

## SECTION 6 EARTHWORKS

### EXCAVATION FOR WATER COURSES, CANALS, INTERCEPTING DITCHES

WATERWAYS UP TO 6M WIDE, 2.5M DEEP

| | | UNIT | L | P | M | £ NETT |
|---|---|---|---|---|---|---|
| Excavate suitable material in new watercourses | | | | | | |
| up to 1.0m deep | hr | m3 | 0.10 | 0.10 | | |
| | £ | | 1.69 | 1.34 | - | 3.03 |
| 1.0 - 2.5m deep | hr | m3 | 0.12 | 0.12 | | |
| | £ | | 2.02 | 1.60 | - | 3.62 |
| Excavate unsuitable material in new watercourses | | | | | | |
| up to 1.0m deep | hr | m3 | 0.11 | 0.11 | | |
| | £ | | 1.85 | 1.47 | - | 3.32 |
| 1.0 - 2.5m deep | hr | m3 | 0.13 | 0.13 | | |
| | £ | | 2.19 | 1.74 | - | 3.93 |
| Excavate suitable material in clearing abandoned watercourses | | | | | | |
| up to 1.0m deep | hr | m3 | 0.09 | 0.09 | | |
| | £ | | 1.52 | 1.20 | - | 2.72 |
| 1.0 - 2.5m deep | hr | m3 | 0.11 | 0.11 | | |
| | £ | | 1.85 | 1.47 | - | 3.32 |
| Excavate unsuitable material in clearing abandoned watercourses | | | | | | |
| up to 1.0m deep | hr | m3 | 0.10 | 0.10 | | |
| | £ | | 1.69 | 1.34 | - | 3.03 |
| 1.0 - 2.5m deep | hr | m3 | 0.123 | 0.123 | | |
| | £ | | 2.07 | 1.64 | - | 3.71 |

**SECTION 6 EARTHWORKS**

**DEPOSITION OF FILL MATERIALS**

| | | | UNIT | L | P | M | £ NETT |
|---|---|---|---|---|---|---|---|
| Deposition in embankments and other areas of fill, other than adjacent to structures | | | | | | | |
| Deposition of suitable material | | | | | | | |
| depths up to 1.0m | | hr | m3 | 0.01 | 0.01 | | |
| | | £ | | **0.14** | **0.26** | - | **0.40** |
| depths 1.0 - 2.0m | | hr | m3 | 0.008 | 0.008 | | |
| | | £ | | **0.12** | **0.20** | - | **0.32** |
| depths 2.0 - 3.0m | | hr | m3 | 0.01 | 0.01 | | |
| | | £ | | **0.14** | **0.26** | - | **0.40** |
| depths 3.0 - 5.0m | | hr | m3 | 0.014 | 0.014 | | |
| | | £ | | **0.20** | **0.36** | - | **0.56** |
| depths 5.0 - 7.0m | | hr | m3 | 0.019 | 0.019 | | |
| | | £ | | **0.28** | **0.48** | - | **0.76** |
| Disposal of unacceptable material offsite (excluding any tips, fees or royalties etc.) | | | | | | | |
| Tipping lorry on or offsite use | | | | | | | |
| ne 10t capacity | ne 1Km | hr | m3 | 0.05 | 0.05 | | |
| | | £ | | **0.40** | **0.64** | - | **1.04** |
| | Add per Km | hr | m3 | 0.025 | 0.025 | | |
| | | £ | | **0.20** | **0.32** | - | **0.52** |
| 10 - 15t capacity | ne 1Km | hr | m3 | 0.04 | 0.04 | | |
| | | £ | | **0.32** | **0.67** | - | **0.99** |
| | Add per Km | hr | m3 | 0.02 | 0.02 | | |
| | | £ | | **0.16** | **0.33** | - | **0.49** |
| 15 - 25t capacity | ne 1Km | hr | m3 | 0.03 | 0.03 | | |
| | | £ | | **0.24** | **0.71** | - | **0.95** |
| | Add per Km | hr | m3 | 0.015 | 0.015 | | |
| | | £ | | **0.12** | **0.35** | - | **0.47** |

## SECTION 6 EARTHWORKS

**IMPORTED FILLS** (Waste/loss allowance 5%) placed density conversion = 1.85 tonnes/m3

| | | UNIT | L | P | M | £ NETT |
|---|---|---|---|---|---|---|
| Imported suitable fill deposited in embankments and other areas of fill other than adjacent to structures (PC £1.90/tonne) | | | | | | |
| depths up to 1.0m | hr | m3 | 0.01 | 0.01 | | |
| | £ | | **0.14** | **0.26** | **3.59** | **3.99** |
| 1.00 - 2.0m | hr | m3 | 0.013 | 0.013 | | |
| | £ | | **0.19** | **0.33** | **3.59** | **4.11** |
| 2.00 - 4.0m | hr | m3 | 0.018 | 0.018 | | |
| | £ | | **0.26** | **0.46** | **3.59** | **4.31** |
| 4.00 - 6.0m | hr | m3 | 0.025 | 0.025 | | |
| | £ | | **0.36** | **0.64** | **3.59** | **4.59** |
| Imported graded granular fill deposited in embankments and other areas of fill other than adjacent to structures (PC £2.38/tonne) | | | | | | |
| depths up to 1.0m | hr | m3 | 0.013 | 0.013 | | |
| | £ | | **0.19** | **0.33** | **4.50** | **5.02** |
| 1.00 - 2.0m | hr | m3 | 0.018 | 0.018 | | |
| | £ | | **0.26** | **0.46** | **4.50** | **5.22** |
| 2.00 - 4.0m | hr | m3 | 0.021 | 0.021 | | |
| | £ | | **0.30** | **0.54** | **4.50** | **5.34** |
| 4.00 - 6.0m | hr | m3 | 0.025 | 0.025 | | |
| | £ | | **0.36** | **0.64** | **4.50** | **5.50** |
| Imported well graded fill deposited in embankments and other areas of fill other than adjacent to structures (PC £3.65/tonne) | | | | | | |
| depths up to 1.0m | hr | m3 | 0.015 | 0.015 | | |
| | £ | | **0.22** | **0.38** | **6.90** | **7.50** |
| 1.00 - 2.0m | hr | m3 | 0.02 | 0.02 | | |
| | £ | | **0.29** | **0.51** | **6.90** | **7.70** |
| Imported well graded fill as a drainage layer (PC £3.80/tonne) | | | | | | |
| depths up to 0.25m | hr | m3 | 0.01 | 0.01 | | |
| | £ | | **0.14** | **0.26** | **7.18** | **7.58** |
| 0.25 - 0.5m | hr | m3 | 0.013 | 0.013 | | |
| | £ | | **0.19** | **0.33** | **7.18** | **7.70** |
| 0.50 - 1.0m | hr | m3 | 0.015 | 0.015 | | |
| | £ | | **0.22** | **0.38** | **7.18** | **7.78** |

**SECTION 6 EARTHWORKS**

**IMPORTED FILLS** (Waste allowance 5%) - continued

| | | UNIT | L | P | M | £ NETT |
|---|---|---|---|---|---|---|
| Imported selected graded fill deposited adjacent to structures (PC £2.60/tonne) | | | | | | |
| depths up to 1.0m | hr | m3 | 0.01 | 0.01 | | |
| | £ | | 0.14 | 0.26 | 4.90 | 5.30 |
| 1.0 - 2.0m | hr | m3 | 0.013 | 0.013 | | |
| | £ | | 0.19 | 0.33 | 4.90 | 5.42 |
| 2.0 - 4.0m | hr | m3 | 0.019 | 0.019 | | |
| | £ | | 0.28 | 0.48 | 4.90 | 5.66 |
| 4.0 - 6.0m | hr | m3 | 0.026 | 0.026 | | |
| | £ | | 0.38 | 0.66 | 4.90 | 5.94 |
| Imported selected graded fill adjacent to reinforced earth structures (PC £4.55/tonne) | | | | | | |
| depths up to 1.0m | hr | m3 | 0.013 | 0.013 | | |
| | £ | | 0.19 | 0.33 | 8.60 | 9.12 |
| 1.0 - 2.0m | hr | m3 | 0.019 | 0.019 | | |
| | £ | | 0.28 | 0.48 | 8.60 | 9.36 |
| 2.0 - 4.0m | hr | m3 | 0.022 | 0.022 | | |
| | £ | | 0.32 | 0.56 | 8.60 | 9.48 |
| 4.0 - 6.0m | hr | m3 | 0.026 | 0.026 | | |
| | £ | | 0.38 | 0.66 | 8.60 | 9.64 |
| Imported well graded fill deposited adjacent to structures (PC £3.65/tonne) | | | | | | |
| depths up to 1.0m | hr | m3 | 0.016 | 0.016 | | |
| | £ | | 0.23 | 0.41 | 6.90 | 7.54 |
| 1.0 - 2.0m | hr | m3 | 0.02 | 0.02 | | |
| | £ | | 0.29 | 0.51 | 6.90 | 7.70 |

**SECTION 6 EARTHWORKS**

**IMPORTED FILLS** (Waste allowance 5%) - continued

| | | UNIT | L | P | M | £ NETT |
|---|---|---|---|---|---|---|
| Selected graded fill deposited around structural foundations (PC £2.60/tonne) | | | | | | |
| depths up to 1.0m | hr | m3 | 0.01 | 0.01 | | |
| | £ | | 0.14 | 0.26 | 4.90 | 5.30 |
| 1.0 - 2.0m | hr | m3 | 0.014 | 0.014 | | |
| | £ | | 0.20 | 0.36 | 4.90 | 5.46 |
| 2.0 - 4.0m | hr | m3 | 0.019 | 0.019 | | |
| | £ | | 0.28 | 0.48 | 4.90 | 5.66 |
| 4.0 - 6.0m | hr | m3 | 0.027 | 0.027 | | |
| | £ | | 0.39 | 0.69 | 4.90 | 5.98 |
| Imported selected graded fill around structural foundations (PC £4.55/tonne) | | | | | | |
| depths up to 1.0m | hr | m3 | 0.014 | 0.014 | | |
| | £ | | 0.20 | 0.36 | 8.60 | 9.16 |
| 1.0 - 2.0m | hr | m3 | 0.019 | 0.019 | | |
| | £ | | 0.28 | 0.48 | 8.60 | 9.36 |
| 2.0 - 4.0m | hr | m3 | 0.023 | 0.023 | | |
| | £ | | 0.33 | 0.59 | 8.60 | 9.52 |
| 4.0 - 6.0m | hr | m3 | 0.027 | 0.027 | | |
| | £ | | 0.39 | 0.69 | 8.60 | 9.68 |
| Imported well graded fill deposited around structural foundations (PC £3.65/tonne) | | | | | | |
| depths up to 1.0m | hr | m3 | 0.016 | 0.016 | | |
| | £ | | 0.23 | 0.41 | 6.90 | 7.54 |
| 1.0 - 2.0m | hr | m3 | 0.02 | 0.02 | | |
| | £ | | 0.29 | 0.51 | 6.90 | 7.70 |

**SECTION 6 EARTHWORKS**

**IMPORTED FILLS** (Waste allowance 5%) - continued

| | | UNIT | L | P | M | £ NETT |
|---|---|---|---|---|---|---|
| Imported rock fill (PC £4.20/tonne) | | | | | | |
| depths up to 1.0m | hr | m3 | 0.015 | 0.015 | | |
| | £ | | **0.22** | **0.38** | **7.94** | **8.54** |
| 1.00 - 1.5m | hr | m3 | 0.018 | 0.018 | | |
| | £ | | **0.26** | **0.46** | **7.94** | **8.66** |
| 1.50 - 2.0m | hr | m3 | 0.02 | 0.02 | | |
| | £ | | **0.29** | **0.51** | **7.94** | **8.74** |
| Imported rock fill into soft areas (rock punching) (PC £4.40/tonne) | | | | | | |
| depths up to 1.0m | hr | m3 | 0.021 | 0.021 | | |
| | £ | | **0.30** | **0.54** | **8.32** | **9.16** |
| 1.00 - 1.5m | hr | m3 | 0.024 | 0.024 | | |
| | £ | | **0.35** | **0.61** | **8.32** | **9.28** |
| 1.50 - 2.0m | hr | m3 | 0.025 | 0.025 | | |
| | £ | | **0.36** | **0.64** | **8.32** | **9.32** |
| Imported clay fill (cohesive material) (PC £4.10/tonne) | | | | | | |
| depths up to 0.25m | hr | m3 | 0.025 | 0.025 | | |
| | £ | | **0.36** | **0.64** | **7.75** | **8.75** |
| 0.25 - 0.5m | hr | m3 | 0.027 | 0.027 | | |
| | £ | | **0.39** | **0.69** | **7.75** | **8.83** |
| 0.50 - 1.0m | hr | m3 | 0.03 | 0.03 | | |
| | £ | | **0.43** | **0.77** | **7.75** | **8.95** |

**SECTION 6 EARTHWORKS**

**IMPORTED FILLS** (Waste allowance 5%) - continued

**(SHORE PROTECTION, SEA DEFENCES)**

| | | UNIT | L | P | M | £ NETT |
|---|---|---|---|---|---|---|
| Imported bedding/free draining materials (PC £3.80/tonne) depths up to 1.0m | hr | m3 | 0.02 | 0.02 | | |
| | £ | | **0.29** | **0.57** | **7.18** | **8.04** |
| Imported rock fill (as a core embankment) (PC £4.40/tonne) Embankment depth up to 1.0m | hr | m3 | 0.01 | 0.01 | | |
| | £ | | **0.14** | **0.28** | **8.32** | **8.74** |
| 1.0 - 2.0m | hr | m3 | 0.021 | 0.021 | | |
| | £ | | **0.30** | **0.60** | **8.32** | **9.22** |
| 2.0 - 4.0m | hr | m3 | 0.033 | 0.033 | | |
| | £ | | **0.48** | **0.94** | **8.32** | **9.74** |

Imported Armour Stones
depth up to 2.0m
(rocks individually handled from stockpile)

(Waste allowance 2%) placed density conversion = 1.6 tonnes/m3

| | | UNIT | L | P | M | £ NETT |
|---|---|---|---|---|---|---|
| individual stones up to 500Kg (PC £9.05/tonne) | hr | m3 | 0.056 | 0.056 | | |
| | £ | | **0.81** | **1.59** | **14.77** | **17.17** |
| up to 1000Kg (PC £12.20/tonne) | hr | m3 | 0.032 | 0.032 | | |
| | £ | | **0.46** | **0.91** | **19.90** | **21.27** |
| up to 3000Kg (PC £19.25/tonne) | hr | m3 | 0.026 | 0.026 | | |
| | £ | | **0.38** | **0.74** | **31.42** | **32.54** |

**SECTION 6 EARTHWORKS**

**SUBGRADES AND CAPPING LAYERS** (Waste allowance 5%) placed density 1.8 tonnes/m3

| | | UNIT | L | P | M | £ NETT |
|---|---|---|---|---|---|---|
| Imported selected fill in capping layer deposited in embankments and other areas of fill other than adjacent to structures (PC £2.75/tonne) | | | | | | |
| depth up to 0.5m | hr | m3 | 0.13 | 0.13 | | |
| | £ | | 1.88 | 3.32 | 5.20 | 10.40 |
| 0.5 - 1.0m | hr | m3 | 0.08 | 0.08 | | |
| | £ | | 1.16 | 2.04 | 5.20 | 8.40 |
| Imported selected fill in capping layer deposited adjacent to structures, excluding around foundations | | | | | | |
| depth up to 0.5m | hr | m3 | 0.14 | 0.14 | | |
| | £ | | 2.03 | 3.57 | 5.20 | 10.80 |
| 0.5 - 1.0m | hr | m3 | 0.095 | 0.095 | | |
| | £ | | 1.37 | 2.42 | 5.20 | 8.99 |

**SECTION 6 EARTHWORKS**

**COMPACTION OF FILL**

| | | UNIT | L | P | M | £ NETT |
|---|---|---|---|---|---|---|
| Compaction of suitable/granular fill material in embankments and other areas of fill other than adjacent to structures | hr | m3 | 0.005 | 0.005 | | |
| | £ | | **0.07** | **0.13** | - | **0.20** |
| Compaction of fill material on sub base or capping layers under verges, central reserves or side slopes | hr | m3 | 0.02 | 0.02 | | |
| | £ | | **0.29** | **0.51** | - | **0.80** |
| Compaction of graded fill material in embankments and other areas of fill other than adjacent to structures | hr | m3 | 0.006 | 0.006 | | |
| | £ | | **0.09** | **0.15** | - | **0.24** |
| Compaction of rock fill materials in embankments and other areas of fill other than adjacent to structures | hr | m3 | 0.009 | 0.009 | | |
| | £ | | **0.13** | **0.23** | - | **0.36** |
| Compaction of clay fill deposited in embankments and other areas of fill other than adjacent to structures | hr | m3 | 0.03 | 0.03 | | |
| | £ | | **0.43** | **0.77** | - | **1.20** |

**SECTION 6 EARTHWORKS**

**COMPACTION OF FILL - continued**

| | | UNIT | L | P | M | £ NETT |
|---|---|---|---|---|---|---|
| Compaction of suitable/granular fill material adjacent to structures | hr | m3 | 0.008 | 0.008 | | |
| | £ | | **0.12** | **0.20** | - | **0.32** |
| Compaction of fill material adjacent to structures | hr | m3 | 0.035 | 0.035 | | |
| | £ | | **0.51** | **0.89** | - | **1.40** |
| Compaction of graded fill material adjacent to structures | hr | m3 | 0.01 | 0.01 | | |
| | £ | | **0.14** | **0.26** | - | **0.40** |
| Compaction of rock fill materials adjacent to structures | hr | m3 | 0.014 | 0.014 | | |
| | £ | | **0.20** | **0.36** | - | **0.56** |
| Compaction of clay fill deposited adjacent to structures | hr | m3 | 0.048 | 0.048 | | |
| | £ | | **0.70** | **1.22** | - | **1.92** |

**SECTION 6 EARTHWORKS**

**COMPACTION OF FILL - continued**

| | | UNIT | L | P | M | £ NETT |
|---|---|---|---|---|---|---|
| Compaction of suitable/granular fill around structural foundations | hr | m3 | 0.012 | 0.012 | | |
| | £ | | **0.17** | **0.31** | - | **0.48** |
| Compaction of fill material around structural foundations | hr | m3 | 0.04 | 0.04 | | |
| | £ | | **0.58** | **1.02** | - | **1.60** |
| Compaction of graded fill material around structural foundations | hr | m3 | 0.015 | 0.015 | | |
| | £ | | **0.22** | **0.38** | - | **0.60** |
| Compaction of rock fill materials around structural foundations | hr | m3 | 0.017 | 0.017 | | |
| | £ | | **0.25** | **0.43** | - | **0.68** |
| Compaction of clay fill deposited around structural foundations | hr | m3 | 0.048 | 0.048 | | |
| | £ | | **0.70** | **1.22** | - | **1.92** |

**SECTION 6 EARTHWORKS**

**COMPLETION OF FORMATION**

| | | UNIT | L | P | M | £ NETT |
|---|---|---|---|---|---|---|
| Formation of suitable material | hr | m2 | 0.007 | 0.007 | | |
| | £ | | **0.10** | **0.18** | - | **0.28** |
| Formation of granular fill | hr | m2 | 0.005 | 0.005 | | |
| | £ | | **0.07** | **0.13** | - | **0.20** |
| Formation of selected fill in capping layers | hr | m2 | 0.006 | 0.006 | | |
| | £ | | **0.09** | **0.15** | - | **0.24** |

**SECTION 6 EARTHWORKS**

**REINFORCED AND ANCHORED EARTH STRUCTURES**

SPECIALIST ADVICE

As each structure is different, it is virtually impossible to give general prices. The prices will vary with the following parameters:-

a) Type of structure
b) Where located (in water, dry condition)
c) Where geographically in the country
d) Type of fill
e) Duration of structure
f) Size of structure, etc.

Consequently, it is easier to give a price indication for a hypothetical structure with the following characteristics:

Location - Good foundation

Area - Midlands - therefore good fill

Wall - Supporting urban highway

Structure - 6m high x 150m in length

Material cost in the region of £75 per square metre of face area including concrete panels, strips, fixings and accessories. The finished structure on a square metre basis will be in the range of £120 to £140. Depending on the size of a structure, the finished price range in UK is from £100 to £180 per square square metre of completed structures.

## SECTION 6 EARTHWORKS

### GEOTEXTILES

The prices below include for preparing surfaces, overlaps and turnups, jointing and sealing, fixing material in place if required and reasonable waste between 5% and 10%.

| | | UNIT | L | P | M | £ NETT |
|---|---|---|---|---|---|---|
| Geotextiles - examples only are included here for a full range refer to Basic Material Prices | | | | | | |
| "Typar" spun bonded polypropylene Geotextile 0 - 90 degrees | hr | m2 | 0.075 | - | | |
| | £ | | 0.47 | - | 1.16 | 1.63 |
| "Filtram" laminated filter Drainage membrane IBI 0 - 90 degrees | hr | m2 | 0.12 | - | | |
| | £ | | 0.74 | - | 3.68 | 4.42 |
| "Lotrak" permeable membrane 10/7 0 - 90 degrees | hr | m2 | 0.035 | - | | |
| | £ | | 0.22 | - | 0.31 | 0.53 |
| "Netlon" geogrid reinforcing textile CE 111 0 - 90 degrees | hr | m2 | 0.08 | - | | |
| | £ | | 0.50 | - | 1.29 | 1.79 |
| "Terram" 2000 permeable membrane 0 - 90 degrees | hr | m2 | 0.033 | - | | |
| | £ | | 0.20 | - | 0.72 | 0.92 |
| "Tensar" SS2 Geogrid 0 - 90 degrees | hr | m2 | 0.07 | - | | |
| | £ | | 0.43 | - | 1.65 | 2.08 |
| "Enkamat" Erosion control matting (Including pegs) 0 - 45 degrees | hr | m2 | 0.13 | - | | |
| | £ | | 0.81 | - | 3.65 | 4.46 |
| "Armoflex" Revetmet matting fixed with duckbill anchors (including nicolon textile undermat and binding layer) 0 - 45 degrees | hr | m2 | 0.82 | 0.075 | | |
| | £ | | 5.08 | 1.43 | 15.94 | 22.45 |

## SECTION 6 EARTHWORKS

### CRIB WALLING

There are a number of products specially designated for large-scale earth control. Crib walling consists of a rigid unit built of rectangular interlocking timber or precast concrete members forming skeletons or cells laid on top of each other and filled with earth or rock. Prices for these items depend on quantity, difficulty of access to the site and availabilty of suitable filling material: estimates should be obtained from the manufacturer when the site conditions have been determined. The following are approximate estimates for guidance only:

| | PC £ | UNIT | SPECIALIST ADVICE £ NETT |
|---|---|---|---|
| Supply and construct crib wall of precast concrete crib units dry jointed; excavation and foundation measured separately: | | | |
| Andacrib | | | |
| Andacrib Maxi; dowelless system | m2 | | |
| 2.5m high | 58.00 | m | 287.55 |
| 4.0m high | 58.00 | m | 453.18 |
| Andacrib Mini; dowelless system | m2 | | |
| 1.0m high | 42.00 | m | 92.82 |
| 1.5m high | 42.00 | m | 133.50 |
| Extra for hand filling with imported topsoil and fertilizer | m3 | | |
| 1.0m high | 10.00 | m2 | 16.75 |
| 1.5m high | 10.00 | m2 | 24.15 |
| 2.5m high | 10.00 | m2 | 38.00 |
| 4.0m high | 10.00 | m2 | 58.15 |
| Supply and construct crib wall of Permacrib Timber crib walling system - prices are ex works. | | | |
| Model 550, 36mm thick for landscape walls up to 2m high | m2<br>32.40 | m2 | 69.70 |
| Model 733, 36mm thick for retaining walls up to 3.6m high | 45.00 | m2 | 95.57 |
| Model 825, 46mm thick for retaining walls up to 4.10m high | 62.30 | m2 | 123.53 |
| Model 1100, 46mm thick for retaining walls up to 5.75m high | 73.64 | m2 | 146.77 |
| Model 1375, 46mm thick for retaining walls up to 6.30m high | 90.00 | m2 | 172.04 |

## SECTION 6 EARTHWORKS

### GABIONS

The rates shown are based upon the following cost elements:

Labour = £ 6.00 per hour
Plant = £ 13.50 per hour
Stone = £ 6.00 per tonne

| | | UNIT | L | P | M | SPECIALIST ADVICE £ NETT |
|---|---|---|---|---|---|---|
| Maccaferri gabions | | | | | | |
| PVC coated galvanised wire mesh | | | | | | |
| In module sizes 1m x 1.0m | hr | m3 | 1.50 | 1.50 | | |
| | £ | | 9.00 | 20.25 | 27.50 | 56.75 |
| 1m x 0.5m | hr | m3 | 2.00 | 2.00 | | |
| | £ | | 12.00 | 27.00 | 36.60 | 75.60 |
| Heavily galvanized woven wire mesh | | | | | | |
| In module sizes 1m x 1.0m | hr | m3 | 1.30 | 1.30 | | |
| | £ | | 7.80 | 17.55 | 24.84 | 50.19 |
| 1m x 0.5m | hr | m3 | 1.60 | 1.60 | | |
| | £ | | 9.60 | 21.60 | 27.60 | 58.80 |
| Tensar gabions (GM4) | | | | | | |
| HDPE | | | | | | |
| 2 x 1 x 1m Retaining structure | hr | m3 | 1.50 | 1.50 | | |
| | £ | | 9.00 | 20.25 | 25.80 | 55.05 |
| 4 x 1 x 1m River banks | hr | m3 | 2.00 | 2.00 | | |
| | £ | | 12.00 | 27.00 | 28.30 | 67.30 |
| HDPE Mattresses (GM4) | | | | | | |
| 6 x 2 x 0.30m reclaimed land/ marine | hr | m3 | 1.30 | 1.30 | | |
| | £ | | 7.80 | 17.55 | 37.80 | 63.15 |
| 6 x 2 x 0.17m retaining structure | hr | m3 | 1.60 | 1.60 | | |
| | £ | | 9.60 | 21.60 | 43.80 | 75.00 |

## SECTION 6 EARTHWORKS

## GRASSING

| | | UNIT | L | P | M | £ NETT |
|---|---|---|---|---|---|---|
| Seeding; PC £2.80/Kg @ 50g/m2 (inc. 2% waste) supply and sow in two operations. | | | | | | |
| 10 degrees or less | hr | m2 | 0.01 | - | | |
| | £ | | 0.12 | - | 0.14 | 0.26 |
| more than 10 degrees | hr | m2 | 0.015 | - | | |
| | £ | | 0.19 | - | 0.14 | 0.33 |
| 45 - 90 degrees to horizontal | hr | m3 | 0.02 | - | | |
| | £ | | 0.25 | - | 0.14 | 0.39 |
| Turfing; PC £1.07/m2 (inc. 2% waste) | | | | | | |
| Surfaces sloping at 10 degrees or less to the horizontal | hr | m3 | 0.12 | - | | |
| | £ | | 1.49 | - | 1.07 | 2.56 |
| Surfaces sloping at more than 10 degrees to the horizontal | hr | m2 | 0.14 | - | | |
| | £ | | 1.74 | - | 1.07 | 2.81 |
| 45 - 90 degrees to horizontal (including pegging and netting) | hr | m2 | 0.19 | - | | |
| | £ | | 2.36 | - | 1.07 | 3.43 |
| 'Hydraulic' mulch grass seeding | | | | | | |
| 10 degrees or less | hr | m2 | 0.005 | 0.005 | | |
| | £ | | 0.06 | 0.05 | 0.05 | 0.16 |
| more than 10 degrees | hr | m2 | 0.005 | 0.005 | | |
| | £ | | 0.06 | 0.05 | 0.05 | 0.16 |
| 45 - 90 degrees to horizontal | hr | m2 | 0.007 | 0.007 | | |
| | £ | | 0.09 | 0.07 | 0.05 | 0.21 |

**SECTION 6 EARTHWORKS**

**LANDSCAPING**

**PLANTING TREES**

The cost of planting semi-mature trees will depend on the size and species, and on the access to the site for tree handling machines. Prices should be obtained for individual trees and planting. (PC = Prime Cost. Total rate = Labour, Plant and Materials, including PC).

| | PC £ | MATERIAL £ | UNIT | TOTAL RATE £ |
|---|---|---|---|---|
| Break up subsoil in treepit to a depth of 200m | - | - | each | 0.72 |
| Supply and plant Extra Heavy Standard trees in prepared pit and backfill with excavated topsoil, minimum 600 mm deep | each 22.50 | 24.35 | each | 43.93 |
| Supply and plant Heavy Standard tree in prepared pit and backfill with excavated topsoil, minimum 600 mm deep | 16.85 | 18.25 | each | 29.42 |
| Supply and plant Selected Standard tree in prepared pit and backfill with excavated topsoil, minimum 600 mm deep | 8.90 | 9.70 | each | 19.72 |
| Supply and plant Standard tree in prepared pit and backfill with excavated topsoil, minimum 600 mm deep | 6.65 | 7.20 | each | 13.38 |
| Supply and plant Light Standard tree in prepared pit and backfill with excavated topsoil, minimum 600 mm deep | 5.15 | 5.52 | each | 9.70 |
| Extra for filling with topsoil from spoil heap not exceeding 100 m travel | - | - | m3 | 12.02 |
| Extra for filling with imported topsoil | m3 11.70 | - | m3 | 23.22 |
| Extra for incorporating manure or compost into topsoil 1 m3:5 m3 | - - | - - | - | +60% |
| Extra for supply and fix tree stake 3.0m long x 100 mm diam. driven 500 mm into firm ground and trimmed to approved height, including two tree ties to approved pattern | each 5.60 | 6.40 | each | 9.88 |
| As above but with tree stakes 2.4m x 100mm diameter | 4.40 | 5.23 | each | 7.73 |
| As above but with two stakes 3.0m x 100mm diameter | 11.20 | 12.13 | each | 19.28 |
| As above but with two stakes 2.4m x 100mm diameter | 8.80 | 9.67 | each | 14.26 |

**SECTION 6 EARTHWORKS**

**LANDSCAPING - continued**

**PLANTING TREES - continued**

| | PC £ | MATERIAL £ | UNIT | TOTAL RATE £ |
|---|---|---|---|---|
| Supply and support tree with three collars and wire guys; including pickets: | | | | |
| galvanised steel 50mm x 600mm | - | 19.69 | each | 45.26 |
| hardwood 75mm x 600mm | - | 19.96 | each | 48.15 |
| Extra for land drain to tree pit 100mm dia clayware agricultural drain including excavating drain trench, laying pipes, and backfilling | - | 2.63 | each | 5.90 |
| Supply and fix standard steel tree guard | - | 15.62 | each | 21.25 |

SECTION 6 EARTHWORKS

LANDSCAPING - continued

PLANTING SHRUBS

| | PC | MATERIAL | RATE 100 | RATE EACH |
|---|---|---|---|---|
| | £ | £ | £ | £ |
| Form planting hole in cultivated area measured elsewhere, supply and plant specified shrub and backfill with excavated material: | 100 rate each | | | |
| shrub 300mm high | 2.07 | - | 235.40 | 2.35 |
| shrub 600mm high | 1.33 | - | 154.10 | 1.55 |
| shrub 900mm high | 1.64 | - | 188.30 | 1.89 |
| shrub 1.0m + high | 2.20 | - | 246.10 | 2.46 |
| Extra for supply and fix shrub stake 1.5m x 75mm including two ties | each 3.32 | 4.10 | - | 6.37 |

| | PC | RATE 100 | RATE EACH |
|---|---|---|---|
| | £ | £ | £ |
| Extra for incorporating manure | m3 | m3 | m3 |
| at 1m3/10m2 | 2.25 | 275.75 | 2.80 |
| at 1m3/15m2 | 2.25 | 266.06 | 2.69 |
| at 1m3/20m2 | 2.25 | 252.51 | 2.54 |
| Extra for incorporating fertiliser | kg | 100 m2 | m2 |
| at 35g/m2 | 1.84 | 17.59 | 0.19 |
| at 50g/m2 | 1.84 | 21.25 | 0.22 |
| at 70g/m2 | 1.84 | 26.82 | 0.28 |
| at 100g/m2 | 1.84 | 33.39 | 0.35 |

PLANTING HERBACEOUS AND GROUNDCOVER PLANTS

| | | | |
|---|---|---|---|
| Form planting hole in cultivated area measured elsewhere, supply and plant s specified herbaceous plant and backfill with excavated material: | 100 | 100m2 | m2 |
| 1 - 3 plants per m2 | 129.00 | 434.90 | 4.42 |
| 4 - 11 plants per m2 | 129.00 | 1534.90 | 15.47 |
| 12 - 25 plants per m2 | 129.00 | 3418.50 | 34.34 |
| 26 - 35 plants per m2 | 129.00 | 4722.30 | 47.70 |
| 36 - 50 plants per m2 | 129.00 | 6656.80 | 67.20 |

**SECTION 6 EARTHWORKS**

**LANDSCAPING - continued**

**PLANTING HERBACEOUS AND GROUNDCOVER PLANTS - continued**

| | PC £ | RATE 100 £ | EACH £ |
|---|---|---|---|
| Extra for incorporating manure | m3 | 100m2 | m2 |
| at 1 m3/10m2 | 2.25 | 275.95 | 2.80 |
| at 1 m3/15m2 | 2.25 | 266.06 | 2.69 |
| at 1 m3/20m2 | 2.25 | 252.51 | 2.59 |
| Extra for incorporating fertilizer | kg | 100m2 | m2 |
| at 35g/m2 | 1.84 | 17.59 | 0.19 |
| at 50g/m2 | 1.84 | 21.25 | 0.22 |
| at 70g/m2 | 1.84 | 26.82 | 0.28 |
| at 100g/m2 | 1.84 | 33.39 | 0.35 |
| Supply and fix Plant Support Netting on 50mm diameter stakes 750mm long driven into ground at 1.5m centres | roll | | |
| green extruded plastic mesh 125mm sq | 300 x 1.5m 37.85 | 1.80 | m2 |

**PLANTING BULBS**

| | PC £ | RATE 100 £ | EACH £ |
|---|---|---|---|
| Form planting hole in cultivated area measured elsewhere, supply and plant bulbs and backfill with excavated material: | 100 | 100 | each |
| small | 11.15 | 30.10 | 0.32 |
| medium | 17.80 | 40.00 | 0.42 |
| large | 21.20 | 49.00 | 0.51 |
| Supply and plant bulbs in grassed area using bulb planter and backfill with screened topsoil or peat and cut turf plug: | | | |
| small | 11.15 | 29.15 | 0.31 |
| medium | 17.80 | 38.80 | 0.42 |
| large | 21.20 | 48.50 | 0.51 |
| Extra to items above for planting in prefabricated or in-situ planters | | | |
| Shrubs | - | - | +20% |
| Herbaceous plants | - | - | +20% |
| Bulbs | - | - | +15% |

**SECTION 6 EARTHWORKS**

**LANDSCAPING - continued**

**OPERATIONS AFTER PLANTING (MAINTENANCE)**

| | PC | HAND 100M | M |
|---|---|---|---|
| | £ | £ | £ |
| Initial cut back to shrubs and hedge plants, clear away all cuttings | - | 31.00 | 0.32 |
| Protect planted areas with windbreak fencing 1m high fixed to stakes 1.50m x 50mm diameter at 2m centres and clear away on completion | m 2.25 | 516.75 | 5.25 |
| Cut out dead and diseased wood, prune, trim and cut to shape treat with sealant | | | |
| shrubs up to 1.0m high | each | 0.43 | |
| 1-2m high | each | 0.95 | |
| 2-3m high | each | 1.69 | |
| Cut and trim hedge to specified profile clean away cuttings | | | |
| 2.0m high | m | 0.95 | |
| | 100m | 93.36 | |
| 2.4m high | m | 1.65 | |
| | 100m | 165.05 | |
| Trim field hedge to specified height and shape | | | |
| using flail | 100m | 310.00 | |
| using cutting bar | 100m | 465.00 | |
| Weed and handfork planted areas including removing and dumping weeds and debris on site | m2 | 0.64 | |
| | 100m2 | 63.00 | |
| Supply and apply selective weed killer at: | | | |
| 35g/m2 | m2 | 0.22 | |
| 70g/m2 | m2 | 0.34 | |
| 100g/m2 | m2 | 0.44 | |

**SECTION 6 EARTHWORKS**

**GROUND ANCHORAGES**

Ground anchorages consist of the installation of a cable or solid bar tendon fixed in the ground by grouting and tensioned to exceed the working load to be carried. Ground anchors may be of a permanent or temporary nature and can be used in conjunction with diaphragm walling or sheet piling to eliminate the use of strutting etc. The following costs are based on the installation of 50 ground anchors.

| | | UNIT | L | P | M | £ NETT |
|---|---|---|---|---|---|---|
| Establishment of standard plant and equipment and removal on completion | hr | sum | - | - | | |
| | £ | | - | - | - | 8000.00 |
| Standing time (equivalent to Gang F Cost) | hr | hr | 1.00 | 1.00 | | |
| | £ | | 35.00 | 65.00 | - | 100.00 |
| Number in material other than rock or artificial hard material (Type of ground) (Load in tonnes) | | | | | | |
| temporary or permanent; maximum depth 15m | | | | | | |
| Rock alluvial or clay 0 - 50 tonnes load | hr | nr | 0.58 | 0.58 | | |
| | £ | | 12.30 | 37.70 | - | 50.00 |
| Rock or alluvial 50 - 90 tonnes load | hr | nr | 0.58 | 0.58 | | |
| | £ | | 22.30 | 37.70 | - | 60.00 |
| Rock only 90 - 150 tonnes load | hr | nr | 0.58 | 0.58 | | |
| | £ | | 32.30 | 37.70 | - | 70.00 |
| Total length of tendons in material other than rock or artificial hard material. | | | | | | |
| temporary | | | | | | |
| Rock alluvial or clay 0 - 50 tonnes load | hr | nr | 0.54 | 0.54 | | |
| | £ | | 4.90 | 35.10 | - | 40.00 |
| Rock or alluvial 50 - 90 tonnes load | hr | nr | 0.58 | 0.58 | | |
| | £ | | 22.30 | 37.70 | - | 60.00 |
| Rock only 90 - 150 tonnes load | hr | nr | 0.65 | 0.65 | | |
| | £ | | 37.75 | 42.25 | - | 80.00 |
| permanent | | | | | | |
| Rock alluvial or clay 0 - 50 tonnes load | hr | nr | 0.81 | 0.81 | | |
| | £ | | 7.35 | 52.65 | - | 60.00 |
| Rock or alluvial 50 - 90 tonnes load | hr | nr | 0.85 | 0.85 | | |
| | £ | | 24.75 | 55.25 | - | 80.00 |
| Rock only 90 - 150 tonnes load | hr | nr | 0.92 | 0.92 | | |
| | £ | | 40.20 | 59.80 | - | 100.00 |

## SECTION 6 EARTHWORKS

### GROUND IMPROVEMENTS

### GROUND CONSOLIDATION - DYNAMIC COMPACTION

Ground consolidation by dynamic compaction is a technique which involves the dropping of a steel or concrete pounder several times in each location on a grid pattern that covers the whole site. For ground compaction up to 10m a 15t pounder from a free fall of 20m would be typical. Several passes over the site are normally required to achieve full compaction. The process is recommended for naturally cohesive soils and is usually uneconomic for areas of less than 4,000m2 for sites with granular or mixed granular cohesive soils and 6,000m2 for a site with weak cohesive soils. The main considerations to be taken into account when using this method of consolidation are:

sufficient area to be viable

proximity and condition of adjacent property and services

need for blanket layer of granular material for a working surface and as backfill to offset induced settlement

water table level

The final bearing capacity and settlement criteria that can be achieved depends on the nature of the material being compacted. Allowable bearing capacity may be increased by up to twice the pre-treated value for the same settlement. Control testing can be by crater volume measurements, site levelling between passes, penetration tests or plate loading tests.

The following range of costs are averages based on treating an area of about 10,000m2 for a 5 - 6m compaction depth. Typical progress would be 1,500 - 2,000m2 per week.

| | | UNIT | L | P | M | £ NETT |
|---|---|---|---|---|---|---|
| Establishment of standard plant and equipment and removal on completion | hr | sum | - | - | | |
| | £ | | - | - | - | 30000.00 |
| Ground treatment | hr | m2 | 0.02 | 0.02 | | |
| | £ | | 1.80 | 4.20 | - | 6.00 |
| Free-draining granular blanket layer as working surface and backfill material (300mm thickness required of filter material) | hr | m2 | - | - | | |
| | £ | | 0.65 | 0.55 | 8.05 | 9.25 |
| Control testing including levelling, piezometers and penetrameter testing | hr | m2 | - | - | | |
| | £ | | - | - | - | 2.50 |
| Kentledge load test | hr | nr | - | - | | |
| | £ | | - | - | - | 8000.00 |

## SECTION 7 PAVEMENTS

### SUB BASES

The following unit costs are generally based on the Department of Transport (DTp) Specification for Highway Works and reference is made throughout this section to clauses within that specification.

| | | | |
|---|---|---|---|
| LABOUR based on Gang A | | | |
| Total hourly cost | = | £ | 33.71 |
| PLANT | | | |
| Grader 130G | | | 22.05 |
| Tractor loader 931 | | | 11.48 |
| Towed roller BW10 | | | 5.79 |
| Total hourly cost | = | £ | 39.32 |

Included in this section are:-

- granular sub bases
- lean concrete sub bases
- hardcore
- wet mix macadam

For information on coverage of sub bases and voids in materials, refer to page 793, material prices are on pages 86 - 87.

**SECTION 7 PAVEMENTS**

**SUB BASES - continued**

| | | UNIT | L | P | M | £ NETT |
|---|---|---|---|---|---|---|
| Granular material DTp specified type 1 (PC = £6.70/tonne) | | | | | | |
| 75mm depth | hr | m3 | 0.035 | 0.035 | | |
| | £ | | 1.18 | 1.38 | 12.25 | 14.81 |
| 100mm depth | hr | m3 | 0.04 | 0.04 | | |
| | £ | | 1.35 | 1.57 | 12.25 | 15.17 |
| 150mm depth | hr | m3 | 0.045 | 0.045 | | |
| | £ | | 1.52 | 1.77 | 12.25 | 15.54 |
| 200mm depth | hr | m3 | 0.05 | 0.05 | | |
| | £ | | 1.69 | 1.97 | 12.25 | 15.91 |
| Granular material DTp specified type 2 (PC = £6.25/tonne) | | | | | | |
| 75mm depth | hr | m3 | 0.035 | 0.035 | | |
| | £ | | 1.18 | 1.38 | 11.42 | 13.98 |
| 100mm depth | hr | m3 | 0.04 | 0.04 | | |
| | £ | | 1.35 | 1.57 | 11.42 | 14.34 |
| 150mm depth | hr | m3 | 0.045 | 0.045 | | |
| | £ | | 1.52 | 1.77 | 11.42 | 14.71 |
| 200mm depth | hr | m3 | 0.05 | 0.05 | | |
| | £ | | 1.69 | 1.97 | 11.42 | 15.08 |
| Lean concrete DTp specified strength mix (1:3:6 - 20mm aggregate) | | | | | | |
| 100mm depth | hr | m2 | 0.016 | 0.016 | | |
| | £ | | 0.54 | 0.63 | 4.22 | 5.39 |
| 200mm depth | hr | m2 | 0.017 | 0.017 | | |
| | £ | | 0.57 | 0.67 | 8.45 | 9.69 |

**SECTION 7 PAVEMENTS**

**SUB BASES - continued**

| | | UNIT | L | P | M | £ NETT |
|---|---|---|---|---|---|---|
| Hardcore | | | | | | |
| 100mm depth | hr | m3 | 0.04 | 0.04 | | |
| | £ | | 1.35 | 1.57 | 2.60 | 5.52 |
| 150mm depth | hr | m3 | 0.045 | 0.045 | | |
| | £ | | 1.52 | 1.77 | 2.60 | 5.89 |
| 200mm depth | hr | m3 | 0.05 | 0.05 | | |
| | £ | | 1.69 | 1.97 | 2.60 | 6.26 |
| Wet mix macadam; DTp clause 808 | | | | | | |
| 75mm depth | hr | m2 | 0.053 | 0.053 | | |
| | £ | | 1.79 | 2.08 | 1.10 | 4.97 |
| 100mm depth | hr | m2 | 0.055 | 0.055 | | |
| | £ | | 1.85 | 2.16 | 1.47 | 5.48 |
| 200mm depth | hr | m2 | 0.058 | 0.058 | | |
| | £ | | 1.96 | 2.28 | 2.93 | 7.17 |

**SECTION 7 PAVEMENTS**

**FLEXIBLE ROAD BASES AND SURFACING**

| | | | |
|---|---|---|---|
| LABOUR based on Gang G | | | |
| Total hourly cost | = | £ | 61.57 |
| | | | |
| PLANT | | | |
| 2 Asphalt pavers BK171 | | | 58.96 |
| 2 107 Rollers (3 point) | | | 11.58 |
| Tractair + Tools | | | 8.85 |
| (Part time) Tar Sprayer | | | 1.15 |
| S.P. Chip spreader | | | 4.21 |
| Heat Iron | | | 0.75 |
| Total hourly cost | = | £ | 85.50 |

Notes: Material prices include delivery via insulated road tippers and include an allowance of 1% for waste and 1% for sundry materials and extra consumables (diesel/propane/fuels etc.) All outputs are based on clear runs without undue delay to two pavers with a 75% utilisation. The outputs can be adjusted as follows to take account of space or time influences on the utilisation.

| | | | |
|---|---|---|---|
| 1 paver @ 75% utilisation | = | x | 2.00 |
| 1 paver @ 100% utilisation | = | x | 1.50 |
| 2 paver @ 100% utilisation | = | x | 0.75 |

**SECTION 7 PAVEMENTS**

**FLEXIBLE ROAD BASES AND SURFACING - continued**
**(All references to 1986 DTp Spec)**

| | | UNIT | L | P | M | £ NETT |
|---|---|---|---|---|---|---|
| Dense Bitumen Macadam | | | | | | |
| Road Base to DTp clause 903 | | | | | | |
| 100mm depth | hr | m2 | 0.007 | 0.007 | | |
| | £ | | **0.43** | **0.60** | **3.85** | **4.88** |
| 150mm depth | hr | m2 | 0.009 | 0.009 | | |
| | £ | | **0.55** | **0.77** | **5.70** | **7.02** |
| 200mm depth | hr | m2 | 0.011 | 0.011 | | |
| | £ | | **0.68** | **0.94** | **7.65** | **9.27** |
| Base Course to DTp clause 906 | | | | | | |
| 50mm depth | hr | m2 | 0.010 | 0.010 | | |
| | £ | | **0.62** | **0.86** | **1.70** | **3.18** |
| 100mm depth | hr | m2 | 0.013 | 0.013 | | |
| | £ | | **0.80** | **1.11** | **2.50** | **4.41** |
| Wearing Course to DTp clause 912 | | | | | | |
| 30mm depth | hr | m2 | 0.004 | 0.004 | | |
| | £ | | **0.25** | **0.34** | **1.50** | **2.09** |
| 50mm depth | hr | m2 | 0.007 | 0.007 | | |
| | £ | | **0.43** | **0.60** | **2.50** | **3.53** |
| Bitumen Macadam | | | | | | |
| Base Course to DTp clause 908 | | | | | | |
| 35mm depth | hr | m2 | 0.004 | 0.004 | | |
| | £ | | **0.25** | **0.34** | **1.50** | **2.09** |
| 70mm depth | hr | m2 | 0.006 | 0.006 | | |
| | £ | | **0.37** | **0.51** | **3.00** | **3.88** |
| Dense Tarmacadam | | | | | | |
| Road Base to DTp clause 902 | | | | | | |
| 50mm depth | hr | m2 | 0.004 | 0.004 | | |
| | £ | | **0.25** | **0.34** | **2.07** | **2.66** |
| 100mm depth | hr | m2 | 0.007 | 0.007 | | |
| | £ | | **0.43** | **0.60** | **4.10** | **5.13** |
| Base Course to DTp clause 907 | | | | | | |
| 60mm depth | hr | m2 | 0.004 | 0.004 | | |
| | £ | | **0.25** | **0.34** | **2.65** | **3.24** |
| 80mm depth | hr | m2 | 0.005 | 0.005 | | |
| | £ | | **0.31** | **0.43** | **3.50** | **4.24** |

**SECTION 7 PAVEMENTS**

**FLEXIBLE ROAD BASES AND SURFACING - continued**

| | | UNIT | L | P | M | £ NETT |
|---|---|---|---|---|---|---|
| Dense Tar Surfacing | | | | | | |
| Wearing Course to DTp clause 913 | | | | | | |
| 30mm depth | hr | m2 | 0.004 | 0.004 | | |
| | £ | | **0.25** | **0.34** | **1.50** | **2.09** |
| 50mm depth | hr | m2 | 0.005 | 0.005 | | |
| | £ | | **0.31** | **0.43** | **2.50** | **3.24** |
| Cold Asphalt | | | | | | |
| Wearing course to DTp clause 914 | | | | | | |
| 15mm depth | hr | m2 | 0.004 | 0.004 | | |
| | £ | | **0.25** | **0.34** | **0.77** | **1.36** |
| 30mm depth | hr | m2 | 0.005 | 0.005 | | |
| | £ | | **0.31** | **0.43** | **1.52** | **2.26** |
| Rolled Asphalt | | | | | | |
| Base Course to DTp clause 905 | | | | | | |
| 60mm depth | hr | m2 | 0.005 | 0.005 | | |
| | £ | | **0.31** | **0.43** | **2.60** | **3.34** |
| 80mm depth | hr | m2 | 0.006 | 0.006 | | |
| | £ | | **0.37** | **0.51** | **3.50** | **4.38** |
| Wearing Course to DTp clause 911 | | | | | | |
| 40mm depth | hr | m2 | 0.006 | 0.006 | | |
| | £ | | **0.37** | **0.51** | **2.40** | **3.28** |
| 60mm depth | hr | m2 | 0.007 | 0.007 | | |
| | £ | | **0.43** | **0.60** | **3.60** | **4.63** |
| Slurry sealing | | | | | | |
| BS434 class K3 clause 918 | | | | | | |
| 3mm depth | hr | m2 | 0.003 | 0.003 | | |
| | £ | | **0.18** | **0.26** | **0.75** | **1.19** |
| 3mm depth | hr | m2 | 0.004 | 0.004 | | |
| | £ | | **0.25** | **0.34** | **1.00** | **1.59** |

**SECTION 7 PAVEMENTS**

**FLEXIBLE ROAD BASES AND SURFACING - continued**

| | | UNIT | L | P | M | £ NETT |
|---|---|---|---|---|---|---|
| Surface dressing | | | | | | |
| coated chippings, 9 - 11 kg/m2 - clause 915 | | | | | | |
| 6mm nominal size | hr | m2 | 0.002 | 0.002 | | |
| | £ | | 0.12 | 0.17 | 0.14 | 0.43 |
| 8mm nominal size | hr | m2 | 0.002 | 0.002 | | |
| | £ | | 0.12 | 0.17 | 0.15 | 0.44 |
| 10mm nominal size | hr | m2 | 0.002 | 0.002 | | |
| | £ | | 0.12 | 0.17 | 0.16 | 0.45 |
| 12mm nominal size | hr | m2 | 0.003 | 0.003 | | |
| | £ | | 0.18 | 0.26 | 0.18 | 0.62 |
| Bituminous spray | | | | | | |
| tack coat: BS434 K1 - 40 clause 920 | | | | | | |
| large area over 20m2 | hr | m2 | - | - | | |
| | £ | | - | - | 0.08 | 0.08 |
| small area below 20m2 | hr | m2 | - | - | | |
| | £ | | - | - | 0.10 | 0.10 |

## SECTION 7 PAVEMENTS

### FLEXIBLE ROAD BASES AND SURFACING - continued

LABOUR based on Gang G

| | | |
|---|---|---|
| Total hourly cost | = £ | 61.57 |

PLANT

| | | | | |
|---|---|---|---|---|
| Cold planer | 28.03 | Heat planer | | 44.00 |
| Wheeled loader (2.30m3) | 13.95 | Wheeled loader | | 13.95 |
| Total hourly cost = £ | 41.98 | | = £ | 57.95 |

| | | UNIT | L | P | M | £ NETT |
|---|---|---|---|---|---|---|
| Removal of flexible surfaces (assumes disposal on site or re-use as fill but excludes transport if required) | | | | | | |
| depth 75mm | hr | m2 | 0.027 | 0.027 | | |
| | £ | | 1.66 | 1.13 | - | 2.79 |
| depth 100mm | hr | m2 | 0.036 | 0.036 | | |
| | £ | | 2.22 | 1.51 | - | 3.73 |
| depth 50mm; scarifying surface | hr | m2 | 0.022 | 0.022 | | |
| | £ | | 1.35 | 0.92 | - | 2.27 |
| depth 75mm; scarifying surface | hr | m2 | 0.037 | 0.037 | | |
| | £ | | 2.28 | 1.55 | - | 3.83 |
| depth 25mm; heat planing for re-use | hr | m2 | 0.032 | 0.032 | | |
| | £ | | 1.97 | 1.85 | - | 3.82 |
| depth 50mm; heat planing for re-use | hr | m2 | 0.056 | 0.056 | | |
| | £ | | 3.45 | 3.25 | - | 6.70 |
| raking over scarified or heat planed material up to 50mm depth and compacting with 10t roller | hr | m2 | 0.005 | 0.005 | | |
| | £ | | 0.31 | 0.29 | - | 0.60 |

**SECTION 7 PAVEMENTS**

**CONCRETE PAVEMENTS**

Note: The following unit costs are for jointed reinforced concrete pavements.

LABOUR based on Gang G

Total hourly cost = £ 61.57

PLANT

Concrete paver train (tracked)
and ancillary equipment

Total hourly cost = £ 66.92

| | | UNIT | L | P | M | £ NETT |
|---|---|---|---|---|---|---|
| Carriageway slabs of Dtp specified paving grade quality concrete grade C30, 20mm aggregate | | | | | | |
| depth 180mm | hr | m2 | 0.0125 | 0.0125 | | |
| | £ | | **0.77** | **0.84** | **8.91** | **10.52** |
| depth 220mm | hr | m2 | 0.015 | 0.015 | | |
| | £ | | **0.92** | **1.00** | **10.89** | **12.81** |
| depth 260mm | hr | m2 | 0.018 | 0.018 | | |
| | £ | | **1.11** | **1.20** | **12.87** | **15.18** |
| depth 300mm | hr | m2 | 0.021 | 0.021 | | |
| | £ | | **1.29** | **1.41** | **14.85** | **17.55** |
| Steel fabric reinforcement to BS4483 | | | | | | |
| nominal mass 2.22kg Ref A142 | hr | m2 | 0.018 | - | | |
| | £ | | **1.11** | **-** | **1.32** | **2.43** |
| nominal mass 3.95kg Ref A252 | hr | m2 | 0.027 | - | | |
| | £ | | **1.66** | **-** | **2.36** | **4.02** |
| nominal mass 4.53kg Ref B385 | hr | m2 | 0.030 | - | | |
| | £ | | **1.85** | **-** | **2.81** | **4.66** |
| nominal mass 5.55kg Ref C636 | hr | m2 | 0.031 | - | | |
| | £ | | **1.91** | **-** | **3.38** | **5.29** |
| nominal mass 5.93kg Ref B503 | hr | m2 | 0.031 | - | | |
| | £ | | **1.91** | **-** | **3.55** | **5.46** |

**SECTION 7 PAVEMENTS**

**CONCRETE PAVEMENTS - continued**

| | | UNIT | L | P | M | £ NETT |
|---|---|---|---|---|---|---|
| Mild Steel bar reinforcement to BS4449 cut and bent (nominal size) | | | | | | |
| 6mm | hr | t | 4.05 | - | | |
| | £ | | 249.36 | - | 552.21 | 801.57 |
| 8mm | hr | t | 3.75 | - | | |
| | £ | | 230.89 | - | 507.78 | 738.67 |
| 10mm | hr | t | 3.15 | - | | |
| | £ | | 193.95 | - | 476.47 | 670.42 |
| 12mm | hr | t | 2.50 | - | | |
| | £ | | 153.93 | - | 451.22 | 605.15 |
| 16mm | hr | t | 1.80 | - | | |
| | £ | | 110.83 | - | 419.91 | 530.74 |
| High yield steel bar reinforcement to BS4449 cut and bent (nominal size) | | | | | | |
| 6mm | hr | t | 4.05 | - | | |
| | £ | | 249.36 | - | 556.26 | 805.62 |
| 8mm | hr | t | 3.75 | - | | |
| | £ | | 230.89 | - | 511.82 | 742.71 |
| 10mm | hr | t | 3.15 | - | | |
| | £ | | 193.95 | - | 448.51 | 642.46 |
| 12mm | hr | t | 2.50 | - | | |
| | £ | | 153.93 | - | 455.26 | 609.19 |
| 16mm | hr | t | 1.80 | - | | |
| | £ | | 110.83 | - | 423.95 | 534.78 |
| Waterproof membranes below concrete pavements (nominal size) plastic sheeting; 1200 grade | | | | | | |
| 1000 gauge | hr | m2 | 0.003 | | | |
| | £ | | 0.18 | - | 0.47 | 0.65 |
| 2000 gauge | hr | m2 | 0.003 | - | | |
| | £ | | 0.18 | - | 0.83 | 1.01 |

**SECTION 7 PAVEMENTS**

**JOINTS IN CONCRETE PAVEMENTS**

| | UNIT | | L | P | M | £ NETT |
|---|---|---|---|---|---|---|
| Longitudinal joints (joint depth) DTp clause 1010 type 2 | | | | | | |
| 180mm | hr | m | 0.13 | 0.001 | | |
| | £ | | 8.00 | 0.07 | 12.82 | 20.89 |
| 220mm | hr | m | 0.13 | 0.001 | | |
| | £ | | 8.00 | 0.07 | 13.67 | 21.74 |
| 260mm | hr | m | 0.14 | 0.001 | | |
| | £ | | 8.62 | 0.07 | 16.25 | 24.94 |
| 300mm | hr | m | 0.14 | 0.001 | | |
| | £ | | 8.62 | 0.07 | 17.24 | 25.93 |
| Expansion joints (joint depth) DTp clause 1009; 13mm wide | | | | | | |
| 180mm | hr | m | 0.18 | 0.001 | | |
| | £ | | 11.08 | 0.07 | 19.81 | 30.96 |
| 220mm | hr | m | 0.19 | 0.001 | | |
| | £ | | 11.70 | 0.07 | 23.24 | 35.01 |
| 260mm | hr | m | 0.20 | 0.001 | | |
| | £ | | 12.31 | 0.07 | 26.84 | 39.22 |
| 300mm | hr | m | 0.21 | 0.001 | | |
| | £ | | 12.93 | 0.07 | 27.54 | 40.54 |
| Contraction joints (joint depth) DTp clause 1009; 10mm wide | | | | | | |
| 180mm | hr | m | 0.12 | 0.001 | | |
| | £ | | 7.39 | 0.07 | 10.61 | 18.07 |
| 220mm | hr | m | 0.13 | 0.001 | | |
| | £ | | 8.00 | 0.07 | 11.47 | 19.54 |
| 260mm | hr | m | 0.14 | 0.001 | | |
| | £ | | 8.62 | 0.07 | 15.04 | 23.73 |
| 300mm | hr | m | 0.14 | 0.001 | | |
| | £ | | 8.62 | 0.07 | 15.04 | 23.73 |

**SECTION 7 PAVEMENTS**

**JOINTS IN CONCRETE PAVEMENTS - continued**

| | | UNIT | L | P | M | £ NETT |
|---|---|---|---|---|---|---|
| Warping joints (joint depth) DTp clause 1009; 10mm wide | | | | | | |
| 180mm | hr | m | 0.14 | 0.001 | | |
| | £ | | 8.62 | 0.07 | 22.90 | 31.59 |
| 220mm | hr | m | 0.15 | 0.001 | | |
| | £ | | 9.24 | 0.07 | 25.10 | 34.41 |
| 260mm | hr | m | 0.16 | 0.002 | | |
| | £ | | 9.85 | 0.13 | 29.00 | 38.98 |
| 300mm | hr | m | 0.17 | 0.002 | | |
| | £ | | 10.47 | 0.13 | 39.50 | 50.10 |
| Construction joints (joint depth) DTp clause 1009 | | | | | | |
| 180mm | hr | m | 0.11 | 0.001 | | |
| | £ | | 6.77 | 0.07 | 6.65 | 13.49 |
| 220mm | hr | m | 0.11 | 0.001 | | |
| | £ | | 6.77 | 0.07 | 7.32 | 14.16 |
| 260mm | hr | m | 0.12 | 0.001 | | |
| | £ | | 7.39 | 0.07 | 8.06 | 15.52 |
| 300mm | hr | m | 0.12 | 0.001 | | |
| | £ | | 7.39 | 0.07 | 8.86 | 16.32 |

## SECTION 11 KERBS, FOOTWAYS AND PAVED AREAS

### KERBS, CHANNELS AND EDGINGS

LABOUR based on Gang E

| | | | |
|---|---|---|---|
| Total hourly cost | = | £ | 27.81 |

PLANT

| | | | |
|---|---|---|---|
| (Part time) JCB 3CX | | | 2.58 |
| Power tools, Roadforms | | | 2.09 |
| Total hourly cost | = | £ | 4.67 |

| | | UNIT | L | P | M | £ NETT |
|---|---|---|---|---|---|---|
| Precast concrete kerbs to BS340 (Bullnosed splayed or half battered) Straight or curved to radius exceeding 12m | | | | | | |
| 150 x 305mm | hr | m | 0.07 | 0.07 | | |
| | £ | | 1.95 | 0.33 | 5.98 | 8.26 |
| 125 x 255mm | hr | m | 0.07 | 0.07 | | |
| | £ | | 1.95 | 0.33 | 4.19 | 6.47 |
| 125 x 150mm | hr | m | 0.06 | 0.06 | | |
| | £ | | 1.67 | 0.28 | 2.99 | 4.94 |
| Curved to radius not exceeding 12m | | | | | | |
| 150 x 305mm | hr | m | 0.075 | 0.075 | | |
| | £ | | 2.09 | 0.35 | 6.90 | 9.34 |
| 125 x 255mm | hr | m | 0.075 | 0.075 | | |
| | £ | | 2.09 | 0.35 | 4.79 | 7.23 |
| 125 x 150mm | hr | m | 0.065 | 0.065 | | |
| | £ | | 1.81 | 0.30 | 3.94 | 6.05 |
| Precast concrete kerbs to BS340 | | | | | | |
| quadrants; 455 x 455 x 255mm | hr | nr | 0.10 | 0.10 | | |
| | £ | | 2.78 | 0.47 | 5.10 | 8.35 |
| quadrants; 305 x 305 x 150mm | hr | nr | 0.08 | 0.08 | | |
| | £ | | 2.22 | 0.37 | 4.33 | 6.92 |
| drop kerbs; 125 x 255mm | hr | m | 0.07 | 0.07 | | |
| | £ | | 1.95 | 0.33 | 4.20 | 6.48 |

**SECTION 11 KERBS, FOOTWAYS AND PAVED AREAS**

**KERBS, CHANNELS AND EDGINGS - continued**

| | | UNIT | L | P | M | £ NETT |
|---|---|---|---|---|---|---|
| drop kerbs; 150 x 305mm | hr | m | 0.07 | 0.07 | | |
| | £ | | 1.95 | 0.33 | 6.90 | 9.18 |
| Precast concrete channels | | | | | | |
| Straight or curved to radius exceeding 12m; 255 x 125mm | hr | m | 0.07 | 0.07 | | |
| | £ | | 1.95 | 0.33 | 4.79 | 7.07 |
| Curved to radius not exceeding 12m; 255 x 125mm | hr | m | 0.07 | 0.07 | | |
| | £ | | 1.95 | 0.33 | 5.69 | 7.97 |
| | | | | | | SPECIALIST ADVICE |
| In-situ kerbing | | | | | | |
| In-situ Asphalt kerbing | | | | | | |
| Straight or curved to radius exceeding 12m (BS5931) | | | | | | |
| Kerb height 75mm | hr | m | - | - | | |
| | £ | | - | - | - | 3.12 |
| Kerb height 100mm | hr | m | - | - | | |
| | £ | | - | - | - | 4.40 |
| Kerb height 125mm | hr | m | - | - | | |
| | £ | | - | - | - | 5.75 |
| Curved to radius not exceeding 12m (BS5931) | | | | | | |
| Kerb height 75mm | hr | m | - | - | | |
| | £ | | - | - | - | 3.30 |
| Kerb height 100mm | hr | m | - | - | | |
| | £ | | - | - | - | 4.55 |
| Kerb height 125mm | hr | m | - | - | | |
| | £ | | - | - | - | 5.92 |
| In-situ Asphalt channelling | | | | | | |
| Straight or curved to radius exceeding 12m (40mm thick) | | | | | | |
| 300mm wide | hr | m | - | - | | |
| | £ | | - | - | - | 8.55 |
| 250mm wide | hr | m | - | - | | |
| | £ | | - | - | - | 6.50 |
| Curved to radius not exceeding 12m | | | | | | |
| 300mm wide | hr | m | - | - | | |
| | £ | | - | - | - | 8.72 |
| 250mm wide | hr | m | - | - | | |
| | £ | | - | - | - | 6.66 |

**SECTION 11 KERBS, FOOTWAYS AND PAVED AREAS**

**KERBS, CHANNELS AND EDGINGS - continued**

| | | | SPECIALIST ADVICE | | | |
|---|---|---|---|---|---|---|
| In-situ concrete kerbing<br>Straight or curved to radius exceeding 12m (BS5931)<br>Kerb height 100mm | hr | m | - | - | | |
| | £ | | - | - | - | 5.20 |
| Kerb height 125mm | hr | m | - | - | | |
| | £ | | - | - | - | 6.12 |
| Curved to radius not exceeding 12m (BS5931)<br>kerb height 100mm | hr | m | - | - | | |
| | £ | | - | - | - | 5.20 |
| Kerb height 125mm | hr | m | - | - | | |
| | £ | | - | - | - | 5.30 |

| | | UNIT | L | P | M | £ NETT |
|---|---|---|---|---|---|---|
| Precast concrete edgings 150 x 50mm | | | | | | |
| Straight or curved to radius exceeding 12m | hr | m | 0.04 | 0.04 | | |
| | £ | | 1.11 | 0.19 | 1.06 | 2.36 |
| Curved to radius not exceeding 12m | hr | m | 0.045 | 0.045 | | |
| | £ | | 1.25 | 0.21 | 1.46 | 2.92 |
| Precast concrete 'Safticurb' | | | | | | |
| 250 x 254 medium duty | hr | m | 0.075 | 0.075 | | |
| | £ | | 2.09 | 0.35 | 11.98 | 14.42 |
| 305 x 305 heavy duty | hr | m | 0.095 | 0.095 | | |
| | £ | | 2.64 | 0.44 | 19.16 | 22.24 |
| Precast concrete 'Ellis Trief' safety kerb laid | | | | | | |
| Straight or curved to radius exceeding 12m (415 x 380mm) | hr | m | 0.225 | 0.225 | | |
| | £ | | 6.26 | 1.05 | 33.40 | 40.71 |
| Curved to radius not exceeding 12m | hr | m | 0.25 | 0.25 | | |
| | £ | | 6.95 | 1.17 | 33.40 | 41.52 |

**SECTION 11 KERBS, FOOTWAYS AND PAVED AREAS**

**KERBS, CHANNELS AND EDGINGS - continued**

| | | UNIT | L | P | M | £ NETT |
|---|---|---|---|---|---|---|
| Precast concrete combined kerb and drainage block (Beany Block System) | hr | m | 0.06 | 0.06 | | |
| | £ | | 1.67 | 0.28 | 38.30 | 40.25 |
| Precast concrete or pressed clay. Block kerbing straight or curved to radius exceeding 12m (natural shades) | hr | m | 0.05 | 0.05 | | |
| | £ | | 1.39 | 0.23 | 17.75 | 19.37 |
| Extra over for additional concrete for kerbing | hr | m3 | 0.02 | 0.02 | | |
| | £ | | 0.56 | 0.09 | 38.50 | 39.15 |
| **FOOTWAYS** | | | | | | |
| Flexible paved footway excluding sub base | | | | | | |
| Bitumen macadam surfacing | | | | | | |
| 40 base course | hr | m2 | 0.04 | 0.04 | | |
| | £ | | 1.11 | 0.19 | 2.47 | 3.77 |
| 20 wearing course | hr | m2 | 0.045 | 0.045 | | |
| | £ | | 1.25 | 0.21 | 2.73 | 4.19 |
| For sub bases refer to Section 7 Pavements | | | | | | |
| Extra over flexible footways for colour tinting additives (including alterations to designs) | | | | | | |
| Red | hr | m2 | - | - | | |
| | £ | | - | - | 1.10 | 1.10 |
| Green | hr | m2 | - | - | | |
| | £ | | - | - | 1.85 | 1.85 |
| Central Reserve Treatments | | | | | | |
| Provide and compact 10mm down limestone 100mm depth, spray twice with cold bituminous emulsion and blind with 6mm Quartzite fine gravels | hr | m2 | - | - | | |
| | £ | | 0.55 | 0.20 | 2.12 | 2.87 |

**SECTION 11 KERBS, FOOTWAYS AND PAVED AREAS**

**KERBS, CHANNELS AND EDGINGS - continued**

| | | UNIT | L | P | M | £ NETT |
|---|---|---|---|---|---|---|
| 50mm thick 'Gold Top' gravels in central reserves (onto sub bases) (Johnstones hot bitumen spray and hot chip process) | hr | m2 | - | - | | |
| | £ | | 0.70 | 0.30 | 3.30 | 4.30 |
| Surface dressing, double layer to clause 919 | hr | m2 | - | - | | |
| | £ | | 0.20 | 0.20 | 1.25 | 1.65 |
| 600 x 900 x 63mm thick precast concrete paved footway on granular sub base material type 2, 100mm thick to surface sloping at 10 degrees or less to the horizontal | hr | m2 | 0.27 | 0.27 | | |
| | £ | | 7.51 | 1.26 | 5.00 | 13.77 |
| 600 x 900 x 63mm thick precast concrete paved footway on 20mm mortar bed on class E concrete 100mm thick to surface sloping at 10 degrees or less to the horizontal | hr | m2 | 0.30 | 0.30 | | |
| | £ | | 8.34 | 1.40 | 8.05 | 17.79 |
| 600 x 600 x 50mm thick precast concrete paved footway on granular sub base material type 2, 100mm thick to surface sloping at 10 degrees or less to the horizontal | hr | m2 | 0.30 | 0.30 | | |
| | £ | | 8.34 | 1.40 | 5.52 | 15.26 |
| 450 x 600 x 50mm thick precast concrete paved footway on granular sub base material type 2, 100mm thick to surface sloping at 10 degrees or less to the horizontal | hr | m2 | 0.31 | 0.31 | | |
| | £ | | 8.62 | 1.45 | 6.05 | 16.12 |
| 80mm thick concrete block paving on type 2 sub base 100mm thick to surfaces sloping at 10 degrees or less to horizontal | hr | m2 | 0.30 | 0.30 | | |
| | £ | | 8.34 | 1.40 | 8.90 | 18.64 |
| 80mm stock flex paver bricks on on type 2 sub base 100mm thick to surfaces sloping at 10 degrees or less to horizontal. (PC 418/1000) | hr | m2 | 0.30 | 0.30 | | |
| | £ | | 8.34 | 1.40 | 15.25 | 24.99 |

**SECTION 11 KERBS, FOOTWAYS AND PAVED AREAS**

**KERBS, CHANNELS AND EDGINGS - continued**

| | | UNIT | L | P | M | £ NETT |
|---|---|---|---|---|---|---|
| Extra over for class E concrete bases to block paving or paver bricks (50mm thickness) | hr | m2 | 0.03 | 0.03 | | |
| | £ | | 0.83 | 0.14 | 2.50 | 3.47 |
| Remove from store and relay random sized granite setts laid to random pattern on 25mm cement mortar on 100mm granular sub base material type 2 | hr | m2 | 0.20 | 0.20 | | |
| | £ | | 5.56 | 0.93 | 2.60 | 9.09 |
| Remove from store and relay 100mm x 100mm granite setts laid to specified pattern on 25mm cement mortar on 100mm granular sub base material type 2 | hr | m2 | 0.30 | 0.30 | | |
| | £ | | 8.34 | 1.40 | 2.60 | 12.34 |
| Extra over cost for new granite setts (200Kg/m2) | hr | m2 | - | - | | |
| | £ | | - | - | 19.80 | 19.80 |
| Extra over any item of kerbing or paving for excavation of rock, masonry, mass concrete and the like | hr | m3 | 0.71 | 0.71 | | |
| | £ | | 19.75 | 3.32 | - | 23.07 |
| Extra for any item of kerbing or paving for excavation for reinforced concrete | hr | m3 | 1.08 | 1.08 | | |
| | £ | | 30.03 | 5.04 | - | 35.07 |

## SECTION 12 TRAFFIC SIGNS AND ROAD MARKINGS

### TRAFFIC SIGNS AND BOLLARDS

| | | |
|---|---|---|
| LABOUR based on Gang I.2 | | |
| Total hourly cost | = £ | 38.51 |
| PLANT | | |
| (Part time) | | |
| Compressor and tools | | 1.04 |
| JCB 3CX | | 1.94 |
| 12T Lorry with hiab lift | | 3.31 |
| Total hourly cost | = £ | 6.29 |

In this section prices will vary depending upon the diagram configurations. The following are average costs of Signs and Bollards. Diagram numbers refer to the Traffic Signs Regulations and General Directions 1981 and the figure numbers refer to the Traffic Signs Manual.

Examples of prime costs for traffic and road signs (ex works) for orders exceeding £500.

| | Class 1 (High Intensity) | Class 2 (Engineering Grade) |
|---|---|---|
| 600mm diameter | 39.68 | 29.19 |
| 600mm Tri | 33.12 | 24.00 |
| 600mm x 450 | 30.85 | 23.27 |
| 1200mm x 400 | 54.02 | 39.92 |
| Anodized Aluminimum Posts | 34.30 | |
| Plastic Coated Steel Posts | 30.49 | |
| Stainless Steel Clips pair | 2.30 | |
| Base Plates for posts | 1.67 | |

## SECTION 12 TRAFFIC SIGNS AND ROAD MARKINGS

### TRAFFIC SIGNS AND BOLLARDS - continuedd

| | | UNIT | L | P | M | £ NETT |
|---|---|---|---|---|---|---|
| Standard reflectorised traffic sign 0.25m2 in area, on Tubular Steel Post (76mm diameter) | hr | nr | 0.28 | 0.28 | | |
| | £ | | 10.78 | 1.76 | 64.89 | 77.43 |
| Standard reflectorised traffic sign 0.25m2 in area, on independently measured or existing column | hr | nr | 0.05 | 0.05 | | |
| | £ | | 1.93 | 0.31 | 20.34 | 22.58 |
| Standard externally illuminated reflectorised traffic sign 0.25m2 in area, on Tubular Steel Post (76mm diameter) | hr | nr | 0.61 | 0.61 | | |
| | £ | | 23.49 | 3.84 | 130.44 | 157.77 |
| Standard externally illuminated reflectorised traffic sign 0.75m2 in area, on Tubular Steel Post (100mm diameter) | hr | nr | 0.86 | 0.86 | | |
| | £ | | 33.12 | 5.41 | 187.87 | 226.40 |
| Standard externally illuminated reflectorised traffic sign 1.50m2 in area on two Tubular Steel Posts (100mm diameter) | hr | nr | 1.675 | 1.675 | | |
| | £ | | 64.50 | 10.54 | 350.00 | 425.04 |
| Standard externally illuminated reflectorised traffic sign 4.00m2 in area on three 100mm diameter Tubular Steel Posts | hr | nr | 2.88 | 2.88 | | |
| | £ | | 110.91 | 18.12 | 561.25 | 690.28 |
| Standard internally illuminated traffic sign 0.25m2 in area on Tubular Steel Posts (76mm diameter) | hr | nr | 0.53 | 0.53 | | |
| | £ | | 20.41 | 3.33 | 121.15 | 144.89 |
| Standard internally illuminated traffic sign 0.75m2 in area on Tubular Steel Posts (100mm diameter) | hr | nr | 0.77 | 0.77 | | |
| | £ | | 29.65 | 4.84 | 170.10 | 204.59 |

**SECTION 12 TRAFFIC SIGNS AND ROAD MARKINGS**

**TRAFFIC SIGNS AND BOLLARDS - continuedd**

| | | UNIT | L | P | M | £ NETT |
|---|---|---|---|---|---|---|
| Standard internally illuminated traffic sign 0.25m2 in area on an independently measured or existing column | hr | nr | 0.31 | 0.31 | | |
| | £ | | 11.94 | 1.95 | 73.25 | 87.14 |
| Standard internally illuminated traffic sign up to 4.00m2 in area, on four Tubular Steel Posts (120mm diameter) | hr | nr | 1.98 | 1.98 | | |
| | £ | | 76.25 | 14.57 | 428.10 | 518.92 |
| Standard internally illuminated traffic bollard approximately 0.25m2 in area | hr | nr | 0.48 | 0.48 | | |
| | £ | | 18.48 | 3.53 | 110.25 | 132.26 |
| Extra over any item of traffic signs for excavation in existing pavement or rock | hr | m3 | - | - | | |
| | £ | | - | - | - | 59.25 |

**SECTION 12 TRAFFIC SIGNS AND ROAD MARKINGS**

**SIGNS ON GANTRIES**

LABOUR based on Gang 1

| | | |
|---|---|---|
| Total hourly cost | = £ | 38.51 |

PLANT

| | | |
|---|---|---|
| (Part time) | | |
| Lorry with hiab | | 3.31 |
| 20t Mobile Crane | | 4.05 |
| Total hourly cost | = £ | 7.36 |

| | | UNIT | L | P | M | £ NETT |
|---|---|---|---|---|---|---|
| Externally illuminated reflectorised signs | | | | | | |
| 0.75m2 | hr | nr | 1.785 | 1.785 | | |
| | £ | | 68.74 | 13.14 | 146.25 | 228.13 |
| 1.00m2 | hr | nr | 2.15 | 2.15 | | |
| | £ | | 82.80 | 15.82 | 192.29 | 290.91 |
| 1.50m2 | hr | nr | 3.07 | 3.07 | | |
| | £ | | 118.23 | 22.60 | 274.70 | 415.53 |
| Internally illuminated reflectorised signs | | | | | | |
| 0.75m2 | hr | nr | 1.56 | 1.56 | | |
| | £ | | 60.08 | 11.48 | 131.05 | 202.61 |
| 1.00m2 | hr | nr | 1.70 | 1.70 | | |
| | £ | | 65.47 | 12.51 | 175.00 | 252.98 |
| 1.50m2 | hr | nr | 2.41 | 2.41 | | |
| | £ | | 92.81 | 17.74 | 248.50 | 359.05 |

## SECTION 12 TRAFFIC SIGNS AND ROAD MARKINGS

## ROAD MARKINGS, ROAD STUDS AND TRAFFIC DELININATORS ETC.

| | | UNIT | SPECIALIST ADVICE £ NETT |
|---|---|---|---|
| ROAD MARKINGS | | | |
| (New road, clean surface closed to traffic) | | | |
| Continuous line in reflectorised white thermoplastic screed or spray | | | |
| 150mm wide | £ | m | 0.54 |
| 200mm wide | £ | m | 0.70 |
| Continuous line in reflectorised yellow thermoplastic screed or spray | | | |
| 100mm wide | £ | m | 0.37 |
| 150mm wide | £ | m | 0.53 |
| Intermittent line in reflectorised white thermoplastic screed or spray | | | |
| 100mm wide with 1000mm line and and 5000mm gap | £ | m | 0.38 |
| 100mm wide with 2000mm line and and 7000mm gap | £ | m | 0.38 |
| 100mm wide with 4000mm line and and 2000mm gap | £ | m | 0.38 |
| 100mm wide with 6000mm line and and 3000mm gap | £ | m | 0.38 |
| 150mm wide with 1000mm line and and 5000mm gap | £ | m | 0.54 |
| 150mm wide with 6000mm line and and 3000mm gap | £ | m | 0.54 |

**SECTION 12 TRAFFIC SIGNS AND ROAD MARKINGS**

**ROAD MARKINGS, ROAD STUDS AND TRAFFIC DELININATORS ETC. - continued**

| | | UNIT | SPECIALIST ADVICE £ NETT |
|---|---|---|---|
| ROAD MARKINGS - continued | | | |
| 150mm wide with 600mm line and and 300mm gap | £ | m | 0.54 |
| 200mm wide with 600mm line and and 300mm gap | £ | m | 0.70 |
| 200mm wide with 1000mm line and and 1000mm gap | £ | m | 0.70 |
| 60mm wide with 600mm line and and 600mm gap | £ | m | 0.30 |
| Ancillary line in reflectorised white thermoplastic screed or spray | | | |
| 200mm wide in hatched areas | £ | m | 0.80 |
| 150mm wide in hatched areas | £ | m | 0.64 |
| Ancillary lines in yellow thermoplastic screed or spray | | | |
| 150mm wide in hatched areas | £ | m | 0.64 |
| Kerb markings 100mm wide | £ | nr | 0.25 |

**SECTION 12 TRAFFIC SIGNS AND ROAD MARKINGS**

**ROAD MARKINGS, ROAD STUDS AND TRAFFIC DELININATORS ETC. - continued**

| | | UNIT | SPECIALIST ADVICE £ NETT |
|---|---|---|---|
| Arrows in reflectorised white thermoplastic screed or spray | | | |
| 4000mm long straight or turning | £ | nr | 7.50 |
| 6000mm long straight or turning | £ | nr | 10.00 |
| 6000mm double headed | £ | nr | 11.00 |
| 8000mm long double headed | £ | nr | 23.00 |
| 16000mm long double headed | £ | nr | 55.00 |
| 32000mm long double headed | £ | nr | 120.00 |
| Triangle or letter 1600mm high | £ | nr | 5.00 |
| Triangle or letter 2800mm high | £ | nr | 7.00 |
| Triangle 3750mm high | £ | nr | 7.50 |
| Letter 3750mm high | £ | nr | 11.00 |
| Removal of existing markings | | | |
| Lines in reflectorised thermoplastic | | | |
| 100mm wide | £ | m | 6.00 |
| 150mm wide | £ | m | 6.00 |
| 200mm wide | £ | m | 6.00 |
| Arrow or letter up to 6.00m long | £ | nr | 10.50 |
| Arrow or letter 6.00 to 16.00m long | £ | nr | 46.26 |

## SECTION 12 TRAFFIC SIGNS AND ROAD MARKINGS

### ROAD MARKINGS, ROAD STUDS AND TRAFFIC DELININATORS ETC. - continued

| | | UNIT | L | P | M | SPECIALIST ADVICE £ NETT |
|---|---|---|---|---|---|---|
| 'Verynyl' strip markings. (Pedestrian crossing and similar locations) | | | | | | |
| Continuous lines 200mm wide | hr | m | - | - | | |
| | £ | | - | - | - | 4.70 |
| 600 x 300 single stud tiles | hr | nr | - | - | | |
| | £ | | - | - | - | 6.50 |
| Corner cube plastic Amber reflecting two way Road Stud size 100 x 100mm | hr | nr | - | - | | |
| | £ | | 2.34 | 0.66 | 3.25 | 6.25 |
| Catseye white reflecting two way Road Stud size 140 x 254mm | hr | nr | - | - | | |
| | £ | | 2.24 | 0.66 | 4.90 | 7.80 |
| Catseye red reflecting one way Road Stud size 140 x 254mm | hr | nr | - | - | | |
| | £ | | 2.19 | 0.66 | 4.95 | 7.80 |
| Catseye green refelecting one way road stud | hr | nr | - | - | | |
| | £ | | 2.19 | 0.66 | 4.95 | 7.80 |
| Catseye road studs without inserts | hr | nr | - | - | | |
| | £ | | 1.18 | 0.66 | 3.96 | 5.80 |
| Remove existing corner cube road stud | hr | nr | - | - | | |
| | £ | | - | - | - | 0.50 |
| Remove existing catseye road stud | hr | nr | - | - | | |
| | £ | | - | - | - | 2.00 |
| Marker Posts type 1,2,3 or 4 | hr | nr | - | - | | |
| | £ | | 2.15 | 0.52 | 6.77 | 9.44 |
| Marker Posts type 5,6,7 or 8 | hr | nr | - | - | | |
| | £ | | 2.15 | 0.52 | 6.11 | 8.78 |
| Linto posts for emergency crossing | hr | nr | - | - | | |
| | £ | | 0.50 | - | 4.45 | 4.95 |
| Standard reflectorised traffic cylinder 1000mm high 125mm diameter mounted in cats eye base (deliniator) | hr | nr | - | - | | |
| | £ | | 0.50 | - | 15.75 | 16.25 |

**SECTION 12 TRAFFIC SIGNS AND ROAD MARKINGS**

**ROAD MARKINGS, ROAD STUDS AND TRAFFIC DELININATORS ETC. - continued**

**TRAFFIC SIGNALS**

| | | UNIT | SPECIALIST ADVICE £ NETT |
|---|---|---|---|
| Traffic signal installation is exclusively carried out by Specialist Contractors. Certain items are dealt with by main Contractors/Sub-contractors. | | | |
| Installation of signal pedestals, loop detector unit pedestals, controller unit boxes and cable connection pillars. | | | |
| Installation of signal pedestals | £ | nr | 20.94 |
| Installation of loop detector unit pedestals | £ | nr | 12.77 |
| Installation of controller unit box | £ | nr | 31.77 |
| Traffic signal cable trench<br>Cable trench depth 1.50m or less | £ | m | 3.91 |
| Excavation in rock, pavement, brickwork, concrete, reinforced concrete, masonry and the like.<br>Excavation in rock | £ | m3 | 21.92 |
| Excavation in pavement, brickwork, concrete, masonry and the like | £ | m3 | 21.92 |
| Saw cutting grooves in pavement for detector loops and feeder cables seal after installation with hot bitumen sealant | | | |
| 25mm deep | £ | m | 3.27 |

**SECTION 12 TRAFFIC SIGNS AND ROAD MARKINGS**

**MISCELLANEOUS STREET FURNITURE**

LABOUR based on Gang I.1

| | | |
|---|---|---|
| Total hourly cost | = £ | 27.23 |

PLANT

| | | |
|---|---|---|
| (Part time) Compressor and tools | | 1.04 |
| JCB 3CX | | 1.94 |
| 12T Lorry with hiab lift | | 3.31 |
| Total hourly cost | = £ | 6.29 |

| | | UNIT | L | P | M | £ NETT |
|---|---|---|---|---|---|---|
| Precast Concrete Bollard 750mm high, minimum diameter 150mm | hr | nr | 0.40 | 0.40 | | |
| | £ | | 10.89 | 2.52 | 43.20 | 56.61 |
| Precast Concrete Bollard 750mm high, minimum diameter 300mm | hr | nr | 0.40 | 0.40 | | |
| | £ | | 10.89 | 2.52 | 23.75 | 37.16 |
| Extra over last item for exposed aggregate finish | hr | nr | - | - | | |
| | £ | | - | - | 7.50 | 7.50 |
| Removable lockable Bollard (Galvanised Steel) | hr | nr | 0.40 | 0.40 | | |
| | £ | | 10.89 | 2.52 | 44.10 | 57.51 |
| Lifting Traffic Barrier, all Galvanised Steel 4.0m wide | hr | nr | 1.20 | 1.20 | | |
| | £ | | 32.68 | 7.55 | 619.40 | 659.63 |
| Concrete Bench Seat 2m long | hr | nr | 0.75 | 0.75 | | |
| | £ | | 20.42 | 4.72 | 92.65 | 117.79 |
| Bench Seat with Concrete end and Timber slats 2m long | hr | nr | 0.75 | 0.75 | | |
| | £ | | 20.42 | 4.72 | 119.73 | 144.87 |
| Timber Bench Seat fixed to Concrete 2m long | hr | nr | 0.45 | 0.45 | | |
| | £ | | 12.25 | 2.83 | 37.08 | 52.16 |
| Metal Bench Seat 2m long | hr | nr | 0.75 | 0.75 | | |
| | £ | | 20.42 | 4.72 | 89.10 | 114.24 |

**SECTION 12 TRAFFIC SIGNS AND ROAD MARKINGS**

**TAKE FROM STORE AND RE-ERECT**

| | | UNIT | L | P | M | £ NETT |
|---|---|---|---|---|---|---|
| Take from store and re-erect existing 4m high Lighting Column complete | hr | nr | 0.69 | 0.69 | | |
| | £ | | 18.79 | 4.34 | 10.92 | 34.05 |
| Take from store and re-erect existing 10m high Lighting Column complete | hr | nr | 0.86 | 0.86 | | |
| | £ | | 23.42 | 5.41 | 19.66 | 48.49 |
| Take from store and re-erect existing Road Sign 3m high | hr | nr | 0.28 | 0.28 | | |
| | £ | | 7.62 | 1.76 | 10.92 | 20.30 |
| Take from store and re-erect existing Road Sign on two posts | hr | nr | 0.50 | 0.50 | | |
| | £ | | 13.62 | 3.15 | 21.84 | 38.61 |
| Take from store and re-erect existing Pedestrian Guard Railing in 3m length. Panels 1m high | hr | nr | 0.18 | 0.18 | | |
| | £ | | 4.90 | 1.13 | 2.91 | 8.94 |

**SECTION 13 ROAD LIGHTING COLUMNS AND BRACKETS**

LABOUR based on Gang I

| | | | |
|---|---|---|---|
| Total hourly cost | = | £ | 38.51 |

| | | | |
|---|---|---|---|
| PLANT (Part time) | | | |
| JCB 3CX | | | 1.94 |
| Use of power tools compressor, dumper etc. | | | 2.75 |
| 15T Mobile Crane | | | 6.71 |
| Pick-up van | | | 6.05 |
| Total hourly cost | = | £ | 17.45 |

This section and 14 following are broken down into columns, brackets, lamps and cabling.

The outputs assume operations are continuous and are based on at least 10 complete units.

**SECTION 13 ROAD LIGHTING COLUMNS AND BRACKETS**

**COLUMNS**

| | | UNIT | L | P | M | £ NETT |
|---|---|---|---|---|---|---|
| Supply and erection of Galvanised Steel Lighting Columns to BS1840 (including all control gear, switching, fuses and internal column cables) | | | | | | |
| 4m nominal height | hr | nr | 0.75 | 0.75 | | |
| | £ | | 28.88 | 13.09 | 60.50 | 102.47 |
| 6m nominal height | hr | nr | 0.80 | 0.80 | | |
| | £ | | 30.81 | 13.96 | 123.00 | 167.77 |
| 8m nominal height | hr | nr | 0.96 | 0.96 | | |
| | £ | | 36.97 | 16.75 | 153.50 | 207.22 |
| 10m nominal height | hr | nr | 1.03 | 1.03 | | |
| | £ | | 39.67 | 17.97 | 182.00 | 239.64 |
| 12m nominal height | hr | nr | 1.10 | 1.10 | | |
| | £ | | 42.36 | 19.20 | 219.50 | 281.06 |
| 15m nominal height | hr | nr | 1.18 | 1.18 | | |
| | £ | | 45.44 | 20.59 | 259.75 | 325.78 |
| 3m cast iron column (pedestrian/landscape area) | hr | nr | 0.70 | 0.76 | | |
| | £ | | 26.96 | 13.26 | 95.00 | 135.22 |
| Supply and erection of Concrete Lighting Columns Columns to BS1308 | | | | | | |
| 5m nominal height | hr | m | 0.75 | 0.75 | | |
| | £ | | 28.88 | 13.09 | 80.38 | 122.35 |
| 10m nominal height | hr | m | 1.03 | 1.03 | | |
| | £ | | 39.67 | 17.97 | 194.25 | 251.89 |

Painting and protection of columns apart from galvanising is not included in the above prices or outputs.

There is no allowance in the above outputs for on site remedial works after erection of columns.

**SECTION 13 ROAD LIGHTING COLUMNS AND BRACKETS**

**BRACKETS**

| | | UNIT | L | P | M | £ NETT |
|---|---|---|---|---|---|---|
| Supply and assemble onto column to BS1840 and BS5649 | | | | | | |
| Bracket arm with 5 degree uplift | | | | | | |
| Galvanised steel | | | | | | |
| 0.5m outreach single arm | hr | nr | 0.12 | 0.12 | | |
| | £ | | 4.62 | 2.09 | 10.75 | 17.46 |
| 1.0m outreach single arm | hr | nr | 0.14 | 0.14 | | |
| | £ | | 5.39 | 2.44 | 23.25 | 31.08 |
| 1.5m outreach single arm | hr | nr | 0.16 | 0.16 | | |
| | £ | | 6.16 | 2.79 | 25.25 | 34.20 |
| 2.0m outreach single arm | hr | nr | 0.16 | 0.16 | | |
| | £ | | 6.16 | 2.79 | 28.10 | 37.05 |
| 1.0m outreach double arm | hr | nr | 0.17 | 0.17 | | |
| | £ | | 6.55 | 2.97 | 38.50 | 48.02 |
| 2.0m outreach double arm | hr | nr | 0.19 | 0.19 | | |
| | £ | | 7.32 | 3.32 | 46.55 | 57.19 |
| Concrete | | | | | | |
| 1.0m outreach single arm | hr | nr | 0.14 | 0.14 | | |
| | £ | | 5.39 | 2.44 | 23.18 | 31.01 |
| 2.0m outreach single arm | hr | nr | 0.16 | 0.16 | | |
| | £ | | 6.16 | 2.79 | 27.95 | 36.90 |
| 1.0m outreach double arm | hr | nr | 0.17 | 0.17 | | |
| | £ | | 6.55 | 2.97 | 38.00 | 47.52 |
| 2.0m outreach double arm | hr | nr | 0.19 | 0.19 | | |
| | £ | | 7.32 | 3.32 | 46.10 | 56.74 |

**SECTION 14 ELECTRICAL WORK FOR ROAD LIGHTING AND TRAFFIC SIGNS**

**LUMINAIRES LAMPS**

| | | |
|---|---|---|
| **LABOUR based on Gang I** | | |
| Total hourly cost | = £ | 38.51 |
| **PLANT (Part time)** | | |
| JCB 3CX | | 1.94 |
| Use of power tools compressor, dumper etc. | | 2.75 |
| Simon Hoist (50ft) | | 5.92 |
| Pick-up van | | 6.05 |
| Total hourly cost | = £ | 16.66 |

| | | **UNIT** | **L** | **P** | **M** | **£ NETT** |
|---|---|---|---|---|---|---|
| Supply assemble and install lamps, lanterns and luminares to suit Columns and Brackets (including integral control gear and Photo-Electric Cell Socket) | | | | | | |
| 55W SON (P226) to suit 4m and 5m columns | hr | nr | 0.40 | 0.40 | | |
| | £ | | 15.40 | 6.66 | 61.98 | 84.04 |
| 70W SON (P236) to suit 5m and 6m columns | hr | nr | 0.40 | 0.40 | | |
| | £ | | 15.40 | 6.66 | 84.98 | 107.04 |
| 250W SON (P426) to suit 8m,10m and 12m columns | hr | nr | 0.50 | 0.50 | | |
| | £ | | 19.26 | 8.33 | 130.40 | 157.99 |
| Sphere 70W SON to suit 3m (P456) | hr | nr | 0.30 | 0.30 | | |
| | £ | | 11.55 | 5.00 | 122.40 | 138.95 |
| 400W SON High pressure sodium to suit 12m and 15m columns | hr | nr | 0.50 | 0.50 | | |
| | £ | | 19.26 | 8.33 | 199.20 | 226.79 |

## SECTION 14 ELECTRICAL WORK FOR ROAD LIGHTING AND TRAFFIC SIGNS

### TRENCHES AND CABLING

LABOUR based on Gang I

| | | | |
|---|---|---|---|
| Total hourly cost | = | £ | 38.51 |

PLANT (Part time)

| | | | |
|---|---|---|---|
| JCB 3CX | | | 3.88 |
| Use of power tools compressor, dumper etc. | | | 2.75 |
| Trench Excavator | | | 6.00 |
| Total hourly cost | = | £ | 12.63 |

600/1000V grade, PVC insulated copper conductor steel wire armoured cable

In trench

| | | Unit | L | P | M | £ NETT |
|---|---|---|---|---|---|---|
| size 2.5mm2 2 core | hr | m | 0.04 | 0.04 | | |
| | £ | | 1.54 | 0.51 | 0.82 | 2.87 |
| 4.0mm2 2 core | hr | m | 0.04 | 0.04 | | |
| | £ | | 1.54 | 0.51 | 0.94 | 2.99 |
| 6.0mm2 2 core | hr | m | 0.04 | 0.04 | | |
| | £ | | 1.54 | 0.51 | 1.13 | 3.18 |
| 10.0mm2 2 core | hr | m | 0.04 | 0.04 | | |
| | £ | | 1.54 | 0.51 | 1.51 | 3.56 |
| 16.0mm2 2 core | hr | m | 0.04 | 0.04 | | |
| | £ | | 1.54 | 0.51 | 2.15 | 4.20 |
| 16.0mm2 4 core | hr | m | 0.034 | 0.034 | | |
| | £ | | 1.31 | 0.43 | 3.50 | 5.24 |
| 25.0mm2 2 core | hr | m | 0.04 | 0.04 | | |
| | £ | | 1.54 | 0.51 | 4.19 | 6.24 |
| 35.0mm2 4 core | hr | m | 0.034 | 0.034 | | |
| | £ | | 1.31 | 0.43 | 5.43 | 7.17 |
| 70.0mm2 4 core | hr | m | 0.08 | 0.08 | | |
| | £ | | 3.08 | 1.01 | 7.28 | 11.37 |

**SECTION 14 ELECTRICAL WORK FOR ROAD LIGHTING AND TRAFFIC SIGNS**

**TRENCHES AND CABLING - continued**

| | | UNIT | L | P | M | £ NETT |
|---|---|---|---|---|---|---|
| 600/1000V grade, PVC insulated steel wire armour cable | | | | | | |
| In ducts or pipe bays or troughs | | | | | | |
| size 2.5mm2 2 core | hr | m | - | - | | |
| | £ | | 0.05 | - | 0.82 | 0.87 |
| 4.0mm2 2 core | hr | m | - | - | | |
| | £ | | 0.05 | - | 0.94 | 0.99 |
| 6.0mm2 2 core | hr | m | - | - | | |
| | £ | | 0.06 | - | 1.13 | 1.19 |
| 10.0mm2 2 core | hr | m | - | - | | |
| | £ | | 0.08 | - | 1.51 | 1.59 |
| 16.0mm2 2 core | hr | m | - | - | | |
| | £ | | 0.08 | - | 2.15 | 2.23 |
| 16.0mm2 4 core | hr | m | - | - | | |
| | £ | | 0.09 | - | 3.50 | 3.59 |
| 25.0mm2 2 core | hr | m | - | - | | |
| | £ | | 0.10 | - | 4.19 | 4.29 |
| 35.0mm2 4 core | hr | m | - | - | | |
| | £ | | 0.10 | - | 5.43 | 5.53 |
| 70.0mm2 4 core | hr | m | - | - | | |
| | £ | | 0.11 | - | 7.28 | 7.39 |
| Extra over for sand backfill to trench | hr | m | - | - | | |
| | £ | | - | - | - | 0.80 |
| Extra over for PVC marker tape | hr | m | - | - | | |
| | £ | | - | - | - | 0.31 |

## SECTION 14 ELECTRICAL WORK FOR ROAD LIGHTING AND TRAFFIC SIGNS

### JOINTS AND TERMINATIONS

| | | UNIT | L | P | M | £ NETT |
|---|---|---|---|---|---|---|
| Branch joints in 600/1000V grade PVC insulated copper conductor steel wire armoured cable | | | | | | |
| size 2.5mm2 2 core | hr | nr | 0.127 | 0.127 | | |
| | £ | | 4.89 | 2.12 | 34.03 | 41.04 |
| 4.0mm2 2 core | hr | nr | 0.127 | 0.127 | | |
| | £ | | 4.89 | 2.12 | 34.03 | 41.04 |
| 6.0mm2 2 core | hr | nr | 0.127 | 0.127 | | |
| | £ | | 4.89 | 2.12 | 34.03 | 41.04 |
| 10.0mm2 2 core | hr | nr | 0.13 | 0.13 | | |
| | £ | | 5.01 | 2.17 | 34.90 | 42.08 |
| 16.0mm2 2 core | hr | nr | 0.133 | 0.133 | | |
| | £ | | 5.12 | 2.22 | 35.73 | 43.07 |
| 16.0mm2 4 core | hr | nr | 0.136 | 0.136 | | |
| | £ | | 5.24 | 2.27 | 36.58 | 44.09 |
| 25.0mm2 2 core | hr | nr | 0.142 | 0.142 | | |
| | £ | | 5.47 | 2.37 | 38.28 | 46.12 |
| 35.0mm2 4 core | hr | nr | 0.15 | 0.15 | | |
| | £ | | 5.78 | 2.50 | 38.98 | 47.26 |
| 70.0mm2 4 core | hr | nr | 0.165 | 0.165 | | |
| | £ | | 6.35 | 2.75 | 44.24 | 53.34 |

**SECTION 14 ELECTRICAL WORK FOR ROAD LIGHTING AND TRAFFIC SIGNS**

**JOINTS AND TERMINATIONS - continued**

| | | UNIT | L | P | M | £ NETT |
|---|---|---|---|---|---|---|
| Terminations of 600/1000V grade PVC insulated copper conductor steel wire armoured cable | | | | | | |
| size 2.5mm2 2 core | hr | nr | 0.023 | 0.023 | | |
| | £ | | **0.89** | **0.38** | 2.93 | 4.20 |
| 4.0mm2 2 core | hr | nr | 0.028 | 0.028 | | |
| | £ | | 1.08 | 0.47 | 3.67 | 5.22 |
| 6.0mm2 2 core | hr | nr | 0.029 | 0.029 | | |
| | £ | | 1.12 | 0.48 | 3.69 | 5.29 |
| 10.0mm2 2 core | hr | nr | 0.03 | 0.03 | | |
| | £ | | 1.16 | 0.50 | 3.95 | 5.61 |
| 16.0mm2 2 core | hr | nr | 0.034 | 0.034 | | |
| | £ | | 1.31 | 0.57 | 4.32 | 6.20 |
| 16.0mm2 4 core | hr | nr | 0.037 | 0.037 | | |
| | £ | | 1.42 | 0.62 | 4.79 | 6.83 |
| 25.0mm2 2 core | hr | nr | 0.039 | 0.039 | | |
| | £ | | 1.50 | 0.65 | 5.01 | 7.16 |
| 35.0mm2 4 core | hr | nr | 0.05 | 0.05 | | |
| | £ | | 1.93 | 0.83 | 6.33 | 9.09 |
| 70.0mm2 4 core | hr | nr | 0.06 | 0.06 | | |
| | £ | | 2.31 | 1.00 | 7.93 | 11.24 |
| Cable terminations, 2 cables add 20% | | | | | | |
| Cable terminations, 3 cables | | | | | | |
| **LIGHTNING PROTECTION** | | | | | | |
| Minimal protection using Earth rods, plates or stops protective tape and joint | | | | | | |
| Suit columns up to 12m | hr | nr | - | - | | |
| | £ | | - | - | - | 166.25 |
| Suit columns up to 15m | hr | nr | - | - | | |
| | £ | | - | - | - | 220.50 |
| Suit Superstructure or Buildings using copper lead conductor (per 230m height) | hr | nr | - | - | | |
| | £ | | - | - | - | 375.00 |

## SECTION 14 ELECTRICAL WORK FOR ROAD LIGHTING AND TRAFFIC SIGNS

### FEEDER PILLARS

| | | UNIT | L | P | M | £ NETT |
|---|---|---|---|---|---|---|
| Supply erect and install feeder pillars (excluding control gear and equipment) | | | | | | |
| size 460mm x 850mm 'Austin Taylor' Lighting Pillar | hr | nr | 4.64 | 4.64 | | |
| | £ | | 178.69 | 77.30 | 27.70 | 283.69 |
| 300mm x 600mm 'Austin Taylor' Lighting Pillar | hr | nr | 4.24 | 4.24 | | |
| | £ | | 163.28 | 70.64 | 25.30 | 259.22 |
| 600mm x 850mm concrete feeder pillar 'Sheffield' type | hr | nr | 4.96 | 4.96 | | |
| | £ | | 191.01 | 82.63 | 29.60 | 303.24 |
| 300mm x 600mm feeder pillar 'DJ Young' type | hr | nr | 4.19 | 4.19 | | |
| | £ | | 161.36 | 69.81 | 25.00 | 256.17 |

### EXTRA OVER ITEMS

| | | UNIT | L | P | M | £ NETT |
|---|---|---|---|---|---|---|
| Extra over any item of street lighting for excavation in Rock or Reinforced concrete | hr | m3 | 1.04 | 1.04 | | |
| | £ | | 40.05 | 17.33 | - | 57.38 |
| Extra over any item of street lighting for excavation of pavement brickwork or mass concrete | hr | m3 | 0.74 | 0.74 | | |
| | £ | | 28.50 | 12.33 | - | 40.83 |
| Abandoned holes, after fully excavating for column roots | hr | nr | 0.37 | 0.37 | | |
| | £ | | 14.25 | 6.16 | - | 20.41 |

**SECTION 15 MOTORWAY COMMUNICATIONS**

**COMMUNICATIONS INCLUDING DETECTOR LOOPS AND FOGWARN**

| | | UNIT | SPECIALIST ADVICE £ NETT |
|---|---|---|---|
| CABLES, TRENCHES AND TERMINATIONS | | | |
| cable trench, 450mm deep, with sand fill surround to cable and marker tape | £ | m | 4.12 |
| Type A1 2 pair 0.9mm armoured multi-pair communication cable | £ | m | 1.99 |
| Type A2 20 pair 0.9mm armoured multi-pair communication cable | £ | m | 3.60 |
| Type A3 30 pair 0.9mm armoured multi-pair communication cable | £ | m | 4.40 |
| Type A4 10mm2 armoured split concentric power cable | £ | m | 3.15 |
| Type A6 50/0.25m2 single core detector feeder cable | £ | m | 2.40 |
| Type A7 50/0.25m2 single core detector feeder cable | £ | m | 2.40 |
| Termination of type A1 cable | £ | nr | 27.95 |
| Termination of type A2 cable | £ | nr | 125.00 |
| Termination of type A3 cable | £ | nr | 179.00 |
| Termination of type A4 cable | £ | nr | 30.05 |
| Termination of type A6 cable | £ | nr | 17.70 |
| Termination of type A7 cable | £ | nr | 17.70 |

**SECTION 15 MOTORWAY COMMUNICATIONS**

**COMMUNICATIONS INCLUDING DETECTOR LOOPS AND FOGWARN**

| | | UNIT | SPECIALIST ADVICE £ NETT |
|---|---|---|---|
| INSTALLATION OF EQUIPMENT AS DESCRIBED | | | |
| Plinth as foundation for cabinets | £ | nr | 82.50 |
| Plinth as foundation for matrix signal posts | £ | nr | 82.50 |
| Communication equipment cabinet 600 type series | £ | nr | 83.58 |
| Terminator type II | £ | nr | 87.95 |
| Emergency telephone post | £ | nr | 44.18 |
| Telephone housing | £ | nr | 13.63 |
| Matrix signal post | £ | nr | 44.18 |
| Motorwarn/Fogwarn | £ | nr | 89.96 |
| Distributor on gantry | £ | nr | 114.98 |
| Isolator switch for gantry | £ | nr | 247.10 |
| Heater unit 65W Henleys type 22501 on gantry | £ | nr | 16.25 |
| Plinth as foundation for CCTV camera | £ | nr | 82.50 |
| Wall mounted brackets (up to 15m) | £ | nr | 27.50 |

NB. ABOVE RATES ARE FOR INSTALLATION ONLY. THEY EXCLUDE THE ACTUAL PROVISION OF EQUIPMENT

## SECTION 15 MOTORWAY COMMUNICATIONS

### COMMUNICATIONS INCLUDING DETECTOR LOOPS AND FOGWARN

| | | UNIT | SPECIALIST ADVICE £ NETT |
|---|---|---|---|
| **TERMINAL BLOCKS** | | | |
| Klippon BK6 | £ | nr | 5.30 |
| Klippon BK12 | £ | nr | 5.95 |
| **GROOVES IN PAVEMENT FOR LOOP DETECTOR CIRCUIT** | | | |
| Grooves in pavement for detector loops and feeders | £ | m | 3.12 |
| Additional cost for sealing with hot bitumen sealant | £ | m | 0.27 |

## SECTION 16 PILING AND DIAPHRAGM WALLING

There are a number of different types of piling which are available for use in differing situations. Selection of the most suitable type of piling for a particular site will depend on a number of factors including the physical conditions likely to be encountered during driving, the loads to be carried, the design of superstructure, etc. The most commonly used systems are included in this section.
It is essential that a thorough and adequate site investigation is carried out to ascertain details of the ground strata and bearing capacities to enable a a proper assessment to be made of the most suitable and economical type of piling to be adopted.
There are so many factors, apart from design considerations, which influence the cost of piling that it is not possible to give more than an approximate indication of costs. To obtain reliable costs for a particular contract advice should be sought from a company specialising in the particular type of piling proposed. Some Specialists Contractors will also provide a design service if required.

The section comprises:
bored in-situ reinforced concrete piling (tripod rig)
bored in-situ reinforced concrete piling (mobile rig)
concrete injected piles
large diameter bored in-situ piles (general note only)
driven in situ reinforced concrete piling
driven timber piles
Frodingham steel sheet piling
Larssen steel sheet piling
steel bearing piles
steel tubular piles

### BORED IN-SITU REINFORCED CONCRETE PILING (TRIPOD RIG)

The following unit costs cover the construction of small diameter bored piling using light and compact tripod rigs requiring no expensive site levelling or levelling or access ways. Piling can be constructed in very restricted headroom or on confined and difficult sites. Standard diameters are between 400 and 600mm with a normal maximum depth of 30m.

The costs are based on installing 100 piles of 500mm nominal diameter, a concrete strength of 20N/mm2 with nominal reinforcement, on a clear site with reasonable access. Disposal of excavated material is not included in the unit costs.

| | | UNIT | L | P | M | £ NETT |
|---|---|---|---|---|---|---|
| Provision of all plant, equipment and labour including transporting to and from the site and establishing and dismantling | hr | sum | - | - | | |
| | £ | | 1600.00 | 2800.00 | - | 4400.00 |
| Set up at each pile position | hr | nr | - | - | | |
| | £ | | 15.00 | 15.00 | - | 30.00 |
| Boring to a depth of 10m and placing concrete and reinforcement | hr | nr | - | - | | |
| | £ | | 210.00 | 450.00 | - | 660.00 |

## SECTION 16 PILING AND DIAPHRAGM WALLING

### BORED IN-SITU REINFORCED CONCRETE PILING (TRIPOD RIG) - continued

| | | UNIT | L | P | M | £ NETT |
|---|---|---|---|---|---|---|
| Add for additional length of pile up to a depth of 15m | hr | m | - | - | | |
| | £ | | - | - | 48.00 | 48.00 |
| Deduct for reduction in length of pile below 10m | hr | m | - | - | | |
| | £ | | 9.00 | 15.00 | - | 24.00 |
| Add for boring through obstructions | hr | hr | - | - | | |
| | £ | | 30.00 | 45.00 | - | 75.00 |
| Standing time | hr | hr | - | - | | |
| | £ | | 30.00 | 45.00 | - | 75.00 |
| Carry out pile test for maximum test load of 600kN on a working pile as reaction | | | | | | |
| first pile | hr | nr | - | - | | |
| | £ | | 1300.00 | 2300.00 | - | 3600.00 |
| subsequent pile | hr | nr | - | - | | |
| | £ | | 1050.00 | 1650.00 | - | 2700.00 |
| Take and test undisturbed soil samples | hr | nr | - | - | | |
| | £ | | 50.00 | 65.00 | - | 115.00 |
| Make, cure and test concrete cubes | hr | nr | - | - | | |
| | £ | | - | - | - | 6.00 |
| Trim down heads of piles | hr | nr | - | - | | |
| | £ | | 16.00 | 24.00 | - | 40.00 |

**SECTION 16 PILING AND DIAPHRAGM WALLING**

**BORED IN-SITU REINFORCED CONCRETE PILING (MOBILE RIG) - continued**

The following unit costs cover the construction of small diameter bored piles using lorry or crawler mounted rotary boring rigs. This type of plant is more mobile and faster in operation than the tripod rigs and is ideal for large contracts in cohesive ground. Construction of piles under bentonite suspension can be carried out to obviate the use of liners. Standard diameters of 450 to 900mm diameter can be constructed to depths of 30m.

The costs are based on installing 100 piles of 500mm nominal diameter, a concrete strength of 20N/mm2 with nominal reinforcement, on a clear site with reasonable access. Disposal of excavated material is not included in the unit costs.

| | | UNIT | L | P | M | £ NETT |
|---|---|---|---|---|---|---|
| Provision of all plant, equipment and labour including transport to and from site and establishing and dismantling | hr | sum | - | - | | |
| | £ | | 2750.00 | 4250.00 | - | 7000.00 |
| Set up at each pile position | hr | nr | - | - | | |
| | £ | | 7.00 | 28.00 | - | 35.00 |
| Boring to a depth of 10m, placing concrete and reinforcement | hr | nr | - | - | | |
| | £ | | - | - | - | 300.00 |
| Add for additional length of pile up to a depth of 15m | hr | m | - | - | | |
| | £ | | - | - | - | 28.00 |
| Deduct for reduction in length of pile below 10m | hr | m | - | - | | |
| | £ | | - | - | - | 15.00 |
| Add for boring through obstructions | hr | hr | - | - | | |
| | £ | | 35.00 | 105.00 | - | 140.00 |
| Standing time | hr | nr | - | - | | |
| | £ | | 50.00 | 90.00 | - | 140.00 |
| Carry out pile test for maximum test load of 600kN on a working pile as reaction | | | | | | |
| first pile | hr | nr | - | - | | |
| | £ | | - | - | - | 3500.00 |

**SECTION 16 PILING AND DIAPHRAGM WALLING**

**BORED IN-SITU REINFORCED CONCRETE PILING (MOBILE RIG) - continued**

| | | UNIT | L | P | M | £ NETT |
|---|---|---|---|---|---|---|
| subsequent piles | hr | nr | - | - | | |
| | £ | | - | - | - | 2500.00 |
| Take and test undisturbed soil samples | hr | nr | - | - | | |
| | £ | | - | - | - | 120.00 |
| Make, cure and test concrete cubes | hr | nr | - | - | | |
| | £ | | - | - | - | 6.00 |
| Trim down heads of piles | hr | nr | - | - | | |
| | £ | | 16.00 | 24.00 | - | 40.00 |

## SECTION 16 PILING AND DIAPHRAGM WALLING

### CONCRETE INJECTED PILES

The following unit costs cover the construction of piles by screwing a continuous flight auger into the ground to the required design depth. Concrete is then pumped through the hollow stem of the auger to the bottom and the pile formed as the auger is withdrawn. Spoil is removed by the auger as it is withdrawn. This is a fast method of construction without causing disturbance or vibration to adjacent ground. No casing is required even in unsuitable soils. Reinforcement can be placed after grouting is complete. The costs are based on installing 100 piles on a clear site with reasonable access. Disposal of excavated material is not included in the unit costs.

| | | UNIT | L | P | M | £ NETT |
|---|---|---|---|---|---|---|
| **Provision of all plant, equipment and labour including transport to and from the site and establishing and dismantling** | | | | | | |
| 450mm/ 650kN | hr | sum | - | - | | |
| | £ | | 2300.00 | 4200.00 | - | 6500.00 |
| 600mm/1400kN | hr | sum | - | - | | |
| | £ | | 2300.00 | 4200.00 | - | 6500.00 |
| 750mm/2200kN | hr | sum | - | - | | |
| | £ | | 2300.00 | 4200.00 | - | 6500.00 |
| **Set up at each pile position** | | | | | | |
| 450mm/ 650kN | hr | nr | - | - | | |
| | £ | | 7.00 | 28.00 | - | 35.00 |
| 600mm/1400kN | hr | nr | - | - | | |
| | £ | | 7.00 | 28.00 | - | 35.00 |
| 750mm/2200kN | hr | nr | - | - | | |
| | £ | | 7.00 | 28.00 | - | 35.00 |
| **Construct pile with 20N/mm2 concrete and reinforce with starter bars** | | | | | | |
| 450mm/ 650kN | hr | m | - | - | | |
| | £ | | 5.00 | 7.00 | 18.00 | 30.00 |
| 600mm/1400kN | hr | m | - | - | | |
| | £ | | 8.00 | 10.50 | 23.50 | 42.00 |
| 750mm/2200kN | hr | m | - | - | | |
| | £ | | 14.00 | 20.25 | 32.75 | 67.00 |

**SECTION 16 PILING AND DIAPHRAGM WALLING**

**CONCRETE INJECTED PILES - continued**

| | | UNIT | L | P | M | £ NETT |
|---|---|---|---|---|---|---|
| Add for additional length of pile over design depth up to a maximum depth of 22m | | | | | | |
| 450mm/ 650kN | hr | m | - | - | | |
| | £ | | 10.00 | 7.00 | 18.00 | 35.00 |
| 600mm/1400kN | hr | m | - | - | | |
| | £ | | 10.00 | 12.50 | 24.50 | 47.00 |
| 750mm/2200kN | hr | m | - | - | | |
| | £ | | 17.00 | 23.25 | 34.75 | 75.00 |
| Deduct for reduction in length of pile under design depth | | | | | | |
| 450mm/ 650kN | hr | m | - | - | | |
| | £ | | 3.50 | 3.50 | 9.00 | 16.00 |
| 600mm/1400kN | hr | m | - | - | | |
| | £ | | 5.00 | 6.00 | 10.00 | 21.00 |
| 750mm/2200kN | hr | m | - | - | | |
| | £ | | 8.00 | 10.00 | 16.00 | 34.00 |
| Standing time | | | | | | |
| 450mm/ 650kN | hr | hr | - | - | | |
| | £ | | 49.00 | 91.00 | - | 140.00 |
| 600mm/1400kN | hr | hr | - | - | | |
| | £ | | 49.00 | 91.00 | - | 140.00 |
| 750mm/2200kN | hr | hr | - | - | | |
| | £ | | 49.00 | 91.00 | - | 140.00 |
| Carry out pile test for maximum load first pile | | | | | | |
| 450mm/ 650kN | hr | nr | - | - | | |
| | £ | | 1300.00 | 2300.00 | - | 3600.00 |
| 600mm/1400kN | hr | nr | - | - | | |
| | £ | | 1630.00 | 2870.00 | - | 4500.00 |
| 750mm/2200kN | hr | nr | - | - | | |
| | £ | | 2000.00 | 3500.00 | - | 5500.00 |

**SECTION 16 PILING AND DIAPHRAGM WALLING**

**CONCRETE INJECTED PILES - continued**

| | UNIT | | L | P | M | £ NETT |
|---|---|---|---|---|---|---|
| subsequent piles | | | | | | |
| 450mm/ 650kN | hr | nr | - | - | | |
| | £ | | 920.00 | 1680.00 | - | 2600.00 |
| 600mm/1400kN | hr | nr | - | - | | |
| | £ | | 1260.00 | 2240.00 | - | 3500.00 |
| 750mm/2200kN | hr | nr | - | - | | |
| | £ | | 1630.00 | 2870.00 | - | 4500.00 |
| Make, cure and test concrete | | | | | | |
| cubes | | | | | | |
| 450mm/ 650kN | hr | nr | - | - | | |
| | £ | | - | - | - | 6.00 |
| 600mm/1400kN | hr | nr | - | - | | |
| | £ | | - | - | - | 6.00 |
| 750mm/2200kN | hr | nr | - | - | | |
| | £ | | - | - | - | 6.00 |
| Trim down heads of piles | | | | | | |
| 450mm/ 650kN | hr | nr | - | - | | |
| | £ | | 15.50 | 24.50 | - | 40.00 |
| 600mm/1400kN | hr | nr | - | - | | |
| | £ | | 28.00 | 42.00 | - | 70.00 |
| 750mm/2200kN | hr | nr | - | - | | |
| | £ | | 47.00 | 63.00 | - | 110.00 |

**SECTION 16 PILING AND DIAPHRAGM WALLING**

**LARGE DIAMETER BORED IN-SITU PILES**

Large diameter bored in-situ piles may be constructed using rotary driven augers mounted on lorry or crawler cranes. Piles may be straight shafted or belled at the bottom to give greater bearing area. In utilising this type of pile it is of great importance to carry out a very thorough site investigation of the strata to be encountered before design proceeds.
These piles may be constructed in nominal diameters from 750 to 2100mm increments of 150mm and underreamed bells may be formed in suitable ground up to three times the diameter of the shaft. Loads of up to 30,000kN can be carried and piles can be constructed to depths up to 33m. Visual inspection of the shaft and bell can be made using special man-riding cages.
Shafts may be lined where necessary or bentonite suspension may be used in straight shafted piles in poor ground.
It is not practical to give average rates for this type of piling and the advice of a Specialist Contractor should be sought.

**DRIVEN IN-SITU REINFORCED CONCRETE PILING**

The following unit costs cover the construction of piles by driving a tube into the ground either by using an internal hammer acting on a gravel or concrete plug or by using an external hammer on a driving helmet at the top of the tube. After driving to the required depth an enlarged base is formed by hammering out successive charges of concrete down the tube. The tube is then filled with concrete which is compacted as the tube is vibrated and withdrawn piles of 350 to 500mm diameter can be constructed with rakes up to 1 in 4 to carry working loads up to 120t per pile.
The costs are based on installing 100 piles of 430mm nominal diameter, a concrete strength of 20N/mm2 suitably reinforced for a working load not exceeding 750kN, on a clear site with reasonable access.

| | | UNIT | L | P | M | £ NETT |
|---|---|---|---|---|---|---|
| Provision of all plant, equipment and labour including transporting to and from the site and establishing and dismantling | | | | | | |
| Bottom driven | hr | sum | - | - | | |
| | £ | | 1710.00 | 3990.00 | - | 5700.00 |
| Top driven | hr | sum | - | - | | |
| | £ | | 1710.00 | 3990.00 | - | 5700.00 |
| Set up at each pile position | | | | | | |
| Bottom driven | hr | nr | - | - | | |
| | £ | | 25.50 | 59.50 | - | 85.00 |
| Top driven | hr | nr | - | - | | |
| | £ | | 19.50 | 45.50 | - | 65.00 |

## SECTION 16 PILING AND DIAPHRAGM WALLING

### DRIVEN IN-SITU REINFORCED CONCRETE PILING - continued

| | | UNIT | L | P | M | £ NETT |
|---|---|---|---|---|---|---|
| Drive shell and form pile shaft including reinforcement to a depth of 10m | | | | | | |
| Bottom driven | hr | m | - | - | | |
| | £ | | - | - | - | 22.00 |
| Top driven | hr | m | - | - | | |
| | £ | | - | - | - | 20.00 |
| Add for additional length exceeding 10m but ne 15m | | | | | | |
| Bottom driven | hr | m | - | - | | |
| | £ | | - | - | - | 23.50 |
| Top driven | hr | m | - | - | | |
| | £ | | - | - | - | 21.25 |
| Add for driving through obstructions where within the capabilities of the normal plant | | | | | | |
| Bottom driven | hr | nr | - | - | | |
| | £ | | - | - | - | 140.00 |
| Top driven | hr | nr | - | - | | |
| | £ | | - | - | - | 140.00 |
| Standing time | | | | | | |
| Bottom driven | hr | nr | - | - | | |
| | £ | | - | - | - | 130.00 |
| Top driven | hr | nr | - | - | | |
| | £ | | - | - | - | 130.00 |
| Carry out test on 430mm diameter pile for maximum test load of 1125kN using non working tension tension piles as reaction first piles | | | | | | |
| Bottom driven | hr | m | - | - | | |
| | £ | | 1005.00 | 2345.00 | - | 3350.00 |
| Top driven | hr | m | - | - | | |
| | £ | | 855.00 | 1995.00 | - | 2850.00 |

**SECTION 16 PILING AND DIAPHRAGM WALLING**

**DRIVEN IN-SITU REINFORCED CONCRETE PILING - continued**

| | | UNIT | L | P | M | £ NETT |
|---|---|---|---|---|---|---|
| subsequent piles | | | | | | |
| Bottom driven | hr | nr | - | - | | |
| | £ | | 690.00 | 1610.00 | - | 2300.00 |
| Top driven | hr | nr | - | - | | |
| | £ | | 690.00 | 1610.00 | - | 2300.00 |
| Make, cure and test concrete cubes | | | | | | |
| Bottom driven | hr | nr | - | - | | |
| | £ | | - | - | - | 5.75 |
| Top driven | hr | nr | - | - | | |
| | £ | | - | - | - | 5.75 |
| Trim down heads of piles | | | | | | |
| Bottom driven | hr | nr | - | - | | |
| | £ | | 7.80 | 18.20 | - | 26.00 |
| Top driven | hr | nr | - | - | | |
| | £ | | 7.80 | 18.20 | - | 26.00 |

**SECTION 16 PILING AND DIAPHRAGM WALLING**

**DRIVEN TIMBER PILES**

The following unit costs are based on installing 100 piles on a clear site with reasonable access

| | | UNIT | L | P | M | £ NETT |
|---|---|---|---|---|---|---|
| Provision of all plant, equipment and labour including transporting to and from the site and establishing and dismantling | hr | sum | - | - | | |
| | £ | | 1500.00 | 3500.00 | - | 5000.00 |
| Standing time | hr | hr | - | - | | |
| | £ | | 60.00 | 140.00 | - | 200.00 |
| Provide timber piles (Pile size) | | | | | | |
| Douglas fir | | | | | | |
| 225 x 225mm | hr | m | - | - | | |
| | £ | | - | - | 16.05 | 16.05 |
| 300 x 300mm | hr | m | - | - | | |
| | £ | | - | - | 24.85 | 24.85 |
| 350 x 350mm | hr | m | - | - | | |
| | £ | | - | - | 31.50 | 31.50 |
| 450 x 450mm | hr | m | - | - | | |
| | £ | | - | - | 50.15 | 50.15 |
| Greenheart (hewn to mean size) | | | | | | |
| 225 x 225mm | hr | m | - | - | | |
| | £ | | - | - | 31.95 | 31.95 |
| 300 x 300mm | hr | m | - | - | | |
| | £ | | - | - | 53.45 | 53.45 |
| 350 x 350mm | hr | m | - | - | | |
| | £ | | - | - | 70.40 | 70.40 |
| 450 x 450mm | hr | m | - | - | | |
| | £ | | - | - | 113.85 | 113.85 |
| Set top at each pile position | | | | | | |
| 225 x 225mm | hr | nr | - | - | | |
| | £ | | - | - | - | 55.00 |
| 300 x 300mm | hr | nr | - | - | | |
| | £ | | - | - | - | 55.00 |
| 350 x 350mm | hr | nr | - | - | | |
| | £ | | - | - | - | 55.00 |
| 450 x 450mm | hr | nr | - | - | | |
| | £ | | - | - | - | 55.00 |

**SECTION 16 PILING AND DIAPHRAGM WALLING**

**DRIVEN TIMBER PILES - continued**

| | UNIT | | L | P | M | £ NETT |
|---|---|---|---|---|---|---|
| Drive piles | | | | | | |
| 225 x 225mm | hr | m | - | - | | |
| | £ | | 6.05 | - | - | 6.05 |
| 300 x 300mm | hr | m | - | - | | |
| | £ | | 7.50 | - | - | 7.50 |
| 350 x 350mm | hr | m | - | - | | |
| | £ | | 8.25 | - | - | 8.25 |
| 450 x 450mm | hr | m | - | - | | |
| | £ | | 9.10 | - | - | 9.10 |
| Cut heads of pile to level | | | | | | |
| Douglas fir | | | | | | |
| 225 x 225mm | hr | nr | - | - | | |
| | £ | | 1.85 | - | - | 1.85 |
| 300 x 300mm | hr | nr | - | - | | |
| | £ | | 3.30 | - | - | 3.30 |
| 350 x 350mm | hr | nr | - | - | | |
| | £ | | 4.24 | - | - | 4.24 |
| 450 x 450mm | hr | nr | - | - | | |
| | £ | | 8.11 | - | - | 8.11 |
| Greenheart | | | | | | |
| 225 x 225mm | hr | nr | - | - | | |
| | £ | | 3.65 | - | - | 3.65 |
| 300 x 300mm | hr | nr | - | - | | |
| | £ | | 10.60 | - | - | 10.60 |
| 350 x 350mm | hr | nr | - | - | | |
| | £ | | 8.50 | - | - | 8.50 |
| 450 x 450mm | hr | nr | - | - | | |
| | £ | | 15.75 | - | - | 15.75 |

## SECTION 16 PILING AND DIAPHRAGM WALLING

### FRODINGHAM STEEL SHEET PILING

The following unit costs are based on driving/extracting 1500m2 of sheet piling on a clear site with reasonable access.

| | | UNIT | L | P | M | £ NETT |
|---|---|---|---|---|---|---|
| Provision of all plant, equipment and labour including transport to and from the site and establishing and dismantling for the driving of sheet piling | hr | sum | - | - | | |
| | £ | | 1300.00 | 3000.00 | - | 4300.00 |
| Standing time | hr | hr | - | - | | |
| | £ | | 60.00 | 140.00 | - | 200.00 |
| Provision of all plant, equipment and labour including transport to and from the site and establishing and dismantling for the extraction of sheet piling | hr | sum | - | - | | |
| | £ | | 1140.00 | 2660.00 | - | 3800.00 |
| Standing time | hr | hr | - | - | | |
| | £ | | 60.00 | 140.00 | - | 200.00 |

**SECTION 16 PILING AND DIAPHRAGM WALLING**

**FRODINGHAM STEEL SHEET PILING - continued**

| | UNIT | | L | P | M | £ NETT |
|---|---|---|---|---|---|---|
| Provide new steel sheet piling to BS4360 Grade 43A in lengths 7 - 14m (Piling section reference) (Weight in kg/m2) | | | | | | |
| 1N 99.10 | hr | m2 | - | - | | |
| | £ | | - | - | 37.46 | 37.46 |
| 2N 112.30 | hr | m2 | - | - | | |
| | £ | | - | - | 42.45 | 42.45 |
| 3NA 137.10 | hr | m2 | - | - | | |
| | £ | | - | - | 51.82 | 51.82 |
| 4N 170.80 | hr | m2 | - | - | | |
| | £ | | - | - | 64.56 | 64.56 |
| 5 236.90 | hr | m2 | - | - | | |
| | £ | | - | - | 89.55 | 89.55 |
| Add to last item for copper content ne 0.35% (Piling section reference) (Weight in kg/m2) | | | | | | |
| 1N 99.10 | hr | m2 | - | - | | |
| | £ | | - | - | 0.99 | 0.99 |
| 2N 112.30 | hr | m2 | - | - | | |
| | £ | | - | - | 1.12 | 1.12 |
| 3NA 137.10 | hr | m2 | - | - | | |
| | £ | | - | - | 1.37 | 1.37 |
| 4N 170.80 | hr | m2 | - | - | | |
| | £ | | - | - | 1.70 | 1.70 |
| 5 236.90 | hr | m2 | - | - | | |
| | £ | | - | - | 2.37 | 2.37 |

**SECTION 16 PILING AND DIAPHRAGM WALLING**

**FRODINGHAM STEEL SHEET PILING - continued**

| | UNIT | | L | P | M | £ NETT |
|---|---|---|---|---|---|---|
| copper content 0.35 - 0.5% (Piling section reference) (Weight in kg/m2) | | | | | | |
| 1N 99.10 | hr | m2 | - | - | | |
| | £ | | - | - | 1.39 | 1.39 |
| 2N 112.30 | hr | m2 | - | - | | |
| | £ | | - | - | 1.57 | 1.57 |
| 3NA 137.10 | hr | m2 | - | - | | |
| | £ | | - | - | 1.92 | 1.92 |
| 4N 170.80 | hr | m2 | - | - | | |
| | £ | | - | - | 2.39 | 2.39 |
| 5 236.90 | hr | m2 | - | - | | |
| | £ | | - | - | 3.32 | 3.32 |
| one coat black tar varnish (Piling section reference) (Weight in kg/m2) | | | | | | |
| 1N 99.10 | hr | m2 | - | - | | |
| | £ | | 2.80 | - | 4.20 | 7.00 |
| 2N 112.30 | hr | m2 | - | - | | |
| | £ | | 2.80 | - | 4.20 | 7.00 |
| 3NA 137.10 | hr | m2 | - | - | | |
| | £ | | 2.80 | - | 4.20 | 7.00 |
| 4N 170.80 | hr | m2 | - | - | | |
| | £ | | 2.80 | - | 4.20 | 7.00 |
| 5 236.90 | hr | m2 | - | - | | |
| | £ | | 2.80 | - | 4.20 | 7.00 |

**SECTION 16 PILING AND DIAPHRAGM WALLING**

**FRODINGHAM STEEL SHEET PILING - continued**

| | UNIT | | L | P | M | £ NETT |
|---|---|---|---|---|---|---|
| fabricated piles (Piling section reference) (Weight in kg/m2) | | | | | | |
| 1N 99.10 | hr | t | - | - | | |
| | £ | | 201.00 | 469.00 | 342.78 | 1012.78 |
| 2N 112.30 | hr | t | - | - | | |
| | £ | | 240.00 | 560.00 | 352.50 | 1152.50 |
| 3NA 137.10 | hr | t | - | - | | |
| | £ | | 300.00 | 700.00 | 352.50 | 1352.50 |
| 4N 170.80 | hr | t | - | - | | |
| | £ | | 360.00 | 840.00 | 352.50 | 1552.50 |
| 5 236.90 | hr | t | - | - | | |
| | £ | | 420.00 | 980.00 | 352.50 | 1752.50 |
| Extra for piles under 4 m long (Piling section reference) (Weight in kg/m2) | | | | | | |
| 1N 99.10 | hr | m2 | - | - | | |
| | £ | | 0.20 | 0.50 | 4.46 | 5.16 |
| 2N 112.30 | hr | m2 | - | - | | |
| | £ | | 0.20 | 0.50 | 4.50 | 5.20 |
| 3NA 137.10 | hr | m2 | - | - | | |
| | £ | | 0.20 | 0.50 | 4.80 | 5.50 |
| 4N 170.80 | hr | m2 | - | - | | |
| | £ | | 0.20 | 0.50 | 4.27 | 4.97 |
| 5 236.90 | hr | m2 | - | - | | |
| | £ | | 0.20 | 0.50 | 4.74 | 5.44 |

**SECTION 16 PILING AND DIAPHRAGM WALLING**

**FRODINGHAM STEEL SHEET PILING - continued**

| | UNIT | | L | P | M | £ NETT |
|---|---|---|---|---|---|---|
| Extra for piles 4 - 7m long (Piling section reference) (Weight in kg/m2) | | | | | | |
| 1N 99.10 | hr | m2 | - | - | | |
| | £ | | 0.09 | 0.23 | 1.98 | 2.30 |
| 2N 112.30 | hr | m2 | - | - | | |
| | £ | | 0.10 | 0.25 | 2.25 | 2.60 |
| 3NA 137.10 | hr | m2 | - | - | | |
| | £ | | 0.09 | 0.22 | 2.06 | 2.37 |
| 4N 170.80 | hr | m2 | - | - | | |
| | £ | | 0.12 | 0.28 | 2.56 | 2.96 |
| 5 236.90 | hr | m2 | - | - | | |
| | £ | | 0.11 | 0.27 | 2.37 | 2.75 |
| Extra for piles 14 - 20m long (Piling section reference) (Weight in kg/m2) | | | | | | |
| 1N 99.10 | hr | m2 | - | - | | |
| | £ | | 0.04 | 0.10 | 1.00 | 1.14 |
| 2N 112.30 | hr | m2 | - | - | | |
| | £ | | 0.05 | 0.18 | 1.12 | 1.35 |
| 3NA 137.10 | hr | m2 | - | - | | |
| | £ | | 0.03 | 0.06 | 0.70 | 0.79 |
| 4N 170.80 | hr | m2 | - | - | | |
| | £ | | 0.04 | 0.14 | 0.85 | 1.03 |
| 5 236.90 | hr | m2 | - | - | | |
| | £ | | 0.06 | 0.13 | 1.18 | 1.37 |

**SECTION 16 PILING AND DIAPHRAGM WALLING**

**FRODINGHAM STEEL SHEET PILING - continued**

| | UNIT | | L | P | M | £ NETT |
|---|---|---|---|---|---|---|
| Extra for piles 20 - 25m long (Piling section reference) (Weight in kg/m2) | | | | | | |
| 1N 99.10 | hr | m2 | - | - | | |
| | £ | | - | - | - | - |
| 2N 112.30 | hr | m2 | - | - | | |
| | £ | | - | - | - | - |
| 3NA 137.10 | hr | m2 | - | - | | |
| | £ | | 0.16 | 0.37 | 3.43 | 3.96 |
| 4N 170.80 | hr | m2 | - | - | | |
| | £ | | 0.16 | 0.37 | 3.42 | 3.95 |
| 5 236.90 | hr | m2 | - | - | | |
| | £ | | 0.17 | 0.38 | 3.55 | 4.10 |
| Extra for piles 25 - 29m long (Piling section reference) (Weight in kg/m2) | | | | | | |
| 1N 99.10 | hr | m2 | - | - | | |
| | £ | | - | - | - | - |
| 2N 112.30 | hr | m2 | - | - | | |
| | £ | | - | - | - | - |
| 3NA 137.10 | hr | m2 | - | - | | |
| | £ | | - | - | - | - |
| 4N 170.80 | hr | m2 | - | - | | |
| | £ | | 0.24 | 0.55 | 5.13 | 5.92 |
| 5 236.90 | hr | m2 | - | - | | |
| | £ | | 0.27 | 0.62 | 5.95 | 6.84 |

**SECTION 16 PILING AND DIAPHRAGM WALLING**

**FRODINGHAM STEEL SHEET PILING - continued**

| | | UNIT | | L | P | M | £ NETT |
|---|---|---|---|---|---|---|---|
| Handle pitch and drive piles (Piling section reference) (Weight in kg/m2) | | | | | | | |
| 1N | 99.10 | hr | m2 | - | - | | |
| | | £ | | 8.24 | 17.52 | - | 25.76 |
| 2N | 112.30 | hr | m2 | - | - | | |
| | | £ | | 7.40 | 16.80 | - | 24.20 |
| 3NA | 137.10 | hr | m2 | - | - | | |
| | | £ | | 6.24 | 15.60 | - | 21.84 |
| 4N | 170.80 | hr | m2 | - | - | | |
| | | £ | | 5.26 | 14.22 | - | 19.48 |
| 5 | 236.90 | hr | m2 | - | - | | |
| | | £ | | 5.78 | 15.62 | - | 21.40 |
| Burn off tops of piles to level (Piling section reference) (Weight in kg/m2) | | | | | | | |
| 1N | 99.10 | hr | m | - | - | | |
| | | £ | | 1.70 | 5.30 | - | 7.00 |
| 2N | 112.30 | hr | m | - | - | | |
| | | £ | | 1.70 | 5.30 | - | 7.00 |
| 3NA | 137.10 | hr | m | - | - | | |
| | | £ | | 1.70 | 5.30 | - | 7.00 |
| 4N | 170.80 | hr | m | - | - | | |
| | | £ | | 1.88 | 5.62 | - | 7.50 |
| 5 | 236.90 | hr | m | - | - | | |
| | | £ | | 2.25 | 6.75 | - | 9.00 |

**SECTION 16 PILING AND DIAPHRAGM WALLING**

**FRODINGHAM STEEL SHEET PILING - continued**

| | | UNIT | | L | P | M | £ NETT |
|---|---|---|---|---|---|---|---|
| Extract piling and stack on site (Piling section reference) (Weight in kg/m2) | | | | | | | |
| 1N | 99.10 | hr | m2 | - | - | | |
| | | £ | | 5.13 | 12.00 | - | 17.13 |
| 2N | 112.30 | hr | m2 | - | - | | |
| | | £ | | 3.95 | 11.25 | - | 15.20 |
| 3NA | 137.10 | hr | m2 | - | - | | |
| | | £ | | 2.92 | 9.24 | - | 12.16 |
| 4N | 170.80 | hr | m2 | - | - | | |
| | | £ | | 2.81 | 8.89 | - | 11.70 |
| 5 | 236.90 | hr | m2 | - | - | | |
| | | £ | | 3.11 | 9.81 | - | 12.92 |

**SECTION 16 PILING AND DIAPHRAGM WALLING**

**LARSSEN STEEL SHEET PILING**

The following unit costs are based on driving/extracting 1500m2 of sheet piling on a clear site with reasonable access.

| | UNIT | | L | P | M | £ NETT |
|---|---|---|---|---|---|---|
| Provision of all plant, equipment and labour including transport to and from the site and establishing and dismantling for the driving of sheet piling | hr | sum | - | - | | |
| | £ | | 1300.00 | 3000.00 | - | 4300.00 |
| Standing time | hr | hr | - | - | | |
| | £ | | 60.00 | 140.00 | - | 200.00 |
| Provision of all plant, equipment and labour including transport to and from the site and establishing and dismantling for the extraction of sheet piling | hr | sum | - | - | | |
| | £ | | 1140.00 | 2660.00 | - | 3800.00 |
| Standing time | hr | hr | - | - | | |
| | £ | | 60.00 | 140.00 | - | 200.00 |
| Provide new steel sheet piling to BS4360 Grade 43A in lengths 7 - 14m (Piling section reference) (Weight in kg/m2) | | | | | | |
| 6W 85.10 | hr | m2 | - | - | | |
| | £ | | - | - | 35.50 | 35.50 |
| 9W 97.10 | hr | m2 | - | - | | |
| | £ | | - | - | 39.50 | 39.50 |
| 12W 115.10 | hr | m2 | - | - | | |
| | £ | | - | - | 45.15 | 45.15 |
| 16W 130.10 | hr | m2 | - | - | | |
| | £ | | - | - | 51.00 | 51.00 |
| 20W 147.20 | hr | m2 | - | - | | |
| | £ | | - | - | 57.70 | 57.70 |

**SECTION 16 PILING AND DIAPHRAGM WALLING**

**LARSSEN STEEL SHEET PILING - continued**

| | | UNIT | L | P | M | £ NETT |
|---|---|---|---|---|---|---|
| Add to last item for copper content ne 0.35% (Piling section reference) (Weight in kg/m2) | | | | | | |
| 6W 85.10 | hr | m2 | - | - | | |
| | £ | | - | - | 0.85 | 0.85 |
| 9W 97.10 | hr | m2 | - | - | | |
| | £ | | - | - | 0.97 | 0.97 |
| 12W 115.10 | hr | m2 | - | - | | |
| | £ | | - | - | 1.15 | 1.15 |
| 16W 130.10 | hr | m2 | - | - | | |
| | £ | | - | - | 1.30 | 1.30 |
| 20W 147.20 | hr | m2 | - | - | | |
| | £ | | - | - | 1.47 | 1.47 |
| copper content 0.35 - 0.5% (Piling section reference) (Weight in kg/m2) | | | | | | |
| 6W 85.10 | hr | m2 | - | - | | |
| | £ | | - | - | 1.19 | 1.19 |
| 9W 97.10 | hr | m2 | - | - | | |
| | £ | | - | - | 1.36 | 1.36 |
| 12W 115.10 | hr | m2 | - | - | | |
| | £ | | - | - | 1.61 | 1.61 |
| 16W 130.10 | hr | m2 | - | - | | |
| | £ | | - | - | 1.82 | 1.82 |
| 20W 147.20 | hr | m2 | - | - | | |
| | £ | | - | - | 2.06 | 2.06 |

**SECTION 16 PILING AND DIAPHRAGM WALLING**

**LARSSEN STEEL SHEET PILING - continued**

| | | UNIT | L | P | M | £ NETT |
|---|---|---|---|---|---|---|
| one coat black tar varnish (Piling section reference) (Weight in kg/m2) | | | | | | |
| 6W 85.10 | hr | m2 | - | - | | |
| | £ | | 2.80 | - | 4.20 | 7.00 |
| 9W 97.10 | hr | m2 | - | - | | |
| | £ | | 2.80 | - | 4.20 | 7.00 |
| 12W 115.10 | hr | m2 | - | - | | |
| | £ | | 2.80 | - | 4.20 | 7.00 |
| 16W 130.10 | hr | m2 | - | - | | |
| | £ | | 2.80 | - | 4.20 | 7.00 |
| 20W 147.20 | hr | m2 | - | - | | |
| | £ | | 2.80 | - | 4.20 | 7.00 |
| fabricated piles (Piling section reference) (Weight in kg/m2) | | | | | | |
| 6W 85.10 | hr | t | - | - | | |
| | £ | | 72.00 | 168.00 | 417.15 | 657.15 |
| 9W 97.10 | hr | t | - | - | | |
| | £ | | 120.00 | 280.00 | 406.85 | 806.85 |
| 12W 115.10 | hr | t | - | - | | |
| | £ | | 180.00 | 420.00 | 391.40 | 991.40 |
| 16W 130.10 | hr | t | - | - | | |
| | £ | | 240.00 | 560.00 | 391.40 | 1191.40 |
| 20W 147.20 | hr | t | - | - | | |
| | £ | | 300.00 | 700.00 | 391.06 | 1391.06 |

**SECTION 16 PILING AND DIAPHRAGM WALLING**

**LARSSEN STEEL SHEET PILING - continued**

| | UNIT | L | P | M | £ NETT |
|---|---|---|---|---|---|
| Extra for piles under 4m long (Piling section reference) (Weight in kg/m2) | | | | | |
| 6W 85.10 | hr m2 | - | - | | |
| | £ | 0.18 | 0.41 | 3.83 | 4.42 |
| 9W 97.10 | hr m2 | - | - | | |
| | £ | 0.18 | 0.42 | 3.88 | 4.48 |
| 12W 115.10 | hr m2 | - | - | | |
| | £ | 0.17 | 0.38 | 3.45 | 4.00 |
| 16W 130.10 | hr m2 | - | - | | |
| | £ | 0.18 | 0.42 | 3.90 | 4.50 |
| 20W 147.20 | hr m2 | - | - | | |
| | £ | 0.17 | 0.40 | 3.68 | 4.25 |
| Extra for piles 4 - 7m long (Piling section reference) (Weight in kg/m2) | | | | | |
| 6W 85.10 | hr m2 | - | - | | |
| | £ | 0.08 | 0.18 | 1.70 | 1.96 |
| 9W 97.10 | hr m2 | - | - | | |
| | £ | 0.09 | 0.22 | 1.94 | 2.25 |
| 12W 115.10 | hr m2 | - | - | | |
| | £ | 0.08 | 0.19 | 1.73 | 2.00 |
| 16W 130.10 | hr m2 | - | - | | |
| | £ | 0.09 | 0.22 | 1.95 | 2.26 |
| 20W 147.20 | hr m2 | - | - | | |
| | £ | 0.10 | 0.24 | 2.21 | 2.55 |

**SECTION 16 PILING AND DIAPHRAGM WALLING**

**LARSSEN STEEL SHEET PILING - continued**

| | UNIT | | L | P | M | £ NETT |
|---|---|---|---|---|---|---|
| piles 14 - 20m long (Piling section reference) (Weight in kg/m2) | | | | | | |
| 6W 85.10 | hr | m2 | - | - | | |
| | £ | | 0.04 | 0.10 | 0.85 | 0.99 |
| 9W 97.10 | hr | m2 | - | - | | |
| | £ | | 0.04 | 0.14 | 0.97 | 1.15 |
| 12W 115.10 | hr | m2 | - | - | | |
| | £ | | 0.05 | 0.18 | 1.15 | 1.38 |
| 16W 130.10 | hr | m2 | - | - | | |
| | £ | | 0.03 | 0.06 | 0.65 | 0.74 |
| 20W 147.20 | hr | m2 | - | - | | |
| | £ | | 0.03 | 0.06 | 0.74 | 0.83 |
| piles 20 - 25m long (Piling section reference) (Weight in kg/m2) | | | | | | |
| 6W 85.10 | hr | m2 | - | - | | |
| | £ | | - | - | - | - |
| 9W 97.10 | hr | m2 | - | - | | |
| | £ | | - | - | - | - |
| 12W 115.10 | hr | m2 | - | - | | |
| | £ | | - | - | - | - |
| 16W 130.10 | hr | m2 | - | - | | |
| | £ | | 0.15 | 0.36 | 3.25 | 3.76 |
| 20W 147.20 | hr | m2 | - | - | | |
| | £ | | 0.17 | 0.40 | 3.68 | 4.25 |

**SECTION 16 PILING AND DIAPHRAGM WALLING**

**LARSSEN STEEL SHEET PILING - continued**

| | | UNIT | L | P | M | £ NETT |
|---|---|---|---|---|---|---|
| piles over 25m long (Piling section reference) (Weight in kg/m2) | | | | | | |
| 6W 85.10 | hr | m2 | - | - | | |
| | £ | | - | - | - | - |
| 9W 97.10 | hr | m2 | - | - | | |
| | £ | | - | - | - | - |
| 12W 115.10 | hr | m2 | - | - | | |
| | £ | | - | - | - | - |
| 16W 130.10 | hr | m2 | - | - | | |
| | £ | | - | - | - | - |
| 20W 147.20 | hr | m2 | - | - | | |
| | £ | | 0.24 | 0.56 | 5.15 | 5.95 |
| Handle pitch and drive piles (Piling section reference) (Weight in kg/m2) | | | | | | |
| 6W 85.10 | hr | m2 | - | - | | |
| | £ | | 8.46 | 17.96 | - | 26.42 |
| 9W 97.10 | hr | m2 | - | - | | |
| | £ | | 6.76 | 15.78 | - | 22.54 |
| 12W 115.10 | hr | m2 | - | - | | |
| | £ | | 5.78 | 14.18 | - | 19.96 |
| 16W 130.10 | hr | m2 | - | - | | |
| | £ | | 4.84 | 13.06 | - | 17.90 |
| 20W 147.20 | hr | m2 | - | - | | |
| | £ | | 3.98 | 12.58 | - | 16.56 |

**SECTION 16 PILING AND DIAPHRAGM WALLING**

**LARSSEN STEEL SHEET PILING - continued**

| | UNIT | | L | P | M | £ NETT |
|---|---|---|---|---|---|---|
| Burn off tops of piles to level (Piling section reference) (Weight in kg/m2) | | | | | | |
| 6W 85.10 | hr | m | - | - | | |
| | £ | | 1.50 | 4.50 | - | 6.00 |
| 9W 97.10 | hr | m | - | - | | |
| | £ | | 1.63 | 4.87 | - | 6.50 |
| 12W 115.10 | hr | m | - | - | | |
| | £ | | 1.68 | 5.32 | - | 7.00 |
| 16W 130.10 | hr | m | - | - | | |
| | £ | | 1.68 | 5.32 | - | 7.00 |
| 20W 147.20 | hr | m | - | - | | |
| | £ | | 1.80 | 5.70 | - | 7.50 |
| Extract piling and stack on site (Piling section reference) (Weight in kg/m2) | | | | | | |
| 6W 85.10 | hr | m2 | - | - | | |
| | £ | | 5.03 | 11.74 | - | 16.77 |
| 9W 97.10 | hr | m2 | - | - | | |
| | £ | | 4.48 | 10.47 | - | 14.95 |
| 12W 115.10 | hr | m2 | - | - | | |
| | £ | | 3.15 | 9.45 | - | 12.60 |
| 16W 130.10 | hr | m2 | - | - | | |
| | £ | | 2.82 | 8.94 | - | 11.76 |
| 20W 147.20 | hr | m2 | - | - | | |
| | £ | | 2.62 | 8.30 | - | 10.92 |

**SECTION 16 PILING AND DIAPHRAGM WALLING**

**LARSSEN STEEL SHEET PILING - continued**

| | | | | | | | £ |
|---|---|---|---|---|---|---|---|
| | | | **UNIT** | **L** | **P** | **M** | **NETT** |
| Provide new steel sheet piling to BS4360 Grade 43A (Piling section reference) (Weight in kg/m2) | | | | | | | |
| 25W | 167.40 | hr | m2 | - | - | | |
| | | £ | | - | - | **65.62** | **65.62** |
| 32W | 197.40 | hr | m2 | - | - | | |
| | | £ | | - | - | **78.37** | **78.37** |
| 3 | 155.50 | hr | m2 | - | - | | |
| | | £ | | - | - | **60.18** | **60.18** |
| 4A | 185.10 | hr | m2 | - | - | | |
| | | £ | | - | - | **71.63** | **71.63** |
| 6/131kg/m | 311.80 | hr | m2 | - | - | | |
| | | £ | | - | - | **122.23** | **122.23** |
| Add to last item for copper content ne 0.35% (Piling section reference) (Weight in kg/m2) | | | | | | | |
| 25W | 167.40 | hr | m2 | - | - | | |
| | | £ | | - | - | **1.68** | **1.68** |
| 32W | 197.40 | hr | m2 | - | - | | |
| | | £ | | - | - | **1.97** | **1.97** |
| 3 | 155.50 | hr | m2 | - | - | | |
| | | £ | | - | - | **1.56** | **1.56** |
| 4A | 185.10 | hr | m2 | - | - | | |
| | | £ | | - | - | **1.85** | **1.85** |
| 6/131kg/m | 311.80 | hr | m2 | - | - | | |
| | | £ | | - | - | **3.12** | **3.12** |

**SECTION 16 PILING AND DIAPHRAGM WALLING**

**LARSSEN STEEL SHEET PILING - continued**

| | | UNIT | L | P | M | £ NETT |
|---|---|---|---|---|---|---|
| copper content 0.35 - 0.5% (Piling section reference) (Weight in kg/m2) | | | | | | |
| 25W | 167.40 | hr m2 | - | - | | |
| | | £ | - | - | 2.35 | 2.35 |
| 32W | 197.40 | hr m2 | - | - | | |
| | | £ | - | - | 2.76 | 2.76 |
| 3 | 155.50 | hr m2 | - | - | | |
| | | £ | - | - | 2.18 | 2.18 |
| 4A | 185.10 | hr m2 | - | - | | |
| | | £ | - | - | 2.60 | 2.60 |
| 6/131kg/m | 311.80 | hr m2 | - | - | | |
| | | £ | - | - | 4.36 | 4.36 |
| one coat black tar varnish (Piling section reference) (Weight in kg/m2) | | | | | | |
| 25W | 167.40 | hr m2 | - | - | | |
| | | £ | 2.80 | - | 4.20 | 7.00 |
| 32W | 197.40 | hr m2 | - | - | | |
| | | £ | 2.80 | - | 4.20 | 7.00 |
| 3 | 155.50 | hr m2 | - | - | | |
| | | £ | 2.80 | - | 4.20 | 7.00 |
| 4A | 185.10 | hr m2 | - | - | | |
| | | £ | 2.80 | - | 4.20 | 7.00 |
| 6/131kg/m | 311.80 | hr m2 | - | - | | |
| | | £ | 2.80 | - | 4.20 | 7.00 |

**SECTION 16 PILING AND DIAPHRAGM WALLING**

**LARSSEN STEEL SHEET PILING - continued**

| | | | UNIT | L | P | M | £ NETT |
|---|---|---|---|---|---|---|---|
| fabricated piles (Piling section reference) (Weight in kg/m2) | | | | | | | |
| 25W | 167.40 | hr | t | - | - | | |
| | | £ | | 90.00 | 210.00 | 398.76 | 698.76 |
| 32W | 197.40 | hr | t | - | - | | |
| | | £ | | 120.00 | 280.00 | 404.61 | 804.61 |
| 3 | 155.50 | hr | t | - | - | | |
| | | £ | | 180.00 | 120.00 | 394.73 | 694.73 |
| 4A | 185.10 | hr | t | - | - | | |
| | | £ | | 240.00 | 560.00 | 394.58 | 1194.58 |
| 6/131kg/m | 311.80 | hr | t | - | - | | |
| | | £ | | 300.00 | 700.00 | 399.36 | 1399.36 |
| piles under 4m long (Piling section reference) (Weight in kg/m2) | | | | | | | |
| 25W | 167.40 | hr | m2 | - | - | | |
| | | £ | | 0.15 | 0.36 | 3.35 | 3.86 |
| 32W | 197.40 | hr | m2 | - | - | | |
| | | £ | | 0.18 | 0.43 | 3.95 | 4.56 |
| 3 | 155.50 | hr | m2 | - | - | | |
| | | £ | | 0.22 | 0.50 | 4.67 | 5.39 |
| 4A | 185.10 | hr | m2 | - | - | | |
| | | £ | | 0.21 | 0.50 | 4.63 | 5.34 |
| 6/131kg/m | 311.80 | hr | m2 | - | - | | |
| | | £ | | 0.22 | 0.51 | 4.67 | 5.40 |

**SECTION 16 PILING AND DIAPHRAGM WALLING**

**LARSSEN STEEL SHEET PILING - continued**

| | | UNIT | | L | P | M | £ NETT |
|---|---|---|---|---|---|---|---|
| piles 4 - 7m long (Piling section reference) (Weight in kg/m2) | | | | | | | |
| 25W | 167.40 | hr | m2 | - | - | | |
| | | £ | | 0.08 | 0.17 | 1.68 | 1.93 |
| 32W | 197.40 | hr | m2 | - | - | | |
| | | £ | | 0.09 | 0.22 | 1.97 | 2.28 |
| 3 | 155.50 | hr | m2 | - | - | | |
| | | £ | | 0.11 | 0.26 | 2.33 | 2.70 |
| 4A | 185.10 | hr | m2 | - | - | | |
| | | £ | | 0.13 | 0.29 | 2.78 | 3.20 |
| 6/131kg/m | 311.80 | hr | m2 | - | - | | |
| | | £ | | 0.14 | 0.34 | 3.12 | 3.60 |
| piles 14 - 20m long (Piling section reference) (Weight in kg/m2) | | | | | | | |
| 25W | 167.40 | hr | m2 | - | - | | |
| | | £ | | 0.04 | 0.09 | 0.84 | 0.97 |
| 32W | 197.40 | hr | m2 | - | - | | |
| | | £ | | 0.05 | 0.10 | 0.99 | 1.14 |
| 3 | 155.50 | hr | m2 | - | - | | |
| | | £ | | 0.04 | 0.08 | 0.78 | 0.90 |
| 4A | 185.10 | hr | m2 | - | - | | |
| | | £ | | 0.04 | 0.10 | 0.93 | 1.07 |
| 6/131kg/m | 311.80 | hr | m2 | - | - | | |
| | | £ | | 0.07 | 0.17 | 1.56 | 1.80 |

**SECTION 16 PILING AND DIAPHRAGM WALLING**

**LARSSEN STEEL SHEET PILING - continued**

| | | | UNIT | L | P | M | £ NETT |
|---|---|---|---|---|---|---|---|
| piles 20 - 25m long<br>(Piling section reference)<br>(Weight in kg/m2) | | | | | | | |
| 25W | 167.40 | hr | m2 | - | - | | |
| | | £ | | 0.15 | 0.36 | 3.35 | 3.86 |
| 32W | 197.40 | hr | m2 | - | - | | |
| | | £ | | 0.18 | 0.42 | 3.94 | 4.54 |
| 3 | 155.50 | hr | m2 | - | - | | |
| | | £ | | 0.22 | 0.50 | 4.67 | 5.39 |
| 4A | 185.10 | hr | m2 | - | - | | |
| | | £ | | 0.21 | 0.50 | 4.63 | 5.34 |
| 6/131kg/m | 311.80 | hr | m2 | - | - | | |
| | | £ | | 0.22 | 0.51 | 4.67 | 5.40 |
| piles over 25m long<br>(Piling section reference)<br>(Weight in kg/m2) | | | | | | | |
| 25W | 167.40 | hr | m2 | - | - | | |
| | | £ | | 0.23 | 0.54 | 5.03 | 5.80 |
| 32W | 197.40 | hr | m2 | - | - | | |
| | | £ | | 0.23 | 0.54 | 4.93 | 5.70 |
| 3 | 155.50 | hr | m2 | - | - | | |
| | | £ | | - | - | - | - |
| 4A | 185.10 | hr | m2 | - | - | | |
| | | £ | | 0.30 | 0.71 | 6.48 | 7.49 |
| 6/131kg/m | 311.80 | hr | m2 | - | - | | |
| | | £ | | 0.36 | 0.85 | 7.79 | 9.00 |

**SECTION 16 PILING AND DIAPHRAGM WALLING**

**LARSSEN STEEL SHEET PILING - continued**

| | | | UNIT | L | P | M | £ NETT |
|---|---|---|---|---|---|---|---|
| Handle pitch and drive piles (Piling section reference) (Weight in kg/m2) | | | | | | | |
| 25W | 167.40 | hr | m2 | - | - | | |
| | | £ | | 3.84 | 10.36 | - | 14.20 |
| 32W | 197.40 | hr | m2 | - | - | | |
| | | £ | | 4.26 | 11.54 | - | 15.80 |
| 3 | 155.50 | hr | m2 | - | - | | |
| | | £ | | 6.30 | 17.02 | - | 23.32 |
| 4A | 185.10 | hr | m2 | - | - | | |
| | | £ | | 6.10 | 16.48 | - | 22.58 |
| 6/131kg/m | 311.80 | hr | m2 | - | - | | |
| | | £ | | 6.96 | 18.82 | - | 25.78 |
| Burn off tops of piles to level (Piling section reference) (Weight in kg/m2) | | | | | | | |
| 25W | 167.40 | hr | m | - | - | | |
| | | £ | | 2.11 | 5.69 | - | 7.80 |
| 32W | 197.40 | hr | m | - | - | | |
| | | £ | | 2.22 | 5.98 | - | 8.20 |
| 3 | 155.50 | hr | m | - | - | | |
| | | £ | | 2.32 | 6.28 | - | 8.60 |
| 4A | 185.10 | hr | m | - | - | | |
| | | £ | | 2.10 | 6.65 | - | 8.75 |
| 6/131kg/m | 311.80 | hr | m | - | - | | |
| | | £ | | 3.06 | 9.19 | - | 12.25 |

**SECTION 16 PILING AND DIAPHRAGM WALLING**

**LARSSEN STEEL SHEET PILING - continued**

| | | | | | | £ |
|---|---|---|---|---|---|---|
| | | **UNIT** | **L** | **P** | **M** | **NETT** |
| Extract piling and stack on site<br>(Piling section reference)<br>(Weight in kg/m2) | | | | | | |
| 25W | 167.40 | hr m2 | - | - | | |
| | | £ | **2.62** | **8.30** | **-** | **10.92** |
| 32W | 197.40 | hr m2 | - | - | | |
| | | £ | **2.62** | **8.30** | **-** | **10.92** |
| 3 | 155.50 | hr m2 | - | - | | |
| | | £ | **4.74** | **11.06** | **-** | **15.80** |
| 4A | 185.10 | hr m2 | - | - | | |
| | | £ | **4.74** | **11.06** | **-** | **15.80** |
| 6/131kg/m | 311.80 | hr m2 | - | - | | |
| | | £ | **5.69** | **13.26** | **-** | **18.95** |

**SECTION 16 PILING AND DIAPHRAGM WALLING**

**STEEL BEARING PILES**

Isolated steel bearing piles.
All 356 x 368 x 109 kg/m
Section Grade 50B Steel

| | | UNIT | L | P | M | £ NETT |
|---|---|---|---|---|---|---|
| Establishment of piling plant for steel bearing piles in main piling based on a max 100 | hr | sum | - | - | | |
| mile radius of base | £ | | - | - | - | 6000.00 |
| Establishment of piling plant for steel bearing piles in main piling based on a max 250 | hr | sum | - | - | | |
| mile radius of base | £ | | - | - | - | 15000.00 |
| Moving piling plant for steel bearing piles in main | hr | nr | - | - | | |
| piling | £ | | - | - | - | 180.00 |
| Steel bearing pile exceeding 15 metres in length but not exceeding 20 metres in main | hr | m | - | - | | |
| piling | £ | | - | - | 48.90 | 48.90 |
| Lengthening piece for steel bearing pile not exceeding 5 metres in length in main | hr | m | - | - | | |
| piling | £ | | - | - | 48.90 | 48.90 |
| Driving vertical steel bearing pile exceeding 15 metres in length but not exceeding 20 metres in main | hr | m | - | - | | |
| piling | £ | | 2.40 | 5.60 | - | 8.00 |
| Driving raking steel bearing pile exceeding 15 metres in length but not exceeding 20 metres in main | hr | m | - | - | | |
| piling | £ | | 2.80 | 7.20 | - | 10.00 |
| Driving lengthened vertical steel bearing pile not exceeding 5 metres in length | hr | m | - | - | | |
| in main piling | £ | | 7.50 | 10.50 | - | 18.00 |
| Driving lengthened raking steel bearing pile not exceeding 5 metres in length | hr | m | - | - | | |
| in main piling | £ | | 5.10 | 15.40 | - | 20.50 |

## SECTION 16 PILING AND DIAPHRAGM WALLING

### STEEL BEARING PILES - continued

| | | UNIT | L | P | M | £ NETT |
|---|---|---|---|---|---|---|
| Welding on lengthening piece to vertical steel bearing pile in main piling | hr | nr | - | - | | |
| | £ | | 86.30 | 50.70 | 3.00 | 140.00 |
| Welding on lengthening piece to raking steel bearing pile in main piling | hr | nr | - | - | | |
| | £ | | 86.30 | 50.70 | 3.00 | 140.00 |
| Cutting or burning off surplus length of vertical steel bearing pipe in main piling | hr | nr | - | - | | |
| | £ | | 4.40 | 3.00 | 0.60 | 8.00 |
| Cutting or burning off surplus length of raking steel bearing pipe in main piling | hr | nr | - | - | | |
| | £ | | 4.40 | 3.00 | 0.60 | 8.00 |

### PILING TESTING

| | | UNIT | L | P | M | £ NETT |
|---|---|---|---|---|---|---|
| Establishment of pile testing equipment for steel bearing pile in trial piling as a separate operation in advance of the main piling. | hr | sum | - | - | | |
| | £ | | - | - | - | 16500.00 |
| Vertical load testing of steel bearing pile in trial piling as a separate operation in advance of the main piling to 108 tonnes | hr | nr | - | - | | |
| | £ | | - | - | - | 6200.00 |

## SECTION 16 PILING AND DIAPHRAGM WALLING

### STEEL TUBULAR PILES

Isolated steel tubular piles. All Grade 43A steel, 12.5mm thickness

| | | UNIT | L | P | M | £ NETT |
|---|---|---|---|---|---|---|
| Establishment of piling plant for steel tubular piles in main piling based on a maximum 100 mile radius of base | hr | sum | - | - | | |
| | £ | | - | - | - | 6000.00 |
| Establishment of piling plant for steel tubular piles in main piling based on a maximum 250 mile radius of base | hr | sum | - | - | | |
| | £ | | - | - | - | 15000.00 |
| Moving piling plant for steel tubular piles in main piling | hr | nr | - | - | | |
| | £ | | - | - | - | 200.00 |
| Steel tubular pile exceeding 15 metres in length but not exceeding 20 metres in length in main piling 508mm diameter | hr | m | - | - | | |
| | £ | | - | - | 61.20 | 61.20 |
| Steel tubular pile exceeding 15 metres in length but not exceeding 20 metres in length in main piling 762mm diameter | hr | m | - | - | | |
| | £ | | - | - | 107.65 | 107.65 |
| Lengthening piece of steel tubular pile not exceeding 5 metres in length in main piling 508mm diameter | hr | m | - | - | | |
| | £ | | - | - | 61.20 | 61.20 |
| Lengthening piece of steel tubular pile not exceeding 5 metres in length in main piling 762mm diameter | hr | m | - | - | | |
| | £ | | - | - | 107.65 | 107.65 |

**SECTION 16 PILING AND DIAPHRAGM WALLING**

**STEEL TUBULAR PILES - continued**

| | | UNIT | L | P | M | £ NETT |
|---|---|---|---|---|---|---|
| Driving vertical steel tubular pile exceeding 15 metres in length but not exceeding 20 metres in main piling 508mm diameter | hr | m | - | - | | |
| | £ | | 1.95 | 4.55 | - | 6.50 |
| Driving vertical steel tubular pile exceeding 15 metres in length but not exceeding 20 metres in main piling 762mm diameter | hr | m | - | - | | |
| | £ | | 2.04 | 4.76 | - | 6.80 |
| Driving lengthened vertical steel tubular pile not exceeding 5 metres in length in main piling 508mm diameter | hr | m | - | - | | |
| | £ | | 3.58 | 5.02 | - | 8.60 |
| Driving lengthened vertical steel tubular pile not exceeding 5 metres in length in main piling 762mm diameter | hr | m | - | - | | |
| | £ | | 3.71 | 5.19 | - | 8.90 |
| Welding on lengthening piece to vertical steel tubular pile in main piling 508mm diameter | hr | nr | - | - | | |
| | £ | | 86.30 | 50.70 | 3.00 | 140.00 |
| Welding on lengthening piece to vertical steel tubular pile in main piling 762mm diameter | hr | nr | - | - | | |
| | £ | | 98.00 | 77.00 | 5.00 | 180.00 |
| Cutting or burning off surplus length of vertical steel tubular pipe in main piling 508mm diameter | hr | nr | - | - | | |
| | £ | | 4.84 | 3.30 | 0.66 | 8.80 |
| Cutting or burning off surplus length of vertical steel tubular pipe in main piling 762mm diameter | hr | nr | - | - | | |
| | £ | | 4.98 | 3.39 | 0.68 | 9.05 |

**SECTION 16 PILING AND DIAPHRAGM WALLING**

**STEEL TUBULAR PILES - continued**

| | | UNIT | L | P | M | £ NETT |
|---|---|---|---|---|---|---|
| **PILING TESTING** | | | | | | |
| Establishment of pile testing equipment for steel tubular pile in trial piling as a separate operation in advance of the main piling 508mm diameter | hr | sum | - | - | | |
| | £ | | - | - | - | 16500.00 |
| Establishment of pile testing equipment for steel tubular pile in trial piling as a separate operation in advance of the main piling 762mm diameter | hr | sum | - | - | | |
| | £ | | - | - | - | 16500.00 |
| Vertical load testing of steel tubular pile in trial piling as a separate operation in advance of the main piling to 108 tonnes 508mm diameter | hr | nr | - | - | | |
| | £ | | - | - | - | 6000.00 |
| Vertical load testing of steel tubular pile in trial piling as a separate operation in advance of the main piling to 108 tonnes 762mm diameter | hr | nr | - | - | | |
| | £ | | - | - | - | 6200.00 |

**SECTION 16 PILING AND DIAPHRAGM WALLING**

Diaphragm walls are the construction of vertical walls, cast in place in a trench excavation. They can be formed in reinforced concrete to provide structural elements for temporary or permanent retaining walls. Wall thicknesses of 500 to 1500m up to 40m deep may be constructed. Special equipment such as the Hydrofraise can construct walls up to 100m deep. Restricted urban sites will significantly increase the costs.

The following costs are based on constructing a diaphragm wall with an excavated volume of 4000m3 using a grab. Typical progress would be up to 500m3 per week. Disposal of excavated material is not included in the unit costs.

| | | UNIT | L | P | M | £ NETT |
|---|---|---|---|---|---|---|
| Establishment of standard plant and equipment and removal on completion | hr | sum | - | - | | |
| | £ | | - | - | - | 86000.00 |
| Standing time | hr | hr | 1.00 | 1.00 | | |
| | £ | | 116.00 | 234.00 | - | 350.00 |
| Excavation in material other than rock or artificial hard material (Maximum depth) | | | | | | |
| 5 - 10m | hr | m3 | 0.29 | 0.29 | | |
| | £ | | 37.14 | 67.86 | - | 105.00 |
| 10 - 15m | hr | m3 | 0.31 | 0.31 | | |
| | £ | | 37.46 | 72.54 | - | 110.00 |
| 15 - 20m | hr | m3 | 0.33 | 0.33 | | |
| | £ | | 40.78 | 77.22 | - | 118.00 |
| Concrete grade C25P; 20mm aggregate (Maximum depth) | hr | m3 | - | - | | |
| | £ | | 17.30 | 18.00 | - | 35.30 |
| Guide walls (twin) | hr | m | 0.42 | 0.42 | | |
| | £ | | 61.72 | 98.28 | 45.70 | 205.70 |

| | | | 12mm | 20mm | 25mm |
|---|---|---|---|---|---|
| Mild steel bar reinforcement | hr | t | | | |
| | £ | | 580.00 | 550.00 | 520.00 |

## SECTION 16 PILING AND DIAPHRAGM WALLING

### GROUND ANCHORAGES

Ground anchorages consist of the installation of a cable or solid bar tendon fixed in the ground by grouting and tensioned to exceed the working load to be carried. Ground anchors may be of a permanent or temporary nature and can be used in conjunction with diaphragm walling or sheet piling to eliminate the use of strutting etc. The following costs are based on the installation of 50 ground anchors.

| | | UNIT | L | P | M | £ NETT |
|---|---|---|---|---|---|---|
| Establishment of standard plant and equipment and removal on completion | hr | sum | - | - | | |
| | £ | | - | - | - | 8000.00 |
| Standing time | hr | hr | 1.00 | 1.00 | | |
| | £ | | 35.00 | 65.00 | - | 100.00 |
| Number in material other than rock or artificial hard material | | | | | | |
| temporary or permanent; maximum depth 15m | | | | | | |
| Rock alluvial or clay | | | | | | |
| 0 - 50 tonnes load | hr | nr | 0.58 | 0.58 | | |
| | £ | | 12.30 | 37.70 | - | 50.00 |
| Rock or alluvial | | | | | | |
| 50 - 90 tonnes load | hr | nr | 0.58 | 0.58 | | |
| | £ | | 22.30 | 37.70 | - | 60.00 |
| Rock only | | | | | | |
| 90 - 150 tonnes load | hr | nr | 0.58 | 0.58 | | |
| | £ | | 32.30 | 37.70 | - | 70.00 |
| Total length of tendons in material other than rock or artificial hard material | | | | | | |
| temporary | | | | | | |
| Rock alluvial or clay | | | | | | |
| 0 - 50 tonnes load | hr | nr | 0.54 | 0.54 | | |
| | £ | | 4.90 | 35.10 | - | 40.00 |
| Rock or alluvial | | | | | | |
| 50 - 90 tonnes load | hr | nr | 0.58 | 0.58 | | |
| | £ | | 22.30 | 37.70 | - | 60.00 |
| Rock only | | | | | | |
| 90 - 150 tonnes load | hr | nr | 0.65 | 0.65 | | |
| | £ | | 37.75 | 42.25 | - | 80.00 |
| permanent | | | | | | |
| Rock alluvial or clay | | | | | | |
| 0 - 50 tonnes load | hr | nr | 0.81 | 0.81 | | |
| | £ | | 7.35 | 52.65 | - | 60.00 |
| Rock or alluvial | | | | | | |
| 50 - 90 tonnes load | hr | nr | 0.85 | 0.85 | | |
| | £ | | 24.75 | 55.25 | - | 80.00 |
| Rock only | | | | | | |
| 90 - 150 tonnes load | hr | nr | 0.92 | 0.92 | | |
| | £ | | 40.20 | 59.80 | - | 100.00 |

## SECTION 17 STRUCTURAL CONCRETE

### CONCRETE

LABOUR based on Gang K

| | | |
|---|---|---|
| Total hourly cost | = £ | 46.93 |

PLANT

| | | |
|---|---|---|
| Compressor 2 tool | | 4.83 |
| Vibrators (4,P54) | | 2.52 |
| (Part time) 22RB Crane | | 7.11 |
| Audex skip | | 0.84 |
| Total hourly cost | = £ | 15.30 |

Refer also to CESMM Class F and G.

| | | UNIT | L | P | M | £ NETT |
|---|---|---|---|---|---|---|
| This section is fundamentally different from the CESMM classes in that provision of concrete of different classes and placing are included in the unit costs. | | | | | | |
| **BLINDING CONCRETE** | | | | | | |
| In-situ concrete Class E in blinding up to 75mm in thickness | hr | m3 | 0.18 | 0.01 | | |
| | £ | | 8.45 | 0.15 | 42.24 | 50.84 |
| In-situ concrete Class E in blinding up to 75mm in thickness placed in narrrow widths up to 1.0m wide or in bottoms of trenches up to 2.5m wide | hr | m3 | 0.20 | 0.20 | | |
| | £ | | 9.39 | 3.06 | 42.24 | 54.69 |
| In-situ concrete class C10P in blinding up to 150mm in thickness (excluding formwork to sides) | hr | m3 | 0.16 | 0.01 | | |
| | £ | | 7.51 | 0.15 | 44.54 | 52.20 |

**SECTION 17 STRUCTURAL CONCRETE**

**BLINDING CONCRETE - continued**

| | | UNIT | L | P | M | £ NETT |
|---|---|---|---|---|---|---|
| In-situ concrete class C10P in blinding up to 150mm in thickness(excluding formwork to sides) placed in narrow widths up to 1.0m wide or in bottoms of trenches up to 2.5m wide | hr | m3 | 0.18 | 0.18 | | |
| | £ | | 8.45 | 2.75 | 44.54 | 55.74 |

**IN-SITU CONCRETE**

| | | UNIT | L | P | M | £ NETT |
|---|---|---|---|---|---|---|
| Bases, footings, pile caps and ground beams (C20/20 concrete) thickness | | | | | | |
| ne 150mm | hr | m3 | 0.20 | 0.20 | | |
| | £ | | 9.39 | 3.06 | 47.18 | 59.63 |
| 150 - 300mm | hr | m3 | 0.17 | 0.17 | | |
| | £ | | 7.98 | 2.60 | 47.18 | 57.76 |
| 300 - 500mm | hr | m3 | 0.15 | 0.15 | | |
| | £ | | 7.04 | 2.30 | 47.18 | 56.52 |
| exceeding 500mm | hr | m3 | 0.14 | 0.14 | | |
| | £ | | 6.57 | 2.14 | 47.18 | 55.89 |
| (C30/20 concrete) thickness | | | | | | |
| ne 150mm | hr | m3 | 0.21 | 0.21 | | |
| | £ | | 9.86 | 3.21 | 52.09 | 65.16 |
| 150 - 300mm | hr | m3 | 0.18 | 0.18 | | |
| | £ | | 8.45 | 2.75 | 52.09 | 63.29 |
| 300 - 500mm | hr | m3 | 0.15 | 0.15 | | |
| | £ | | 7.04 | 2.30 | 52.09 | 61.43 |
| exceeding 500mm | hr | m3 | 0.14 | 0.14 | | |
| | £ | | 6.57 | 2.14 | 52.09 | 60.80 |
| Walls; (C20/20 concrete) thickness | | | | | | |
| ne 150mm | hr | m3 | 0.21 | 0.21 | | |
| | £ | | 9.86 | 3.21 | 47.18 | 60.25 |
| 150 - 300mm | hr | m3 | 0.15 | 0.15 | | |
| | £ | | 7.04 | 2.30 | 47.18 | 56.52 |
| 300 - 500mm | hr | m3 | 0.13 | 0.13 | | |
| | £ | | 6.10 | 1.99 | 47.18 | 55.27 |
| exceeding 500mm | hr | m3 | 0.12 | 0.12 | | |
| | £ | | 5.63 | 1.84 | 47.18 | 54.65 |

**SECTION 17 STRUCTURAL CONCRETE**

**IN-SITU CONCRETE - continued**

| | | UNIT | L | P | M | £ NETT |
|---|---|---|---|---|---|---|
| Walls; (C30/20 concrete) thickness | | | | | | |
| ne 150mm | hr | m3 | 0.22 | 0.22 | | |
| | £ | | 10.32 | 3.37 | 52.09 | 65.78 |
| 150 - 300mm | hr | m3 | 0.16 | 0.16 | | |
| | £ | | 7.51 | 2.45 | 52.09 | 62.05 |
| 300 - 500mm | hr | m3 | 0.13 | 0.13 | | |
| | £ | | 6.10 | 1.99 | 52.09 | 60.18 |
| exceeding 500mm | hr | m3 | 0.12 | 0.12 | | |
| | £ | | 5.63 | 1.84 | 52.09 | 59.56 |
| Suspended slabs; (C20/20 concrete) thickness | | | | | | |
| ne 150mm | hr | m3 | 0.27 | 0.27 | | |
| | £ | | 12.67 | 4.13 | 47.18 | 63.98 |
| 150 - 300mm | hr | m3 | 0.21 | 0.21 | | |
| | £ | | 9.86 | 3.21 | 47.18 | 60.25 |
| 300 - 500mm | hr | m3 | 0.19 | 0.19 | | |
| | £ | | 8.92 | 2.91 | 47.18 | 59.01 |
| exceeding 500mm | hr | m3 | 0.19 | 0.19 | | |
| | £ | | 8.92 | 2.91 | 47.18 | 59.01 |
| (C30/20 concrete) thickness | | | | | | |
| ne 150mm | hr | m3 | 0.28 | 0.28 | | |
| | £ | | 13.14 | 4.28 | 52.07 | 69.49 |
| 150 - 300mm | hr | m3 | 0.22 | 0.22 | | |
| | £ | | 10.32 | 3.37 | 52.07 | 65.76 |
| 300 - 500mm | hr | m3 | 0.19 | 0.19 | | |
| | £ | | 8.92 | 2.91 | 52.07 | 63.90 |
| exceeding 500mm | hr | m3 | 0.19 | 0.19 | | |
| | £ | | 8.92 | 2.91 | 52.07 | 63.90 |

**SECTION 17 STRUCTURAL CONCRETE**

**IN-SITU CONCRETE - continued**

| | | UNIT | L | P | M | £ NETT |
|---|---|---|---|---|---|---|
| Columns, piers and beams; (C20/20 concrete) thickness | | | | | | |
| ne 0.03m2 | hr | m3 | 0.50 | 0.50 | | |
| | £ | | 23.47 | 7.65 | 47.18 | 78.30 |
| 0.03 - 0.10m2 | hr | m3 | 0.40 | 0.40 | | |
| | £ | | 18.77 | 6.12 | 47.18 | 72.07 |
| 0.10 - 0.25m2 | hr | m3 | 0.35 | 0.35 | | |
| | £ | | 16.43 | 5.36 | 47.18 | 68.97 |
| 0.25 - 1.00m2 | hr | m3 | 0.35 | 0.35 | | |
| | £ | | 16.43 | 5.36 | 47.18 | 68.97 |
| exceeding 1m2 | hr | m3 | 0.28 | 0.28 | | |
| | £ | | 13.14 | 4.28 | 47.18 | 64.60 |
| (C30/20 concrete) thickness | | | | | | |
| ne 0.03m2 | hr | m3 | 0.53 | 0.53 | | |
| | £ | | 24.87 | 8.11 | 52.09 | 85.07 |
| 0.03 - 0.10m2 | hr | m3 | 0.42 | 0.42 | | |
| | £ | | 19.71 | 6.43 | 52.09 | 78.23 |
| 0.10 - 0.25m2 | hr | m3 | 0.36 | 0.36 | | |
| | £ | | 16.89 | 5.51 | 52.09 | 74.49 |
| 0.25 - 1.00m2 | hr | m3 | 0.35 | 0.35 | | |
| | £ | | 16.43 | 5.36 | 52.09 | 73.88 |
| exceeding 1m2 | hr | m3 | 0.28 | 0.28 | | |
| | £ | | 13.14 | 4.28 | 52.09 | 69.51 |

Add to the above prices for

| | UNIT | £ |
|---|---|---|
| Sulphur resisting cement C20 | m3 | 1.68 |
| Sulphur resisting cement C30 | m3 | 2.23 |
| Air entrained concrete | m3 | 0.51 |
| Water repellent concrete | m3 | 1.80 |

**SECTION 17 STRUCTURAL CONCRETE**

**PRECAST CONCRETE**

The cost of precast concrete item is very much dependent on the complexity of the moulds, the number of units to be cast from each mould and the size and weight of the unit to be handled. The unit rates below are for standard precast items that are often to be found on a Civil Engineering project. It would be misleading to quote for indicative costs for tailor-made precast concrete units and it is advisable to contact Specialist Manufacturers for guide prices.

**PRESTRESSED PRE-TENSIONED BEAMS**

| | | UNIT | L | P | M | £ NETT |
|---|---|---|---|---|---|---|
| Concrete mix C20 | | | | | | |
| size 100 x 65 x 1050mm long | hr | nr | - | - | | |
| | £ | | 2.40 | - | 4.76 | 7.16 |
| size 265 x 65 x 1800mm long | hr | nr | - | - | | |
| | £ | | 2.29 | 2.18 | 19.05 | 23.52 |
| Inverted 'T' beams, flange width 495mm | | | | | | |
| Section T4; length 12m and depth 575mm | hr | nr | - | - | | |
| | £ | | 78.00 | 90.00 | 635.00 | 803.00 |
| Section T7; length 12m and depth 695mm | hr | nr | - | - | | |
| | £ | | 105.00 | 122.00 | 760.00 | 987.00 |
| Section T9; length 16m and depth 775mm | hr | nr | - | - | | |
| | £ | | 138.00 | 160.00 | 1125.00 | 1423.00 |
| Section T10; length 18m and depth 815mm | hr | nr | - | - | | |
| | £ | | 162.00 | 189.00 | 1330.00 | 1681.00 |
| 'M' beams, flange width 970mm | | | | | | |
| Section M3; length 18m and depth 800mm | hr | nr | - | - | | |
| | £ | | 250.00 | 291.00 | 2890.00 | 3431.00 |
| Section M6; length 22m and depth 1040mm | hr | nr | - | - | | |
| | £ | | 340.00 | 397.00 | 3920.00 | 4657.00 |
| Section M8; length 25m and depth 1200mm | hr | nr | - | - | | |
| | £ | | 444.00 | 518.00 | 4725.00 | 5687.00 |

**SECTION 17 STRUCTURAL CONCRETE**

**PRECAST CONCRETE - continued**

| | | UNIT | L | P | M | £ NETT |
|---|---|---|---|---|---|---|
| 'U' beams, base width 970mm | | | | | | |
| Section U3; length 16m and depth 900mm | hr | nr | - | - | | |
| | £ | | 460.00 | 536.00 | 4400.00 | 5396.00 |
| Section U5; length 20m and depth 1000mm | hr | nr | - | - | | |
| | £ | | 587.00 | 684.00 | 5860.00 | 7131.00 |
| Section U8; length 24m and depth 1200mm | hr | nr | - | - | | |
| | £ | | 786.00 | 918.00 | 8200.00 | 9904.00 |
| Section U12; length 30m and depth 1600mm | hr | nr | - | - | | |
| | £ | | 1255.00 | 1460.00 | 13150.00 | 15865.00 |

| | | UNIT | L | P | M | £ NETT |
|---|---|---|---|---|---|---|
| Precast concrete culverts, cattle creeps, and subway units, with rebated joints. | | | | | | |
| Rectangular section | | | | | | |
| 500mm high x 1250 wide | hr | m | - | - | | |
| | £ | | 3.40 | 4.25 | 135.85 | 143.50 |
| 1250mm high x 1500 wide | hr | m | - | - | | |
| | £ | | 7.65 | 9.75 | 246.95 | 264.35 |
| 1250mm high x 1750 wide | hr | m | - | - | | |
| | £ | | 13.85 | 17.50 | 266.20 | 297.55 |
| 2000mm high x 2750 wide | hr | m | - | - | | |
| | £ | | 25.90 | 27.85 | 704.00 | 757.75 |
| 2750mm high x 3000 wide | hr | m | - | - | | |
| | £ | | 34.25 | 55.90 | 927.30 | 1017.45 |
| Extra over for curved on plan less than 20m radius | hr | m | - | - | | |
| | £ | | - | - | - | 260.00 |

**SECTION 17 STRUCTURAL CONCRETE**

**FORMWORK**

| | | | |
|---|---|---|---|
| LABOUR based on Gang C | | | |
| Large area | | | |
| Total hourly cost | = | £ | 35.21 |
| PLANT | | | |
| (Part time) | | | |
| 20T Mobile Crane | | | 4.04 |
| Saw Bench | | | 1.01 |
| Small Power Tools | | | 0.50 |
| Total hourly cost | = | £ | 5.55 |

These unit costs are based upon those outputs and prices detailed in Class G but are referenced to The Specification for Highway Works, clause 1708.

The following rates do not include for formwork outside the payline and are based on an optimum of a minumum 8 uses.

ADD to formwork costs generally for number of uses

1 90-170%
2 50-80%
3 15-30%
6 5-10%

F1 Formwork is rough finish
F2 Formwork is fair finish
F3 Formwork is extra smooth finish

Materials wasteage allowance = 5 - 7%
Refer also to CESMM Class G.

**SECTION 17 STRUCTURAL CONCRETE**

**FORMWORK - continued**

| | | UNIT | L | P | M | £ NETT |
|---|---|---|---|---|---|---|
| Formwork 300mm wide or less at any inclination Class Fl | hr | m2 | 0.716 | 0.716 | | |
| | £ | | 25.21 | 3.97 | 3.52 | 32.70 |
| Formwork 300mm wide or less at any inclination Class Fl | hr | m | 0.20 | 0.20 | | |
| | £ | | 7.04 | 1.11 | 0.70 | 8.85 |
| Formwork more than 300mm wide at any inclination up to and including 5 degrees to the horizontal Class Fl | hr | m2 | 0.52 | 0.52 | | |
| | £ | | 18.31 | 2.89 | 2.80 | 24.00 |
| Formwork more than 300mm wide at any inclination more than 5 degrees up to and including 85 degrees to the horizontal Class Fl | hr | m2 | 0.55 | 0.55 | | |
| | £ | | 19.37 | 3.05 | 3.23 | 25.65 |
| Formwork more than 300mm wide at any inclination more than 85 degrees up to and including 90 degrees to the horizontal Class Fl | hr | m2 | 0.61 | 0.61 | | |
| | £ | | 21.48 | 3.39 | 3.32 | 28.19 |
| Formwork 300mm wide or less, at any inclination. Class F2 | hr | m2 | 0.745 | 0.745 | | |
| | £ | | 26.23 | 4.13 | 4.77 | 35.13 |
| Formwork 300mm wide or less, at any inclination. Class F2 | hr | m | 0.20 | 0.20 | | |
| | £ | | 7.04 | 1.11 | 0.83 | 8.98 |
| Formwork more than 300mm wide at any inclination up to and including 5 degrees to the horizontal. Class F2 | hr | m2 | 0.54 | 0.54 | | |
| | £ | | 19.01 | 3.00 | 3.64 | 25.65 |
| Formwork more than 300mm wide at any inclination more than 5 degrees up to and including 85 degrees to the horizontal. Class F2 | hr | m2 | 0.57 | 0.57 | | |
| | £ | | 20.07 | 3.16 | 4.24 | 27.47 |
| Formwork more than 300mm wide at any inclination more than 85 degrees up to and including 90 degrees to the horizontal. Class F2 | hr | m2 | 0.63 | 0.63 | | |
| | £ | | 22.18 | 3.50 | 4.31 | 29.99 |

**SECTION 17 STRUCTURAL CONCRETE**

**FORMWORK - continued**

| | | UNIT | L | P | M | £ NETT |
|---|---|---|---|---|---|---|
| Formwork 300mm wide or less at any inclination. Class F3 | hr | m2 | 0.761 | 0.761 | | |
| | £ | | 26.79 | 4.22 | 5.83 | 36.84 |
| Formwork 300mm wide or less at any inclination. Class F3 | hr | m | 0.20 | 0.20 | | |
| | £ | | 7.04 | 1.11 | 1.02 | 9.17 |
| Formwork more than 300mm wide at any inclination up to and including 5 degrees to the horizontal. Class F3 | hr | m2 | 0.553 | 0.553 | | |
| | £ | | 19.47 | 3.07 | 4.48 | 27.02 |
| Formwork more than 300mm wide at any inclination more than 5 degrees up to and including 85 degrees to the horizontal. Class F3 | hr | m2 | 0.585 | 0.585 | | |
| | £ | | 20.60 | 3.25 | 5.18 | 29.03 |
| Formwork more than 300mm wide at any inclination more than 85 degrees up to and including 90 degrees to the horizontal. Class F3 | hr | m2 | 0.65 | 0.65 | | |
| | £ | | 22.89 | 3.61 | 5.30 | 31.80 |
| Curved formwork 1m radius or greater | | | | | | |
| Curved of both girth and width more than 300mm at any inclination. Class F1 | hr | m2 | 0.72 | 0.72 | | |
| | £ | | 25.35 | 4.00 | 3.63 | 32.98 |
| Curved of both girth or width more than 300mm or less at any inclination. Class F1 | hr | m2 | 0.95 | 0.95 | | |
| | £ | | 33.45 | 5.27 | 4.04 | 42.76 |
| Curved of both girth and width more than 300mm at any inclination. Class F2 | hr | m2 | 0.75 | 0.75 | | |
| | £ | | 26.41 | 4.16 | 4.71 | 35.28 |
| Curved of both girth or width more than 300mm or at any inclination. Class F2 | hr | m2 | 0.98 | 0.98 | | |
| | £ | | 34.51 | 5.44 | 5.28 | 45.23 |

**SECTION 17 STRUCTURAL CONCRETE**

**FORMWORK - continued**

| | | UNIT | L | P | M | £ NETT |
|---|---|---|---|---|---|---|
| Curved of both girth and width more than 300mm at any inclination. Class F3 | hr | m2 | 0.77 | 0.77 | | |
| | £ | | 27.11 | 4.27 | 5.80 | 37.18 |
| Curved of both girth or width of 300mm or less at any inclination. Class F3 | hr | m2 | 0.99 | 0.99 | | |
| | £ | | 34.86 | 5.49 | 6.48 | 46.83 |

SURFACE FINISHES

| | | UNIT | L | P | M | £ NETT |
|---|---|---|---|---|---|---|
| Class U1 | hr | m2 | 0.012 | 0.012 | | |
| | £ | | 0.42 | 0.07 | - | 0.49 |
| U2 | hr | m2 | 0.013 | 0.013 | | |
| | £ | | 0.46 | 0.07 | - | 0.53 |
| U3 | hr | m2 | 0.02 | 0.02 | | |
| | £ | | 0.70 | 0.11 | - | 0.81 |
| U4 | hr | m2 | 0.033 | 0.033 | | |
| | £ | | 1.16 | 0.18 | - | 1.34 |
| U5 | hr | m2 | 0.034 | 0.034 | | |
| | £ | | 1.20 | 0.19 | - | 1.39 |
| Add for additional craneage and rub up (where required) | £ | m2 | 1.16 | 1.26 | 0.02 | 2.44 |
| Add for additional craneage and rub up (where required) | £ | m | 3.88 | 4.20 | 0.08 | 8.16 |

**SECTION 17 STRUCTURAL CONCRETE**

**REINFORCEMENT**

| | | | |
|---|---|---|---|
| LABOUR based on Gang D | | | |
| Total hourly cost | = | £ | 41.41 |

| | | | |
|---|---|---|---|
| PLANT<br>(Part time)<br>Crane 20T Mobile | | | 4.04 |
| Bar Cropper | | | 0.50 |
| Small Power Tools | | | 0.15 |
| Tirfors, kentlidge, etc. | | | 0.50 |
| Total hourly cost | = | £ | 5.19 |

| | | UNIT | L | P | M | £ NETT |
|---|---|---|---|---|---|---|
| Refer also to CESMM2 Class F and G. | | | | | | |
| Straight bars | | | | | | |
| nominal size 6mm | | | | | | |
| Mild steel to BS4449 | hr | t | 8.00 | 8.00 | | |
| | £ | | 331.28 | 41.52 | 473.29 | 846.09 |
| High yield steel to BS4449 or BS4461 | hr | t | 8.00 | 8.00 | | |
| | £ | | 331.28 | 41.52 | 477.39 | 850.19 |
| nominal size 8mm | | | | | | |
| Mild steel to BS4449 | hr | t | 6.74 | 6.74 | | |
| | £ | | 279.10 | 34.98 | 433.32 | 747.40 |
| High yield steel to BS4449 or BS4461 | hr | t | 6.74 | 6.74 | | |
| | £ | | 279.10 | 34.98 | 437.42 | 751.50 |
| nominal size 10mm | | | | | | |
| Mild steel to BS4449 | hr | t | 6.74 | 6.74 | | |
| | £ | | 279.10 | 34.98 | 422.04 | 736.12 |
| High yield steel to BS4449 or BS4461 | hr | t | 6.74 | 6.74 | | |
| | £ | | 279.10 | 34.98 | 427.39 | 741.47 |

**SECTION 17 STRUCTURAL CONCRETE**

**REINFORCEMENT - continued**

| | | UNIT | L | P | M | £ NETT |
|---|---|---|---|---|---|---|
| Straight bars | | | | | | |
| nominal size 12mm | | | | | | |
| Mild steel to BS4449 | hr | t | 6.74 | 6.74 | | |
| | £ | | 279.10 | 34.98 | 411.79 | 725.87 |
| High yield steel to BS4449 or BS4461 | hr | t | 6.74 | 6.74 | | |
| | £ | | 279.10 | 34.98 | 415.89 | 729.97 |
| nominal size 16mm | | | | | | |
| Mild steel to BS4449 | hr | t | 6.15 | 6.15 | | |
| | £ | | 254.67 | 31.92 | 390.27 | 676.86 |
| High yield steel to BS4449 or BS4461 | hr | t | 6.15 | 6.15 | | |
| | £ | | 254.67 | 31.92 | 394.37 | 680.96 |
| nominal size 20mm | | | | | | |
| Mild steel to BS4449 | hr | t | 4.44 | 4.44 | | |
| | £ | | 183.86 | 23.04 | 389.24 | 596.14 |
| High yield steel to BS4449 or BS4461 | hr | t | 4.44 | 4.44 | | |
| | £ | | 183.86 | 23.04 | 393.34 | 600.24 |
| nominal size 25mm | | | | | | |
| Mild steel to BS4449 | hr | t | 4.44 | 4.44 | | |
| | £ | | 183.86 | 23.04 | 392.32 | 599.22 |
| High yield steel to BS4449 or BS4461 | hr | t | 4.44 | 4.44 | | |
| | £ | | 183.86 | 23.04 | 396.42 | 603.32 |
| nominal size 32mm | | | | | | |
| Mild steel to BS4449 | hr | t | 4.44 | 4.44 | | |
| | £ | | 183.86 | 23.04 | 399.49 | 606.39 |
| High yield steel to BS4449 or BS4461 | hr | t | 4.44 | 4.44 | | |
| | £ | | 183.86 | 23.04 | 403.59 | 610.49 |
| nominal size 40mm | | | | | | |
| Mild steel to BS4449 | hr | t | 4.44 | 4.44 | | |
| | £ | | 183.86 | 23.04 | 411.79 | 618.69 |
| High yield steel to BS4449 or BS4461 | hr | t | 4.44 | 4.44 | | |
| | £ | | 183.86 | 23.04 | 415.89 | 622.79 |

**SECTION 17 STRUCTURAL CONCRETE**

**REINFORCEMENT - continued**

| | | UNIT | L | P | M | £ NETT |
|---|---|---|---|---|---|---|
| Bent bars | | | | | | |
| nominal size 6mm | | | | | | |
| Mild steel to BS4449 | hr | t | 8.00 | 8.00 | | |
| | £ | | 331.28 | 41.52 | 552.21 | 925.01 |
| High yield steel to BS4449 or BS4461 | hr | t | 8.00 | 8.00 | | |
| | £ | | 331.28 | 41.52 | 556.26 | 929.06 |
| nominal size 8mm | | | | | | |
| Mild steel to BS4449 | hr | t | 6.74 | 6.74 | | |
| | £ | | 279.10 | 34.98 | 507.78 | 821.86 |
| High yield steel to BS4449 or BS4461 | hr | t | 6.74 | 6.74 | | |
| | £ | | 279.10 | 34.98 | 511.82 | 825.90 |
| nominal size 10mm | | | | | | |
| Mild steel to BS4449 | hr | t | 6.74 | 6.74 | | |
| | £ | | 279.10 | 34.98 | 476.47 | 790.55 |
| High yield steel to BS4449 or BS4461 | hr | t | 6.74 | 6.74 | | |
| | £ | | 279.10 | 34.98 | 480.51 | 794.59 |
| nominal size 12mm | | | | | | |
| Mild steel to BS4449 | hr | t | 6.74 | 6.74 | | |
| | £ | | 279.10 | 34.98 | 451.22 | 765.30 |
| High yield steel to BS4449 or BS4461 | hr | t | 6.74 | 6.74 | | |
| | £ | | 279.10 | 34.98 | 455.26 | 769.34 |
| nominal size 16mm | | | | | | |
| Mild steel to BS4449 | hr | t | 6.15 | 6.15 | | |
| | £ | | 254.67 | 31.92 | 419.91 | 706.50 |
| High yield steel to BS4449 or BS4461 | hr | t | 6.15 | 6.15 | | |
| | £ | | 254.67 | 31.92 | 423.95 | 710.54 |
| nominal size 20mm | | | | | | |
| Mild steel to BS4449 | hr | t | 4.44 | 4.44 | | |
| | £ | | 183.86 | 23.04 | 417.89 | 624.79 |
| High yield steel to BS4449 or BS4461 | hr | t | 4.44 | 4.44 | | |
| | £ | | 183.86 | 23.04 | 417.89 | 624.79 |

**SECTION 17 STRUCTURAL CONCRETE**

**REINFORCEMENT - continued**

| | | UNIT | L | P | M | £ NETT |
|---|---|---|---|---|---|---|
| Bent bars | | | | | | |
| nominal size 25mm | | | | | | |
| Mild steel to BS4449 | hr | t | 4.44 | 4.44 | | |
| | £ | | 183.86 | 23.04 | 416.88 | 623.78 |
| High yield steel to BS4449 or BS4461 | hr | t | 4.44 | 4.44 | | |
| | £ | | 183.86 | 23.04 | 420.92 | 627.82 |
| nominal size 32mm | | | | | | |
| Mild steel to BS4449 | hr | t | 4.44 | 4.44 | | |
| | £ | | 183.86 | 23.04 | 418.90 | 625.80 |
| High yield steel to BS4449 or BS4461 | hr | t | 4.44 | 4.44 | | |
| | £ | | 183.86 | 23.04 | 422.93 | 629.83 |
| nominal size 40mm | | | | | | |
| Mild steel to BS4449 | hr | t | 4.44 | 4.44 | | |
| | £ | | 183.86 | 23.04 | 431.02 | 637.92 |
| High yield steel to BS4449 or BS4461 | hr | t | 4.44 | 4.44 | | |
| | £ | | 183.86 | 23.04 | 435.06 | 641.96 |

**SECTION 17 STRUCTURAL CONCRETE**

**REINFORCEMENT - continued**

| | | UNIT | L | P | M | £ NETT |
|---|---|---|---|---|---|---|
| ADD to the above for bars over 12 - 15m long | | | | | | |
| Mild steel to BS4449 | hr | t | - | - | | |
| | £ | | - | - | - | 14.00 |
| High yield steel to BS4449 or BS4461 | hr | t | - | - | | |
| | £ | | - | - | - | 14.00 |
| ADD to the above for bars over 15m long, per 500mm increment | | | | | | |
| Mild steel to BS4449 | hr | t | - | - | | |
| | £ | | - | - | - | 3.05 |
| High yield steel to BS4449 or BS4461 | hr | t | - | - | | |
| | £ | | - | - | - | 3.05 |
| High yield steel fabric to BS4483 | | | | | | |
| nominal mass ne 2kg/m2; ref D49 | hr | m2 | 0.02 | - | | |
| | £ | | 0.83 | - | 0.84 | 1.67 |
| nominal mass ne 2kg/m2; ref A98 | hr | m2 | 0.03 | - | | |
| | £ | | 1.24 | - | 1.04 | 2.28 |
| nominal mass 2-3 kg/m2; ref A142 | hr | m2 | 0.03 | - | | |
| | £ | | 1.24 | - | 1.32 | 2.56 |
| nominal mass 2-3 kg/m2; ref C283 | hr | m2 | 0.03 | - | | |
| | £ | | 1.24 | - | 1.57 | 2.81 |
| nominal mass 3-4 kg/m2; ref A193 | hr | m2 | 0.04 | - | | |
| | £ | | 1.66 | - | 1.79 | 3.45 |
| nominal mass 3-4 kg/m2; ref B196 | hr | m2 | 0.04 | - | | |
| | £ | | 1.66 | - | 1.96 | 3.62 |
| nominal mass 3-4 kg/m2; ref C385 | hr | m2 | 0.04 | - | | |
| | £ | | 1.66 | - | 2.16 | 3.82 |
| nominal mass 3-4 kg/m2; ref B283 | hr | m2 | 0.04 | - | | |
| | £ | | 1.66 | - | 2.34 | 4.00 |
| nominal mass 3-4 kg/m2; ref A252 | hr | m2 | 0.04 | - | | |
| | £ | | 1.66 | - | 2.36 | 4.02 |

## SECTION 17 STRUCTURAL CONCRETE

**REINFORCEMENT - continued**

| | | UNIT | L | P | M | £ NETT |
|---|---|---|---|---|---|---|
| nominal mass 4-5 kg/m2; ref C503 | hr | m2 | 0.05 | - | | |
| | £ | | 2.07 | - | 2.62 | 4.69 |
| nominal mass 4-5 kg/m2; ref B385 | hr | m2 | 0.05 | - | | |
| | £ | | 2.07 | - | 2.81 | 4.88 |
| nominal mass 5-6 kg/m2; ref C636 | hr | m2 | 0.05 | - | | |
| | £ | | 2.07 | - | 3.38 | 5.45 |
| nominal mass 5-6 kg/m2; ref B503 | hr | m2 | 0.05 | - | | |
| | £ | | 2.07 | - | 3.55 | 5.62 |
| nominal mass 6-7 kg/m2; ref A393 | hr | m2 | 0.07 | - | | |
| | £ | | 2.90 | - | 3.66 | 6.56 |
| nominal mass 6-7 kg/m2; ref C785 | hr | m2 | 0.07 | - | | |
| | £ | | 2.90 | - | 4.09 | 6.99 |
| nominal mass 8.14 kg/m2; ref B785 | hr | m2 | 0.08 | - | | |
| | £ | | 3.31 | - | 4.85 | 8.16 |
| nominal mass 10.90 kg/m2; ref B1131 | hr | m2 | 0.09 | - | | |
| | £ | | 3.73 | - | 6.56 | 10.29 |
| Stainless steel bars bent and cut lengths | | | | | | |
| 8mm | hr | t | 6.74 | 6.74 | | |
| | £ | | 279.10 | 34.98 | 4100.91 | 4414.99 |
| 10mm | hr | t | 6.74 | 6.74 | | |
| | £ | | 279.10 | 34.98 | 3860.00 | 4174.08 |
| 12mm | hr | t | 6.74 | 6.74 | | |
| | £ | | 279.10 | 34.98 | 3457.92 | 3772.00 |
| 16mm | hr | t | 6.15 | 6.15 | | |
| | £ | | 254.67 | 31.92 | 3141.11 | 3427.70 |
| 20mm | hr | t | 4.44 | 4.44 | | |
| | £ | | 183.86 | 23.04 | 3019.14 | 3226.04 |
| 25mm | hr | t | 4.44 | 4.44 | | |
| | £ | | 183.86 | 23.04 | 2992.89 | 3199.79 |

## SECTION 18 STEELWORK FOR STRUCTURES

| | UNIT | SPECIALIST ADVICE £ NETT |
|---|---|---|
| Fabrication of members<br>Steel grade 43A | | |
| Columns | t | 850.00 |
| Beams | | |
| universal beams; straight on plan | t | 700.00 |
| universal beams; curved on plan | t | 1,250.00 |
| Trusses and built up girders | | |
| straight on plan | t | 950.00 |
| curved on plan | t | 1,500.00 |
| Masts, towers, pylons | t | 1,250.00 |
| Trestles and frames | t | 1,200.00 |
| Fabrication of main members<br>Steel grade 43A | | |
| Box girder | t | 1,150.00 |
| Plate girder | t | 950.00 |
| Rolled sections | t | 920.00 |
| **ERECTION OF STEELWORK** | | |
| Permanent erection of columns beams and frames | t | 180.00 |
| Permanent erection of superstructure steelwork | t | 205.00 |
| Permanent erection of subsidiary superstructure steelwork | t | 210.00 |

**SECTION 19 PROTECTION OF STEELWORK AGAINST CORROSION**

Refer also to CESMM - Class V

LABOUR based on Gang I

| | | |
|---|---|---|
| Total hourly cost | = £ | 38.51 |

PLANT
(Part time)

| | | |
|---|---|---|
| Power tools | | 2.09 |
| Access scaffolding, trestles and ladders | | 5.08 |
| Transport, van | | 3.28 |
| Total hourly cost | = £ | 10.45 |

To new works in Inland 'B' exposed environment.

Typically applied to metal parapets and fencing, lighting columns, brackets. Metal coating - Galvanised to BS729.

1st coat: Mordant T wash
2nd coat: Zinc phosphate, CR/Alkyd Undercoat
3rd coat: Zinc phosphate, CR/Alkyd Undercoat
4th coat: MIO CR Undercoat
5th coat: CR finish. Colour to be specified

First 3 coats at works. Coats 4 and 5 on site.

| | | UNIT | L | P | M | £ NETT |
|---|---|---|---|---|---|---|
| Brush or airless spray dry film thickness 250 um | hr | m2 | 0.075 | 0.075 | | |
| | £ | | 2.89 | 0.78 | 6.35 | 10.02 |

Typically applied to subsidiary steelwork, interior finishes. Blast clean to BS4232 (surface preparation).

1st coat: Zinc Chromate, Red Oxide Blast primer
2nd coat: Zinc Phosphate, Epoxy Ester Undercoat
3rd coat: MIO Undercoat
4th coat: MIO finished colour

First 3 coats at works. Coat 4 on site.

| | | UNIT | L | P | M | £ NETT |
|---|---|---|---|---|---|---|
| Brush or airless spray dry film thickness 175um | hr | m2 | 0.062 | 0.062 | | |
| | £ | | 2.39 | 0.65 | 4.56 | 7.60 |

**SECTION 19 PROTECTION OF STEELWORK AGAINST CORROSION**

Typically applied to Main Steel members. Blast clean to BS4232 (surface preparation). Metal coating - Aluminium spray, applied at works.

1st coat: Zinc Chromate Etch primer (2 pack)
2nd coat: Zinc Phosphate, CR/Alkyd Undercoat
3rd coat: Zinc Phosphate, CR/Alkyd Undercoat
4th coat: M10 CR Undercoat
5th coat: CR finish

First 3 coats at works Coats 4 and 5 on site.

| | | **UNIT** | **L** | **P** | **M** | **£ NETT** |
|---|---|---|---|---|---|---|
| Brush or airless spray dry film thickness 250um | hr | m2 | 0.067 | 0.067 | | |
| | £ | | 2.58 | 0.70 | 5.18 | 8.46 |

Typically applied to Internal Steel members. Blast clean to BS4232 (2nd quality surface preparation). Remove all surface defects to BS4360.

1st coat: Zinc rich primer (2 pack)
2nd coat: Epoxy High Build M10 (2 pack)
3rd coat: Polyurethane Undercoat (2 pack)
4th coat: Finish coat polyurethane (2 pack)

Coats 1 and 2 at works. Coats 3 and 4 on site.

| | | **UNIT** | **L** | **P** | **M** | **£ NETT** |
|---|---|---|---|---|---|---|
| Brush or airless spray | hr | m2 | 0.062 | 0.062 | | |
| | £ | | 2.39 | 0.65 | 2.00 | 5.04 |

Other surface treatments

| | **UNIT** | **£** |
|---|---|---|
| Galvanising (Hot dip) to BS729 | tonne | 400.00 approx |
| Shot blasting (at works) | m2 | 1.00 approx |
| Grit blasting (at works) | m2 | 1.85 approx |
| Sand blasting (at works) | m2 | 3.00 approx |
| Shot blasting (on site) | m2 | 1.40 approx |
| Grit blasting (on site) | m2 | 2.50 approx |
| Sand blasting (on site) | m2 | 4.10 approx |

## SECTION 20 WATERPROOFING FOR STRUCTURES

Refer also to CESMM Class W

**LABOUR**

| | | |
|---|---|---|
| Ganger | | 6.82 |
| Skilled labour | | 6.54 |
| Unskilled labour | | 6.20 |
| Total hourly cost | = £ | 19.56 |

**PLANT**

| | | |
|---|---|---|
| Power tools | | 4.18 |
| Tar boiler | | 0.96 |
| (Part time) Compressor, tools, dumper | | 2.02 |
| Total hourly cost | = £ | 7.16 |

This section is based around the installation of proprietory systems to new/recently completed works as part of major scheme, for minor works/repairs ouputs will be many times more. Outputs are also based on use of skilled labour therefore efficiency is high and wastage low (5 - 7% only allowed) excepting laps where required.

**SECTION 20 WATERPROOFING FOR STRUCTURES**

| | | UNIT | L | P | M | £ NETT |
|---|---|---|---|---|---|---|
| | | | **SPECIALIST ADVICE** | | | |
| Proprietary system - Bitu-thene | | | | | | |
| '1000 grade' more than 300mm wide up to 45 degrees to horizontal | hr | m2 | 0.30 | 0.30 | | |
| | £ | | 5.87 | 2.15 | 4.89 | 12.91 |
| '1000 grade' more than 300mm wide 45 - 90 degrees to the horizontal | hr | m2 | 0.33 | 0.33 | | |
| | £ | | 6.45 | 2.36 | 5.03 | 13.84 |
| '1000 grade' 300mm wide or less | hr | m2 | 0.38 | 0.38 | | |
| | £ | | 7.43 | 2.72 | 5.97 | 16.12 |
| 'Heavy duty grade' more than 300mm wide up to 45 degrees to horizontal | hr | m2 | 0.30 | 0.30 | | |
| | £ | | 5.87 | 2.15 | 5.14 | 13.16 |
| 'Heavy duty grade' more than 300mm wide 45-90 degrees to the horizontal | hr | m2 | 0.33 | 0.33 | | |
| | £ | | 6.45 | 2.36 | 5.28 | 14.09 |
| 'Heavy duty grade' 300mm wide or less | hr | m2 | 0.38 | 0.38 | | |
| | £ | | 7.43 | 2.72 | 6.12 | 16.27 |

## SECTION 20 WATERPROOFING FOR STRUCTURES

| | | UNIT | L | P | M | £ NETT |
|---|---|---|---|---|---|---|
| Refer also to CESMM Class W. | | | | | | |
| Proprietory system - Bitu-thene with Bitu-shield Green protective membranes | | | | | | |
| more than 300mm wide up to 45 degrees to the horizontal | hr | m2 | 0.32 | 0.32 | | |
| | £ | | 6.26 | 2.29 | 10.90 | 19.45 |
| more than 300mm wide 45 - 90 degrees to the horizontal | hr | m2 | 0.35 | 0.35 | | |
| | £ | | 6.85 | 2.51 | 11.12 | 20.48 |
| 300mm wide or less | hr | m2 | 0.40 | 0.40 | | |
| | £ | | 7.82 | 2.86 | 12.97 | 23.65 |
| Proprietory system - 'Famguard' (Hot applied) with Fam-primer | | | | | | |
| more than 300mm wide up to 45 degrees to the horizontal | hr | m2 | 0.32 | 0.32 | | |
| | £ | | 6.26 | 2.29 | 6.80 | 15.35 |
| more than 300mm wide 45 - 90 degrees to the horizontal | hr | m2 | 0.35 | 0.35 | | |
| | £ | | 6.85 | 2.51 | 6.94 | 16.30 |
| 300mm wide or less | hr | m2 | 0.40 | 0.40 | | |
| | £ | | 7.82 | 2.86 | 8.09 | 18.77 |
| Proprietory system - 'Famflex' (Hot applied) with Fam-primer | | | | | | |
| more than 300mm wide up to 45 degrees to the horizontal | hr | m2 | 0.32 | 0.32 | | |
| | £ | | 6.26 | 2.29 | 5.29 | 13.84 |
| more than 300mm wide 45 - 90 degrees to the horizontal | hr | m2 | 0.34 | 0.34 | | |
| | £ | | 6.65 | 2.43 | 5.40 | 14.48 |
| 300mm wide or less | hr | m2 | 0.38 | 0.38 | | |
| | £ | | 7.43 | 2.72 | 6.30 | 16.45 |

**SECTION 20 WATERPROOFING FOR STRUCTURES**

| | | UNIT | L | P | M | £ NETT |
|---|---|---|---|---|---|---|
| Two coats of RIW liquid asphaltic composition sprayed or brushed on, any inclination | hr | m2 | 0.07 | 0.07 | | |
| | £ | | 1.37 | 0.50 | 1.90 | 3.77 |
| Two coats of 'Mulseal' sprayed or brushed on, any inclination | hr | m2 | 0.07 | 0.07 | | |
| | £ | | 1.37 | 0.50 | 1.26 | 3.13 |
| 20mm thick red tinted sand asphalt layer onto Bridge Deck | hr | m2 | - | - | | |
| | £ | | 0.45 | 1.05 | 4.75 | 6.25 |

**SECTION 21 BRIDGE BEARINGS**

Bridge bearings are manufactured and installed to individual specifications. The following are for comparative costs for different sizes of simple Bridge Bearings. If requirements are known then Specialist Manufacturers such as CCL or Mageba will be pleased to quote.

| LABOUR | | |
|---|---|---|
| Skilled operative | | 6.54 |
| 2 unskilled operatives | | 12.40 |
| Total hourly cost | = £ | 18.94 |

| | | UNIT | L | P | M | £ NETT |
|---|---|---|---|---|---|---|
| Supply plain rubber bearings approx size 300 x 150 x 20mm | hr | nr | - | - | | |
| | £ | | - | - | 2.93 | 2.93 |
| Supply laminated rubber bearings size 305 x 152 x 18mm | hr | nr | - | - | | |
| | £ | | - | - | 31.11 | 31.11 |
| Supply plain rubber bearings approx size 10,000 x 152 x 25mm | hr | nr | - | - | | |
| | £ | | - | - | 242.93 | 242.93 |
| Install plain rubber bearings 305 x 152 x 18mm | hr | nr | 0.12 | - | | |
| | £ | | 2.27 | - | 0.51 | 2.78 |
| Install laminated rubber bearings 10,000mm x 152 x 25mm | hr | nr | 0.15 | - | | |
| | £ | | 2.84 | - | 0.51 | 3.35 |
| Install plain rubber bearings 10,000mm x 152 x 25mm | hr | nr | 0.74 | - | | |
| | £ | | 14.02 | - | 4.82 | 18.84 |
| Supply elastomeric bearing size 300 x 150 x 18mm | hr | nr | - | - | | |
| | £ | | - | - | 34.84 | 34.84 |
| Supply elastomeric bearing size 450 x 200 x 20mm | hr | nr | - | - | | |
| | £ | | - | - | 62.04 | 62.04 |

## SECTION 21 BRIDGE BEARINGS

| | | UNIT | L | P | M | £ NETT |
|---|---|---|---|---|---|---|
| Supply elastomeric bearing size 600 x 300 x 20mm | hr | nr | - | - | | |
| | £ | | - | - | 70.68 | 70.68 |
| Supply mechanical guide to suit 600 x 300 x 20mm bearing | hr | nr | - | - | | |
| | £ | | - | - | 179.38 | 179.38 |
| Supply mechanical anchor to suit 600 x 300 x 20mm bearing | hr | nr | - | - | | |
| | £ | | - | - | 223.02 | 223.02 |

If there is a requirement for testing bridge Bearings prior to installation then the tests should be enumerated separately. Again specialist advice should be sought once details are known.

| | | UNIT | L | P | M | £ NETT |
|---|---|---|---|---|---|---|
| Compression test for elastomeric bearings | hr | nr | - | - | | |
| | £ | | - | - | 107.63 | 107.63 |
| Shear test for elastomeric bearing 300 x 150 x 18mm | hr | nr | - | - | | |
| | £ | | - | - | 133.25 | 133.25 |
| Bond test for elastomeric bearing 300 x 150 x 18mm | hr | nr | - | - | | |
| | £ | | - | - | 133.25 | 133.25 |
| Bond test for elastomeric bearing 450 x 200 x 20mm | hr | nr | - | - | | |
| | £ | | - | - | 157.65 | 157.65 |
| Bond test for elastomeric bearing 600 x 300 x 20mm | hr | nr | - | - | | |
| | £ | | - | - | 165.54 | 165.54 |

**SECTION 21 BRIDGE BEARINGS**

**APPROXIMATE COSTS OF INSTALLING MANUFACTURED BEARINGS**

| | | UNIT | L | P | M | £ NETT |
|---|---|---|---|---|---|---|
| Installation of elastomeric bearing 300 x 150 x 18mm | hr | nr | 1.40 | - | | |
| | £ | | 26.52 | - | - | 26.52 |
| Installation of elastomeric bearing 450 x 200 x 20mm | hr | nr | 1.42 | - | | |
| | £ | | 26.89 | - | - | 26.89 |
| Installation of elastomeric bearing 600 x 300 x 20mm | hr | nr | 1.44 | - | | |
| | £ | | 27.27 | - | - | 27.27 |
| Installation of mechanical guide to suit 600 x 300 x 20mm bearing | hr | nr | 4.26 | - | | |
| | £ | | 80.68 | - | - | 80.68 |
| Installation of mechanical anchor to suit 600 x 300 x 20mm bearing | hr | nr | 3.36 | - | | |
| | £ | | 63.64 | - | - | 63.64 |

**SECTION 22 PARAPETS**

LABOUR based on Gang I

| | | | |
|---|---|---|---|
| Total hourly cost | = | £ | 38.51 |
| **PLANT** | | | |
| Tractor and tools | | | 8.03 |
| (Part time) Lorry with hiab. | | | 6.62 |
| Power tools | | | 2.03 |
| Total hourly cost | = | £ | 16.68 |

The heights of the following parapets are in accordance with the Standard Designs and DTp requirements. The rates include for all anchorages and fixings and in the case of steel, galvanising at works and painting four coat paint system on site, together with etching the galvanised surface, as necessary.

**STEEL PARAPETS**

| | | **UNIT** | **L** | **P** | **M** | **£ NETT** |
|---|---|---|---|---|---|---|
| Fabrication of group P1 parapet (113 Kph) (1.0m high) comprising steel yielding posts and steel horizontal rails | hr | m | - | - | | |
| | £ | | - | - | **70.00** | **70.00** |
| Installation of group P1 parapet (113 Kph) (1.0m high) comprising steel yielding posts and steel horizontal rails | hr | m | 0.65 | 0.65 | | |
| | £ | | 25.03 | 10.84 | **1.95** | **37.82** |
| Fabrication of group P2 parapet (113 Kph) (1.0m high) comprising steel yielding posts and steel horizontal rails with mesh infill | hr | m | - | - | | |
| | £ | | - | - | **86.00** | **86.00** |
| Installation of group P2 parapet (113 Kph) (1.0m high) comprising steel yielding posts and steel horizontal rails with mesh infill | hr | m | 0.69 | 0.69 | | |
| | £ | | 26.57 | 11.51 | **1.95** | **40.03** |
| Fabrication of group P2 parapet (48 Kph) (1.0m high) comprising steel yielding posts and steel horizontal rails with mesh infill | hr | m | - | - | | |
| | £ | | - | - | **125.00** | **125.00** |

SECTION 22 PARAPETS

STEEL PARAPETS

| | | UNIT | L | P | M | £ NETT |
|---|---|---|---|---|---|---|
| Installation of group P2 parapet (48 Kph) (1.0m high) comprising steel yielding posts and steel horizontal rails with mesh infill | hr | m | 0.67 | 0.67 | | |
| | £ | | 25.80 | 11.18 | 1.95 | 38.93 |
| Fabrication of group P4 parapet (1.1m high) comprising steel yielding posts and steel horizontal rails | hr | m | - | - | | |
| | £ | | - | - | 75.00 | 75.00 |
| Installation of group P4 parapet (1.1m high) comprising steel yielding posts and steel horizontal rails | hr | m | 0.33 | 0.33 | | |
| | £ | | 12.71 | 5.50 | 0.72 | 18.93 |
| Fabrication of group P5 parapet (1.5m high) (113kph) comprising steel yielding posts and steel horizontal rails with solid sheet infill | hr | m | - | - | | |
| | £ | | - | - | 200.00 | 200.00 |
| Installation of group P5 parapet (1.5m high) (113kph) comprising steel yielding posts and steel horizontal rails with solid sheet infill | hr | m | 0.90 | 0.90 | | |
| | £ | | 34.66 | 15.01 | 1.95 | 51.62 |
| Fabrication of group P6 parapet (1.5m high) comprising steel yielding posts and steel horizontal rails mesh or sheet infill | hr | m | - | - | | |
| | £ | | - | - | 360.00 | 360.00 |
| Installation of group P6 parapet (1.5m high) comprising steel yielding posts and steel horizontal rails | hr | m | 1.25 | 1.25 | | |
| | £ | | 48.14 | 20.85 | 1.95 | 70.94 |
| Colebrand 50 tonne structural shock transmission unit to P6 parapet | hr | nr | - | - | | |
| | £ | | - | - | 4000.00 | 4000.00 |

## SECTION 22 PARAPETS

### ALUMINIUM PARAPETS

| | UNIT | | L | P | M | £ NETT |
|---|---|---|---|---|---|---|
| Fabrication of group P1 parapet (1.0m high) comprising aluminium yielding posts and aluminium horizontal rails | hr | m | - | - | | |
| | £ | | - | - | 130.00 | 130.00 |
| Installation of group P1 parapet (1.0m high) comprising aluminium yielding posts and aluminium horizontal rails | hr | m | 0.30 | 0.30 | | |
| | £ | | 11.55 | 5.00 | 1.33 | 17.88 |
| Fabrication of group P2 parapet (113 Kph) (1.0m high) comprising aluminium yielding posts and aluminium horizontal rails with mesh infill | hr | m | - | - | | |
| | £ | | - | - | 150.00 | 150.00 |
| Installation of group P2 parapet (113 Kph)(1.0m high) comprising aluminium yielding posts and aluminium horizontal rails with mesh infill | hr | m | 0.33 | 0.33 | | |
| | £ | | 12.71 | 5.50 | 1.33 | 19.54 |
| Fabrication of group P4 parapet (1.1m high) comprising aluminium yielding posts and aluminium horizontal rails | hr | m | - | - | | |
| | £ | | - | - | 114.00 | 114.00 |
| Installation of group P4 parapet (1.1m high) comprising aluminium yielding posts and aluminium horizontal rails | hr | m | 0.22 | 0.22 | | |
| | £ | | 8.47 | 3.67 | 1.33 | 13.47 |
| Fabrication of group P5 parapet (1.5m high) (113 Kph) comprising aluminium yielding posts and aluminium horizontal rails with solid sheet infill | hr | m | - | - | | |
| | £ | | - | - | 245.00 | 245.00 |
| Installation of group P5 parapet (1.5m high) (113 Kph) comprising aluminium yielding posts and aluminium horizontal rails with solid sheet infill | hr | m | 0.65 | 0.65 | | |
| | £ | | 25.03 | 10.84 | 1.33 | 37.20 |

**SECTION 23 EXPANSION JOINTS AND SEALING OF GAPS**

| | | | |
|---|---|---|---|
| LABOUR | | | |
| Ganger | | | 6.82 |
| Skilled labour | | | 6.54 |
| Unskilled labour | | | 6.20 |
| Total hourly cost | = | £ | 19.56 |
| | | | |
| PLANT | | | |
| Power tools (saws, etc.) | | | 4.18 |
| Tar Boiler | | | 0.96 |
| (Part time)<br>Compressor and tools,<br>dumper, etc. | | | 2.02 |
| Total hourly cost | = | £ | 7.16 |

## SECTION 23 EXPANSION JOINTS AND SEALING OF GAPS

### MOVEMENT JOINTS TO BRIDGE AND VIADUCT DECK

Major movement joints to bridge and viaduct decks are manufactured and installed to individual specifications determined by the type of structure location in the deck, amount of movement to be expected, and many other variables. The following are unit rates for other types of movement joints found in structures.

### MOVEMENT JOINTS OTHER THAN TO BRIDGE AND VIADUCT DECK

| | | UNIT | L | P | M | £ NETT |
|---|---|---|---|---|---|---|
| Plain open surface including scabbling concrete for subsequent pour: | | | | | | |
| 300mm wide or less | hr | m2 | 0.06 | 0.06 | | |
| | £ | | 1.17 | 0.43 | - | 1.60 |
| exceeding 300mm wide | hr | m2 | 0.05 | 0.05 | | |
| | £ | | 0.98 | 0.36 | - | 1.34 |
| Open surface with 10mm flexcell joint filler any width | hr | m2 | 0.06 | 0.06 | | |
| | £ | | 1.17 | 0.43 | 3.62 | 5.22 |
| Formed surface with filler (excluding Formwork) | | | | | | |
| 300mm wide or less 10mm flexcell joint | hr | m2 | 0.16 | - | | |
| | £ | | 3.13 | - | 2.94 | 6.07 |
| 20mm flexcell joint | hr | m2 | 0.16 | - | | |
| | £ | | 3.13 | - | 5.65 | 8.78 |
| 25mm flexcell joint | hr | m2 | 0.16 | - | | |
| | £ | | 3.13 | - | 6.85 | 9.98 |

## SECTION 23 EXPANSION JOINTS AND SEALING OF GAPS

### MOVEMENT JOINTS OTHER THAN TO BRIDGE AND VIADUCT DECK - continued

| | | UNIT | L | P | M | £ NETT |
|---|---|---|---|---|---|---|
| exceeding 300mm wide | | | | | | |
| 10mm flexcell joint | hr | m2 | 0.08 | - | | |
| | £ | | 1.56 | - | 2.94 | 4.50 |
| 20mm flexcell joint | hr | m2 | 0.08 | - | | |
| | £ | | 1.56 | - | 5.65 | 7.21 |
| 25mm flexcell joint | hr | m2 | 0.08 | - | | |
| | £ | | 1.56 | - | 6.85 | 8.41 |
| Sealing of gaps, grooves and rebates | | | | | | |
| 10mm x 20mm hot poured rubber bitumen sealant | hr | m | 0.033 | - | | |
| | £ | | 0.65 | - | 0.27 | 0.92 |
| 20mm x 20mm hot poured rubber bitumen sealant | hr | m | 0.04 | - | | |
| | £ | | 0.78 | - | 0.50 | 1.28 |
| 15mm x 25mm hot poured rubber bitumen sealant | hr | m | 0.065 | - | | |
| | £ | | 1.27 | - | 0.47 | 1.74 |
| 25mm x 25mm hot poured rubber bitumen sealant | hr | m | 0.073 | - | | |
| | £ | | 1.43 | - | 0.76 | 2.19 |
| Cold applied polysulphide joint sealant 20mm x 20mm | hr | m | 0.065 | - | | |
| | £ | | 1.27 | - | 3.20 | 4.47 |
| Gun grade cold applied elastomeric joint sealant | | | | | | |
| 25mm x 25mm on 3mm foam strip | hr | m | 0.065 | - | | |
| | £ | | 1.27 | - | 6.03 | 7.30 |
| 50mm x 25mm on 3mm foam strip | hr | m | 0.09 | - | | |
| | £ | | 1.76 | - | 11.96 | 13.72 |

**SECTION 23 EXPANSION JOINTS AND SEALING OF GAPS**

**MOVEMENT JOINTS OTHER THAN TO BRIDGE AND VIADUCT DECK - continued**

| | | UNIT | L | P | M | £ NETT |
|---|---|---|---|---|---|---|
| Building paper slip joint to abutment toe | hr | m2 | 0.011 | - | | |
| | £ | | 0.22 | - | 0.75 | 0.97 |
| Bond breaking agent | hr | m2 | 0.025 | - | | |
| | £ | | 0.49 | - | 2.50 | 2.99 |
| PVC centre bulb waterstop | | | | | | |
| 160mm wide | hr | m | 0.08 | - | | |
| | £ | | 1.56 | - | 4.26 | 5.82 |
| 210mm wide | hr | m | 0.09 | - | | |
| | £ | | 1.76 | - | 6.32 | 8.08 |
| 260mm wide | hr | m | 0.11 | - | | |
| | £ | | 2.15 | - | 9.00 | 11.15 |
| Rubber centre bulb waterstop | | | | | | |
| 150mm wide | hr | m | 0.08 | - | | |
| | £ | | 1.56 | - | 15.13 | 16.69 |
| 230mm wide | hr | m | 0.10 | - | | |
| | £ | | 1.96 | - | 22.60 | 24.56 |
| 305mm wide | hr | m | 0.12 | - | | |
| | £ | | 2.35 | - | 37.27 | 39.62 |
| Dowels, plain or greased | | | | | | |
| 12mm mild steel 450mm long | hr | m | 0.04 | - | | |
| | £ | | 0.78 | - | 0.28 | 1.06 |
| 16mm mild steel 750mm long | hr | m | 0.045 | - | | |
| | £ | | 0.88 | - | 0.62 | 1.50 |
| 16mm mild steel 750mm long with debonding agent for 375mm | hr | m | 0.053 | - | | |
| | £ | | 1.04 | - | 0.74 | 1.78 |
| Dowels sleeved or capped | | | | | | |
| 12mm mild steel 450mm long with debonding agent for 225mm and PVC dowel cap | hr | m | 0.045 | - | | |
| | £ | | 0.88 | - | 0.50 | 1.38 |

**SECTION 24 BRICKWORK BLOCKWORK AND STONEWORK**

LABOUR based on Gang J

| | | | |
|---|---|---|---|
| Total hourly cost | = | £ | 50.51 |

PLANT

| | | | |
|---|---|---|---|
| (Part time) | | | |
| Mixer | | | 0.83 |
| Dumper | | | 1.15 |
| Power tools | | | 2.09 |
| Scaffold etc | | | 0.90 |
| Total hourly cost | = | £ | 4.97 |

| | | UNIT | L | P | M | £ NETT |
|---|---|---|---|---|---|---|
| **IN WALLS** | | | | | | |
| Half brick thick vertical straight walls | | | | | | |
| Commons PC £120/1000 | hr | m2 | 0.23 | 0.23 | | |
| | £ | | 11.62 | 1.14 | 8.56 | 21.32 |
| Facings PC £350/1000 | hr | m2 | 0.34 | 0.34 | | |
| | £ | | 17.17 | 1.69 | 22.72 | 41.58 |
| Engineering PC £300/1000 | hr | m2 | 0.29 | 0.29 | | |
| | £ | | 14.65 | 1.44 | 19.49 | 35.58 |
| One brick thick vertical straight walls | | | | | | |
| Commons PC £120/1000 | hr | m2 | 0.44 | 0.44 | | |
| | £ | | 22.22 | 2.19 | 17.17 | 41.58 |
| Facings PC £350/1000 | hr | m2 | 0.574 | 0.574 | | |
| | £ | | 28.99 | 2.85 | 45.53 | 77.37 |
| Engineering PC £300/1000 | hr | m2 | 0.485 | 0.485 | | |
| | £ | | 24.50 | 2.41 | 39.03 | 65.94 |

SECTION 24 BRICKWORK BLOCKWORK AND STONEWORK

| | | UNIT | L | P | M | £ NETT |
|---|---|---|---|---|---|---|
| Two brick thick vertical straight walls | | | | | | |
| Commons PC £120/1000 | hr | m2 | 0.78 | 0.78 | | |
| | £ | | 39.40 | 3.88 | 34.85 | 78.13 |
| Facings PC £350/1000 (commons in centre) | hr | m2 | 1.08 | 1.08 | | |
| | £ | | 54.55 | 5.37 | 70.04 | 129.96 |
| Engineering PC £300/1000 | hr | m2 | 0.905 | 0.905 | | |
| | £ | | 45.71 | 4.50 | 78.68 | 128.89 |
| COPINGS, STRING COURSES AND THE LIKE | | | | | | |
| Copings; brick-on-edge 215mm wide | | | | | | |
| Commons PC £120/1000 | hr | m | 0.105 | 0.105 | | |
| | £ | | 5.30 | 0.52 | 1.97 | 7.79 |
| Facings PC £350/1000 | hr | m | 0.134 | 0.134 | | |
| | £ | | 6.77 | 0.67 | 5.22 | 12.66 |
| Engineering PC £300/1000 | hr | m | 0.134 | 0.134 | | |
| | £ | | 6.77 | 0.67 | 4.48 | 11.92 |
| Half brick thick vertical curved walls | | | | | | |
| Commons PC £120/1000 | hr | m2 | 0.304 | 0.304 | | |
| | £ | | 15.36 | 1.51 | 8.56 | 25.43 |
| Facings PC £350/1000 | hr | m2 | 0.45 | 0.45 | | |
| | £ | | 22.73 | 2.24 | 22.72 | 47.69 |
| Engineering PC £300/1000 | hr | m2 | 0.416 | 0.416 | | |
| | £ | | 21.01 | 2.07 | 19.49 | 42.57 |
| One brick thick vertical curved walls | | | | | | |
| Commons PC £120/1000 | hr | m2 | 0.59 | 0.59 | | |
| | £ | | 29.80 | 2.93 | 17.17 | 49.90 |
| Facings PC £350/1000 | hr | m2 | 0.84 | 0.84 | | |
| | £ | | 42.43 | 4.17 | 45.53 | 92.13 |
| Engineering PC £300/1000 | hr | m2 | 0.78 | 0.78 | | |
| | £ | | 39.40 | 3.88 | 39.03 | 82.31 |

## SECTION 24 BRICKWORK BLOCKWORK AND STONEWORK

| | | UNIT | L | P | M | £ NETT |
|---|---|---|---|---|---|---|
| **Two brick thick vertical curved walls** | | | | | | |
| Commons PC £120/1000 | hr | m2 | 1.04 | 1.04 | | |
| | £ | | 52.53 | 5.17 | 34.85 | 92.55 |
| Facings PC £350/1000 (commons in centre) | hr | m2 | 1.39 | 1.39 | | |
| | £ | | 70.21 | 6.91 | 70.04 | 147.16 |
| Engineering PC £300/1000 | hr | m2 | 1.30 | 1.30 | | |
| | £ | | 65.66 | 6.46 | 78.68 | 150.80 |

IN SPECIFIED FORM

| | | UNIT | L | P | M | £ NETT |
|---|---|---|---|---|---|---|
| **Half brick thick walls built to battered face** | | | | | | |
| Commons PC £120/1000 | hr | m2 | 0.304 | 0.304 | | |
| | £ | | 15.36 | 1.51 | 8.56 | 25.43 |
| Facings PC £350/1000 | hr | m2 | 0.45 | 0.45 | | |
| | £ | | 22.73 | 2.24 | 22.72 | 47.69 |
| Engineering PC £300/1000 | hr | m2 | 0.416 | 0.416 | | |
| | £ | | 21.01 | 2.07 | 19.49 | 42.57 |
| **One brick thick walls built to battered face** | | | | | | |
| Commons PC £120/1000 | hr | m2 | 0.59 | 0.59 | | |
| | £ | | 29.80 | 2.93 | 17.17 | 49.90 |
| Facings PC £350/1000 | hr | m2 | 0.84 | 0.84 | | |
| | £ | | 42.43 | 4.17 | 45.53 | 92.13 |
| Engineering PC £300/1000 | hr | m2 | 0.78 | 0.78 | | |
| | £ | | 39.40 | 3.88 | 39.03 | 82.31 |

## SECTION 24 BRICKWORK BLOCKWORK AND STONEWORK

| | | UNIT | L | P | M | £ NETT |
|---|---|---|---|---|---|---|
| Two brick thick walls built to battered face | | | | | | |
| Commons PC £120/1000 | hr | m2 | 1.04 | 1.04 | | |
| | £ | | 52.53 | 5.17 | 34.85 | 92.55 |
| Facings PC £350/1000 | hr | m2 | 1.39 | 1.39 | | |
| | £ | | 70.21 | 6.91 | 70.04 | 147.16 |
| Engineering PC £300/1000 | hr | m2 | 1.30 | 1.30 | | |
| | £ | | 65.66 | 6.46 | 78.68 | 150.80 |
| Half brick thick vertical facing to concrete | | | | | | |
| Commons PC £120/1000 | hr | m2 | 0.262 | 0.262 | | |
| | £ | | 13.23 | 1.30 | 8.56 | 23.09 |
| Facings PC £350/1000 | hr | m2 | 0.368 | 0.368 | | |
| | £ | | 18.59 | 1.83 | 22.72 | 43.14 |
| Engineering PC £300/1000 | hr | m2 | 0.318 | 0.318 | | |
| | £ | | 16.06 | 1.58 | 19.49 | 37.13 |
| One brick thick vertical facing to concrete | | | | | | |
| Commons PC £120/1000 | hr | m2 | 0.51 | 0.51 | | |
| | £ | | 25.76 | 2.53 | 17.17 | 45.46 |
| Facings PC £350/1000 | hr | m2 | 0.66 | 0.66 | | |
| | £ | | 33.34 | 3.28 | 45.53 | 82.15 |
| Engineering PC £300/1000 | hr | m2 | 0.56 | 0.56 | | |
| | £ | | 28.29 | 2.78 | 39.03 | 70.10 |

## SECTION 24 BRICKWORK BLOCKWORK AND STONEWORK

| | | UNIT | L | P | M | £ NETT |
|---|---|---|---|---|---|---|
| **IN ARCHES** | | | | | | |
| Flat arches; 215mm high on face; facings PC £350/1000; | | | | | | |
| 102mm wide exposed soffit | hr | m | 0.21 | 0.21 | | |
| | £ | | 10.61 | 1.04 | 5.26 | 16.91 |
| 215mm wide exposed soffit | hr | m | 0.315 | 0.315 | | |
| | £ | | 15.91 | 1.57 | 10.40 | 27.88 |
| Flat arches; 215mm high on face; bullnosed specials PC £110/100; | | | | | | |
| 102mm wide exposed soffit | hr | m | 0.22 | 0.22 | | |
| | £ | | 11.11 | 1.09 | 15.69 | 27.89 |
| 215mm wide exposed soffit | hr | m | 0.33 | 0.33 | | |
| | £ | | 16.67 | 1.64 | 31.26 | 49.57 |
| Segmental arches; one ring; 215mm high on face; facings PC £350/1000; | | | | | | |
| 102mm wide exposed soffit | hr | m | 0.37 | 0.37 | | |
| | £ | | 18.69 | 1.84 | 7.75 | 28.28 |
| 215mm wide exposed soffit | hr | m | 0.55 | 0.55 | | |
| | £ | | 27.78 | 2.73 | 15.19 | 45.70 |
| Segmental arches; two ring; 215mm high on face; facings PC £350/1000; | | | | | | |
| 102mm wide exposed soffit | hr | m | 0.49 | 0.49 | | |
| | £ | | 24.75 | 2.44 | 7.75 | 34.94 |
| 215mm wide exposed soffit | hr | m | 0.74 | 0.74 | | |
| | £ | | 37.38 | 3.68 | 15.19 | 56.25 |
| Segmental arches; 215mm high on face; cut voussoirs PC £230/100; | | | | | | |
| 102mm wide exposed soffit | hr | m | 0.39 | 0.39 | | |
| | £ | | 19.70 | 1.94 | 32.49 | 54.13 |
| 215mm wide exposed soffit | hr | m | 0.58 | 0.58 | | |
| | £ | | 29.30 | 2.88 | 64.85 | 97.03 |

**SECTION 24 BRICKWORK BLOCKWORK AND STONEWORK**

| | | UNIT | L | P | M | £ NETT |
|---|---|---|---|---|---|---|
| Segmental arches; one and a half ring; 320mm high on face; cut voussoirs PC £230/100 | | | | | | |
| 102mm wide exposed soffit | hr | m | 0.55 | 0.55 | | |
| | £ | | 27.78 | 2.73 | 63.98 | 94.49 |
| 215mm wide exposed soffit | hr | m | 0.83 | 0.83 | | |
| | £ | | 41.92 | 4.13 | 127.52 | 173.57 |
| **IN ALTERATION WORKS** | | | | | | |
| Common bricks; including bonding ends to existing | | | | | | |
| Half brick thick | hr | m | 0.07 | 0.07 | | |
| | £ | | 3.54 | 0.35 | 0.68 | 4.57 |
| One brick thick | hr | m | 0.102 | 0.102 | | |
| | £ | | 5.15 | 0.51 | 1.31 | 6.97 |
| One and a half brick thick | hr | m | 0.153 | 0.153 | | |
| | £ | | 7.73 | 0.76 | 1.99 | 10.48 |
| Two brick thick | hr | m | 0.225 | 0.225 | | |
| | £ | | 11.36 | 1.12 | 2.62 | 15.10 |
| Engineering bricks; including bonding ends to existing | | | | | | |
| Half brick thick | hr | m | 0.08 | 0.08 | | |
| | £ | | 4.04 | 0.40 | 1.50 | 5.94 |
| One brick thick | hr | m | 0.112 | 0.112 | | |
| | £ | | 5.66 | 0.56 | 2.93 | 9.15 |
| One and a half brick thick | hr | m | 0.174 | 0.174 | | |
| | £ | | 8.79 | 0.86 | 4.42 | 14.07 |
| Two brick thick | hr | m | 0.245 | 0.245 | | |
| | £ | | 12.37 | 1.22 | 5.85 | 19.44 |

## SECTION 24 BRICKWORK BLOCKWORK AND STONEWORK

| | | | UNIT | L | P | M | £ NETT |
|---|---|---|---|---|---|---|---|
| **RUBBLE MASONRY** (based on 2.0 tonnes of stone per m3 of wall) | | | | | | | |
| Random rubble walling (PC £38.40 per tonne) | | | | | | | |
| Thickness of wall | 300mm | hr | m2 | 0.40 | 0.40 | | |
| | | £ | | 20.20 | 1.99 | 25.48 | 47.67 |
| | 450mm | hr | m2 | 0.55 | 0.55 | | |
| | | £ | | 27.78 | 2.73 | 38.23 | 68.74 |
| | 600mm | hr | m2 | 0.75 | 0.75 | | |
| | | £ | | 37.88 | 3.73 | 50.98 | 92.59 |
| Squared rubble walling (as a facing) (PC £48.00 per tonne) | | | | | | | |
| Thickness of facing | 150mm | hr | m2 | 0.35 | 0.35 | | |
| | | £ | | 17.68 | 1.74 | 15.93 | 35.35 |
| | 225mm | hr | m2 | 0.40 | 0.40 | | |
| | | £ | | 20.20 | 1.99 | 23.89 | 46.08 |
| | 300mm | hr | m2 | 0.45 | 0.45 | | |
| | | £ | | 22.73 | 2.24 | 31.85 | 56.82 |
| Drystone walling (PC £38.40 per tonne) | | | | | | | |
| Thickness of wall | 300mm | hr | m2 | 0.27 | 0.27 | | |
| | | £ | | 13.64 | 1.34 | 23.04 | 38.02 |
| | 450mm | hr | m2 | 0.35 | 0.35 | | |
| | | £ | | 17.68 | 1.74 | 34.60 | 54.02 |
| | 600mm | hr | m2 | 0.70 | 0.70 | | |
| | | £ | | 35.36 | 3.48 | 46.08 | 84.92 |

Apply the following multipliers to Labour and Plant for stone walls

| | |
|---|---|
| Height up to 2m | 1.00 |
| Height 2m to 5m | 1.21 |
| Height 5m to 10m | 1.37 |
| Wall to radius small | 1.75 |
| Wall to radius large | 1.50 |
| Wall to rake or batter | 1.15 |
| Wall in piers and stanchion | 1.50 |
| Wall in butteress | 1.15 |

## SECTION 27 ACCOMMODATION WORKS AND WORKS FOR STATUTORY UNDERTAKERS

Items for Accommodation Works are usually measured in accordance with the various sections 2 - 24.

The following items are those that may be specifically included in Bills of Quantities for Works for statutory undertakers.

| | UNIT | £ |
|---|---|---|
| **Demolitions** | | |
| Demolish to ground level and dispose off site (in small quantities) | | |
| - Brick walls | m3 | 12.00 |
| - Masonry walls | m3 | 13.25 |
| - Concrete walls | m3 | 18.00 |
| **Excavations** | | |
| Excavate Trial Holes in material other than rock, approx. size 1m x 2m | | |
| depth ne 1m | nr | 12.35 |
| depth 1 - 2m | nr | 23.96 |
| Excavate Trial Holes in material which includes rock approx. size 1m x 2m | | |
| depth ne 1m | nr | 14.27 |
| depth 1 - 2m | nr | 26.00 |
| Excavate trenches for pipes, ducts, conduits or services width up to 450mm | | |
| depth 0 - 300mm | m3 | 4.00 |
| 300 - 750mm | m3 | 4.20 |
| 750 - 1500mm | m3 | 4.50 |
| width up to 600mm | | |
| depth 0 - 300mm | m3 | 5.40 |
| 300 - 750mm | m3 | 5.70 |
| 750 - 1500mm | m3 | 6.00 |
| Excavate rock or Brickwork in trench | m3 | 33.00 |

## SECTION 27 ACCOMMODATION WORKS AND WORKS FOR STATUTORY UNDERTAKERS

**SPECIFIC ITEMS - continued**

| | UNIT | £ |
|---|---|---|
| Backfill trenches or pipe bays (in layers, including compaction) | | |
| with sand up to 750mm deep | m3 | 32.85 |
| with pea gravel up to 750mm deep | m3 | 59.20 |
| with excavated materials up to 750mm deep | m3 | 9.20 |
| concrete surround to ducts or pipes | m3 | 107.00 |
| **SERVICES** | | |
| Water | | |
| Water mains in uPVC to BS3505 in trench or conduit | | |
| 2" pipework with joints in the running length | m | 6.10 |
| 3" pipework with joints in the running length | m | 6.60 |
| joint to existing supply (PVC - PVC) | nr | 8.00 |
| joint to meter | nr | 6.00 |
| Water mains in ductile iron pipework to BS4772 in trench | | |
| 150mm pipework with joints in the running length | m | 21.50 |
| 90 degree bends | nr | 46.00 |
| 45 degree bends | nr | 40.00 |
| 250mm pipework with joints in the running length | m | 47.00 |
| 90 degree bends | nr | 148.00 |
| 45 degree bends | nr | 99.00 |
| joint to existing supply (DI - DI) | nr | 33.00 |
| Water standpipe fixed to wall inside locking box | nr | 70.00 |

## SECTION 27 ACCOMMODATION WORKS AND WORKS FOR STATUTORY UNDERTAKERS

### SPECIFIC ITEMS - continued

### SERVICES - continued

| | UNIT | £ |
|---|---|---|
| ELECTRIC | | |
| 600/1000V cables 2 core copper to BS6346 | | |
| 6mm2 laid in trench | m | 2.60 |
| 6mm2 drawn into duct | m | 2.75 |
| 16mm2 laid in trench | m | 3.00 |
| 16mm2 drawn into duct | m | 3.20 |
| 25mm2 laid in trench | m | 4.20 |
| 25mm2 drawn into duct | m | 4.43 |
| 600/1000V cables 4 core copper to BS6346 | | |
| 25mm2 laid in trench | m | 4.70 |
| 25mm2 drawn into duct | m | 4.95 |
| 70mm2 laid in trench | m | 7.70 |
| 70mm2 drawn into duct | m | 7.95 |
| PVC marker tape 150mm wide laid into trench | m | 0.31 |
| 300/500V cables PVC/copper flat twin earth | | |
| 2.5mm2 in conduit | m | 0.95 |
| 2.5mm2 externally | m | 0.90 |
| GAS - not included | | |
| CABLE DUCTS - Eg. BRITISH TELECOM | | |
| Vitrified clay (Hepduct) including draw rope flexible couplings | | |
| 100mm pipe in trench | m | 4.57 |
| bends | m | 2.09 |
| plugs | m | 1.10 |
| 150mm pipe in trench | m | 8.80 |
| bends | m | 4.50 |
| plugs | m | 1.75 |

**SECTION 27 ACCOMMODATION WORKS AND WORKS FOR STATUTORY UNDERTAKERS**

**SPECIFIC ITEMS - continued**

**SERVICES - continued**

| | UNIT | £ |
|---|---|---|
| uPVC ducts (Marley) including draw rope, ring seal sockets. | | |
| 82mm pipe in trench | m | 6.55 |
| 82mm pipe on supports | m | 6.85 |
| 110mm pipe in trench | m | 6.75 |
| 110mm pipe on supports | m | 7.10 |

CHAMBERS

| | | |
|---|---|---|
| Water meter chambers (430 x 280) | nr | 138.00 |
| Fire hydrant chambers (430 x 280) | nr | 167.00 |
| In-situ concrete B.T. joint box (600 x 600) | nr | 302.00 |
| In situ concrete draw pit (900 x 600) up to 1m deep with BT cover and frame | nr | 640.00 |
| Pre-cast concrete soakaway 900 diameter up to 2m deep | nr | 510.00 |
| Vitrified clay yard gully 225 diameter | nr | 90.00 |

## SECTION 27 ACCOMMODATION WORKS AND WORKS FOR STATUTORY UNDERTAKERS

### EQUIPMENT BASES AND SUPPORTS

| | UNIT | £ |
|---|---|---|
| Cabinet bases (for control cabinet feeder pillar equipment housing etc.) In-situ concrete 150mm deep | | |
| size 450 x 450 | nr | 51.50 |
| 600 x 600 | nr | 62.10 |
| 900 x 900 | nr | 82.00 |
| Message sign bases | | |
| 300 x 300 x 300 | nr | 57.50 |
| 900 x 600 x 300 | nr | 112.00 |
| CCTV bases | nr | 108.00 |
| Services gantry foundations/structure | | |
| Concrete substructure | m3 | 110.00 |
| Fabricate and erect steelwork Superstructure | tonne | 2365.00 |

PART 6

# Oncosts and Profit

*In Part 1 of this book, it is stressed that the cost information given in Parts 2, 3 and 4 leads to a cost estimate that requires further adjustment before it is submitted as a tender. This part deals with those adjustments and includes a worked example of a Tender Summary.*

## RISK/OPPORTUNITY

The factors to be taken into account when gauging the possibility of the Estimator's prediction of cost being inadequate or excessive are given in Part 2. Clearly it is considered in parallel with profit and it is not possible to give any indicative guidance on the level of adjustment that might result. For the purpose of a preliminary estimate, it is suggested that no adjustment is made to the costs generated by the other parts of this book.

At the same time as making a general appraisal of risk/opportunity, management will look at major quantities and may suggest amendments to the unit rates proposed.

## HEAD OFFICE OVERHEADS

An addition needs to be made to the nett estimate to cover all costs incurred in operating the central services provided by head office. Apart from general management and accountancy, this will normally include the departments dealing with:

- estimating
- planning and design
- purchasing
- surveying
- insurance
- wages and bonus
- site safety.

The appropriate addition varies with the extent of services provided centrally, rather than on site, and with the size of organisation, but a range of 4% to 8% on turnover would cover most circumstances.

Some organisations would include finance costs with head office overheads, as a general charge to the company, but for the purposes of this book finance costs are treated separately (see below).

## PROFIT

Obviously, the level of profit is governed by the degree of competition applicable to the job - which is in turn a function of the industry's current workload. Again, the appropriate addition is highly variable, but for the purposes of a preliminary estimate an addition of 2% to 5% onto nett turnover is suggested.

## FINANCE COSTS - ASSESSMENT OF CONTRACT INVESTMENT

The following procedure may be followed to give an indication of the average amount of capital investment required to finance the contract. It must be emphasised that this method will not give an accurate investment as this can only be done by preparing a detailed cash flow related to the programme for the contract. The example is based on the same theoretical contract used for the worked example in Part 2 and should be read in conjunction with the Tender Summary that follows.

**FINANCE COSTS - ASSESSMENT OF CONTRACT INVESTMENT**

The average monthly income must first be assessed. This is done by deducting from the Tender total the contingency items and those items for which immediate payment is necessary.

| | £ | £ |
|---|---|---|
| Tender total (excluding finance at £68 030 + 5% employer's contingencies) | | 5 850 500 |
| Deduct | | |
| Subcontractors | 1 070 000 | |
| Prime cost sums | 21 000 | |
| Employer's contingencies (excluding 5% on finance) | 283 090 | 1 374 090 |
| Amount to be financed | | £4 476 410 |

The average monthly income is this sum (£4 382 982) divided by the contract period (12 months), that is, £365 248

The average contract investment may now be calculated as follows :

| | £ | £ |
|---|---|---|
| Plant and equipment to be purchased | | 75 000 |
| Non time related | | |
| Contractor | 118 760 | |
| Employer | 5 850 | |
| Other services, charges and fees | NIL | |
| | (1) | |
| Subtotal | 124 610 | |
| Take 50% as an average | x50% | 62 305 |
| Stores and unfixed materials on site | | 10 000 |
| Work done but not paid for | | |
| 2.5 months at £338 455 (2) | 846 150 | |
| Less retention at 5% (3) | 42 300 | 803 850 |
| Retention (5% with limit of 3%) | | |
| Average retention (4) 3% of £4 476 410 | | 134 290 |
| | | 1 085 445 |
| Subtotal | | |
| Deduct | | |
| Advance payment by client | NIL | |
| Bill loading (5) | 150 000 | |
| Creditors (suppliers) | 360 000 | 510 000 |
| Average contract investment | | £575 445 |

The interest charges that must be added to the Tender price (or absorbed from profit if capital needs to be borrowed) are therefore :

£575 445 x say 14% x 1 year = say £ 80 562

**FINANCE COSTS - ASSESSMENT OF CONTRACT INVESTMENT**

Notes

(1) These non time related oncosts and services are incurred as lump sums during the period of the contract and, therefore, only 50% of such costs are taken for investment purposes.
(2) This period depends on the terms of payment set out in the contract.
(3) Retention is deducted as full retention is taken into account later.
(4) Average retention will depend on the retention condition set out in the contract, taking into account any partial completion dates.
(5) The contractor assesses here any financial advantage he may obtain by varying his spread on bill items.

**VALUE ADDED TAX**

All of the figures quoted in this book exclude value added tax, which the conditions of contract normally make the subject of a separate invoicing procedure between the contractor and the employer.

The reader's attention is also drawn to a ruling of the European Court on 21 June 1988. As from 1 April 1989, Value Added Tax will be chargeable at the standard rate, currently 15%, on supplies of services in the course of:

1. The construction of a non-domestic building
2. The construction or demolition of a civil engineering work
3. The demolition of any building, and
4. The approved alteration of a non-domestic protected building

**TENDER SUMMARY**

This summary sets out a suggested method of collecting together the various costs and other items and sums which, together, make up the total Tender sum for the example contract.

| | | |
|---|---|---|
| Preliminaries and General Items (from Part 2) | £ | £ |
| Contractor's site oncosts - time related | 359 804 | |
| Contractor's site oncosts - non time related | 117 111 | |
| Employer's requirements - time related | 59 350 | |
| Employer's requirements - non time related | 5 780 | |
| Other services, charges and fees | NIL | |
| Temporary works not included in unit costs | 34 505 | |
| Plant not included in unit costs | 130 701 | 707 251 |
| Estimated nett cost of measured work, priced at unit costs | | 2 843 547 |
| Subtotal | | 3 550 798 |
| Allowance for fixed price contract | £ | |
| 6% on labour (assumed to be £1 050 000) | 63 000 | |
| 4% on materials (assumed to be £1 900 000) | 76 000 | |
| 4% on plant (assumed to be £300 000) | 12 000 | |
| 5% on staff, overheads etc. (assumed to be £300 000) | 15 000 | 166 000 |
| Carried forward | | £3 716 798 |

**TENDER SUMMARY**

| | £ | £ |
|---|---|---|
| Brought forward | | 3 716 798 |
| Sub-contractors (nett) | | 1 070 000 |
| Prime cost sums | | 21 000 |
| Adjustments made at Management Appraisal | | |
| price adjustments, add say | 25 000 | |
| risk evaluation, add say | 10 000 | 35 000 |
| Subtotal | | 4 842 798 |
| Head office overheads and profit at 9% | | 435 852 |
| Finance costs (from page 557) | | 80 562 |
| Provisional Sums | | 65 000 |
| Dayworks Bill | | 65 000 |
| | | 5 489 212 |
| Subtotal | | |
| Employer's contingencies add 5% | | 274 461 |
| Tender total | | £5 763 673 |

PART 7

# Cost and Tender Price Indices

*The purpose of this part is to present historic changes in Civil Engineering costs and tender prices. It gives published and constructed indices and diagrammatic comparisons between building and Civil Engineering costs and tender prices and the retail price index, and will provide a basis for updating historical cost or tender price information.*

## INTRODUCTION

It is important to distinguish between costs and tender prices; Civil Engineering costs are the costs incurred by a Contractor in the course of his business; Civil Engineering tender prices are the prices for which a Contractor undertakes work. Tender prices will be based on Contractor's costs but will also take into account market considerations such as the availability of labour and materials and the prevailing workload for Civil Engineering Contractors. This can mean that in a period when work is scarce tender prices may fall as costs are rising while when there is plenty of work prices will tend to increase at a faster rate than costs. This section comprises published Civil Engineering cost and tender indices, a constructed Civil Engineering cost index and comparisons of these with building cost and tender indices and with the retail price index.

## PRICE ADJUSTMENT FORMULA INDICES

Cost indices for labour, plant and materials in Civil Engineering work are compiled and maintained by the Property Services Agency as Technical Secretariat to the Working Group. They are published in a monthly bulletin by HMSO and are reproduced here with their permission. These indices were formerly known as NEDO or 'Baxter' indices. They comprise 14 indices derived from Government sources. Two of them, DERV fuel (index 8) and light re-rolled bars and sections (index 11b), are little used and a further two, indices 12 and 13, represent materials and labour specifically for structural steelwork. Thus there are 10 current indices for general Civil Engineering work; the reference numbers and titles are listed below and quarterly values, with a base of 1970 = 100 unless otherwise stated, are tabulated overleaf:

| Index No | Title | Previous Title |
|---|---|---|
| 1 | Labour and supervision in Civil Engineering | Cost of labour in Civil Engineering construction |
| 2 | Plant and road vehicles: provision and maintenance | Cost of providing and maintaining constructional plant & equipment |
| 3 | Aggregates | Sand and gravel ex pit or works |
| 4 | Bricks and clay products | Engineering bricks delivered |
| 5 | Cements | Calcareous cement (other than clinker) delivered |
| 6 | Cast iron products | Cast & spun iron pipes & fittings |
| 7 | Coated roadstone for road pavement & bituminous products | Coated limestone roadstone |
| 9 | Gas oil fuel | |
| 10 | Timber | Imported softwood |
| 11a | Steel for reinforcement | Steel for reinforcement (cut, bent and delivered) |

*Quarterly values of price adjustment formula indices base:* 1970 = 100
Except for index 11a which has a base date of July 1976.
*Source* derived from HMSO

**PRICE ADJUSTMENT FORMULA INDICES**

| Year | Q | 1 | 2 | 3 | 4 | 5 | 6 | 7 | 9 | 10 | 11a |
|---|---|---|---|---|---|---|---|---|---|---|---|
| 1977 | 1 | 245 | 262 | 271 | 315 | 277 | 315 | 357 | 429 | 383 | 104 |
| | 2 | 250 | 272 | 295 | 324 | 289 | 331 | 372 | 461 | 388 | 106 |
| | 3 | 263 | 281 | 313 | 338 | 315 | 337 | 377 | 467 | 386 | 112 |
| | 4 | 264 | 288 | 330 | 339 | 315 | 338 | 384 | 467 | 363 | 116 |
| 1978 | 1 | 264 | 293 | 335 | 357 | 319 | 354 | 388 | 467 | 349 | 117 |
| | 2 | 269 | 300 | 346 | 374 | 335 | 362 | 399 | 467 | 347 | 120 |
| | 3 | 291 | 313 | 352 | 379 | 346 | 364 | 406 | 467 | 356 | 120 |
| | 4 | 294 | 318 | 366 | 383 | 346 | 385 | 415 | 467 | 362 | 126 |
| 1979 | 1 | 294 | 323 | 382 | 396 | 369 | 395 | 425 | 478 | 379 | 128 |
| | 2 | 297 | 331 | 402 | 401 | 384 | 428 | 458 | 548 | 399 | 137 |
| | 3 | 337 | 349 | 431 | 460 | 417 | 445 | 502 | 659 | 411 | 149 |
| | 4 | 340 | 357 | 451 | 476 | 434 | 449 | 528 | 666 | 424 | 149 |
| 1980 | 1 | 340 | 367 | 473 | 541 | 468 | 488 | 574 | 742 | 436 | 150 |
| | 2 | 342 | 374 | 502 | 557 | 538 | 491 | 619 | 796 | 460 | 153 |
| | 3 | 410 | 399 | 530 | 563 | 538 | 504 | 647 | 822 | 466 | 153 |
| | 4 | 411 | 402 | 534 | 563 | 538 | 508 | 665 | 822 | 467 | 153 |
| 1981 | 1 | 411 | 405 | 547 | 585 | 559 | 519 | 687 | 878 | 467 | 153 |
| | 2 | 412 | 410 | 577 | 592 | 601 | 522 | 729 | 903 | 459 | 153 |
| | 3 | 435 | 426 | 581 | 598 | 601 | 529 | 754 | 966 | 439 | 157 |
| | 4 | 451 | 435 | 613 | 623 | 601 | 552 | 768 | 1051 | 439 | 159 |
| 1982 | 1 | 455 | 442 | 633 | 656 | 646 | 563 | 788 | 1068 | 436 | 170 |
| | 2 | 455 | 446 | 655 | 611 | 646 | 587 | 802 | 1059 | 450 | 170 |
| | 3 | 480 | 452 | 667 | 668 | 646 | 587 | 823 | 1120 | 456 | 171 |
| | 4 | 480 | 454 | 685 | 686 | 646 | 591 | 856 | 1249 | 459 | 171 |
| 1983 | 1 | 474 | 455 | 700 | 720 | 646 | 593 | 863 | 1299 | 471 | 171 |
| | 2 | 481 | 462 | 731 | 723 | 646 | 587 | 879 | 1297 | 510 | 181 |
| | 3 | 507 | 472 | 735 | 730 | 646 | 595 | 906 | 1323 | 531 | 181 |
| | 4 | 507 | 473 | 755 | 753 | 646 | 621 | 911 | 1206 | 547 | 181 |
| 1984 | 1 | 507 | 478 | 766 | 789 | 646 | 624 | 933 | 1253 | 562 | 190 |
| | 2 | 509 | 483 | 785 | 797 | 646 | 670 | 978 | 1286 | 582 | 200 |
| | 3 | 533 | 494 | 783 | 806 | 646 | 676 | 988 | 1318 | 595 | 202 |
| | 4 | 529 | 499 | 792 | 837 | 646 | 729 | 1011 | 1451 | 608 | 205 |
| 1985 | 1 | 530 | 507 | 803 | 849 | 646 | 739 | 1048 | 1622 | 638 | 215 |
| | 2 | 532 | 512 | 830 | 861 | 655 | 787 | 1091 | 1516 | 636 | 218 |
| | 3 | 557 | 524 | 840 | 900 | 674 | 785 | 1095 | 1561 | 612 | 218 |
| | 4 | 557 | 526 | 243 | 907 | 674 | 783 | 1085 | 1501 | 592 | 218 |
| 1986 | 1 | 558 | 531 | 865 | 943 | 674 | 777 | 1048 | 1197 | 580 | 217 |
| | 2 | 559 | 534 | 885 | 943 | 674 | 785 | 1020 | 1102 | 590 | 218 |
| | 3 | 558 | 546 | 885 | 972 | 674 | 785 | 991 | 976 | 609 | 210 |
| | 4 | 588 | 547 | 881 | 972 | 674 | 783 | 1000 | 1100 | 641 | 207 |
| 1987 | 1 | 588 | 557 | 904 | 973 | 674 | 783 | 1018 | 1108 | 645 | 207 |
| | 2 | 584 | 559 | 929 | 1022 | 674 | 787 | 1047 | 1120 | 647 | 206 |
| | 3 | 618 | 572 | 924 | 1022 | 662 | 787 | 1066 | 1205 | 661 | 206 |
| | 4 | 618 | 573 | 939 | 1025 | 662 | 811 | 1073 | 1221 | 666 | 206 |
| 1988 | 1 | 619 | 578 | 987 | 1047 | 662 | 811 | 1074 | 1116 | 676 | 210 |
| | 2 | 623 | 590 | 1002 | 1071 | 662 | 821 | 1076 | 1125 | 678 | 231 |
| | 3 | 655 | 602 | 1012 | 1091 | 702 | 821 | 1081 | 1127 | 687 | 253 |
| | 4 | 655 | 606 | 1002 | 1091 | 702 | 853 | 1098 | 1134 | 701 | 253 |
| 1989 | 1 | 655 | 612 | 1031 | 1157 | 917 | 855 | 1052 | 1105 | 714 | 253 |
| | 2 | 651 | 618 | 1090 | 1166 | 750 | 861 | 1099 | 1140 | 727 | 266 |
| | 3 | 704 | 637 | 1089 | 1166 | 750 | 860 | 1134 | 1149 | 755 | 266 |
| | 4 | 705 | 640 | 1097 | 1166 | 750 | 874 | 1153 | 1323 | 782 | 264 |
| 1990 | 1 | 705* | 647* | 1114* | 1166* | 763 | 876 | 1193 | 1416 | 806 | 263 |
| | 2 | | | | | | | | | | |
| | 3 | | | | | | | | | | |
| | 4 | | | | | | | | | | |

* Provisional

**A CONSTRUCTED COST INDEX BASED ON THE PRICE ADJUSTMENT FORMULA INDICES**

Although the above indices are prepared and published in order to provide a common basis for calculating reimbursement of increased costs during the course of a contract, they also present time series of cost indices for the main components of Civil Engineering work. They can therefore be used as the basis of an index for Civil Engineering work. The method used here is to construct a composite index by allocating weightings to each of the 10 indices, the weightings being established from an analysis of actual projects. The composite index is calculated by applying these weightings to the appropriate price adjustment formula indices and totalling the results; this index is again presented with a base date of 1970.

*Constructed Civil Engineering Cost Index* base: 1970 = 100

| *Year* | *First quarter* | *Second quarter* | *Third quarter* | *Fourth quarter* | *Annual average* |
|---|---|---|---|---|---|
| 1975 | 198 | 209 | 220 | 227 | 213 |
| 1976 | 241 | 250 | 269 | 279 | 260 |
| 1977 | 291 | 303 | 314 | 319 | 307 |
| 1978 | 322 | 330 | 341 | 348 | 335 |
| 1979 | 357 | 373 | 409 | 420 | 390 |
| 1980 | 440 | 462 | 494 | 498 | 474 |
| 1981 | 509 | 524 | 541 | 558 | 533 |
| 1982 | 572 | 578 | 594 | 608 | 588 |
| 1983 | 613 | 625 | 640 | 639 | 629 |
| 1984 | 649 | 663 | 676 | 671 | 665 |
| 1985 | 690 | 698 | 712 | 710 | 703 |
| 1986 | 698 | 689 | 685 | 693 | 691 |
| 1987 | 701 | 709 | 728 | 730 | 717 |
| 1988 | 736 | 743 | 760 | 763 | 751 |
| 1989 | 767 | 786 | 809* | 825* | 797* |

*Source*: DL&E based on HMSO figures.

The figure on page 164 illustrates graphically the movement of the constructed index, the DL&E index of building costs and the retail price index.

**A CONSTRUCTED COST INDEX BASED ON THE PRICE ADJUSTMENT FORMULA INDICES**

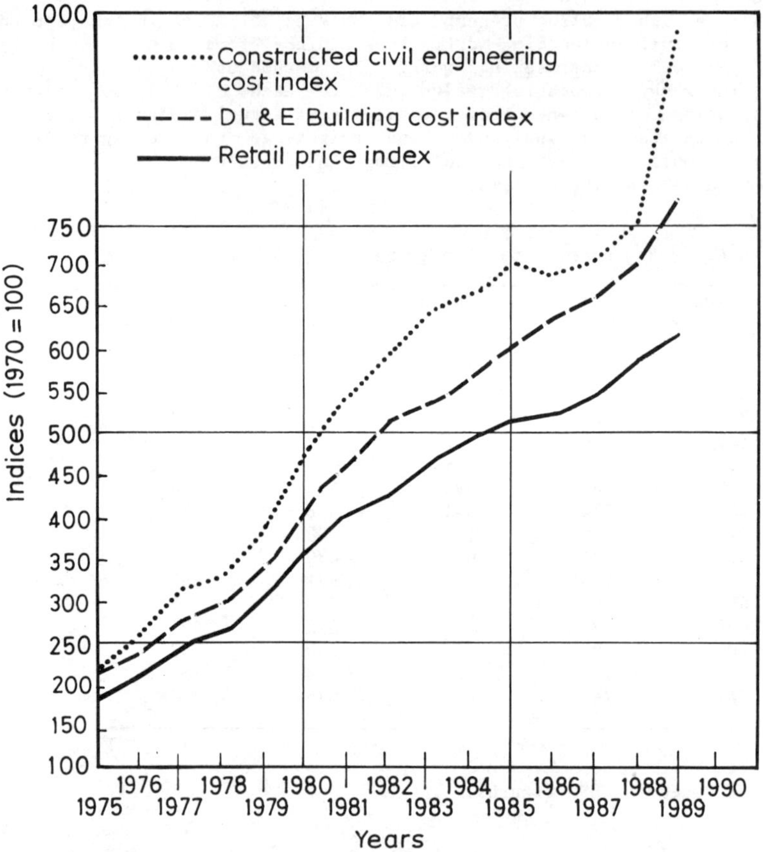

**THE ROAD CONSTRUCTION PRICE INDEX**

Civil Engineering work generally does not lend itself easily to the preparation of tender price indices in the same way as building work. There is, however, a published index for road construction tender prices (available on subscription from the Department of the Environment) and this is reproduced below with the permission of HMSO. The index is produced by the Department of the Environment and is intended to indicate the movement in tender prices for road construction contracts. It is currently based on contracts of £250 000 or more and is published with a base of 1980 = 100.

*Road Construction Price Index* base: 1980 = 100

| *Year* | *First quarter* | *Second quarter* | *Third quarter* | *Fourth quarter* | *Annual average* |
|---|---|---|---|---|---|
| 1970 | | | | | 21 |
| 1971 | 22 | 22 | 23 | 23 | 22 |
| 1972 | 23 | 23 | 24 | 24 | 24 |
| 1973 | 26 | 27 | 29 | 33 | 29 |
| 1974 | 35 | 38 | 41 | 44 | 39 |
| 1975 | 45 | 45 | 44 | 42 | 44 |
| 1976 | 42 | 41 | 41 | 42 | 42 |
| 1977 | 46 | 48 | 51 | 54 | 50 |
| 1978 | 58 | 61 | 62 | 65 | 62 |
| 1979 | 68 | 73 | 77 | 89 | 77 |
| 1980 | 97 | 102 | 103 | 98 | 100 |
| 1981 | 95 | 92 | 90 | 88 | 91 |
| 1982 | 89 | 92 | 96 | 98 | 94 |
| 1983 | 89 | 96 | 95 | 96 | 97 |
| 1984 | 96 | 97 | 97 | 101 | 98 |
| 1985 | 98 | 100 | 99 | 102 | 100 |
| 1986 | 97 | 93 | 94 | 95 | 95 |
| 1987 | 101 | 106 | 107 | 118 | 109 |
| 1988 | 110 | 114 | 117 | 116 | 114 |
| 1989 | 150 | 125* | 129* | | |

Note: * Provisional
Source: HMSO

The figure on page 166 illustrates graphically the movement of the road construction price index, the DL&E building tender price index and the retail price index.

**THE ROAD CONSTRUCTION PRICE INDEX**

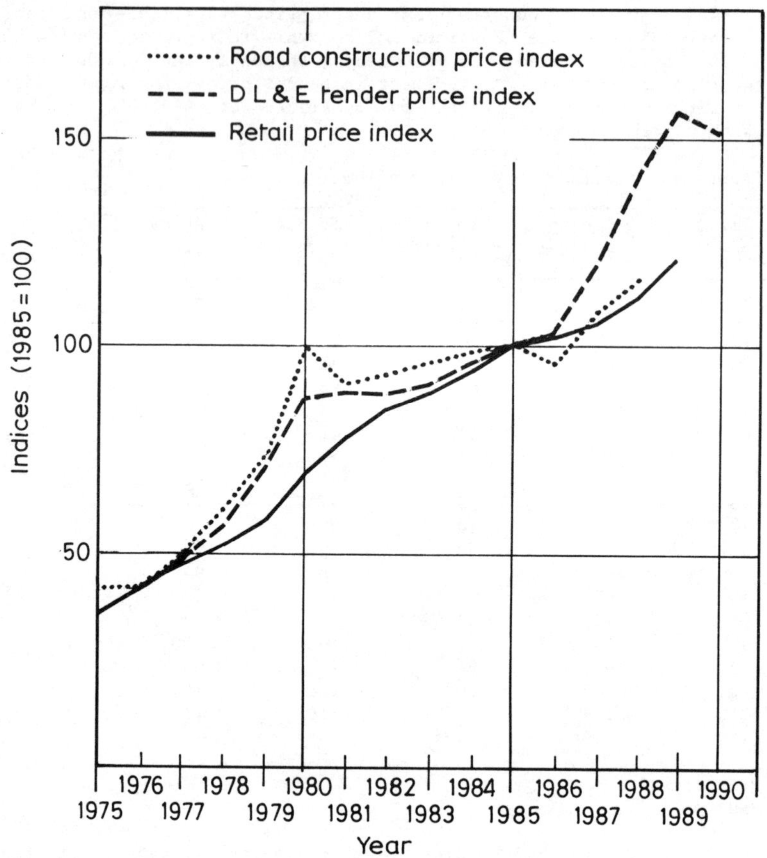

PART 8

# Daywork

## Keep your figures up to date, free of charge

This section, and most of the other information in this Price Book, is brought up to date every three months in the *Price Book Update.*

The *Update* is available free to all Price Book purchasers.

To ensure you receive your copy, simply complete the reply card from the centre of the book and return it to us.

CIVIL ENGINEERING INDUSTRY

**DAYWORK IN CONNECTION WITH CIVIL ENGINEERING WORKS**

**INTRODUCTION**

Daywork ordered by the Engineer pursuant to clause 52(3) of the ICE Conditions of Contract is normally paid for at the rates and prices and under the conditions contained in the "Schedule of Dayworks carried out incidental to contract work" issued by the Federation of Civil Engineering Contractors, current at the date of execution of the Daywork.

The most recent schedules are those dated 22 January 1990 and are published by the Federation of Civil Engineering Contractors. Copies can be obtained from:

THE FEDERATION OF CIVIL ENGINEERING CONTRACTORS.
Cowdray House
6 Portugal Street
London
WC2A 2HH

These schedules identify a number of items that are excluded from the rates and percentages quoted, but should be recovered by the Contractor when valuing his Daywork account. Under normal circumstances it is impractical to accurately value these items for each Daywork event, although certain items can be allowed for by means of a percentage addition. Suggested methods of calculating such additions are includedin this section.

In Summary the Daywork schedule allows for an addition:-

of 148% to wages paid to workmen

of 88% to labour only sub-contractors and hired plant drivers

of 12.5% to subsistence allowances and travel paid to workmen

of 12.5% to materials used on dayworks

of 12.5% to full cost of plant hired for dayworks

of 12.5% to cost of operating welfare facilities.

Part 3 of this book (Resources - Plant) details references to the plant section if the 1990 daywork schedule.

CIVIL ENGINEERING INDUSTRY

1. **GENERAL SERVICING OF PLANT**

Schedule 3(1) specifically excludes time spent on general servicing from the Hire Rates.

General servicing in this context can be assumed to mean:

Checking, replenishing (or changing, if applicable)

Engine lubrication
Transmission lubrications
General greasing
Coolants
Hydraulic oils and brake systems
Filters
Tyres

Inspecting special items

eg. Buckets
Blades
Hoses/airlines
Shank protecters
Cables/ropes/hawsers
Rippers
Blades, steels, etc

Assuming an 8 hour working day these operations would take an operator on average:

20 minutes for a large machine per day
(equates to 1 hour for each 24 worked)

10 minutes for a medium sized machine per day
(equates to 1 hour for each 48 worked)

5 minutes for a small machine
(equates to 1 hour for each 96 worked)

These times expressed as a percentage of the *schedule hire rates*

Examples:

| | |
|---|---|
| D8 Dozer | = 8.92 - (24 x 86.25) = **0.5%** |
| JCB 3CX | = 7.95 - (48 x 15.36) = **1.1%** |
| 1 tonne dumper | = 6.20 - (96 x 5.82) = **1.1%** |

Considering the range of these sizes of plant which are normally deployed on site, the following would provide a reasonable average percentage addition

0.5 x 2 = 1.0
1.1 x 3 = 3.3
1.1 x 6 = 6.6
10.9
- 11
1.0 %

Information from Plant Hire firms

CIVIL ENGINEERING INDUSTRY

2. **FUEL DISTRIBUTION**

Schedule 3 (2) allows for charging for fuel distribution.

i. A heavy item of plant, for example a Cat D8N Tractor Bulldozer with a 212kW engine, is quoted as using 5 gallons of fuel (Gas Oil) per working hour, as a maximum fuel consumption. A site utilisation rate for this machine; ie the percentage of time the machine operates at its maximum fuel consumption, is 75%.

Therefore, the machine average is 5 x 0.75 = 3.75 gallons per hour

Fuel tank capacity of this machine is 45 gallons, so it would require filling after 12 hours operating (45 - 3.75).

Assume this is done with a towed fuel bowser behind a farm type tractor with driver/labourer in attendance, the operation would take 30 minutes to visit, service, and return or continue onto the next machine.

Machine: hourly rate (ref 32/7) = £86.25 x 12 = £1035.00

divided into

Attendance: hourly rate tractor/bowser/driver
= £10.09 + £1.16 + £6.20 = £17.45 x 0.5 = £8.73

So percentage addition for fuelling the machine = **0.9%**

ii. A medium sized item of plant, a JCB 3CX using 25 gallons per hour maximum site utilisation at 75% has an average consumption therefore of 1.88 gallons per hour.

Fuel capacity is 20 gallons

Requires filling after 10.6 working hours

Tractor/bowser service taking 10 minutes

Machine: hourly rate (ref 10/37)
£15.36 x £10.6 = £162.82

divided into

Attendance: hourly rate tractor/bowser
£17.45 x 0.17 = £2.97

So percentage addition for fuelling the machine = **1.8%**

CIVIL ENGINEERING INDUSTRY

2. **FUEL DISTRIBUTION**

iii. A small item of plant, a 2 tonne dumper, using 1.75 gallons per hour maximum, site utilisation at 80%, average consumption then is 1.40 gallons per hour.

Fuel capacity is 8 gallons

Requires refuelling each 5.75 working hours

Tractor/bowser service taking 5 minutes

Machine: hourly rate (ref 9/5)
£5.82 x 5.75 = £33.47

divided into

Attendance: hourly rate tractor/bowser
£17.45 x 0.08 = £1.40

So percentage additional for fuelling the machine = **4.2%**

Considering the range of these categories of plant which are normally deployed on site, the following would provide a reasonable average percentage addition

| | | | |
|---|---|---|---|
| **0.9 x 2** | = | **1.8** |
| **1.8 x 3** | = | **5.4** |
| **4.2 x 6** | = | **25.2** |
| | | 32.4 |
| | - | 11 |
| | | 3.0% |

CIVIL ENGINEERING INDUSTRY

### 3. WELFARE FACILITIES

Schedule 4 (5) allows for the net cost of operating these facilities plus 12.5%

This is an example of how these costs may be charged to the Daywork account on an 18 month, £2.5m Civil Engineering scheme, where the total labour element is £200,000

| Facility | | Weekly Cost |
|---|---|---|
| | | £ |
| Jack Leg toilet unit (2) | | 27.40 |
| Wheeled cabin 22' | | 11.24 |
| Wheeled cabin 12' | | 6.40 |
| Labour to clean, maintain and make tea, etc | | |
| 1 man, 2 hours per day, 6 days x £6.20 | | 74.40 |
| Consumables (heat, light, soap, disinfectant, etc) | say | 25.00 |
| Rates, insurance, taxes, etc (add 2%) | | 2.90 |
| Total weekly cost | | 147.34 |
| x 80 weeks (construction period) | | |
| plus | | |
| 6 weeks (maintenance period) | | |
| Total cost to contract | | 12671.24 |

| | | |
|---|---|---|
| This cost as a percentage of the Labour element of the contract = $\frac{12,671}{200,000}$ | = | 6.3% |
| add, as schedule 4 (5) 12.5% | = | 0.8% |
| **Percentage addition for facilities** | = | 7.2% |

CIVIL ENGINEERING INDUSTRY

4. **HANDLING AND OFFLOADING MATERIALS**

Schedule 2 (3) states that an allowance for handling materials, and an allowance for unloading or storage including wastage should be an additional charge.

Example, for an 18 month, £2.5m Civil Engineering scheme where the total cost of materials that require handling (excluding Ready Mix concrete imported fills fuels and similar items is £

The following gang is employed (part time) throughout the contract for offloading and handling of materials.

| | (net cost) £ |
|---|---|
| 2 labourers | 12.40 |
| Lorry (8T) with Hiab lift | 11.52 |
| | £23.92/hour |

allow an average of 5 hours per week over 80 weeks

Total cost = 400 x £23.92 = £9,568.00

This cost as a percentage of the materials element of the contract

$= \frac{9,568}{375,000} = 2.5\%$

Material waste should be added direct to the cost of materials used for each particular Daywork items as an appropriate percentage.

# SPON'S
# CONTRACTORS' HANDBOOKS
# ORDER FORM

## Please send me:

____ copy/ies of **MINOR WORKS, ALTERATIONS, REPAIRS AND MAINTENANCE** 4th Edition @ £17.95

____ copy/ies of **ROOFING** 3rd Edition @ £16.95

____ copy/ies of **PAINTING AND DECORATING** 4th Edition @ £15.95

____ copy/ies of **FLOOR, WALL AND CEILING FINISHINGS** 3rd Edition @ £17.95

____ copy/ies of **PLUMBING AND DOMESTIC HEATING** 3rd Edition @£16.95

____ copy/ies of **ELECTRICAL INSTALLATION** 2nd Edition @ £16.95

## Please tick as appropriate:

☐ I enclose a cheque/PO for £________ payable to E & F N Spon

☐ Please debit my Access/Diners/Visa/AmEx card

Card Number______________________________

Expiry Date_______________________________

☐ I enclose an official order, please send invoice with books

Name____________________________________________________________

Address__________________________________________________________

____________________________________________Postcode______________

Signature_______________________________________Date_______________

Return to: ***The Promotion Dept., E & F N Spon, 2-6 Boundary Row, London SE1 8HN***

PART 9

# Professional Fees

CONSULTING ENGINEERS' FEES

## INTRODUCTION

The following scale of professional charges for consulting engineering services in connection with civil engineering works is as published by the Association of Consulting Engineers in 1981, 1983 printing and incorporating 1987 amendments.

## AGREEMENT 1.

### Introduction

The following paragraphs describe the scope of professional services provided by the Consulting Engineer and give general advice about his appointment.

The Association of Consulting Engineers has drawn up standard Conditions of Engagement to form the basis of the agreement between the Client and the Consulting Engineer, for five different types of appointment as hereafter described. Each of the standard Conditions of Engagement is accompanied by a recommended Memorandum of Agreement. These two documents, taken together, constitute the recommended form of Agreement in each case.

When the standard Conditions of Engagement are not so used, it is in the interests of both the Client and the Consulting Engineer that there should be an exchange of letters defining the duties which the Consulting Engineer is to perform and the terms of payment.

### Report and Advisory Work

For reports, and for advisory work, the services required from the Consulting Engineer will usually comprise one or more of the following:

(a) investigating and advising on a project and submitting a report thereon. The Consulting Engineer may be asked to examine alternatives; review all technical aspects; make an economic appraisal of costs and benefits; draw conclusions and make recommendations.

(b) inspecting existing works (e.g. a reservoir or a building or an installation) and reporting thereon. If the Client requires continuing advice on maintenance or operation of an existing project, the Consulting Engineer may be appointed to make periodic visits.

(c) making a special investigation of an engineering problem and reporting thereon.

(d) making valuations of plant and undertakings.

Payment for the services provided under Items (a) to (d) above should normally be on a time basis or, where the duration and extent of the services can be defined clearly, by an agreed lump sum. It is recommended that the conditions governing an appointment for these services should be based on:

**Agreement 1** - Conditions of Engagement for Report and Advisory Work.

CONSULTING ENGINEERS' FEES

**AGREEMENT 1.**

**Design and Supervision of Construction**

When the Client has decided to proceed with the construction of engineering works or the installation of plant, it is normal practice for the Consulting Engineer who prepared or assisted in the preparation of any report to be appointed for the subsequent stages of the work or the relevant part thereof. The Consulting Engineer will assist the Client in obtaining the requisite approvals, then prepare the designs and tender documents to enable competitive tenders to be obtained or orders to be placed, and will be responsible for the technical control and administration of the construction of the Works.

It is recommended that the agreement for this type of appointment should be based on the most appropriate of the following:

**Agreement 2** Conditions of Engagement for Civil, Mechanical and Electrical Work and for Structural Engineering Work where an Architect is not appointed by the Client.

**Agreement 3** Conditions of Engagement for Structural Engineering Work where an Architect is appointed by the Client.

**Agreement 4A** Conditions of Engagement for Engineering Services in relation to Sub-contract Works. For convenience Agreement 4A is published in three separate books

(i) Full Duties
(ii) Abridged Duties
(iii) Performance Duties

**Agreement 4B** Conditions of Engagement for Engineering Services in relation to Director Contract Works.

**Terms of Payment**

The Association considers that, in normal circumstances, the level of remuneration represented by the scales of percentage fees and hourly charging rates set out in the ACE Conditions of Engagement is such as to ensure the provision by a Consulting Engineer to his Client of a full, competent and reliable standard of service.

These scales and rates are not mandatory but solely guidelines, and may by negotiation be adjusted upwards or downwards to take account of the abnormal complexity or simplicity of design, increase or diminution of the extent of services to be provided, long-standing client relationships or other circumstances.

However, the Association strongly advised clients to satisfy themselves that the level and quality of services they want will be covered if they make appointments based on charges which are appreciably lower than those shown.

CONSULTING ENGINEERS' FEES

**AGREEMENT 1.**

### Direct Labour Works

The ACE Conditions of Engagement require to be modified and supplemented when the Client intends to have the Works constructed wholly or partly by direct labour under the control of the Consulting Engineer.

The additional services and substantially greater responsibilities undertaken by the Consulting Engineer in connection with work carried out by direct labour usually entitle him to a higher level of remuneration than that payable in respect of Works carried out by contract.

### Partial Services

When the Client wishes to appoint the Consulting Engineer for partial services only, it is important that both parties recognise the limitation which such an appointment places upon the responsibility of the Consulting Engineer who cannot be held liable for matters that are outside his control.

The terms of references for the appointment should be carefully drawn up and the relevant ACE Conditions of Engagement should be adapted to suit the scope of services required.

Professional charges for partial services are usually best calculated on a time basis, but may, in suitable cases, be a commensurate part of the percentage fee for normal services shown in the standard Conditions of Engagement.

### Inspection Services

The inspection of materials and plant during manufacture or on site is usually required during the construction stage of a project. Consequently, the arrangements for this service are described in the standard Conditions of Engagement. If, however, the Client wishes to engage the Consulting Engineer to provide only inspection services, the charges of the Consulting Engineer may be either on a time basis or a percentage of the cost of the materials to be inspected.

### Acting as Arbitrator, Umpire or Expert Witness

A consulting Engineer may be appointed to act as Arbitrator or Umpire, or be required to attend as an Expert Witness at Parliamentary Committees, Courts of Law, Arbitrations or Official Inquiries.

Payment for any of these services should be on the basis of a lump sum retainer plus time charges, not less than three hours per day being chargeable for attendance, however short, either before or after a mid-day adjournment.

CONSULTING ENGINEERS' FEES

**AGREEMENT 1.**

**Payment on a Time Basis**

When it is not possible to estimate in advance the duration and extent of the Consulting Engineer's services, neither a lump sum payment alone nor a percentage of the estimated construction cost would normally be a fair basis of remuneration. The most satisfactory and equitable method of payment in these cases makes allowance for the actual time occupied in providing the services required, and comprises the following elements, as applicable:

(1) A charge in the form of hourly rate(s) for the services of a Partner or a Consultant of the firm. The hourly rate will depend upon his standing, the nature of the work and any special circumstances. Alternatively a lump sum fee may be charged instead of the said hourly rate(s).

(2) A charge which covers technical and supporting staff salary and other payroll costs actually incurred by the Consulting Engineer, together with a fair proportion of his overhead costs, plus an element of profit. This charge is most conveniently calculated by applying a multiplier to the salary cost and then adding the net amount of other payroll costs. The major part of the multiplier is attributable to the Consulting Engineer's overheads which may include, inter alia, the following indirect costs and expenses:

    (a) rent, rates and other expenses of upkeep of his office, its furnishings, equipment and supplies.
    (b) insurance premiums other than those recovered in the payroll costs;
    (c) administrative, accounting, secretarial and financing costs;
    (d) the expense of keeping abreast of advances in engineering;
    (e) the expense of preliminary arrangements for new or prospective projects;
    (f) loss of productive time of technical staff between assignments.

(3) A charge for use of a computer or other special equipment.

When calculating amounts chargeable on a time basis, a Consulting Engineer is entitled to include time spent by Partners, Consultants and technical and supporting staff in travelling in connection with the performance of the services. The time spent by secretarial staff or by staff engaged on general accountancy or administration duties in the Consulting Engineer's office is not chargeable unless otherwise agreed.

**Disbursements and Out of Pocket Expenses**

In addition to the charges and percentage fees referred to in the preceding Sections, the Consulting Engineer is entitled to recover from the Client all disbursements and out of pocket expenses incurred in performing his services.

CONSULTING ENGINEERS' FEES

**AGREEMENT 1.**

**Appointments Outside the United Kingdom**

The standard ACE Conditions of Engagement are suitable for appointments in the United Kingdom. For work overseas, it is impracticable to make definite recommendations as the conditions vary widely from country to country. There are added complications in documentation relating to import customs, conditions of payment, insurance, freight, etc. Furthermore, it is necessary to arrange for site visits to be undertaken by partners or senior staff whose absence abroad during such periods represents a serious reduction of their earning power.

The additional duties, responsibilities and non-recoverable costs involved, and the extra work on general co-ordination, therefore justify higher fees in such cases. Special arrangements are also necessary to cover travelling and other out of pocket expenses in excess of those normally incurred on similar work in the United Kingdom - including such matters as local cost-of-living allowances and the cost of providing home-leave facilities to expatriate staff.

CONSULTING ENGINEERS' FEES

**AGREEMENT 2.**

1. **Definitions**

In construing this Agreement the following expressions shall have the meanings hereby assigned to them except where the context otherwise requires:

'The Consulting Engineer' means the person or firm named in the Memorandum of Agreement and shall include any other person or persons taken into partnership by such person or firm during the currency of this Agreement and the surviving member or members of any such partnership.

'The Works' means the Works in connection with which the Client has engaged the Consulting Engineer to perform professional services.

'Contractor' means any person or persons, firm or company under contract to the Client to perform work and/or supply goods in connection with the Works.

'Salary Cost' means the annual salary and other emoluments paid by the Consulting Engineer to any person employed by him divided by 1600 (being deemed to be the average annual total of effective working hours of an employee) and multiplied by the number of working hours spent by such person in performing any of the services in respect of which payment under this Agreement is to be made to the Consulting Engineer upon the basis of Salary Cost. For the Purposes of this definition the annual salary of a person employed by the Consulting Engineer for a period less than a full year shall be calculated pro rata to such person's salary and other emoluments for such lesser period.

'Other Payroll Cost' means the annual amount of all contributions and payments made by the Consulting Engineer on behalf of or in respect of a person employed by him for staff pension and life assurance schemes, and also for National Insurance Contributions and for any other tax, charge, levy impost or payment of any kind whatsoever which the Consulting Engineer at any time during the performance of this Agreement is obliged by law to make on behalf of or in respect of such person, divided by 1600 (being deemed to be the average annual total of effective working hours of an employee) and multiplied by the number of working hours spent by such person in performing any of the services in respect of which payment under this Agreement is to be made to the Consulting Engineer upon the basis of Other Payroll Cost. For the purposes of this definition the annual amount of all contributions and payments made by the Consulting Engineer on behalf of or in respect of a person employed by him for a period less than a full year shall be calculated pro rata to the amount of such contributions and payments for such lesser period.

Words printed in italics are explanatory of the terms of these conditions and do not form part thereof.

Words importing the singular include the plural and vice versa where the context requires.

CONSULTING ENGINEERS' FEES

**AGREEMENT 2.**

2. **Duration of Engagement**

2.1. The appointment of the Consulting Engineer shall commence from the date stated in the Memorandum of Agreement or from the time when the Consulting Engineer shall have begun to perform for the Client any of the services specified in Clauses 6 and 7 of this Agreement, whichever is the earlier.

2.2. The Consulting Engineer shall not, without the consent of the Client, assign the benefit or in any way transfer the obligations of this Agreement or any part thereof.

2.3. If at any time the Client decides to postpone or abandon the Works, he may thereupon by notice in writing to the Consulting Engineer forthwith terminate this Agreement, provided that, in any case in which the Consulting Engineer is paid for his services under Clause 6 otherwise than in accordance with Clause 10.2 the Client may, in lieu of so terminating, by notice in writing to the Consulting Engineer require him to suspend the carrying out of his services under this Agreement for the time being.

2.4. If at any time the Client decides to postpone or abandon any part of the Works, he may thereupon by notice in writing to the Consulting Engineer seek to vary this Agreement either by excluding the services to be performed by the Consulting Engineer in relation to such part of the Works, or by suspending performance of the same (unless the Consulting Engineer is to be paid for his services under Clause 6 in accordance with Clause 10.2) and in such notice, the Client shall specify the services affected. The Consulting Engineer shall forthwith comply with the Client's notice and the Client shall pay to the Consulting Engineer a sum calculated in accordance with the provisions of Clause 17.2 in respect of such compliance.

2.5. If the Client shall not have required the Consulting Engineer to resume the performance of services in respect of the whole or any part of the Works suspended under Clause 2.3 or Clause 2.4 hereof within a period of 12 months from the date of the Client's notice:

(i) In the case of a suspension under Clause 2.3 this Agreement shall forthwith automatically terminate; or

(ii) In the case of a suspension under Clause 2.4 the suspended services shall be deemed to have been excluded from the services to be performed by the Consulting Engineer under this Agreement.

CONSULTING ENGINEERS' FEES

**AGREEMENT 2.**

2. **Duration of Engagement** - *continued*

2.6. In the event of the failure of the Client to comply with any of his obligations under this Agreement, or upon the occurrence of any circumstances beyond the control of the Consulting Engineer which are such as to delay for a period of more than 12 months or prevent or unreasonably impede the carrying out by the Consulting Engineer of his services under this Agreement, the Consulting Engineer may upon not less than 60 days' notice in writing to the Client terminate his appointment, under this Agreement, provided that in lieu of so terminating his appointment the Consulting Engineer may:

(a) forthwith upon any such failure or the occurrence of any such circumstances suspend the carrying out of his services hereunder for a period of 60 days (provided that he shall as soon as practicable inform the Client in writing of such suspension and the reasons therefore), and
(b) at the expiry of such period of suspension either continue with the carrying out of his services under this Agreement or else, if any of the reasons for the suspension then remain, forthwith in writing to the Client terminate his appointment under this Agreement.

2.7. The Consulting Engineer shall, upon receipt of any notice in accordance with Clause 2.3 or in the event of the termination by him of his appointment in pursuance of Clause 2.6, proceed in an orderly manner but with all reasonable speed and economy to take such steps as are necessary to bring to an end his services under this Agreement.

2.8. Unless terminated under this Clause the Consulting Engineer's appointment under this Agreement shall terminate when the Consulting Engineer shall make and issue the certificate authorising the final payment to the Contractor.

2.9. Any termination of the Consulting Engineer's appointment under this Agreement shall not prejudice or affect the accrued rights or claims of either party to this Agreement.

CONSULTING ENGINEERS' FEES

**AGREEMENT 2.**

3. **Ownership of Documents and Copyright**

3.1. The copyright in all drawings, reports, specifications, bills of quantities, calculations and other documents provided by the Consulting Engineer in connection with the Works shall remain vested in the Consulting Engineer, but the Client shall have a licence to use such drawings and other documents for the purpose of constructing the Works. Save as aforesaid, the Client shall not make copies of such drawings or other documents nor shall he use the same in connection with the making or improvement of any works other than those comprised in the Works without the prior written approval of the Consulting Engineer and upon such terms as may be agreed between the Client and the Consulting Engineer.

3.2. The Consulting Engineer may with the consent of the Client, which consent shall not be unreasonably withheld, publish alone or in conjunction with any other person any articles, photographs or other illustrations relating to the Works.

4. **Settlement of Disputes**

Any dispute or difference arising out of this Agreement shall be referred to the arbitration of a person to be agreed upon between the Client and the Consulting Engineer or, failing agreement, nominated by the President for the time being of the Chartered Institute of Arbitrators.

**OBLIGATIONS OF THE CONSULTING ENGINEER**

5. **Care and Diligence**

5.1. The Consulting Engineer shall exercise all reasonable skill, care and diligence in the discharge of the services agreed to be performed by him. If in the performance of his services the Consulting Engineer has a discretion exercisable as between the Client and the Contractor, the Consulting Engineer shall exercise his discretion fairly.

5.2. The Consulting Engineer may recommend that specialist suppliers and/or contractors should design and execute certain part or parts of the Works in which circumstances the Consulting Engineer shall co-ordinate and integrate the design of such part or parts with the overall design of the Works but he shall be relieved of all responsibility for the design, manufacture, installation and performance of any such part or parts of the Works. Where any person or persons are engaged in accordance with Clause 7.4, the Consulting Engineer shall not be liable for acts of negligence, default or omission by such person or persons.

CONSULTING ENGINEERS' FEES

**AGREEMENT 2.**

6. **Normal services**

6.1. Design Stage I

The services to be provided by the Consulting Engineer at this stage shall comprise all or any of the following as may be necessary in the particular case:

(a) Investigating data and information relevant to the Works which are reasonably accessible to the Consulting Engineer, and considering any reports relating to the Works which have either been previously prepared by the Consulting Engineer or prepared by others and made available to the Consulting Engineer by the Client.

(b) Making any normal topographical survey of the proposed site of the Works which may be necessary to supplement the topographical information already available to the Consulting Engineer.

(c) Advising the Client on the need to carry out any geotechnical investigations which may be necessary to supplement the geotechnical information already available to the Consulting Engineer, arranging for such investigation when authorised by the Client, certifying the amount of any payments to be made by the Client to the persons or firms carrying out such investigations under the Consulting Engineer's direction, and advising the Client on the results of such investigations.

(d) Advising the Client on the need for arrangements to be made, in accordance with Clause 7, for the carrying out of special surveys, special investigations or model tests and advising the Client of the results of any such surveys, investigations or tests carried out.

(e) Consulting any architect appointed by the Client in connection with the architectural treatment of the Works.

(f) Preparing such documents as are reasonably necessary to enable the Client to consider the Consulting Engineer's general proposals for the construction of the Works in the light of the investigations carried out by him at this stage, and to enable the Client to apply for approval in principle of the execution of the Works in accordance with such proposals.

CONSULTING ENGINEERS' FEES

**AGREEMENT 2.**

6.2. Design Stage II

The services to be provided by the Consulting Engineer at this stage shall comprise all or any of the following as may be necessary in the particular case:

(a) Preparing designs and tender drawings in connection with the Works.

(b) Advising as to the appropriate conditions of contract to be incorporated in any contract to be made between the Client and a Contractor.

(c) Preparing such specifications, schedules and bills of quantities as may be necessary to enable the Client to obtain tenders or otherwise award a contract for carrying out the Works.

(d) Advising the Client as to the suitability for carrying out the Works of persons and firms tendering and further as to the relative merits of tenders, prices and estimates received for carrying out the Works.

As soon as the Consulting Engineer shall have submitted advice to the Client upon tenders, his services at this stage shall be complete.

6.3. Construction Stage

The Consulting Engineer shall not accept any tender in respect of the Works unless the Client gives him instructions in writing to do so, and any acceptance so made by the Consulting Engineer on the instructions of the Client shall be on behalf of the Client. The services to be provided by the Consulting Engineer at this stage shall include all or any of the following as may be necessary in the particular case:

(a) Advising on the preparation of formal contract documents relating to accepted tenders for carrying out the Works or any part thereof.

(b) Inspecting and testing during manufacture and installation such electrical and mechanical materials, machinery and plant supplied for incorporation in the Works as are usually inspected and tested by Consulting Engineers, and arranging and witnessing acceptance tests.

(c) Advising the Client on the need for special inspection or testing other than that referred to in Clause 6.3(b).

(d) Advising the Client on the appointment of site staff in accordance with Clause 8.

(e) Preparing bar bending schedules and any further designs and drawings which may be necessary.

CONSULTING ENGINEERS' FEES

**AGREEMENT 2.**

6. **Normal Services** - *continued*

(f) Examining the Contractor's proposals.

(g) Making such visits to site as the Consulting Engineer shall consider necessary to satisfy himself as to the performance of any site staff appointed pursuant to Clause 8, and to satisfy himself that the Works are executed generally according to the contract and otherwise in accordance with good engineering practice.

(h) Giving all necessary instructions to the Contractor, provided that the Consulting Engineer shall not without the prior approval of the Client give any instructions which in the opinion of the Consulting Engineer are likely substantially to increase the cost of the Works unless it is not in the circumstances practicable for the Consulting Engineer to obtain such prior approval.

(j) Issuing certificates for payment to the Contractor.

(k) Performing any services which the Consulting Engineer may be required to carry out under any contract for the execution of the Works, including where appropriate the supervision of any specified tests and of the commissioning of the Works, provided that the Consulting Engineer may decline to perform any services specified in a contract the terms of which have not initially been approved by the Consulting Engineer.

(l) Delivering to the Client on the completion of the Works such records and manufacturers' manuals as are reasonably necessary to enable the Client to operate and maintain the Works.

(m) Deciding any dispute or difference arising between the Client and the Contractor and submitted to the Consulting Engineer for his decision, provided that this service shall not extend to advising the Client following the taking of any step in or towards any arbitration or litigation in connection with the Works.

6.4. General

Without prejudice to the preceding provisions of this clause, the Consulting Engineer shall from time to time as may be necessary advise the Client as to the need for the Client to be provided with additional services in accordance with Clause 7.

CONSULTING ENGINEERS' FEES

**AGREEMENT 2.**

7. **Additional Services Not Included in Normal Services**

7.1. As services additional to those specified in Clause 6, the Consulting Engineer shall, if so requested by the Client, provide any of the services specified in Clause 7.2 and provide or take all reasonable steps to arrange for the provision by others, of any of the services specified in Clause 7.3.

7.2. (a) Preparing any report or additional contract documents required for consideration of proposals for the carrying out of alternative works.

(b) Carrying out work consequent upon a decision by the Client to seek parliamentary powers.

(c) Carrying out work in connection with any application by the Client for any order, sanction, licence, permit or other consent, approval or authorisation necessary to enable the Works to proceed.

(d) Carrying out work arising from the failure of the Client to award a contract in due time.

(e) Preparing details for shop fabrication of ductwork, metal or plastic frameworks.

(f) Carrying out work consequent upon any assignment of a contract by the Contractor or upon the failure of the Contractor properly to perform any contract or upon delay by the Client in fulfilling his obligations under Clause 9 or in taking any other step necessary for the due performance of the Works.

(g) Advising the Client and carrying out work following the taking of any step in or towards any litigation or arbitration relating to the Works.

(h) Carrying out work in conjunction with others employed to provide any of the services specified in Clause 7.3.

(j) Carrying out such other additional services, if any, as are specified in Article 7 of the Memorandum of Agreement.

7.3. (a) Specialist technical advice on any abnormal aspects of the Works.

(b) Architectural, legal, financial and other professional services.

(c) Services in connection with the valuation, purchase, sale or leasing of lands and the obtaining of wayleaves.

CONSULTING ENGINEERS' FEES

**AGREEMENT 2.**

7. **Additional Services Not Included in Normal Services** - *continued*

(d) The carrying out of marine and air surveys, and land surveys other than those referred to in Clause 6, and the making of model tests or special investigations.

(e) The carrying out of special inspections or tests advised by the Consulting Engineer under Clause 6.3(c).

7.4. The Consulting Engineer shall obtain the prior agreement of the Client to the arrangements which he proposes to make on the Client's behalf for the provision of any of the services specified in Clause 7.3 The Client shall be responsible to any person or persons providing such services for the cost thereof.

CONSULTING ENGINEERS' FEES

**AGREEMENT 2.**

8. **Supervision on Site**

8.1. If in the opinion of the Consulting Engineer the nature of the Works, including the carrying out of any geotechnical investigation pursuant to Clause 6.1 warrants full-time or part-time engineering supervision on site, the Client shall agree to the appointment of such suitably qualified technical and clerical site-staff as the Consulting Engineer shall consider reasonably necessary to enable such supervision to be carried out.

8.2. Persons appointed pursuant to Clause 8.1 shall be employed either by the Consulting Engineer or, if the Client and the Consulting Engineer shall so agree, by the Client directly, provided that the Client shall not employ any person as a member of the site-staff unless the Consulting Engineer has first selected or approved such person as suitable for employment.

8.3. The terms of service of all site-staff to be employed by the Consulting Engineer shall be subject to the approval of the Client, which approval shall not be unreasonably withheld.

8.4. The Client shall procure that the contracts of employment of site-staff employed by the Client shall stipulate that the staff so employed shall in no circumstances take or act upon instructions other than those of the Consulting Engineer.

8.5. Where any supervision on site is performed by staff employed other than by the Consulting Engineer, the consulting Engineer shall not be responsible for any failure on the part of such staff properly to comply with any instructions given by the Consulting Engineer.

CONSULTING ENGINEERS' FEES

**AGREEMENT 2.**

**OBLIGATIONS OF THE CLIENT**

9. **Information to be Supplied to the Consulting Engineer**

9.1. The Client shall supply to the Consulting Engineer without charge and in such reasonable time as not to delay or disrupt the performance by the Consulting Engineer of his services under this Agreement, all necessary and relevant data and information in the possession of the Client or the Contractors or their respective servants, agents or sub-contractors and the Client shall give and shall procure that such persons give such assistance as shall reasonably be required by the Consulting Engineer in the performance of his services under this Agreement.

9.2. The Client shall give his decision on all sketches, drawings, reports, recommendations, tender documents and other matters properly referred to him for decision by the Consulting Engineer in such reasonable time as not to delay or disrupt the performance by the Consulting Engineer of his services under his Agreement.

CONSULTING ENGINEERS' FEES

**AGREEMENT 2.**

10. **Payment for Normal Services**

*Three different methods of payment for services carried out under Clause 6 are detailed in the three succeeding Clauses 10.1, 10.2 and 10.3. The method of payment which the Client and the Consulting Engineer agree to adopt must be specified in the Memorandum of Agreement. It should be noted that the method of payment specified in Clause 10.3 extends also to services other than those carried out under Clause 6.*

10.1. Payment depending upon the actual cost of the Works.

*This method of payment is recommended for use in normal circumstances.*

10.1.1. The sum payable by the Client to the Consulting Engineer for his services under Clause 6 shall be calculated as follows:

(a) The Works shall first be classified into one or more of the following classes as shall be appropriate:

Class A Civil and structural engineering including Class D work incidental thereto and geotechnical investigations.

Class B Buildings including Class D work incidental thereto and building and engineering services associated with buildings.

Class C Mechanical and electrical plant and equipment.

Class D Structural work in reinforced concrete, prestressed concrete, steel and other metals.

(b) The cost of each class of work shall next be calculated and

(c) The sum payable by the Client to the Consulting Engineers shall then be determined and shall be an amount or the sum of the amounts calculated in respect of each relevant class of work in accordance with the percentage stated in Article 4(a) of the Memorandum of Agreement.

10.1.2. The cost of each class of work shall be calculated in accordance with Clause 19. Where the Works have been classified in accordance with Clause 10.1.1 into more than one class, there shall be attributed to each class an appropriate portion of any general and preliminary items included in the total cost of the Works so that the total cost of all work in Classes A, B and C shall equal the total cost of the Works.

CONSULTING ENGINEERS' FEES

**AGREEMENT 2.**

10. **Payment for Normal Services** - *continued*

Class D work shall be taken to include concrete, reinforcement, prestressing tendons and anchorages, formwork, inserts and all labours associated therewith together with the appropriate portion of any general and preliminary items.

10.1.3. Where it is stated in the Memorandum of Agreement that the Works are to be constructed in more than one phase, then the percentage fees stated for each phase in Article 4 (d) and the provisions of this Agreement shall apply separately to each phase as if the expression 'the Works' meant, in the case of each phase, the work comprised in that phase.

10.2. Payment of a fixed sum

*This method of payment may be used when*

*(a) the nature of the Works and the services to be provided can be closely defined and will not change, and*

*(b) the Client and the Consulting Engineer have both approved an itemised estimate of the cost of the Works, and*

*(c) there are good prospects of the construction stage following the design stages without delay.*

*Ordinarily the fixed sum will not be less than that deducible from Clause 10.1 using, instead of the actual cost of each separate class of work, the estimates thereof approved by the Client and the Consulting Engineer at the time of the Consulting Engineer's appointment.*

The sum payable by the client to the Consulting Engineer for his services under Clause 6 shall be the sum specified in Article 4(b) of the Memorandum of Agreement, provided that the Consulting Engineer shall, in addition to the said sum, be paid in accordance with Clause 11.2 for any services of the kind specified in Clause 6.3 (m) which it is necessary for him to provide.

10.3. Payment at hourly rates plus fixed fee

*This method of payment may be used when*

*(a) the extent and duration of services to be provided by the Consulting Engineer are difficult to foresee at the time of appointment, or*

*(b) the services are likely to be substantially different from those described in Clause 6, or*

*(c) the time likely to be devoted by the Consulting Engineer bears little relation to the value of the Works.*

CONSULTING ENGINEERS' FEES

**AGREEMENT 2.**

10. **Payment for Normal Services** - *continued*

The fee referred to in Clause 10.3.1(b) may be agreed as a rate per month payable for the duration of this Agreement or alternatively as a lump sum.

10.3.1. In respect of services provided by the Consulting Engineer under the following clauses:-

| | |
|---|---|
| Clause 6 | Normal Services |
| Clause 7 | Additional Services |
| Clause 12 | Computer, etc. |
| Clause 13.3 | Site Visits |
| Clause 15 | Alterations, etc. |
| Clause 16 | Works Damaged, etc. |

the Client shall pay the Consulting Engineer

(a) For technical and supporting staff: Salary Cost times the multiplier specified in Article 4(c) of the Memorandum of Agreement, plus Other Payroll Cost.

(b) The fee specified in Article 4(d) of the Memorandum of Agreement, which shall be deemed to cover the services of Partners and Consultants of the firm but not their expenses which are reimbursable separately under Clause 14.

(c) Any charge for the use of a computer or other special equipment payable under Clause 12(d).

10.3.2. Time spent by technical and supporting staff in travelling in connection with the Works shall be chargeable on the above basis.

10.3.3. Unless otherwise agreed between the Client and the Consulting Engineer, the Consulting Engineer shall not be entitled to any payment in respect of time spent by secretarial staff or by staff engaged on general accountancy or administration duties in the Consulting Engineer's office.

10.3.4. The Consulting Engineeer shall submit to the Client at the time of submission of the monthly accounts referred to in Clause 20 such supporting data as may be agreed between the Client and the Consulting Engineer.

CONSULTING ENGINEERS' FEES

**AGREEMENT 2.**

11. **Payment for Additional Services**

11.1. In respect of additional services provided by the Consulting Engineer under Clause 7, the Client shall, subject to Clause 10.3 pay the Consulting Engineer in accordance with the Scale of Charges set out in Clause 11.2.

11.2. Scale of Charges:

(a) Partners and Consultant: At the hourly rate or rates specified in Article 4 (e) of the Memorandum of Agreement.

(b) Technical and supporting staff: Salary Cost times the multiplier specified in Article 4(c) of the Memorandum of Agreement, plus Other Payroll Costs.

(c) Time spent by Partners, Consultants, technical and supporting staff in travelling in connection with the Works shall be chargeable on the above basis.

(d) Unless otherwise agreed between the Client and the Consulting Engineer, the Consulting Engineer shall not be entitled to any payment in respect of time spent by secretarial staff or by staff engaged on general accountancy or administration duties in the Consulting Engineer's office.

12. **Payment for Use of Computer or Other Special Equipment**

Where the Client has agreed to pay the Consulting Engineer

(a) in accordance with Clause 10.3 and the Consulting Engineer decides to use a computer or other special equipment in the carrying out of any of his services, or

(b) for his services under Clause 6 in accordance with Clause 10.1 or 10.2 and the Consulting Engineer decides to use a computer or other special equipment in carrying out any additional services in accordance with Clause 7 or is expressly required by the Client to use a computer or other special equipment in the carrying out of his services under Clause 6,

the Client shall, unless otherwise agreed between the Client and the Consulting Engineer, pay the Consulting Engineer

(c) for the time spent in connection with the use of a computer or other special equipment, including the development and writing of programmes and the operation of the computer in trial and final runs, in accordance with Clause 10.3 when applicable and otherwise in accordance with the Scale of Charges set out in Clause 11.2 and

(d) a reasonable charge for the use of the computer or other special equipment, which charge shall be agreed between the Client and the Consulting Engineer before the work is put in hand.

CONSULTING ENGINEERS' FEES

**AGREEMENT 2.**

13. **Payment for Site Supervision**

13.1. In addition to any other payment to be made by the Client to the Consulting Engineer under this Agreement, the Client shall

(a) reimburse the Consulting Engineer in respect of all salary and wage payments made by the Consulting Engineer to site staff employed by the Consulting Engineer pursuant to Clause 8 and in respect of all other expenditure incurred by the Consulting Engineer in connection with the selection, engagement and employment of site staff, and

(b) pay to the Consulting Engineer an increment on the amount due under Clause 13.1(a) calculated at the percentage specified in Article 4(f) of the Memorandum of Agreement in respect of head office overhead costs incurred on site staff administration

provided that in lieu of payments under (a) and (b) above the Client and the Consulting Engineer may agree upon inclusive monthly or other rates to be paid by the Client to the Consulting Engineer for each member of site staff employed by the Consulting Engineer.

13.2. The Client shall also in all cases be responsible for the cost of providing such local office accommodation, furniture, telephones, equipment and transport as shall be reasonably necessary for the use of site staff appointed pursuant to Clause 8, and for the reasonable running costs of such necessary local office accommodation and other facilities, including those of stationery, telephone calls, telex, telegrams and postage. Unless otherwise agreed between the Client and the Consulting Engineer, the Consulting Engineer shall arrange, whether through the Contractor or otherwise, for the provision of local office accommodation and facilities for the use of site staff.

13.3. In cases where the Consulting Engineer has thought it proper that site staff should not be appointed, or where the necessary site staff is not available at site due to sickness or any other cause, the Consulting Engineer shall, subject to Clause 10.3, be paid in accordance with the Scale of Charges set out in Clause 11.2 for site visits which would have been unnecessary but for the absence or nonavailability of site staff.

CONSULTING ENGINEERS' FEES

**AGREEMENT 2.**

14. **Disbursements**

14.1. The Client shall in all cases reimburse the Consulting Engineer in respect of all the Consulting Engineer's disbursements properly made in connection with:

(a) Printing, reproduction and purchase of all documents, drawings, maps, records and photographs.

(b) Telegrams, telex and telephone calls (other than local telephone calls).

(c) Postage and similar delivery charges except in the case of items weighing less than 250 grammes sent by the ordinary inland post.

(d) Travelling, hotel expenses and other similar disbursements.

(e) Advertising for tenders and site staff.

(f) The provision of additional services to the Client pursuant to Clause 7.4.

(g) Professional Indemnity Insurance taken out by the Consulting Engineer to accord with the wishes of the client and as set out in the Memorandum of Agreement.

14.2. The Client, by agreement with the Consulting Engineer and in satisfaction of his liability to the Consulting Engineer in respect of these disbursements, may make to the Consulting Engineer a lump sum payment or alternatively a sum calculated as a percentage of the fees and charges falling due under Clause 10, 11 and 12 as specified in Article 4(g) of the Memorandum of Agreement.

CONSULTING ENGINEERS' FEES

**AGREEMENT 2.**

15. **Payment for Alteration or Modification to Design**

15.1. Subject to Clauses 15.2 and 15.3, if, at any time after the commencement of the Consulting Engineer's appointment, any design whether completed or in progress or any specification, drawing or other document prepared in whole or in part by the Consulting Engineer shall require to be modified or revised by reason of instructions received by the Consulting Engineer from or on behalf of the Client, or by reason of circumstances which could not reasonably have been foreseen by the Consulting Engineer, then the Client shall make additional payment to the Consulting Engineer for making any necessary modifications or revisions and for any consequential reproduction of documents. Subject to Clause 10.3, and unless otherwise agreed between the Client and the Consulting Engineer the additional sum to be paid to the Consulting Engineer shall be calculated in accordance with the Scale of Charges set out in Clause 11.2, and shall also include any appropriate reimbursements in accordance with Clause 14.1 provided always that the Consulting Engineer shall give the Client written notice as soon as it is reasonably apparent that additional payments will arise under this Clause.

15.2. Subject to Clause 15.3, where in the Consulting Engineer's opinion the Client's instruction necessitates a fundamental redesign of the part or parts of the Works affected by the instruction such that designs, specifications, drawings and other documents prepared by the Consulting Engineer cannot be modified or revised to take account thereof or where the modification or revision instructed by the Client results in a reduction in the cost of the part or parts of the Works affected thereby as contained in the Consulting Engineer's most recent estimate under Clause 20.1(a) by 10% or more, then such part or parts of the Works shall be deemed to have been abandoned and the Consulting Engineer shall be paid therefore in accordance with Clause 17.2. The Consulting Engineer shall carry out such further work and shall produce such further designs, specifications, drawings and other documents as may be necessary to comply with the Client's instructions and the Consulting Engineer shall be paid therefore in accordance with the provisions of this Agreement.

15.3. In any case in which the Consulting Engineer is paid for his services under Clause 6 in accordance with Clause 10.2 any instruction from or on behalf of the Client under Clause 15.1 or 15.2 shall be subject to the prior written agreement of the Consulting Engineer and the Client and the Consulting Engineer shall agree the appropriate adjustment to the sum payable under Clause 10.2.

CONSULTING ENGINEERS' FEES

**AGREEMENT 2.**

16. **Payment when Works are Damaged or Destroyed**

If at any time before completion of the Works any part of the Works or any materials, plant or equipment whether incorporated in the Works or not shall be damaged or destroyed, the Client shall make additional payment to the Consulting Engineer in respect of any expenses incurred or additional work required to be carried out by the Consulting Engineer as a result of such damage or destruction. Subject to Clause 10.3, the amount of such additional payment shall be calculated in accordance with the Scale of Charges set out in Clause 11.2, and shall also include any appropriate reimbursements in accordance with Clause 14.

17. **Payment following Termination or Suspension by the Client**

17.1. Upon a termination or suspension by the Client in pursuance of Clause 2.3, the Client shall pay to the Consulting Engineer the sums specified in (a), (b) and (c) below less the amount of payments previously made to the Consulting Engineer under the terms of this Agreement.

(a) A fair and reasonable proportion of the sum which would have been payable to the Consulting Engineer under Clause 10 if no such termination or suspension had taken place. In the assessment of such proportion, the services carried out by the Consulting Engineer up to the date of termination or suspension and in pursuance of Clause 2.7 shall be compared with a reasonable assessment of the services which the Consulting Engineer would have carried out but for the termination or suspension. In any case in which it is necessary to assess the payment to be made to the Consulting Engineer in accordance with this sub-clause by reference to the cost of the Works, then to the extent that such cost is not known the assessment shall be made upon the basis of the Consulting Engineer's best estimate of cost.

(b) Amounts due to the Consulting Engineer under any other clauses of this Agreement.

(c) A disruption charge equal to one-sixth of the difference between the sum which would have been payable to the Consulting Engineer under Clause 10, but for the termination or suspension, and the sum payable under (a).

17.2. Upon a termination or suspension by the Client in pursuance of Clause 2.4, the Client shall pay to the Consulting Engineer the sums specified in (a) and (b) below.

(a) A fair and reasonable proportion of the sum which would have been payable to the Consulting Engineer under Clause 10 in respect of the services affected if no such termination or suspension had taken place. The proportion shall be calculated in accordance with the provisions of Clause 17.1(a); and

**AGREEMENT 2.**

(b) A disruption charge, to be calculated in accordance with the provision of Clause 17.1(c).

17.3. In any case in which the Client has required the Consulting Engineer to suspend the carrying out of the Consulting Engineer's services in pursuance of Clause 2.3 or 2.4, the Client may at any time within the period of 12 months from the date of the Client's notice require the Consulting Engineer in writing to resume the performance of such services. In such event:

(a) the Consulting Engineer shall within a reasonable time of receipt by him of the Client's said requirement in writing resume the performance of services in accordance with this Agreement, the payment made under Clause 17.1(a) or 17.2(a) as the case may be, ranking as a payment on account towards the total sum payable to the Consulting Engineer under Clause 10, but,

(b) notwithstanding such resumption, the Consulting Engineer shall be entitled to retain or receive as an additional payment due in accordance with this Agreement the disruption charge referred to in Clause 17.1(c) or 17.2(b) as the case may be.

17.4. If the Consulting Engineer shall need to perform any additional services in connection with the resumption of his services in accordance with Clause 17.3, the Client shall pay the Consulting Engineer in respect of the performance of such additional services in accordance with Clause 10.3 or the Scale of Charges set out in Clause 11, as the case may be, and any appropriate reimbursements in accordance with Clause 14.

18. **Payment following Termination by the Consulting Engineer**

Upon a termination by the Consulting Engineer in pursuance of Clause 2.6, the Client shall pay to the Consulting Engineer the sums specified in Clauses 17.1(a) and (b) less the amount of payments previously made to the Consulting Engineer under the terms of this Agreement. Upon the payment of such sums, the Consulting Engineer shall deliver to the Client such completed drawings, specifications and other similar documents relevant to the Works as are in his possession. The Consulting Engineer shall be permitted to retain copies of any documents so delivered to the Client. The provisions of this Clause are without prejudice to any other rights and remedies which the Consulting Engineer may possess.

**AGREEMENT 2.**

19. **Cost of the Works**

19.1. The cost of the Works or any part thereof shall be deemed to include:

(a) The cost to the Client, of the Works however incurred, without deduction of any liquidated damages or penalties payable by the Contractor to the Client but including any payments made by the Client to the Contractor by way of bonus, incentive or ex-gratia payments, or in settlement of claims.

(b) A fair valuation of any labour, materials manufactured goods, machinery or other facilities provided by the Client, and of the full benefit accruing to the Contractor from the use of construction plant and equipment belonging to the Client which the Client has required to be used in the execution of the Works.

(c) The market value, as if purchased new, of any second-hand materials, manufactured goods and machinery incorporated in the Works.

(d) The cost of geotechnical investigations.

(e) A fair proportion of the total cost to the Client of any work in connection with the provision or diversion of public utilities systems which is carried out, other than by the Contractor, under arrangements made by the Consulting Engineer. The said fair proportion shall be assessed with reference to the costs incurred by the Consulting Engineer in making such arrangements.

19.2. The cost of the Works shall not include:

(a) Administration expenses incurred by the Client.

(b) Costs incurred by the Client under this Agreement.

(c) Interest on capital during construction, and the cost of raising moneys required for carrying out the construction of the Works.

(d) Cost of land and wayleaves.

CONSULTING ENGINEERS' FEES

**AGREEMENT 2.**

20. **Payment of Accounts**

20.1. Unless otherwise agreed between the Client and the Consulting Engineer from time to time:

(a) The sum payable to the Consulting Engineer under Clause 10.1 shall, until the cost of the Work is known, be paid by the Client to the Consulting Engineer in instalments and shall be calculated by reference to the Consulting Engineer's most recent estimate of the cost of the Works taking into account any accepted tender or tenders when available and any fluctuations clauses contained therein. Such instalments shall be paid during each of the several stages of the Consulting Engineer's services in equal amounts at the intervals specified in Article 5 of the Memorandum of Agreement so that by the end of each stage the cumulative total of all instalments then paid shall amount to the relevant proportion, as specified in Article 5 of the Memorandum of Agreement, of the estimated total sum payable to the Consulting Engineer under this Agreement;

(b) the sum payable to the Consulting Engineer under Clause 10.2 or Clause 10.3.1(b) as the case may be shall be paid by the Client in the instalments and at the intervals stated in Article 6 of the Memorandum of Agreement and the lump sum payable under Clause 14.2 shall be paid as stated in Article 4(g)(2);

(c) all other sums due to the Consulting Engineer shall be paid by the Client on accounts rendered monthly by the Consulting Engineer.

20.2. Instalments paid by the Client to the Consulting Engineer in accordance with Clause 20.1(a) shall constitute no more than payments on account. A statement of the total sum due to the Consulting Engineer shall be prepared when the cost of the Works is fully known. Such statement, after giving credit to the Client for all instalments previously paid, shall state the balance (if any) due from the Client to the Consulting Engineer or from the Consulting Engineer to the Client, as the case may be, which balance shall be paid to or by the Consulting Engineer as the case may require.

20.3. All sums due from the Client to the Consulting Engineer in accordance with the terms of this Agreement shall be paid within 40 days of the submission by the Consulting Engineer of his accounts therefore to the Client, and any sums remaining unpaid at the expiry of such period of 40 days shall bear interest thereafter; such interest to accrue from day to day at the rate of 2% per annum above the Base Rate of the Engineer's Principal Bank as stated in Article 4(h) of the Memorandum of Agreement.

**AGREEMENT 2.**

20.4. If any item or part of an item of an account rendered by the Consulting Engineer is disputed or subject to question by the Client, the payment by the Client of the remainder of that account shall not be withheld on those grounds and the provisions of Clause 20.3 shall apply to such remainder and also to the disputed or questioned item to the extent that it shall subsequently be agreed or determined to have been due to the Consulting Engineer.

20.5. All fees set out in this Agreement are exclusive of Value Added Tax, the amount of which, at the rate and in the manner prescribed by law, shall be paid by the Client to the Consulting Engineer.

CONSULTING ENGINEERS' FEES

## MEMORANDUM OF AGREEMENT

BETWEEN CLIENT AND CONSULTING ENGINEER FOR CIVIL, MECHANICAL AND ELECTRICAL WORKS AND FOR STRUCTURAL ENGINEERING WORK WHERE AN ARCHITECT IS NOT APPOINTED BY THE CLIENT

MEMORANDUM OF AGREEMENT made the ..............................................

day of ............................ 19 ..........................................

BETWEEN

..................................................................................

.............................................. (hereinafter called 'the Client')

of the one part and ..................................................................

..................................................................................

(hereinafter called 'the Consulting Engineer') of the other part.

WHEREAS the Client has considered and approved the general proposals recommended in

a Report dated ............................................. submitted by the Consulting Engineer and/or intends to proceed with ............................

..................................................................................

..................................................................................

..................................................................................

..................................................................................

and has requested the Consulting Engineer to provide professional services in connection with

..................................................................................

..................................................................................

................................ (referred to in this Agreement as 'the Works')

CONSULTING ENGINEERS' FEES

**MEMORANDUM OF AGREEMENT** - *continued*

NOW IT IS HEREBY AGREED as follows:

1. The Client agrees to engage the Consulting Engineer subject to and in accordance with the Conditions of Engagement attached hereto and the Consulting Engineer agrees to provide professional services subject to and in accordance with the said Conditions of Engagement.

2. This Memorandum of Agreement and the said Conditions of Engagement shall together constitute the Agreement between the Client and the Consulting Engineer.

3. The method of payment for services under clause 6 of the said Conditions of Engagement shall be that described in clause 10.1*, 10.2*, 10.3* thereof.

4. In the said Conditions of Engagement:-

(a) the fees referred to in Clause 10.1 shall be the following percentages of the cost of each class of work

* Class A work ..................................................%

* Class B work ..................................................%

* Class C work ..................................................%

* Class D work ..................................................%

(b) the fixed sum referred to in Clause 10.2 shall be

....................................................(£ )*

(c) the multiplier referred to in Clause 10.3.1(a) and 11.2(b) shall be

..................................................................

(d) the fee referred to in Clause 10.3.1(b) shall be

....................................................(£ )*

payable in ....................... equal monthly instalments.*

(e) the rate or rates referred to in Clause 11.2(a) shall be

..................................................................

..................................................................

(f) the percentages to be applied to the cost of site staff referred to in Clause 13.1(b) shall be ............... % for specifically recruited staff and ...................... % for seconded staff.

CONSULTING ENGINEERS' FEES

**MEMORANDUM OF AGREEMENT** - *continued*

(g) the payment for disbursements referred to in Clause 14 shall be

(1) reimbursed at cost*

(2) a lump sum of ..............................................

............(£ ) payable in ............ equal monthly

instalments*

(3) ............. % of the fees and charges due to the Consulting Engineer for all services, as specified in Clause 14.2.*

(h) The Engineer's Principal Bank referred to in Clause 20.3 shall be

..............................................................

..............................................................

5. The intervals for the payment of instalments under Clause 20.1(a) of the said Conditions of Engagement shall be monthly/quarterly* reckoned from the commencement of the Consulting Engineer's appointment, and the cumulative proportions referred to in the said Clause shall be as follows:-

| | | |
|---|---|---|
| On completion of Design Stage I | ....... | per cent of the total fee |
| On completion of Design Stage II | ....... | per cent of the total fee |
| On completion of Construction Stage | 100 | per cent of the total fee |

6. The sum payable under Clause 20.1(b) of the said Conditions of Engagement shall be paid in equal monthly/quarterly* instalments.

7. The additional services to be carried out in accordance with Clause

7.2(j) of the said Conditions of Engagement shall be

..............................................................

..............................................................

..............................................................

..............................................................

..............................................................

..............................................................

CONSULTING ENGINEERS' FEES

**MEMORANDUM OF AGREEMENT** - *continued*

8. Professional Indemnity Insurance

The amount of professional indemnity insurance referred to in Clause 14.1 of the said Conditions of Engagement shall be

.............................................................. pounds (£ ) for any one occurrence or series of occurrences arising out of this engagement.

This professional indemnity insurance shall be maintained for a period of ..............years from the date of this Memorandum of Agreement, unless such insurance cover ceases to be available in which event the Consulting Engineer will notify the client immediately. The sum payable by the Client to the Consulting Engineer as a contribution to the additional cost of the professional indemnity insurance thus provided shall be

.............................................................. pounds (£ ) and such an amount shall become due to the Consulting Engineer immediately upon acceptance by or on behalf of the Client of any tender in respect of the Works or any part thereof/immediately upon submission by the Consulting Engineer to the Client of his report in relation to the Task.*

9. LIMITATION OF LIABILITY

Notwithstanding anything to the contrary contained elsewhere in this Agreement, the total liability of the Consulting Engineer under or in connection with this Agreement, whether in contract, in tort, for breach or statutory duty or otherwise, shall not exceed

.............................................................. pounds (£ ). The Client shall indemnify and keep indemnified the Consulting Engineer from and against all claims, demands, proceedings, damages, costs, charges and expenses arising out of or in connection with this Agreement, the Works and/or the Project in excess thereof.

AS WITNESS the hands of the parties the day and year first above written.

Duly Authorised
Representative
of the Client ........................ Consulting Engineer ..........................

Witness ........................ Witness ..........................

* *Delete as appropriate*

CONSULTING ENGINEERS' FEES

**MEMORANDUM OF AGREEMENT** - *continued*

*Basis for completing Memorandum of Agreement*

1. *The works shall be comprehensively described in the Memorandum of Agreement. If the Works are to be carried out in more than one phase it is important that the work included in each phase be defined.*

2. *In Article 4(a) the percentages for fee calculations are fixed for the duration of the Agreement. Where the Works are to be carried out in phases, specific percentages should be stated for each phase utilising additional columns as necessary.*

3. *For works of average complexity the percentages for fee calculation may be determined from the graph and notes on page . The estimated cost of work in each class shall be agreed with the Client and the costs shall be divided by the current Output Price Index (1975 = 100) published by the Department of the Environment. From the figures so obtained the percentage fee appropriate for each class of work can be determined from the graph.*

   *e.g. If the cost of Class A work is estimated to be £200 000 and the Output Price Index is 200, the category would be 1000 and the percentage would be 8.25.*

4. *Articles 4(c) and 4(e) must be completed in all cases.*

5. *In Article 4(c) the normal multiplier is 2.5.*

6. *In Article 4(f) the percentage to be added to the direct costs of providing site supervisory staff is to cover all overheads and head office charges incurred in administering such staff and should include for administration of payroll, finance costs and an allowance for potential redundancy payments. The normal figure is 15% for staff recruited on a contract basis. For seconded staff the terms are to be negotiated between the Client and the Consulting Engineer.*

7. *In Article 5, the following table, provided for guidance only, is typical of the proportionate amount of the total services performed at the end of each stage:*

| | *Option 1* | *Option 2* | *Option 3* |
|---|---|---|---|
| *On completion of Design Stage I* | *20%* | *20%* | *20%* |
| *On completion of Design Stage II* | *70%* | *60%* | *50%* |
| *On completion of Construction Stage* | *100%* | *100%* | *100%* |

*Option 1 is appropriate for civil engineering work and buildings where the majority of drawings are prepared before the construction stage is reached.*

*Option 2 is appropriate for civil engineering work and buildings where the majority of drawings are prepared during the construction stage.*

*Option 3 is appropriate for mechanical and electrical plant.*

*Note the current Price Index referred to above was 270 at 3RD Quarter 1988.*

CONSULTING ENGINEERS' FEES

**MEMORANDUM OF AGREEMENT** - *continued*

Graph showing relationship of percentage fee with size of project

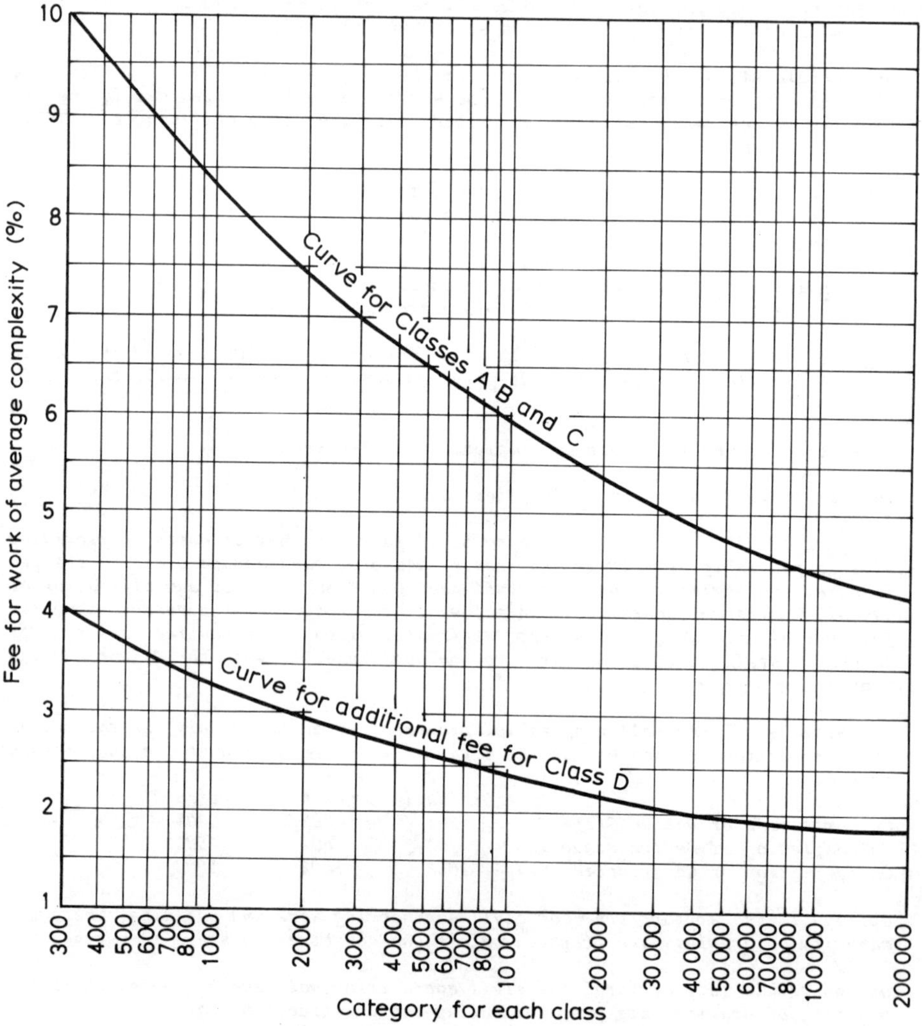


Cost of works in each class in £ divided by output price index (1975=100)

Notes

For projects in categories below 300, time charges should be used

For projects in categories above 200 000, the fee for Classes A, B and C should not be less than 4% and the additional fee for Class D should not be less than 1.6%

QUANTITY SURVEYORS' FEES

**INTRODUCTION**

The following scale of professional charges for quantity surveying services in connection with civil engineering works is as published by the Royal Institution of Chartered Surveyors and is effective from 29 July 1988. The scale has been determined by the Fees Committee appointed by the Quantity Surveyors Divisional Council of the Royal Institution of Chartered Surveyors. The scale is recommended and not mandatory.

Please note this scale is currently under review in the light of recent Government policy changes concerning restrictive practices.

SCALE No 38. ITEMISED SCALE OF PROFESSIONAL CHARGES FOR QUANTITY SURVEYING SERVICES FOR CIVIL ENGINEERING WORKS

1.0. **Generally**

1.1. This Scale of professional charges is applicable where the contract provides for the bills of quantities and final account to be based on measurements prepared in accordance with or based on the principles of the Standard Method of Measurement of Civil Engineering Quantities issued by the Institution of Civil Engineers.

1.2. The fees are in all cases exclusive of travelling and other expenses (for which the actual disbursement is recoverable unless there is some prior arrangement for such charges) and of the cost of reproduction of bills of quantities and other documents, which are chargeable in addition at net cost.

1.3. The fees are in all cases exclusive of services in connection with the allocation of the cost of the works for purposes of calculating value added tax for which there shall be an additional fee based on the time involved (see Section 8.0).

1.4. If any of the materials used in the works are supplied by the employer or charged at a preferential rate, then the actual or estimated market value thereof shall be included in the amounts upon which fees are to be calculated.

1.5. The fees are in all cases exclusive of preparing a specification of the materials to be used and the works to be done.

1.6. If the quantity surveyor incurs additional costs due to exceptional delays in construction operations or any other cause beyond the control of the quantity surveyor then the fees may be adjusted by agreement between the employer and the quantity surveyor.

1.7. If the works are substantially varied at any stage or if the quantity surveyor is involved in abortive work there shall be an additional fee based on the time involved (see section 8.0).

1.8. The fees and charges are in all cases exclusive of value added tax which will be applied in accordance with legislation.

QUANTITY SURVEYORS' FEES

1.0. **Generally** - *continued*

1.9. This scale is not intended to apply to works of a civil engineering nature which form a subsidiary part of a building contract or to buildings which are ancillary to a civil engineering contract. In these cases the fees to be charged for quantity surveying services shall be in accordance with the scales applicable to building works.

1.10. When works of both categories I and II are included in one contract the fee to be charged shall be calculated by taking the total value of the sections of work in each of the categories and applying the appropriate scale from the beginning in each case. General items such as preliminaries (and in the case of post contract fees contract price fluctuations) or sections of works which cannot be specifically allocated to either category shall be apportioned pro rata to the values of the other sections of the works and added thereto in order to ascertain the total value of works in each category.

1.11. When a project is the subject of a number of contracts then, for the purpose of calculating fees, the value of such contracts shall not be aggregated but each contract shall be taken separately and the scale of charges applied as appropriate.

1.12. Roads, railways, earthworks and dredging which are ancillary only to any Category II work shall be regarded as Category II work. Works or sections of works of Category I which incorporate piled construction shall be regarded as Category II works.

1.13. No addition to the fees given hereunder shall be made in respect of works of alteration or repair where such works are incidental to the new works. If the work covered by a single contract is mainly one of alteration or repair then an additional fee shall be negotiated.

1.14. In the absence of agreement to the contrary, payments to the quantity surveyor shall be made by instalments by arrangement between the employer and the quantity surveyor.

1.15. Copyright in bills of quantities and other documents prepared by the quantity surveyor is reserved to the quantity surveyor.

QUANTITY SURVEYORS' FEES

## 2.0. Contracts based on Bills of Quantities: Pre-contract Services

### Estimated Quantities

2.1. *Basic Scale*: For preparing bills of quantities giving estimated quantities of the work required, examining tenders received and reporting thereon.

(a) Category I. Works or sections of works such as monolithic walls for quays, jetties, dams and reservoirs; caissons; tunnels; airport runways and tracks; roads; railways; and earthworks and dredging.

| Value of work | | Category I fee | |
|---|---|---|---|
| £ | £ | £ | £ |
| Up to | 300,000 | 0.40% (minimum fee £800) | |
| 300,000 - | 600,000 | 1,200 + 0.38% on balance over | 300,000 |
| 600,000 - | 1,500,000 | 2,340 + 0.30% on balance over | 600,000 |
| 1,500,000 - | 3,000,000 | 5,040 + 0.25% on balance over | 1,500,000 |
| 3,000,000 - | 10,000,000 | 8,790 + 0.20% on balance over | 3,000,000 |
| 10,000,000 - | 20,000,000 | 22,790 + 0.15% on balance over | 10,000,000 |
| Over | 20,000,000 | 37,790 + 0.14% on balance over | 20,000,000 |

(b) Category II. Works or sections of works such as piled quay walls; suspended jetties and quays; bridges and their abutments; culverts; sewers; pipe lines; electric mains; storage and treatment tanks; water cooling towers; and structures for housing heavy industrial and public utility plant, e.g., furnace houses and rolling mills to steel works; and boiler houses, reactor blocks and turbine halls to electricity generating stations.

| Value of work | | Category II fee | |
|---|---|---|---|
| £ | £ | £ | £ |
| Up to | 300,000 | 0.75% (minimum fee £800) | |
| 300,000 - | 600,000 | 2,250 + 0.70% on balance over | 300,000 |
| 600,000 - | 1,500,000 | 4,350 + 0.50% on balance over | 600,000 |
| 1,500,000 - | 3,000,000 | 8,850 + 0.40% on balance over | 1,500,000 |
| 3,000,000 - | 10,000,000 | 14,850 + 0.35% on balance over | 3,000,000 |
| 10,000,000 - | 20,000,000 | 39,350 + 0.30% on balance over | 10,000,000 |
| Over | 20,000,000 | 69,350 + 0.28% on balance over | 20,000,000 |

QUANTITY SURVEYORS' FEES

2.2. Mechanical and electrical installations

(a) When bills of quantities giving estimated quantities of the work required are prepared by the quantity surveyor for the mechanical and electrical services installed in structures for domestic or environmental purposes common to buildings generally (i.e. purposes other than industrial processes) there shall be a fee for these services (which shall include examining tenders received and reporting thereon) in addition to the fee calculated in accordance with paragraph 2.1 as follows:-

| Value of work | | Additional fee | |
|---|---|---|---|
| £ | £ | £ | £ |
| Up to | 120,000 | 1.80% | |
| 120,000 - | 240,000 | 2,160 + 1.55% on balance over | 120,000 |
| 240,000 - | 480,000 | 4,020 + 1.30% on balance over | 240,000 |
| 480,000- | 750,000 | 7,140 + 1.20% on balance over | 480,000 |
| 750,000 - | 1,000,000 | 10,380 + 1.00% on balance over | 750,000 |
| 1,000,000 - | 4,000,000 | 12,880 + 0.85% on balance over | 1,000,000 |
| Over | 4,000,000 | 38,380 + 0.82% on balance over | 4,000,000 |

QUANTITY SURVEYORS' FEES

2.0. **Contracts based on Bills of Quantities: Pre-contract Services** - *continued*

2.2 (b) When bills of quantities giving estimated quantities of the work required are prepared for machinery and other permanent mechanical and electrical installation or plant in connection with industrial processes other than those included in paragraph 2.1 then a fee in addition to that envisaged under paragraph 2.1 shall be negotiated.

(c) Fees shall be calculated upon the accepted tender for each individual contract.

Firm Quantities

2.3 For preparing bills of quantities giving firm quantities for works or sections of works (i.e. not intended to be subject to remeasurement), fees shall be calculated in accordance with paragraphs 2.1 and 2.2 and the amounts of fees so obtained shall be increased by 37.5%

Consultative Services, Negotiating Tenders and Pricing Bills of Quantities

2.4 When the quantity surveyor is appointed to provide cost consultative services or prepare approximate estimates, feasibility studies or submissions for the approval of financial grants, negotiate tenders price bills of quantities to provide an estimate comparable with tenders or similar services, then the fee shall be based on the time involved (see Section 8.0) or alternatively on a lump sum or percentage basis agreed between the employer and the quantity surveyor.

3.0. **Qualifications in connection with fees for Pre-contract Services**

3.1 The scales of fees for preparing bills of quantities (paragraphs 2.1, 2.2 and 2.3) shall take into account the following:-

(a) The scales of fees are overall scales based upon the inclusion of all provisional and prime cost items. When work, normally included in construction works is the subject of a separate contract for which the quantity surveyor has not been paid fees under any other paragraph hereof, the value of such work shall be included in the amount upon which fees are charged.

(b) Fees shall be calculated upon the accepted tender for the whole of the work subject to the provision of paragraph 2.2(c). In the event of no tender being accepted, fees shall be calculated upon the basis of the lowest original bona fide tender received. In the event of no such tender being received the fees shall be calculated upon a reasonable valuation of the works based upon the original bills of quantities.

Note: In the foregoing context 'bona fide tender' shall be deemed to mean a tender submitted in good faith without major errors of computation and not subsequently withdrawn by the tenderer.

QUANTITY SURVEYORS' FEES

3.0. **Qualifications in connection with fees for Pre-contract Services** - *continued*

3.1 (c) In calculating the amount upon which fees are charged the total of any credits and the totals of any alternative bills shall be aggregated and added to the amount described above. The value of any omission or addition forming part of an alternative bill shall not be added unless measurement or abstraction from the original dimension sheets was necessary.

(d) If a tender, when received, is reduced before acceptance, fees are to be calculated upon the amount of the unreduced tender. When the preparation of bills of reduction is required the fee shall be based upon the time involved (see Section 8.0) or, alternatively, on a lump sum or percentage basis agreed between the employer and the quantity surveyor.

3.2 Fees calculated in accordance with paragraphs 2.1, 2.2 and 2.3 include for the preparation of bills of quantities on a normal trade basis. If the employer requires additional information to be provided in the bills of quantities or the bills to be prepared in an elemental, operational or similar form, then the fee may be adjusted by agreement between the employer and the quantity surveyor.

QUANTITY SURVEYORS' FEES

## 4.0. **Contracts based on Bills of Quantities: Post-contract Services**

Estimated Quantities

4.1. Basic Scale: For taking particulars and reporting valuations for interim certificates for payments on account to the contractor and measuring for and preparing the final account for work executed, including pricing and agreeing final account totals with the contractor.

(a) Category I. Works or sections of works such as monolithic walls for quays, jetties, dams and reservoirs; caissons; tunnels; airport runways and tracks; roads; railways; and earthworks and dredging.

| Value of work | | Category I fee | |
|---|---|---|---|
| £ | £ | £ | £ |
| Up to | 300,000 | 1.25% (minimum fee £2,500) | |
| 300,000 - | 600,000 | 3,750 + 1.10% on balance over | 300,000 |
| 600,000 - | 1,500,000 | 7,050 + 0.80% on balance over | 600,000 |
| 1,500,000- | 3,000,000 | 14,250 + 0.65% on balance over | 1,500,000 |
| 3,000,000 - | 10,000,000 | 24,000 + 0.50% on balance over | 3,000,000 |
| 10,000,000 - | 20,000,000 | 59,000 + 0.45% on balance over | 10,000,000 |
| Over | 20,000,000 | 104,000 + 0.42% on balance over | 20,000,000 |

(b) Category II. Works or sections of works such as piled quay walls, suspended jetties and quays, bridges and their abutments; culverts; sewers; pipe lines; electric mains; storage and treatment tanks; water cooling towers; and structures for housing heavy industrial and public utility plant, e.g., furnace houses and rolling mills to steel works; and boiler houses, reactor blocks and turbine halls to electricity generating stations.

QUANTITY SURVEYORS' FEES

4.0. **Contracts based on Bills of Quantities:Post-contract Services** - *continued*

| Value of work £ | £ | Category II fee £ | | £ |
|---|---|---|---|---|
| Up to | 300,000 | 2.25% (minimum fee £2,500) | | |
| 300,000 - | 600,000 | 6,750 + 2.00% | on balance over | 300,000 |
| 600,000 - | 1,500,000 | 12,750 + 1.50% | on balance over | 600,000 |
| 1,500,000- | 3,000,000 | 26,250 + 1.20% | on balance over | 1,500,000 |
| 3,000,000 - | 10,000,000 | 44,250 + 1.05% | on balance over | 3,000,000 |
| 10,000,000 - | 20,000,000 | 117,750 + 0.90% | on balance over | 10,000,000 |
| Over | 20,000,000 | 207,750 + 0.84% | on balance over | 20,000,000 |

4.2. *Mechanical and electrical installations*

(a) When final accounts are prepared by the quantity surveyor for mechanical and electrical services installed in structures for domestic or environmental purposes common to buildings generally (i.e. purpose other than industrial processes) there shall be a fee for these services in addition to the fee calculated in accordance with paragraph 4.1 as follows.

| Value of work £ | £ | Category II fee £ | | £ |
|---|---|---|---|---|
| Up to | 120,000 | 2.70% | | |
| 120,000 - | 240,000 | 3,240 + 2.30% | on balance over | 120,000 |
| 240,000 - | 480,000 | 6,000 + 1.95% | on balance over | 240,000 |
| 480,000- | 750,000 | 10,680 + 1.80% | on balance over | 480,000 |
| 750,000 - | 1,000,000 | 15,540 + 1.50% | on balance over | 750,000 |
| 1,000,000 - | 4,000,000 | 19,290 + 1.30% | on balance over | 1,000,000 |
| Over | 4,000,000 | 58,290 + 1.23% | on balance over | 4,000,000 |

(b) When final accounts are required to be prepared by the quantity surveyor for machinery and other permanent mechanical and electrical installations or plant in connection with industrial processes other than those included in paragraph 4.1 then a fee in addition to that envisaged under paragraph 4.1 shall be negotiated.

(c) Fees shall be calculated upon the basis of the accounts for each individual contract for which the final account has been prepared by the quantity surveyor.

**Firm Quantities**

4.3. For taking particulars and reporting valuations for interim certificates for payments on account to the contractor and measuring and making up bills of variations including pricing and agreeing totals with the contractor, fees shall be calculated in accordance with paragraphs 4.1 and 4.2 and the amounts of fees so obtained shall be decreased by 12.5% in respect of paragraph 4.1 and 25% in respect of paragraph 4.2.

QUANTITY SURVEYORS' FEES

5.0. **Fluctuations**

5.1. When the conditions of contract provide for contract price fluctuations by the use of a price adjustment formula method post contract fees shall be charged on the account for the whole of the work including the amount of the fluctuations.

5.2. When the conditions of contract provide for contract price fluctuations calculated other than by the use of a price adjustment formula method then an additional fee shall be charged based upon the time involved (see Section 8.0) or alternatively on a lump sum or percentage basis agreed between the employer and the quantity surveyor. Post contract fees shall be charged on the account for the whole of the work including the amount of the fluctuations.

6.0. **Qualifications in connection with fees for Post-contract Services**

6.1. The scales of fees for post-contract services shall take into account the following:

(a) The scales of fees are overall scales based upon the inclusion of all nominated sub-contractors' and nominated suppliers' accounts. When work normally included in construction works is the subject of a separate contract for which the quantity surveyor has not been paid fees under any other paragraph hereof, the value of such work shall be included in the amount on which fees are charged.

(b) Fees shall be calculated upon the basis of the account for the whole of the work, subject to the provisions of paragraph 4.2(c) and 7.2.

(c) In calculating the amount on which fees are charged the total of any credits is to be added to the amount described above.

(d) When the quantity surveyor is required to prepare valuations of materials or goods off site, an additional fee shall be charged based on the time involved (see Section 8.0).

(e) When the employer specifically requests a cost monitoring service an additional fee shall be charged based on the time involved (see Section 8.0) or, alternatively, on a lump sum or percentage basis agreed between the employer and the quantity surveyor.

(f) If the quantity surveyor appointed to carry out the post-contract services did not prepare the bills of quantities then the fees in paragraphs 4.1 and 4.2 may be increased to cover the additional services undertaken by the quantity surveyor.

QUANTITY SURVEYORS' FEES

7.0. **Additional Services**

7.1. For additional services not normally necessary, such as those arising as a result of the termination of a contract before completion, liquidation, fire or other damage to the works, services in connection with arbitration, litigation, grant and taxation matters and all similar services where the employer specifically instructs the quantity surveyor, the charges shall be in accordance with Section 8.0.

7.2. When the quantity surveyor is required to provide services in connection with establishing the validity of a contractor's claim and consequential valuation and negotiation an additional fee shall be charged based on the time involved (see Section 8.0). When a fee has been charged under this paragraph the amount of the claim shall be deducted from the total of the final account on which fees would otherwise be charged.

8.0. **Time Charges**

8.1. (a) For consultancy and other services performed by a principal, a fee by arrangement according to the circumstances including the professional status and qualifications of the quantity surveyor.

(b) When a principal does work which would normally be done by a member of staff, the charge shall be calculated as paragraphs 8.2 and 8.3 below.

8.2. (a) For services by a member of staff, the charges for which are to be based on the time involved, such charges shall be calculated on the hourly cost of the individual involved plus 145%.

(b) A member of staff shall include a principal doing work normally done by an employee (as paragraph 8.1 (b) above), technical and supporting staff, but shall exclude secretarial staff or staff engaged upon general administration.

(c) For the purpose of paragraph 8.2(b) above, a principal's time shall be taken at the rate applicable to a senior assistant in the firm.

(d) The supervisory duties of a principal shall be deemed to be included in the addition of 145% as paragraph 8.2(a) above and shall not be charged separately.

(e) The hourly cost to the employer shall be calculated by taking the sum of the annual cost of the member of staff of:-

i) salary and bonus but excluding expenses;
ii) employer's contributions payable under any pension and life assurance schemes;

QUANTITY SURVEYORS' FEES

8.0. **Time Charges** - *continued*

iii) employer's contributions made under the National Insurance Acts, the Redundancy Payment Act and any other payments made in respect of the employee by virtue of any statutory requirements; and

iv) any other payments or benefits made or granted by the employer in pursuance of the terms of employment of the member of staff;

and dividing by 1650.

8.3. The foregoing time charges under paragraph 8.2 are intended for use where other sections of the Scale (not related to time charges) form a significant proportion of the overall fee. In all other cases an increased time charge may be agreed.

PART 10

# Approximate Estimates

## 1. INDUSTRIAL AND COMMERCIAL BUILDINGS AND CIVIL ENGINEERING FACILITIES

Prices given under this heading are average prices on a 'fluctuating basis' for typical buildings during the third quarter 1990 with a tender price level index of 330 (1976 = 100). Unless otherwise stated, prices do not allow for external works other than those adjacent to the building, furniture, loose or special equipment and are, of course, exclusive of fees for professional services.

Prices are based upon the total floor area of all storeys, measured between external walls and without deduction for internal walls, columns, stairwells, liftwells and the like.

As in previous editions it is emphasised that the prices must be treated with reserve in that they represent the average of prices from our records and cannot provide more than a rough guide to the probable cost of a building or structure.

In many instances normal commercial pressures together with a limited range of available specifications ensure that a single rate is sufficient to indicate the prevailing average price. However, where such restrictions do not apply a range has been given; this is not to suggest that figures outside this range will not be encountered, but simply that the calibre of such a type of building can itself vary significantly.

As elsewhere in this edition, prices do not include Value Added Tax, which should be applied at the current rate to all non-domestic building.

| | Square metre excluding VAT £ |
|---|---|
| Surface car parking | 35 to 50 |
| Multi-storey car parks | |
| split level | **190 to 210** |
| flat slab | **200 to 250** |
| warped | **220 to 265** |
| Underground car parks | |
| partially underground under buildings | **325 to 390** |
| completely underground under buildings | **390 to 475** |
| completely underground with landscaped roof | **430 to 515** |
| Railway stations | **1200 to 2000** |
| Bus and coach stations | **600 to 1000** |
| Bus garages | **575 to 625** |
| Garage showrooms | **750 to 1100** |
| Garages, domestic | **450 to 700** |
| Airport passenger terminal buildings (excluding aprons) | |
| national standard | **1000 to 1200** |
| international standard | **1400 to 1800** |
| Airport facility buildings | |
| large hangars | **1000 to 1200** |
| workshops and small hangars | **630 to 720** |
| TV, radio and video studios | **725 to 1175** |
| Telephone exchanges | **600 to 950** |
| Telephone engineering centres | **500 to 625** |
| Branch Post Offices | **625 to 850** |

| | Square metre excluding VAT £ |
|---|---|
| **Industrial facilities** | |
| Sub-stations | 800 to 950 |
| Agricultural storage buildings | 300 to 400 |
| Factories | |
| for letting (incoming services only) | 215 to 295 |
| for letting (including lighting, power and heating) | 285 to 390 |
| nursery units (including lighting, power and heating) | 355 to 515 |
| workshops | 400 to 515 |
| maintenance/motor transport workshops | 400 to 675 |
| owner occupation - for light industrial use | 360 to 515 |
| owner occupation - for heavy industrial use | 675 to 765 |
| Factory/office buildings - high technology production | |
| for letting (shell and core only) | 390 to 515 |
| for letting (ground floor shell, first floor offices) | 630 to 810 |
| for owner occupation (controlled environment, fully finished) | 810 to 1080 |
| Warehouses | |
| low bay (6-8 m high) for letting (no heating) | 215 to 260 |
| low bay for owner occupation (including heating) | 300 to 380 |
| high bay (9-18 m high) for owner occupation (including heating) | 400 to 515 |
| Cold stores, refrigerated stores | 450 to 575 |

**Water and Treatment Facilities**

| | Cubic metre excluding VAT |
|---|---|
| Reinforced Concrete tanks: including all excavation; fill; structural work; valves; penstocks; pipeworks per $m^3$ of concrete in structure | |
| treatment tanks | 270 to 305 |
| fire ponds and lagoons | 240 to 275 |
| reservoirs | 245 to 280 |
| Reinforced Concrete river weirs, quay and wave walls: including temporary dams; caissons overpumping, anchorages, all structural works. | 195 to 300 |
| Reinforced Concrete dams (upto 12 m high): | |
| arch dam, including excavation anchorages and structural work only. $m^3$ of completed structure | 650 to 950 |
| flat slab dam including excavation anchorages and structural work only. $m^3$ of completed structure | 495 to 775 |
| Earth dams: | |
| rock fill with concrete core $m^3$ of completed structure | 130 to 450 |
| hydraulic fill embankment dam $m^3$ of completed structure | 70 to 200 |

## 2. FOUNDATIONS FOR STRUCTURES

In-situ concrete foundations for steel portal framed buildings:

| | gross floor area | Unit | £ |
|---|---|---|---|
| depth up to 1m | up to 100 $m^2$ | $m^3$ | 265 |
| | 100 to 500 $m^2$ | $m^3$ | 210 |
| | 500 to 1000 $m^2$ | $m^3$ | 165 |
| | 1000 to 2500 $m^2$ | $m^3$ | 125 |
| | over 2500 $m^2$ | $m^3$ | 120 |
| depth 1 - 2m | up to 100 $m^2$ | $m^3$ | 295 |
| | 100 to 500 $m^2$ | $m^3$ | 260 |
| | 500 to 1000 $m^2$ | $m^3$ | 215 |
| | 1000 to 2500 $m^2$ | $m^3$ | 165 |
| | over 2500 $m^2$ | $m^3$ | 120 |
| depth over 2m | | | |
| | up to 100 $m^2$ | $m^3$ | 325 |
| | 100 to 500 $m^2$ | $m^3$ | 280 |
| | 500 to 1000 $m^2$ | $m^3$ | 215 |
| | 1000 to 2500 $m^2$ | $m^3$ | 165 |
| | over 2500 $m^2$ | $m^3$ | 120 |

Reinforced Concrete slabs for internal or external:

reinforced concrete slab C30 concrete, reinforced with A393 fabric laid in bays size 20 m x 10 m

slabs on sub-base, 250 mm granular material including all excavation (250 mm depth) formwork and formed expansion joints with filler

| | | Unit | £ |
|---|---|---|---|
| Tamped finish | 200 mm thick | $m^2$ | 37.00 |
| | 250 mm thick | $m^2$ | 44.00 |
| | 275 mm thick | $m^2$ | 44.50 |
| | 300 mm thick | $m^2$ | 46.50 |
| extra over cost for hand trowel finish | | $m^2$ | 2.50 |
| for power float finish | | $m^2$ | 1.85 |
| for wire brush finish | | $m^2$ | 1.15 |
| extra over cost for additional sub base depth (including excavation/disposal) | | | |
| | 50 mm | $m^2$ | 2.25 |
| | 100 mm | $m^2$ | 4.30 |
| | 150 mm | $m^2$ | 6.50 |

The prices following have been assembled from the relevant items in the unit costs section. They are intended to give a broad overall price or help in comparisons between a number of basic construction procedures. These approximate estimates are for construction only. They do not include for preliminaries, design/supervision costs, land purchase or OH&P etc.

Prices in this section are based on the same information and outputs as used in the unit costs section. Costs per $m^2$ or $m^3$ of completed structure.

| 3. EARTH RETENTION AND STABILISATION | Unit | £ |
|---|---|---|
| Reinforced in-situ concrete retaining wall: including excavation; reinforcement formwork; expansion joints and granular backfill and profiled formwork finish to one side (typical retaining wall, allowing for profiling finishes) | | |
| 1 m high | $m^3$ | **185** |
| 3 m high | $m^3$ | **209** |
| 6 m high | $m^3$ | **279** |
| 9 m high | $m^3$ | **395** |
| Precast concrete block, earth retaining wall: including granular fill; earth anchors and proprietary units or concrete panels, strips, fixings and accessories (Reinforced Earth) | | |
| 1.50 m high | $m^2$ | **120** |
| 3.0 m high | $m^2$ | **140** |
| 6.0 m high | $m^2$ | **180** |
| Precast, reinforced concrete unit retaining wall (in-situ foundation): including excavation and fill; reinforced concrete foundation; pre-cast concrete units, joints | | |
| 1.00 high | $m^3$ | **222** |
| 2.00 high | $m^3$ | **250** |
| 3.00 high | $m^3$ | **265** |
| Precast concrete crib wall including excavation; stabilisation and foundation work | | |
| up to 2.5 m high | $m^2$ | **123** |
| up to 4.0 m high | $m^2$ | **108** |
| Timber crib walling | | |
| up to 2.0 m high | $m^2$ | **90** |
| up to 3.6 m high | $m^2$ | **120** |
| up to 5.7 m high | $m^2$ | **190** |
| Rock gabions including preparation and anchoring | | |
| up to 2 m high | $m^3$ | **58** |
| up to 6 m high | $m^3$ | **74** |

## 4. BRIDGEWORKS

The following prices are based on recovered data and information from approximately 50 separate structures completed as part of recent projects.

| | | Unit | £ |
|---|---|---|---|
| **Road Bridges** | | | |
| Reinforced in-situ concrete viaduct: including excavation; reinforcement; formwork; concrete; bearings; expansion joints; deck waterproofing; deck finishings; P1 parapet. | | | |
| per $m^2$ of deck maximum span between piers or abutments | 15 m | $m^2$ | 800 to 900 |
| | 20 m | $m^2$ | 760 to 800 |
| | 25 m | $m^2$ | 740 to 760 |
| Reinforced concrete bridge with precast beams: including excavation; reinforcement; formwork; concrete; bearings; expansion joints; deck waterproofing; deck finishings; P1 parapet. | | | |
| per $m^2$ of deck maximum span between piers or abutments | 12 m | $m^2$ | 785 to 815 |
| | 17 m | $m^2$ | 735 to 760 |
| | 22 m | $m^2$ | 700 to 720 |
| | 27 m | $m^2$ | 685 to 725 |
| Reinforced concrete bridge with pre-fabricated steel beams: including excavation; reinforcement; formwork; concrete; bearings; expansion joints; deck waterproofing; deck finishings; P1 parapet. | | | |
| per $m^2$ of deck maximum span between piers or abutments | 20 m | $m^2$ | 780 to 880 |
| | 30 m | $m^2$ | 730 to 750 |
| | 40 m | $m^2$ | 710 to 730 |
| **Footbridges** | | | |
| Reinforced in-situ concrete with precast beams: including excavation; reinforcement; formwork; concrete; bearings; expansion joints; deck waterproofing; deck finishings; P6 parapet. | | | |
| widths up to 6 m wide per $m^2$ of deck maximum span | 5 m | $m^2$ | 700 to 725 |
| | 10 m | $m^2$ | 650 to 675 |
| | 20 m | $m^2$ | 675 to 700 |
| Structural steel bridge with concrete foundations, | | | |
| width up to 4 m wide per $m^2$ of deck maximum span | 10 m | $m^2$ | 630 to 660 |
| | 12 m | $m^2$ | 635 to 665 |
| | 16 m | $m^2$ | 640 to 670 |
| | 20 m | $m^2$ | 660 to 680 |
| Timber footbridge (stress graded with concrete piers) | | | |
| per $m^2$ of deck maximum span | 12 m | $m^2$ | 580 to 600 |
| | 18 m | $m^2$ | 600 to 620 |

## 5. HIGHWAY WORKS

The following prices are the approximate costs per metre run of carriageway as described, and are based on information from a number of sources including engineers estimates, tenders, final account values etc on approximately 20 various recent road contracts.

The price include for earthworks, structures, drainage, pavements, footways, signs, lighting, fencing and barrier works as well as allowance for accommodation works, statutory undertakings and landscaping as appropriate to the type and location of the carriageway. The earthworks elements can be adjusted by reference to factors detailed at the end of this sub-section.

| **Highway Type** | Unit | £ range |
|---|---|---|
| Rural Motorways | | |
| Dual 2 lane Motorway (each carriageway is 7.3 m wide) | m | **2025 to 2550** |
| Dual 3 lane Motorway (each carriageway is 11.0 m wide) | m | **2300 to 3600** |
| Dual 4 lane Motorway (each carriageway is 14.6 m wide) | m | **3250 to 5600** |
| Rural All-Purpose Roads | | |
| Dual 2 lane all-purpose road (each carriageway is 7.3 m wide) | m | **1750 to 2350** |
| Dual 3 lane all-purpose road (each carriageway is 11.0 m wide) | m | **2025 to 3100** |
| Single carriageway all-purpose road (carriagway is 7.3 m wide) | m | **1120 to 1530** |
| Wide single carriageway all-purpose road (carriageway is 10.0 m wide) | m | **1420 to 1850** |
| Rural Link Roads | | |
| Two lane link road (carriageway is 7.3 m wide) | m | **1245 to 1600** |
| Single lane link road (carriageway is 3.7 m wide) | m | **800 to 975** |
| Rural Motorway or Dual Carriageway Slip Roads | | |
| Single carriageway slip road (carriageway width is 5.0 m) | m | **920 to 1280** |
| Urban Motorways | | |
| Dual 2 lane Motorway (each carriageway is 7.3 m wide) | m | **2355 to 2900** |
| Dual 3 lane Motorway (each carriageway is 11.0 m wide) | m | **2750 to 4200** |
| Dual 4 lane Motorway (each carriageway is 14.6 m wide) | m | **3900 to 6875** |
| Urban Motorway or Dual Carriageway Sliproads | | |
| Single carriageway slip road (carriageway is 6.0 m wide) | m | **1090 to 1480** |

## 5. HIGHWAY WORKS

| | Unit | £ range |
|---|---|---|
| **Urban All-Purpose Roads** | | |
| Dual 2 lane all-purpose road - no footpaths (each carriageway is 7.3 m wide) | m | **1900 to 2650** |
| Dual 3 lane all-purpose road - no footpaths (each carriageway is 11.0 m wide) | m | **2255 to 3150** |
| Dual 2 lane all-purpose road with verge or footway each side (each carriageway is 7.3 m wide, each footway is 3.0 m wide) | m | **2025 to 2550** |
| Dual 3 lane all-purpose road with verge or footway each side (each carriageway is 11.0 m wide, each footway is 3.0 m wide) | m | **2375 to 3500** |
| Single carriageway all-purpose road with footway 1 side (Carriageway is 7.3 m wide footway is 3.0 m wide) | m | **1240 to 1650** |
| Wide single carriageway all purpose road with footway each side (carriageway is 10.0 m wide each footway is 3.0 m wide) | m | **1580 to 1990** |
| Cost of nom 3.0 m cycle track to all-purpose roads | m | **325 to 500** |
| **Urban Link Roads** | | |
| Two lane link road (carriageway is 7.3 m wide) | m | **1075 to 1390** |
| Single lane link road (carriageway is 3.7 m wide) | m | **960 to 1190** |

The average Earthworks proportion of the above prices is 33%

Use the following tables of multipliers to adjust this percentage for depths and ground conditions.

| Depths (m) Cut | Fill |
|---|---|
| 0-2 x 1.00 | 0-1 x 1.00 |
| 2-3 x 1.17 | 1-2 x 1.20 |
| 3-4 x 1.25 | 2-3 x 1.45 |
| 4-5 x 1.37 | 3-4 x 1.75 |
| | 4-5 x 2.15 |

| Ground Conditions | |
|---|---|
| Sandy clay or marl | x 1.00 |
| Hard clay/soft chalk | x 1.20 |
| Soft rock/mudstones | x 1.25 |
| medium rock/existing pavements | x 1.70 |
| Hard rock | x 2.00 (min) |
| Running sands | x 1.50 (min) |
| Swamp/water logged ground | x 2.25 (min) |

As a very general guide to current Highway Construction Costs, in £ million.

| | | |
|---|---|---|
| Rural Motorways | **£3.105 per km** | **(£5.000 per mile)** |
| Urban Motorways | **£3.700 per km** | **(£5.955 per mile)** |
| Motorway re-construction | **£1.250 per km** | **(£2.000 per mile)** |
| All purpose dual carriageway | **£2.400 per km** | **(£3.850 per mile)** |
| All purpose single carriageway | **£1.545 per km** | **(£2.480 per mile)** |
| Light duty single carriageway up to 5 m wide | **£0.950 per km** | **(£1.525 per mile)** |

## 5. HIGHWAY WORKS

The following are approximate costs for the installation of roads and drains to serve as part of the development of infrastructure for Housing, Retail or Industrial development. The density (ie percentage of area used for Roads etc is also given to enable adjustments) NB, excludes car parking.

Type A Construction = medium duty carriageway consisting of 100 mm surfacing, roadbase up to 115 mm subbase of 150 mm.

Type B Construction = heavy duty carriageway consisting of reinforced concrete slab 225 mm thick subbase 130 mm thick capping layer 280 mm thick.

| Density of Facility (House/Outlet/Unit) | | per unit £ | | per hectare £ | | Density of road per hectare |
|---|---|---|---|---|---|---|
| per hectare | (per acre) | Type A | Type B | Type A | Type B | |
| 5 | (2) | 6408 | 7000 | 32040 | 35000 | 2% |
| 15 | (6) | 3600 | 3933 | 54000 | 59000 | 3.5% |
| 20 | (8) | 2975 | 3260 | 59500 | 65200 | 3.8 |
| 25 | (10) | 2620 | 2842 | 65500 | 71050 | 4.1% |
| 30 | (12) | 2368 | 2606 | 71050 | 78200 | 4.5 |
| 37 | (15) | 2221 | 2461 | 82175 | 91050 | 5.2% |
| 50 | (20) | 2013 | 2250 | 100650 | 112500 | 6.5% |

| | | £ |
|---|---|---|
| Turning or passing bay = 35 $m^2$ overall area suitable cars, vans | | |
| Type A construction | each | 5775.00 |
| Type B construction | each | 6320.00 |
| Turning or passing bay = 100 $m^2$ overall area suitable semi trailer | | |
| Type A construction | each | 16500.00 |
| Type B construction | each | 18000.00 |
| Bus lay-by = 40 $m^2$ overall area | | |
| Type B construction | each | 7200.00 |
| Parking lay-by = 200 $m^2$ overall area | | |
| Type A construction | each | 33000.00 |
| Type B construction | each | 36000.00 |
| Vehicle crossing verge/footway/central reserve | $m^2$ | 170.00 |
| Footway construction (bit-mac plus edgings) | $m^2$ | 19.00 |

## 6. CIVIL ENGINEERING WORKS SITE UTILITIES AND INFRASTRUCTURE

The following prices have been compiled for various services and provide average costs for site utilities and infrastructure works.

| | Unit | To estate road £ | To major road £ |
|---|---|---|---|
| **Underpasses** | | | |
| Provision of underpasses to new roads, constructed as part of a road building programme | | | |
| Pedestrian underpass 3 m wide x 2.5 m high | m | 2900 | 3500 |
| Vehicle underpass 7 m wide x 5 m high | m | 14700 | 18000 |
| Vehicle underpass 14 m wide x 5 m high | m | - | 36000 |

**Road Crossing**

Costs include road markings, beacons, lights, signs, advance danger signs etc

| | Unit | £ |
|---|---|---|
| 4 way traffic signal installation | each | 33000 - 45000 |
| zebra crossing | each | 3500 - 4000 |
| pelican crossing | each | 14000 - 15000 |
| pedestrian guard railing | m | 80 - 100 |

**Street furniture**

There is an almost infinite variety of items with which local authorities and private developers can enhance the roadside and public areas, including statutory requirements such as lighting. It is therefore impossible to price all the different street furniture available. The following however gives a selection of the more common items:-

| | £ each |
|---|---|
| reflectorised traffic signs 0.25-0.75 $m^2$ area on steel post | 78 to 157 |
| internally illuminated traffic signs (dependent on area) | 160 to 210 |
| externally illuminated traffic signs (up to 4 $m^2$) | 175 to 750 |
| lighting to pedestrian areas and estates roads on 4-6 m columns with up to 70 W lamps | 190 to 275 |
| lighting to main roads on 10-12 m columns with 250 W lamps | 415 to 470 |
| lighting to main roads on 12-15 m columns with 400 W high pressure sodium lighting | 540 to 590 |
| benches - hardwood and precast concrete | 120 to 150 |
| litter bins: precast concrete | 150 to 170 |
| hardwood slatted | 60 to 80 |
| cast iron | 250 |
| large aluminium | 450 |
| concrete bollards | 40 to 70 |
| steel bollards | 35 to 60 |
| bus stops | 275 |
| bus stops inc basic shelter | 625 |
| pillar box on post | 225 |
| telephone box | 2500 |
| Hi mast radio/beacon - 60 m | 17000 |
| automatic barrier equipment | 12000 |

## 6. CIVIL ENGINEERING WORKS SITE UTILITIES AND INFRASTRUCTURE

### Footpaths and Pavings

Costs include excavation, base course as necessary and precast concrete edgings on foundations to one or both sides

| | Unit | £ |
|---|---|---|
| bitumen macadam footpath | $m^2$ | 19 |
| precast concrete paving flags | $m^2$ | 28 |
| precast concrete block pavings | $m^2$ | 35 |
| clay brick paving | $m^2$ | 43 |
| granite setts | $m^2$ | 51 |
| cobbled paving | $m^2$ | 60 |

### Surface Car Parks

Surface car parking is the cheapest way of providing car parking. The cost of rooftop or basement car parking can be 10 to 15 times the cost of the surface car parking. The vogue in car parking for all except industrial schemes is for concrete block pavers, with different colours marking out parking bays and zones. Grasscrete is an alternative but is not usually used for main parking areas.

Guideline figures for parking requirements are shown below:-

| | one car space per |
|---|---|
| offices | 22 - 25 $m^2$ GFA |
| industrial - factories | 45 - 55 $m^2$ GFA |
| - warehouses | 200 $m^2$ GFA |
| shops | 20 - 25 $m^2$ GFA |
| superstores | 10 $m^2$ GFA |
| cinemas, theatres | 3 - 5 seats |
| hospital | 3 hospital beds |
| residential | 1 - 2 dwellings |

(GFA - Ground Floor Area)

For surface level car parking the area required per car will not generally show much variation. Site shape, position of the building on the site, and parking configuration will be the main determinants of the area to be allowed per car, however, a fairly low range of 20-23 $m^2$ per car will usually be sufficient.

Typical costs for surface car parking including lighting and drainage are illustrated below:-

| | Cost/$m^2$ | Cost per car |
|---|---|---|
| tarmacadam finished, marked out with thermoplastic road paint | **38** | **760 - 880** |
| interlocking or herringbone concrete block paving marked out with coloured blocks | **42** | **840 - 960** |
| grasscrete precast concrete units filled with top soil and grass seed | **45** | **900 - 1050** |

Note: Costs include forecourts, aprons and access areas but not approach roads.

## 6. CIVIL ENGINEERING WORKS SITE UTILITIES AND INFRASTRUCTURE

### Services

Costs of services to a site are built up of connection charges and service runs. The calculation of connection charges to a site is complicated by the possibility of changes in the responsibilities of the statutory bodies supplying the services in the future. For instance the privatisation of telephone, water and gas and the future privatisation of electricity could complicate the assessment of contributions.

However, typical service charges for an estate of 200 houses might be as follows:-

| | Charge per house £ | |
|---|---|---|
| Water | 400 to 800 | |
| Electric | 200 if all electric<br>400 if gas/electric | ) Plus cost of<br>) substation<br>) £12,000 -<br>) £17,000 |
| Gas | 400 - 500 plus cost of governing station £11,000 | |
| Telephone | 130 | |
| Sewerage | 300 - 400 | |

Mains laid trenches including excavation and filling.

| | Unit | £ |
|---|---|---|
| Water main | | |
| 75 mm upvc main in 150 mm ductile iron pipe as duct | m | 47 |
| Electric main | | |
| 600/1000 volt cables. Two core 25 mm cable inc 100 mm clayware duct | m | 20 |
| Gas main | | |
| 150 mm ductile or cast iron gas pipe | m | 37 |
| Telephone | | |
| British Telecom installation in 110 mm upvc duct | m | 15 |
| Drainage | | |
| 100/150 mm clay or upvc pipes on granular bed and surround up to 3 m deep | m | 18 |
| 100/150 mm cast iron pipes on concrete beds | m | 50 |
| 375 mm clay drain pipe up to 3 m deep on granular bed and surround | m | 46 |
| 450 mm concrete pipe on concrete beds | m | 52 |
| 900 mm diameter concrete sewage pipe on granular bed up to 3.5 m deep | m | 120 |
| Brick manholes in commons, rendered internally, clay channel with three branches, concrete cover slab, cast iron cover and frame 900 - 1500 mm deep | each | 675 |
| Precast concrete manholes with precast concrete rings up to 1500 mm diameter, channel with three branches, concrete cover slab and frame 900-1500 mm deep | each | 595 |
| Vitrified clay or precast concrete road gully 450 x 900 mm deep with concrete surround, brick seating and cast iron grating | each | 152 |

## 7. LANDSCAPING

Landscaping as a subject matter is sufficiently large to fill a book (see *Spons Landscaping and External Works Price Book*). However, there are certain items that arise on a majority of projects and other discrete items for which indicative costs can be produced. The following rates include for normal surface excavation but exclude bulk excavation, levelling or earth shifting and land drain provision.

Soft landscaping

| | Unit | £ |
|---|---|---|
| cultivate ground, remove rubbish and plant with grass seed | $m^2$ | 4 - 6 |
| ditto and turf | $m^2$ | 6 - 10 |
| shrubbed areas | $m^2$ | 17 - 35 |
| shrubbed areas including allowance for small trees | $m^2$ | 25 - 60 |
| standard tree in tree pit including stake | nr | 38 |
| ditto but with tree guard and precast tree grid slabs | nr | 120 |

Sports pitches

The provision of sports facilities will involve different techniques of earth shifting and cultivation and usually will be carried out by specialist landscaping contractors.

Costs include for cultivating ground, bringing to appropriate levels for the specified game, applying fertiliser, weedkiller, seeding and rolling and white line marking with nets, posts etc as required.

| | £ |
|---|---|
| football pitch (114 x 72 m) | **13,000 each** |
| cricket outfield (160 x 142 m) | **44,000 each** |
| cricket square (20 x 20 m) inc imported marl or clay loam bringing to accurate levels, seeding with cricket square type grass | **4,000 each** |
| bowling green (38 x 38 m) rink, inc french drain and gravel path on four sides | **16,000 each** |
| grass tennis courts 1 court (35 x 17 m) inc bringing to accurate levels, chain link perimeter fence and gate, tennis posts and net | **16,200 each** |
| two grass tennis courts (35 x 32 m) ditto | **27,500 pair** |
| artificial surface tennis courts (35 x 17 m) inc chain link fencing, gate, tennis posts and net | **13,000 each** |
| two courts (45 x 32 m) ditto | **24,000 pair** |
| artificial football pitch inc sub-base, bitumen macadam open textured base and heavy duty astro-turf carpet | **250,000 each** |
| golf putting green | **1,300/hole** |
| pitch and put course | **4000-6000/hole** |
| full length golf course, full specifications inc watering system | **16,000 - 30,000/hole** |
| championship course | **up to 100,000 per hole** |

## 7. LANDSCAPING

### Parklands

As with all sports pitches, parklands will involve different techniques of earth shifting and cultivation.

The following rates include for normal surface excavation.

| | £ |
|---|---|
| parklands, inc cultivating ground applying fertiliser etc seeding with parks type grass | **12,300/ha** |
| general sportsfield | **14,800/ha** |
| lakes inc excavation up to 10 m deep, laying 1.5 mm thick butyl rubber sheet and spreading top soil evenly on top to depth of 300 m | |
| under 1 ha in area | **270,000/ha** |
| between 1 and 5 ha in area | **250,000/ha** |
| extra for planting aquatic plants in lake top soil | **40/m2** |

### Playground Equipment

| | £ each |
|---|---|
| modern swings with flat rubber safety seats: four seats, two bays | **1000** |
| stainless steel slide, 3.40 m long | **1100** |
| climbing frame - igloo type 3.20 x 3.75 m on plan x 2.00 m high | **1150** |
| seesaw comprising timber plank on sealed ball bearings 3960 x 230 x 70 mm thick | **800** |

| | Unit | £ |
|---|---|---|
| wicksteed tumbleguard type safety surfacing around play equipment | $m^2$ | **66** |
| bark particles type safety surfacing 150 mm thick on hardcore bed | $m^2$ | **9** |

### Land Drainage

The above rates exclude provision of any land drainage. If land drainage is required on a project, the propensity of the land to flood will decide the spacing of the land drains. However, some indicative figures can be given for land drainage.

Costs include for excavation and backfilling of trenches and laying agricultural clay drain pipes with 75 mm diameter lateral runs average 600 mm deep and 100 mm diameter main runs average 750 mm deep.

| | £ |
|---|---|
| land drainage to parkland with laterals at 30 m centres and main runs at 100 m centres | **2,500/ha** |
| land drainage to sportsfields with laterals at 10 m centres and main runs at 33 mm centres | **7,500/ha** |

## 8. TEMPORARY WORKS

Installation of temporary Bailey bridges:

including temporary concrete abutments; erection maintenance; dismantling

span up to 10 m

| | £ | £ |
|---|---|---|
| Hire costs (for 52 weeks) | 7550.00 | |
| Delivery/collect | 560.00 | |
| Erect/dismantle | 2750.00 | |
| Concrete abutments, inc | 2232.00 | |
| Demolish after dismantling | 390.00 | |
| Allowance for maintenance, etc | 1200.00 = | 14682.00 |
| span 15 m | | 17082.00 |
| span 20 m | | 22235.00 |
| span 25 m | | 24627.00 |

Installation of cofferdams: £ range

(based on driven steel sections with recovery value)
including all plant, for installation and dismantling; loss of materials; pumping and maintainance excluding excavation and disposal of material - backfilling on completion.

cost range based on 12 weeks installation

| | | £ range |
|---|---|---|
| depth of drive | up to 5 m | |
| diameter | up to 2 m | 1720.00 - 1885.00 |
| or sides | up to 10 m | 8500.00 - 9420.00 |
| | up to 20 m | 15010.00 - 18850.00 |
| depth of drive | 5 - 10 m | |
| diameter | up to 2 m | 2500.00 - 3510.00 |
| or sides | up to 10 m | 11050.00 - 17595.00 |
| | up to 20 m | 20350.00 - 35190.00 |
| depth of drive | 10 - 15 m | |
| diameter | up to 2 m | 3500.00 - 4900.00 |
| or sides | up to 10 m | 17000.00 - 24500.00 |
| | up to 20 m | 31075.00 - 49000.00 |
| depth of drive | 15 - 20 m | |
| diameter | up to 2 m | 6000.00 - 7550.00 |
| or sides | up to 10 m | 17050.00 - 30165.00 |
| | up to 20 m | 42800.00 - 60300.00 |

Above based on soft-medium ground conditions

add for medium-hard + 20%
hard but not rock + 33%

## 8. TEMPORARY WORKS

Access Scaffolding

The following are guideline costs for the hire and erection of proprietary scaffold systems, (tube and coupling) costs are very general and are based on a minimum area of 360 $m^2$ at 1.80m deep.

| | | | £ |
|---|---|---|---|
| 1. | Approximate hire (supply and fix) of patent scaffold per sq. metre per 4 week (cost dependant upon length of hire, quality of system, number of toe boards handrails etc.)<br><br>Based on this a typical cost of a 60 x 10m (600 $m^2$) area for 8 weeks would be about £4,200. | = | 3.50 - 4.70 |
| 2. | Approximate hire (supply and fix of mobile access towers per sq. metre per 4 week hire<br><br>(refer also to plant hire in Section 3) | = | 12.00 - 18.00 |
| 3. | Additional costs of Pole ladder access, per 4.0m high, per 4 week | = | 4.50 - 5.10 |
| 4. | Additional cost of stair towers extra over the cost of scaffold system per 2m rise, per 4 week | = | 50.00 |
| 5. | Additional cost of hoarding around base perimeter (using multi use ply sheeting) per $m^2$ per 4 week | = | 0.70 - 1.25 |
| 6. | Additional cost of polythene debris netting (no re-use) per $m^2$ | = | 0.12 |
| 7. | Additional cost of Monaflex 'T' plus weatherproof sheeting including anchors and straps (based on 3 uses) per $m^2$ | = | 0.69 |

Erection of scaffolding system is based on 3 men erecting a 16 $m^2$ bay in about 1 hour and dismantling the same in 20 minutes. (i.e. experienced scaffolders)

Note: Although scaffolding is essentially plant hire, allowance must be made for inevitable loss and damage to fittings, for consumables used and for maintenance during hire periods.

## 8. TEMPORARY WORKS

Earthwork Support

The following are comparative costs for Earthwork and trench support based on the hire of modular systems (trench box hydraulic) with an allowance for consumable materials (maximum 5 day hire allowance).

| | £/m$^2$ |
|---|---|
| Earthwork support not exceeding 1m deep, distance between opposing faces not exceeding 2m | 2.60 |
| Earthwork support not exceeding 1m deep distance between opposing faces 2-4m | 4.40 |
| Earthwork support not exceeding 2m deep distance between opposing faces not exceeding 2m | 4.90 |
| Earthwork support not exceeding 2m deep distance between opposing faces 2-4m | 8.75 |
| Earthwork support not exceeding 4m deep distance between opposing faces not exceeding 2m | 12.50 |
| Earthwork support not exceeding 4m deep distance between opposing face 2-4m | 18.00 |

For larger excavations requiring Earthwork support refer to cofferdam estimates; or sheet piling within unit cost sections.

The following are approximate weekly hire costs for a range of basic support equipment used on site (see also page 129 plant section).

| | | £ per week |
|---|---|---|
| Steel Sheet piling | | |
| Frodingham or Larssen section | per tonne | 12.00 - 15.00 |
| Trench Sheeting | | |
| Standard overlapping sheets | per m$^2$ | 0.55 - 0.65 |
| Interlocking type | per m$^2$ | 1.10 - 1.50 |
| Heavy duty overlapping sheets | | |
| 6 mm thick | per m$^2$ | 1.75 - 1.90 |
| 8 mm thick | per m$^2$ | 1.85 - 2.00 |
| driving cap | each | 5.00 |
| extraction clamp | each | 10.00 |
| Trench box with hydraulic walings - as above | | |
| Trech struts | | |
| No 0 0.3 - 0.40 m | each | 0.23 |
| No 1 0.5 - 0.75 m | each | 0.26 |
| No 2 0.7 - 1.14 m | each | 0.27 |
| No 3 1.0 - 1.75 | each | 0.32 |

PART 11

# Outputs

**HYDRAULIC BACKACTERS - Production rates (90° swing)**

| Weight (t) (Heaped Capacity) | Power (kw) | Material | Cycle Time (sec) | Operating Co-eff | Bucket Co-eff | Hourly Standard Prod $m^3$/h |
|---|---|---|---|---|---|---|
| 6.2 (0.25$m^3$) | 36 | Soil | Av 13 10-16 | 0.83 (50/60 min) | 1.0 | Av 55 45-75 |
| 11.5 (0.40$m^3$) | 53 | Soil | Av 13 11-15 | 0.83 | 1.0 | Av 90 80-110 |
| 13.25 (0.45$m^3$) | 63 | Soil | Av 13 11-15 | 0.83 | 1.0 | Av 100 90-120 |
| 16 (0.55$^3$) | 68 | Soil | Av 14 12-16 | 0.83 | 1.0 | Av 120 100-140 |
| 19 (0.70$m^3$) | 85 | Soil | Av 14 12-16 | 0.83 | 1.0 | Av 150 130-170 |
| 21 (1.00$m^3$) | 95 | Soil | Av 17 15-19 | 0.83 | 1.0 | Av 175 160-200 |
| 23.5 (0.90$m^3$) | 110 | Soil | Av 16 14-18 | 0.83 | 1.0 | Av 170 150-190 |
| 28.5 (1.20$m^3$) | 136 | Soil | Av 16 14-18 | 0.83 | 1.0 | Av 220 200-260 |
| 35.7 (1.20$m^3$) | 150 | Soil | Av 18 16-20 | 0.83 | 1.0 | Av 200 180-225 |
| 41 (1.60$m^3$) | 184 | Soil | Av 19 17-21 | 0.83 | 1.0 | Av 250 230-280 |
| 41 (1.60$m^3$) | 184 | Rock: well blasted | Av 21 19-23 | 0.83 | 0.9 | Av 205 190-230 |
| 41 (1.60$m^3$) | 184 | Rock: poorly blasted | Av 25 23-27 | 0.83 | 0.9 | Av 170 160-190 |
| 51 (2.00$m^3$) | 220 | Soil | Av 21 18-22 | 0.83 | 1.0 | Av 280 240-330 |
| 51 (2.00$m^3$) | 220 | Rock: well blasted | Av 24 22-26 | 0.83 | 0.9 | Av 220 200-240 |
| 51 (2.00$m^3$) | 220 | Rock: poorly blasted | Av 29 27-31 | 0.83 | 0.9 | Av 180 170-200 |
| 20 (0.8$m^3$) (Long reach excavator VC20) | 90 | Soil | Av 24 23-25 | 0.83 | 1.0 | Av 85 80-90 |

**BACKACTER IDENTIFICATION CHART**

| **MACHINE TYPE** | **GROUP 1** | **GROUP 1A** (wheeled) | **GROUP 2** | **GROUP 3** |
|---|---|---|---|---|
| Excavator weight group in tonnes | 8 to 13 | 8 to 13 | 13 to 18 | 18 to 21 |
| Akerman 'H' range | H3 | - | 7C | 10B |
| Atlas 'D' range | 1104 | 1204 | 1304, 1404 | 1604 |
| Caterpillar range | E70, 110, 120, 205 | - | E140, 211, 213 | E180, 215 |
| Daewoo range | DH040, 110, 130 | - | DH120M | DH180 |
| Demag range | - | - | H15 | H21 |
| Dresser range | 625E | | | 635E |
| Fiat/Allis range | FE12 | - | FE16, 18 | FE20 |
| Ford/Richier range | H42 | - | H45 | H48 |
| Hannomag/M-F range | - | - | HC150, 170 | HC190 |
| Hitachi 'EX' range | EX90, 100, 120 | - | EX150 | EX200 |
| Hymac range | 580 | - | - | 181, 211 |
| International - 1H range | 620 | - | 630, 635B | 640 |
| JCB range | - | - | 812, 814 | 817, 818, 820 |
| John Deere range | - | - | - | 690 |
| Komatsu 'PC' range | 100, 120 | - | 150 | 180 |
| Liebherr range | - | R992 | R900B, 902 | R912, 922 |
| Macmoter Spa range | M5, 7, 9 | - | - | - |
| Mitsubishi range | 110 | - | 140 | 180 |
| O & K 'RH' range | 2, 4 | - | 4 Plus | 5, 6 |
| Poclain range | 61 | - | 81 | 90 |
| Priestman Mustang range | 2 - 12 | - | 2 - 15 | 2 - 18, VC 15 |

| <u>GROUP 4</u> | <u>GROUP 5</u> | <u>GROUP 6</u> | <u>GROUP 7</u> | <u>GROUP 8</u> |
|---|---|---|---|---|
| <u>21 to 28</u> | <u>28 to 34</u> | <u>34 to 45</u> | <u>45 to 55</u> | <u>55 to 65</u> |
| 12B | 14B | 16D | | 25C |
| 1704, 1804 | 1902 | 2202 | - | - |
| E240, 225B | 229, E300 | - | - | - |
| DH280 | - | FE40 | - | - |
| - | - | H40 | H55 | - |
| 645 | 650B | - | - | - |
| FE24 | FE28 | FE40 | - | - |
| H50 | H52 | - | - | - |
| - | - | - | - | - |
| EX220 | EX300 | - | - | - |
| 251 | 301 | - | - | - |
| 645 | 650 | 650B | - | - |
| - | - | - | - | - |
| - | - | 890 | - | - |
| 210, 240 | 280 | 360 | 400 | - |
| R932 | R942 | R952 | R962 | R974 |
| - | - | - | - | - |
| 230 | 280 | 380 | - | 580 |
| 8, 9 | 11,12 | 20 | 30 | 30C |
| - | - | 220 | 220 | 350 |
| - | VC20 | - | - | - |

**DISPOSAL OF EXCAVATED MATERIALS**

OUTPUTS PER HUNDRED (100) CUBIC METRES

| Tipper Capacity and Length of Haul | Outputs Average Speed 10 mph | | | Outputs Average Speed 15 mph | | | Outputs Average Speed 20 mph | | |
|---|---|---|---|---|---|---|---|---|---|
| | Driver | Labour | Total | Driver | Labour | Total | Driver | Labour | Total |
| 3 Cu M tipper: | | | | | | | | | |
| 1 km haul | 15.0 | 2.8 | 17.8 | 11.6 | 2.8 | 14.4 | 10.5 | 2.8 | 13.3 |
| 5 km haul | 21.8 | 2.8 | 24.6 | 16.2 | 2.8 | 19.0 | 14.0 | 2.8 | 16.8 |
| 9 km haul | 28.2 | 3.0 | 31.2 | 20.6 | 3.0 | 23.6 | 17.3 | 3.0 | 20.3 |
| 12 km haul | 36.0 | 3.0 | 39.0 | 26.8 | 3.0 | 29.8 | 21.0 | 3.0 | 24.0 |
| 15 km haul | 41.7 | 2.5 | 44.2 | - | - | - | - | - | - |
| 8 Cu M tipper: | | | | | | | | | |
| 1 km haul | 11.3 | 2.1 | 13.4 | 8.8 | 2.0 | 10.8 | 7.9 | 2.1 | 9.0 |
| 5 km haul | 16.2 | 2.1 | 18.3 | 12.0 | 2.0 | 14.0 | 10.4 | 2.1 | 12.5 |
| 9 km haul | 21.6 | 2.0 | 23.6 | 15.8 | 2.3 | 18.1 | 13.2 | 2.2 | 15.4 |
| 12 km haul | 26.4 | 2.0 | 28.4 | 18.7 | 2.3 | 21.0 | 15.6 | 2.2 | 17.8 |
| 15 km haul | 31.3 | 1.3 | 32.6 | - | - | - | - | - | - |
| 12 Cu M tipper: | | | | | | | | | |
| 1 km haul | 9.0 | 1.7 | 10.7 | 7.0 | 1.7 | 8.7 | 6.3 | 1.6 | 7.9 |
| 5 km haul | 13.0 | 1.7 | 14.7 | 9.7 | 1.7 | 11.4 | 8.3 | 1.7 | 10.0 |
| 9 km haul | 17.1 | 1.8 | 18.9 | 12.3 | 1.8 | 14.1 | 10.4 | 1.7 | 12.1 |
| 12 km haul | 21.0 | 2.0 | 23.0 | 15.0 | 2.0 | 17.0 | 12.4 | 1.7 | 14.1 |
| 15 km haul | 25.0 | 1.7 | 26.7 | - | - | - | - | - | - |
| 15 Cu M tipper: | | | | | | | | | |
| 1 km haul | 5.6 | 1.0 | 6.6 | 4.8 | 1.0 | 5.8 | 4.0 | 1.0 | 5.0 |
| 5 km haul | 8.2 | 1.0 | 9.2 | 6.0 | 1.0 | 7.0 | 5.2 | 1.0 | 6.2 |
| 9 km haul | 10.5 | 1.1 | 11.6 | 7.8 | 1.1 | 8.9 | 6.5 | 1.0 | 7.5 |
| 12 km haul | 13.2 | 1.1 | 14.3 | 9.2 | 1.1 | 10.3 | 7.6 | 1.0 | 8.6 |
| 15 km haul | 15.6 | 1.3 | 16.9 | - | - | - | - | - | - |

Manhours *include* round trip for tipper and driver, unloading and labour for positioning during loading or tipping.

Manhours *do not include* for excavation or loading of trucks.

**BREAKING OUT PAVEMENT, BRICKWORK, CONCRETE & MASONRY BY HAND & PNEUMATIC BREAKER**

| Description | Unit | Break Out By Hand Using Picks, Shovels & Points Labour | $7m^3$ Compressor (2 Tool) Compressor | Labour | $10m^3$ Compressor (3 Tool) Compressor | Labour |
|---|---|---|---|---|---|---|
| Break out bit mac surfaces on sub base or hardcore: | | | | | | |
| 75mm thick | $m^2$/hr | 1 | 25 | 13 | 50 | 17 |
| 100mm thick | $m^2$/hr | 1 | 20 | 10 | 33 | 11 |
| Break out asphalt roads on hardcore: | | | | | | |
| 150mm thick | $m^2$/hr | 0.60 | 8 | 4 | 11 | 4 |
| 225mm thick | $m^2$/hr | 0.50 | 6 | 3 | 8 | 3 |
| 300mm thick | $m^2$/hr | 0.40 | 4 | 2 | 7 | 2 |
| Remove existing sett paving | $m^2$/hr | 0.80 | 10 | 5 | 17 | 6 |
| Break out brickwork in cement mortar: | | | | | | |
| 200mm thick | $m^2$/hr | 0.20 | 3 | 2 | 5 | 2 |
| Break out concrete in areas: | | | | | | |
| 100mm thick | $m^2$/hr | 0.90 | 9 | 5 | 14 | 5 |
| 150mm thick | $m^2$/hr | 0.50 | 7 | 4 | 10 | 4 |
| 225mm thick | $m^2$/hr | 0.30 | 4 | 2 | 6 | 2 |
| 300mm thick | $m^2$/hr | 0.20 | 3 | 1 | 4 | 1 |
| Break out reinforced concrete | $m^3$/hr | 0.02 | 0.40 | 0.20 | 0.60 | 0.20 |
| Break out sandstone | $m^3$/hr | 0.03 | 0.50 | 0.30 | 0.80 | 0.30 |

**LOADING LOOSE MATERIALS AND ITEMS BY HAND**

| Material | Unit | Loading into Vehicles Tonne | Cubic Metre |
|---|---|---|---|
| Bricks | hr | 1.7 | 2.9 |
| Concrete, batches | hr | 1.4 | 1.2 |
| Gully grates and frames | hr | 1.0 | |
| Kerb | hr | 1.1 | |
| Paving slabs | hr | 0.9 | |
| Pipes, concrete and clayware | hr | 0.9 | |
| Precast concrete items | hr | 0.9 | |
| Soil | hr | 1.4 | 2.0 |
| Steel reinforcement | hr | 0.8 | |
| Steel sections, etc | hr | 1.0 | |
| Stone and aggregates: | | | |
| bedding material | hr | 1.5 | 1.8 |
| filter/subbase | hr | 1.3 | 1.4 |
| rock fill (6" down) | hr | 1.3 | 1.3 |
| Trench planking and shoring | hr | 0.9 | 2.9 |

**LAYING AND JOINTING, FLEXIBLE JOINTED CLAYWARE PIPES**

| Diameter of Pipe in mm | Unit | In Trench not exceeding 1.5m Drainage Gang | In Trench not exceeding 3m Drainage Gang | In Trench 3-4.5m Drainage Gang |
|---|---|---|---|---|
| 100 | m/hr | 10 | 8 | 7 |
| 150 | m/hr | 7 | 6 | 5 |
| 225 | m/hr | 5 | 4 | 3 |
| 300 | m/hr | 4 | 3 | 2 |
| 375 | m/hr | 3 | 2 | 2 |
| 450 | m/hr | 2 | 2 | 1 |
| **Bends** | | | | |
| 100 | m/hr | 20 | 17 | 14 |
| 150 | m/hr | 17 | 14 | 13 |
| 225 | m/hr | 13 | 10 | 8 |
| 300 | m/hr | 10 | 8 | 7 |
| 375 | m/hr | 7 | 6 | 5 |
| 450 | m/hr | 3 | 3 | 2 |
| **Single junctions** | | | | |
| 100 | m/hr | 13 | 10 | 8 |
| 150 | m/hr | 7 | 6 | 5 |
| 225 | m/hr | 6 | 5 | 4 |
| 300 | m/hr | 4 | 3 | 3 |
| 375 | m/hr | 3 | 2 | 2 |
| 450 | m/hr | 2 | 2 | 1 |

**PRECAST CONCRETE MANHOLES IN SECTIONS**

| Description | Unit | Pipelayer | Labourer |
|---|---|---|---|
| Place 675mm dia shaft rings | m/hr | 1 | 0.30 |
| Place 900mm manhole rings | m/hr | 0.60 | 0.20 |
| Place 1200mm manhole rings | m/hr | 0.40 | 0.10 |
| Place 1500mm manhole rings | m/hr | 0.30 | 0.10 |
| Place 900/675 mm tapers | nr/hr | 1 | 0.30 |
| Place 1200/675mm tapers | nr/hr | 0.60 | 0.20 |
| Place 1500/675mm tapers | nr/hr | 0.50 | 0.20 |
| Place cover slabs to 675mm rings | nr/hr | 2.50 | 0.80 |
| Place cover slabs to 900mm rings | nr/hr | 2 | 0.70 |
| Place cover slabs to 1200mm rings | nr/hr | 1.40 | 0.50 |
| Place cover slabs to 1500mm rings | nr/hr | 0.90 | 0.30 |
| Build in pipes and make good base: | | | |
| 150mm diameter | nr/hr | 4.0 | - |
| 300mm diameter | nr/hr | 2 | - |
| 450mm diameter | nr/hr | 1 | - |
| Benching 150mm thick | $m^2$/hr | 1.20 | 1.20 |
| Benching 300mm thick | $m^2$/hr | 0.60 | 0.60 |
| Render Benching 25mm thick | $m^2$/hr | 1.10 | 1.10 |
| Fix Manhole covers frame | nr/hr | 1.3 | 1.3 |

**READY MIXED CONCRETE (PLACING IN THE WORKS)**

| Description | Labour Gang ($m^3$ per hour) |
|---|---|
| **Mass Concrete** | |
| Blinding: | |
| 150mm thick | 5.50 |
| 150mm - 300mm thick | 6.25 |
| 300mm - 500mm thick | 7.00 |
| Bases and oversite concrete: | |
| not exceeding 150mm thick | 5.00 |
| not exceeding 300mm thick | 5.75 |
| not exceeding 500mm thick | 6.75 |
| exceeding 500mm thick | 7.00 |
| **Reinforced Concrete** | |
| Bases: | |
| not exceeding 300mm thick | 5.50 |
| not exceeding 300mm - 500mm thick | 6.25 |
| exceeding 500mm thick | 6.75 |
| Suspended slabs not exceeding 3m above pavement level: | |
| 150mm thick | 3.75 |
| 150mm - 300mm thick | 4.75 |
| exceeding 300mm thick | 5.25 |
| Walls and stems not exceeding 3m above pavement: | |
| thickness not exceeding 150mm | 3.50 |
| thickness not exceeding 300mm | 4.50 |
| thickness exceeding 300mm | 5.00 |
| Beams, columns and piers not exceeding 3m above pavement: | |
| sectional area not exceeding $0.03m^2$ | 2.00 |
| sectional area $0.03\text{-}1.0m^2$ | 2.50 |
| sectional area exceeding $1.0m^2$ | 3.50 |

**FIXING BAR REINFORCEMENT**

All bar delivered to site, cut and bent and marked. Including craneage and hoisting (maximum height 5m).

| Description (fix only) | Unit | up to 6 mm steel fixer | labour | 7 to 12 mm steel fixer | labour | 13 to 19 mm steel fixer | labour | > 19 mm steel fixer | labour |
|---|---|---|---|---|---|---|---|---|---|
| **Straight Round Bars** | | | | | | | | | |
| To beams, floors, roofs and walls | tonne /hour | 0.03 | 0.03 | 0.04 | 0.04 | 0.06 | 0.06 | 0.08 | 0.08 |
| To braces, columns, sloping roofs and battered walls | tonne /hour | 0.01 | 0.01 | 0.02 | 0.02 | 0.03 | 0.03 | 0.05 | 0.05 |
| **Bent Round Bars** | | | | | | | | | |
| To beams, floors, roofs and walls | tonne /hour | 0.02 | 0.02 | 0.03 | 0.03 | 0.03 | 0.03 | 0.04 | 0.04 |
| To braces, columns, sloping roofs and battered walls | tonne /hour | 0.01 | 0.01 | 0.01 | 0.01 | 0.02 | 0.02 | 0.03 | 0.03 |
| **Straight, Indented or Square Bars** | | | | | | | | | |
| To beams, floors, roofs and walls | tonne /hour | 0.02 | 0.02 | 0.04 | 0.04 | 0.05 | 0.05 | 0.07 | 0.07 |
| To braces, columns, sloping roofs and battered walls | tonne /hour | 0.01 | 0.01 | 0.02 | 0.02 | 0.02 | 0.02 | 0.04 | 0.04 |
| **Bent, Indented or Square Bars** | | | | | | | | | |
| To beams, floors, roofs and walls | tonne /hour | 0.02 | 0.02 | 0.02 | 0.02 | 0.03 | 0.03 | 0.03 | 0.03 |
| To braces, columns, sloping roofs and battered walls | tonne /hour | 0.01 | 0.01 | 0.01 | 0.01 | 0.02 | 0.02 | 0.02 | 0.02 |
| (Average based on Gang D) | | 0.125 | | 0.148 | | 0.163 | | 0.225 | |

**FORMWORK TO BEAMS AND WALLS**

| Erect and strike formwork (first fix) | | Unit | Joiner | Labourer |
|---|---|---|---|---|
| Walls - vertical face: | up to 1.5m | $m^2$/hr | 1.7 | 0.8 |
| | 1.5 to 3.0m | $m^2$/hr | 1.4 | 0.7 |
| | 3.0 to 4.5m | $m^2$/hr | 1.2 | 0.6 |
| | 4.5 to 6.0m | $m^2$/hr | 1.0 | 0.5 |

**FORMWORK TO SLABS**

| Erect and strike formwork (first fix) | | Unit | Joiner | Labourer |
|---|---|---|---|---|
| Horizontal flat formwork at heights: | up to 3.0m | $m^2$/hr | 1.11 | 1.11 |
| | 3.0 to 3.6m | $m^2$/hr | 1.05 | 1.05 |
| | 3.6 to 4.2m | $m^2$/hr | 1.00 | 1.00 |
| | 4.2 to 4.8m | $m^2$/hr | 0.95 | 0.95 |
| | 4.8 to 5.4m | $m^2$/hr | 0.90 | 0.90 |
| | 5.4 to 6.0m | $m^2$/hr | 0.83 | 0.83 |

**MULTIPLIERS FOR FORMWORK**

| Description | Multiplier |
|---|---|
| Walls built to batter | 1.20 |
| Walls built circular to large radius | 1.70 |
| Walls built circular to small radius | 2.10 |
| Formwork used once | 1.00 |
| Formwork used twice, per use | 0.85 |
| Formwork used three times, per use | 0.75 |
| Formwork used four times, per use | 0.72 |
| Formwork used five times, per use | 0.68 |
| Formwork used six times, per use | 0.66 |
| Formwork used seven or more times, per use | 0.63 |
| Formwork to slope not exceeding $45^{o}$ from horizontal | 1.25 |

PART 12

# Tables and Memoranda

**QUICK REFERENCE CONVERSION TABLES**

1. LINEAR

| | | | | | | |
|---|---|---|---|---|---|---|
| 0.039 | in | --- | 1 | --- | mm | 25.4 |
| 3.281 | ft | --- | 1 | --- | metre | 0.305 |
| 1.094 | yd | --- | 1 | --- | metre | 0.914 |

2. WEIGHT

| | | | | | | |
|---|---|---|---|---|---|---|
| 0.020 | cwt | --- | 1 | --- | kg | 50.802 |
| 0.984 | ton | --- | 1 | --- | tonne | 1.016 |
| 2.205 | lb | --- | 1 | --- | kg | 0.454 |

3. CAPACITY

| | | | | | | |
|---|---|---|---|---|---|---|
| 1.760 | pint | --- | 1 | --- | litre | 0.568 |
| 0.220 | gal | --- | 1 | --- | litre | 4.546 |

4. AREA

| | | | | | | |
|---|---|---|---|---|---|---|
| 0.002 | $in^2$ | --- | 1 | --- | $mm^2$ | 645.16 |
| 10.764 | $ft^2$ | --- | 1 | --- | $m^2$ | 0.093 |
| 1.196 | $yd^2$ | --- | 1 | --- | $m^2$ | 0.836 |
| 2.471 | acre | --- | 1 | --- | ha | 0.405 |
| 0.386 | $mile^2$ | --- | 1 | --- | $km^2$ | 2.59 |

5. VOLUME

| | | | | | | |
|---|---|---|---|---|---|---|
| 0.061 | $in^3$ | --- | 1 | --- | $cm^3$ | 16.387 |
| 35.315 | $ft^3$ | --- | 1 | --- | $m^3$ | 0.028 |
| 1.308 | $yd^3$ | --- | 1 | --- | $m^3$ | 0.765 |

6. POWER

| | | | | | | |
|---|---|---|---|---|---|---|
| 1.310 | HP | --- | 1 | --- | kW | 0.746 |

**CONVERSION FACTORS**

| Multiply Metric Unit | By | To obtain English Unit |
|---|---|---|
| kilometre (km) | .6214 | mile |
| metre (m) | 1.0936 | yard |
| centimetre (cm) | .0328 | foot |
| millimetre (mm) | .03937 | inch |
| sq kilometre ($km^2$) | .3861 | square mile |
| hectare (ha) | 2.471 | acre |
| sq metre ($m^2$) | 10.764 | square foot |
| sq metre ($m^2$) | 1550 | square inch |
| sq centimetre ($cm^2$) | .1550 | square inch |
| cu centimetre ($cm^3$) | .061 | cubic inch |
| cu metre ($m^3$) | 1.308 | cubic yard |
| litre (L) | 61.02 | cubic inch |
| litre (L) | .001308 | cubic yard |
| km/h | .621 | mph |
| litre (L) | .2642 | US gallon |
| litre (L) | .22 | Imperial gallon |
| metric tonne (t) | .984 | long ton |
| metric tonne (t) | 1.102 | short ton |
| kilogram (kg) | 2.205 | pound, avdp |
| gram (g or gr) | .0353 | ounce, avdp |
| kilonewton (kN) | 225 | pound |
| newton (N) | .225 | pound |
| cu centimetre ($cm^3$) | .0338 | fluid ounce |
| kilograms/cu metre | 1.686 | pounds/cu yd |
| kilograms/cu metre | .062 | pounds/cu ft |
| kilograms/sq cm ($kg/cm^2$) | 14.225 | pounds/sq in |
| kilocalorie (kcal) | 3.968 | Btu |
| kilogram-metre (kg.m) | 7.233 | foot-pound |
| metre-kilogram (m.kg) | 7.233 | pound-foot |
| metric horsepower (CV) | .9863 | hp |
| kilowatt (kW) | 1.341 | hp |
| kilopascal (kPa) | .145 | psi |
| bar | 14.5 | psi |
| tonnes/$m^3$ | 1692 | pounds/cu yd |

**CONVERSION FACTORS**

| Multiply English Unit | By | To obtain Metric Unit |
|---|---|---|
| mile statute (m) | 1.609 | kilometre |
| yard (yd) | .9144 | metre |
| foot (ft) | .3048 | metre |
| inch (in) | 25.4 | millimetre |
| sq mile ($mile^2$) | 2.590 | sq kilometre |
| acre | .4047 | hectare |
| sq foot ($ft^2$) | .0929 | sq metre |
| sq inch ($in^2$) | .000645 | sq metre |
| cu yard ($yd^3$) | .7645 | cu metre |
| cu inch ($in^3$) | 16.387 | cu centimetre |
| cu foot ($ft^3$) | .0283 | cu metre |
| cu inch ($in^3$) | .0164 | litre |
| cubic yard ($yd^3$) | 764.55 | litre |
| mph | 1.61 | km/h |
| Ton - mph | 1.459 | tkm/h |
| US gallon (US Gal) | 3.785 | litre |
| US gallon | .833 | Imperial gallon |
| long ton (lg ton) | 1.016 | metric ton |
| short ton (sh ton) | .907 | metric ton |
| pound (lb) | .4536 | kilogram |
| ounce (oz) | 28.35 | gram |
| pound (lb) | .00445 | kilonewton |
| pound (lb) | 4.45 | newton |
| fluid oz (fl oz) | 29.57 | cu centimetre |
| lb/cu ft ($lb/ft^3$) | 16.018 | kg/cu metre |
| lb/cu yd ($lb/yd^3$) | .5933 | kg/cu metre |
| pounds/sq in | .0703 | kilogram/sq cm |
| psi | .0689 | bar |
| psi | 6.89 | kilopascal |
| Btu | .2520 | kilogram-calorie |
| foot-pound (ft-lb) | .1383 | kilogram metre |
| horsepower (hp) | 1.014 | metric horsepower |
| horsepower (hp) | .7457 | kilowatt |
| pounds/cu yd | .0005928 | $tons/m^3$ |
| pounds (No 2 diesel fuel) | .1413 | U S gallon |

**CONVERSION TABLES**

Conversion factors

**Length**

| | Unit | | | | | | |
|---|---|---|---|---|---|---|---|
| Centimetre | cm | 1 in | = | 2.54 | cm | 1 cm | = 0.3937 in |
| Metre | m | 1 ft | = | 0.3048 | m | 1 m | = 3.2808 ft |
| | | 1 yd | = | 0.9144 | m | 1 m | = 1.0936 yd |
| Kilometre | km | 1 mile | = | 1.6093 | km | 1 km | = 0.6214 mile |
| Note: 1 cm = 10 | mm | 1 ft | = | 12 | in | | |
| 1 m = 100 | cm | 1 yd | = | 3 | ft | | |
| 1 km = 1 000 | m | 1 mile | = | 1 760 | yd | | |

**Area**

| | | | | | | | |
|---|---|---|---|---|---|---|---|
| Square Millimetre | $mm^2$ | 1 $in^2$ | = | 645.2 | $mm^2$ | 1 $mm^2$ | = 0.0016 $in^2$ |
| Square Centimetre | $cm^2$ | 1 $in^2$ | = | 6.4516 | $cm^2$ | 1 $cm^2$ | = 1.1550 $in^2$ |
| Square Metre | $m^2$ | 1 $ft^2$ | = | 0.0929 | $m^2$ | 1 $m^2$ | = 10.764 $ft^2$ |
| | | 1 $yd^2$ | = | 0.8361 | $m^2$ | 1 $m^2$ | = 1.1960 $yd^2$ |
| Square Kilometre | $km^2$ | 1 $mile^2$ | = | 2.590 | $km^2$ | 1 $km^2$ | = 0.3861 $mile^2$ |
| Note: 1 $cm^2$ = 100 | $mm^2$ | 1 $ft^2$ | = | 144 | $in^2$ | | |
| 1 $m^2$ = 10 000 | $cm^2$ | 1 $yd^2$ | = | 9 | $ft^2$ | | |
| | | 1 $km^2$ | = | 100 | hectares | 1 acre | = 4 840 $yd^2$ |
| | | 1 $mile^2$ | = | 640 | acres | | |

**Volume**

| | | | | | | | |
|---|---|---|---|---|---|---|---|
| Cubic Centimetre | $cm^3$ | 1 $cm^3$ | = | 0.0610 | $in^3$ | 1 $in^3$ | = 16.387 $cm^3$ |
| Cubic Decimetre | $dm^3$ | 1 $dm^3$ | = | 0.0353 | $ft^3$ | 1 $ft^3$ | = 28.329 $dm^3$ |
| Cubic metre | $m^3$ | 1 $m^3$ | = | 35.3147 | $ft^3$ | 1 $ft^3$ | = 0.0283 $m^3$ |
| | | 1 $m^3$ | = | 1.3080 | $yd^3$ | 1 $yd^3$ | = 0.7646 $m^3$ |
| Litre | L | 1 L | = | 1.76 | pint | 1 pint | = 0.5683 L |
| | | | = | 2.113 | US pt | | = 0.4733 US L |
| Note: 1 $dm^3$ = 1 000 | $cm^3$ | 1 $ft^3$ | = | 1.728 | $in^3$ | | |
| 1 $m^3$ = 1 000 | $dm^3$ | 1 $yd^3$ | = | 27 | $ft^3$ | | |
| 1 L = 1 | $dm^3$ | 1 pint | = | 20 | fl oz | | |
| 1 HL = 100 | L | 1 gal | = | 8 | pints | | |

**Mass**

| | | | | | | | |
|---|---|---|---|---|---|---|---|
| Milligram | mg | 1 mg | = | 0.0154 | grain | 1 grain | = 64.935 mg |
| Gram | g | 1 g | = | 0.0353 | oz | 1 oz | = 28.35 g |
| Kilogram | kg | 1 kg | = | 2.2046 | lb | 1 lb | = 0.4536 kg |
| Tonne | t | 1 t | = | 0.9842 | ton | 1 ton | = 1.016 t |
| Note: 1 g = 1000 | mg | 1 oz | = | 437.5 | grains | | |
| 1 kg = 1000 | g | 1 lb | = | 16 | oz | | |
| 1 t = 1000 | kg | 1 stone | = | 14 | lb | | |
| | | 1 cwt | = | 112 | lb | | |
| | | 1 ton | = | 20 | cwt | | |

**CONVERSION TABLES**

| | Unit | Conversion factors | | | | |
|---|---|---|---|---|---|---|
| **Force** | | | | | | |
| Newton | N | 1 lb f | = | 4.448 N | 1 kg f | = 9.807 N |
| Kilonewton | kN | 1 lb f | = | 0.004448 kN | 1 ton f | = 9.964 kN |
| Meganewton | mN | 100 ton f | = | 0.9964 mN | | |
| **Pressure and stress** | | | | | | |
| Kilonewton per square metre | $kN/m^2$ | 1 lb $f/in^2$ | = | 6.895 $kN/m^2$ | | |
| | | 1 bar | = | 100 $kN/m^2$ | | |
| Meganewton per square metre | $mN/m^2$ | 1 ton $f/ft^2$ | = | 107.3 $kN/m^2$ | = 0.1073 $mN/m^2$ | |
| | | 1 kg $f/cm^2$ | = | 98.07 $kN/m^2$ | | |
| | | 1 lb $f/ft^2$ | = | 0.04788 $kN/m^2$ | | |
| **Coefficient of consolidation (Cv) or swelling** | | | | | | |
| Square metre per year | $m^2$/year | 1 $cm^2/s$ | = | 3.154 $m^2$/year | | |
| | | 1 $ft^2$/year | = | 0.0929 $m^2$/year | | |
| **Coefficient of permeability** | | | | | | |
| Metre per second | m/s | 1 cm/s | = | 0.01 m/s | | |
| Metre per year | m/year | 1 ft/year | = | 0.3048 m/year | | |
| | | | = | 0.9651 x (10)8 m/s | | |
| **Temperature** | | | | | | |
| Degree celcius | C | °C = $^5/_9$ (°F - 32°) | | °F = $^9/_5$ (°C + 32°) | | |

**Metric Equivalents**

| | | |
|---|---|---|
| 1 km | = | 1000 m |
| 1 m | = | 100 cm |
| 1 cm | = | 10 mm |
| 1 $km^2$ | = | 100 ha |
| 1 ha | = | 10,000 $m^2$ |
| 1 $m^2$ | = | 10,000 $cm^2$ |
| 1 $cm^2$ | = | 100 mm |
| 1 $m^3$ | = | 1000 litres |
| 1 litre | = | 1000 $cm^3$ |
| 1 metric tonne | = | 1000 kg |
| 1 quintal | = | 100 kg |
| 1 N | = | 0.10197 kg |
| 1 kg | = | 1000 g |
| 1 g | = | 1000 mg |
| 1 bar | = | 14.504 psi |
| 1 cal | = | 427 kg.m |
| | = | 0.0016 cv.h |
| | = | 0.00116 kw.h |
| torque unit | | |
| 1 CV | = | 75 kg.m/s |
| 1 $kg/cm^2$ | = | 0.97 atmosph |

## CONVERSION TABLES

### English Unit Equivalents

| | | |
|---|---|---|
| 1 mile | = | 1760 yd |
| 1 yd | = | 3 ft |
| 1 ft | = | 12 in |
| 1 sq mile | = | 640 acres |
| 1 acre | = | 43,560 sq ft |
| 1 sq ft | = | 144 sq in |
| 1 cu ft | = | 7.48 gal liq |
| 1 gal | = | 231 cu in |
| | = | 4 quarts liq |
| 1 quart | = | 32 fl oz |
| 1 fl oz | = | 1.80 cu in |
| 1 sh ton | = | 2000 lb |
| 1 lg ton | = | 2240 lb |
| 1 lb | = | 16 oz, avdp |
| 1 Btu | = | 778 ft lb |
| | = | 0.000393 hph |
| | = | 0.000293 kwh |
| 1 hp | = | 550 ft-lb/sec |
| 1 atmosph | = | 14.7 lb/in$^2$ |

### Power Units

| | | | | |
|---|---|---|---|---|
| kW | = | Kilowatt | | |
| HP | = | Horsepower | | |
| CV | = | Cheval Vapeur (Steam Horsepower) | | |
| | = | French designation for **Metric** Horsepower | | |
| PS | = | Pferderstarke (Horsepower) | | |
| | = | German designation of **Metric** Horsepower | | |
| 1 HP | = | 1.014 CV | = | 1.014 PS |
| | = | .7457 kW | | |
| 1 PS | = | 1 CV | = | .9863 HP |
| | = | .7355 kW | | |
| 1 kW | = | 1.341 HP | | |
| | = | 1.359 CV | | |
| | = | 1.359 PS | | |

## FORMULAE

### Two dimensional figures

| Figure | Area |
|---|---|
| Triangle | 0.5 x base x height<br>or N(s(s - a)(s-b)(s-c)) where s = 0.5 x the sum of the three sides and a, b and c are the lengths of the three sides,<br>or $a^2 = b^2 + c^2 - 2$ x bc x COS A where A is the angle opposite side a |
| Hexagon | 2.6 x (side) sq |
| Octagon | 4.83 x (side) sq |
| Trapezoid | height x 0.5(base + top) |
| Circle | 3.142 x radius sq or 0.7854 x diameter sq<br>(circumference = 2 x 3.142 x radius or 3.142 x diameter) |
| Sector of a circle | 0.5 x length of arc x radius |
| Segment of a circle | area of sector - area of triangle |
| Ellipse | 3.142 x AB<br>(where A = 0.5 x height and B = 0.5 x length) |
| Spandrel | $^3/_{14}$ x radius |

### Three dimensional figures

| Figure | Volume | Surface Area |
|---|---|---|
| Prism | Area of base x height | circumference of base x height |
| Cube | (side) cubed | 6 x (side) sq |
| Cylinder | 3.142 x radius sq x height | 2 x 3.142 x radius x (height - radius) |
| Sphere | $^4/_3$ x 3.142 x radius cubed | 4 x 3.142 x radius sq |
| Segment of a sphere | [(3.142 x h)] $/_6$ x (3 x r2 + h2)] | [2 x 3.142 x r x h] |
| Pyramid | $^1/_3$ of area of base x height | 0.5 x circumstance of base x slant height |
| Cone | $^1/_3$ x 3.142 x radius sq x h | 3.142 x radius x slant height |

**FORMULAE**

**Three dimensional figures**

| **Figure** | **Volume** | **Surface Area** |
|---|---|---|
| Frustrum of a pyramid | 0.33 x height [A + B + sq root (AB)] where A is the area of the large end and B is the area of the small end | 0.5 mean circumference x slant height |
| Frustrum of a cone | (0.33 x 3.142 x height (R sq + r sq + R x r)) where R is the radius of the large end and r is the radius of the small end | (3.142 x slant height x (R + r)) |

**Other formulae**

| **Formula** | **Description** |
|---|---|
| Pythagoras theorem | $A^2 = B^2 + C^2$ where A is the hypotenuse of a right-angled triangle and B and C are the two adjacent sides |
| Simpsons Rule | The Area is divided into an even number of strips of equal length, and therefore has an odd number of ordinates at the division points<br><br>Area = $\frac{h}{3}$ (A + 2B + 4C)<br><br>where h = common interval<br>A = sum of first and last ordinates<br>B = sum of remaining odd ordinates<br>C = sum of the even ordinates |
| Trapezoidal Rule | (0.16 x [Total length of trench] x [area of first section x 4 times area of middle section + area of last section]) |

Note: Both Simpsons and Trapezoidal Rule are useful in calculating the volume of an irregular solid accurately.

**WEIGHTS OF TYPICAL MATERIALS HANDLED BY EXCAVATORS**

| Material | $lb/yd^3$ Solid | $lb/yd^3$ Loose | $kg/m^3$ Solid | $kg/m^3$ Loose |
|---|---|---|---|---|
| Ashes | | 945 | | 560 |
| Ballast and sand: dry | | 2565-2970 | | 1520-1760 |
| Basalt | 4725 | 2700 | 2800 | 1600 |
| Bauxite: alum ore | | 2300 | | 1365 |
| Carnotite | 3700 | 2750 | 2200 | 1250 |
| Cement | | 2430 | | 1440 |
| Chalk (hard) | 4050 | 2300 | 2400 | 1365 |
| Cinders | | 1215 | | 720 |
| Clay: dry | 2970 | 2430 | 1760 | 1440 |
| Clay: wet | 3500 | 2970 | 2075 | 1760 |
| Coal: heavy | 2700 | 1620 | 1600 | 960 |
| Coal: light | 2025 | 1215 | 1200 | 720 |
| Coke | | 810 | | 480 |
| Copper ore: low grade | 4400 | 2700 | 2620 | 1600 |
| Crushed stone | 4500 | 2700 | 2670 | 1600 |
| Decomposed rock | | | | |
| 75% rock 25% earth | 4705 | 3305 | 2790 | 1960 |
| 50% rock 50% earth | 3845 | 2900 | 2280 | 1720 |
| 25% rock 75% earth | 3305 | 2650 | 1960 | 1570 |
| Dolomite | | 2700 | | 1600 |
| Earth: dry | 2565 | 2050 | 1520 | 1215 |
| Earth: moist | 2700 | 2160 | 1600 | 1200 |
| Earth: wet | 3100 | 2920 | 1835 | 1730 |
| Flint | 4375 | | 2595 | |
| Granite | 4725 | 2700 | 2800 | 1600 |
| Gravel | 3240 | 2970 | 1920 | 1760 |
| Gypsum | | 2700 | | 1600 |
| Iron punchings | | 7290 | | 4320 |
| Iron turnings | | 4725 | | 2800 |
| Iron ore 60% | 8100 | 4800 | 4800 | 2845 |
| Iron ore 50% | 6750 | 4000 | 4000 | 2370 |
| Iron ore 40% | 5400 | 3200 | 3200 | 1895 |
| Iron ore 25% | 4500 | 2700 | 2670 | 1600 |
| Lignite broken | | 1400 | | 8030 |
| Limestone | 4450 | 2565 | 2640 | 1520 |
| Limestone - crushed | 4385 | 2600 | 2610 | 1540 |
| - broken | 4385 | 2600 | 2610 | 1540 |
| Manganese ore | | 3350 | | 1985 |
| Magnesite | | 2835 | | 1680 |
| Peat: dry | 810 | | 480 | |
| Peat: moist | 1350 | | 800 | |
| Peat: wet | 1890 | | 1120 | |
| Phosphate: rock | | 2970 | | 1760 |
| Ragstone: Kent | 4450 | 2780 | 2640 | 1650 |
| Sand: dry | 2700 | 2430 | 1600 | 1440 |
| Sand: wet | 3240 | 2970 | 1920 | 1760 |
| Sand and gravel - dry | 3255 | 2900 | 1930 | 1720 |
| - wet | 3760 | 3405 | 2230 | 2020 |
| Sandstone | 4320 | 2260 | 2560 | 1340 |
| Shale | 4320 | 2430 | 2560 | 1440 |
| Slag (blast) | 3375 | 2600 | 2000 | 1540 |
| Slate | 4860 | | 2880 | |
| Taconite | 4215 | 2865 | 2500 | 1700 |
| Topsoil | 2310 | 1600 | 1370 | 950 |
| Traprock | 4590 | 2835 | 2720 | 1680 |
| Whinstone | 4725 | 2835 | 2800 | 1680 |

## WEIGHTS OF CONSTRUCTIONAL MATERIALS

### Miscellaneous Materials

| | $kg/m^3$ | $lb/ft^3$ |
|---|---|---|
| Tarmacadam | 2306 | 144 |
| Macadam (waterbound) | 2563 | 160 |
| Snow - compact | 100 to 816 | 15 to 50 |
| - loose | 82 to 194 | 5 to 12 |
| Vermiculite (aggregate) | 82 | 5 |
| Terracotta | 2122 | 132 |
| Glass | 2723 | 170 |
| Cork - compressed | 388 | 24 |
| Water | 1000 | 62 |
| | $kg/m^2$ | $lb/ft^2$ |
| Clay floor tiles | 59 | 12 |
| Pavement lights | 122 | 25 |
| Damp proof course | 5 | 1 |
| | $kg/m^2$ per mm thickness | $lb/ft^2$ per in thickness |
| Felt (insulating) | 0.20 | 1 |
| Paving Slabs (stone) | 2.70 | 14 |
| Granite setts | 2.88 | 15 |
| Asphalt | 2.30 | 12 |
| Rubber paving | 1.54 | 8 |
| Polyvinylchloride | 1.94 (avg) | 10 (avg) |
| Glass-fibre (forms) | 0.20 | 1 |

### Timber

| | $kg/m^3$ | $lb/ft^3$ |
|---|---|---|
| General | 806 (avg) | 50 (avg) |
| Douglas fir | 479 | 30 |
| Yellow pine, spruce | 479 | 30 |
| Pitch pine | 673 | 42 |
| Larch, elm | 561 | 35 |
| Oak (English) | 724 to 959 | 45 to 60 |
| Teak | 643 to 877 | 40 to 55 |
| Jarrah | 959 | 60 |
| Greenheart | 1040 to 1204 | 65 to 75 |
| Quebracho | 1285 | 80 |
| | $kg/m^2$ per mm | $lb/ft^2$ per in |
| Wooden boarding and blocks | | |
| Softwood | 0.48 | 2.5 |
| Hardwood | 0.76 | 4 |
| Hardboard | 1.06 | 5.5 |
| Chipboard | 0.76 | 4 |
| Plywood | 0.62 | 3.25 |
| Blockboard | 0.48 | 2.5 |
| Fibreboard | 0.29 | 1.5 |
| Wood-wool | 0.58 | 3 |
| Plasterboard | 0.96 | 5 |
| Weather boarding | 0.39 | 2 |

**WEIGHTS OF CONSTRUCTIONAL MATERIALS** - *continued*

**Stone and other materials**

| | $kg/m^3$ | $lb/ft^3$ |
|---|---|---|
| Natural stone (solid) | | |
| Granite | 2560 to 2927 | 160 to 183 |
| Limestone - Bath stone | 2081 | 130 |
| - Marble | 2723 | 170 |
| - Portland stone | 2244 | 140 |
| Sandstone | 2244 to 2407 | 140 to 150 |
| Slate | 2880 | 180 |
| Stone rubble (packed) | 2244 | 140 |
| Quarry waste | 1438 | 90 |
| Hardcore (consolidated) | 1928 | 120 |
| All-in aggregate | 1999 | 125 |

**Metals, steel construction, etc**

| | | |
|---|---|---|
| Iron - cast | 7207 | 450 |
| - wrought | 7687 | 480 |
| - ore - general | 2407 | 150 |
| - (crushed) Swedish | 3682 | 230 |
| Steel | 7854 | 490 |
| Copper - cast | 8731 | 545 |
| - wrought | 8945 | 558 |
| Brass | 8497 | 530 |
| Bronze | 8945 | 558 |
| Aluminium | 2774 | 173 |
| Lead | 11322 | 707 |
| Zinc (rolled) | 7140 | 446 |

| | $g/mm^2$ per metre | $lb/ft^2$ per foot |
|---|---|---|
| Steel bars | 7.85 | 3.4 |

| | |
|---|---|
| Structural steelwork | Net weight of member @ 7854 $kg/m^3$ |
| rivetted | + 10% for cleats, rivets, bolts, etc |
| welded | + 1.25% to 2.5% for welds, etc |
| Rolled sections - beams | + 2.5% |
| - stanchions | + 5% (extra for caps and bases) |
| Plate-web girders | + 10% for rivets or welds, stiffeners, etc |
| Roof trusses and wall framing | see separate table |

| | kg/m | lb/ft |
|---|---|---|
| Steel stairs : industrial type 1 m or 3ft wide | 84 | 56 |
| Steel tubes 50 mm or 2 in bore | 5 to 6 | 3 to 4 |
| Gas piping 20 mm or 3/4 in | 2 | 1 1/4 |

**WEIGHTS OF CONSTRUCTIONAL MATERIALS** - *continued*

**Rail tracks**

| | | kg/m of track | lb/ft of track |
|---|---|---|---|
| **standard guage** | Bull-head rails, chairs, transverse timber (softwood) sleepers, etc. | 245 | 165 |
| **main lines** | Flat-bottom rails, transverse prestressed concrete sleepers, etc | 418 | 280 |
| | Add for electric third rail | 51 | 35 |
| | Add for crushed stone ballast | 2600 | 1750 |
| | | $kg/m^2$ | $lb/ft^2$ |
| | Overall average weight - rails connections, sleepers, ballast, etc | 733 | 150 |
| | | kg/m of rail | lb/ft of rail |
| | Bridge rails, longitudinal timber sleepers, etc | 112 | 75 |

## WEIGHTS OF CONCRETE AND CONCRETE ELEMENTS

### Ordinary concrete (dense aggregates)

| | | | $kg/m^3$ | $lb/ft^3$ |
|---|---|---|---|---|
| Non-reinforced plain or mass concrete | Nominal weight | | 2305 | 144 |
| | Aggregate - limestone | | 2162 to 2407 | 135 to 150 |
| | - gravel | | 2244 to 2407 | 140 to 150 |
| | - broken brick | | 2000 (av) | 125 (av) |
| | - other crushed stone | | 2326 to 2489 | 145 to 155 |
| Reinforced Concrete | Nominal weight | | 2407 | 150 |
| | Reinforcement - 1% | | 2305 to 2468 | 144 to 154 |
| | - 2% | | 2356 to 2519 | 147 to 157 |
| | - 4% | | 2448 to 2703 | 153 to 163 |
| | | | $kg/m^2$ | $lb/ft^2$ |
| | Solid slabs (floors, walls etc) | Thickness | | |
| | | 75 mm or 3 in | 184 | 37.5 |
| | | 100 mm or 4 in | 245 | 50 |
| | | 150 mm or 6 in | 378 | 75 |
| | | 250 mm or 10 in | 612 | 125 |
| | | 300 mm or 12 in | 734 | 150 |
| | Ribbed slabs | 125 mm or 5 in | 204 | 42 |
| | | 150 mm or 6 in | 219 | 45 |
| | | 225 mm or 9 in | 281 | 57 |
| | | 300 mm or 12 in | 342 | 70 |

### Special concretes, etc

| | | $kg/m^3$ | $lb/ft^3$ |
|---|---|---|---|
| Heavy concrete | Aggregates - barytes, magnetite | 3210 (min) | 200 (min) |
| | steel shot, punchings | 5280 | 330 |
| Lean mixes | Dry-lean (gravel aggregate) | 2244 | 140 |
| | Soil-cement (normal mix) | 1601 | 100 |
| | | $kg/m^2$ per mm thick | $lb/ft^2$ per in thick |
| Finishes etc | Rendering, screed etc ) | | |
| | Granolithic, terrazzo ) | 1928 to 2401 | 10 to 12.5 |
| | Glass-block (hollow) concrete | 1734 (approx) | 9 (approx) |

Prestressed concrete — Weights as for reinforced concrete (upper limits)

Air-entrained concrete — Weights as for plain or reinforced concrete

## WEIGHTS OF ROOFS AND WALLS

### Roofs

| | | | | $kg/m^2$ | $lb/ft^2$ |
|---|---|---|---|---|---|
| Cladding | (Weights per $m^2$ or per $ft^2$ of slope of roof) | | | $kg/m^2$ | $lb/ft^2$ |
| | Patent glazing (with lead covered astragals) | | | 30 | 6 |
| | ditto including steel purlins, etc | | | 39 | 8 |
| | Slates or tiles, battens, steel purlins, etc | | | 68 to 88 | 14 to 18 |
| | ditto with boarding, felt, etc | | | 93 to 112 | 17 to 23 |
| | Corrugated asbestos or steel sheeting, steel purlins, etc | | | 39 to 49 | 8 to 10 |
| Roof Trusses | Spacing of trusses | | | 3.0 m | 4.5 m |
| | Approximate weights of steel roof trusses in kg/m2 of plan area of roof | Span of trusses | 7.5 m | 9690 | 7344 |
| | | | 9 m | 12240 | 7344 |
| | | | 12 m | 13464 | 8568 |
| | | | 15 m | 14688 | 11016 |
| | | | 18 m | 20706 | 14688 |
| | | | 25 m | 24378 | 17136 |

### Walls and components of walls

| | | | |
|---|---|---|---|
| | | $kg/m^2$ per mm thick | $lb/ft^2$ per in thick |
| Brickwork and blockwork | Hollow clay blocks | 1.15 (av) | 6 (av) |
| | Common clay blocks | 1.90 | 10 |
| | Engineering clay bricks | 2.30 | 12 |
| | Refactory bricks | 1.15 | 6 |
| | Sand-lime (and similar) bricks | 2.02 | 10.5 |
| | | $kg/m^2$ | $lb/ft^2$ |
| Plaster | Gypsum - two coat 12 mm or 0.5 in thick | 22.00 | 4.5 |
| | - plasterboard ditto | 11.00 | 2.25 |
| | Lath and plaster (two-faced inc studding) | 48.93 | 10 |
| Sheeting, etc | Corrugated steel or asbestos-cement sheeting (inc bolts, sheeting, rails etc) | 44.00 | 9 |
| | Steel wall framing (for sheeting or brick panels) | 24.50-34.15 | 5 to 7 |
| | ditto with brick panels and windows | 244.65 | 50 (av) |
| | ditto with steel or asbestos-cement sheeting | 73.40 | 15 (av) |
| | Windows (industrial type - metal or wooden frames) | 24.50 | 5 (av) |
| | Doors (ordinary industrial type - wooden) | 38.75 | 8 |

## EXCAVATION AND EARTHWORK

### Transport capacities

| Type of vehicle | Capacity of vehicle Payload (kg) | Heaped capacity ($m^3$) |
|---|---|---|
| Wheelbarrow | 150 | 0.10 |
| 1 tonne dumper | 1250 | 1.00 |
| 2.5 tonne dumper | 4000 | 2.50 |
| Articulated dump truck (Volvo A20 6x4) | 18500 | 11.00 |
| Articulated dump truck (Volvo A35 6x6) | 32000 | 19.00 |
| Large capacity rear dumper (Euclid R35) | 35000 | 22.00 |
| Large capacity rear dumper (Euclid R85) | 85000 | 50.00 |

### Earthwork support

Maximum depth of excavation in various soils without the use of earthwork support

| Ground conditions | Feet (ft) | Metres (m) |
|---|---|---|
| Compact soil | 12 | 3.66 |
| Drained loam | 6 | 1.83 |
| Dry sand | 1 | 0.30 |
| Gravelly earth | 2 | 0.61 |
| Ordinary earth | 3 | 0.91 |
| Stiff clay | 10 | 3.05 |

It is important to note that the above table should only be used as a guide. Each case must be taken on its merits and, as the limited distances given above are approached, careful watch must be kept for the slightest signs of caving in.

### Bulkage of soils after excavation

| | Approximate bulk of $1m^3$ after excavation |
|---|---|
| Vegetable soil and loam | 25 - 30% |
| Soft clay | 30 - 40% |
| Stiff clay | 10 - 20% |
| Gravel | 20 - 25% |
| Sand | 40 - 50% |
| Chalk | 40 - 50% |
| Rock, weathered | 30 - 40% |
| Rock, unweathered | 50 - 60% |

### Shrinkage of materials on being deposited

| | Approximate shrinkage of $1m^3$ after excavation |
|---|---|
| Clay | 10% |
| Gravel | 8% |
| Gravel and sand | 9% |
| Loam and light sandy soils | 12% |
| Loose vegetable soils | 15% |

## EXCAVATION AND EARTHWORK

### Materials: angles of repose

| | |
|---|---|
| Clay (Dry) | 30° |
| Clay (Damp, well drained) | 45° |
| Clay (Wet) | 15-20° |
| Earth (Dry) | 30° |
| Earth (Damp) | 45° |
| Gravel (Moist) | 48° |
| Sand (Dry or moist) | 35° |
| Sand (Wet) | 25° |
| Loam | 40° |

### Slopes and Angles

| Ratio of base to height | Angle in degrees |
|---|---|
| 5 : 1 | 11 |
| 4 : 1 | 14 |
| 3 : 1 | 18 |
| 2 : 1 | 26 |
| 1.5 : 1 | 33 |
| 1 : 1 | 45 |
| 1 : 1.5 | 56 |
| 1 : 2 | 63 |
| 1 : 3 | 72 |
| 1 : 4 | 76 |
| 1 : 5 | 78 |

### Grade in Degrees and Percents

| DEGREES | PERCENT | DEGREES | PERCENT |
|---|---|---|---|
| 1 | 1.8 | 24 | 44.5 |
| 2 | 3.5 | 25 | 46.6 |
| 3 | 5.2 | 26 | 48.8 |
| 4 | 7.0 | 27 | 51.0 |
| 5 | 8.8 | 28 | 53.2 |
| 6 | 10.5 | 29 | 55.4 |
| 7 | 12.3 | 30 | 57.7 |
| 8 | 14.0 | 31 | 60.0 |
| 9 | 15.8 | 32 | 62.5 |
| 10 | 17.6 | 33 | 64.9 |
| 11 | 19.4 | 34 | 67.4 |
| 12 | 21.3 | 35 | 70.0 |
| 13 | 23.1 | 36 | 72.7 |
| 14 | 24.9 | 37 | 75.4 |
| 15 | 26.8 | 38 | 78.1 |
| 16 | 28.7 | 39 | 81.0 |
| 17 | 30.6 | 40 | 83.9 |
| 18 | 32.5 | 41 | 86.9 |
| 19 | 34.4 | 42 | 90.0 |
| 20 | 36.4 | 43 | 93.3 |
| 21 | 38.4 | 44 | 96.6 |
| 22 | 40.4 | 45 | 100.0 |
| 23 | 42.4 | | |

## EXCAVATION AND EARTHWORKS

### Bearing powers

| GROUND/CONDITIONS | BEARING POWER | | |
|---|---|---|---|
| | km/m$^2$ | lb/in$^2$ | metric t/m$^2$ |
| Rock (broken) | 483 | 70 | 50 |
| Rock (solid) | 2415 | 350 | 240 |
| Clay, dry or hard | 380 | 55 | 40 |
| medium dry | 190 | 27 | 20 |
| soft or wet | 100 | 14 | 10 |
| Gravel, cemented | 760 | 110 | 80 |
| Sand, compacted | 380 | 55 | 40 |
| clean dry | 190 | 27 | 20 |
| Swamp and alluvial soils etc | 48 | 7 | 5 |

## CONCRETE WORK

### Approximate average weights of aggregates

| Materials | Percentage of voids (%) | Weight per $m^3$ (kg) |
|---|---|---|
| Sand | 39 | 1660 |
| Gravel 10 - 20 mm | 45 | 1440 |
| Gravel 35 - 75 mm | 42 | 1555 |
| Crushed stone | 50 | 1330 |
| Crushed granite (over 15 mm) | 50 | 1345 |
| (n.e. 15 mm) | 47 | 1440 |
| 'All-in' ballast | 32 | 1800 - 2000 |

### Common mixes for various types of work per $m^3$

| Recommended mix | Class of work suitable for:- | Cement (kg) | Sand (kg) | Course Aggregate (kg) | No of 50kg bags of cement per $m^3$ of combined aggregate |
|---|---|---|---|---|---|
| 1:3:6 | Roughest type of mass concrete such as footings, road haunching over 300 mm thick. | 208 | 905 | 1509 | 4.00 |
| 1:2.5:5 | Mass concrete of better class than 1:3:6 such as bases for machinery, walls below ground etc. | 249 | 881 | 1474 | 5.00 |
| 1:2:4 | Most ordinary uses of concrete, such as mass walls above ground, road slabs etc. and general reinforced concrete work. | 304 | 889 | 1431 | 6.00 |
| 1:1.5:3 | Watertight floors, pavements and walls, tanks, pits, steps, paths, surface of 2 course roads, reinforced concrete where extra strength is required. | 371 | 801 | 1336 | 7.50 |
| 1:1:2 | Work of thin section such as fence posts and small precast work. | 511 | 720 | 1206 | 10.50 |

## CONCRETE WORK

### Prescribed mixes for ordinary structural concrete

Weights of cement and total dry aggregates in kg to produce approximately one cubic metre of fully compacted concrete together with the percentages by weight of fine aggregate in total dry aggregates.

| Concrete grade | Nominal max size of aggregate (mm) | 40 | | 20 | | 14 | | 10 | |
|---|---|---|---|---|---|---|---|---|---|
| | Workability | Medium | High | Medium | High | Medium | High | Medium | High |
| | Limits to slump that may be expected (mm) | 50-100 | 100-150 | 25-75 | 75-125 | 10-50 | 50-100 | 10-25 | 25-50 |
| 7 | Cement (kg) | 180 | 200 | 210 | 230 | - | - | - | - |
| | Total aggregate (kg) | 1950 | 1850 | 1900 | 1800 | - | - | - | - |
| | Fine aggregate (%) | 30-45 | 30-45 | 35-50 | 35-50 | - | - | - | - |
| 10 | Cement (kg) | 210 | 230 | 240 | 260 | - | - | - | - |
| | Total aggregate (kg) | 1900 | 1850 | 1850 | 1800 | - | - | - | - |
| | Fine aggregate (%) | 30-45 | 30-45 | 35-50 | 35-50 | - | - | - | - |
| 15 | Cement (kg) | 250 | 270 | 280 | 310 | - | - | - | - |
| | Total aggregate (kg) | 1850 | 1800 | 1800 | 1750 | - | - | - | - |
| | Fine aggregate (%) | 30-45 | 30-45 | 35-50 | 35-50 | - | - | - | - |
| 20 | Cement (kg) | 300 | 320 | 320 | 350 | 340 | 380 | 360 | 410 |
| | Total aggregate (kg) | 1850 | 1750 | 1800 | 1750 | 1750 | 1700 | 1750 | 1650 |
| | Sand | | | | | | | | |
| | Zone 1 (%) | 35 | 40 | 40 | 45 | 45 | 50 | 50 | 55 |
| | Zone 2 (%) | 30 | 35 | 35 | 40 | 40 | 45 | 45 | 50 |
| | Zone 3 (%) | 30 | 30 | 30 | 35 | 35 | 40 | 40 | 45 |
| 25 | Cement (kg) | 340 | 360 | 360 | 390 | 380 | 420 | 400 | 450 |
| | Total aggregate (kg) | 1800 | 1750 | 1750 | 1700 | 1700 | 1650 | 1700 | 1600 |
| | Sand | | | | | | | | |
| | Zone 1 (%) | 35 | 40 | 40 | 45 | 45 | 50 | 50 | 55 |
| | Zone 2 (%) | 30 | 35 | 35 | 40 | 40 | 45 | 45 | 50 |
| | Zone 3 (%) | 30 | 30 | 30 | 35 | 35 | 40 | 40 | 45 |
| 30 | Cement (kg) | 370 | 390 | 400 | 430 | 430 | 470 | 460 | 510 |
| | Total aggregate (kg) | 1750 | 1700 | 1700 | 1650 | 1700 | 1600 | 1650 | 1550 |
| | Sand | | | | | | | | |
| | Zone 1 (%) | 35 | 40 | 40 | 45 | 45 | 50 | 50 | 55 |
| | Zone 2 (%) | 30 | 35 | 35 | 40 | 40 | 45 | 45 | 50 |
| | Zone 3 (%) | 30 | 30 | 30 | 35 | 35 | 40 | 40 | 45 |

## CONCRETE WORK

### Bar reinforcement

Cross-sectional area and mass

| Nominal sizes (mm) | Cross-sectional area ($mm^2$) | Mass per metre mm (kg) | Length of bar (m) per tonne |
|---|---|---|---|
| 6* | 28.3 | 0.222 | 4505 |
| 8 | 50.3 | 0.395 | 2532 |
| 10 | 78.5 | 0.616 | 1623 |
| 12 | 113.1 | 0.888 | 1126 |
| 16 | 201.1 | 1.579 | 633 |
| 20 | 314.2 | 2.466 | 406 |
| 25 | 490.9 | 3.854 | 259 |
| 32 | 804.2 | 6.313 | 158 |
| 40 | 1256.6 | 9.864 | 101 |
| 50* | 1963.5 | 15.413 | 65 |

Weights of bars at specific spacings

Weights of metric (millimetre) bars in kilogrammes per square metre

| Size (mm) | Spacing of bars in millimetres | | | | | | | | | |
|---|---|---|---|---|---|---|---|---|---|---|
| | 75 | 100 | 125 | 150 | 175 | 200 | 225 | 275 | 300 | |
| 6* | 2.960 | 2.220 | 1.776 | 1.480 | 1.269 | 1.110 | 0.987 | 0.888 | 0.807 | 0.740 |
| 8 | 5.267 | 3.950 | 3.160 | 2.633 | 2.257 | 1.975 | 1.756 | 1.580 | 1.436 | 1.317 |
| 10 | 8.213 | 6.160 | 4.928 | 4.107 | 3.520 | 3.080 | 2.738 | 2.464 | 2.240 | 2.053 |
| 12 | 11.84 | 8.880 | 7.104 | 5.920 | 5.074 | 4.440 | 3.947 | 3.552 | 3.229 | 2.960 |
| 16 | 21.05 | 15.79 | 12.63 | 10.53 | 9.023 | 7.895 | 7.018 | 6.316 | 5.742 | 5.263 |
| 20 | 32.88 | 24.66 | 19.73 | 16.44 | 14.09 | 12.33 | 10.96 | 9.864 | 8.967 | 8.220 |
| 25 | 51.39 | 38.54 | 30.83 | 25.69 | 22.02 | 19.27 | 17.13 | 15.42 | 14.01 | 12.85 |
| 32 | - | 63.13 | 50.50 | 42.09 | 36.07 | 31.57 | 28.06 | 25.25 | 22.96 | 21.04 |
| 40 | - | - | 78.91 | 65.76 | 56.37 | 49.32 | 43.84 | 39.46 | 35.87 | 32.88 |
| 50* | - | - | - | 102.75 | 88.07 | 77.07 | 68.50 | 61.65 | 56.05 | 51.38 |

Basic weight = 0.00785 $kg/mm^2/m$
Weight per metre = 0.006165 $\#^2$ (kg)
Weight per $mm^2$ at spacing s (mm) = 6.165 $\#^2/s$ (kg)
# = diameter of bar in millimetres

* Where a bar larger than 40 mm is to be used the recommended size is 50 mm. Where a bar smaller than 8 mm is to be used the recommended size is 6 mm.

**CONCRETE WORK**

**Fabric reinforcement**

Preferred range of designated fabric types and stock sheet sizes

| **Fabric reference** | **Longitudinal wires** | | **Cross wires** | | |
|---|---|---|---|---|---|
| | Nominal wire size (mm) | Pitch (mm) | Nominal wire size (mm) | Pitch (mm) | Mass ($kg/m^2/m$) |
| Square mesh | | | | | |
| A393 | 10 | 200 | 10 | 200 | 6.16 |
| A252 | 8 | 200 | 8 | 200 | 3.95 |
| A193 | 7 | 200 | 7 | 200 | 3.02 |
| A142 | 6 | 200 | 6 | 200 | 2.22 |
| A98 | 5 | 200 | 5 | 200 | 1.54 |
| Structural mesh | | | | | |
| B1131 | 12 | 100 | 8 | 200 | 10.90 |
| B785 | 10 | 100 | 8 | 200 | 8.14 |
| B503 | 8 | 100 | 8 | 200 | 5.93 |
| B385 | 7 | 100 | 7 | 200 | 4.53 |
| B283 | 6 | 100 | 7 | 200 | 3.73 |
| B196 | 5 | 100 | 7 | 200 | 3.05 |
| Long mesh | | | | | |
| C785 | 10 | 100 | 6 | 400 | 6.72 |
| C636 | 9 | 100 | 6 | 400 | 5.55 |
| C503 | 8 | 100 | 5 | 400 | 4.34 |
| C385 | 7 | 100 | 5 | 400 | 3.41 |
| C283 | 6 | 100 | 5 | 400 | 2.61 |
| Wrapping mesh | | | | | |
| D98 | 5 | 200 | 5 | 200 | 1.54 |
| D49 | 2.5 | 100 | 2.5 | 100 | 0.77 |

Stock sheet size Length 4.8 m Width 2.4 m Sheet area $11.52m^2$

**CONCRETE WORK**

**Wire**

| SWG | 6g | 5g | 4g | 3g | 2g | 1g | 1/0g | 2/0g | 3/0g | 4/0g | 5/0g |
|---|---|---|---|---|---|---|---|---|---|---|---|
| dia | | | | | | | | | | | |
| in | 0.192 | 0.212 | 0.232 | 0.252 | 0.276 | 0.300 | 0.324 | 0.348 | 0.372 | 0.400 | 0.432 |
| mm | 4.9 | 5.4 | 5.9 | 6.4 | 7.0 | 7.6 | 8.2 | 8.8 | 9.5 | 10.2 | 11.0 |
| area | | | | | | | | | | | |
| $in^2$ | 0.029 | 0.035 | 0.042 | 0.050 | 0.060 | 0.071 | 0.082 | 0.095 | 0.109 | 0.126 | 0.146 |
| $mm^2$ | 19 | 23 | 27 | 32 | 39 | 46 | 53 | 61 | 70 | 81 | 95 |

**Average weight $kg/m^3$ of steelwork reinforcement in concrete for various building elements**

| Substructure | | $kg/m^3$ concrete |
|---|---|---|
| Pile caps | 110 - 150 | |
| Tie beams | 130 - 170 | |
| Ground beams | | 230 - 330 |
| Bases | 90 - 130 | |
| Footings | 70 - 110 | |
| Retaining walls | | 110 - 150 |
| **Superstructure** | | |
| Slabs - one way | | 75 - 125 |
| Slabs - two way | | 65 - 135 |
| Plate slab | 95 - 135 | |
| Cant slab | 90 - 130 | |
| Ribbed floors | | 80 - 120 |
| Columns | 200 - 300 | |
| Beams | | 250 - 350 |
| Stairs | | 130 - 170 |
| Walls - normal | | 30 - 70 |
| Walls - wind | | 50 - 90 |

Note: For exposed elements add the following % :

Walls 50%, Beams 100%, Columns 15%

## CONCRETE WORK

### Formwork stripping times

Normal Curing Periods

| Conditions under which concrete is maturing | Minimum periods of protection for different types of cement: Number of days (where the average surface temperature of the concrete exceeds $10^{\circ}C$ during the whole period) | | | Equivalent maturity (degree hours) calculated as the age of the concrete in hours multiplied by the number of degrees Celsius by which the average surface temperature of the concrete exceeds $-10^{\circ}C$ | | |
|---|---|---|---|---|---|---|
| | Other* | SRPC | OPC or RHPC | Other* | SRPC | OPC or RHPC |
| 1. Hot weather or drying winds | 7 | 4 | 3 | 3500 | 2000 | 1500 |
| 2. Conditions not covered by 1 | 4 | 3 | 2 | 2000 | 1500 | 1000 |

KEY: OPC - Ordinary Portland Cement
RHPC - Rapid-hardening Portland cement
SRPC - Sulphate-resisting Portland cement

Minimum period before striking formwork

| Type of Formwork | Minimum period before striking — Surface Temperature of concrete: $16^{\circ}C$ | $7^{\circ}C$ | $t^{\circ}C$ (any temperature between $0^{\circ}$ and $25^{\circ}C$) |
|---|---|---|---|
| Vertical formwork to columns, walls and large beams | 12 hours | 18 hours | $\frac{300}{t+10}$ hours |
| Soffit formwork to slabs | 4 days | 6 days | $\frac{100}{t+10}$ days |
| Props to slabs | 10 days | 15 days | $\frac{250}{t+10}$ days |
| Soffit formwork to beams | 9 days | 14 days | $\frac{230}{t+10}$ days |
| Props to beams | 14 days | 21 days | $\frac{360}{t+10}$ days |

**DRAINAGE**

**Width required for trenches for various diameters of pipes**

| Pipe diameter (mm) | Trench not exceeding 1.5 m deep | Trench over 1.5 m deep |
|---|---|---|
| not exceeding 100 mm | 450 mm | 600 mm |
| 100 - 150 mm | 500 mm | 650 mm |
| 150 - 225 mm | 600 mm | 750 mm |
| 225 - 300 mm | 650 mm | 800 mm |
| 300 - 400 mm | 750 mm | 900 mm |
| 400 - 450 mm | 900 mm | 1050 mm |
| 450 - 600 mm | 1100 mm | 1300 mm |

**Weights and dimensions of typically sized uPVC pipes**

| Nominal size | Mean outside diameter (mm) | | Wall thickness | Weight kg per metre |
|---|---|---|---|---|
| Standard pipes | min | max | | |
| 82.4 | 82.4 | 82.7 | 3.2 | 1.2 |
| 110.0 | 110.0 | 110.4 | 3.2 | 1.6 |
| 160.0 | 160.0 | 160.6 | 4.1 | 3.0 |
| 200.0 | 200.0 | 200.6 | 4.9 | 4.6 |
| 250.0 | 250.0 | 250.7 | 6.1 | 7.2 |
| Perforated pipes | | | | |
| Heavy grade as above | | | | |
| Thin wall | | | | |
| 82.4 | 82.4 | 82.7 | 1.7 | - |
| 110.0 | 110.0 | 110.4 | 2.2 | - |
| 160.0 | 160.0 | 160.6 | 3.2 | - |

**Cubic metres of concrete drain beds per lineal metre**

| 150 mm Bed | Diameter of Pipe mm | | | | | |
|---|---|---|---|---|---|---|
| | 100 | 150 | 225 | 300 | 450 | 600 |
| Haunched | 0.10 | 0.12 | 0.14 | 0.18 | 0.25 | 0.32 |
| Surrounded | 0.15 | 0.17 | 0.23 | 0.30 | 0.43 | 0.54 |
| *Deduct* if 100 mm Bed | 0.017 | 0.025 | 0.025 | 0.034 | 0.042 | 0.05 |

In all cases trench is taken as 300mm wider than pipe

**DRAINAGE**

Vitrified clay pipes

| Product | Nominal diameter (mm) | Effective pipe length (mm) | BS 65 limits of bore tolerance min (mm) | max (mm) | Crushing strength (kN/m) | Weight kg/pipe | kg/m |
|---|---|---|---|---|---|---|---|
| Supersleve | 100 | 1600 | 96 | 105 | 35.00 | 14.71 | 9.19 |
| | 150 | 1750 | 146 | 158 | 35.00 | 29.24 | 16.71 |
| Hepsleve | 225 | 1850 | 221 | 236 | 28.00 | 84.03 | 45.42 |
| | 300 | 2500 | 295 | 313 | 34.00 | 193.05 | 77.22 |
| Hepseal | 150 | 1500 | 146 | 158 | 22.00 | 37.04 | 24.69 |
| | 225 | 1750 | 221 | 236 | 28.00 | 85.47 | 48.84 |
| | 300 | 2500 | 295 | 313 | 34.00 | 204.08 | 81.63 |
| | 400 | 2500 | 394 | 414 | 44.00 | 357.14 | 142.86 |
| | 450 | 2500 | 444 | 464 | 44.00 | 454.55 | 181.63 |
| | 500 | 2500 | 494 | 514 | 48.00 | 555.56 | 222.22 |
| | 600 | 2500 | 591 | 615 | 57.00 | 796.23 | 307.69 |
| | 700 | 3000 | 689 | 719 | 67.00 | 1111.11 | 370.45 |
| | 800 | 3000 | 788 | 822 | 72.00 | 1351.35 | 450.45 |
| Hepline | 100 | 1600 | 95 | 107 | 22.00 | 14.71 | 9.19 |
| | 150 | 1750 | 145 | 160 | 22.00 | 29.24 | 16.71) |
| | 225 | 1850 | 219 | 239 | 28.00 | 84.03 | 45.42 |
| | 300 | 1850 | 292 | 317 | 34.00 | 142.86 | 77.22 |
| Hepduct | 90 | 1500 | - | - | 28.00 | 12.05 | 8.03 |
| (Conduit) | 100 | 1600 | - | - | 28.00 | 14.71 | 9.19 |
| | 125 | 1750 | - | - | 28.00 | 20.73 | 11.84 |
| | 150 | 1750 | - | - | 28.00 | 29.24 | 16.71 |
| | 225 | 1850 | - | - | 28.00 | 84.03 | 45.42 |
| | 300 | 1850 | - | - | 34.00 | 142.86 | 77.22 |

**Concrete pipes**

| Nom Intl Diam mm | Nom Wall thickness | Approx weight kg/m | Nom Intl Diam mm | Nom Wall thickness | Approx weight kg/m |
|---|---|---|---|---|---|
| 150 | 25 | 45 | 750 | 54 | 454 |
| 225 | 29 | 71 | 900 | 60 | 616 |
| 300 | 32 | 122 | 1200 | 76 | 912 |
| 375 | 35 | 162 | 1500 | 89 | 1458 |
| 450 | 38 | 191 | 1800 | 102 | 1884 |
| 600 | 48 | 317 | 2100 | 127 | 2619 |

*Wall thickness, weights and pipe lengths vary, depending on type of pipe required.*

The particulars shown above represent a selection of available diameters and are applicable to strength class 1 pipes with flexible rubber ring joints.

Tubes with Ogee joints are also available.

**STRUCTURAL STEEL SECTIONS**

**UNIVERSAL BEAMS**

(Based on 7854 kg/m$^3$)

| Designation | | Depth of Section | Width of Section | Thickness | |
|---|---|---|---|---|---|
| Serial Size | Mass per unit length | | | Web | Flange |
| mm | kg/m | mm | mm | mm | mm |
| 914 x 419 | 388 | 920.5 | 420.5 | 21.5 | 36.6 |
| | 343 | 911.4 | 418.5 | 19.4 | 32.0 |
| 914 x 305 | 289 | 926.6 | 307.8 | 19.6 | 32.0 |
| | 253 | 918.5 | 305.5 | 17.3 | 27.9 |
| | 224 | 910.3 | 304.1 | 15.9 | 23.9 |
| | 201 | 903.0 | 303.4 | 15.2 | 20.2 |
| 838 x 292 | 226 | 850.9 | 293.8 | 16.1 | 26.8 |
| | 194 | 840.7 | 292.4 | 14.7 | 21.7 |
| | 176 | 834.9 | 291.6 | 14.0 | 18.8 |
| 762 x 267 | 197 | 269.6 | 268.0 | 15.6 | 25.4 |
| | 173 | 762.0 | 266.7 | 14.3 | 21.6 |
| | 147 | 753.9 | 265.3 | 12.9 | 17.5 |
| 686 x 254 | 170 | 692.9 | 255.8 | 14.5 | 23.7 |
| | 152 | 687.6 | 254.5 | 13.2 | 21.0 |
| | 140 | 683.5 | 253.7 | 12.4 | 19.0 |
| | 125 | 677.9 | 253.0 | 11.7 | 16.2 |
| 610 x 305 | 238 | 633.0 | 311.5 | 18.6 | 31.4 |
| | 179 | 617.5 | 307.0 | 14.1 | 23.6 |
| | 149 | 609.6 | 304.8 | 11.9 | 19.7 |
| 610 x 229 | 140 | 617.0 | 230.1 | 13.1 | 22.1 |
| | 125 | 611.9 | 229.0 | 11.9 | 19.6 |
| | 113 | 607.3 | 228.2 | 11.2 | 17.3 |
| | 101 | 602.2 | 227.6 | 10.6 | 14.8 |
| 533 x 210 | 122 | 544.6 | 211.9 | 12.8 | 21.3 |
| | 109 | 539.5 | 210.7 | 11.6 | 18.8 |
| | 101 | 536.7 | 210.1 | 10.9 | 17.4 |
| | 92 | 533.1 | 209.3 | 10.2 | 15.6 |
| | 82 | 528.3 | 208.7 | 9.6 | 13.2 |
| 457 x 191 | 98 | 467.4 | 192.8 | 11.4 | 19.6 |
| | 89 | 463.6 | 192.0 | 10.6 | 17.7 |
| | 82 | 460.2 | 191.3 | 9.9 | 16.0 |
| | 74 | 457.2 | 190.5 | 9.1 | 14.5 |
| | 67 | 453.6 | 189.9 | 8.5 | 12.7 |
| 457 x 152 | 82 | 465.1 | 153.5 | 10.7 | 18.9 |
| | 74 | 461.3 | 152.7 | 9.9 | 17.0 |
| | 67 | 457.2 | 151.9 | 9.1 | 15.0 |
| | 60 | 454.7 | 152.9 | 8.0 | 13.3 |
| | 52 | 449.8 | 152.4 | 7.6 | 10.9 |

**STRUCTURAL STEEL SECTIONS**

**UNIVERSAL BEAMS**

(Based on 7854 kg/m$^3$)

| Designation | | Depth of Section | Width of Section | Thickness | |
|---|---|---|---|---|---|
| Serial Size | Mass per unit length | | | Web | Flange |
| mm | kg/m | mm | mm | mm | mm |
| 406 x 178 | 74 | 412.8 | 179.7 | 9.7 | 16.0 |
| | 67 | 409.4 | 178.8 | 8.8 | 14.3 |
| | 60 | 406.4 | 177.8 | 7.8 | 12.8 |
| | 54 | 402.6 | 177.6 | 7.6 | 10.9 |
| 406 x 140 | 46 | 402.3 | 142.4 | 6.9 | 11.2 |
| | 39 | 397.3 | 141.8 | 6.3 | 8.6 |
| 356 x 171 | 67 | 364.0 | 173.2 | 9 1 | 15 7 |
| | 57 | 358.6 | 172.1 | 8.0 | 13.0 |
| | 51 | 355.6 | 171.5 | 7.3 | 11.5 |
| | 45 | 352.0 | 171.0 | 6.9 | 9.7 |
| 356 x 127 | 39 | 352.8 | 126.0 | 6.5 | 10.7 |
| | 33 | 348.5 | 125.4 | 5.9 | 8.5 |
| 305 x 165 | 54 | 310.9 | 166.8 | 7.7 | 13.7 |
| | 46 | 307.1 | 165.7 | 6.7 | 11.8 |
| | 40 | 303.8 | 165.1 | 6.1 | 10.2 |
| 305 x 127 | 48 | 310.4 | 125.2 | 8.9 | 14.0 |
| | 42 | 306.6 | 124.3 | 8.0 | 12.1 |
| | 37 | 303.8 | 123.5 | 7.2 | 10.7 |
| 305 x 102 | 33 | 312.7 | 102.4 | 6.6 | 10.8 |
| | 28 | 308.9 | 101.9 | 6.1 | 8.9 |
| | 25 | 304.8 | 101.6 | 5.8 | 6.8 |
| 254 x 146 | 43 | 259.6 | 147.3 | 7.3 | 12.7 |
| | 37 | 256.0 | 146.4 | 6.4 | 10.9 |
| | 31 | 251.5 | 146.1 | 6.1 | 8.6 |
| 254 x 102 | 28 | 260.4 | 102.1 | 6.4 | 10.0 |
| | 25 | 257.0 | 101.9 | 6.1 | 8.4 |
| | 22 | 254.0 | 101.9 | 5.8 | 6.8 |
| 203 x 133 | 30 | 206.8 | 133.8 | 6.3 | 9.6 |
| | 25 | 203.2 | 133.4 | 5.8 | 7.8 |
| 203 x 102 | 23 | 203.2 | 101.6 | 5.2 | 9.3 |
| 178 x 102 | 19 | 177.8 | 101.6 | 4.7 | 7.9 |
| 152 x 89 | 16 | 152.4 | 88.9 | 4.6 | 7.7 |
| 406 x 140 | 13 | 127.0 | 76.2 | 4.2 | 7.6 |

**STRUCTURAL STEEL SECTIONS**

**UNIVERSAL COLUMNS**

| Designation | | Depth of Section | Width of Section | Thickness | |
|---|---|---|---|---|---|
| Serial Size | Mass per unit length | | | Web | Flange |
| mm | kg/m | mm | mm | mm | mm |
| 356 x 406 | 634 | 474.7 | 424.1 | 47.6 | 77.0 |
| | 551 | 455.7 | 418.5 | 42.0 | 67.5 |
| | 467 | 436.6 | 412.4 | 35.9 | 58.0 |
| | 393 | 419.1 | 407.0 | 30.6 | 49.2 |
| | 340 | 406.4 | 403.0 | 26.5 | 42.9 |
| | 287 | 393.7 | 399.0 | 22.6 | 36.5 |
| | 235 | 381.0 | 395.0 | 18.5 | 30.2 |
| Column Core | 477 | 427.0 | 424.4 | 48.0 | 53.2 |
| 356 x 368 | 202 | 374.7 | 374.4 | 16.8 | 27.0 |
| | 177 | 368.3 | 372.1 | 14.5 | 23.8 |
| | 153 | 362.0 | 370.2 | 12.6 | 20.7 |
| | 129 | 355.6 | 368.3 | 10.7 | 17.5 |
| 305 x 305 | 283 | 365.3 | 321.8 | 26.9 | 44.1 |
| | 240 | 352.6 | 317.9 | 23.0 | 37.7 |
| | 198 | 339.9 | 314.1 | 19.2 | 31.4 |
| | 158 | 327.2 | 310.6 | 15.7 | 25.0 |
| | 137 | 320.5 | 308.7 | 13.8 | 21.7 |
| | 118 | 314.5 | 306.8 | 11.9 | 18.7 |
| | 97 | 307.8 | 304.8 | 9.9 | 15.4 |
| 254 x 254 | 167 | 289.1 | 264.5 | 19.2 | 31.7 |
| | 132 | 276.4 | 261.0 | 15.6 | 25.3 |
| | 107 | 266.7 | 258.3 | 13.0 | 20.5 |
| | 89 | 260.4 | 255.9 | 10.5 | 17.3 |
| | 73 | 254.0 | 254.0 | 8.6 | 14.2 |
| 203 x 203 | 86 | 222.3 | 208.8 | 13.0 | 20.5 |
| | 71 | 215.9 | 206.2 | 10.3 | 17.3 |
| | 60 | 209.6 | 205.2 | 9.3 | 14.2 |
| | 52 | 206.2 | 203.3 | 8.0 | 12.5 |
| | 46 | 203.2 | 203.2 | 7.3 | 11.0 |
| 152 x 152 | 37 | 161.8 | 154.4 | 8.1 | 11.5 |
| | 30 | 157.5 | 152.9 | 6.6 | 9.4 |
| | 23 | 152.4 | 152.4 | 6.1 | 6.8 |

**STRUCTURAL STEEL SECTIONS**

**UNIVERSAL BEARING PILES**

| Designation | | Depth of Section | Width of Section | Thickness | |
|---|---|---|---|---|---|
| Serial Size | Mass per unit length | | | Web | Flange |
| mm | kg/m | mm | mm | mm | mm |
| 356 x 368 | 174 | 361.5 | 378.1 | 20.4 | 20.4 |
| | 152 | 256.4 | 375.5 | 17.9 | 17.9 |
| | 133 | 351.9 | 373.3 | 15.6 | 15.6 |
| | 109 | 346.4 | 370.5 | 12.9 | 12.9 |
| 305 x 305 | 223 | 338.0 | 325.4 | 30.5 | 30.5 |
| | 186 | 328.3 | 320.5 | 25.6 | 25.6 |
| | 149 | 318.5 | 315.6 | 20 7 | 20 7 |
| | 126 | 312.6 | 312.5 | 17.7 | 17.7 |
| | 110 | 307.9 | 310.3 | 15.4 | 15.4 |
| | 95 | 303.8 | 308.3 | 13.4 | 13.4 |
| | 88 | 301.7 | 307.2 | 12.3 | 12.3 |
| | 79 | 299.2 | 306.0 | 11.1 | 11.1 |
| 254 x 254 | 85 | 254.3 | 259.7 | 14.3 | 14.3 |
| | 77 | 249.9 | 257.5 | 12.1 | 12.1 |
| | 63 | 246.9 | 256.0 | 10.6 | 10.6 |
| 203 x 203 | 54 | 203.9 | 207.2 | 11.3 | 11.3 |
| | 45 | 200.2 | 205.4 | 9.5 | 9.5 |

**STRUCTURAL STEEL SECTIONS**

MASSES OF ANGLES

IN KILOGRAMMES PER LINEAR METRE

| EQUAL ANGLES | | | | UNEQUAL ANGLES | | | |
|---|---|---|---|---|---|---|---|
| Nominal Size mm | Mass per metre in kg | Nominal Size mm | Mass per metre in kg | Nominal Size mm | Mass per metre in kg | Nominal Size mm | Mass per metre in kg |
| 200 x 200 x 24 | 71.1 | 70 x 70 x 10 | 10.3 | 200 x 150 x 18 | 47.1 | 100 x 65 x 10 | 12.3 |
| 20 | 59.9 | 8 | 8.36 | 15 | 39.6 | 8 | 9.94 |
| 18 | 54.2 | 6 | 6.38 | 12 | 32.0 | 7 | 8.77 |
| 18 | 48.6 | | | | | | |
| 150 x 150 x 18 | 40.1 | 60 x 60 x 10 | 8.69 | 200 x 100 x 15 | 33.7 | 80 x 69 x 8 | 8.34 |
| 15 | 33.8 | 8 | 7.09 | 12 | 27.3 | 7 | 7.36 |
| 12 | 27.3 | 6 | 5.42 | 10 | 23.0 | 6 | 6.37 |
| 10 | 23.0 | 5 | 4.57 | | | | |
| 120 x 120 x 15 | 26.6 | 50 x 50 x 8 | 5.82 | 150 x 90 x 15 | 26.6 | 75 x 50 x 8 | 7.39 |
| 12 | 21.6 | 6 | 4.47 | 12 | 21.6 | 6 | 5.65 |
| 10 | 18.2 | 5 | 3.77 | 10 | 18.2 | | |
| 8 | 14.7 | | | | | | |
| 100 x 100 x 15 | 21.9 | 45 x 45 x 6 | 4.00 | 150 x 75 x 15 | 24.8 | 65 x 50 x 8 | 6.75 |
| 12 | 17.8 | 5 | 3.38 | 12 | 20.2 | 6 | 5.16 |
| 8 | 12.2 | 4 | 2.74 | 10 | 17.0 | 5 | 4 35 |
| 90 x 90 x 12 | 15.9 | 45 x 40 x 6 | 3.52 | 125 x 75 x 12 | 17.8 | 60 x 30 x 6 | 3.99 |
| 10 | 13.4 | 5 | 2.97 | 10 | 15.0 | 5 | 3.37 |
| 8 | 10.9 | 4 | 2.42 | 8 | 12.2 | | |
| 6 | 8.30 | | | | | | |
| 80 x 80 x 10 | 11.9 | 30 x 30 x 5 | 2.18 | 100 x 75 x 12 | 15.4 | 40 x 25 x 4 | 1.93 |
| 8 | 9.63 | 4 | 1.78 | 10 | 13.0 | | |
| 6 | 7.34 | 3 | 1.36 | 8 | 10.6 | | |
| | | 25 x 25 x 5 | 1.78 | | | | |
| | | 4 | 1.45 | | | | |
| | | 3 | 1.12 | | | | |

**STRUCTURAL STEEL SECTIONS**

SPACING OF HOLES IN ANGLES

| Nominal Leg Length | Spacing of Holes | | | | | | Maximum Diameter of Bolt or Rivet | | |
|---|---|---|---|---|---|---|---|---|---|
| | mm | mm | mm | mm | mm | mm | mm | mm | mm |
| 200 | | 75 | 75 | 55 | 55 | 55 | | 30 | 20 |
| 150 | | 55 | 55 | | | | | 20 | |
| 125 | | 45 | 60 | | | | | 20 | |
| 120 | | | | | | | | | |
| 100 | 55 | | | | | | 24 | | |
| 90 | 50 | | | | | | 24 | | |
| 80 | 45 | | | | | | 20 | | |
| 75 | 45 | | | | | | 20 | | |
| 70 | 40 | | | | | | 20 | | |
| 65 | 35 | | | | | | 20 | | |
| 60 | 35 | | | | | | 16 | | |
| 50 | 28 | | | | | | 12 | | |
| 45 | 25 | | | | | | | | |
| 40 | 23 | | | | | | | | |
| 30 | 20 | | | | | | | | |
| 25 | 15 | | | | | | | | |

## STRUCTURAL STEELWORK

### Surface Areas

| Size (mm) | Weight (kg/m) | Area ($m^2/m$) |
|---|---|---|
| Universal beams | | |
| 914 x 419 | 388 | 3.404 |
| | 343 | 3.382 |
| 914 x 305 | 289 | 2.988 |
| | 253 | 2.967 |
| | 224 | 2.948 |
| | 201 | 2.932 |
| 838 x 292 | 226 | 2.791 |
| | 194 | 2.767 |
| | 176 | 2.754 |
| 762 x 267 | 197 | 2.530 |
| | 173 | 2.512 |
| | 147 | 2.493 |
| 686 x 254 | 170 | 2.333 |
| | 152 | 2.320 |
| | 140 | 2.310 |
| | 125 | 2.298 |
| 610 x 305 | 238 | 2.421 |
| | 179 | 2.381 |
| | 149 | 2.361 |
| 610 x 229 | 140 | 2.088 |
| | 125 | 2.075 |
| | 113 | 2.064 |
| | 101 | 2.053 |
| 533 x 210 | 122 | 1.872 |
| | 109 | 1.860 |
| | 101 | 1.853 |
| | 92 | 1.844 |
| | 82 | 1.833 |
| 457 x 191 | 98 | 1.650 |
| | 89 | 1.641 |
| | 82 | 1.633 |
| | 74 | 1.625 |
| | 67 | 1.617 |
| 457 x 152 | 82 | 1.493 |
| | 74 | 1.484 |
| | 67 | 1.474 |
| 547 x 152 | 60 | 1.487 |
| | 52 | 1.476 |
| 406 x 178 | 74 | 1.493 |
| | 67 | 1.484 |
| | 60 | 1.476 |
| | 54 | 1.468 |
| 406 x 140 | 46 | 1.332 |
| | 39 | 1.320 |
| 356 x 171 | 67 | 1.371 |
| | 57 | 1.358 |
| | 51 | 1.351 |
| | 45 | 1.343 |
| 356 x 127 | 39 | 1.169 |
| | 33 | 1.160 |
| 305 x 165 | 54 | 1.245 |
| | 46 | 1.235 |
| | 40 | 1.227 |
| 305 x 127 | 48 | 1.079 |
| | 42 | 1.069 |
| | 37 | 1.062 |
| 305 x 102 | 33 | 1.006 |
| | 28 | 0.997 |
| | 25 | 0.988 |
| 254 x 146 | 43 | 1.069 |
| | 37 | 1.060 |
| | 31 | 1.050 |
| 254 x 102 | 28 | 0.900 |
| | 25 | 0.893 |
| | 22 | 0.887 |
| 203 x 133 | 30 | 0.912 |
| | 25 | 0.904 |
| 203 x 102 | 23 | 0.789 |
| 78 x 102 | 19 | 0.740 |
| 152 x 89 | 16 | 0.638 |
| 127 x 76 | 13 | 0.537 |

**STRUCTURAL STEELWORK**

| Size (mm) | Weight (kg/m) | Area ($m^2/m$) | Size (mm) | Weight (kg/m) | Area ($m^2/m$) |
|---|---|---|---|---|---|
| Universal columns | | | | | |
| 356 x 406 | 634 | 2.525 | 254 x 254 | 167 | 1.576 |
| | 551 | 2.475 | | 132 | 1.543 |
| | 467 | 2.425 | | 107 | 1.519 |
| | 393 | 2.379 | | 89 | 1.502 |
| | 340 | 2.346 | | 73 | 1.485 |
| | 287 | 2.132 | | | |
| | 235 | 2.279 | 203 x 203 | 86 | 1.236 |
| | | | | 71 | 1.218 |
| 356 x 368 | 202 | 2.187 | | 60 | 1.204 |
| | 177 | 2.170 | | 52 | 1.194 |
| | 153 | 2.154 | | 46 | 1.187 |
| | 129 | 2.137 | | | |
| | | | 152 x 152 | 37 | 0.912 |
| 305 x 305 | 283 | 1.938 | | 30 | 0.900 |
| | 240 | 1.905 | | 23 | 0.889 |
| | 198 | 1.872 | | | |
| | 158 | 1.839 | | | |
| | 137 | 1.822 | | | |
| | 118 | 1.806 | | | |
| | 97 | 1.789 | | | |
| Joists | | | | | |
| 254 x 203 | 81.85 | 1.213 | 127 x 114 | 26.79 | 0.635 |
| 254 x 114 | 37.20 | 0.898 | 114 x 114 | 26.79 | 0.600 |
| 203 x 152 | 52.09 | 0.931 | 102 x 102 | 23.07 | 0.528 |
| 152 x 127 | 37.20 | 0.722 | 89 x 89 | 19.35 | 0.460 |
| 127 x 114 | 29.76 | 0.635 | 76 x 76 | 12.65 | 0.403 |
| Channels | | | | | |
| 432 x 102 | 65.54 | 1.217 | 203 x 89 | 29.78 | 0.720 |
| 381 x 102 | 55.10 | 1.118 | 203 x 76 | 23.82 | 0.675 |
| 305 x 102 | 46.18 | 0.966 | 178 x 89 | 26.81 | 0.671 |
| 305 x 102 | 41.69 | 0.920 | 178 x 76 | 20.84 | 0.625 |
| 254 x 89 | 35.74 | 0.820 | 152 x 89 | 23.84 | 0.621 |
| 254 x 76 | 28.29 | 0.774 | 152 x 76 | 17.88 | 0.575 |
| 229 x 89 | 32.76 | 0.770 | 127 x 64 | 14.90 | 0.476 |
| 229 x 76 | 26.06 | 0.725 | 102 x 51 | 10.42 | 0.379 |

**STRUCTURAL STEELWORK**

**Surface Areas** - continued

| Size (mm) | Weight (kg/m) | Area ($m^2$/m) | Size (mm) | Weight (kg/m) | Area ($m^2$/m) |
|---|---|---|---|---|---|
| Rectangular hollow sections (square) | | | Rectangular hollow sections (square) | | |
| 20 x 20 | 1.12 | 0.076 | 150 x 150 | 22.70 | 0.589 |
| | 1.39 | 0.074 | | 28.30 | 0.586 |
| | | | | 35.40 | 0.583 |
| 30 x 30 | 2.21 | 0.114 | | 43.60 | 0.579 |
| | 2.65 | 0.113 | | 66.40 | 0.566 |
| 40 x 40 | 3.03 | 0.154 | 180 x 180 | 34.20 | 0.706 |
| | 3.66 | 0.153 | | 43.00 | 0.703 |
| | 4.46 | 0.151 | | 53.00 | 0.699 |
| | | | | 65.20 | 0.693 |
| 50 x 50 | 4.66 | 0.193 | | 81.40 | 0.686 |
| | 5.72 | 0.191 | | | |
| | 6.97 | 0.189 | 200 x 200 | 38.20 | 0.786 |
| | | | | 48.00 | 0.783 |
| 60 x 60 | 5.67 | 0.233 | | 59.30 | 0.779 |
| | 6.97 | 0.231 | | 73.00 | 0.773 |
| | 8.54 | 0.229 | | 91.50 | 0.766 |
| 70 x 70 | 7.46 | 0.272 | 250 x 250 | 48.10 | 0.986 |
| | 10.10 | 0.269 | | 60.50 | 0.983 |
| | | | | 75.00 | 0.979 |
| 80 x 80 | 8.59 | 0.312 | | 92.60 | 0.973 |
| | 11.70 | 0.309 | | 117.00 | 0.966 |
| | 14.40 | 0.306 | | | |
| 90 x 90 | 9.72 | 0.352 | 300 x 300 | 90.70 | 1.18 |
| | 13.30 | 0.349 | | 112.00 | 1.17 |
| | | | | 142.00 | 1.17 |
| | 16.40 | 0.346 | | | |
| | | | 350 x 350 | 106.00 | 1.38 |
| 100 x 100 | 12.00 | 0.391 | | 132.00 | 1.37 |
| | 14.80 | 0.389 | | 167.00 | 1.37 |
| | 18.40 | 0.386 | | | |
| | 22.90 | 0.383 | 400 X 400 | 122.00 | 1.58 |
| | 27.90 | 0.379 | | 152.00 | 1.57 |
| 120 x 120 | 18.00 | 0.469 | | | |
| | 22.30 | 0.466 | | | |
| | 27.90 | 0.463 | | | |
| | 34.20 | 0.459 | | | |

STRUCTURAL STEELWORK

**Surface Areas** - continued

| Size (mm) | Weight (kg/m) | Surface Area ($m^2/m$) | Size (mm) | Weight (kg/m) | Surface Area ($m^2/m$) |
|---|---|---|---|---|---|
| Rectangular hollow sections - (size mm) | | | | | |
| 50 x 30 | 3.03 | 0.154 | 160 x 80 | 18.00 | 0.469 |
| | 3.66 | 0.153 | | 22.30 | 0.466 |
| | | | | 27.90 | 0.463 |
| 60 x 40 | 4.66 | 0.193 | | 34.20 | 0.459 |
| | 5.72 | 0.191 | | | |
| | | | 200 x 100 | 22.70 | 0.589 |
| 80 x 40 | 5.67 | 0.233 | | 28.30 | 0.586 |
| | 6.97 | 0.231 | | 35.40 | 0.583 |
| | | | | 43.60 | 0.579 |
| 90 x 50 | 7.46 | 0.272 | | 53.40 | 0.573 |
| | 10.10 | 0.269 | | | |
| | | | 250 x 150 | 38.20 | 0.786 |
| 100 x 50 | 7.18 | 0.293 | | 48.00 | 0.783 |
| | 10.90 | 0.289 | | 59.30 | 0.779 |
| | 8.86 | 0.291 | | 73.00 | 0.773 |
| | | | | 91.50 | 0.766 |
| 100 x 60 | 8.59 | 0.312 | | | |
| | 11.70 | 0.309 | 300 x 200 | 48.10 | 0.986 |
| | 14.40 | 0.306 | | 60.50 | 0.983 |
| | | | | 75.00 | 0.979 |
| 120 x 60 | 9.72 | 0.352 | | 92.60 | 0.973 |
| | 13.30 | 0.349 | | 117.00 | 0.966 |
| | 16.40 | 0.346 | | | |
| | | | 400 x 200 | 90.70 | 1.18 |
| 120 x 80 | 14.80 | 0.389 | | 112.00 | 1.17 |
| | 18.40 | 0.386 | | 142.00 | 1.17 |
| | 22.90 | 0.383 | | | |
| | 27.90 | 0.379 | 450 x 250 | 106.00 | 1.38 |
| | | | | 132.00 | 1.37 |
| 150 x 100 | 18.70 | 0.489 | | 1.67 | 1.37 |
| | 23.30 | 0.486 | | | |
| | 29.10 | 0.483 | | | |
| | 35.70 | 0.479 | | | |

| Size (mm) | Surface area $m^2/m$ | Size (mm) | Surface area $m^2/m$ |
|---|---|---|---|
| Circular hollow sections - outside dia (mm) | | | |
| 21.3 | 0.067 | 168.3 | 0.529 |
| 26.9 | 0.085 | 193.7 | 0.609 |
| 33.7 | 0.106 | 219.1 | 0.688 |
| 42.4 | 0.133 | 244.5 | 0.768 |
| 48.3 | 0.152 | 273.0 | 0.858 |
| 60.3 | 0.189 | 323.9 | 1.020 |
| 76.1 | 0.239 | 355.6 | 1.117 |
| 88.9 | 0.279 | 406.4 | 1.280 |
| 114.3 | 0.359 | 457 | 1.436 |
| 139.7 | 0.439 | | |

## BRICKWORK AND BLOCKWORK

### Common Brickwork

### Quantities of Bricks and Mortar

| Thickness | Materials per $m^2$ No. of Bricks | Mortar $m^3$ |
|---|---|---|
| Half brick (112.5 mm) | 58 | 0.022 |
| One brick (225 mm) | 116 | 0.055 |
| Cavity, both skins (275 mm) | 116 | 0.045 |
| 1.5 brick (337 mm) | 174 | 0.074 |
| Mass brickwork per $m^3$ | 464 | 0.360 |

### Mortar mixes: Quantities of Dry Materials

| | Imperial cu yd | | | metric $m^3$ | | |
|---|---|---|---|---|---|---|
| Mix | Cement cwts | Lime cwts | Sand cu yds | Cement tonnes | Lime tonnes | Sand cu m |
| 1:3 | 7.0 | - | 1.04 | 0.54 | - | 1.10 |
| 1:4 | 6.3 | - | 1.10 | 0.40 | - | 1.20 |
| 1:1:6 | 3.9 | 1.6 | 1.10 | 0.27 | 0.13 | 1.10 |
| 1:2:9 | 2.6 | 2.1 | 1.10 | 0.20 | 0.15 | 1.20 |
| 0:1:3 | - | 3.3 | 1.10 | - | 0.27 | 1.00 |

### Mortar mixes for various uses

1:3 Construction designed to withstand heavy loads in all seasons

1:1:6 Normal construction *not* designed for heavy loads. Sheltered and moderate conditions in spring and summer. Work above d:p:c - sand, lime bricks, clay blocks etc.

1:2:9 Internal partitions with blocks which have high drying shrinkage, pumice blocks, etc. any periods.

0:1:3 Hydraulic lime only should be used in this mix and may be used for construction not designed for heavy loads and above d:p:c spring and summer.

## BRICKWORK AND BLOCKWORK

### Quantities of bricks and mortar required per m$^2$ of walling

| Description | Unit | No of bricks required | Mortar required (cubic metres) No frogs | Single frogs | Double frogs |
|---|---|---|---|---|---|
| Standard bricks | | | | | |
| Brick size 215 x 102.5 x 50 mm | | | | | |
| half brick wall (103 mm) | m$^2$ | 72 | 0.022 | 0.027 | 0.032 |
| 2 x half brick cavity wall (270 mm) | m$^2$ | 144 | 0.044 | 0.054 | 0.064 |
| one brick wall (215 mm) | m$^2$ | 144 | 0.052 | 0.064 | 0.076 |
| one and a half brick wall (328 mm) | m$^2$ | 216 | 0.073 | 0.091 | 0.108 |
| Mass brickwork | m$^3$ | 576 | 0.347 | 0.413 | 0.480 |
| Brick size 215 x 102.5 x 65 mm | | | | | |
| half brick wall (103 mm) | m$^2$ | 58 | 0.019 | 0.022 | 0.026 |
| 2 x half brick cavity wall (270 mm) | m$^2$ | 116 | 0.038 | 0.045 | 0.055 |
| one brick wall (215 mm) | m$^2$ | 116 | 0.046 | 0.055 | 0.064 |
| one and a half brick wall (328 mm) | m$^2$ | 174 | 0.063 | 0.074 | 0.088 |
| Mass brickwork | m$^3$ | 464 | 0.307 | 0.360 | 0.413 |
| Metric modular bricks | | | Perforated | | |
| Brick co-ordinating size 200 x 100 x 75 mm | | | | | |
| 90 mm thick | m$^2$ | 67 | 0.016 | 0.019 | |
| 190 mm thick | m$^2$ | 133 | 0.042 | 0.048 | |
| 290 mm thick | m$^2$ | 200 | 0.068 | 0.078 | |
| Brick co-ordinating size 200 x 100 x 100 mm | | | | | |
| 90 mm thick | m$^2$ | 50 | 0.013 | 0.016 | |
| 190 mm thick | m$^2$ | 100 | 0.036 | 0.041 | |
| 290 mm thick | m$^2$ | 150 | 0.059 | 0.067 | |
| Brick co-ordinating size 300 x 100 x 75 mm | | | | | |
| 90 mm thick | m$^2$ | 33 | - | 0.015 | |
| Brick co-ordinating size 300 x 100 x 100 mm | | | | | |
| 90 mm thick | m$^2$ | 44 | 0.015 | 0.018 | |

Note: Assuming 10 mm deep joints.

## WOODWORK

### Conversion tables (for sawn timber only)

| Inches | Millimetres | Feet | Metres | Inches | Millimetres | Feet | Metres |
|---|---|---|---|---|---|---|---|
| 1 | 25 | 1 | 0.300 | 13 | 325 | 13 | 3.900 |
| 2 | 50 | 2 | 0.600 | 14 | 350 | 14 | 4.200 |
| 3 | 75 | 3 | 0.900 | 15 | 375 | 15 | 4.500 |
| 4 | 100 | 4 | 1.200 | 16 | 400 | 16 | 4.800 |
| 5 | 125 | 5 | 1.500 | 17 | 425 | 17 | 5.100 |
| 6 | 150 | 6 | 1.800 | 18 | 450 | 18 | 5.400 |
| 7 | 175 | 7 | 2.100 | 19 | 475 | 19 | 5.700 |
| 8 | 200 | 8 | 2.400 | 20 | 500 | 20 | 6.000 |
| 9 | 225 | 9 | 2.700 | 21 | 525 | 21 | 6.300 |
| 10 | 250 | 10 | 3.000 | 22 | 550 | 22 | 6.600 |
| 11 | 275 | 11 | 3.300 | 23 | 575 | 23 | 6.900 |
| 12 | 300 | 12 | 3.600 | 24 | 600 | 24 | 7.200 |

### Planed softwood

The finished end section size of planed timber is usually 3/16" less than the original size from which it is produced. This however varies slightly dependant upon availability of material and origin of species used.

Standard (timber) to cubic metres and cubic metres to standards (timber)

| Cubic metres | Cubic metres standards | Standard |
|---|---|---|
| 4.672 | 1 | 0.214 |
| 9.344 | 2 | 0.428 |
| 14.017 | 3 | 0.642 |
| 18.689 | 4 | 0.856 |
| 23.361 | 5 | 1.070 |
| 28.033 | 6 | 1.284 |
| 32.706 | 7 | 1.498 |
| 37.378 | 8 | 1.712 |
| 42.050 | 9 | 1.926 |
| 46.722 | 10 | 2.140 |
| 93.445 | 20 | 4.281 |
| 140.167 | 30 | 6.421 |
| 186.890 | 40 | 8.561 |
| 233.612 | 50 | 10.702 |
| 280.335 | 60 | 12.842 |
| 327.057 | 70 | 14.982 |
| 373.779 | 80 | 17.122 |
| 420.502 | 90 | 19.263 |
| 467.224 | 100 | 21.403 |

Standards (timber) to cubic metres and cubic metres to standards (timber)

1 cu metre = 35.3148 cu ft = 0.21403 std
1 cu ft = 0.028317 cu metres
1 std = 4.67227 cu metres

## PAVING AND SURFACING

### Weight table

| Description of Item | Size | Quantity per Tonne |
|---|---|---|
| Paving 50 mm thick | 900 x 600 mm | 15 |
| Paving 50 mm thick | 750 x 600 mm | 18 |
| Paving 50 mm thick | 600 x 600 mm | 23 |
| Paving 50 mm thick | 450 x 600 mm | 30 |
| Paving 38 mm thick | 600 x 600 mm | 30 |
| Path edging | 914 x 50 x 150 mm | 60 |
| Kerb (including radius and tapers) | 125 x 254 x 914 mm | 15 |
| Kerb (including radius and tapers) | 125 x 150 x 914 mm | 25 |
| Square channel | 125 x 254 x 914 mm | 15 |
| Dished channel | 125 x 254 x 914 mm | 15 |
| Quadrants | 300 x 300 x 254 mm | 19 |
| Quadrants | 450 x 450 x 254 mm | 12 |
| Quadrants | 300 x 300 x 150 mm | 30 |
| Internal angles | 300 x 300 x 254 mm | 30 |
| Fluted pavement channel | 255 x 75 x 914 mm | 25 |
| Corner stones | 300 x 300 mm | 80 |
| Corner stones | 360 x 360 mm | 60 |
| Cable covers | 914 x 175 mm | 55 |
| Gulley kerbs | 220 x 220 x 150 mm | 60 |
| Gulley kerbs | 220 x 200 x 75 mm | 120 |

### Voids in material used as sub bases or beddings

| Material | $m^3$ of voids per $m^3$ | Material | $m^3$ of voids per $m^3$ |
|---|---|---|---|
| Alluvium | 0.37 | Gravel | 0.39 |
| River grit | 0.29 | Broken stone | 0.45 |
| Quarry sand | 0.24 | Broken bricks | 0.42 |
| Shingle | 0.37 | | |

### Coverage ($m^2$) per cubic metre of materials used as sub bases or capping layers

| Consolidated thickness laid in (mm) | Square metre coverage Gravel | Sand | Hardcore |
|---|---|---|---|
| 50 | 15.80 | 16.50 | - |
| 75 | 10.50 | 11.00 | - |
| 100 | 7.92 | 8.20 | 7.42 |
| 125 | 6.34 | 6.60 | 5.90 |
| 150 | 5.28 | 5.50 | 4.95 |
| 175 | - | - | 4.23 |
| 200 | - | - | 3.71 |
| 225 | - | - | 3.30 |
| 300 | - | - | 2.47 |

## PAVING AND SURFACING

### Approximate rate of spreads

| Average Thickness of Course | Description | Approximate Rate of Spread: Open Textured | | Dense, Medium & Fine Textured | |
|---|---|---|---|---|---|
| | | kg/$m^2$ | $m^2$/t | kg/$m^2$ | $m^2$/t |
| mm | | | | | |
| 35 | 14 mm open textured or dense wearing course | 60-75 | 13-17 | 70-85 | 12-14 |
| 40 | 20 mm open textured or dense base course | 70-85 | 12-14 | 80-100 | 10-12 |
| 45 | 20 mm open textured or dense base course | 80-100 | 10-12 | 95-100 | 9-10 |
| 50 | 20 mm open textured or dense, or 28 mm dense base course | 85-110 | 9-12 | 110-120 | 8-9 |
| 60 | 28 mm dense base course, 40 mm open textured or dense base course or 40 mm single course as base course | 95-125 | 8-10 | 130-150 | 7-8 |
| 65 | 28 mm dense base course, 40 mm open textured or dense base course or 40 mm single course | 100-135 | 7-10 | 140-160 | 6-7 |
| 75 | 40 mm single course, 40 mm open textured or dense base course, 40 mm dense roadbase | 120-150 | 7-8 | 165-185 | 5-6 |
| 100 | 40 mm dense base course or roadbase | - | - | 220-240 | 4-4.5 |

### Surface Dressing Roads: Coverage ($m^2$) per tonne of Material

| Size in mm | Sand | Granite chips | Gravel | Limestone Chips |
|---|---|---|---|---|
| Sand | 168 | - | - | - |
| 3 | - | 148 | 152 | 165 |
| 6 | - | 130 | 133 | 144 |
| 9 | - | 111 | 114 | 123 |
| 13 | - | 85 | 87 | 95 |
| 19 | - | 68 | 71 | 78 |

**USEFUL ADDRESSES FOR FURTHER INFORMATION**

ACOUSTICAL INVESTIGATION & RESEARCH ORGANISATION LTD
Duxon's Turn, Maylands Avenue,
Hemel Hempstead, Herts HP2 4SB
Hemel Hempstead (0442) 47146/7d

AGGREGATE CONCRETE BLOCK ASSOCIATION
60 Charles Street, Leicester LE1 1FB
Leicester (0533) 536161

ALUMINIUM EXTRUDERS ASSOCIATION
Broadway House, Calthorpe Road,
Five Ways, Birmingham B15 1TN
021-455 0311
Telex BIRCOM G 338024 ALFED

ALUMINIUM FEDERATION LTD
Broadway House, Calthorpe Road,
Five Ways, Birmingham B15 1TN
021-455 0311
Telex BIRCOM G 338024 ALFED

ALUMINIUM WINDOW ASSOCIATION
Ivy Bank House, Ivy Bank Park,
Entry Hill, Bath BA2 5NF
Bath (0225) 835811

AMERICAN PLYWOOD ASSOCIATION
101 Wigmore Street, London W1H 9AB
01-629 3437/8
Telex 296009 USAGOF G

ARBORICULTURAL ADVISORY & INFORMATION SERVICE
Forest Research Station,
Alice Holt Lodge, Farnham
Surrey GU10 4LH
Bentley (0420) 22255

ARBORICULTURAL ASSOCIATION
Ampfield House, Ampfield,
Romsey, Hants SO5 9PA
Braishfield (0794) 68717

ARCHITECTURAL ASSOCIATION
34/36 Bedford Square, London
WC1B 3ES
071-636 0974

ARMS (ASSOCIATION OF ROOFING AND MATERIALS SUPPLIERS)
c/o Allan Harris & Sons Ltd,
Station Rd, Georges,
Weston-Super-Mare, Avon
BS22 OXN
Weston-Super-Mare (0934) 511166
Alloa (0259) 721010
Darwen (0254) 771722
London 071-485 1791

ASBESTOS INFORMATION CENTRE LTD
St Andrews House, 22-28 High
Epsom, Surrey KT18 8AH
Epsom (037 27) 42055
Telex 21120 Ref 2526

ASBESTOS REMOVAL CONTRACTORS ASSOCIATION
45 Sheen Lane, London SW14 8AB
01-876 4415/6
Telex 927298

BRITISH AGGREGATE CONSTRUCTION MATERIALS INDUSTRIES LTD
156 Buckingham Palace Road, London
SW1W 9TR
071-730 8194

BRITISH AIRPORTS AUTHORITY
Head Office, Gatwick Airport
Gatwick, West Sussex RH6 OHZ
Gatwick (0293) 517755
Telex 877995 BAA HQG

BRITISH ANODISING ASSOCIATION
Broadway House, Calthorpe Road,
Five Ways, Birmingham B15 1TN

BRITISH ARCHITECTURAL LIBRARY
RIBA, 66 Portland Place, London
W1N 4AD
071-580 5533

BRITISH ASSOCIATION OF LANDSCAPE INDUSTRIES
9 Henry Street, Keighley, West
Yorkshire
BD21 3DR

**USEFUL ADDRESSES FOR FURTHER INFORMATION**

BRITISH CLAYWARE LAND DRAIN INDUSTRY
Federation House, Station Rd,
Stoke-on-Trent ST4 2TJ
Stoke-on-Trent (0782) 416256
Telex 367446

BRITISH COMBUSTION EQUIPMENT MANUFACTURERS ASSOCIATION
The Fernery, Market Place, Midhurst.
West Sussex GU29 9DP
Midhurst (073 081) 2782

BRITISH CONCRETE MASONARY ASSOCIATION
St John's Works, Bedford MK42 0DL
Bedford (022234) 63171/9

BRITISH CONCRETE PUMPING ASSOCIATION LTD
1 Highline Cottage, Lodge Hill, Newton,
Nr Southampton, Hants
Fareham (0329) 232220

BRITISH CONSTRUCTIONAL STEELWORK ASSOCIATION LTD
35 Old Queen St, London SW1H 9HZ
071-222 2254

BRITISH EFFLUENT & WATER ASSOCIATION
51 Castle Street, High Wycombe, Bucks
HP13 6RN
High Wycombe (0494) 444544

BRITISH ELECTRICAL & ALLIED MANUFACTURERS ASSOCIATION
Leicester House, 8 Leicester St,
London WC2H 7BN
071-437 0678

BRITISH ELECTRICAL SYSTEM ASSOCIATION (BESA)
Granville Chambers, 2 Radford St,
Stone, Staffs ST15 8DA
Stone (0785) 812426

BRITISH FIRE PROTECTION SYSTEMS ASSOCIATION LTD
48a Eden Street, Kingston-upon-Thames,
Surrey KT1 1EE
Kingston-upon-Thames 081-549 5855

BRITISH FIRE SERVICES ASSOCIATION
88 London Road, Leicester LE2 0QR
Leicester (0533) 5428279

BRITISH FLAT ROOFING COUNCIL
PO Box 125, Haywards Heath, West Sussex
RH16 3TJ
Haywards Heath (0444) 416681/2

BRITISH FLUE & CHIMNEY MANUFACTURERS ASSOCIATION
Nicholson House, High St, Maidenhead,
Berks SL6 1LF
Maidenhead (0628) 34667

BRITISH INDUSTRIAL FASTNERS FEDERATION
Queen's House, Queen's Rd, Coventry
CV1 3EG
Coventry (0203) 22325
Telex 311650

**USEFUL ADDRESSES FOR FURTHER INFORMATION**

BRITISH LEAD MANUFACTURERS ASSOCIATION
68 High St, Weybridge, Surrey KT13 8BL
Weybridge (0932) 56621

BRITISH LIBRARY LENDING DIVISION
Boston Spa, Wetherby, West Yorks LS23 7BQ
Boston Spa (0937) 843434
Telex 557381

BRITISH LIBRARY, SCIENCE REFERENCE LIBRARY
25 Southampton Buildings, Chancery Lane, London WC2A 1AW
01-405 8721, ex 3344/5
Telex 266959

BRITISH LOCK MANUFACTURERS ASSOCIATION
5 Greenfield Crescent, Edgbaston
Birmingham B15 3BE
021-454 2177
Telex 336006

BRITISH MALLEABLE TUBE FITTINGS ASSOCIATION
105 Meadow View Road, Catford, London SE6 3NJ
081-698 8856

BRITISH NON-FERROUS METALS FEDERATION
Crest House, 7 Highfield Road, Edgbaston, Birmingham B15 3ED
021-454 7766
Telex 339161

BRITISH NON-FERROUS METALS TECHNICAL CENTRE
Grove Laboratories, Detchworth Rd, Wantage, Oxon OX12 9BJ
Wantage (023 57) 2992
Telex 837166

BRITISH PLASTICS FEDERATION
5 Belgrave Square, London SW1X 8PH
071-235 9483
Information 071-235 9888
Telex 895 1528

BRITISH PRECAST CONCRETE FEDERATION
60 Charles St, Leicester LE1 1FB
Leicester (0533) 536161

BRITISH PUMP MANUFACTURERS ASSOCIATION
3 Pannells Court, Chertsey St, Guildford, Surrey GU1 4EU
Guildford (0483) 37997/8

BRITISH READY MIXED CONCRETE ASSOCIATION
Bramber Court
Bramber Road
London W14
071-381 6582

BRITISH REINFORCEMENT MANUFACTURERS ASSOCIATION (BRMA)
15 Tooks Court, London EC4 1LA
071-831 7581

BRITISH ROAD FEDERATION
Cowdray House, 6 Portugal St, London WC2A 2HG
071-242 1285

BRITISH RUBBER MANUFACTURERS ASSOCIATION LTD
90/91 Tottenham Court Road. London W1P 0BR
071-580 2794
Telex 8813271 GECOMS G

BRITISH STANDARDS INSTITUTION
2 Park Street, London W1A 2BS
071-629 9000
Enquiries: Milton Keynes (0908) 320066

BRITISH STEEL CORPORATION
Swinden Laboratories, Moorgate
Rotherham S60 3AR
Rotherham (0709) 60166

BRITISH VALVE MANUFACTURERS ASSOCIATION LTD
3 Pannells Court, Chertsey Street, Guildford, Surrey GU1 4EU
Guildford (0483) 37379

BRITISH WOODWORK FEDERATION
82 New Cavendish St, London W1M 8AD
071-580 5588
Telex 265763

BUILDING CENTRE: BRISTOL
The Building Centre, Stonebridge House, Colston Avenue, The Centre, Bristol BS1 4TW
Management Bristol: (0272) 22953
Information Bristol: (0272) 277003

## USEFUL ADDRESSES FOR FURTHER INFORMATION

BUILDING CENTRE: COVENTRY
Coventry Building Information Centre, Dept of Architecture & Planning, Tower Block, Council Offices, Much Park St, Coventry CV1 5RT
Coventry (0203) 25555, ext 2512
Telex 31469

BUILDING CENTRE: DURHAM
Northern Counties Building Information Centre, c/o NFBTE
Green Lane, Durham DH1 3JI
Durham (0385) 62611
Telex 537544

BUILDING CENTRE: LONDON
The Building Centre, 26 Store St, London WC1E 7BT
Administration: 01-637 1022
Telex 261507 Ref 3324
Bookshop: 071-637 3151
Information Service: (0344) 884999

BUILDING CENTRE: MANCHESTER
The Building Centre, 113-115 Portland Street, Manchester M1 6FB
061-236 6933/9802

BUILDING CENTRE: PETERBOROUGH
Building Materials Information Service, 22 Broadway, Peterborough PE1 1RU
Peterborough (0733) 314239

BUILDING CENTRE: SCOTLAND
The Building Centre Scotland 1971 Ltd, Macdata Unit, 47 High Street, Paisley PA1 2AL
041-840 1199

BUILDING CENTRE: STOKE-ON-TRENT
The Building Information Centre, Stoke-on-Trent Cauldon College, The Concourse, Stoke Rd, Shelton, Stoke-on-Trent ST4 2DG
Stoke-on-Trent (0782) 29561

BUILDING EMPLOYERS CONFEDERATION
82 New Cavendish Street, London W1M 8AD
071-580 5588

BUILDING MAINTENANCE INFORMATION
85/87 Clarence Street, Kingston-upon-Thames, Surrey KT1 1RB
Kingston-upon-Thames 081-546 7555

BUILDING RESEARCH ESTABLISHMENT: SCOTLAND
Kelvin Rd, East Killbridge, Glasgow G75 0RX
East Killbridge (035 52) 33001

BRITISH SERVICES RESEARCH AND INFORMATION ASSOCIATION
Old Bracknell Lane West, Bracknell Berks RG12 4AH
Bracknell (0344) 426511

CALCIUM SILICATE BRICK ASSOCIATION
24 Fearnley Rd, Welwyn Garden City, Herts AL8 0HW
Welwyn Garden (070 73) 2538

CEMENT ADMIXTURES ASSOCIATION
2a High St, Hythe, Southampton SO4 6YW
Southampton (0703) 842765

CEMENT AND CONCRETE ASSOCIATION
Wexham Springs, Slough, Berks SL3 6PL
Fulmer (028 16) 2727

CEMENT MAKERS' FEDERATION
Terminal House, 52 Grosvenor Garden, London SW1W 0AH
01-730 2148
Telex 261700 CEMFED

CHARTERED INSTITUTE OF ARBITRATORS
75 Cannon Street, London EC4N 5BH
01-236 8761
Telex 893466 CIARB G

CHARTERED INSTITUTE OF BUILDING
Englemere, Kings Ride, Ascot, Berks SL5 8BJ
Ascot (0990) 23355

CHARTERED INSTITUTION OF BUILDING SERVICES
Delta House, 222 Balham High Road, London SW12 9BS
081-675 5211

CLAY PIPE DEVELOPMENT ASSOCIATION
Drayton House, 30 Gordon Street London WC1H 0AN
081-388 0025/6

**USEFUL ADDRESSES FOR FURTHER INFORMATION**

COLD ROLLED SECTIONS ASSOCIATIONS OF SPECIALIST ENGINEERING CONTRACTORS
ESCA House, 34 Palace Court, Bayswater, London W2 4JG
071-229 2488
Telex 27929

CONCRETE BRICK MANUFACTURERS ASSOCIATION
c/o British Precast Concrete Federation, 60 Charles St, Leicester LE1 1FB
Leicester (0533) 536161

CONCRETE PIPE ASSOCIATION OF GREAT BRITAIN
60 Charles Street, Leicester LE1 1FB
Leicester (0533) 536161

CONCRETE SOCIETY
Devon House, 12-15 Dartmouth Street, London SW1H 9BL
071-222 1822

CONFEDERATION OF BRITISH INDUSTRY
Centre Point, 103 New Oxford Street,
071-379 7400
Telex 21352

CONSTRADO (CONSTRUCTURAL STEEL RESEARCH AND DEVELOPMENT ORGANISATION)
NLA Tower, 12 Addiscombe Road, Croydon CR9 3JH
Croydon 081-688 2688, 081-686 0366

CONTRACT FLOORING ASSOCIATION
23 Chippenham Mews, London W9 2AN
071-286 4499

CONTRACTORS MECHANICAL PLANT ENGINEERS
20 Knave Wood Rd, Kemsing, Sevenoaks Kent TN15 6RH
Otford (09592) 2628

COPPER CYLINDER AND BOILER MANUFACTURERS ASSOCIATION
56 Oxford St, Manchester M1 6U
061-236 0384

DRY STONE WALLING ASSOCIATION
YFC Office, National Agricultural Centre, Kenilworth, Warwickshire CV8 2LG
Coventry (0203) 56131

DUCTILE IRON PRODUCERS ASSOCIATION
8th Floor, Bridge House, 121 Smallbrook, Queensway, Birmingham B5 5JP
021-643 3377

ELECTRIC CABLE MAKERS CONFEDERATION
56 Palace Road, East Molesey, Surrey KT8 9DW
East Molesey 081-941 4079
Telex 24893

ELECTRICAL CONTRACTORS ASSOCIATION (ECA)
Eca House, 34 Palace Court, Bayswater, London W2 4HY
081-229 1266
Telex 27929

ELECTRICAL CONTRACTORS ASSOCIATION OF SCOTLAND
23 Heriot Row, Edinburgh EH3 6EW
031-225 7221

ELECTRICAL INSTALLATION EQUIPMENT MANUFACTURERS ASSOCIATION
Leicester House, 8 Leicester St, London WC2H 7BN
071-437 0678

ENERGY EFFICIENCY OFFICE
Room 1312, Department of Energy, Thames House South, Millbank, London SW1P 4QJ

ENERGY SYSTEM TRADE ASSOCIATION LTD (ESTA)
P O Box 16, Stroud, Glos GL5 5EB
Amberket (045 387) 3568

EUROBUILD
26 Rue la Perouse, 75116 Paris
(010 33) 7201020
Telex 611975 F

FARM BUILDING ASSOCIATION
National Agricultural Centre, Stoneleigh, Kenilworth, Warwickshire CV8 2LG
Coventry (0203) 23345

**USEFUL ADDRESSES FOR FURTHER INFORMATION**

FEDERATION OF BUILDING & CIVIL ENGINEERING CONTRACTORS (NI) LTD
143 Malone Rd, Belfast BT9 6SU
Belfast (0232) 22345

FEDERATION OF CIVIL ENGINEERING CONTRACTORS
Cowdray House, 6 Portugal Street, London WC2A 2HH
071-404 4020

FEDERATION OF RESIN FORMULATORS AND APPLICATORS LTD (FERFA)
16 Courtmoor Avenue, Fleet, Aldershot, Hampshire
GU13 9UF
Fleet (025 144) 6936

FEDERATION OF MANUFACTURERS OF CONSTRUCTION EQUIPMENT & CRANES
7-15 Lansdowne Rd, Croydon, Surrey CR9 2P1
Croydon 081-688 4422
Telex 917857 BINDER G

FEDERATION OF MASTER BUILDERS
33 John Street, London WC1N 2BB
071-242 7583

FEDERATION OF PILING SPECIALISTS
Dickens House, 15 Tooks Court, London EC4A 1LA
071-831 7581
Telex 23485

FLAT ROOFING CONTRACTORS ADVISORY BOARD
Maxwelton House, 41/43 Boltro Road, Haywards Heath, West Sussex RH16 1BJ
Haywards Heath (0444) 451835/6

FENCING CONTRACTORS ASSOCIATION
St Johns House, 23 St Johns Rd, Watford WD1 1PY
Watford (0923) 27236

FIBRE CEMENT MANUFACTURERS ASSOCIATION LTD (FCMA)
PO Box 92, Elmswell, Bury St Edmunds, Suffolk IP30 9HS
Elmswell (0359) 40963

FINNISH PLYWOOD INTERNATIONAL
PO Box 99, Welwyn Garden City, Herts AL6 0HS
Bulls Green (043 879) 746

FLEXIBLE ROOFING ASSOCIATION
125 Queen's Road, Brighton, East Sussex BN1 3YW
Brighton (0273) 33322

GLASSFIBRE REINFORCEMENT
5 Upper Bar, Newport, Shropshire
TF10 7EH
Newport (0952) 811397

INDUSTRIAL BUILDING BUREAU
33 Upper St, London N1 0PN
01-359 9877

INSTITUTE OF ASPHALT TECHNOLOGY
Unit 18, Central Trading Estate
Staines, Middlesex TW18 4XE
Staines (0784) 65387

INSTITUTE OF CONCRETE TECHNOLOGY
PO Box 52, Slough SL3 6PL
Postal enquires only

INSTITUTE OF SHEET METAL ENGINEERS
Queensway House, 2 Queensway, Redhill, Surrey RH1 1QS
Redhill (0737) 68611

INSTITUTE OF WASTES MANAGEMENT
2 Albion Place, Northampton NN1 1UD
Northampton (0604) 20426

INSTITUTE OF WATER POLLUTION CONTROL
Ledstone House, 53 London Road
Maidstone, Kent ME16 8JH
Maidstone (0622) 601399

INSTITUTE OF CIVIL ENGINEERS
1-7 Great George St, London SW1P 3AA
071-222 7722

INSTITUTION OF ELECTRICAL ENGINEERS
Savoy Place, London WC2H 0BL
071-240 1871
Telex 261176 IEELDN G

**USEFUL ADDRESSES FOR FURTHER INFORMATION**

JOINT CONTRACTS TRIBUNAL
66 Portland Place, London W1N 4AD
071-580 5533, 01-580 5588

LEAD DEVELOPMENT ASSOCIATION
34 Berkeley Square
071-499 8422

NATIONAL PAVING AND KERB ASSOCIATION
60 Charles St, Leicester LE1 6AJ
071-499 8422

PIPELINE INDUSTRIES GUILD
17 Grosvenor Crescent, London SW1X 7ES
071-235 7938

PITCH FIBRE PIPE ASSOCIATION OF GREAT BRITAIN
c/o Croda Hydrocarbons Ltd, Pipes Division, PO Box 16, Weeland Rd, Knottingley, West Yorkshire WF11 8DZ
Knottingley (0977) 87161
Telex 8814171

PLASTICS PIPE MANUFACTURERS SOCIETY
89 Cornwall Street, Birmingham B3 3BY
021-236 1866

POST TENSIONING ASSOCIATION
CCL Systems Ltd, Cabca House, 296 Ewell Road, Surbiton, Surrey, KT6 7AH
Surbiton 081-390 1122

PRECAST CONCRETE FRAME ASSOCIATION
60 Charles St, Leicester LE1 1FB
Leicester (0533) 53161

PREFABRICATED BUILDING MANUFACTURERS ASSOCIATION OF GREAT BRITAIN
Westgate House, Chalk Lane, Epsom, Surrey KT18 7AJ
Epsom (037 27) 40044

PRESTRESSED CONCRETE ASSOCIATION
60 Charles St, Leicester LE1 1FB
Leicester (0533) 536161

PROPERTY SERVICES AGENCY
DEPARTMENT OF THE ENVIRONMENT
Whitgift Centre, Wellesley Road, Croydon CR9 3LY
Croydon 081-686 8710
Architectural Services: 081-686 8710
Civil Accommodation/Estate Surveying Services: 071-928 7999
Defence Services 1 (Army, Overseas): 081-397 5266
Defence Services 2 (Air/Navy): 081-686 8710
Diplomatic/Post Offices Services: 081-686 5622
Engineering/QS Services: 081 686 3499

ROYAL INSTITUTE OF BRITISH STRUCTURAL INSULATION ASSOCIATION
45 Sheen Lane, London SW14 8AB
081-876 4415/6

SAND AND GRAVEL ASSOCIATION
3rd Floor, 32-36 Fleet Lane
London EC4M 4YA

SCOTTISH BUILDING EMPLOYERS FEDERATION
13 Woodside Crescent, Glasgow G3 7UP
0141-332 7144
Aberdeen (0224) 643838
Inverness (0463) 237626

SCOTTISH DEVELOPMENT AGENCY
120 Bothwell Street, Glasgow G2 7JP
041-248 2700
Telex 777600

SCOTTISH PRECAST CONCRETE MANUFACTURERS ASSOCIATION
9 Princes Street, Falkirk FK1 1LS
Falkirk (0324) 22088

SCOTTISH WIREWORK MANUFACTURERS ASSOCIATION
36 Renfield Street, Glasgow G2 18D
041-248 6161

SOCIETY OF CHAIN LINK FENCING MANUFACTURERS
PO Box 121, The Fountain Precinct, 1 Balm Green, Sheffield S1 3AF
Shefield (0742) 751234

**USEFUL ADDRESSES FOR FURTHER INFORMATION**

STEEL CASTINGS RESEARCH & TRADE ASSOCIATION
East Bank Road, Sheffield S2 3PT
Sheffield (0742) 28647

STEEL WINDOW ASSOCIATION
26 Store Street, London WC1E 7JR
071-637 3571/2

STONE FEDERATION
82 New Cavendish Street, London W1M 8AD
071-580 5588
Telex 265763

STRUCTURAL INSULATION ASSOCIATION
45 Sheen Lane, London SW14 8AB
081-876 4415/6
Telex 927298 ALLEN g

SWEDISH FINNISH TIMBER COUNCIL
21 Carolgate, Retford, Notts DN22 6BZ
Retford (0777) 706616/7

SWIMMING POOL AND ALLIED TRADES ASSOCIATION (SPATA)
Faraday House, 17 Essendene Road, Caterham, Surrey CR3 5PB
Caterham (0883) 40110

TAR INDUSTRIES SERVICES
Mill Lane, Wingerworth, Chesterfield
Derbyshire S42 6NG
Chesterfield (0246) 76823
Telex 547061

THERMAL INSULATION CONTRACTORS ASSOCIATION
Kenway House, 388 High Rd, Ilford, Essex IG1 1TL
Ilford 081-514 2120

TIMBER TRADE FEDERATION OF THE UNITED KINGDOM
Clareville House, 26-27 Oxenden St, London SW1Y 4EL

TOWN AND COUNTRY PLANNING ASSOCIATION
17 Carlton House Terrace London, SW1 5AS 071-930 8392

TRADA
Stocking Lane, Huhenden Valley, High Wycombe, Bucks HP14 4ND
Naphill (024 024) 3091
Telex 21252

WELDING INSTITUTE
Abingdon Hall, Abingdon
Cambridge CB1 6AL
Cambridge (0223) 892588
Fax (0223) 891162
Telex 81183

WESTERN PRODUCTS ASSOCIATION (USA)
69 Wigmore St, London W1H 9LG
071-486 7488/9

WIRE GOODS MANUFACTURERS ASSOCIATION
Kensington House, 136 Suffolk St, Queensway, Birmingham B1 1LL
021-643 4488
Telex 338876 GRBHAMG

WOOD WOOL SLAB MANUFACTURERS ASSOCIATION
10 Great George St, London SW1P 3AE
071-222 5315

ZINC DEVELOPMENT ASSOCIATION
34 Berkeley Square, London W1X 6AJ
071-499 6636
Telex 261286

PART 13

# Updates

**INTRODUCTION**

This section has been complied as an aid in making adjustments to the rates an prices contained in Parts 3, 4, 5 and 10 of the Book

Users are advised to mark up the relevant pro-forma in line with any adjustments to the rates prices and costs which will be reported in the free quarterly updates.

This will allow the reader to obtain maximum benefit of the cost research that has been used to compile the data contained in the Book.

**LABOUR COSTS**

Revisions to labour costs as detailed on page 39.

| Categories of labour | 1 | 2 | 3 | 4 | 5 |
|---|---|---|---|---|---|
| Revised cost per hour<br>Update 1 | | | | | |
| 2 | | | | | |
| 3 | | | | | |

**MATERIAL PRICES**

This section details market prices of materials incorporating any appropriate trade discounts and delivery to site (except where noted), although distribution after delivery is not included.

| MATERIAL PRICES | Price changes by Update | | |
|---|---|---|---|
| | End Oct | End Dec | End Mar |
| Materials generally | | | |
| aluminium cladding | | | |
| bar and fabric reinforcement | | | |
| brickwork/blockwork | | | |
| cast iron pipes and fittings | | | |
| clayware pipes and fittings | | | |
| concrete and cement | | | |
| concrete manholes | | | |

**MATERIAL PRICES**

| MATERIAL PRICES | Price changes by Update End Oct | End Dec | End Mar |
|---|---|---|---|
| concrete pipes and fittings | | | |
| consumable stores | | | |
| contractors site equipment | | | |
| culverts | | | |
| earth retention | | | |
| fencing (site boundary fencing) | | | |
| geotextiles | | | |
| gully grates and frames to BS 497 | | | |
| instrumentation (soil) | | | |
| interlocking steel sheet piling and bearing piles | | | |
| joint fillers and waterstops | | | |
| landscaping/soiling | | | |
| manhole covers & frames to BS 497 | | | |
| paint/stains/protective coatings | | | |
| polymer channels and fittings | | | |
| precast concrete kerbs, etc | | | |
| quarry products, aggregates etc | | | |
| scour and erosion protection | | | |
| septic tanks, cesspools etc | | | |
| shuttering, timber and nails etc | | | |
| steel pipes and fittings | | | |
| structural steelwork | | | |
| timber | | | |
| upvc drain pipes and fittings | | | |
| upvc raineater gutters and pipes | | | |
| upvc subsoil drainage systems | | | |
| miscellaneous and minor items | | | |

**PLANT COSTS**

| PLANT | Price changes by Update | | |
|---|---|---|---|
| | End Oct | End Dec | End Mar |
| access platforms | | | |
| agriculture type tractors | | | |
| asphalt/road construction | | | |
| compaction | | | |
| cleaners sweepers | | | |
| compressors | | | |
| compressor tools | | | |
| concrete equipment | | | |
| concrete pumps | | | |
| concrete skips | | | |
| cranes | | | |
| diesel generators | | | |
| excavators | | | |
| excavator equipment | | | |
| hoists and lifting | | | |
| piling plant | | | |
| portable accommodation | | | |
| rollers | | | |
| scaffolding and staging | | | |
| tractors (dozers, loaders and scrapers) | | | |
| transport (tippers and dumpers) | | | |
| transport (vans etc) | | | |
| trench sheets | | | |
| pumping and dewatering | | | |
| miscellaneous | | | |
| power tools | | | |
| site tools | | | |

**UNIT COSTS**

| UNIT COSTS | Price changes by Update | | |
|---|---|---|---|
| | End Oct | End Dec | End Mar |
| Ground investigation, Geotechnical and other specialist processes | | | |
| Demolition and Site Clearance | | | |
| Fencing, Hedges etc. | | | |
| Safety Fencing, Parapets | | | |
| Drainage Work | | | |
| Earthworks - excavation | | | |
| - Filling (imported) | | | |
| - Landscaping | | | |
| Concrete and Reinforcement | | | |
| Pre-cast items | | | |
| Formwork | | | |
| Pavements - Surfacing flexible | | | |
| - Surfacing concrete | | | |
| Kerbs and footways | | | |
| Signs and lighting columns | | | |
| - Trenches | | | |
| - Electrical Work | | | |
| Piling Works | | | |
| Structual Steelwork | | | |
| Timber Work | | | |
| Painting | | | |
| Rail Track | | | |
| Tunnels | | | |
| Waterproofing | | | |
| Roofing and Cladding | | | |
| Bridge Bearings | | | |
| Expansion joints | | | |
| Brickwork, Blockwork, Stonework | | | |

**APPROXIMATE ESTIMATES**

| APPROXIMATE ESTIMATES | Price changes by Update | | |
|---|---|---|---|
| | End Oct | End Dec | End Mar |
| Industrial and Commercial Facilities | | | |
| Water & Treatment facilities | | | |
| Foundations for structures | | | |
| Concrete slabs for structures | | | |
| Earth retention and stabilisation | | | |
| - Concrete retaining walls | | | |
| - Crib Walls | | | |
| - Giobions | | | |
| Bridgeworks | | | |
| - Concrete | | | |
| - Steel | | | |
| - Timber | | | |
| Highway Works | | | |
| Civil Engineering Utilities and Infrastructure | | | |
| - Roadworks etc | | | |
| - Services | | | |
| - Lanscaping | | | |
| Temporary Works | | | |
| Bailey Bridges | | | |
| Piled Cofferdams | | | |
| Trench Supports | | | |
| Scaffolding | | | |

# Index to Advertisers

# Index

NOTES

NOTES

NOTES

NOTES

NOTES

NOTES

NOTES

NOTES